Landscape Restoration
H A N D B O O K
Second Edition

Landscape Restoration
H A N D B O O K
Second Edition

by

Donald Harker • Gary Libby
Kay Harker • Sherri Evans
Marc Evans

CRC Press
Taylor & Francis Group
Boca Raton London New York

CRC Press is an imprint of the
Taylor & Francis Group, an **informa** business

CRC Press
Taylor & Francis Group
6000 Broken Sound Parkway NW, Suite 300
Boca Raton, FL 33487-2742

First issued in paperback 2019

ISBN-13: 978-1-56670-175-4 (hbk)
ISBN-13: 978-0-367-39994-8 (pbk)
Library of Congress Card Number 98-46072

Library of Congress Cataloging-in-Publication Data

Landscape restoration handbook / Donald Harker ... [et al.] ; United
    States Golf Association, New York Audubon Society. -- 2nd ed.
        p.   cm.
    Includes bibliographical references (p.   ).
    ISBN 1-56670-175-9 (alk. paper)
    1. Natural landscaping. 2.Restoration ecology. 3. Natural
landscaping--United States. 4. Restoration ecology--United States.
5. Native plants for cultivation--United States. 6. Plant
communities--United States. 7. Phytogeography--United States.
I. Harker, Donald F. II. United States Golf Association. III. New
York Audubon Society.
SB439.L35 1999
719'.0973—dc21   98-46072
                                                                    CIP

Visit the Taylor & Francis Web site at
http://www.taylorandfrancis.com

and the CRC Press Web site at
http://www.crcpress.com

Foreword

Cooperation and communication are the keys to improving the quality of our environment. Cooperation opens the door to a strong, unified effort to advance a progressive environmental agenda. Communication is integral to cooperation. We need to talk about our concerns for the environment and our common responsibility for taking care of our planet. We need to discuss our concerns and responsibilities, as well as our goals and objectives, in order to enhance our ability to actively work together.

In 1887, Frank Chapman, the first president of The Audubon Society of New York State, spoke with women and encouraged them to sign pledge cards that stated their intention to never again wear bird feathers on their hats. The fashion of adorning hats and clothes with feathers nearly caused the extinction of several bird species. By communicating the importance of bird protection to women and by offering them the opportunity to work positively and cooperatively toward protecting birds, Chapman was able to generate a forceful, unified effort toward changing prevalent environmental attitudes. It was from this effort that the Audubon movement and the early stages of today's international bird conservation efforts evolved.

Today, in that same tradition, The Audubon Society of New York State continues its communication efforts by maintaining that every one of us is responsible for taking care of the earth; every one of us should do what we can where we live, work, and recreate to protect the environment. The Audubon Society of New York State offers a cooperative approach to actively care for and protect our environment—The Audubon Cooperative Sanctuary System. The Cooperative Sanctuary System is divided into four programs (Individual, Golf, School, and Corporate and Business) and is designed to encourage landowners and managers to become actively involved in conservation and wildlife enhancement, and to publicly recognize those who are involved in conservation and wildlife enhancement activities.

Over the twenty-odd years that I have been involved in the environmental movement, people have said over and over, "Quit telling me everything you're opposed to and tell me what you're in favor of." Well... here you go. The concepts in the *Landscape Restoration Handbook* are what New York Audubon is for. The United States Golf Association is cooperating with us, along with hundreds of existing golf courses across the country, and many individuals, corporations, and schools. I'd be pretty surprised if any one of the people managing these facilities ever got up in the morning and said, "Let me see. Today I think I'll pollute the environment." My experience has been that what they do say with increasing regularity is "Give me some information, some direction. Tell me how I can help wildlife and the environment."

The *Landscape Restoration Handbook* is our response. It is a cooperative effort between New York Audubon and the United States Golf Association and its purpose is to make information about enhancing the environment available to landowners and land managers. The *Landscape Restoration Handbook* is an important component of The Audubon Cooperative Sanctuary System. But, in reality, **you** are the most important part of The Cooperative System. You, the homeowner, the golf course superintendent, the school official, the corporate land manager, the architect, or the concerned citizen who has this book in your hands; you hold the key to conserving our environment. The key is not this book, though the book will open many doors by making available essential information. It is, in fact, your land, your home, your golf course, your business property, or school that is the key to a healthy environment. The *Landscape Restoration Handbook* is the best information that we could gather to help you achieve both your goal and ours—managing your land with the best interests of the environment in mind.

Thank you for the opportunity to share our thoughts and ideas with you.

Ronald G. Dodson
President
The Audubon Society of New York State

About the Authors

Donald F. Harker, president of the Mountain Association for Community Economic Development, received his B.S. degree in biology from Austin Peay State University, Clarksville, Tennessee and his M.S. degree in biology from the University of Notre Dame, South Bend, Indiana, where he studied "killer" bees in Brazil. He is the recent co-founder of Communities by Choice, an organization dedicated to sustainable community development.

Don has been a naturalist, farmer, and environmental consultant. He spent over ten years in Kentucky state government as Director of the Kentucky State Nature Preserves Commission, Director of the Kentucky Division of Water, and Director of the Kentucky Division of Waste Management. Don's consulting and research have taken him to Alaska, Mexico, Venezuela, Ecuador, Brazil, Costa Rica, Bahamas, and throughout the United States.

Gary Libby, ecologist with EcoTech, Inc. of Frankfort, Kentucky, received his B.A. in biology from Berea College, Berea, Kentucky and is completing a M.S. in biology and applied ecology at Eastern Kentucky University, Richmond, Kentucky. Gary worked as a botanist with the Kentucky State Nature Preserves Commission for two years.

Kay Harker, manager of the Planning and Program Coordination branch of the Kentucky Department for Environmental Protection, received her B.S. and M.S. degrees in biology from Austin Peay State University, Clarksville, Tennessee, and has completed all the course work toward a Ph.D. in biology at the University of Kentucky, Lexington, Kentucky.

Since 1985 Kay has worked for the Kentucky Department for Environmental Protection on solid waste, hazardous waste, and water issues. Kay has also been a naturalist, college teacher, and high school teacher. She is an avid gardener.

Sherri Evans, ecologist with the Kentucky Department of Fish and Wildlife and owner and operator of Shooting Star Nursery, Frankfort, Kentucky, received her B.A. in biology and M.A. in zoology from Southern Illinois University, Carbondale, Illinois. She has worked as a wildlife researcher and environmental consultant, and has spent over seven years in state government natural resource management programs.

Shooting Star Nursery was established in 1988, the first nursery in Kentucky to specialize in the propagation and use of native plants for landscaping and restoration. The nursery offers consulting services in landscape and restoration planning and has worked with parks, golf courses, national forests, utility companies, city planners, museums, nature preserves, businesses, and others to create attractive, self-sustaining, native plant landscapes.

Marc Evans, Botanist/Ecologist for the Kentucky State Nature Preserves Commission, received his B.A. and M.S. in botany from Southern Illinois University in Carbondale, Illinois.

Marc has been a botanist/ecologist with the Kentucky State Nature Preserves Commission for fifteen years. He has worked as a biological consultant, researcher, and park naturalist. Marc is co-owner with Sherri of Shooting Star Nursery.

Acknowledgments

The authors are deeply indebted to many people who helped with this book and wish to thank the following:

The United States Golf Association who generously supported this work, and especially the USGA Environmental Research Committee for their support, ideas, and review of the manuscript.

Ron Dodson, President of the New York Audubon Society, for his support of this effort from its conception to its completion. He edited the manuscript, shared ideas, and generously wrote the Foreword. His good cheer and enthusiasm help keep us all going.

Mike Kenna, Research Director of the Greens Section of USGA, for his support, encouragement, and enthusiasm. His editing improved the manuscript and he generously shared ideas that improved this book.

David Snyder, Austin Peay State University, who could not pass up the opportunity to critique two former students and teach them just a little more. David critically read portions of the manuscript.

Stan Beikmann, former Director of Fernwood, Inc., who has taught many about gardens and enthusiastically supported the Fernwood Prairie restoration project. Stan critically read portions of the manuscript.

Susan Kiely, who researched, edited, typed, and managed much of the production of this book for the year we all worked on it. Susan has now finished her Masters degree in landscape design.

Cathy Guthrie for her research on the ecological community descriptions and species lists.

Van Fritts for editing, research, typing, and helping us tend to all of the details. Lou Martin for proofing and editing the final manuscript.

Deborah White for editing and research on plant species and ecological communities.

Liz Natter for critically reading the manuscript and offering many helpful suggestions.

Hal Bryan, Eco-Tech, Inc., for sharing his vast experience in wetland restoration with us and for critically reading the manuscript and offering many helpful suggestions.

John A. Bacone, Bill Carr, Chris Chappell, William Dick–Peddie, Bruce Hoagland, Chris Lea, Darla Lenz, John Logan, Max Medley, Bob Moseley, Linda Pearsall, John Pearson, Mike Penskar, John Sawyer, Alfred Schotz, Timothy E. Smith, Mary Kay Solecki, and Elizabeth Thompson for providing valuable information and suggestions on the ecological communities and species lists.

Leslie Sauer, Andropogon Associates, for reviewing the manuscript and offering helpful suggestions that improved the completeness of this book.

Thomas Heilbron, founder/developer of The Champions golf course, for enthusiasm about the concept and allowing us to use The Champions for the conceptual design work. Karen Bess for help on The Champions conceptual plan.

Bill Martin, Eastern Kentucky University, for sharing ideas, experience, and his library.

Brian Lewis for support, encouragement, and understanding.

April Smith who did the final layout and design of this book.

Others who helped with research, editing, and/or typing include Ambika Chawla, Steve Cooper, Jim Fries, Hayden Harker, Heather Harker, Marilyn Wrenn Harrell, Francis Harty, Glenn Humphress, John Kartesz, and Jean Mackay .

PlanGraphics, Inc., especially John Antenucci, Kaye Brothers, Leann Rodgers, and Jani Sivills for computer assistance.

MACED's GIS lab for preparing the maps in this book, especially Shepard McAninch and Jennifer Hromyak.

And last, but not least, to our many friends and family members who always had an encouraging word when we needed it. The authors take responsibility for any errors in this book.

Dedication

This book is dedicated to our children, Heather Harker, Hayden Harker, Porter Libby, and Daniel Evans, and the children of their generation, who soon must care for the earth for the next generation. This is how humanity survives.

Contents

Inside Back Cover Pocket—Natural Regions of the Conterminous United States (Color Map)

CHAPTER 1

Naturalizing the Managed Landscape

Introduction

Since the beginning of recorded time, human beings have felt both a part of nature and, at the same time, apart from nature. Sometimes we are able to simply enjoy the beauty that nature has to offer. Sometimes we use natural forces to our advantage. At other times, we are at the mercy of nature. The spring rains that moisten the soil and foster lush, green growth can become a deluge, washing the hopes of farmers and homeowners into the nearest stream or river. When challenged by nature's forces, one can work with them or attempt to control them.

Today, we understand that working with nature, and not against it, makes both environmental and economic sense. Working with nature does not necessarily mean just letting nature take its own course. It means making sound decisions about how to manage the land. It means finding out what will work with the land, given its physical and chemical characteristics. When we view the land, whether it's our own backyard, a forest, a meadow, a park, or a golf course, we each have some vision of how we think that land "ought" to look. Weaving nature into our vision for a landscape means seeing the beauty in a field of wildflowers, instead of thinking it looks like unkempt weeds.

This book is based upon the concept that the managed landscape *can* be naturalized and that the process of naturalizing the landscape makes long-term environmental and economic sense. Naturalizing the landscape means using native plants to restore and beautify the landscape. It is both restoring or creating an ecological community, using as

full a complement of native species as possible, and landscaping with native plants. It takes hundreds or even thousands of years for an ecological community to develop. It may not be possible to completely recreate such a community and, certainly not in a short time. We can, however, begin the process by educating ourselves about what may have been present before the land was altered and then developing a long-term plan to restore the land, as much as possible, to its former natural condition. Since nature is dynamic, the most we can ever do is establish the conditions for natural processes to work.

The naturalization process can be either in the form of natural landscaping or ecological restoration. This book provides guidance on both of these. For example, if one chooses a site to create a forest and plants five species of trees from the long list of the oak–hickory forest type, that can be either ecological restoration or natural landscaping. It depends to a large degree upon the goal and outcome of the project. If the intention is to maintain a five–species complex or planting, then it is natural landscaping. If, however, the intention is to add trees, shrubs, and herbaceous species of the oak–hickory forest, and to allow changes that result in the succession of these plants into a more natural ecological community, then it is ecological restoration. Both natural landscaping and ecological restoration have value. Given the size or shape of the site, the surrounding land uses, or unchangeable cultural or developed features, complete restoration of an ecological community may not be possible. This should not keep the land manager from utilizing natural landscaping on the property.

The greenspaces of the earth's landscape mosaic are diminishing. Parks, farms, yards, greenways, wooded river corridors, woodlots, and golf courses are all part of the greenspace patchwork where people seek escape from the stress and noise of cities and highways. All of these areas are, or have the potential to be, important parts of the naturalized landscape. As the natural components of these areas increase, the connection between the natural world and the human–influenced world increases. When naturalized, greenspaces provide birdsong, the flicker of orange or blue from a winging butterfly, prairie grasses blowing in the wind, or a chipmunk scurrying through the leaves—sounds and sights that remind us of our role, our place, in nature.

Golf courses are a kind of managed landscape where "nature by design" is an integral part of the game and the landscape. Over the years, however, particularly in the United States, the human vision of how a golf course "should" look has overshadowed nature's design. The popularity of golf has increased steadily in the United States since 1946. According to the National Golf Foundation (1997b), from 1985 to 1996, the number of golfers increased from 17.5 to 25 million. In this same period, the number of rounds of golf played annually increased by 17.6 percent. Between 1968 and 1996, the number of golf facilities increased from 9,600 to 14,341, 70% of which allow public access (National Golf Foundation 1997a). Today, it is estimated that over 12 percent of the United States population over the age of twelve plays golf. Assuming an average size of 124 to 180 acres per facility, there are from 1.8 to 2.6 million acres of land dedicated to golf facilities. Opportunities to use natural landscaping or practice ecological restoration abound on each of these facilities. Out-of-play areas, stream corridors, wetland areas, and woodlots can all play important roles for wildlife and environmental conservation. By practicing the principles of natural landscaping and ecological restoration and passing on information concerning those efforts to golfers and other users of managed areas, we will spread the seeds to encourage individual participation in naturalizing and restoring the landscape beyond those managed areas.

Naturalizing and restoring the land are parts of an ecological renaissance and plan of action for the future. Human progress is now being measured in terms such as "sustainable," "environmentally friendly," and "quality of life." Living within sensible ecological boundaries and restoring parts of the earth to a more natural state are integral parts of that ecological renaissance.

Why Naturalness?

There is broad scientific agreement that the world's biological resources (biodiversity) are being significantly diminished. Protecting these resources and restoring naturalness to human-altered landscapes are important to protecting the world's biodiversity. "Naturalness," simply defined, is "the degree to which the present community of plants and animals resembles the community that existed before human intervention." Restoring naturalness can mean planting a native tree or planting a forest. The act of becoming knowledgeable about and appreciating our natural heritage can lead to a recognition of our common destiny with the natural world and a desire to both protect and restore the earth's resources.

Some have asked whether we should spend our resources on restoring the earth when so much protection effort is needed to save the natural areas that remain. John Berger (1990), executive director of Restoring the Earth, speaks to the issue. "I have been told that we should not publicize or even propose restoration because society will use it to excuse new assaults on the environment. Yet the technology of resource restoration is developing rapidly and cannot be wished away. Just as our understanding of atomic energy cannot be unlearned, so, too, the genie of restoration is 'out of the bottle.' Our task is to see that restoration is used properly — to repair past damage, not to legitimize new disruption. An epochal development has clearly begun: For the first time in human history, masses of people now realize not only that we must stop abusing the earth, but that we also must restore it to ecological health. We must all work cooperatively toward that goal, with the help of restoration science and technology."

Much of the landscape has already been altered from its natural state. Clearly naturalization will, of necessity, play an increasingly significant role in efforts worldwide to preserve the planet's biodiversity.

By increasing the naturalness of our landscape, we become a positive force in contributing to a sustainable world. Landscape restoration affects the world in a variety of ways.

- creates a healthier, sustainable mosaic of land uses on the landscape
- maintains a diversity of plants and animals
- maintains the gene pool of particular plant and animal species, promoting hardiness, disease resistance, and adaptability
- protects ecosystems and ecological communities
- improves water quality
- minimizes erosion
- creates positive, progressive, and constructive attitudes about the natural world
- promotes the concept that natural is beautiful and desirable
- creates lower maintenance landscapes, reducing our dependency on water and the production and use of chemicals

The *Landscape Restoration Handbook* is an "ecological call to action." It is a call to those who manage yards, farms, corporate land, parks, school yards, roadsides, and golf courses to consider a more natural vision for the human-managed landscape. We are facing ever-increasing demands and impacts on the landscape. The cumulative impacts of our burgeoning population over time are nearly incalculable. Naturalizing and restoring the earth must become the business of everyone. Every attempt at restoration, from planting a tree to restoring a forest, matters, for each one shows that our reconciliation with the earth is desired and is progressing.

There is a story (original teller unknown) about a big storm that came and washed thousands of starfish onto the beach. A little girl came down to the beach and was terribly upset by all of the stranded starfish that would soon die. She began picking up starfish, running down to the ocean, and throwing them back into the sea. A man came along and reasoned with the child that she was wasting her time. There were so many starfish on the beach that her efforts just would not matter. The little girl grabbed up a starfish, ran to the sea, and threw it into the sea. She returned to the man and, with a smile, said "It mattered to that one."

Photo. 1.1. Golf course with a naturalized landscape.

Using This Book

This book is divided into several chapters that outline principles for developing a comprehensive naturalization program. A three-part "Greenlinks" program is discussed in Chapter 2, outlining a broad–based program for education, regional planning, and naturalization planning. Also discussed in the book are principles and guidelines for increasing biological diversity (Chapter 3), natural landscaping (Chapter 4), and ecological restoration (Chapter 5). These three chapters provide a basic background for understanding conceptual and detailed naturalization plans. The diversity of landscape conditions, such as soil, slope, and available moisture, demands site specific analyses in order to develop detailed naturalization plans. While such analyses are beyond the scope of this book, the principles in this book can be generally applied by the land manager in developing a naturalization plan.

In addition to the Greenlinks program and principles and guidelines for naturalization, the dominant ecological communities are described for thirty natural regions of the continental United States (Chapter 6 and Appendix A). These descriptions provide a framework for determining which common ecological communities occur or occurred in each region of the country. A list of plant species associated with each ecological community (Appendix A) is included to help the landscaper or restorationist select appropriate species. Chapter 7 and Appendix B describe 27 Ecological Restoration Types. Each of these types combine one or more Dominant Ecological Communities for the purpose of describing generally applicable restoration techniques. Appendix B contains a description of site conditions, establishment, and maintenance and management for each Restoration Type. Characteristics of each species are presented in a woody species matrix and a herbaceous species matrix

(Appendix C) to help in the selection of species appropriate for the planting conditions and design specifications of a naturalized planting. The matrices include, for each plant, information on its type, environmental tolerances, aesthetic values, wildlife values, flower color, bloom time, and landscape uses. The matrices are organized alphabetically by scientific name.

Appendix D is a list of nurseries which carry native plants that can be used for either natural landscaping or ecological restoration. The nurseries are organized by state for the Continental U.S.

The plant lists used in this book have come from a variety of sources. Because of the diverse sources and the age of some sources, all scientific and common names have been synonymized with *Synonymized Checklist of the Vascular Flora of the United States* (Kartesz 1991) and *Common Names for the North America Flora* (Kartesz 1990).

Appendix E lists resources important to ecological restoration including professionals, organizations, Natural Heritage programs, and web sites.

Appendix F contains a brief outline of the regulatory considerations relating to restoring wetlands. Since regulations change, we suggest that the most current information be obtained from regulators before restoring wetlands.

This book also contains a glossary and bibliography. The bibliography contains more than literature cited.

This book can be read from front to back or used as a reference. Those interested in landscaping or ecological restoration can go directly to the appropriate chapter. For a broader view of landscape restoration, however, a full reading can stimulate imagination and creativity in approaching the naturalization of our human-managed landscapes. In addition, it is hoped that this book will help stimulate a renewed sense of interest in taking active steps to become environmentally sensitive caretakers of planet earth.

CHAPTER 2

Greenlinks

Introduction

Greenlinks is the name given to a comprehensive program for naturalizing the landscape. It involves both planning and education. A Greenlinks program is divided into three basic parts: (1) introducing a target audience to the idea of naturalizing the landscape through a strong education program; (2) working with adjacent and regional landowners to link more and more greenspaces and natural areas together in a regional context; and (3) developing a detailed naturalization plan for the managed site.

An overall plan for naturalizing managed landscapes generally includes a combination of ecological restoration and natural landscaping. A variety of goals and objectives can be developed for a site. They include improving wildlife habitat, creating more aesthetically pleasing surroundings, restoring natural communities, protecting local rare plants or animals, or preserving the local gene pool of a particular species or natural community.

It is not possible to offer detailed prescriptions of how to restore or landscape every type of site. To develop a detailed site plan requires applying the principles for maintaining natural diversity, ecological restoration, and natural landscaping to specific parcels of landscape. This chapter offers guidelines for developing an educational program, regional overview, and naturalization plan.

Greenlinks Education Program

Any naturalization or ecological restoration project benefits from, and is more successful with, a strong education program. Naturalizing the landscape provides opportunities for education about a region's natural heritage, the value of natural land-

scaping, and restoring natural communities. We recommend that the opportunity created by a naturalization project be used to educate owners, managers, members, visitors to managed areas, and public officials. Also, a cooperative effort with surrounding landowners can lead to regional naturalization programs that increase biological diversity. This will be especially important if landowners of large natural areas or important corridors become involved. Leslie Sauer (1998) states that "the most important conditions for a restoration project are that it be community-based and that it be science-based."

As development increasingly encroaches on the few remaining natural lands, opportunities to learn about nature are diminished. Restoring ecological communities and establishing natural landscapes play important roles in preserving the unique vegetational character that distinguishes each region of the continent. Naturalization projects on managed landscapes provide an opportunity for the public, members, visitors, and students to learn about the natural heritage of the region. This educational experience will increase awareness and appreciation of efforts to preserve that heritage. Leslie Sauer (1998) says "we cannot talk about restoring particular habitats without addressing the larger issues driving change within the regional ecosystem, including human behavior and values. Our attitudes toward the forest must also change, and we must come to know the landscapes around us. Indeed, the act of restoration itself is one of the most powerful vehicles for fostering awareness of place and environment." What people learn can also be transferred to other areas, such as their yards and the property where they work.

Ecological restoration and natural landscaping

Photo. 2.1. Children at Fernwood Prairie getting a lesson about a disappearing ecological community.

serve as excellent examples of how to create less intensively managed landscapes which help reduce maintenance costs, conserve natural resources, increase biological diversity, and benefit wildlife. Keeping the public well informed about the establishment process is very important for maintaining continuous support for these projects. This is especially important since it may take several years for a restoration project or natural landscape to mature and achieve its full aesthetic potential. During the interim, the landscape may appear "weedy" or unkempt to the uninformed viewer. Natural landscapes and ecological communities go through a dynamic process of change as they become established and mature. These changes demonstrate how the forces of nature shape the natural environment and are exciting and educational to observe.

Education occurs when one has a new awareness or experience, learns the name of a plant or bird, or has his or her consciousness raised about an issue. A comprehensive Greenlinks education program will seek to bring people into the overall planning and development of the program. Various kinds of educational approaches can be used. The following are a few recommended elements in a comprehensive Greenlinks education program.

■ Solicit the help of local nature centers, native plant societies, botanical gardens, and local people with expertise in native plants and restoration.

■ Involve the public, members and supporters of the managed area, local officials, and students in planning and implementing the naturalization program.

■ Invite adjacent and surrounding landowners to meet and discuss the regional concept of a Greenlinks program. Involve them in developing a regional plan.

■ Sponsor educational forums in the community to demonstrate examples and highlight the positive components of Greenlinks.

■ Join the Audubon Cooperative Sanctuary System or Audubon Signature Program. These programs are sponsored by the Audubon International which provides an advisory information service promoting the protection and enhancement of wildlife habitats and water resources.

■ When a plan is completed, develop informational brochures to provide background on the goals and processes of establishing natural landscapes.

■ Be sure your plan includes self-interpreting walking paths with interpretive signs at strategic locations, such as trailheads and in front of natural gardens, to help the public learn to recognize native species of plants, the wildlife they attract, and other benefits of using native plants.

Regional Overview Plan

One of the first recommendations of the New York Audubon Cooperative Sanctuary Program is to create a resource committee. A principal task for that committee can be to obtain aerial photographs, soil maps, and topographic maps and use them to examine the regional ecological context of the particular site. Attention should be given to the habitats surrounding the site, the location of streams and rivers, and the current and potential land uses of the area. If a regional Greenlinks program is attempted, this exercise will provide a good idea of the opportunities available for linking patches of natural communities and greenspaces together. Regional landscape approaches are necessary to maintain plant and animal diversity. It cannot be done by creating many small patches of habitat on one small site.

"The regional landscape approach to preservation should therefore recognize the importance of broad corridors connecting habitat islands. Fence rows and shelterbelts should be widened whenever possible. Regional planners should draw corridors of natural habitat onto their blueprints. Park planners might connect significant patches of habitat within a given park, which would minimize island effects while still permitting development of considerable land area. Stream corridors, which can be effective avenues of dispersal for terrestrial as well as aquatic organisms, particularly if they are wide and contain some upland habitat, should be protected wherever possible" (Noss 1983).

There is also a temporal dimension to any site. In other words, one must consider what will happen to the area over time. The concern is not just what will happen to the site, but what will happen to the surrounding landscape. If development is occurring in a particular adjoining area, then visual screening may be appropriate. Effective screening may take years to develop. Activities upstream affect the quality and quantity of water downstream. Wildlife corridors can be broken by many types of activity. Working with adjacent landowners on a regional plan can pay great dividends in the future.

Figure 2.1 shows the region around The Champions golf course in the central Bluegrass region of Kentucky. This area is heavily impacted by agriculture, horse farms, and development, leaving very few natural areas or high quality greenspaces. The regional context suggests that corridors to natural areas in the region can be established through a naturalized stream corridor network, wooded fencerows, and wooded patches. The Kentucky River is a primary corridor that remains wooded. It runs from the Eastern Kentucky mountains to the Ohio River. Providing links to this area will increase the wildlife for a site. There are also some large natural areas (wooded, but impacted) in the Sinking Creek area. These are important to the regional biota and should be part of a Greenlinks program. A comprehensive Greenlinks program will require developing these ideas and bringing together the landowners, citizens, and local officials who can help make such an idea a reality.

The Naturalization Plan

Basic to the concept of naturalizing a managed site is the development of the naturalization plan. The four main components of the plan are (1) a statement of objectives, (2) a synthesis and analysis of baseline information on existing site conditions, (3) a conceptual design plan identifying potentials for natural landscaping and restoration, and (4) a program for establishment and maintenance that includes a reasonable timetable and an estimate of costs. Component four is part of a detailed site plan and beyond the scope of this book.

Site Objectives

An important first step in the planning process is to define the objectives for naturalizing the managed site. In many cases there will be multiple objectives. They may be functional, such as controlling erosion; aesthetic, such as visually screening an unpleasant view or creating a colorful wildflower garden; economic, such as reducing dependency upon chemicals used for landscape maintenance; or biological, such as increasing diversity or protecting a rare species. Many of these objectives

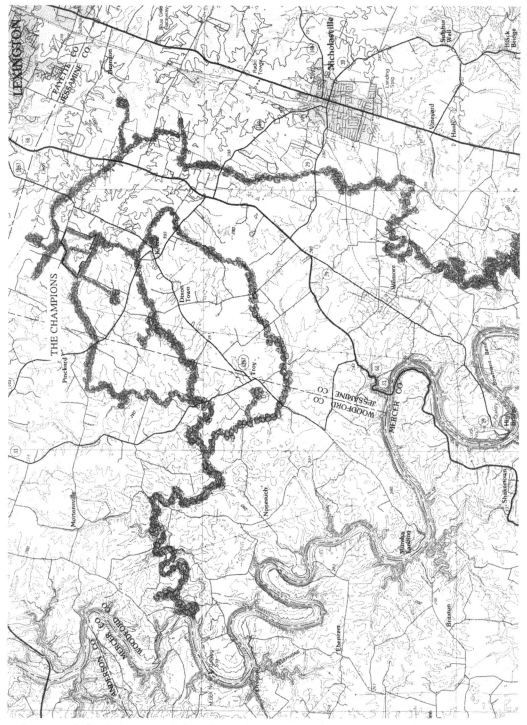

Figure. 2.1. This figure shows one proposal for using wooded rivers, streams, fencerows, railroad rights-of-way, and roads to connect woodland patches throughout the landscape.

are interrelated, and multiple objectives are often attainable on a given site when each is considered fully and early in the planning process. For example, many of the most showy or fragrant flowers suitable for landscaping are also excellent nectar-producing plants that will attract wildlife. Many native trees and shrubs with showy flowers or brilliant autumn color also provide valuable cover or food for wildlife as well as welcome shade on a hot summer day.

Baseline Information

Synthesis and analysis of baseline information help guide decisions concerning the possibilities and limitations for naturalizing the site. The end result is the delineation of naturalized landscapes to be created, restored, or enhanced. A site survey should be conducted that includes the biota and an analysis of the light, moisture, soils, slope, altitude, wind, and microclimate. This information is then used to select appropriate ecological communities as landscape models. The plant and animal data are especially useful to help identify the potential of existing communities for preservation, restoration, or enhancement.

Baseline information forms the framework of the naturalization plan. Baseline data is usually presented on a base map which provides a visual record of the plan. The size of the planning site dictates the scale of the base map, but a scale of one–quarter inch to the foot is generally the minimum used for effective planning (Weddle 1983). Some types of baseline information to be included on the base map are features such as property boundaries; easement boundaries; above and below ground utility and communication rights–of–way; existing trees, shrub borders, meadows, water bodies, and other natural features; and buildings, paved roads, and other permanent structures, with access and service corridors noted. Symbols may be used to identify orientation, predominant seasonal wind patterns, and seasonal changes in the angle of sunlight. Information on existing local ordinances governing vegetation, minimum setbacks, and other restrictions should be included. A series of transparent overlays can be used to record physical, biotic, and visual survey data.

A series of landscape surveys may be needed to obtain accurate baseline data, especially when knowledge of measurable quantities and qualities of the site is desired. Surveys are often a compilation of existing data from architectural drawings, maps, and aerial photographs, combined with on-site inventories of physical, biotic, and visual characteristics. Survey results are presented as drawings, photographs, and written documentation.

Physical survey data on soils, geology, topography, and hydrology can often be compiled from U.S. Geological Survey topographic and geologic maps, and county-based soil surveys. Soil survey data can be misleading in areas disturbed by past construction activities, in which case samples from test holes can be used to plot topsoil and subsoil layers across a site. Soil tests, available through local Cooperative Extension Service offices, analyze soil fertility and acidity.

A biotic or ecological survey is often needed to assess the quality of existing natural communities. The survey should produce a complete record of plant and animal life, including the identification of rare or sensitive species requiring special consideration during site planning and development. The relative abundance of nonnative species and a recommended eradication program should be included as part of the report. At the very least, an assessment of the relative quality of existing natural communities within a regional context should be included. The species composition of the plant communities can be used as an indicator of local environmental conditions.

A viewshed survey may be desirable to qualitatively evaluate the visual character of the landscape. The visual survey may encompass visual horizons, water, vegetation patterns, and visual attributes and intrusions of adjacent land. The beyond-site viewsheds are judged subjectively as attractive or intrusive, often as viewed from one or more preselected locations on the site. Recommendations are made as to the potential to open or reduce viewsheds from various areas. Photographic documentation, particularly with a panoramic series, provides a permanent record for later reference.

The Conceptual Design Plan

The landscape can be considered as patches within patches. One can look at a large piece of landscape and see fields, streams, and forest. At that scale, the forest or field may seem homogeneous. A closer look reveals that each is, in reality, many smaller patches. Each separate patch is considered a site or planning unit for developing a naturalization plan.

The conceptual plan identifies the patches or planning units potentially available for restoration or natural landscaping. These planning units are discrete parcels of similar physical makeup that will receive similar treatment. A planning unit can be a portion of hillside with the same soil type, exposure, and moisture regime on which the same ecological community will be restored.

A conceptual design has been developed to illustrate some of the restoration and landscaping principles that can be applied to an existing golf course. The design was created for The Champions golf course (Figure. 2.2) located in the Bluegrass region of Kentucky near Lexington. Photos 2.2 through 2.13 show how the course currently looks.

The concepts illustrated in the design created for The Champions golf course include the following.

Ecological Restoration

- Upland Ecological Community Restoration—The central Kentucky Bluegrass region had a variety of upland plant communities. They included the oak–ash savanna, oak–hickory forest, and canebrakes. All of these communities are used in the concept plan.
- Wetland Ecological Community Restoration—Two natural wetland communities are recommended. Below the spillway of the large impoundment is a good location for a marsh. This marsh will increase the plant and animal diversity and act as a water quality improvement system for runoff coming into the lake and out through the spillway.

  The other wetland is a wooded area at the head of the impoundment. It is recommended that this area be expanded. Part of this planning unit is currently forested and part is old field with an existing pond located on adjacent property under the same management. This pond should be left and surrounded by woods. This forested wetland area can serve as a buffer to the impoundment by trapping sediment and agricultural runoff from the surrounding area.
- Corridors—The concept plan maximizes the idea of connecting patches of forest, wildflower meadows, canebrakes, wetlands, and savannas. These corridors will allow animals to move throughout the area more readily.

The corridors also connect patches on the golf course with areas off the course. This will bring more species onto the course from outside. It actually creates a two-way path, with animals moving both ways.

- Vegetative Structure—The types and availability of habitat are significantly increased with increased vegetative structure. This means tree, shrub, and herbaceous layers should be planned and allowed to develop corridors. Dead and fallen logs and leaf litter should not be removed.

Natural Landscaping

- Enhancement Zones—Certain edge areas with high visibility have been selected as enhancement zones. These zones are considered landscaping because they will have a high concentration of showy prairie wildflowers rather than the normal dominance of grasses. It will cost more than native savanna to develop these areas so they are restricted in size.
- Screening—The entire perimeter of the site will be screened from the road and surrounding area. This screening increases privacy, reduces noise from traffic, and provides more habitat.
- Special Features—Three special features were identified on the site. They are generally described as follows:

  The first special feature is a natural outcropping of limestone. Fortunately, a number of mature trees were left standing in this area (see Photo 2.11). It is recommended that a bench be placed here. Recommended plantings include native flowering shrubs and herbs.

  The second special feature is a spring head that has a stone spring house built around it. Unfortunately, it occurs in an open area; however, it will be much more attractive with a backdrop of trees (see Photo 2.12).

  The third special feature is a natural seep in an area with some topographic relief. The area was left with mature trees. It will be very attractive with a planting of large native ferns such as cinnamon fern and royal fern and a diversity of wildflowers adapted to moist soils (see Photo 2.13).

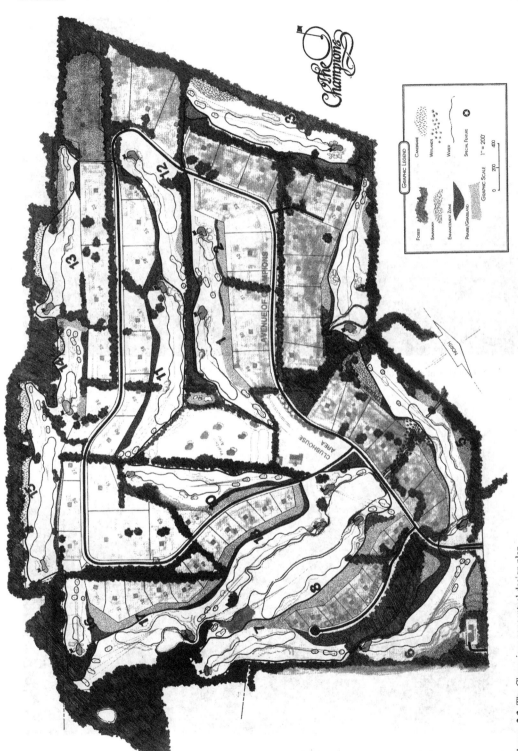

Figure. 2.2. The Champions conceptual design plan.

Photo. 2.2. Overview of portion of course.

Photo. 2.3. Fairway with patches of natural woodland left on either side.

Photo. 2.4. Golf cart path goes through edge of natural woods.

Photo. 2.5. View from golf cart path into natural woods.

Photo. 2.6. Row of trees currently raked and mowed.

Photo. 2.7. Row of trees with shrub layers, herbaceous layers, and leaf litter left on the ground.

Photo. 2.8. Woodland that can be naturalized to screen a road.

Photo. 2.9. Open area available for naturalization including shrub and herbaceous layers.

Photo. 2.10. Open area available for native grassland restoration and enhancement zone.

Photo. 2.11. Special feature No. 1 is a natural limestone outcropping.

Photo. 2.12. Special feature No. 2 is a natural spring with a stone springhouse.

Photo. 2.13. Special feature No. 3 is a natural spring.

CHAPTER 3

Principles for Maintaining and Restoring Natural Diversity

Introduction

An effective naturalization plan is based on the application of principles for maintaining natural diversity in a managed landscape. Diversity, as used here, is not the diversity of a zoo, but is diversity as seen in natural communities and ecosystems. The aim is not to plant as many different kinds of trees as possible on a particular site, but to restore a natural community on a site with as full a complement of native species as possible. Restoration of a natural plant community provides a variety of habitats valuable to a variety of animals. These principles are not prescriptive; they are designed to help people think about the use of a particular parcel of land. Application of the principles will enable managers of golf courses, parks, recreation areas, farms, schools, and other parcels of land to increase the diversity of plants and animals on their land. If applied widely within a region, these principles can result in a new landscape mosaic that increases overall naturalness and regional diversity of native plant and animal species. Restoring many small areas to natural communities contributes to increased individual, population, and community diversity.

Thorne (1993) offers that "the fundamental goal of ecological design should be to maintain ecological integrity, also referred to as ecological health. Ecological integrity is characterized by (1) natural levels of plant (primary) productivity, (2) a high level of native biological diversity, (3) natural (usually very low) rates of soil erosion and nutrient loss, and (4) clean water and healthy aquatic communities."

Maintaining and restoring natural diversity to our increasingly fragmented landscape is our challenge. Thorne (1993) states "the landscape is a particularly useful unit for understanding broadscale issues of ecological integrity. Landscapes typically include a mixture of both human and natural features and contain numerous interacting ecosystems, such as forests, fields, waterways, and human settlements. By understanding the nature of interactions across landscapes, it is possible to address systemic ecological issues over the long term." A number of studies address this issue. For example, as forests in the northeastern United States were converted to farmland in the late 1800s and early 1900s, the number of species of migratory songbirds using these areas declined. But, as small forest patches were reestablished in the latter half of this century, the number of bird species increased. A study of small urban and suburban forest preserves of less than 247 acres suggests that populations of many species may not be self-sustaining in a small preserve, but are instead an important part of a larger, regional population (Askins and Philbrick 1987). The study illustrates that diversity can increase in an area even when the creation of a large preserve is not possible. When created, small preserves can play an important role in restoring our natural heritage without diminishing the intended human use of the land. To preserve the regional diversity of native plants and animals, it is important to restore many small sites.

Small areas will not satisfy all the needs of all species that occur within a region. Robinson et al. (1995) pointed out that many neotropical migrants are declining. Declines are attributed to the loss of breeding, wintering, and migration stopover habitats. Forest fragmentation appears to be a major cause. Nesting success is reduced below replacement levels in some fragmented landscapes. Brood parasitism by brown-headed cowbirds increased with fragmentation in nine midwestern study areas. In landscapes with less than 55 percent forest cover, most wood thrush nests were parasitized. In heavily forested landscapes, nest parasitism by cowbirds was not significant. Robinson et al. (1995) suggest that "conservation strategies should consider preservation and restoration of large, unfragmented 'core' areas in each region."

By analyzing specific characteristics of an individual area and consulting the lists of native plants for that ecological community, the landscaper can determine the types of plants that will best survive and prosper at a specific site. In restoring native plant communities, the landscaper provides an opportunity for the animal species associated with those communities to find suitable shelter, food, and areas for raising young. Although only a few species may be present on any given restored site, the total of all restored sites will preserve a large number of species and create diversity within a region.

This chapter provides principles, guidelines, and design suggestions for increasing and maintaining a diversity of plants and animals on parcels of managed land. Each parcel, whether a golf course, school, yard, farm, park, or unused corporate land, can play an important role in maintaining regional diversity. The principles and designs suggested in this chapter can be used to increase species diversity, increase the populations of particular species, and assist in restoration of specific ecosystems. These principles can be applied during the establishment of a new managed area or on an area currently in use.

What Is Diversity?

Biological diversity, or biodiversity, refers to the variety of life in all its forms and levels of organization (Hunter 1990). Deliberations on biological diversity include discussions of genetic diversity of individuals within a species, diversity of populations, diversity of communities, and diversity of ecosystems and regional vegetation types called biomes. This chapter discusses four levels of biological organization: individuals, populations, communities, and ecosystems.

Scientists identify all individual plants or animals by a two-part scientific name. The first word is the genus to which the species belongs and the second word is the specific epithet, or species name. Each species is adapted to survive in a certain range of conditions. All individuals of a species differ slightly from one another in their abilities to adjust and respond to a particular range of environmental conditions. These slight differences represent diversity within the species. Since all environments are continually changing, this diversity is important in the long-term survival of species.

A population is defined as a group of individuals of a single species that interact, interbreed, and live in a given area. The individuals in a population are genetically diverse, and a viable population has sufficient genetic diversity to withstand changes in the environment. All environments change. In genetically diverse populations, some individuals may die as a result of changes, but many survive and the population survives. Usually a large population contains more genetic diversity than a small population and thus has a better chance of surviving environmental change. Density, or the number of organisms per unit area, is one measure used by scientists to characterize a population.

An ecological community is an interrelated assemblage of plants and animals associated with the abiotic or physical environment. A community type repeats itself, with variations, under similar conditions across the local landscape (or within an ecoregion). Through time, species within a community have developed complex relationships and interactions. Competition among similar species for scarce food supplies is a common occurrence in communities. Predation of one species upon another affects the number of different kinds of species in a community. Studies of competition, predation, decomposition, and many other relationships among species within communities suggest that preservation of a single species isolated from its community is difficult, and often impossible. Because of the many varied relationships, interactions, and requirements of species, studies of the needs of a given species are often incomplete and are always time consuming to conduct. Although autecological studies (studies of individual species' needs) may be very successful in preserving a species, time and financial constraints make such studies of all species within a community impractical. An effective and less expensive means of preserving a species is usually through preserving or restoring its natural community. Restoration of natural communities does not require knowledge of all requirements of all species.

An ecosystem is a complex system of biotic and abiotic components and relationships. It includes not only the plant and animal communities, but also the microorganisms, climate, sunlight, water, and soil that interact with communities and with each other. Nutrients, water, and energy cycle through the system. Animals use other animals or plants as food. Slightly different soils, slopes, and many other factors make for subtle differences in each locale or habitat within a given ecosystem. An ecosystem usually contains, or can include, many different ecological communities and may cover a large area.

The place that a plant or animal lives is its habitat. Plants and animals of a given species have specific requirements which must be met in order to survive. Each site or parcel of landscape has characteristic soil, exposure, moisture, and temperature ranges which meet the requirements of a given assemblage of plants. Each community of plants provides habitat for a particular group of animals. Wetland and upland trees may be found in the same geographic area, but the two types of trees are not found in the same habitat. Each has different requirements for soil type and moisture. It is better to restore or preserve an upland site or a wetland site than attempt to artificially recreate such a site where it did not exist.

The intricate interdependencies of living things dictate that restoration and conservation efforts be focused on the habitat and community levels. The north-facing and south-facing slopes of a hill in the eastern deciduous forest provide different habitats and thus different communities of plants and animals. These differences result from different amounts of direct sunlight that fall on the hillsides which, in turn, affect the moisture and temperature regimes in the soil. These two communities in the same ecosystem are different.

Since the 1800s, ecologists have attempted to define criteria that determine the numbers and kinds of organisms that live in any given ecosystem. In one woodland ecosystem in England, scientists have attempted to list all species of plants and animals. This effort has taken several years, does not include microorganisms, and is not complete (Hunter 1990). Because ecosystems are so complex and are continually changing, ecologists frequently study a particular group of organisms such as rodents, birds, trees, or

grasses, or an individual species. From these studies they attempt to develop general rules of ecosystem function or preservation.

Scientists measure and evaluate biological diversity in several ways. One concept of diversity is the number of different types or species of organisms within an ecosystem or over a defined parcel of landscape. In order to compare different sites and different studies, ecologists have developed ways to measure diversity. One of the simplest measurements of diversity is the number of species that dwell in a habitat. The number of species is referred to as species richness. Habitats with more species are considered more diverse. But simple numbers of species often do not tell the entire story of diversity. A forest with 90 percent of one tree species and 5 percent each of two other species is different from a community containing the same three tree species but at densities of 25, 35, and 40 percent. Each forest has the same richness with three species of trees, but the forest with the more even distribution is considered more diverse. The distribution of abundance among different species or the relative abundance of species is referred to as evenness. Forests with high evenness are considered more diverse than forests with low evenness. Ecologists have developed several different equations combining richness and evenness to measure diversity. In these equations, richness or number of species is a more important factor than evenness. Frequently, ecologists use only the number of species, or richness, when studying an area.

Diversity can be increased to very high levels. But diversity is not our only goal. We would like also to maintain or restore the natural communities that occurred in a region. Restoration of natural communities is suggested for two main reasons. First, by using plants that originally occurred in an area, the manager will also be more likely to attract and reestablish native animals within the area. Trees, for example, attract fewer bird species when planted in areas in which they are not native. Exotic pine trees in France (Constant et al. 1973) and Australia (Disney and Stokes 1976) contain fewer species of birds than are found in pine plantations in their native countries. Eucalyptus trees are not native to California, Sardinia, or Chile (Smith 1974) and plantations in these areas contain few bird species. Eucalyptus trees, however, support a diverse community of birds in their native Australia.

The implications of these studies are clear. Native populations of plants will maximize the diversity of native animal species. A second reason to restore natural communities is that current scientific methods do not provide the means to understand all relationships and individual species' requirements necessary to create a community. Diversity may be increased without understanding all the reasons for that increase. Restoration of a natural community should include as many of the components of the community as possible, to ensure that the requirements of individual species are met.

Principles of Diversity

Landscape restoration is based on an understanding of how ecosystems work. Studies of species' requirements, maintenance of natural areas and biological diversity, island biogeography, landscape ecology, and ecosystems have led to the

recognition of specific principles that can be applied to managed parcels of land to maintain and restore natural diversity.

Ecology is the study of how organisms interact with each other and with their environment. Ecologists are often satisfied with general explanations of why certain things happen. However, managers of golf courses, recreation areas, or parks want to predict the outcome of a particular action under specified conditions. Because of the complex relationships within an ecosystem, the predictive powers of ecology are limited. But some ecological principles are understood well enough to be applied, resulting in increased diversity. Application of these principles will increase the diversity on small tracts of land and at the same time increase the aesthetic value and preserve the functionality of the tract.

"The key to planning the management for all species of wildlife is to know the species' habitat requirements and provide a variety of habitat components in a desirable combination that will meet the needs of as many species as possible" (Schneegas 1975). To increase the diversity of native plants requires the evaluation of existing site conditions and the application of ecological and horticultural principles in using native plants. Most plants, being rooted in the soil, are totally dependent on the site conditions. For animals, we may analyze why certain species or combinations of species are found in certain places, and recreate the conditions specifically related to those species. Alternatively, we may recreate the native plant community and rely upon immigration of animal species (Figure. 3.1). Animals require food, shelter from predators and the weather, and breeding habitats to ensure the best chances of securing mates, nesting, and raising young. The challenge with managed areas is to restore or retain sufficient characteristics of the natural community to maintain a high diversity of plants and animals and at the same time allow human use of the area.

The principles outlined in this book are to some degree hierarchical and overlapping. For example, the theoretical best restoration of a disturbed site is a large, circular (maximize interior, minimize edge)

Figure. 3.1. Changes in density and number of species of breeding birds with advance of community development from abandoned agricultural fields to climax forest. Ecological age is in approximate years (after Johnston and Odum 1956).

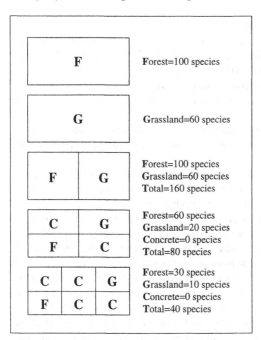

Forest=100 species

Grassland=60 species

Forest=100 species
Grassland=60 species
Total=160 species

Forest=60 species
Grassland=20 species
Concrete=0 species
Total=80 species

Forest=30 species
Grassland=10 species
Concrete=0 species
Total=40 species

Figure. 3.2. Theoretical effects of fragmentation on species on a parcel of land. Numbers of species are for illustrative purposes only. The total number of individuals within each species will also diminish as the degree of fragmentation

preserve that represents the original community found on that site. The minimum size will depend upon whether it is a forest or grassland community. If the site is to have a variety of uses, however, then perhaps the original community cannot be restored. An alternative is to restore a few smaller patches of different habitat communities, none of which may contain a full complement of species.

Some of the principles have a threshold of applicability. If a 500-acre forest exists, most species that normally inhabit a forest, including most of those that inhabit deep forest, will be found in this area. If half of that 500 acres were changed to native grasslands, both the forest species and grassland species would be found on that same 500 acres. If the forest acreage is reduced to below the size and shape that will support the entire complement of forest species, then a reduction in the number of species occurs as the parcel is further subdivided (Figure. 3.2). Knowing the limitations of when and how the principles are applied requires some understanding of natural communities. Discussion with

ecologists, botanists, and zoologists familiar with the local communities may aid in determining which principles to use in any given area.

Beginning in the late 1800s, scientists noted that islands contain fewer species than continental land areas of equal size. In 1967, MacArthur and Wilson formally developed the theory of island biogeography. Much of today's landscape is divided into island-like habitat patches that are decreasing in size, becoming more isolated, and disappearing. Habitat patches display some similarities with islands. Which species can reach a patch? How diverse is each patch? What is the relationship between patch size and its diversity? These are concepts that have analogous island concepts.

Scientists continue to debate specifics of the theory of island biogeography and other ecological concepts, but a few principles for explaining diversity have emerged: smaller islands contain fewer species; islands further from the mainland contain fewer species; boundaries (edges) between different vegetation types contain more species of both plants and animals; more extinctions of species occur in smaller areas; and reductions in the pool of species for immigration will reduce the number of species and the relative abundance of species in a community. Although many authors apply the theory of island biogeography to the design of natural reserves in order to maintain plant or animal communities, there is not agreement on all management strategies.

Golf courses, recreation areas, farms, schools, and parks may be managed to increase diversity by applying ecological principles. The following is a synopsis of ecological principles that represent current scientific thinking on how diversity is maintained, and suggestions for applying these principles to preserve, recreate, or restore natural areas. We have divided the principles into those relating to the spatial aspects of ecosystems on the landscape (Table 3.1) and those relating to the community or biological aspects of ecosystems (Table 3.2). These principles summarize concepts and ideas derived from scientific studies. A more detailed discussion of each spatial and community principle comprises the remainder of this chapter. The reader may elect to skip the detailed discussions of the principles that follow or read about the principles that are of particular interest.

Table 3.1. Spatial Principles and Guidelines

■ **Large areas of natural communities sustain more species than small areas**—Preserve as many large natural communities as possible in single tracts for each ecosystem or increase the size of existing patches to the minimum size needed to sustain viable wildlife populations.

■ **Many small patches of natural communities in an area will help sustain regional diversity**—Where there is no opportunity to preserve, increase, or create large natural community patches, increase the number of small community patches.

■ **The shape of a natural community patch is as important as the size**—Modify or design the shape of natural community patches to create more interior habitat. If space is limited, a circular area will maximize interior habitat.

■ **Fragmentation of habitats, communities, and ecosystems reduces diversity**—Avoid fragmentation of large patches of natural vegetation. Even a narrow access road through a forest can be a barrier to movement of small organisms, eliminate interior habitat, and introduce unwanted species.

■ **Isolated patches of natural communities sustain fewer species than closely associated patches**—Minimize the isolation of patches. Corridors and an increased number of patches can prevent isolation.

■ **Species diversity in patches of natural communities connected by corridors is greater than that of disconnected patches**—Maintain or develop many corridors of similar vegetation to connect isolated patches of the same or similar community types. Opportunities exist along roadways, rivers and streams, urban ravines, fencerows, hedgerows, railroad rights-of-way, to name a few. Wider corridors provide more wildlife benefits and protect water quality better than narrower ones. Breaks in the corridor should be avoided.

■ **A heterogeneous mosaic of natural community types sustains more species and is more likely to support rare species than a single homogeneous community**—On large parcels, mosaics of natural communities should be restored as the diversity of the landscape allows. Smaller parcels should be evaluated within a regional context with the goal of developing such mosaics on the landscape.

■ **Ecotones between natural communities are natural and support a variety of species from both communities and species specific to the ecotone**—Ecotones (transition zones between communities) should be allowed to naturally develop between adjacent communities. The amount of area in ecotones can be increased by increasing the interspersion of community types on a given parcel, but this should not be done at the expense of reducing interior habitat.

Table 3.2. Community Principles and Guidelines

■ **Full restoration of native plant communities sustains diverse wildlife populations**—The more fully restored natural community has a higher diversity. This means introducing as many components of the natural community as possible.

■ **An increase in the structural diversity of vegetation increases species diversity**—The vegetational structure of a community can be enhanced by restoring tree, shrub, and herbaceous layers that are reduced or lacking. Dead logs and litter should also be left.

■ **A high diversity of plant species assures a year-round food supply for the greatest diversity of wildlife**—Introduce as many species known to be part of the natural community as possible. Also retain dead, standing, and fallen trees as they provide important nesting sites for many cavity nesting species and a source of food for other species.

■ **Species survival depends on maintaining minimum population levels**—Different species will have different minimum population requirements. The minimum population in a particular parcel will depend upon factors listed above, such as how connected patches are.

■ **Low intensity land management sustains more species and costs less than high intensity management**—The maintenance costs and environmental impacts associated with landscape management can often be reduced by reducing management intensity. Management intensity can be reduced by converting areas to native vegetation adapted to site conditions. Natural forest, grassland, and wetland communities are low intensity landscapes.

Spatial Principles

The landscape is a mosaic of land uses, such as fields, wetlands, woods, and asphalt. Patches are areas of habitat different from the surrounding area (called a matrix). The study of natural patches provides insight into the ecological mechanisms controlling which species inhabit them. Remnants are isolated patches of persistent, regenerated, or restored native ecosystems. In many areas of the United States, previously disturbed patches may regenerate to the original ecosystem. This regeneration is the regrowth of a natural community through natural succession. The ability of a patch to regenerate will depend on its size, degree of isolation, and how severely it has been disturbed. Restoration can recreate natural ecosystems in areas where regeneration is not possible. Additionally, restoration can reestablish ecosystems more rapidly than regeneration. Principles of restoration and succession are discussed in Chapter 4.

Principle One—Large areas of natural communities sustain more species than small areas

Species richness within communities follows a consistent pattern of an increasing number of species with increasing size of the area. One method of showing the relationship between number of species and area size is the species-area curve. Scientists plot the number of species on one axis of a graph and the area on the other axis. The number of species in any given ecosystem will increase rapidly as the tract size increases up to a certain point, and then increases less rapidly (Dunn and Loehle, 1988). The species-area curve is different for different vegetation types and for their associated animals. A first estimation of the number of species expected in an area of a certain size can come from studies of species-area curves. To achieve maximum diversity for different types of organisms in different regions, areas of different minimum sizes are required, as illustrated by the species-area curves in Figure 3.3.

The optimum method to maintain maximum diversity is to preserve many large areas. The larger the area, the more species preserved. Very large areas ensure the preservation of rare species and large predators. Predators such as hawks and owls require larger patches for survival than do songbirds.

Different geographic regions exhibit different numbers of total species on the same size area. Climate is one of the most important factors affecting geographic variability in species richness. Benign climates have greater species richness than harsh climates (Figure. 3.4). Stable climates have a larger number of species than variable climates (Currie 1991). To protect all or most of the species representative of a particular community or habitat, an area of minimum size must be established. The minimum size area may be different in different ecosystems and the total number of species that can be expected will differ in different areas of the country.

To preserve all species in a forest, large tracts must be preserved. Forest interior species will not occur in small forests. Freemark and Merriam (1986) concluded that large tracts of forest are needed to provide habitat for forest interior birds. Bird species richness increases significantly through a forest island size of approximately sixty acres and is likely to continue increasing significantly at forest sizes greater than sixty acres. Forest interior bird species began appearing in two acre forests (Galli et al. 1976). The interior structure of grassland can be achieved in a smaller area than is required for the establishment of a forest interior. A smaller area can thus provide sufficient habitat for grassland interior birds.

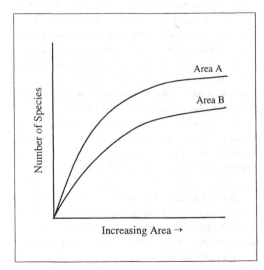

Figure. 3.3. Generalized species-area curve. The number of species increases as the area of a patch or island increases. The actual shape of the curve varies with the type of animal.

(a) Birds, West Coast

(b) Mammals, West Coast

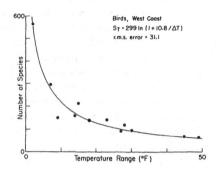

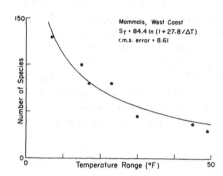

Figure. 3.4. Relationship between temperature fluctuations and species richness in west coast birds from Panama to Nome, Alaska, (a) and mammals from southern Mexico to Nome, Alaska (b). Areas with larger temperature ranges have fewer species. Point for Panama is taken from *Geographical Ecology* (MacArthur 1972, page 135). Mammal data are from Simpson 1964 (reprinted with permission from MacArthur 1975).

Principle Two—Many small patches of natural communities in an area will help sustain regional diversity

The scientific literature contains many articles debating whether a single large or several small preserves of equivalent total area will contain more species. No clear, general answer on either theoretical or empirical grounds has emerged. Even though it is desirable to preserve large areas, fragmentation of natural communities and the mounting pressures from expanding human populations preclude the establishment of large tracts of natural communities in many areas. When preservation of large areas is not possible, the next best option is to preserve many small areas.

The number of species in any area is determined chiefly by the number of different habitats (Forman and Godron 1986). Increasing the size of an area increases subtle habitat differences. Differences in soil, slope, moisture, and associated species affect the composition of the community. Even if sites are randomly chosen, rather than selected to maximize diversity, it is likely that groups of separate sites will encompass various habitats. Each small area will, because of the subtle habitat differences, contain different species. Any single small area may not contain as many species as a large tract, but the total number of species from all smaller tracts may equal or exceed the number of species in a single large tract. When making comparisons, only native species should be counted. Small reserves may contain a higher percentage of exotic species.

By maintaining several similar patches, the likelihood of all individuals of a species being destroyed is reduced. Random events such as epidemics, parasitic infections, or fires may exterminate a species in a patch. If one large patch containing all the individuals of the species is lost, the species will become extinct. But if several small patches contain the species, it is unlikely that an epidemic or fire will spread to all patches, so some individuals will survive. The last population of Heath Hens was found on an island. A large fire and an epidemic among the hens occurred. These combined natural disasters exterminated this last population (Hunter 1990). Had there been populations on other islands or tracts of mainland, some probably would have survived and could have functioned as a source of immigrants for reestablishing the population on the devastated island.

At any moment several small patches may contain different species due to local extinctions. If the region as a whole is made up of many different patches, then each of the small reserves may support a different group of species and the total number of species conserved might exceed that in a large reserve containing a single or few habitat types. Additionally, the different patches serve as a source of species to emigrate to new patches. Studies summarized by Simberloff and Abele (1982) indicate that for plants in England, birds and mammals on mountaintops in the American Great Basin, and lizard species in Australia, several small reserves contain more species than one large reserve. They theorize that the increase is due to subtle habitat variations. Simberloff and Abele (1984) conclude that the ideal strategy is to have many large reserves, but that smaller reserves are useful. For low density species such as predators, however, large reserves are clearly essential.

Blake and Karr (1984) presented data showing that two smaller forest tracts are more likely to have a greater number of bird species but a single large tract will retain more long distance migrants and forest interior species that are dependent on large forest tracts. A single small patch will not preserve as many species as a larger patch. To maximize species diversity, several small patches are needed.

Principle Three—The shape of a natural community patch is as important as the size

Patches of natural habitat occur in many shapes. Depending on the patch, the shape may affect its utility in preserving and creating diversity. Long, thin reserves are optimal for catching immigrants, and functioning as corridors to help direct emigrants to suitable larger reserves; but they do not allow development of distinct interior habitats. A circle is the shape with the least edge per unit area. Use a circular area if space is limited and interior habitat is desired.

Most natural areas are surrounded by human activity that can affect the natural area. Human activities produce disturbances that reduce the number of species. The further the natural area is from human activities, the less likely it is that the system will experience a reduction in the number of species. The edge of a natural area is impacted by surrounding human activities. Oblong or irregularly shaped patches contain a larger portion of edge than round patches or squares with the same total area (Figure. 3.5). Thus round areas have a larger area that experiences less disturbance and will be more likely to have a higher species richness.

When habitat with a distinct interior is to be preserved or created, the optimal shape to use is a circle. If paths are to be created in a natural area, they should be placed as near the edge of the patch as possible and still allow people using the path to experience the natural communities. Placing paths close to the edge of the natural area leaves a larger central area with less disturbance.

Principle Four—Fragmentation of habitats, communities, and ecosystems reduces diversity

Fragmentation can be viewed in several ways. Morrison, et al. (1992) define it as the increase in isolation and decrease in size of resource patches. For the purposes of this book we define fragmentation as the process of altering the landscape, eventually leading to the creation of isolated remnants of natural communities that had once covered the entire landscape. Fragmentation ranges from put-

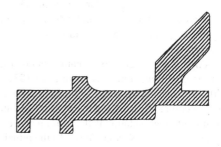

Total area: 96.4 acres
Core area: 0 acres
Species sensitive to fragmentation: 0/16

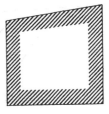

Total area: 116.1 acres
Core area: 49.4 acres
Species sensitive to fragmentation: 6/16

Figure. 3.5. The shape of the patch dictates the amount of edge and interior (Temple 1986).

ting a road through a forest to the elimination of a forest. For the purposes of emphasis and explanation we have specifically chosen to separate the closely related concepts of isolation, corridor development, and size, shape, and number of patches from the overall concept of fragmentation.

Much research has focused on fragments and their relationship to islands, and the application of island biogeographic theory to fragments. Porosity—the measure of density of patches—decreases as fragmentation increases. This means the distance from one patch to another increases. Reduction of species richness within fragments is a generally accepted phenomenon. Agricultural or urban landscapes that surround native habitat patches are radically but not totally different from the native habitat. Some of the native plants and animals still exist in the agricultural or urban landscape, yet the diversity is greatly decreased and they exist isolated from a natural community.

Long-term fragmentation of natural landscapes has been documented in Wisconsin from 1831 to 1950 and in England between 1759 and 1978 (Shafer 1990). A commonly accepted theory is that habitat loss through fragmentation is the leading cause of species extinctions (Norton 1986). As the remaining patches of natural habitat become smaller and smaller, the diversity of animals and plants decreases (Figure. 3.6). Fragmentation of natural communities is occurring on every continent. In some parts of the world, few fragments of natural areas remain. The pattern of decreasing size and increasing modification of fragments is similar on every continent except Antarctica.

Ecological communities and habitats are lost as a result of fragmentation. Fragmentation may result in the complete loss of some habitats and their associated species. The small patches that result from fragmentation may not contain a minimum viable population of a species (i.e., the number of individuals of a given species sufficient to sustain the population). If individuals can move from patch to patch and breed with individuals from other patches, genetic diversity, which is necessary for the long-term survival of a species is maintained. If individuals cannot move among patches, the population may be reduced below the minimum size needed to preserve the population or the genetic diversity may be lowered to a harmful level. As fragmentation progresses, the patches of original habitat become further apart. The increasing distance between similar patches decreases the likelihood that species can migrate among patches. As species are exterminated within a given patch, new individuals cannot immigrate and diversity is decreased.

Rare species and species with patchy distributions are more susceptible to extinction as a result of habitat loss through fragmentation than are common species with an even distribution. As populations in habitat patches are exterminated, the likelihood of extermination of other populations dependent on that species increases.

Effects of fragmentation can be reduced by maintaining or establishing corridors to connect patches. From earlier discussions of eastern woodlots, we have seen that the effects of fragmentation can be reversed as patches of woods are reestablished in farmlands. If large tracts of natural communities exist, every effort should be made to prevent fragmentation. Paths, service corridors, rights-of-way,

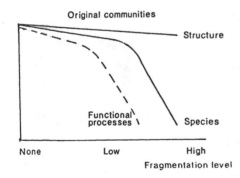

Figure. 3.6. Predicted changes in functional process (e.g., predation, pollination, herbivore), vegetative structure (e.g., stem density) and species numbers (Hansson and Angelstam 1991).

and roads should be placed on the margins of natural areas. Buildings and areas used by humans should be clustered near the edges of natural communities.

Principle Five—Isolated patches of natural communities sustain fewer species than closely associated patches

Insularization is the process by which fragments of the original ecosystem become more and more isolated from one another thus becoming more like islands. Insularization can decrease or even stop colonization of patches from outside areas. The patches become too far apart to allow individuals to move from one patch to another. Random events such as epidemics may cause the death of all individuals in a patch. Without the possibility of immigration, the diversity is thus permanently reduced. Insularization can remove resources that species in the reserve depend on for survival (Wilcox 1980). As the patches become so small that species requirements are not met, extinction occurs. Extinction of a species in turn affects other species and may result in a further decrease in species richness (Balser et al. 1981).

In the book *Wildlife—Habitat Relationships* (Morrison et al. 1992), the authors state that "over time, individuals in local patches or groups of patches might become extinct from chance variations in survivorship and recruitment or from catastrophic or systematic declines in the resource base." Figure 3.7 (a-f) is taken from Morrison as an example of this point.

Patch size and degree of isolation affect both the species composition of bird communities (Butcher et al. 1981, Whitcomb et al. 1981) and the local abundance of birds (Lynch and Whigham 1984). Patch size, acreage of nearby forests, and the distance to extensive forest tracts affect the number of bird species (Opdam et al. 1985). In Maryland, forest fragments of at least three acres will permit most forest interior bird species to breed if the tract is near enough to a larger forest fragment to allow recruitment of individuals (Whitcomb 1977, Whitcomb et al. 1977). Although patches this small might not ensure survival of forest interior bird species over long periods of time (Whitcomb et al 1981), they do contribute to genetic diversity, reduce the effects of random events, and allow more

individuals to produce young each year.

Morrison et al. (1992) states that "connectivity of patches and permeability of edges vary according to a species' body size, habitat specificity, and area of home range. What acts as habitat entirely suitable in type and amount for sustaining a small-bodied, habitat-specific species such as the red tree vole *(Arborimus longicaudus)*, a specialist on Douglas-fir, also might act, at a much wider scale, merely as a dispersal steppingstone for a larger-bodied, less habitat-specific species such as the mountain lion *(Felis concolor)*."

Columbian ground squirrels may move between habitat patches, but they do not colonize some new patches because emigrants only settle near other squirrels of the same species rather than in all vacant patches. Movement into a new patch is related to distance from a source of squirrels, but not to the size of the patch (Weddell 1991). Thus, for some mammal species, location of patches near colonizing sources is important for establishing new populations. Isolation will reduce colonization in these species.

If large tracts of land must be subdivided or fragmented, the resulting smaller areas should be kept close together or connected by corridors to prevent isolation.

Principle Six—Species diversity in patches of natural communities connected by corridors is greater than that of disconnected patches

Corridors are narrow strips of habitat that allow movement between patches of similar habitat (Figure. 3.8). The concept of corridors resulted from studies of species on islands, peninsulas, and mainland fragments. Mainland fragments typically contain more species than islands of the same size. Yet peninsulas or islands that are part of an island group, with some islands near the mainland, have more species than isolated islands. As species become extinct in fragments, corridors allow new species to move into the fragments. Since random events commonly cause local extinctions, corridors are important in maintenance of diversity.

Even small corridors may be effective in providing a path for movement of small mammals, amphibians, and reptiles that travel on the ground and have limited mobility. Some exterminations are inevitable in small reserves or islands. Coloni-

Figure. 3.7 (a–f). Schematic representation of landscape dynamics of patch colonization and occupancy (Morrison et al. 1992).

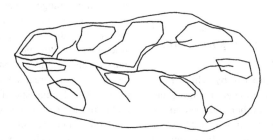

3.7a. Occurrence of ten patches of old-growth forest within a watershed.

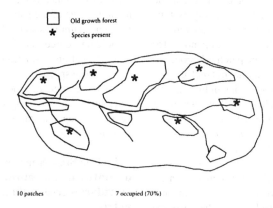

3.7b. Occurrence of old-growth obligate vetebrate in seven of the ten patches.

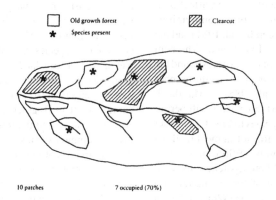

3.7c. Selection of three patches for clear-cut timber harvesting.

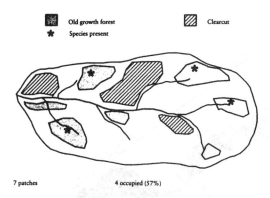

3.7d. Immediate result of harvest disturbance: loss of the species in three patches.

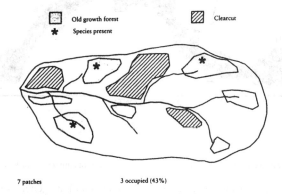

3.7e. Later loss of the species in a distant, isolated, smaller forest patch (faunal relaxation).

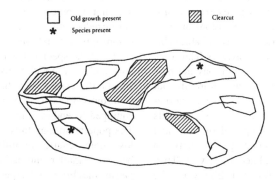

3.7f. Still later loss of the species in a larger forest patch recently isolated.

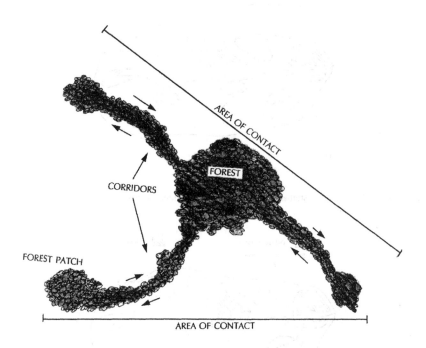

Figure. 3.8. Corridors facilitate movement between patches and broaden the area of contact for species moving across a landscape.

zation, the movement of a species into an area not previously inhabited, may enable a small patch to maintain its diversity in spite of exterminations. The maintenance of species diversity is highly dependent on the ability of individuals to recolonize habitat patches after local exterminations. Corridors increase the chances for colonization.

Steppingstones, or closely spaced suitable habitats, may function similarly to corridors. Thus very small patches of natural vegetation may not provide a habitat suitable for a species entire life, but they may serve as important steppingstones allowing emigration into suitable patches. Shelterbelts of the Great Plains are composed of trees and shrubs and may serve as steppingstones between habitats containing trees adjacent to rivers. They are especially effective for such highly mobile species as birds.

In addition to providing a means to reestablish populations that have been exterminated in a patch,

corridors facilitate genetic exchange among small populations. Such exchange is important in maintaining the viability of small populations. The literature includes many articles discussing minimum population size necessary to ensure that a species does not become extinct. Many events can cause drastic reductions in population size. Reduced populations typically exhibit a loss in the genetic variability of the species. Reduced genetic variability may endanger populations because the reduced genetic diversity lowers the chances that at least some individuals in a population will be genetically equipped to endure changes in the environment. The population may thus be reduced further, and lose more genetic variability. As the population becomes smaller, inbreeding occurs, further impairing the ability of the species to survive. Inbreeding of closely related individuals increases the probability that harmful genetic traits will be expressed in their offspring. Movement of individuals among patches reduces inbreeding. Many animals disperse long distances from their parents before they themselves begin breeding. Corridors increase movement of adults among patches and dispersal of young. Both increase genetic diversity.

Abandoned ravines in urban settings, riparian (river edge) areas, utility rights-of-way, roadsides, railroad rights-of-way, hedgerows, and fence lines may function as corridors. For some species it is not important that corridors be of the same vegetation type as the natural areas they connect, but only that they be different from the surrounding croplands or urban areas (Merriam 1984). In designing corridors, connections to existing corridors as well as patches should be incorporated into the design of a site. In urban areas habitat islands are often small, isolated from natural habitats, experience much disturbance, and thus experience high extinction rates (Davis and Glick 1978). It is important to create corridors that connect with existing corridors and patches to insure their maximum effectiveness.

In arid regions, areas along rivers (riparian habitats) frequently contain trees not found in adjacent grasslands or deserts. Riparian woodlands maintained along streams in arid regions function as corridors for movement. However, as pointed out by Simberloff et al. (1992) some communities, like riparian communities, are very threatened and should be protected independent of their value as corridors.

Forested riparian corridors are also important in agricultural and urban settings. For example, in the cultivated landscape of Kentucky's Bluegrass region several species of salamanders and mammals occur only along the "mainland" wooded corridor of the Kentucky River. Bryan (1991) found smoky shrews (*Sorex fumeus*) only in the deep forest litter of the wooded ravines of the river valley.

Forested corridors are often important resting and feeding areas for migrating birds (Sprunt 1975). They help maintain suitable water temperatures for aquatic life, improve water quality, and preserve stream integrity. Forested areas along streams should be preserved during development of golf courses and other managed areas. To protect water quality, forest filter strips bordering streams in flat terrains should be at least twenty-five feet wide and increase two feet in width for each one percent of slope (Trimble 1959). In municipal settings the filter strip should be doubled in width (Trimble and

Sartz 1957). Filter strips for wildlife may need to be even wider. Stauffer and Best (1980) studied riparian communities of breeding birds in the Central Plains and found that for wooded study plots to contain the maximum diversity of twenty species, a minimum width of 660 feet was needed. A strip 297 feet wide had only thirteen of the twenty species and strips less than twenty-three feet wide contained seven of the twenty.

Two characteristics of corridors—breaks and nodes—have significant implications in increasing and maintaining diversity (Figure. 3.9). Breaks in corridors are long or short areas in which the corridor community is interrupted by the surrounding habitat. Corridors of trees and shrubs around agricultural fields are often interrupted in areas where gates or connections between fields occur. Animal movement along corridors may be reduced or eliminated by such breaks (Forman and Godron 1986). If breaks are necessary, they should be as small as possible. Corridor breaks as small as 50–100 meters may be significant to birds, bats, and other small mammals. Exposure to predation may be a primary reason for avoidance of such openings.

Nodes are areas in which the corridors are wider. Nodes often occur at the junction of two corridors or, in the case of riparian corridors, in the inside of river bends. Increases in species diversity are documented for nodes along fence rows although the nodes show relatively small increases in width compared to the corridor (Forman and Godron 1986).

Opportunities for corridors along roads are often excellent. A slightly wider right-of-way to accommodate native grasses, shrubs, and trees creates an effective corridor. Areas around buildings can serve as screening as well as corridors.

Principle Seven—A heterogenous mosaic of natural community types sustains more species and is more likely to support rare species than a single homogenous community

Sampson and Knopf (1982), in discussing forest management, stressed that managers should strive for a diversity of ecological communities not maximum diversity within one community. Maximum diversity means conserving the greatest number of species possible on one tract of land. Preserving a diversity of ecological communities means preserving many different habitat types on many

communities, we saw this natural mosaic effect. Several distinct communities and habitats may occur in an area of hills with north-facing slopes, south-facing slopes, and wet areas such as springs, seeps, and streams at the bases of the slopes. Disturbed areas and old fields form other patches. Each species has very specific requirements that are satisfied in a particular habitat. The goal should be to preserve different habitats rather than attempt to create a habitat that does not naturally occur in an area. Restoration of an existing south-facing slope community would be easier than attempting to create a north-facing slope community in an area that does not possess the appropriate exposure to sunlight, soils, or hydrology. Restoration of the community that originally occurred in an area will more likely be successful, will require less management, and will exhibit more long-term viability. By analyzing an area and restoring the natural communities as much as possible, we will restore the natural mosaic of communities.

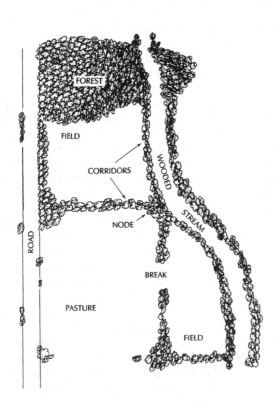

Figure. 3.9. The trees and shrubs along the fence rows, streams, and road connect two forest patches. Nodes and breaks are labeled.

Principle Eight—Ecotones between natural communities are natural and support a variety of species from both communities and species specific to the ecotone

Ecotones are the transition zones or boundaries between two habitats, community types, or ecological systems. The boundary may be between a hay field and a forest or between a natural grassland and a forest (see Figure. 3.10). Typically the edges of the two ecosystems are not distinct lines, but the transition from one ecosystem to another occurs over some area. This area of transition is called an ecotone. An ecotone may be as abrupt as the edge between a lake and the forest on its shore. Larger ecotones exist in the transition from forest type to

different tracts of land. Each tract will contain slightly different environmental factors and thus slightly different communities of plants and animals. Each individual tract may not represent its highest potential diversity, yet a diverse array of species associated with each tract will be preserved. Preserving many tracts of land in many different locations takes advantage of slightly different habitats at each location. Specialized habitats that occur on only a few locations are thus more likely to be preserved and rare species that are found in only these habitats will survive.

In nature, many communities form a natural mosaic of small patches over the landscape. Using the earlier example of north- and south-facing slope

forest type as latitude or altitude increases. More species exist in ecotones than in the individual adjacent communities (Forman and Godron 1986). Certain species from each of the two bordering ecosystems, as well as species that exist only in edges, are found in the ecotone, and thus a greater number of species occur in the ecotone than in either ecosystem alone. However, interior species from the two ecosystems are not found in ecotones. Ecotones are transition zones and not all species can live in transition zones.

Ecotones consist of a mosaic of plants from the two adjacent ecosystems, as well as obligate ecotone species.

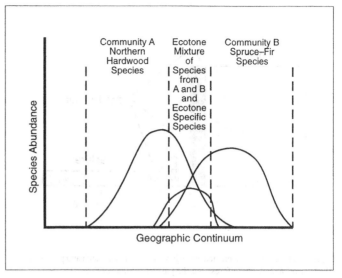

Figure. 3.10. The ecotone between two natural forest communities will include species from both and species that occur only in the ecotone.

Freemark and Merriam (1986) found that the mosaic of different habitats in ecotones is important in the maintenance of populations of migratory birds. Migratory birds that spend their winters in the tropics form the majority of species in the ecotone between the boreal and eastern deciduous forests (Clark et al. 1983). Creating a mosaic of habitats generates ecotones that increase species diversity.

The edge effect or the increase in species richness in ecotones (Figure. 3.11) is important in design consideration but cannot be separated from the area effect. If only a small piece of land is available, a forest interior cannot be created and species limited to forest interior cannot be established. But trees, shrubs, and herbs can be established that will increase the number of species that are not dependent upon interior forests. For eastern deciduous forests, sixty-two acres contain 75 percent of bird species and represent the point on the species-area curve at which rapid increases in species ceases (Tilghman 1987).

A circular reserve design provides the highest probability of maintaining or establishing an interior forest and reduces the edge effect to a minimum. Shafer (1990) suggests that shape is not important as long as reserves are not very thin (e.g., 600 feet), in which case edge effect becomes overwhelming. In natural settings ecotones increase diversity. Fragmentation, over most of the world, has reduced natural communities to small isolated patches. Edges are abundant between forest and fields because of agriculture. The challenge is to protect, restore, or recreate tracts that contain enough area to provide interior habitats and to allow the natural ecotones between communities to develop over larger areas of landscape. Natural ecotones such as where a wetland meets a lowland forest or where an upland forest joins wooded, streamside bottomlands are more desirable than human-created edge.

Community Principles

Principle One—Full restoration of native plant communities sustains diverse wildlife populations

The physical conditions of a particular site determine what community of plants will survive for the long term without special care. When confronted with a previously farmed or otherwise disturbed area, it may be difficult to determine the vegetation history of the site. The soil series on a site is one of the best clues to the original vegetation. The techniques described in Chapter 5 can be used to analyze the site. Restoring the fullest possible complement of those plant species that would most likely have originally occupied an area, while retaining the characteristics of the site needed to

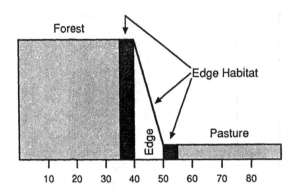

Figure. 3.11. Edge habitat created at the junction of a forest patch and pasture. Edge effects develop in both the forest and the pasture (Luken 1990).

maintain its human uses, results in an increase in native animals (Figure. 3.12).

Principle Two—An increase in the structural diversity of vegetation increases species diversity

Physiognomy, or vegetation structure, is important since it affects the variety of habitats available for animals and plants. A forest with a shrub understory has more structure and more habitat types than one without a shrub layer. Most species are habitat specific. Some ecosystems naturally have a lower diversity of habitats and species than others. Birds that inhabit grasslands are usually quite characteristic of, and restricted to, the grassland vegetation type. Both diversity and density of bird species are low in grasslands compared to other habitats (Cody 1985a).

When migratory birds return to the forests of the eastern United States to breed, they frequently return to the same area year after year. They need food sources, perches from which they sing to attract mates, and nest sites, all of which are affected by vegetation structure. Components of the vegetation that correlate with song posts and nest sites include canopy tree volume, tree cover, tree height and size, number of trees, ground cover, and number and type of shrubs (Clark et al. 1983). This and similar studies in other forested communities suggest that the natural association of trees, shrubs, and ground cover is needed to attract and maintain

a diverse bird population in eastern forests. Manicured forests with a diminished shrub layer and reduced ground cover and litter will attract fewer birds and provide less protective cover for small mammals, reptiles, insects, and amphibians. One study found that the development of a good ground cover resulted in the addition of one or two bird species; a shrub layer added one to four species. A tree layer added approximately twelve species, with the addition of three more species as the trees developed (Willson 1974).

In the Northern Great Plains, the reduced number of sparrows on unreclaimed mine spoils was shown to be related to the loss of sagebrush habitat. Sparrows used the sagebrush as singing posts, for nesting, and for perches from which to feed on grass seeds. A second factor associated with the abundance of sparrows was the presence of litter (Schaid et al. 1983). Litter should be left in place in both forest and shrub communities. Preserving or establishing shrubs along roads, as screening around maintenance buildings, and between fairways can increase the populations of sparrows in mixed grass prairie regions.

In shelterbelts in the Great Plains, multiple rows of trees and shrubs provide more structure than single rows. They provide more winter cover and, therefore, more bird species than single row shelterbelts (Schroeder 1986). Multirow shelterbelts also provide cover for deer and other mammals. The most effective configuration of plants in shelterbelts is tall trees in the middle rows and lower shrubs in the outer rows of the belt. Maximum species richness occurred in shelterbelts of eight or more rows and ninety or more feet in width. In the Central Plains of Iowa, Stauffer and Best (1980) showed that vertical stratification of vegetation, sapling and tree species richness, and sapling and tree size all correlated positively with bird species richness. In this same study it was shown that removal of the shrub and sapling layer adversely affected eleven species and benefited four species. Maximum numbers of bird species occur when three distance layers exist: herbaceous, shrub, and trees over twenty-five feet tall (MacArthur and MacArthur 1961).

Although small mammal populations increase in response to seed production in arid areas and after forest clearing, vegetation cover is the single most important factor affecting small mammal species richness in fence rows of the eastern United States (Asher and Thomas 1985). Structure is also critical to the small mammal community. The biomass and diversity of mice, voles, and shrews are related to the lower strata of vegetation and the depth and friability of the litter and soil humus layers (Bryan pers. com.).

Vegetation structure is increased by establishing several layers of vegetation including herbs, shrubs, understory trees, and canopy trees. Litter and dead logs also increase structure and should not be removed. Horizontal logs retain leaves and build litter layers which enhance habitat for small mammals, amphibians, and reptiles.

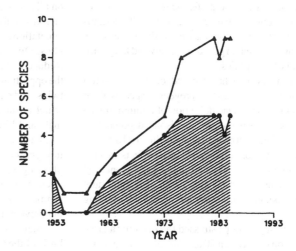

Figure. 3.12. This figure shows the return of interior edge birds (unshaded) and forest interior birds (shaded) in an old field that was allowed to return to natural forest (Askins and Philbrick 1987).

Principle Three—A high diversity of plant species assures a year-round food supply for the greatest diversity of wildlife

Seeds are energy rich and are frequently used by animals as food. Management of areas should include a system to allow seed production. This is especially true for arid areas and grasslands.

Grassland environments exhibit large, unpredictable changes from year to year (Collins and Glenn 1991). A high degree of annual fluctuation in temperature and precipitation occurs with no apparent trend. The amount and timing of precipitation determine the production of seed crops upon which grassland birds depend. Thus, it is not surprising that grassland birds are opportunistic. Unlike birds of the eastern forests, they do not return to the same site each year for nesting, but instead seek the most suitable site for reproduction each year. Establishment of natural vegetation will provide song posts, food supplies, cover, and nesting materials and sites. Since the essentials for survival and reproduction exist at these

sites, they will be selected by birds. Grassland birds often feed on seeds and insects. Management of grassland should include allowing grasses to produce seeds. The grass stems should be left standing to provide perches from which the birds hunt for insects.

Increases in small mammal populations are often related to increased supplies of food such as seeds. In arid regions, small mammal population increases are correlated with increased seed production. Forest clearings or disturbed areas exhibit an increase in deer mice resulting from increased supplies of seeds and insects (Ahlgren 1966). Butterflies, especially their caterpillars, can have very specific food requirements. This requires the planting of particular plant species if those butterflies are desired or considered part of the particular community being restored. Berries, nuts, nectar, leaves, and twigs may all be used as food sources. Some species feed on different food sources during different times of the year. A diversity of plants assures a year-round supply of food for different species.

Principle Four—Species survival depends upon maintaining minimum population levels

The size of a reserve affects the number of individuals of any species that can live in that reserve. Every individual of a population requires a certain amount of space. The amount of space is dependent upon food sources and preferences, species interactions, and reproductive needs. Additionally, a minimum number of individuals of a given species is necessary to maintain genetic diversity and long-term survival of the species.

Much research has focused on the minimum population size needed to prevent a species from becoming extinct. One definition of a minimum viable population is the smallest isolated population having a 99 percent chance of persisting for 100 years in spite of random effects. Random effects include demographic, environmental, and genetic events and natural catastrophes (Shafer 1990). The minimum viable population size thus obviously varies among species.

Early suggestions stressed the need for a large population of between 50 and 500 animals to maintain sufficient within-population variation to prevent negative genetic effects. However, since not all individuals in a population are part of the mating system, more than 50 to 500 individuals may be needed (Boecklen and Simberloff 1986). Recent papers have focused on the subdivision of populations as a means of maintaining genetic diversity rather than trying to determine the minimum population size. This recalls the idea of many small reserves linked by corridors to allow for migration among populations.

In nature, plants frequently exist in small populations. Higher plants, being rooted to one spot, are often highly site-specific. The most favorable sites are a few environmentally heterogeneous reserves of sufficient size to minimize edge effects. Some plants have the ability to propagate asexually. The majority of higher plants are bisexual. Asexual reproduction and bisexuality implies that minimum population sizes for maintenance of genetic diversity can be half those of populations requiring two individuals to reproduce (Ashton 1988). For these reasons plants often need less space than animals to support a viable population.

Principle Five—Low intensity land management sustains more species and costs less than high intensity management

Managed areas in the United States receive a range of treatments, but are often kept neat and tidy. Leaves, dead limbs, fallen trees and limbs, and debris are immediately removed. Grass is mowed on all accessible spots. The resulting communities lack diversity of habitats, do not exhibit complex vegetation structure, and contain reduced food supplies. Diversity of plants and animals is reduced. Reduced management and maintenance provide the opportunity for increased diversity. Dead leaves, twigs, and other debris are a food source for insects which are an important food source for small mammals and birds. Dead limbs provide nest cavities and harbor insects as food supplies for birds. Mature grass provides seed supplies and a more complex vegetation structure. Decreased management intensity increases complex structure of the environment and food supplies, and costs less than intensive management.

Stauffer and Best (1980), in a study of breeding birds of the Central Plains, found that snags (dead, standing trees) were more important than live trees or dead limbs as nesting sites for the ten bird species requiring cavities for nests. Snags or fallen logs also provide food and shelter for many bird species. Mammals, amphibians, and reptiles may also use cavities in snags and logs for shelter. In addition to providing food and shelter, dead logs and snags function in nutrient and energy recycling. Larger species such as the pileated woodpecker can use only trees of a certain minimum size, but small species can use large or small snags. Hunter (1990) suggests that a minimum size of snag to provide for larger species is a thirty foot tall snag with a diameter of twenty inches. Additionally, snags should be distributed throughout the forest since birds establish territories and cannot be crowded into small, localized areas. Explicit equations have been developed to estimate numbers of snags needed to support the maximum number of some woodpeckers. But the autecology of all cavity nesting woodpeckers, much less all species, is too poorly understood to allow the determination of precise numbers of snags needed. Foresters and wildlife managers have generally recommended two to four snags per acre. For North American species, this is supported by some empirical data (Hunter 1990).

Conclusion

Forests in Massachusetts represent a transition zone between the hardwood forests of northern New England and the oak–hickory forests of southern New England. Studies of thirty-two woodlands ranging in size from 2 to 170 acres showed that the size of the woodland was the primary influence on diversity. Forests with manicured understories were not included in the study. The number of species of birds rapidly increased as the size of the woodlots increased from two to sixty-two acres, but increased more slowly thereafter. At sixty-two acres, 75 percent of the bird species were present. Management conclusions from this study include the following (Tilghman 1987):

- establish or maintain woodlands greater than sixty-two acres;
- maintain natural vegetation in the shrub layer;
- include a variety of microhabitats such as small scattered openings;
- create some form of water in, or adjacent to, the woods;
- include patches of conifers or wetland;
- eliminate buildings adjacent to the woods; and,
- limit the number of trails.

Similar recommendations exist for prairie communities (Verner 1975):

- preserve as large an area as possible;
- provide a heterogeneous mosaic of grasses and forbs rather than a uniform stand of either;

- provide cattails of intermediate stem density, rather than bulrushes, for nesting;
- increase the structural complexity of the vegetation by increasing the vertical profile and increasing the percent vegetation cover and total volume of vegetation;
- make every effort to avoid single species plant communities or at least to reduce monocultures to small, intermingled patches;
- provide song posts;
- provide nest cavities; and,
- allow enough vegetation to provide cover for nests.

The studies cited above support the principles outlined in this chapter and the idea that restored or existing natural areas should be allowed to remain in their natural state. Grasses should not be mowed. Limbs, dead trees, litter, and debris should be left in place to produce a varied array of habitats.

The principles in this chapter will not all be used on any one site. However, for each site the use of even one or two of the principles will increase diversity. During development of a site, the creative incorporation of these principles, when appropriate, will increase diversity.

CHAPTER 4

Principles and Practices of Natural Landscaping

Introduction

Natural landscaping as used in this book is defined as the art of capturing the character and spirit of nature in a designed landscape. The objective of natural landscaping is to restore the natural beauty of the landscape by utilizing native plants in a community context. Unlike ecological restoration, which attempts the replication of an ecological community with a full complement of species, natural landscaping uses nature as a model for landscape designs. Once the local site conditions are analyzed and appropriate ecological communities are chosen as models, the landscaper is free to choose any number of species which characterize those models. Species selection often depends as much on the landscaper's personal vision of natural beauty as on other factors.

The term "natural" landscape is commonly used when a more accurate term might be "naturalized" (or "naturalistic") landscape. The designed landscape may resemble a natural one, but it requires planning, site preparation, installation, and maintenance. Natural landscaping should be viewed as a long-term process that ultimately results in a self-sustaining landscape, but you are not trying to recreate the complexities and balance of ecological systems.

The planning unit size often dictates whether ecological restoration or natural landscaping is more appropriate to achieve planning objectives. Both forms of naturalizing use the ecological community as the basis for design. Well-developed ecological communities are highly expressive of the land and its history, and have achieved a balance with the natural forces that shape the landscape (Diekelmann and Schuster 1982). They are not only aesthetically appealing, but are relatively stable and self-perpetuating. The species composition of each community is indicative of the local light, moisture, soil, and other environmental factors. The plants that characterize a particular community are compatible, both in terms of their growing requirements, and their ability to compete with each other for available resources. Such communities ideally serve as models for plantings, as well as indicators of local site conditions. As is true in ecological restoration, care should be taken to model the planting after a community having minimal past disturbance from logging, grazing, fire, construction, or other kinds of development. State natural heritage agencies, native plant societies, university and local botanists, and regional and local offices of The Nature Conservancy can often identify high quality ecological communities to serve as local models.

One approach to designing a naturalized landscape is to first select a desirable ecological community model and then attempt to create site conditions favorable to developing that community. While this method may sometimes work, it can require considerable effort and expense with no guarantee of success. A more effective approach is to determine the existing site conditions and select one or more natural communities adapted to these conditions as landscaping models. This requires a thorough analysis of site conditions and a working knowledge of locally adapted ecological communities. The ecological community descriptions in Appendix A and the species matrices in Appendix C are designed to guide the landscaper in selecting the community models and native species appropriate to the planting site.

The species matrices present information on the aesthetic attributes, wildlife values, environmental tolerances, and suggested landscape uses for plant species listed for each ecological community. This information is particularly helpful to the landscaper for selecting species compatible with design criteria. The list of nursery sources in Appendix D can be used to locate the nearest possible sources of nursery-propagated plants and seeds of selected species.

Landscaping With Native Species

Intrinsic to natural landscaping is the use of native or indigenous species of plants adapted to local site conditions. The term "native" refers to a species' place or region of origin. A native species is one that occurs in a particular region as a result of natural forces without known or suspected human cause or influence. The natural range of a species, in which it is considered native, is the area in which it grows naturally without human intervention. Natural range can vary considerably from species to species. For example, white oak is native to a large area of the eastern United States, while willow oak is native to only the southeastern United States. Both of these oaks, if planted outside their natural range, are considered nonnative or exotic. According to Dr. David Northington, director of the National Wildflower Research Center, a native plant is one that was here, growing in the wild, before European settlement began 300 years ago. Native plants are part of the natural history of a region. They form naturally diverse communities and are well adapted to the climate, soils, and other biotic and abiotic components of their region of origin. An indigenous plant is one native to a particular range. A plant is exotic if it is growing out of its natural range. A naturalized plant is an exotic that has been introduced to an area and is able to perpetuate on its own. Queen Anne's lace is an example, as is, unfortunately, kudzu. Botanists strongly discourage the use of exotics for fear that successful introductions will displace natives.

When incorporating native plants into a designed landscape, it is preferable to use material from a source as close to the planting site as possible to assure it is hardy and adaptable to local growing conditions. While some ecologists believe that source materials obtained from outside a 50- or 100-mile radius of the planting site could contaminate the local gene pool, most would agree that use of such material is preferable to the use of cultivars or nonnative species for restoration projects. Compelling reasons for using native plants in designed landscapes are the desire to have nature's beauty close at hand, conserve natural resources, reduce maintenance costs, preserve biological diversity, and prevent species extinction.

Most state roadside wildflower programs are beginning to rely more on using wildflowers native to the planting region than they have in the past. The sentiments of Mike Creel, a public information director with the South Carolina Department of Wildlife and Marine Resources, concerning the use of native species for roadside beautification, is applicable to managed landscapes in general: "In beautifying [South Carolina's] roadsides, our basic rule should be to follow nature's lead. We must recognize and inventory existing stands of wildflowers, shrubs, and trees that are already growing along our highways. We must consider ways of conserving these existing stands, ways of improving their growing conditions, and ways of expanding their acreage. The native flora and natural habitats of South Carolina are part of our heritage. Our region's special diversity paints a natural tapestry unlike anywhere else in the world. Those interested in conservation and nature should strive to attain a ["South Carolina]-scape" in their wild acreage..." (Shealy 1989).

Approximately 20,000 species of flowering plants are native to North America (Xerces Society 1990). Just a few years ago the scarcity of native plant materials available from reputable nurseries was a barrier to creating an attractive naturalized landscape. As interest in natural landscaping has "blossomed," so have the number and diversity of mail-order nurseries catering to this demand. Today, it is possible to select from a wider offering of native trees, shrubs, vines, herbs, and grasses than ever before, and to select species that are indigenous to each region of the country. This presents exciting, new opportunities to design with nature and to bring some of nature's inspiration and beauty closer to home.

Cumulative environmental impacts (i.e., from

the use of chemicals or mowing) and the costs associated with grounds maintenance can be reduced by using native plants in designed landscapes. Americans are increasingly concerned about environmental quality and are searching for effective alternatives to turf grass and other intensively managed landscapes that consume resources and contribute to environmental pollution. Native plants are adapted to the climatic and environmental conditions of their indigenous regions. When used in appropriate growing conditions, they thrive with minimal need for water, mowing, and chemicals such as fertilizers and pesticides.

The importance of biological diversity to maintaining ecological balance is a motivating reason to use native plants in the naturalized landscape. The loss of native plant communities is a principle cause of species extinction. In the United States alone, some 220 acres of habitat are lost each hour to highway and urban development. About 15 percent of the native plants of North America are presently threatened with extinction. Some biologists estimate there is one plant species for every ten or more animal species; therefore, the loss of even one plant species can potentially affect the ecology of a number of animals (Xerces Society 1990). Many plants, in turn, depend upon animals for such biological services as pollination and seed dispersal. The value of natural landscaping to preserving biological diversity depends upon the type and size of the landscape and its relation to other kinds of land uses. Even on a small scale, however, such landscapes not only protect plant species but also provide food and cover for many kinds of beneficial insects, songbirds, and other animals that are increasingly impacted by habitat loss.

The philosophy behind preserving biodiversity through landscaping works only if we are careful not to contribute to the loss of wild plants. By using only nursery-propagated plants and seeds we can add to the existing flora. Removing plants from the wild disturbs their natural habitats and results in a net loss of plants since only a percentage survive transplanting. Care must also be exercised to avoid introducing native plants into new regions. Many introductions do not adapt, while others become aggressive and invasive into natural habitats to the detriment of native plants and animals. The list in Appendix D can be used to locate the nearest sources of nursery-propagated native plants and seeds. Before ordering plants, be sure to ask if the nursery propagates their plants or collects them from the wild. Do not order from nurseries that collect from the wild.

Opportunities for Natural Landscaping

Opportunities for natural landscaping abound in all types of managed parcels, from small urban lots to extensive national parks. An urban lot is typically a high-intensity management system that relies on a steady regime of mowing, pruning, irrigating, fertilizing, and controlling pests for maintenance. Undeveloped parklands occur at the opposite end of the spectrum, representing low intensity management systems in which maintenance may be limited to provision of access and other low level amenities. On any single managed parcel, there may be one to many levels of management intensity depending upon the objectives and uses of different planning units. One goal of using natural landscaping is to reduce management intensity on as many units as possible. Every time we can lower the intensity from high to moderate or low, we reduce the costs and environmental impacts associated with landscape maintenance while increasing aesthetic, wildlife, and interpretive values.

Some examples of places where natural landscaping can be effectively implemented include:

- Entranceways to a parcel, around signs and kiosks, trailheads.
- Lodge, club house, visitor center, meeting, restaurant, and maintenance facilities.
- Roadsides, foot paths, and hiking trails.
- Picnic and camping sites.
- Parking areas.
- Borders around other kinds of gardens (e.g., vegetable).
- Transition areas between mowed and undeveloped areas.
- Lake, pond, stream, and wetland margins.
- Drainage ditches and septic cells.
- Areas too steep, isolated, or otherwise difficult to mow.

A golf course is an excellent example of varying degrees of landscape management intensity on a single site. The intensity is greatest on the greens, progressively less on the tees and fairways, and minimal in areas of rough (Green and Marshall 1987). The latter areas can comprise more than 50 percent of the total area of a course and afford numerous opportunities for natural landscaping. The degree of difficulty of a course is often related to the extent and severity of hazards such as the rough (Green and Marshall 1987). More difficult courses may provide greater opportunities for natural landscaping to achieve multiple objectives of aesthetics, conservation, and economics. Natural wetlands are being incorporated into the fairways and roughs of golf courses.

Many opportunities exist to use natural landscaping around fairways. Port Ludlow of Washington, ranked in the top one percent of the nation's best designed golf courses by the American Society of Golf Course Architects, uses wildflowers to accent out-of-play areas, knolls between greens, and doglegs on fairways. Larger areas near entrances, the back of the driving range, and around tee boxes are other suitable locations (Stroud 1989). In intervening areas between fairways natural landscaping can be used to create a sense of mystery and anticipation as the player moves between fairways or to visually distinguish one or a series of fairways. Where long-distance views must remain unbroken or where tall vegetation may interfere with play, establishing meadow landscapes may be desirable; elsewhere, wooded landscapes may be more appropriate. Audubon International (Selkirk, New York) has developed an environmentally friendly golf course design program.

Often, when assessing a managed site for landscaping potential, the planner can identify areas that have a low level of use but are in a high intensity management program. Such areas can often be converted to natural landscapes and be maintained at a fraction of the cost (Photo. 4.1). Golf course superin-

tendents have reported savings of 25 to 30 percent compared to turf, largely resulting from reduced mowing, irrigating, and fertilizing. States that have implemented aggressive roadside wildflower programs have reported similar savings. The Georgia Department of Transportation (DOT) reported a 25 to 30 percent savings in maintenance costs for areas previously mowed (Corley 1989). In Florida, Gary Henry, a DOT landscape architect, explains "In the past it would have been seven mowings a year, and now that is reduced to two or three" (Weathers and Hunter 1990). One annual mowing is all that may be needed in some locations.

In addition to reduced maintenance costs, such programs elicit extremely positive public response, increasing tourism and visitor enjoyment. Golf course

Photo. 4.1. Naturalized landscapes add beauty and lower maintenance costs.

superintendents view wildflower plantings as one of their biggest marketing tools, noting that the positive response from guests has been "overwhelming" (Stroud 1989). One South Carolina DOT landscaper reported receiving "about twenty calls and letters a day" from motorists enthused about roadside wildflower plantings. The North Carolina DOT program was described by landscaper Bill Johnson as "the most popular program instituted in DOT." Again, the response was described as overwhelming (Weathers and Hunter 1990). The Texas roadside wildflower program has been around a long time, but has really exploded in the past five years. With renewed efforts beginning in 1980, the statewide mowing budget declined from $33 million annually to $21 million, saving $12 million (Weathers and Hunter 1990). Comparable savings can be realized on any size site, from a corner suburban lot to extensive rural acreage.

An increasing number of government programs are available to help land managers convert managed lands to natural vegetation. For highway projects, monetary support is mandated by the 1987 Surface Transportation and Uniform Relocation Act, which requires that twenty-five cents out of every $100 allocated for landscaping federally funded highways be spent on wildflower plantings. The U.S. Department of Agriculture's Farm Services Agency has a relatively new program to provide cost sharing to landowners establishing native vegetation. Many state agencies now have habitat improvement and forest stewardship programs that offer technical assistance with restoration planning and cost sharing to implement revegetation programs. Additional assistance is sometimes available through local Cooperative Extension Service and Soil and Water Conservation District offices. Local restoration experts are often the most knowledgeable about the assistance programs available in their area.

Landscape Planning

Site Analysis

The process used to develop a plan for naturalizing is presented in Chapter 2. An important part of that process is assembling baseline information about the site and analyzing that information to develop specific design criteria. Physical survey data are used to identify microclimatic effects and characterize topography (and exposures to light and wind), hydrology, and soils. This information is used to select appropriate ecological communities as landscaping models. Biotic survey data are used to characterize vegetation and identify the potential of existing ecological communities for preservation, restoration, or enhancement. This section describes some of the factors to consider when analyzing a site for natural landscaping.

Climate

The effects of climate and existing physical features on establishing or enhancing the naturalized landscape should be evaluated. Large-scale patterns of temperature and precipitation influence the predominant vegetation of a region, but climate can also reflect local influences. Heat reflected off pavement in urban areas, for example, can greatly increase air temperatures. Local climate data should be compared to regional data to determine whether local differences may affect vegetation patterns. The creation of a favorable microclimate may be an essential part of the design, enabling the establishment of certain plant species which do not presently exist on the site. In all regions, the south side of a building is warmer than the north side, but seasonal changes in the length and angle of sunlight and their effects in each region should also be considered (Diekelmann and Schuster 1982).

In the northeast, winds are predominantly from the southwest in summer and the west and northwest in winter, but local influences can drastically alter this pattern. Tall buildings and certain topographic features can influence wind velocity; winds passing over water may be cooler or warmer, depending upon the water body, but are generally more humid. Shelterbelt plantings are often used to modify the patterns and drying effects of wind on a site, as well as to reduce exposure to sunlight and to screen views. The protected microclimate of a woodland environment can be created or enhanced by planting overstory trees to reduce exposures to sunlight and wind and modify the effects of these elements on soil moisture and air and soil temperatures. Use of understory trees, shrubs, and vines can further these effects and create additional microclimatic effects at various vertical above-ground heights. Conversely, pruning or removing woody vegetation can have the opposite effect on a site (Diekelmann and Schuster 1982).

Topography

The existing contours of a site often dictate the placement of roads and other structural components (Weddle 1983). Proper conservation practices dictate that roads be routed along the contours of a slope to prevent soil erosion as well as to minimize long-term maintenance costs. Topography also has a direct effect on the microclimate of a site and can affect the selection of an appropriate plant community type. North- and east-facing slopes generally have less exposure to wind and sunlight, and being more mesic than south- and west-facing slopes support a greater diversity of plant and animal species. Slopes are better drained than level bottoms and terraces, which may be subject to periodic or prolonged flooding or saturated soil conditions.

Topography can be used to screen an undesirable view or as a focal point to display ornamental plantings. Where the topography is relatively flat, construction of berms can add vertical diversity and mystery to the landscape by hiding or revealing different landscape features.

Hydrology

Aquatic features such as streams, springs, and ponds contribute to the aesthetic and biological diversity of a landscape, and should be retained and enhanced wherever possible. Water adds a refreshing element to the landscape, both visually and audibly, and can be the focus of a landscape design. Even small, shallow pools and muddy depressions can present opportunities to establish different kinds of vegetation and are valuable to wildlife. The potential for creating new water bodies on a site should be carefully evaluated. It is often possible to increase the water-retention capabilities of a site simply by modifying existing drainage practices or structures. The benefits of creating new water bodies by damming or otherwise altering natural hydrology, however, should be carefully evaluated in terms of the objectives delineated in the landscape plan. Water bodies impacted by pollution should be examined to determine the potential for improving water quality and establishing natural vegetation in the water and along the shorelines.

Soils

The condition of the soil at various locations within a planning site may affect the potential for establishing different types of natural communities. Areas with naturally hydric soils suggest high water retention and good potential for the establishment of a wet prairie or even an aquatic community. Fertile soils with good drainage can support many kinds of forest and prairie communities in the eastern region. Excessively well-drained soils with low fertility will support a prairie or barrens community in the east, or a desert community in the southwest. Modern soil improvement techniques make it possible to grow plants on naturally poor or highly degraded sites, however, the costs of replenishing soil structure, moisture retention capability, fertility, and other qualities should be considered (Green and Marshall 1987).

Vegetation

To promote aesthetic and biological diversity, good quality representatives of existing ecological communities should be retained and protected to the extent possible during all phases of planning and implementing a naturalization plan. Such communities should be considered as site amenities that influence the overall landscape design. Retaining existing desirable vegetation can reduce the overall costs of implementing the plan. Trees are a long-term investment in the landscape, and healthy trees should always be retained where possible. Even undesirable trees can often be retained to provide needed shelter for new plantings until they are well established.

Landscape Design Considerations

The presence and quality of existing natural communities, as well as the prevailing natural character of the surrounding landscape will, in large part, influence the design. Knowledge of how the predeveloped landscape appeared is helpful in developed areas. Existing ecological communities may be the focal point of a landscape plan, and high quality communities, in general, require minimum effort to achieve multiple objectives for aesthetics, wildlife, and economics. Plantings of ornamental native species with value as wildlife food and cover may be desired to enhance the aesthetics of a high quality community. More disturbed or degraded communities may require more aggressive programs of weed eradication combined with extensive plantings.

Many of the principles outlined in the chapter on ecological restoration (Chapter 5) may be applied toward the enhancement of an existing natural community to satisfy natural landscaping objectives. Where natural communities are lacking, design options should reflect the objectives as stated in the landscape plan.

Principles for Designing With Native Plants

Using native plants in natural landscaping presents unique design opportunities and challenges. Unlike most cultivars, the adaptive quality and natural character of native plants have not been sacrificed to produce larger flowers or prolonged blooming periods. The aesthetic qualities of native plants are varied and often subtle. While some native plants produce large, showy blossoms, the most striking flower displays in nature often result from a tendency for a species to grow in groups or colonies. This clustering tendency can be recreated in the garden by imitating these natural patterns of arrangement. This clustering technique can also be used to display other attributes to their best advantage, such as attractive bark, twig, and foliage colors or textures.

With some exceptions, native plants tend to bloom for relatively brief periods of time. Designing a naturalized landscape with continuous bloom necessitates the use of a diversity of species having successive blooming periods. When a diversity of wildflowers is used, the naturalized landscape changes throughout the season. This dynamic quality is fascinating to observe, as is the variety of wildlife attracted to each successive production of flowers and seeds.

In eastern deciduous forests, many early-blooming wildflowers go completely dormant by summer. Interplantings of such species among other species that emerge later and persist well into the season is an efficient use of available space and keeps the planting interesting. For example, ferns are suitable for interplanting among early bloomers

with early dormancy, such as Dutchman's breeches, Virginia bluebells, shooting star, and dwarf larkspur. It should be kept in mind that a designed landscape of native plants will evolve over time as the plants self-propagate, filling in spaces, and rearranging their positions in the spirit of competition. We feel that this is a desirable progression, but in confined spaces, it is often desirable to control this rearrangement process, either by heavy mulching which discourages self-propagation, or by hand pulling seedlings that appear in undesired locations.

Wildflowers often display a general progression of plant height corresponding to season of bloom. Early spring blooming plants are often low to medium in height, while later blooming species are generally medium to tall. In the design of an island or border planting, early blooming plants are placed in the forefront and middleground, with later blooming plants constituting middle and background areas. Most native plant nurseries provide information on plant height and period of bloom for species they offer (see Appendices B and C).

In addition to the plant component, a naturalized landscape is composed of other natural materials, both organic and inorganic. The addition of materials such as rocks and logs in a design can serve both as substrates for plants to grow on and as interesting features in themselves. There are a great variety of woodland mosses that colonize rocks and woody debris, many of which are extremely attractive and delicately textured. Their bright greens contrast pleasingly with the darker fronds of Christmas and leatherwood ferns, and lend a sense of age and establishment to a landscape. Rocks can form the basis of a rocky glade landscape or be arranged in linear fashion to emulate a dry stream bed, with some boulders buried to one third of their mass to appear more natural than just scattered (Beikmann pers. com.). Large and small boulders arranged along a stream bank can create an illusion of swiftly flowing whitewater or become the structural framework for designed waterfalls. One large boulder with a concave center filled with water can serve as a birdbath or drinking hole.

Natural materials can effectively serve as pathways through the naturalized landscape. Stone pathways made of native bedrock slabs blend well with naturalized plantings and are virtually indestructible. On wet sites, they should be set into a bed of gravel at least four inches deep (Diekelmann and Schuster 1982). Other materials that make satisfactory pathways are gravel, wood chips, and mosses on moist, shady sites. The use of path rush or poverty rush (*Juncus tenuis*), a small plant that often becomes established naturally on foot trails, as a self-sustaining pathway material for low traffic areas deserves investigation.

Landscape Models

Designing a naturalized landscape is an attempt to capture the character of the ecological community serving as the model, whether it is a dry (xeric) forest of stunted, gnarled trees or an alpine meadow with sweeping wildflower vistas. Even on a small scale, naturalized landscapes can suggest the patterns of light, colors, shapes, and textures reminiscent of their natural counterparts. The tranquil, reflective quality of a southern swamp can be captured in a pond garden. A patch of prairie wildflowers and grasses recalls the expansive vistas of the midwestern tallgrass prairies; a rocky xeriscape of cactus and other succulents mimics the textures and hues of the southwestern deserts.

The organization of the naturalized landscape may appear to be random, but it should reflect an order imposed by natural forces such as gravity, sunlight, wind, freezing, and thawing (Cox 1991). Knowing how these forces operate within ecological systems leads to more natural-appearing designs. The contours of a naturalized landscape should not be defined by straight lines or right angles, but should flow in response to a natural or planned scheme that harmonizes with the character of the surrounding landscape. Contours may follow the irregular patterns of soil, water, bedrock, topography, or other natural features, or be laid out to camouflage or create a harmony with rigid features imposed by property boundaries, buildings, and other human-made intrusions.

Naturalized landscapes can be divided into two general categories to assist the landscaper in developing broad design criteria: woodland (shady) landscapes and meadow (sunny) landscapes. Shady landscapes are modeled after forest and woodland communities, while sunny landscapes are based on prairie and meadow communities. Woodland or meadow landscapes can be dry, wet, or gradients in between; desert communities occur on the dry extreme, while wetlands occur on the wet extreme. Theme gardens may be considered as variations in either category. For example, a water garden may be modeled after a wet forest or an open bog; a butterfly garden may be modeled after a dry, mesic, or wet prairie community with special attention given to selecting plants used for food and cover by a diversity (or selected species) of butterflies.

Using the ecological community as a model takes much of the guesswork out of selecting compatible species adapted to site-specific conditions. Plants should be selected first for their fitness for the total planting, and only then for color, shape, and texture (Diekelmann and Schuster 1982). A well-planned landscape is self-perpetuating with minimal maintenance. Perennials, which return each year from the original rootstock, often comprise the bulk of a planting. Annuals, which produce flowers and seeds and die in one season, provide quick cover and color, and usually contribute an abundance of seed for wildlife.

Woodland Landscapes

Forest communities serve as models for shady landscapes, landscapes dominated by trees and shrubs. Forest communities are complex environments with a great diversity of colors, forms, textures, odors, and sounds that appeal to all of our senses. Visitors travel great distances to experience the breathtaking spring wildflowers in the Great Smoky Mountains, the blazing autumn colors in New England, the languid mystery of the southern swamp, and the towering majesty of the giant Sequoias. On a more functional level, such landscapes offer shelter from the hot summer sun and cold winter wind, as well as a psychological sense of solitude and privacy. A desire to duplicate these qualities in our designed landscapes is a strong incentive for choosing the forest community as a landscape model.

The character of the forest is largely shaped by the trees and shrubs, which, by virtue of their numbers and size, influence localized environmental factors such as light, moisture, and wind. The trees in a healthy forest appear randomly spaced and exhibit many shapes and sizes. Unlike the spreading habit or growth pattern of open-grown trees, competition for light causes forest trees to grow straight upward before branching. In a mature forest, the overstory trees form a nearly continuous canopy of foliage, beneath which are several layers of foliage (at different levels) created by the understory trees (many of which are younger age classes of the overstory species), shrubs, and climbing vines. The amount of sunlight reaching the ground is filtered through the many layers of foliage and can be quite subdued. Near the ground is another layer of vegetation comprised of herbaceous plants such as grasses, wildflowers, and ferns, as well as mushrooms, liverworts, and mosses which may also grow part way up tree trunks and branches. The forest floor is frequently littered with pieces of wood and bark, leaves, and other organic debris in varying stages of decomposition. This debris and humus provides sustenance for plants as well as many kinds of insects and other organisms, and makes a major contribution to the rich, loamy soil characteristic of many forest environments.

Existing site conditions, the size of the planting, and overall objectives for the site largely influence the design process. Where a woodland already exists, the objective of the design may simply be to enhance the aesthetic and wildlife qualities by removing undesirable plants and adding desirable species of trees, shrubs, and wildflowers. Trees and shrubs add vertical diversity to the woodland garden and form a structural framework for the wildflowers, ferns, and grasses. Flowering trees and shrubs and wildflowers can be concentrated along access routes and resting areas for maximum visual impact.

An area containing tall shade trees with a grassy groundcover presents greater design opportunities. The bright shade found beneath such trees can be ideal for growing a diversity of woodland plants. The entire area of trees may be developed into a woodland garden or plantings may be established as "islands" or borders centered around clustered groups of trees, with existing grass maintained as mowed pathways between the plantings. Designing a woodland landscape on open, unshaded sites, requires the greatest amount of time and effort, but offers the greatest opportunities for expression. Establishing a tree cover is the first priority. The types and numbers of species should reflect the model community, but may be reduced to one or two species on small sites with an underplanting of wildflowers.

Woodland wildflowers and grasses add seasonal color and textural interest to the woodland garden. Most woodland wildflowers bloom in the spring and can be naturalized in large numbers beneath trees to create masses of solid color. Many wildflowers are grown not only for their spring flowers, but for their attractive foliage or colorful berries. Others, such as shooting star and Virginia bluebells, go completely dormant after setting seed and disappear from the landscape until the following spring.

Ferns are often used as the matrix or background of a woodland garden. Ferns make a handsome ground cover and the more aggressive species, such as the hay-scented, fragile, sensitive, and wood ferns, quickly form a dense cover in mesic forest, even in heavy shade. The shallow rooted species are especially compatible beneath deep-rooted trees and shrubs because they do not compete for moisture and nutrients (Cobb 1963). Pleasing effects may be created by choosing ferns with fronds of a shape, form, and texture that contrast with the growth forms of woody plants. An underplanting of fine-leaved maidenhair or beech ferns, for example, contrasts nicely with the solid formality of native evergreen rhododendrons and laurel (Cobb 1963). The taller species make an excellent backdrop for the colorful spring display of woodland wildflowers and their many shades of green provide a cool respite from the intense light and heat of summer. Like most woodland wildflowers, ferns prefer a light, well-drained soil high in organic content. Cinnamon and royal ferns prefer a constantly moist soil in sun or very bright shade and will grow even in saturated soils.

Meadow Landscapes

Prairie communities serve as the models for meadow landscapes, sunny herbaceous landscapes dominated by grasses and forbs. The name "prairie" brings to mind expansive vistas of tall grasses and brightly colored flowers waving in the summer breeze. The prairie community is best characterized in the great plains and midwestern regions, where it formerly occupied vast areas. In the predominantly forested eastern regions, such communities often occurred as openings of various sizes in the forest, sometimes called meadows or glades.

A prairie community achieves its most dramatic appeal when recreated or restored on a scale that emulates the expansive quality so characteristic of the prairie regions. Even on a smaller scale, a well-designed wildflower meadow or garden possesses all the elements that make the prairie such a desirable landscape system: an open visual field, a diversity of colors, shapes, and textures that change throughout the seasons and from year to year, low maintenance requirements, and high wildlife value.

A prairie or meadow landscape is quicker to establish than a woodland landscape, but requires just as much attention to planning, site preparation, establishment, and maintenance. The prairie community develops under conditions of maximum sunlight. Although many species of prairie plants tolerate shade for part of the day, they may produce fewer flowers and not successfully compete with more shade-tolerant species on sites receiving less than six hours of sun each day. Within forested regions, prairies are more common on southern and western slopes, which are hotter and drier. However, like forest communities, prairies occupy all moisture gradients from wet to dry. Differences in moisture influence species composition.

Bird and Butterfly Gardens

Designing a garden to attract and feed wildlife is similar to other kinds of naturalizing, except that the plants are selected as much for their food and cover value as for their aesthetic appeal. The key to attracting the greatest variety of wildlife is to supply a diverse complement of food and cover plants in all seasons, as well as water. In a small garden, an understory tree or shrub and three or more herbaceous species may be adequate; nest boxes and feeding structures can be used to supply additional niches for wildlife to use. Small gardens will rarely satisfy the life requirements of feeding, nesting, and resting for numerous wildlife species, but they can serve as feeding oases in the midst of developed landscapes to attract and benefit the more mobile species that venture a distance from their forest or grassland homes.

Providing a reliable source of food benefits wildlife since they expend less energy searching for food; this can be especially critical for parents feeding young. The nesting and reproductive success of wildlife is greatest when dependable, season-long sources of food and cover are available. Many species of birds will return to the same locale each year where these requirements are consistently met.

Many wildflowers provide nectar sources for a variety of birds and insects, while others are adapted to only certain species of either group. Of the nectar feeders, hummingbirds are more attracted to tubular flowers in the warm color ranges of red, orange-red, and pink; bees prefer flowers that are blue, yellow, violet, or in the ultraviolet range; butterfly flowers may be purple, lavender, yellow, orange, white, or red and are usually tubular and fragrant.

Hummingbird Gardens

There are twenty-one species of hummingbirds that reach North America, sixteen of which breed here. They are present during some portion of the year in every region except the far north and treeless plains (Arbuckle and Crocker 1991). The ruby-throated hummingbird and a number of western species winter along the Gulf coast in ever-increasing numbers. A number of hummingbird species are year-round residents in the west.

Gardens designed to attract hummingbirds can be of any size, including patio or porch gardens in pots or window boxes. No matter what size, the hummingbird garden should have a combination of food or nectar plants as well as plants that provide cover for resting and nesting. Certain "hummingbird flowers" are specifically adapted for pollination by hummingbirds. Such flowers are generally tubular, produce abundant nectar, have warm colors (often red, orange, or pink), lack fragrance, and are more accessible to hummingbirds than insects. The flowers are often pendant or positioned toward the outside of the plant where hummingbirds can feed without touching the foliage. In addition to nectar, hummingbirds obtain protein food by eating small insects such as beetles, flies, mosquitoes, aphids, and spiders (Terres 1980).

Hummingbirds spend about four-fifths of each day perched on branches near their preferred food sources. At night they roost in dense cover and select trees or shrubs, often in or near the edge of woodland, as nest sites. Water can be an added attraction for drinking and bathing. Water can be supplied in the form of nectar or tree sap, but like other birds, hummingbirds will drink water from a variety of sources. They will bathe in shallow water, but the water in most birdbaths is too deep. Often, they bathe while flying through a fine spray of water from a waterfall or fountain. After bathing, the birds typically perch on a nearby branch to shake off the water and preen their feathers.

Plants should be chosen to provide nectar sources in the garden in all seasons that hummingbirds are present. Wildflowers generally offer more nectar than the cultivars derived from them and should constitute the bulk of a hummingbird planting.

Food can be provided entirely by growing preferred food plants or by the addition of an artificial feeding structure filled with a sugar solution of one part white (sucrose) sugar to four parts water. The latter can be especially important when drought or other factors reduce natural food supplies.

Some of the best native plants that attract hummingbirds in each season are listed below.

Early spring: red columbine, wild blue phlox, fire pink, red buckeye, scarlet larkspur

Mid–late spring: beardtongues, New Jersey tea, coralbells, azalea, evening primrose, Canadian lousewort, ocotillo, black locust, hawthorns

Summer: bee balm, scarlet sage, blazing star, gilia, summer phlox, trumpet honeysuckle, trumpet-creeper, crossvine, rose mallow, royal catchfly, coral berry, wild jewelweeds, wild lilies, bushbean, Indian paintbrush, agave, lupine, crimson monkeyflower, currant, bouvardia, scarlet figwort, butterfly milkweed, scarlet penstemon

Late summer–fall: cardinal flower, California fuschia

Year-round (southern regions): lantana

Songbird Gardens

Songbirds delight us with their songs, brilliant colors, and playful antics. Because of their aerial habits, they tend to be more visible than other kinds of wildlife, and we are able to observe much of their daily activities. While some species of birds, most of them not indigenous, are troublesome pests in our cropland and orchards, songbirds are generally pleasant and interesting, and we find their presence in our gardens desirable.

Most birds require a variety of insect food as well as seeds. During the breeding season, songbird diets are often higher in protein-rich animal foods. A garden that attracts insects will often attract a variety of birds that feed on them. In fall and winter, berries and seeds become important in the bird diet. In winter, birds often feed on the "cones" of alder, birch, and sweetgum (Wilson 1987). Good sources of seed are the native sunflowers, which are very attractive to goldfinch, sparrows, and other finches. Other good sources of seed which are also attractive garden plants include asters, coneflowers, blazing stars, tickseeds, partridge pea, coreopsis, blanket flower, ox-eye sunflower, clovers, wild bergamot, beardtongues, phlox, black-eyed susans, Juneberry, serviceberry, and native blueberry. Some good berry-producing shrubs and small trees that make attractive additions to the garden include dogwood, red mulberry, black cherry, wild plum, elderberry, sumac, red cedar, winterberry, deciduous holly, wild rose, red chokeberry, viburnum, and hawthorn. Some shrubs that have been touted for their berry production but which are extremely invasive into natural habitats and should be avoided include the nonnative honeysuckle, multiflora rose, autumn and Russian olive, nonnative buckthorn, and nonnative barberries (see Chapter 5 for more information on nonnative pests).

Butterfly Gardens

Humans are naturally attracted to the brilliant kaleidoscope of colors and patterns of butterflies, and to their delicate flight which captures the imagination. They inspire curiosity rather than revulsion because they are without the means to sting, bite, or otherwise inflict human injury (Schneck 1990). Many butterfly species have wide ranges while others are found in only certain regions. As a first step in designing a butterfly garden, the landscaper should learn which species occur in the planting region, and then attempt to provide plant species which afford food and cover for all life stages.

Butterflies need food plants in all stages of life; the larval, or caterpillar stage feeds on specific plant hosts, and the adult butterfly feeds on nectar and other materials (Arbuckle and Crocker 1991, Newsom-Brighton 1987, Schneck 1990). Planting many kinds of nectar-producing flowers provides continuous bloom and attracts a diversity of butterflies all season. Since caterpillars are often very specific in the plant species they use for food and cover, it is possible to attract particular butterfly species by planting their preferred food plants.

Since they are cold blooded, butterflies require sunshine to warm up and become active. They also require protection from wind. A background planting or border of trees, shrubs, taller herbaceous plants, or dense vines over a wall, fence, or trellis, can help create a sheltered feeding area and warmer conditions favorable to butterfly movement. The planting should be located on the upwind side of the garden for best protection. Some suitable cover plants include passion flower and pipevine for vine cover; New Jersey tea and leadplant for low shrub cover; hydrangea and sweetshrub for taller shrub cover; and black cherry, mulberry, and hackberry for tree cover.

Butterflies cannot drink directly from open water, but make regular visits to and even congregate in large numbers at puddles of wet sand or earth (Schneck 1990). In addition to water, they imbibe salts and trace minerals. Scattered depressions in the ground with stones and sticks placed on top will further enhance a garden's attractiveness to butterflies.

Water Gardens

Water in the garden has a particular fascination, adding dimensions of light, depth, and movement to the naturalized landscape. Water gardens are often used as an added attraction within other types of landscapes and can become a visual focal point for other garden themes such as bird, butterfly, and rock gardens. Larger ponds may attract a variety of wildlife, including waterfowl, shorebirds, muskrat, and beaver. Even small ponds can support an abundance of life such as fish, frogs, and snails.

Wetland (palustrine) communities serve as models for water gardens and landscapes characterized by emergent aquatic plants. These landscapes can be designed as part of a natural wetland, in ponds or pools constructed for this purpose, or even in water-holding containers of varying sizes. The relatively new innovations in pool liners and preformed pools combined with a greater understanding of how to maintain the balance of life in a pool, has caused a revolution in water gardening.

Methods of constructing a water garden are varied and depend upon the type of materials used. Construction techniques are beyond the scope of this book, but there are many references on the subject, and nurseries and garden centers often have valuable expertise to offer. Existing ponds and lakes can be visually enhanced by planting emergent aquatic plants along the shores. Abrupt or steep shorelines may need to be graded to reduce the slope before plants are installed. Gradual or graduated shorelines are ideally suited to many plant species and allow better accessibility for wildlife.

Any container capable of holding water is a potential water garden. Vinegar and wine barrels cut in half are popular, but wood containers that contained oil, tar, or wood preservative should not be used as an unsightly scum may form on the water surface and the high concentration of arsenic used in most pressure treated materials can be highly toxic to people and wildlife (Beikmann pers. com.). Any container should be thoroughly scrubbed before use, but soaps and detergents should be avoided. Fountains, waterfalls, and other structures help move water through the garden. A good filtration system, including mechanical, bacterial, and plant components, is essential for maintaining clean water in constructed gardens.

Newly constructed water gardens should be carefully sited. Most aquatic plants prefer a good deal of sun but should be given protection from wind. A natural water garden may be located in the lowest part of the landscape as would occur in nature, or at least made to appear lower than the adjacent land (Xerces Society 1990). This effect can be achieved by mounding soil behind the pool, which creates a nice stage for a backdrop planting of wildflowers and ferns to reflect in the pool's surface. The mound can be sited so as to provide protection from wind. If placed in a low-lying area, run off from adjacent land should be diverted around the pool to prevent contamination by fertilizers, pesticides, silt, and decaying matter (Roth 1988). Low-lying areas should be avoided when locating a pool that is artificially lined. The soil should be well-drained to prevent heaving, cracking, and breaking, and the pooling of run-off which can displace a plastic liner. Soil which is very sandy or crumbly may require a retaining wall to prevent soil from caving in.

Most aquatic plants prefer six or more hours of

sunlight each day for extensive blooms (Roth 1988), so water gardens should be located in open areas away from trees, if the objective is to grow water lilies and other open water plants. A water garden may be located in the shade and a variety of plants can be used around the margins, but smaller pools may have to be frequently cleaned of fallen leaves during the autumn to maintain ecological balance.

The selection of plant species depends upon the size and depth of the water garden. Certain emergent plants, such as cattails and water lotus, can become aggressive in shallow water, and in time, may cover the entire water surface. Such plants should be used only around the margins of deep (three feet or deeper) pools if maintaining open water is desired. Around smaller pools, overhanging vegetation should be avoided; leaves, fruits, and other plant parts can accumulate in the water and become toxic to fish. All sizes of water gardens benefit from the inclusion of oxygenating plants. These are usually submerged plants whose function is to maintain healthy, well-oxygenated water for fish, snails, and other aquatic life. By consuming mineral salts, they compete with the green algae which can otherwise become overly abundant, unsightly, and deplete the oxygen supply. Some excellent oxygenator plants include pondweeds, cabomba, coontail, and water celery.

Water plants may be planted directly into the natural substrate of lakes and ponds, or into a substrate of gravel, soil, or mulch placed in the bottom of artificial ponds or pools. Water plants can also be grown in containers placed at the appropriate water depth, usually six to twelve inches below the water surface, in either natural or artificial ponds. Use of containers is often preferred for more aggressive water plants which spread rapidly if planted in substrate (e.g., cattail, lotus). Containerized plants will need to be divided periodically. In regions with cold winters, potted plants can be removed before pools freeze over and stored in a protected location.

Containerized water plants should be planted in good, clean, heavy soil topped off with a generous layer of pea gravel to hold the soil in place. Commercial potting soils are not recommended since they can become "hot" when saturated for extended times. Slow-release fertilizer tablets designed for water plants can be inserted directly into the pot (caution: some fertilizers may release ammonia into the water, which can be toxic to fish).

Different plant species have different requirements for water depth and should be placed at the appropriate depth recommended by the nursery source. Water gardens are often constructed with shelves at different depths to accommodate varying depth requirements. Pots can also be placed on variously sized rocks or other materials set within the water garden.

Artificial Animal Structures

The wildlife value of a landscape can often be improved with the addition of artificial shelters such as bird and bat houses. In a natural forest, there are normally numerous dead and dying standing trees that provide cavities used by woodpeckers and other bird species. Artificial cavities placed within or along the transition zones of young woodlands will attract cavity nesters such as wren, chickadee, screech owl, woodpecker, raccoon, flying squirrel, and tree squirrel. Wood ducks will use nest boxes placed in woods along streams and swamps, and artificial houses provide shelter for beneficial insect-eating bats. Recent information indicates that larger, colonial houses are more likely to attract bats. Bluebirds and kestrels prefer nest boxes placed in or near an open meadow. A roost box, which usually has the opening near the bottom of the box, provides shelter and warmth for smaller birds in winter. Warmer air is trapped at the upper end of the box. Nesting platforms are used by the robin, barn swallow, and other "edge nesters." Many kinds of nesting and roosting structures are available at garden centers and companies which offer restoration services often have specifications for making many kinds of wildlife structures.

Establishment and Maintenance

Establishment Techniques

A naturalized landscape may be established using seed, nursery stock, or a combination of both. No matter how a natural landscape is established, special attention must be given to site preparation, installation, and maintenance during the establishment period. Design considerations, the type of community being established, availability of the desired species of seed or stock, budget constraints, and establishment time influence the technique used.

Establishing a naturalized landscape from seed can be more economical on larger sites (usually 1,000 square feet or larger), but normally requires a longer establishment time than when nursery stock is used. Seed mixes are often used to establish meadow landscapes on larger sites and although it can take several years to mature, observing the successional process can be rewarding. Establishment using nursery stock is often preferable on smaller sites. While initially more expensive, this technique has the shortest establishment time and allows for complete control over the placement of species by height, color, and other design specifications. Nursery stock may also be preferred where weeds are a problem because it is easier to control weeds by mulching around established plants than among unevenly distributed seedlings. Another option is to establish a meadow planting by seed, supplemented with nursery stock to fill in gaps in the planting or to highlight visually sensitive areas with flowers of a preferred color or height. Using a wildflower seed mix can be a cost-effective way to establish a meadow landscape on larger sites. It is a common misconception, however, that it is easy to grow a lush meadow of beautiful wildflowers simply by throwing a seed mixture on unprepared ground.

Site Preparation

Site preparation begins with delineating the boundaries of the planting site on the ground, followed by eliminating competing vegetation and preparing the planting bed for seeds or nursery stock. Selecting appropriate plant species for existing site conditions reduces the amount of effort and cost associated with preparing the planting site.

Soil Improvement

Selecting a natural community adapted to existing site conditions will, in most instances, eliminate the need for extensive soil improvement. Soils with naturally poor drainage will often support a wetland or bog community, while dry, excessively drained sites may support xeric forest, glade, or dry prairie or desert communities. The need for improving the soil will be indicated by survey data or may be ascertained from the condition of the veg-

etation growing on the site. A site which presently supports a lush growth of vegetation probably does not need much improvement. Sites which do not presently support a healthy stand of vegetation will probably require soil improvement before planting or seeding. Sites disturbed by construction may be compacted and have little or no topsoil. Amending severely degraded soils requires the application of large amounts of topsoil and other organic matter and may take several years to prepare. Subsoiling may be necessary on soil compacted by heavy equipment.

Organic matter may need to be increased if the site has been under cultivation and is being restored to woodland. Woodland plants generally prefer a soil high in organic content. Woodland herbs adapted to a mesic site, for example, may require soil organic content ranging anywhere from 10 to 33 percent of volume, or even more in clay or sandy soils (Diekelmann and Schuster 1982). Many prairie plants will tolerate a well-drained soil with average fertility, and tend to become "leggy" and topple in a very rich soil. Fertile soils with poor drainage can be improved with the addition of coarse, quarried sand or gravel.

A variety of soil amendments may be used to improve the organic content, fertility, aeration, water retention, drainage, and structure of a substandard soil. Composted materials such as leaves, grass clippings, manure, yard waste, and kitchen waste can be used. Commercial products such as packaged topsoil, compost, humus, peat, and shredded bark can also be used, and are sometimes available by the truckload; keep in mind that unpackaged materials sometimes contain many weed seeds.

Most native plants have a broad tolerance for soil pH (e.g., the degree of acidity or alkalinity of a soil). Soil pH may be lowered (more acidic) using copper sulfate, peat, shredded pine bark or needles, or other available materials, or raised using lime, which also helps break up heavy clay soil into more friable particles (Beikmann, pers. com.) (follow label directions for appropriate application rates).

Weed Elimination and Control

Whether the planting area is to be established by seed or nursery stock, the soil needs to be loosened and cleared of sod, weeds, and other competing vegetation. When establishing an area from seed, it is particularly important to eliminate weeds prior to seeding, as post-emergent weed control is difficult and marginally effective. Even with effective weed control prior to seeding, a variety of annual, biennial, and perennial weeds can be expected to appear in the planting. Thorough preplanting weed control gives native plant seedlings a head start in competing with weeds during the establishment process.

Sod can be manually removed with a sod cutter (available at most rental centers) and stockpiled for up to several weeks for use at another location. Sod can be killed under a thick layer of grass clippings or other kind of mulch, but this usually requires a period of several months to be effective. Or, sod can be killed with a tilling/herbicide program as described below for perennial weeds.

Weeds exist in the form of actively growing vegetation and seeds. Weed seeds often remain dormant in the soil until favorable conditions of light and moisture promote their germination. Tilling and other removal of surface vegetation can expose weed seeds to improved light conditions and promote germination of new weed crops. A deep tilling often exposes even more weed seeds and compounds weed problems. A single, shallow tilling (to a maximum depth of one to four inches) may be expected to expose fewer weed seeds than repeated or deep tillings, but tilling alone is often insufficient to kill perennial weeds. If use of chemicals is not desired, repeated tilling, often for a full year, is required to eliminate weeds. Planting seeds with a seed drill eliminates the need to till and is an excellent method of seeding on slopes and other erosion-prone sites.

An effective weed control program for sites with actively growing weeds combines shallow tilling with use of a powerful broad-spectrum herbicide such as a glyphosate. Many of the most persistent perennial weeds and grasses such as Johnson grass, Bermuda grass, and fescue have strong root systems which will vigorously resprout if the plants are simply cut or tilled under. While tilling may be used to break up sod and kill shallow rooted plants, it may need to be followed by one or more herbicide treatments to kill resprouting perennial weeds. After the first herbicide treatment, a waiting period of ten days will allow for the re-sprouting of weeds, and keeping the soil moist will hasten the process. Raking treated vegetation away from the planting site will help prevent resprouting. If new growth appears, a second herbicide treatment is warranted, followed by another ten-day waiting period.

Herbicides are most effective on actively growing vegetation. After mowing, new growth should be well underway before treatment. Herbicides are potentially toxic to fish and wildlife and should be used only in accordance with the manufacturer's instructions and under conditions of no wind or precipitation. Oil- or ester-based herbicides should be avoided near streams and wetlands. Using large spray droplets and low pressure in the sprayer helps prevent wind drift. Wick application of herbicides is one way to prevent drift, which often occurs with spraying. Herbicides which quickly break down in the soil will minimize environmental hazards. Seed can generally be planted seven to ten days after using such products. Areas previously in cropland that were treated with long-term residuals such as atrazine should not be seeded for one or two years following treatment; a soil test can be taken to identify residue levels or a small test plot can be seeded to test for germination success.

On smaller planting sites, it may be practical to use solarization to kill weeds. After tilling, the site is covered with clear plastic which is well anchored to the ground. With the heat of the sun, moisture trapped beneath the plastic turns to steam, and the heat and steam act to kill both weeds and weed seeds. The process requires from one to two months, but usually eliminates the need for herbicides.

The common horticultural practice of mulching, using various materials to prevent weeds and conserve soil moisture, can be used to prevent weeds around planted nursery stock and is effective in preventing frost heaving in fall plantings. Mulch should be applied after the ground freezes to prevent rodents from feeding on plant parts over winter. But mulching, if applied too heavily, can also discourage natural propagation. Mulch should be used sparingly, or not at all, if self-propagation is desired, or used heavily to discourage natural spreading. Annual native plants, which live only a single season, typically produce large numbers of seed to assure their perpetuation in the wild. The seed must make good contact with the soil to germinate. Biennials usually spend their first season in a vegetative condition, dying the second season after producing flowers and seed. Similar to the annuals, biennials primarily rely upon seed production for perpetuation in the wild. Heavy mulching can prevent seeds from germinating. Perennial plants are typically long-lived from a perennial root system and may not bloom until the second or third season. Many perennial species multiply by sending out rhizomes or stolons from the parent plant, with new plants produced along their lengths. Heavy mulching will reduce light availability and discourage the production of new plants.

Establishment from Seed

Many native species can be established from seed, but it can take several years for a planting from seed to become well established. Wildflowers often have complex seed dormancy mechanisms that prevent germination of seed during unfavorable conditions. Some species must be subjected to a period of moist cold, called moist stratification. Such species should be sown before winter for spring germination, or can be stored in a sealed container with moist sand, vermiculite, or other inert material for one to two months to simulate a natural winter chilling; containers should be examined weekly to adjust moisture levels and to prevent molding. Seeds with hard coats may require scarification to promote germination. This consists of breaking down the seed coat by soaking in water or acid, nicking with a knife, or by abrasion with sandpaper. Seed of some species (e.g., bloodroot, celandine poppy), must be sown fresh or germination can take years or may never occur. Legume seeds benefit from inoculation with a species-specific bacterium. Most seed companies provide directions for the best germination methods for each wildflower species. Wildflowers require specific soil, temperature, and moisture conditions for successful seed germination and seedling growth. Attention must be given to site selection, species selection, seedbed preparation, maintaining favorable soil moisture, and controlling weeds throughout the establishment process.

The development of a meadow planting can be an interesting and educational process when expectations are realistic and informed. Many environmental factors can affect the success of a wild-

flower seeding. Adverse weather conditions such as drought, hail, or excessive winds may seriously affect successful germination and seedling growth. Over-wintering seeds may be subject to animal predation, rot, and erosional forces which can displace seeds from the site. Use of a cover crop or erosion netting can help reduce seed loss. Vinyl netting can capture and kill both birds and snakes and should be avoided. Jute or other material that quickly biodegrades is a good substitute.

Possibly the most important factor in a successful seeding is patience. Many wildflower plantings from seed have been plowed under because the growers were uninformed about the establishment process. It is unreasonable to expect to produce in one or two years what nature has taken hundreds of years to achieve. Establishing a wildflower meadow from seed requires several years and during the establishment period, a wildflower planting might appear weedy. The creation of a sustainable landscape necessitates the use of perennial species, many of which are slow to establish and mature to blooming size. Many perennial species will not germinate until exposed to a period of moist cold; such species may not bloom until the third season following a spring planting. During the first growing season, 75 percent of the growth of native warm season grasses can be devoted to the development of an extensive root system. A diversity of native species should be apparent by the second season.

Many factors must be considered when formulating a wildflower seed mix, including flower color, number of blooms per plant, period of bloom, seed availability and price, regional adaptivity, and number of seeds per pound. While it may be tempting to custom formulate a seed mix to satisfy certain design objectives, it is often easier, less costly, and more successful to purchase a commercial wildflower seed mix. Commercial mixes vary considerably as to quality and care should be taken to choose a mix with the greatest potential for successful establishment in the planting area. Researchers at the National Wildflower Research Center (1989) offer the following guidelines for selecting wildflower seed mixes.

■ Select a mix with a high percentage of species native to the region where the seed will be planted. The Center warns against the "shotgun approach" to seed mix formulation, in which a mix contains a variety of species adapted to different regions in an attempt to cover a wide geographic area. A high percentage of the species in such a mix will not establish in one region or another, the result being money spent on thousands of seeds that will not germinate or on nonnative annuals that fail to persist after the first year or two. Shotgun mixes may also include inexpensive, bulky filler seed of common cultivated flowers (not wildflowers) or cool season grasses such as fescue. Often this filler seed is present in such large amounts that it competes with the native wildflowers! The Center instead recommends using a regionally adapted mix containing a high percentage of species native to the marketing area. Such mixes are usually produced by small, local nurseries which obtain seed through local field collection and propagation. The advantage of such mixes, when used in the appropriate area, is a much greater establishment rate. While some larger seed companies are now marketing "regionalized" wildflower seed mixes, they generally include a high percentage of nonnative, annual wildflowers which fail to reseed, and are therefore not true regional mixes. When used to provide temporary cover and color, annuals should not be so abundant as to impair the successful establishment of the perennials in the mix.

- Wildflower mixes should be comprised of species which bloom in the spring, summer, and fall to produce a long season of flower display. Commercial mixes often fail to include asters, goldenrods, and sunflowers which extend the blooming period well into the fall months.

- Indigenous native grasses, which are a dominant feature of the natural prairie or meadow community, should comprise 50 to 80 percent of the volume of a seed mix. Native grasses lend structural support to tall wildflowers, stabilize the soils and minimize erosion with their extensive root systems, help prevent invasion by weeds, add late season color and texture to the landscape, and provide valuable food and cover for wildlife. In contrast to sod-forming grasses

such as many of the fescues, bunch-forming native grasses provide adequate spaces in which wildflowers can become established.

Seeding Rates

The seeding rate for wildflowers and grasses will vary depending upon the species or type of seed mix used and the sowing method. The National Wildflower Research Center (1989) suggests seeding rates for establishing pure stands for a number of wildflower species; the rates will change when species are combined. For landscapes that are to be viewed up close, a seeding rate of six to ten seeds per square foot is recommended; for distance viewing, four to five seeds per square foot may be adequate. Seed mixes are often sold in bags of four to six ounces to cover about 1,000 square feet, or in pounds for larger sites. A rate of eight to ten pounds of grass/forb seed mix per acre is often used, but this rate may be reduced or as much as doubled depending upon site conditions and species used. Hydromulching can require a rate three times greater than normal. Seed companies will often recommend a suitable rate for selected species based upon a description of site conditions. Quick-growing, bunch-forming grasses are often included in a mix for temporary cover or as a 'nurse crop' while the perennial grasses are becoming established. Canada ryegrass is often used in northern states, while sheep fescue and hard fescue are used in the west. Annual cover crops such as spring oats and annual rye have been successfully used, but aggressive perennials such as perennial ryegrass and Kentucky 31 fescue should be avoided.

Timing

Most reputable seed companies will recommend the best time to sow wildflower seeds in their marketing region. In the northern and northeastern regions where harsh winters occur, an early spring seeding will promote fast germination of the native grasses, tender annuals, and many species of perennials. A late fall seeding, however, can improve germination rates for many perennial wildflower species which require a period of moist cold to break seed dormancy (e.g., including many kinds of asters, goldenrods, coneflowers, and blazing stars). In the southern and far western regions, fall (September through December) is a favorable time for seeding. This timing takes advantage of warm soil temperatures and soil moisture to promote rapid germination and root growth prior to dormancy. An early spring seeding is also possible, but the early onset of hot, dry weather may necessitate supplemental watering.

Sowing Methods

Prior to sowing, wildflower seed should be mixed with a carrier to increase volume and promote more even seed distribution over the site; sand, sawdust, vermiculite, perlite, and commercial potting soil have all been successfully used. Wildflower seed mixes contain a variety of seed sizes and shapes, and uneven seed distribution can result from the settling of smaller, heavier, or smoother seeds during sowing. Continuous agitation of the mix during sowing, either by hand or mechanically, keeps seed more evenly distributed in the mix and on the ground.

Wildflower seed may be sown by any of the normal seeding methods, but on smaller areas, hand broadcasting is often most effective. Walking two sets of transects across the planting area results in more even coverage. Half of the seed is sown walking a set of parallel transects; the second half is sown while walking a set of transects perpendicular to the first, resulting in a grid pattern of seeding (Figure. 4.1).

Mechanical seed broadcasters and seed drills are often used on larger planting sites. Use of a seed drill adapted to handle warm season grasses can be especially effective on slopes and other erosion-prone sites where tilling is not practical; such drills are not readily available in the east but are more

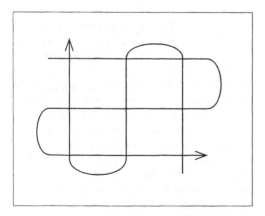

Figure. 4.1. Grid pattern for hand-sown seeds.

common in the west. Hydroseeding, a process where seed is mixed with a liquid slurry and sprayed on a site, is often used on sloping or otherwise difficult-to-seed sites. A tacky or sticky material can be added to the slurry which helps the seed adhere to the ground surface and may eliminate the need for mulching. A disadvantage is that it is difficult to adjust for an accurate seeding rate when using a wildflower seed mix and the seeding rate may need to be tripled to compensate. Hydromulching can be used after broadcast seeding to spray a tacky material on seeded ground in place of mulch.

Good soil contact is essential to anchor seeds and promote successful germination and seedling growth. To achieve good soil contact, surface sown seed can be lightly raked, pressed, or rolled into the soil. Walking on a seeded site is often sufficient to achieve good soil contact. Care should be taken to avoid planting the seed more than one-eighth inch deep; some seed should remain visible on the surface. Small seeds may require light to germinate and do not have enough food stored for a lengthy period of below-surface growth.

Establishment From Nursery Stock

Seed of many wildflower species is commercially available, but seed of woodland species is often available only in limited quantities. Establishing a woodland from seed can be a very rewarding process, but can take generations to achieve results. Woodland community types and smaller meadow gardens are more commonly established using nursery stock.

When examining nursery sources of native plants, be sure the stock is nursery-propagated rather than collected from the wild. The increasing market demand for wildflowers has resulted in a profitable industry involving collection of plants from the wild. Collection of most wild plants, by individuals as well as by nurseries, is essentially unregulated and has the potential to result in the depletion of local populations, and possibly of certain species throughout their range. Part of the incentive behind landscaping with native plants is the preservation of the natural heritage unique to each region of the country. With so many nurseries now offering a wide selection of nursery-propagated natives at affordable prices, strong conservation ethics do not support the use of wild-collected plants in our designed landscapes.

A wide variety of native species are available from mail order nurseries. Perennials can be purchased as bare-root or containerized material. Bare-root material is often less expensive but more difficult to transplant successfully. Container-grown plants should have well-developed root systems that make transplanting into the landscape easy and generally more successful. Many species of native trees and shrubs are available as bare-root stock up to six feet in height. Such stock generally has high transplant success and is often preferred for larger woodland landscapes to hold down initial and replacement costs. Bare-root stock used in northern growing zones (U.S.D.A. hardiness zones six and up) is usually planted during dormancy for greater transplant success. Some species of trees and shrubs are available in larger landscaping sizes and are usually balled and burlapped. Such material can be planted almost anytime and generally has high survivability.

Planting Density

The spacing of nursery stock should in most cases follow the suggestions provided by the nursery where the plants were obtained. One outstanding feature of native plants is their ability to self-propagate both sexually (seed production) and asexually (vegetatively). This ability can be used to advantage in certain situations to reduce plant material and installation costs. The initial planting density can be lower than natural to allow for dispersal by seed or other means. Species which spread by rhizomes or stolons will often form colonies around each parent plant; those which spread by seed can produce hundreds of new plants each season. Within a few years, the intervening spaces will be filled, the rate depending upon the species used and on soil and other characteristics of the planting site. On erosion-prone sites, it may be necessary to install plants at a relatively high density to quickly stabilize soils.

Trees appear randomly spaced in a natural forest. The trees in a woodland design should be spaced close enough to encourage upright growth rather than the spreading habit characteristic of open-grown trees. State forestry offices can often provide guidelines as to suitable planting density for local forest conditions. Keep in mind, however, that commercial timber plantings are often established at rates denser than those designed for aesthetic and wildlife purposes, and a dense planting may require successive thinnings as it matures.

Timing

Planting should be timed to take advantage of natural patterns of precipitation. If supplemental irrigation is possible, plants may be installed anytime except during the hottest months. In the eastern regions, rainfall is most reliable during late winter and spring and thereafter occurs mostly in the form of short, sporadic, often heavy thundershowers. Nursery stock can be planted in early spring as soon as the ground is workable, but young plants should be mulched if danger of hard frost has not passed. An early fall planting in warm soil allows plant roots time to grow and provide an anchoring system which prevents frost-heaving during winter.

Bare-root materials are normally planted during dormancy, i.e., before new growth is initiated in the spring. Bare-root material may be removed from the growing fields during winter, stored under refrigeration, and shipped later in the season before the weather gets hot. Balled and burlapped material can be planted almost anytime, but will require irrigation if natural rainfall is inadequate to sustain soil moisture during the establishment period.

Establishing Tree Cover

Tree cover may be established using larger, landscape-size material for quicker results, but this can be cost prohibitive on large sites and the diversity of species available as large material may be limited. Bare-root material is often used to establish tree cover on large sites, advantages being less initial expense and the availability of a greater variety of species. Larger materials (five or six feet in height) may need staking the first year to prevent root damage from wind (Beikmann pers. com.).

Trees may be planted into an existing groundcover of grass or other cover, but an area the diameter of the tree crown should be cleared of vegetation around each tree planted. This area should be mulched after planting. On dry sites a raised border of mulch should be made around each tree to trap water. A greater increase in trunk diameter will be attained if seedlings are not pruned, particularly the lower branches of balled and burlapped and containerized trees (Beikmann, pers.

com.). Understory plants should not be installed until the tree cover is sufficient to provide protective shade.

Natural Landscape Maintenance

Maintenance of a naturalized landscape initially focuses on maintaining adequate soil moisture and controlling weeds. It is important to keep soil evenly moist, not wet, during the establishment period. During extended dry periods in the first year, water may be necessary in response to wilting leaves. Once established, and if species are sited appropriately, irrigation should not be necessary except in cases of extreme environmental stress.

Post-germination weed control in a landscape established from seed is often difficult because weed seedlings may be hard to identify before they grow large, and herbicides are usually not selective enough to use on weeds growing among young wildflower and grass seedlings. On smaller plantings, manual removal of weeds is the most effective technique but is relatively labor-intensive. Weed control on larger areas is usually accomplished by mowing, spot-spraying with narrow spectrum herbicides, and prescribed burning.

Mowing weeds early in the spring before the natives emerge benefits native plantings by improving sun exposure and reducing weed vigor. Mowing, to a height of six inches, can only be used until the native plants attain this height. Wick application of herbicides is effective on weeds such as thistle when they are taller than the native plants. Spot spraying of herbicides is usually reserved for areas where weed growth is so heavy as to prevent effective establishment of native plants. Such areas may have to be reseeded when the weed problem is under control. Prescribed burning is often used to control weeds and improve vigor in an established planting. The burn should be conducted when cool season weeds have initiated new growth, which is in early spring throughout much of the midwestern and eastern regions. Burning has the advantage of quickly returning nutrients to the soil as well as reducing weed vigor.

Fertilizing a planting is not normally part of maintenance requirements when the planting site is properly prepared. Fertilizer often promotes weed growth at the expense of the native species. In cases where severe nutrient deficiency is suspected of inhibiting a planting, an all-purpose fertilizer can be used at half the strength recommended on the label. Acidifying fertilizers can be used on azaleas and other acid-loving plants.

Water Garden Maintenance

An ecologically balanced water garden will stay healthy and attractive with minimum maintenance. Regular maintenance will be needed to keep the garden clean, but usually a few minutes a week is all that is needed (Roth 1988). During summer, water lost to evaporation will need to be replaced, and regular maintenance of pump and filter systems will assure constantly clean water. Like most gardens, some weeding, thinning, and hand pruning of vegetation may be required, and potted plants may need dividing or transplanting into larger containers. Plants should be inspected each week for insect damage; most pests can be washed off or removed by hand; pesticides should be avoided because they can harm the ecological balance of pond life. Water lilies can be fertilized through December in warmer climates; other plants will not need fertilizer after September. Fish will need to be fed until water temperatures fall below forty-five degrees (F). In autumn, falling leaves and other debris should be removed with a siphon or plastic rake to prevent build up of decaying matter which can be toxic to aquatic animals.

Winter preparation can include a thorough cleaning after leaves have fallen, and pumps and filters are removed and stored. Water lilies should be stored in a location that will not freeze, but other plants can remain in the bottom of in-ground pools; a sheet of white plastic secured over the top will moderate temperatures. Most plastic pots and pond materials will not crack or become damaged during freezing and thawing.

Fish should be over-wintered in an indoor aquarium if the pool is likely to completely freeze or remain covered with ice for an extended period. Ice can be prevented in milder winters by placing a circulating pump on the pond bottom. Small pools can be protected by placing a PVC-covered frame over part of the water. In more severe climates, a thermostatically controlled deicer may be needed. For occasional ice, slowly melt a section using a pan of hot water (Roth 1988).

Algae Control

One of the main concerns in water gardening is controlling the growth of algae, a green plant that can affect the health and beauty of a water garden. A combination of sunlight, warm temperatures, and excessive amounts of nutrients such as nitrates and phosphorous promote algae growth. The green 'bloom' of algae common in natural ponds and small lakes in spring is often the result of an influx of nutrients transported via runoff from nearby land containing fertilizer or animal waste. These nutrients, combined with ample sunlight and warm water temperatures, cause an algae population explosion which turns the water an unattractive, murky green. In addition to its appearance, too much algae depletes the oxygen in the water, which adversely affects fish and other aquatic creatures. The result is an imbalance in the pond's ecology.

Several practices can help minimize algae growth in a pond or pool garden by helping establish and maintain ecological balance.

- Floating plants such as water lilies can be installed to shade the water surface, thus reducing sunlight and water temperature. A single mature water lily can cover an area of eight square feet or more, and smaller-leaved species can be used in small pools or tub gardens. Planting overhanging vegetation along pond margins can have the same effect.

- The amount of oxygen in the water can be increased by using oxygenating plants such as water celery and coontail or by installing fountains, waterfalls, or cascades which aerate the water. This promotes healthy fish populations which will feed on algae and insect larvae and help control mosquitos.

- Fish produce ammonia which can be toxic. In an artificial fish pond, a biological filter will promote the growth of bacteria that break down the ammonia into nitrates which promote the growth of algae, which in turn is eaten by fish.

- Avoid using copper salts and other chemicals or algicides, some of which can kill aquatic wildlife if used at high concentrations. There are some new products on the market which are advertised as safe for fish, but use of chemicals should be a last resort.

- Pond filters help rid the water of algae and debris, thus promoting a cleaner pond environment.

CHAPTER 5

Principles and Practices of Ecological Restoration

Introduction

Restoration as a general term means returning some degraded portion of landscape to an improved and more natural preexisting condition. It means reestablishing a healthy ecosystem which Aldo Leopold (1949) defined as "the capacity of the land for self-renewal." John Berger (1990), author of *Restoring the Earth*, called restoration "an effort to imitate nature in all its artistry and complexity by taking a degraded system and making it more diverse and productive." It can involve the reestablishment of an ecosystem, the control of air pollution, the prevention of acid rain, or the protection of habitat for animals where they nest, rest, and spend the winter. Restoration can mean making something grow on bare soil or on damaged, inverted soils (e.g., mine spoil). Restoration also means replanting a forest in a field. Sometimes it is the practice of an art (things work for reasons not known) and sometimes it is science (things work and we know why). Because this book deals broadly with the restoration of landscapes and managed lands, we decided to use the term ecological restoration to mean returning a specific area to its predisturbance condition, including both functional and structural characteristics. In reality, some of the projects and goals outlined in this book are habitat creations. The uncertainty about the extent and composition of the original community may make habitat creation both practical and necessary for overall ecological restoration. In the limited areas available for golf course construction, for example, restoration of a hydrologically altered wetland may not be possible, but creation of a fully functioning marsh or swamp may be possible and desirable.

What is the difference between ecological restoration and natural landscaping? To some extent, restoration encompasses all plantings designed to improve a site. Natural landscaping can be considered a form of restoration. For the purposes of this book, ecological restoration is distinquished from natural landscaping. If the goal is to restore a native plant community and a complement of native species from that community has been used in a planting scheme, then it is ecological restoration. Natural landscaping is the planting of a group of native species to meet some specific aesthetic, management, or design goals. There is generally a difference of intent and scale between these two actions. Ecological restoration is large and allows a community to evolve and natural succession to occur. Natural landscaping is conducted on a smaller scale and tends to remain at a particular managed level or state.

The natural world without human intervention is the cumulative, dynamic response to millions of years of reacting to itself and the physical world around it. Humans do not understand all of the intricacies of ecosystems. We channelize a stream to drain a wetland and in the simplicity of our desires and understanding, destroy what the wetland is to the stream—a source of water in summer, a filter of chemicals and sediment, a refuge for young fish, and a sponge to soak up and hold heavy rains. Destroying the results of millions of years of nature's work is infinitely easier than restoring it.

Ecological restoration is the art and science of recreating viable natural or ecological communities. It goes beyond just doing plantings for the purpose of stabilizing areas that are eroding. It goes beyond the idea of landscaping with native plants.

Habitat Characteristic	Habitat Elements	No. of Species
1. Forest Cover 0–9 Years Old	Forage, open areas, high stem densities, dense cover	49
2. Old Growth–Forest Cover >100 Years Old	Large diameter, mast, snags, dens, logs, layered vegetation	270
3. Oak and Oak–Pine Forest >50 Years Old	Mast, cavity trees, and snags	80
4. Pole and Sawtimber with Crown Closure over 80%	Forest cover with closed canopies	41
5. Sawtimber with 20–30% Ground Cover	Sawtimber with moderate to open understory	61
6. Oak forest >50 Years Old With a Dense Understory	Sawtimber with moderate to dense understory	74
7. Open and Semi-Open (nonforested) Habitats	Forage, dense cover, some mast, cavities	200
8. Permanent Water Sources	Water for drinking, breeding, feeding, roosting sites	——
9./10. Den Trees/Snags	Cavities for nesting, roosting, song and observation perches, feeding, shelter, and eventually fallen woody material	155

Table 5.1. Representative habitat characteristics identified for the Mark Twain National Forest (Putnam 1988).

It is an attempt to recreate nature. Nature, of course, is a lot of things. It is weather and climate, soils, water, slope, aspect, and altitude. It is also a squirrel that buries an acorn and a bird that eats a seed. Nature is mushrooms, lichens, wildflowers, and trees. To restore a truly natural system is beyond the capacity and knowledge of humans. We can, however, bring together the basic components and characteristic plants and animals of an area (Table 5.1). Nature is assisted and directed by ecological restoration. Natural processes will take over and other components of the natural system will naturally invade the restored system. Of course, some systems are so devastated and the sources of plants and animals available for reinvasion are so remote that nearly everything must be provided by the restorationist.

Don Falk (1990) of the Center for Plant Conservation says in ecological restoration we seek not to "preserve" a static entity but to protect and nurture its capacity for change. "Restoration thus uses the past not as a goal but as a reference point for the future. If we seek to recreate the temperate forests, tallgrass savannas, or desert communities of centuries past, it is not to turn back the evolutionary clock but to set it ticking again."

John Cairns (1988b) questions whether restoration of ecological communities to their original condition may be practical or possible. He offers the following observations:

■ Information about the original system may be inadequate. For example, there is no adequate description of the preindustrial Ohio River that includes a detailed species inventory, along with detailed descriptions of the spatial relationships, trophic dynamics, and functional attributes of the system.

■ An adequate source of species for recolonization may not be available because of the uniqueness of the damaged community or because the remaining communities of this type would be damaged by removing organisms for recolonization elsewhere.

■ It may be impossible to put a halt to some of the factors causing the damage. The notorious example is acid rain, which may originate many miles away in a political jurisdiction over which the restorationist has no influence.

■ The original ecosystem may have developed as a result of a sequence of meteorological events that are unlikely to be repeated and may be difficult or impossible to reproduce.

■ The sheer complexity of duplicating the sequencing of species introductions is overwhelming. To recreate the original community, it is not only necessary that species colonize at the appropriate time, but also that some decolonize (disappear) at the appropriate time.

The restoration efforts envisioned by this book can be considered only partial restorations of natural communities, designed to set the "evolutionary clock ticking again." A repertoire of target ecological communities is compiled as Appendix A. They represent over 200 different dominant communities found in 30 natural regions of the continental United States (see Chapter 6). These communities can be targets for more restoration research as well as starting points for initiating restoration efforts. The species lists will, we hope, encourage nurseries to propagate these plant materials for restoring these basic ecological communities.

The type of restoration program undertaken depends upon the goals for a site or landscape. Goals can include aesthetics, reestablishment of natural conditions, maintenance of a diversity of plant and animal species, protection of the local gene pool of particular species, reduction of grounds maintenance costs and environmental impacts, and creation of representative local ecosystems.

Often more than one goal is possible for any given restoration project. While all the reasons listed above are significant, goals may be quite limited. For instance, restoration to increase the carrying capacity for a given plant or animal is a specific desired outcome. However, in the process of restoring a specific habitat for a species, habitat for other species will also result. A number of spin-off benefits may result from any restoration project, but the reported success or failure of the effort is almost entirely based on a single criterion. A more comprehensive, idealized goal would be to emulate a healthy, ecologically robust, natural, self-regulating system that is integrated with the ecological landscape in which it is situated (Cairns 1991).

The focus of this handbook will be restoring presettlement, predisturbance, and/or natural conditions. Presettlement conditions are defined as conditions encountered by the first European settlers to describe the North American landscape.

Although early historical accounts of the first Europeans are sketchy, we have a general idea of the broad vegetation types present during the 1700s and 1800s. Native Americans had inhabited North America for centuries and through the pattern and timing of harvests, as well as through the burning, pruning, weeding, and planting of some sites, certain mixtures and frequencies of plants and animals were favored (for examples, see Boyd 1986; Whitney 1994; Anderson 1993, 1996). Native American land uses sustained native landscapes rather than severely altering or degrading them. The mutual existence of large herds of bison and Native Americans contrasted with the rapid decline in bison herds as European settlers moved westward supports this idea. In fact, our understanding of indigenous methods and practices may lead to improved restoration methods and approaches to restoration.

Predisturbance is the approximate condition of the site prior to a major land degradation event. Sometimes neither presettlement nor predisturbance conditions are possible as a restoration goal. Information regarding species composition and structure may be lacking or the climate or another important physical parameter may not exert the same influence as it did in presettlement or predisturbance times. The behavior of an ecological system depends to some degree upon its unique past, specific spatial setting, and current influences (Pickett and Parker 1994). All ecosystems are the product of climatological and biogeochemical sequences that probably are unique (Cairns 1991). This is why restoration projects must be very site-specific and flexible in terms of preferred goals.

Restoration to predisturbance condition may not even be as appropriate as restoration to a present natural condition. The difference between predisturbance and natural conditions is that predisturbance is a past condition, whether it is a few years or a few decades. Predisturbance condition is our best estimation of how an area appeared before a human disturbance occurred. Natural condition may be indistinguishable from predisturbance condition, especially when one considers our lack of detailed accounts of many ecological communities. The difference in these conditions is that the natural condition is based on a present day existing model system (such as an adjacent intact natural area) while predisturbance (as well as presettlement) conditions are based on conceived previous or historic accounts. Species in ecosystems are influenced by both short- and long-term events, and the inability to recreate a special sequence may mean that attempts to restore to presettlement or predisturbance conditions are unlikely to succeed because of this fact alone (Cairns 1989). That is exactly why restoration of natural conditions representative of an existing local ecosystem (comparable undisturbed system) is often more realistic. As Dan Willard (quoted in Primack et al. [1995]) stated, "restore to a system that will work, not necessarily the system that you wish would work." Some other types of restoration include rehabilitation of selected attributes (e.g., forested) as well as creation of an alternative ecosystem (Magnuson et al. 1980).

Pickett and Parker (1994) stated that one of the "pitfalls" of restoration ecology is "the assumption that there is one reference state or system that can inform restoration." To assume that there is only one ecologically legitimate or ideal system for a site is a trap. However, Aronson et al. (1995) argue that for the purposes of project design and evaluation, it is desirable to establish at the outset some standard of comparison and evaluation, even if it is arbitrary and imperfect. Aronson et al. (1993a, 1993b) call this the "ecosystem of reference." This standard is followed (in the broad sense) in the present handbook by the use of Ecological Restoration Types and Dominant Ecological Communities. A standard of comparison and evaluation allows the restorationist to track progress and to determine success or evaluate the need to select another ecosystem type for restoration on a particular site. The creation of habitats in places they may never have occupied is an approach with great potential for sites that are so degraded that restoration to even a semblance of the original habitat type is impractical. This approach is valuable for landfilled sites, surface mined areas, contaminated sites, or regionally rare communities.

Matsui (1996) described an interesting project from 20th-century Japan in which the goal was to create a majestic, tranquil forest and landscape that would be able to regenerate and survive almost forever, not for economic reasons, but because this was to be a holy place. Although the plan for the project (Meji Shrine) was modeled to some extent on natural forests of the area, no attempt was made to create a natural ecosystem in the modern sense.

Instead, they assembled species from various areas that they felt would grow well on the site. The primary goal was for the forest to last forever with little or no maintenance. While modeling a restoration project after a natural community is typically how restoration projects are planned, value (ecological and cultural) also exists in creating habitats for other purposes.

The best manner in which restoration projects should be designed and evaluated is often not clearly defined (Cairns 1989, Simberloff 1990, Sprugel 1991). Many ecological and societal reasons influence the choice of certain reference states, including aesthetics, commodity production, ecosystem services, and species protection, among others (Pickett and Parker 1994). Aronson et al. (1995) asked the question, "If no reference or 'control' is selected, how can the project be evaluated?" For the purposes of this handbook, ecological restoration means recreating both functional and structural characteristics of a damaged ecosystem. While this book concentrates mostly on vegetation, discussions of other important considerations also are presented.

A full-scale model need not exist; all else failing, at least a few square feet or tens of square feet may usually be found as a vestige of former vegetation, and this can serve as a micro-model, reference, and inspiration all at the same time (Aronson et al. 1995). No criteria or guidelines that provide ecological predictive capability are found in the professional literature (Cairns 1989). Some efforts will be so far from expectation that the project should be terminated and a fresh start made using a new design. Even an ecologically unsuccessful project may appear successful to laymen because a devegetated area has been revegetated or because they may not understand the management effort required to sustain the ecosystem (ibid.). All restoration projects should have an explicit statement of the product to be ultimately produced and the condition at intermediate stages toward the eventual goal. Each restoration project should have several alternative standby plans so that if Plan A fails, Plan B can be implemented. Communicating that the outcome is uncertain is extremely important. Incorporating an alternative course of action into the restoration plan is equally important. In any restoration project, the goals and the consequences of meeting those goals must be part of the planning process (Cairns 1991).

We now prefer to think in terms of many equally possible ecosystem trajectories and of guiding or piloting the ecosystem under study in one direction or another (Aronson et al. 1995). Luken (1990) used the phrase "directing ecological succession" in order to manage for a particular successional stage.

Ecological Restoration Planning

Many steps are required to conduct a successful restoration project. The most important component for success is understanding that restoration takes time. Actually, restoration, like other aspects of looking after the earth, is an eternal vigil. As part of that vigil, planting trees is an act of hope and optimism. If one plants small trees and waits for a canopy to plant understory shrubs, many years will pass. Even a restored prairie of grasses and forbs will take years before it resembles a natural community. The level of maintenance required to look after a forest while it is developing is, however, relatively low compared to the meticulous, incessant care of a golf green or a lawn. The long-term maintenance of a restored natural ecosystem is negligible.

The Natural Resources Management Staff of the McHenry County (Illinois) Conservation District (1996) have broken the process of natural community restoration into three distinct phases: structural, functional, and compositional. During the **structural phase** a community's original structure is recreated, as invasive brush, fencelines, roadways, buildings, and other human modifications to the site are removed. The **functional phase** involves the reintroduction of ecological factors such as fire, grazing, and hydrological processes that shaped the site's ecological communities prior to Euro-American settlement. Finally, in the **compositional phase**, plant and animal species are reintroduced and exotic species removed in order to restore the community as closely as possible to its historic composition. Planning a restoration project in phases such as these improves the success rate.

Jackson et al. (1995) stated that while every restoration project is unique, all include the following elements:

■ **Judgement of need.** The process of ecological restoration begins with a judgement that an ecosystem is damaged by humans to the point that it will not regain its former characteristic properties in the near term (two generations, or about fifty years), and that continued degradation may occur.

■ **Ecological approach.** Ecological restoration implies that we wish to restore organisms and their interactions with one another and the physical environment. Ecological restoration concentrates on processes such as persistence of species through natural recruitment and survival; functional food webs; system-wide nutrient conservation via relationships among plants, animals, and the decomposer (detritivore) community; the integrity of watersheds; and abiotic processes that shape the community such as periodic floods and fires. Human beings are a part of nature. By working with the full diversity and dynamics of ecosystems, we restore sustainable relationships between nature and culture.

■ **Goal establishment and evaluation process.** Ecological restoration is a deliberate intervention that requires carefully set goals and objective evaluation of the success of restoration activities.

■ **Limitations of ecological restoration.** Ecological restoration in its purest sense is not always possible; it depends upon four interrelated social and biological conditions: (1) how nature is valued by society, (2) the extent of social commitment to ecological restoration, (3) the ecological circumstances under which restoration is attempted, and (4) the quality of restorationists' judgements about how to accomplish restoration. Without optimum conditions in all four areas, complete ecological restoration is not possible.

In addition to Jackson's four elements of every restoration project, there are some crucial steps that should be addressed in the planning stage. These crucial steps are

■ **Seek help from experts and use a multidisciplinary team in planning.** This is especially true of wetland projects that need hydrologic information. Biologists, ecologists, horticulturalists, foresters, agronomists, engineers, geologists, soil scientists, and hydrogeologists all have expertise to offer restorationists. Volunteers, while they are not paid professionals, have been very helpful in many restoration projects. Working with volunteers is also an excellent opportunity for the restorationist to educate the general public about ecological restoration. The team should be well versed on safety issues for a particular site including site hazards, equipment use, poisonous plants, and infectious insects (Sauer 1998).

■ **Clearly define objectives, goals, and measurements**. An ecosystem is complex and dynamic. A successful restoration will evolve through many stages before the end result is achieved. Therefore, it is important to establish site-specific goals for each project. These goals can relate to the number and composition of plant species restored, size of area, structure of vegetation, reinvasion or establishment of animal species, density of exotics, functions restored to the site, and aesthetics. Both short- and long-term goals are needed to ensure that the various stages result in the desired final goal. While many restoration projects are evaluated based on the natural functions established or degree of similarity to a reference ecological community, the majority of these projects do not achieve functional equivalency (at least not for many years, possibly decades). However, restoration projects that do not achieve functional equivalency should not be considered failures, especially if some of the stated goals are met and the site is in an improved condition from that prior to restoration.

■ **Conduct a site-specific analysis of the hydrological, geological, and biological variables at the site**. This can include soil types, basic geology, disturbance history, biological inventory, and natural seed sources of desirable and undesirable species. The county soil survey, published by the U.S.D.A. Natural Resources Conservation Service (formerly the Soil Conservation Service), is available at no cost from the District Conservationists in most county seats in the nation. It is an excellent source of information on soils, hydrology, and potential vegetation. The sources and connections to water supplies should also be identified.

■ **Develop a detailed site plan**. The detailed site plan should include specific information about the characteristics of that site and which areas of a site will be restored to what type of community. Clearly outline the boundaries of the site and different communities within the site on a site plan map. A useful way to do this is to divide a site into appropriate planning units and to use different scales and/or layers of maps and information in the plan. These might include topographic, hydrologic, geologic, soil, man-made structures (i.e., roads, powerlines, underground pipelines, etc.), and cover type/land use maps showing vegetation types. Boundaries of the site must be well known and marked on maps (and possibly in the field) and adjacent land uses will likely need to be considered.

■ **Reflect the local biota in species composition and source of material**. Every regional population of plants and animals has a local gene pool that has evolved. Restoration success is enhanced when local sources of material are used. Many sources suggest using plant material collected within a fifty to sixty mile radius of the restoration site. An adjacent existing community may be the source of material, either through actual seed collecting or through passive dispersal.

■ **Select species carefully**. Use this book, literature sources, local experts, and an analysis of similar communities in the selection of species for use in the restoration project. The Ecological Restoration Types list many literature sources that may be especially helpful in the development of such a list. The species list should be site specific, exclude problematic nonnative species, and include only native species known from that area.

■ **Create a detailed design for each different community including a spatial and temporal planting plan**. Experts can provide valuable information concerning species specific requirements that impact both the location and sequence of plantings. The design should minimize maintenance requirements. Ecological communities are very dynamic and often require plant succession, or seral stages, to achieve the desired condition. For example, if restoring a forest, the plan may be to first establish a canopy before planting the desired understory shrubs and herbaceous plants that require a certain amount of shade to survive.

- **Identify sources for the species to be used in the project**. This may include seed or nursery stock, plants grown in one's own greenhouse, or stock taken from other sites. Taking plants from the wild is viewed as destructive and unethical except in certain circumstances (e.g., road construction site where the plants will otherwise be destroyed). In a restoration project this consideration is of utmost importance as desired native plant species (especially of local stock) are difficult if not impossible to obtain. Often restorationists contract in advance with a nursery, horticulturist, or even local farmers to produce the needed plant materials.
- **Prepare the site**. This can include physical changes such as grading. More often, it will involve decisions concerning whether to add fertilizer, water, and other soil treatments. It can also involve management decisions concerning existing vegetation on the site.
- **Supervise project implementation**. Bulldozer operators and workers removing vegetation or applying chemicals or fertilizer need guidance to ensure that the work done is consistent with the plan. Be sure planting, either with seeds or plants, is conducted correctly and with the guidance and supervision of someone familiar with the site plan and planting techniques.
- **Control exotic and undesirable species**. Monitoring for exotics that may be problems in your area is crucial. Control of exotics may include manual removal or the use of chemicals. A discussion of exotic plants is presented later in this chapter and weed control sections are included in all Ecological Restoration Types.
- **Establish a plan for feedback and midcourse corrections**. Many things can change throughout a restoration. A process should be created for making assessments of conditions and responding at any time. Plants or seed availability, replanting needs, water level changes, substrate differences, and exotic invasions can necessitate mid-course corrections in order to achieve success.
- **Develop a plan for long-term monitoring, management, and maintenance**. This plan should very specifically address the goals of the project and how to measure the success

of reaching those goals. A restored ecosystem should be similar to the original ecosystem in both structure and function. To restore a natural tallgrass prairie with at least fifty native species from the local gene pool is an example of a measurable goal.

Ecological Restoration Design: Considerations

Why is the site degraded?

The primary consideration in designing a restoration plan must address the reason(s) the site became initially degraded/damaged. If the reasons for site degradation are not eliminated or reduced, restoration efforts may not be successful. Several reasons for degradation are grazing, mining, alteration of rivers and streams, urbanization, acid rain, pollution, and logging. Primack et al. (1995) suggest removing the immediate impact, restoring the system the best you can, making continued improvements, and thinking about sustainability.

How is the site degraded?

It is better not to disturb or destroy in the first place. Only the areas with the least disturbance recover (and then only slowly) to the point where differences from natural ecosystems are undetectable. The degree of degradation must be assessed in the restoration planning period to determine fundamental site preparation procedures. For example, is the soil severely degraded so that soil will have to be brought into the system or improved prior to planting? Has the geology or substrate been altered? Is the hydrology altered to the point that it may not support the intended restoration goals? Are seed sources intact or nearby? Are the necessary microorganisms in the soil? These site characteristics must be assessed and managed in all restoration projects. Fortunately, the data found in the large and rapidly growing base of restoration research can be used as tools or benchmarks in an attempt to remedy degradation. As House (1996) proclaimed, "it is impossible to spend long hours under difficult conditions trying to reestablish critical elements of damaged habitats without being acutely aware of land-use practices that led to the damage being repaired."

Early restoration consisted primarily of removing the stress from ecosystems and letting nature take its course (e.g., Gameson and Wheeler 1977,

Cairns et al. 1971). If the soils, substrate, or hydrology have not been severely degraded, restoration may proceed. In these cases, existing desirable species already established on site should be allowed to persist, while exotics and undesirable species should be removed. Ecological succession should be assisted and the area managed, but full-scale restoration as discussed herein may not be required. The need to actively restore an area and the degree to which it should be restored are completely subjective and best determined by a consensus of several experts. A crucial point that we would like to emphasize is that humans are a part of many ecosystems. Human interaction with particular areas must be addressed in designing a restoration project.

What are the roles of groups other than plants?

The structure and function of insect, bird, fish, mammal, fungi, and/or soil invertebrate communities and the role these communities play in restoration design may need to be evaluated and monitored. Functional groups must be considered in light of the balance of trophic levels, predator/prey relationships, pollinators, and seed dispersers.

The structure and function of different groups of organisms (aside from vascular plants) is an important consideration in any restoration project. Complete restoration of the vegetative structure of a site does not necessarily mean that the ecological community will function as a natural one. A good example is found in Zedler (1992). Extensive scale insect *(Haliaspis spartina)* attacks on cordgrass *(Spartina foliosa)* were discovered at San Diego's dredge spoil island. Where predatory beetles *(Coleomegilla fuscilabris)* were abundant, i.e., on the mainland surrounding the Bay, scale insects are kept in check. It was then found that beetles were lacking on the dredge spoil island, thus allowing an unchecked scale insect outbreak (Zedler 1992). Disconnected habitats can reduce or eliminate natural movements of animals, even flying insects.

Mattoni (1989) summarized an attempt to restore habitat for the El Segundo blue butterfly *(Euphilotes bernardino allyni)* in California. Common buckwheat *(Eriogonum fasiculatum)* was introduced as a foodplant for this endangered butterfly. However, populations did not increase as common buckwheat was established. It was later found that the El Segundo blue butterfly was being parasitized by two moth larvae that shared the flowerhead

food resource of the butterfly. The common buckwheat blooms a month earlier than the native coastal buckwheat *(Eriogonum latifolium)*. The moths, with multiple broods each year, produced an extra generation that expanded their populations and essentially overwhelmed the El Segundo blue butterfly. Consequently, removal of the densest common buckwheat stands led to a fourfold increase in the El Segundo blue butterfly populations.

Animals play a crucial role in a number of important processes within an ecosystem, such as organic matter decomposition and pollination. Nearly all the animals required for these processes are sufficiently mobile to appear without assistance, for instance soil fauna (Neumann 1973). However, some species, particularly earthworms, are relatively immobile and it may be valuable to introduce them (Bradshaw 1983). Many animals will colonize without assistance if the conditions are appropriate. In some cases, specific habitat requirements must be provided for particular species.

Terrestrial ecosystems also have a belowground component (Clark 1975). The microbial community in the soil is responsible for tying up and releasing nutrients and for converting them into more available forms. The relationship microbial organisms have with plants is the determining factor for proper ecosystem function (Miller 1985). Some plants do not form mycorrhizal associations, such as mustards (Brassicaceae), sedges (Cyperaceae), rushes (Juncaceae), and many goosefoots (Chenopodiaceae). Many other plants must form mycorrhizal associations to prosper or survive.

Although many schemes for classifying mycorrhizal associations have been proposed, in general, these associations can be divided into three major groups: the sheathing or ectomycorrhizae; nonsheathing, vesicular arbuscular (VA) or endomycorrhizae; and ericaceous or ectendomycorrhizae (ibid.). The kinds of mycorrhizae present in a community may be influenced by the harshness of the site and the types of vegetation present. For example, in the organic soils of tundra and alpine regions, ericoid mycorrhizal associations predominate; in boreal forests ectomycorrhizae are most common; while in grasslands and shrublands, VA mycorrhizae dominate (ibid.). So, the types of mycorrhizae in community vary depending upon the nature of the soil, the vegetation, and the microclimate. For additional information concerning mycorrhizae see Allen (1991, 1992), Harley and Smith (1983), Powell and Bagyaraj (1984), and Safir (1985). New technology has been developed to introduce mycorrhizae to degraded sites and improve the success of planting.

How long will a restoration project take?

Pickett and Parker's 1994) second "pitfall" of restoration ecology (for the first "pitfall" see Introduction) is to think of restoration as a discrete event. Leslie Sauer states "the most difficult aspect of all is that restoration is a long-term effort requiring a high degree of expertise and commitment rather than a quick fix." Because the natural world is an evolving and dynamic collection of systems, restoration should be seen as intervention into an ongoing process rather than a lasting patch or repair (ibid.). For the full benefit to be derived from restoration, the experimental results need to be analyzed over the long term and the models and/or the site adjusted.

In order to evaluate a project objectively, specific criteria for success should be established that reflect the ecological approach described above. Such criteria might include the presence, cover, and distribution of a plant species; the ability of the vegetation to respond to normal disturbance regimes and climatic fluctuations; use of the ecosystem by particular animal species; soil condition and colonization by particular invertebrates, fungi, and bacteria; characteristics of nutrient cycling; and hydrologic regime (Jackson et al. 1995). If one includes self-maintenance as a critical criterion for restoration success, it will be difficult to make such a judgement in a time span of interest to most humans, particularly politicians (Cairns 1991): Restoration exhibiting self-maintenance will take years at a minimum, but may take hundreds of years for selected ecosystems.

An underlying assumption in ecological restoration is that the outcome of certain actions results in a desired condition. Harper (1982) stated "ecology has tended to be highly descriptive in nature, but has thus far made little progress toward reaching maturity as a rigorous experimental and predictive science." Diamond (1987) pointed out that the traditional approach of ecology is reductionist and that restoration provides new knowledge not offered by traditional approaches because it is synthetic. Thus, the prospects for predicting the out-

come of rehabilitation or restoration efforts with any reasonable degree of certainty are not good (Cairns 1986, 1987).

Plant materials, heavy equipment and operators, or any other materials or service may not be available when you would like them to be. This is why a comprehensive restoration plan including these considerations is crucial to success. And a restoration plan is only a beginning. A long-term monitoring plan (including adaptive management practices) is a crucial component of any restoration plan. Restorationists should try to plan projects not only to get the job done, but to answer questions and provide information that will help to do the job better next time. Failures are often the best teachers of techniques. While it may be difficult or expensive, assessments of restoration projects should measure not only community structure or composition, but its functional levels.

It is recommended that project managers develop a timeline for restoration projects, including planning and research, site preparation, implementation, short-term management, long-term expectations, and monitoring.

What impacts does the larger landscape have on restoration success?

The ecologic compatibility of any restoration project within the larger landscape mosaic can be crucial. Ecosystems and ecological communities are not isolated, but are interconnected subsystems of a larger landscape. Conflicts can arise when subsystems are restored at the expense of connected subsystems. For example, stabilization of lake levels or stream flows may affect adjacent wetlands that depend upon variable water levels (Cairns 1991).

If surrounding land use and the dynamics of natural systems are not taken into consideration, some restoration efforts may be in vain. Cairns (1991) gives two examples: (1) the recreation of wetlands will not be possible if the hydrologic cycle is modified in an attempt to prevent flooding of lowland areas; and (2) natural systems evolve and river channels do not remain in a fixed location, nor do barrier islands. Attempts to force natural systems to do so may seriously impair a restoration effort.

The presence and quality of existing natural communities, as well as the prevailing natural character of the surrounding landscape will, in large part, influence the design. Knowledge of how the predeveloped landscape appeared is helpful in developed areas. Existing ecological communities may be the focal point of a landscape plan, and high quality communities, in general, require minimum effort to achieve multiple objectives for aesthetics, wildlife, and economics.

While ecology textbooks that focus on elememts of the landscape (as opposed to communities or ecosystems) are uncommon, a few such texts have been published: Forman and Godron (1986), Urban et al. (1987), Turner (1989), Naveh and Lieberman (1990), Satterlund and Adams (1992), Harker et al. (1993), Hobbs and Saunders (1993), Ehrenfield (1995d), Forman (1995).

Can plant succession alone progress to the desired community?

If disturbance causes a regression to an earlier successional stage and the successional process then follows the same track it did before, one might expect the system to eventually achieve complete recovery (Cairns 1989). This is not always the case since: (1) the species representing the earlier successional stages may no longer be present for reinvasion, and (2) because other species, including exotics, not present during the earlier developmental period may be the post-disturbance colonizers (ibid.). Thus, one has a species composition not characteristic of an earlier successional stage that is likely to influence subsequent stages. And given more time, it does not seem likely that a particular later seral stage will be established.

A particular stable, seral community or successional stage is generally the desired goal of restoration, so it is crucial that we incorporate the principles of plant succession and the development of an ecological community into our restoration plans. This might require a disturbance that was once natural but now must be carried out by humans, i.e., burning, flooding. Vegetation is very dynamic and does not always "succeed" to a predictable and/or stable climax state. The reasons for this include both man-made and natural phenomena that alter the structure of an ecological community so that the natural successional pathway is different than it was when that particular community developed.

As restorationists, this means that we may have to, for example, introduce certain species of trees, shrubs, or herbaceous plants into a restored forest at a particular point in time, such as after a canopy has formed, to accommodate survival of shade tolerant species. Conversely, shade intolerant species (which are unlikely to survive in a mature forest, except in natural openings and tree gaps) will not persist when a canopy develops. Species belonging to later successional stages may be limited by nitrogen accumulation (Leisman 1957, Bradshaw 1983). The principles of plant succession and the concept of a climax condition have implications for restoration that involve temporal and spatial considerations. We can assist ecosystem development by treating the factors limiting development (see Bradshaw 1983). We must accept the view that climax condition is essentially a dynamic flux affected by many factors operating at different scales of time and space.

Some valuable resources that provide discussions of plant succession are Drury and Nisbet (1973), Shugart (1984), Gray et al. (1987), Luken (1990), and Sauer (1998).

Forest Restoration

When we think of recreating a forest, large trees come to mind. But a forest is more than large trees. A forest contains an understory of trees and shrubs, a ground layer of herbaceous plants, and leaf litter. Birds, squirrels, bugs, mushrooms, and a list of living organisms too long to name inhabit the forest. Forests are remarkably complex, with great differences in composition and character, even within the same region. Site-specific strategies must be developed for a restoration project. How do we know when we have restored a forest? The recreation of an upland forest can be evaluated in at least two ways. Evelyn Howell (1986) described it this way.

"One emphasizes community structure and species composition, and judges the success of a restoration effort by asking how closely the resulting community resembles the natural, or 'model,' community with respect to characteristics such as the relative abundances, age-class structure, spacing, and distribution of particular species. An assumption underlying this 'compositional' approach is that if these species groupings are fairly accurately reproduced, then the dynamics and functions of the communities will also resemble those of the model community.

The second way of establishing goals and evaluating the success of restoration projects empha-

sizes ecosystem functions, often with little or no reference to species composition. From this point of view, for example, the presence or absence of a particular species is less important than the provision of functions and processes such as nutrient cycling, erosion control, or biomass production."

Many techniques for restoring a forest have been tried and new procedures are regularly being introduced. The creation of canopy is the most critical issue because of its influence on the microclimate of the community. Starting from a treeless situation, forest restorations use at least six approaches (Howell 1986).

■ Plant canopy trees in ultimately desired densities and proportions; mulch the ground beneath and around the trees; immediately plant desired midstory and understory species.

■ Plant and mulch canopy trees as in the above approach, but plant light-loving ground cover initially (or let "weeds" grow), and add (or encourage natural invasion by) woodland understory and midstory later as shade develops.

■ Plant trees in savanna distribution patterns (less than ultimately desired densities), with savanna understory (prairie plants). Then, as shade develops, gradually plant additional trees; and. finally, plant (and/or manage for the natural invasion of) desired understory and midstory species.

■ Plant trees in greater than ultimately desired densities and either thin or allow self-thinning as the canopy develops. Add midstory and understory species later and/or manage for natural invasion.

■ Plant short-lived, fast-growing trees (aspen) or tall shrubs (pagoda dogwood, alder) as a cover crop and as this canopy develops, underplant with slow-growing, shade-tolerant, long-lived trees that will become the site dominants. Upgrade the understory as the canopy progresses, thinning the cover crop species as necessary to reduce competition with the eventual dominants.

■ Do no planting; allow woody species to invade and selectively remove those which are not desired; treat understory and midstory in a similar fashion.

In most of these methods, the herbaceous layer is added after the formation of canopy. Transplants and potted seedlings are recommended; however,

seeds have been used successfully. "In general, success is closely related to the care the plantings receive during the first year, watering at regular intervals being especially important. Habitat matches are also especially important, factors most likely to be crucial being light, humidity, and depth of litter" (Howell 1986). In some situations, acorn planting can be used to establish forests at lower costs than tree planting. They can be collected locally with minimal impact upon natural ecosystems. About 1000 per acre is a desired stocking rate.

A combination of stresses including drought, low soil fertility, excessive salinity, and herbivory impact desert systems (Virginia 1990). Many woody, legume species of desert trees develop deep root systems and therefore can serve to provide moisture and nutrients (especially by nitrogen fixation) for a herbaceous layer. Techniques used to improve establishment of plants in arid regions include deep ripping, augering holes to improve water infiltration, soil imprinting, careful seed and plant selection, and appropriate symbiont inoculation (Virginia 1990, Dixon 1990, Bainbridge 1990, Bainbridge, et al. 1995).

The important role of ectomycorrhizal fungi in the establishment of forest is now widely recognized. Donald Marx (1991a) reports that ectomycorrhizae occur on about 10 percent of the world flora, and that 2,100 species of fungi form ectomycorrhizae with forest trees in North America. The most common fungus and one now used extensively for inoculating tree roots is *Pisolithus tinctorius*. The fruiting body of this fungus is the familiar puffball. Ectomycorrhizae of *P. tinctorius,* or Pt for short, have formed in many species of *Abies, Carya, Picea, Pinus, Pseudotsuga, Quercus, Castanea, Fagus,* and *Salix.* Because trees with ectomycorrhizae have a more active area for nutrient and water absorption, they are able to absorb and accumulate nitrogen, phosphorus, potassium, and calcium more rapidly and for longer periods than trees without ectomycorrhizae (Marx 1991b). Marx further reports that "ectomycorrhizae also appear to increase the trees' tolerance to drought, high soil temperatures, organic and inorganic soil toxins, and extremes of soil acidity caused by high levels of sulfur, manganese, or aluminum." Private and government nurseries are now inoculating tree stock. Restorationists should look for and use inoculated nursery stock. Three methods have been suggested (Perry and Amaranthus 1990) for inoculating seedlings: (1) using whole soil from established plant communities, (2) using pieces of chopped up root, and (3) using spores of the fungi. The duff inoculation method that has been used in nursery beds can be used in the field. Forest duff, collected from mature stands of target species can be broadcast and plowed in at a particular site (St. John 1990). "Because of the important role of mycorrhizal fungi and other rhizoshpere organisms in creating favorable soil structure, sites with very sandy or very clayey soils may be especially susceptible to improvement" (Perry and Amaranthus 1990).

Sauer (1998) offers a variety of recommendations for building soil systems including building populations of soil fungi, protecting relatively undisturbed native soil, minimizing "working the soil," and reevaluating additives and soil amendments such as nitrogen, mulch, and organic matter.

Grassland Restoration

Modern ecological restoration began in the 1930s at the University of Wisconsin Arboretum when Aldo Leopold began restoring a site to native prairie. Prairie may now be the community type restorationists know best. The recreation of prairie has been accomplished many times over. Also, the management technique of using fire is understood. Fire is so widely used as a management tool that conservation organizations like The Nature Conservancy conduct fire schools for staff.

The restoration of grasslands follows the basic steps outlined for all restoration projects. In addition, the following should be considered:

■ Planting can be accomplished from either seeds or greenhouse grown seedlings.

■ Many prairie forb seeds require stratification (cold treatment); some (i.e., those with hard seed coats) also require scarification (nicking or roughing up the seed coat). Stratification should occur soon after collection to prevent drying and dormancy.

■ Controlled burning after the third year is the most effective management option. Early spring mowing inhibits weeds and woody species.

■ Collecting prairie seeds takes considerable effort. The seeds of prairie species mature at different seasons of the year, at different times at different locations, or a few at a time throughout the season. This means that collecting must be done a number of times at the same site.

■ Legumes need to be inoculated with appropriate rhizobia bacteria for the nitrogen fixing capability to develop. Other mycorrhizal fungi need to be evaluated for improving specific species functioning.

Wetland Restoration

Wetlands are America's most maligned environment. This is ironic because of the many essential functions of wetlands. Wetlands moderate the effects of floods and serve as wildlife habitat, fish propagation and nursery areas, flood storage, groundwater recharge, and filters that prevent sediment and chemicals from entering streams (Mitsch and Gosselink 1993). Wetland restoration is a very active field today. This is primarily due to federal regulations promulgated as a result of extensive wetland losses suffered in the United States. Overall, the United States has lost over fifty percent of its wetlands and in some states, the loss is ninety percent (National Research Council 1992). Wetland restorations are undertaken for a variety of reasons and often fall short of restoring all the values of original wetlands. Some of the reasons for restoration projects include creation of wildlife habitat, improvement of water quality, storage of water, and reduction of flooding.

The success of restoring wetlands is higher for sites that were originally wetlands but modified for some reason such as agriculture. If the hydrologic regime has been destroyed on a site, then it must be restored if the wetland is to survive. Success has occurred most often with the revegetation of coastal, estuarine, and freshwater marshes (Kusler and Kentula 1990). Prior to modification of a site, it is very important to conduct an investigation for jurisdictional wetlands. The U. S. Army Corps of Engineers (ACE) has to concur with any wetland delineation to make it official. It is important to remember that most wetlands may not be disturbed without a permit from at least the ACE and often local or state agencies (see Appendix F for wetland permitting process). It may require wetland replacement or mitigation to obtain a permit, but often the best and cheapest mitigation is to avoid disturbing jurisdictional wetlands. These natural wetlands can often be incorporated into your site plan and increase habitat diversity.

Wetland communities include riparian forests, swamps, bogs, wet meadows, and bottomland hardwood forests. Mitigation of wetland losses, while required by law, is not the only motivating factor for recommending restoration. Because of the value of wetlands and the significant loss of this community type, we recommend restoring these communities whenever appropriate sites are available. If a golf course is going to be developed from an old farm, we would encourage close examination to determine the existence of suitable areas that can be restored to wetland. In many cases there are sites, such as along an intermittent stream, where a small wetland may be created that never existed before.

In addition to all of the general steps that must be followed for any restoration project, there are a few specific minimum requirements that wetland restorationist Hal Bryan (pers. com.) recommends when considering a wetland creation or restoration plan. Three main areas must be addressed: hydrology, soils, and vegetation.

Hydrology

- If you get hydrology right, you will get the wetland right. Evaluate the hydrologic conditions for a site, such as water level elevations, velocity, hydroperiod, salinity, nutrient and chemical levels, and sedimentation rates.
- Source of water can be surface flooding from streams, groundwater, or rainwater.
- Watershed/wetland size ratio should be considered. Usually the larger, the better. Smaller ratios are possible to restore with more poorly drained soils.
- Periodicity of inundation or saturation determines the type of community restored or created. For example, a bottomland hardwood forest may require twenty to thirty days of continuous inundation in the growing season and a period of drydown in the late summer. A shrub swamp may need twice as much inundation with little drydown.
- Restoration of hydrology may require the removal of berms, the plugging of human-made ditches, the renovation of stream meanders and flood regimes, or the destruction of subsurface drainage tiles.
- Creation of shallow ponds for habitat diversity will increase the number of kinds of plants and animals. Use clay subsoil from ponds for ditch plugs or berms.
- The use of water control structures, where possible, will provide some ability to manipulate water levels, which is especially important during the early periods of plant establishment.

Soils

- Select areas of hydric soils for restoration. A list and soil maps can be obtained from the U.S.D.A. Natural Resources Conservation Service. Be sure you are not restoring in an existing jurisdictional wetland without a permit.
- Store and replace topsoil during the construction phase.
- Use topsoil or muck from impacted wetlands, if possible. Muck is a seed source and also provides many microorganisms characteristic of the wetland but not commercially available.
- Be concerned with erosion possibilities. Slopes should be very gentle and stable even without vegetative cover.
- Use small equipment on site; large machinery can compact subsoil and restrict root penetration.

Vegetation

- When possible, select potential restoration locations adjacent to existing wetlands to allow natural regeneration.
- Trees can often be reestablished using bare-root seedlings (usually on ten foot centers) or acorns (1000 per acre).
- Container-grown stock is often the most successful for the price.
- Adjacent wetlands can be used as a seed source.
- Herbaceous wetland plants can be installed in small wetlands or in those wetlands where quick establishment is essential.
- In some cases, seeds of wetland plants are used in a temporary matrix of grass like redtop or rye (nonnatives that will die out in standing water), or switchgrass. Fescue should be avoided except in the most difficult-to-establish areas.
- Carefully select species according to targeted depth of water.

Additional Suggestions

- Use nest boxes for wood ducks, bats, or other animals, if appropriate. Put in perches (dead snags) for birds to use.
- Money should be set aside to monitor and maintain the restored wetland. Be prepared to respond quickly to problems such as erosion or dying trees.
- Plan for buffer areas to protect the wetland from sedimentation, pollution, and other human impacts.

Managing Natural Plant Succession and Pest Species

Plant communities change over time. They go through a process known as succession. This process creates a temporal series of plant communities known as "seres." This basic ecological concept is complex in nature and even though it has been studied since the early 1900s, it is far from understood. Robert McIntosh (1980) calls it "one of the oldest, most basic, yet in some ways, most confounded of ecological concepts." Plant community changes involve "species replacements, shifts in population structure, and changes in availability of resources such as light and soil nutrients" (Luken 1990). Succession must be mentioned in the context of restoration because one option for a particular site is to leave it alone and allow the natural course of plant and animal invasions and extinctions to occur. Succession can also be managed to achieve desired results. James Luken (1990) has dealt with this subject in considerable depth in his book *Directing Ecological Succession*. Table 5.2 from Luken offers a list of management problems where managing succession can achieve management goals.

A number of principles can be gleaned from Luken's book. They include the following:

- All plant communities show some form of succession at all times.
- Management activities modify the rate and direction of succession rather than the state of vegetation.
- The initial or early pool of plants on a site is critical to the succession of that site.
- The "vital attributes" of particular species will determine the course of succession. This involves shade tolerance, pH tolerance, and moisture requirements.
- The depth and duration of standing water is the most critical determining factor for vegetation succession in wetlands.
- Soil nutrient availability can be an important determinant of species composition. In old fields and disturbed soil, the primary limiting element is nitrogen. This can affect which species can invade and grow.
- Grasses are usually favored by fertilization. Woody plants and forbs are more competitive in nutrient-limited situations.

Luken writes that "succession will continue to function as a repair process following human disturbance just as it has done for the last 5000 years. In many situations this may be adequate resource management. In the majority of situations, however, we need to refine and augment this repair process to better preserve and use our dwindling natural resources."

Luken proposes a three-component succession management model. The three components are designed disturbance, controlled colonization, and controlled species performance. "Designed disturbance includes activities initiated to create or eliminate site availability. Controlled colonization includes methods used to decrease or enhance availability and establishment of specific plant species." Colonization is controlled by two factors, "the propagule pool (seeds, spores, rootstocks, bulbs, stumps, rhizomes, plant fragments, and entire plants) and the initial floristics of a site." Controlled species performance includes methods used to decrease or enhance growth and reproduction of specific plant species. When removing undesirable species, care should be taken to avoid harm to the rest of the community. Techniques include mulching, burning, cutting, managing grazing animals, and adjusting water levels. Selective use of herbicides is also used to remove species.

In several places in this book we refer to the need to control pest species. West (1984) defines a pest species as any species not native to a site or management unit that poses a threat to the integrity of natural communities within the site. In addition, native species can sometimes become pests by occurring in large numbers or out-competing desired species.

Controlling pest plants can take a variety of forms.

- **Prevention**—This includes such measures as removing livestock from prairies and not introducing exotics (such as autumn olive) through wildlife enhancement programs.
- **Monitoring**—Sites being restored must be monitored for exotic invaders.
- **Integrated Pest Management (IPM)**—Using biological controls and minimum pesticides is highly desirable.
- **Herbicides**—Foliar application of certain short-lived herbicides can be effective.
- **Mowing**—This practice inhibits woody plants and agricultural weeds.
- **Cutting**—This practice inhibits woody plants from maturing.
- **Burning**—This practice inhibits woody plants and kills some herbaceous pests (usually annual exotics).
- **Hand pulling**—This practice is especially useful in wetlands.
- **Water level manipulation**—Controlling pests in wetlands can be achieved with this method.

The introduction of exotic or nonnative species is still encouraged by a variety of government and private agencies. These exotics are recommended for wildlife purposes, erosion control, ease of establishment, disease resistance, reduced susceptibility to insect damage, and tradition. Recent examples of intentional introductions for the purpose of marsh restoration that are now of urgent concern include the spread of three *Spartina* species (*Spartina alterniflora, S. patens,* and *S. townsendii/anglica*) along the West coast of North America (National Research Council 1992). *Cornus mas,* Cornelian cherry dogwood, is recommended by the National Resources Conservation Service even though there are "at least eleven species of shrubby dogwoods native to the eastern United States that could be substituted instead, with no ecological risks" (Harty 1991). An exotic is not just a species from Europe or Asia. It is any species introduced outside of its natural range. The planting of exotic species should be discouraged if not eliminated.

The concept of substituting native species for exotic species is compelling. Bratton (1982) reports that exotic plants have a variety of impacts. For example, in arid regions, the deep-rooted tamarisk (*Tamarix* spp.) can lower the water table which displaces native plants and eliminates water for wildlife. Exotic species can also change the nutrient balance in the soil, acidify soils, change soil structure, serve as reservoirs for parasites, and produce allelopathic chemicals that retard the growth of other species. In many urban woodlands, exotic shrubs such as privet and honeysuckle are displacing natural trees and shrubs. The overstory trees are not replacing themselves and the forests will be completely gone when canopy residents succumb to old age (Hal Bryan per. com.).

Agencies and organizations can recommend native plants rather than exotic species and achieve the same resource man-

---

Conserving rare or endangered species
Conserving and restoring communities
Manipulating the diversity of plant and animal communities
Creating relatively stable plant communities on rights-of-way
Revegetating drastically disturbed lands
Minimizing the impact and spread of introduced species
Maximizing wood production from forests
Minimizing adverse environmental effects of forestry
Predicting fuel buildup and fire hazards
Increasing animal populations for recreation and aesthetics
Developing multiple-use plans for parks and nature reserves
Determining the minimum size of nature reserves
Minimizing the impact of roads, parking areas, trails, and campsites
Preserving scenic vistas in parks
Minimizing the cost of grounds maintenance on public lands
Controlling water pollution
Minimizing erosion
Maintaining high quality forage production
Minimizing the cost of crop production in agricultural communities
Developing vegetation in wetlands or on the edges of reservoirs

---

Table 5.2. Some resource management problems where succession can be manipulated to achieve management goals (Luken 1990).

agement goals. Harty (1991) reports that the Illinois Department of Conservation nurseries now produce sixty-seven species of native trees and shrubs to use in wildlife habitat, reclamation projects, and community restorations. They also produce seven species of prairie grasses and thirty-seven species of prairie forbs for prairie restoration projects.

To list all known pest species is impossible. The U. S. Congress Office of Technology Assessment stated that 4,500 exotic species have established themselves in the U. S. (Collard 1996). California, Florida, Tennessee, and other states have established Exotic Pest Plant Councils and are part of the National Coalition of Exotic Pest Plant Councils (Appendix E). These Councils have de-

veloped state lists and should be contacted for current information and complete lists. John Randall of The Nature Conservancy polled many managers from their preserves and came up with a list of 177 pest plant species (Cheater 1992). Table 5.3 is a list of pest species compiled from Bratton (1982), Cole (1991), Ebinger (1983, 1991), Ebinger et al. (1984), Glass (1992), Harrington and Howell (1990), Harty (1991), Heidorn (1992), Hester (1991), Hutchison (1992), Kennay and Fell (1992), Myers and Ewel (1990), National Research Council (1992), Szafoni (1991), and West (1984). Restorationists and natural landscapers should avoid using these plants outside of their natural ranges. Some species, such as cattails, should be used carefully even within their natural range.

Table 5.3 Pest Plants
(Scientific names are as they appeared in publications)

Scientific Name	Common Name	Natural Range
Acacia auriculiformis	ear-leaf wattle	Australia
Acacia retinodes	water wattle	Australia
Acer ginnala	Amur maple	E. Asia
Acer platanoides	Norway maple	Europe
Agropyron cristatum	crested wheat grass	Eurasia
Agropyron desertorum	clustered wheat grass	Eurasia
Ailanthus altissima	tree-of-heaven	Asia
Albizzia julibrissin	mimosa	Asia/Africa
Alliaria petiolata	garlic mustard	Europe
Allium vineale	crow garlic	Europe
Ammophila arenaria	European beach grass	Europe
Ampelopsis brevipedunculata	amur peppervine	Asia
Anthoxanthum odoratum	large vernal sweet grass	Europe
Arctium minus	lesser burdock	Eurasia
Artemisia vulgaris	common wormwood	Europe
Arthraxon hispidus	small carp grass	Asia
Ardisia elliptica	shoebutton	India
Arundo donax	giant-reed	Eurasia
Atriplex patula	halberd-leaf orache	Eurasia
Avena fatua	wild oat	Europe
Berberis thunbergii	Japanese barberry	Japan
Bischofia javanica	Javanese bishopwood	tropical Asia
Bromus inermis	Hungarian brome	Europe
Bromus japonicus	Japanese brome	Eurasia
Bromus tectorum	cheat grass	Europe
Broussonetia papyrifera	paper mulberry	China
Buddleia davidii	orange-eye butterfly-bush	China

Carduus acanthoides	spiny plumeless-thistle	Europe
Carduus nutans	musk thistle	Europe
Carpobrotus edulis	common hottentot fig	South Africa
Casuarina equisetifolia	Australian pine	Australia
Celastrus orbiculata	Asian bittersweet	E. Asia
Centaurea spp.	knapweeds and star-thistles	Asia/Europe/Mediterranean
Cerastium spp.	exotic chickweeds	Eurasia
Chondrilla juncea	hogbite	Eurasia
Cichorium intybus	chicory	Mediterranean
Cirsium arvense	"Canada" thistle	Europe
Cirsium vulgare	bull thistle	Eurasia
Conium maculatum	poison-hemlock	Eurasia
Convolvulus arvensis	field bindweed	Europe
Coronilla varia	crown vetch	Europe
Cortaderia jubata	Uruguayan pampas grass	Argentina
Cortaderia selloana	Selloa pampas grass	Argentina
Cotoneaster spp.	exotic cotoneasters	Eurasia
Crataegus monogyna	one-seed hawthorn	Europe
Cupaniopsis anacardioides	carrot wood	Australia
Cyanara cardunculus	cardoon	Mediterranean
Cyperus spp.	nutsedges	Asia/United States/tropical America
Cytisus scoparius	Scotch broom	Europe/Asia
Dactylis glomerata	orchard grass	Europe
Daucus carota	Queen Anne's lace	Europe
Descurainia spp.	exotic tansy-mustards	Europe
Digitalis purpurea	purple foxglove	Mediterranean
Digitaria ssp.	exotic crab grasses	Eurasia
Dipsacus laciniatus	cut-leaved teasel	Europe
Dipsacus sylvestris	wild teasel	Europe
Eichhornia crassipes	water hyacinth	S. America
Elaeagnus angustifolia	Russian olive	Eurasia
Elaeagnus umbellata	autumn olive	E. Asia
Elymus arenarius	European lyme grass	Eurasia
Elytrigia repens	creeping quack grass	Eurasia
Eucalyptus spp.	exotic gums	Australia
Euonymus alata	winged wahoo	E. Asia
Euonymus fortunei	winter creeper	Asia
Euphorbia esula	leafy spurge	Europe
Festuca arundinacea	tall fescue	Europe
Festuca pratensis	tall fescue	Europe
Ficus spp.	exotic figs	Eurasia
Genista monspessulana	French-broom	Mediterranean
Glechoma hederacea	ground ivy	Europe
Gypsophila paniculata	tall baby's breath	Europe
Halogeton glomeratus	saltlover	Eurasia
Hedera helix	English ivy	Europe
Hemerocallis fulva	orange day-lily	temperate Eurasia
Hesperis matronalis	mother-of-the-evening	Europe
Holus lanatus	common velvet grass	Europe
Hydrilla verticillata	water-thyme	Europe

Hypericum perforatum	common St. John's-wort	Europe
Ilex aquifolium	English holly	Europe/Africa
Imperata cylindrica	cogon grass	Pantropical
Jasminum dichotomum	Gold Coast jasmine	tropical Africa
Jasminum fluminense	jazmin-de-trapo	Africa
Juniperus virginiana	red cedar	United States
Kummerowia stipulacea	Korean clover	Asia
Kummerowia striata	Japanese clover	Asia
Lamium spp.	henbits	Eurasia
Lepidium latifolium	broad-leaf pepperwort	Eurasia
Lespedeza cuneata	sericea lespedeza	E. Asia
Leucanthemum vulgare	oxeye daisy	Eurasia
Ligustrum obtusifolium	blunt-leaved privet	Japan
Ligustrum spp.	privets	Eurasia
Ligustrum vulgare	privet	Europe
Lonicera japonica	Japanese honeysuckle	Asia
Lonicera maackii	amur honeysuckle	Eurasia
Lonicera morrowii	Morrow's honeysuckle	Japan
Lonicera tatarica	tartarian honeysuckle	Eurasia
Lotus corniculatus	garden bird's-foot-trefoil	Europe
Lygodium japonicum	Japanese climbing fern	Japan
Lysimachia vulgaris	garden loosestrife	Europe
Lythrum salicaria	purple loosestrife	Europe
Maclura pomifera	osage orange	United States
Melaleuca quinquenervia	cajeput tree	Australia
Melia azedarach	chinaberry	Asia
Melilotus alba	white sweet clover	Europe
Melilotus officinalis	yellow sweet clover	Europe
Merremia tuberosa	Spanish arborvine	tropical America
Mesembryanthemum ssp.	ice plant	Africa
Microstegium vimineum	Nepalese browntop	tropical Asia
Miscanthus sinensis	Chinese silver grass	China
Morus alba	white mulberry	Asia
Myoporum laetum	ngaio tree	New Zealand
Myriophyllum brasiliense	water feather	S. America
Nandina domestica	sacred-bamboo	India/Asia
Narcissus spp.	exotic daffodils	Europe
Nasturtium officinale	watercress	Europe
Pastinaca sativa	wild parsnip	Europe
Paulownia tomentosa	princess tree	E. Asia
Phalaris arundinacea	reed canary grass	Eurasia
Phragmites communis	reed	Eurasia/N. America
Pinus nigra	Austrian pine	Europe
Pinus sylvestris	Scotch pine	Europe
Pinus thunbergii	Japanese black pine	East Asia
Poa compressa	"Canada" bluegrass	Eurasia
Poa pratensis	"Kentucky" bluegrass	Eurasia
Polygonum convolvulus	black bindweed	Europe
Polygonum cuspidatum	Japanese knotweed	E. Asia
Populus alba	white poplar	Europe

Potamogeton crispus	pondweed	Europe
Potentilla recta	sulphur cinquefoil	Eurasia
Prunus mahaleb	perfumed cherry	Europe
Ranunculus spp.	exotic buttercups	Europe
Pueraria lobata	kudzu vine	E. Asia
Rhamnus cathartica	common buckthorn	Europe
Rhamnus davurica	Dahurian buckthorn	E. Asia
Rhamnus frangula	alder buckthorn	Europe
Rhodomyrtus tomentosus	downy myrtle	E. Asia
Robinia pseudoacacia	black locust	United States
Rosa multiflora	multiflora rose	E. Asia
Rubus spp.	exotic brambles	Eurasia
Rumex spp.	exotic docks	Eurasia
Salsola collina	slender Russian-thistle	Eurasia
Sapium sebiferum	Chinese tallowtree	China
Scaevola sericea	beach naupaka	Indian and Pacific Ocean islands and coasts
Schefflera actinophylla	octopus-tree	Australia
Schinus terebinthifolius	peppertree	S. America
Scirpus mucronatus	bog bulrush	Europe
Senecio jacobea	tansy ragwort	Europe
Senecio mikanioides	cape ivy	Europe
Senecio sylvaticus	woodland ragwort	Europe
Senecio vulgaris	old-man-in-the-spring	Eurasia
Sesbania exaltata	peatree	Central America
Setaria faberi	Japanese bristle grass	Asia
Solanum dulcamara	climbing nightshade	Eurasia
Solanum viarum	tropical nightshade	South America
Sonchus spp.	exotic sow-thistles	Europe
Spirea japonica	Japanese meadowsweet	Japan
Sorghum halepense	Johnson grass	Eurasia
Stellaria spp.	exotic chickweeds	Eurasia
Tamarix gallica	tamarisk	S. Europe
Tamarix spp.	exotic tamarisks	Eurasia
Tanacetum vulgare	common tansy	Eurasia
Taraxacum officinale	common dandelion	Eurasia
Taxus cuspidata	Japanese yew	Asia
Tragopogon spp.	exotic goat's-beard	Europe
Trifolium pratense	red clover	Europe
Trifolium repens	white clover	Eurasia
Typha angustifolia	narrow-leaved cattail	United States/Eurasia
Ulex europaeus	common gorse	Europe
Ulmus procera	English elm	Europe
Ulmus pumila	dwarf elm	Asia
Verbascum thapsis	common mullein	Europe
Viburnum lantana	wayfaringtree	Europe
Viburnum opulus	guelder-rose	Europe
Vinca major	large periwinkle	Europe
Vinca minor	common periwinkle	Europe

Ecological Communities

Selecting Model Communities

An important step in any restoration project is determining which natural or ecological community is the appropriate model for restoration. The ecological community descriptions (Appendix A) may be used as a general guide to the species composition for these communities. They are not meant to define the specific species composition at a point on the ground. For that level of detail, one should consult local experts in the state Natural Heritage Programs, native plant societies, environmental organizations, and universities, as well as restorationists and native plant consultants (Appendix E).

For example, suppose individuals in Elizabethtown, Kentucky wanted to restore or create a natural community on their property, but had no idea what types of communities occurred there or which ones would do well. First, they would locate their property on the natural regions map in this book. They find they are located in the Ozark/ Interior Plateaus Natural Region. Turning to that section of Appendix A, they will find a list of Dominant Ecological Communities which occur (or occurred) within that region.

They will then need to determine which of those communities occurred or could be restored on their particular site. To do this, a detailed site analysis will be required to find out certain baseline data such as soil type, geologic substrate, topography, aspect, current vegetation, and land use. Once this is accomplished, a community type or types can be selected. For example, the site may be a gently rolling upland with well drained and moderately fertile soil over limestone bedrock. This eliminates the wetland communities and a selection can be made from the terrestrial communities.

At this point local expertise is needed. Since the Elizabethtown area contained forest, woodland, wetland, and prairie communities at the time of settlement, any of these could be selected for restoration or creation purposes. However, a local expert knows that the oak-pine community did not occur in the area and that mesophytic forest, which does occur in the area, probably did not occur on the site because it is too dry. There-fore the selection comes down to oak/oak-hickory forest, oak barrens, or bluestem prairie as appropriate for this site.

Even though detailed site planning is beyond the scope of this book, two worksheets are provided (Figures. 5.1 and 5.2) to help organize the ecological community information. The site planning checklist can be used for each specific planning unit on the site. The checklist provides the opportunity to assess the basic conditions of each planning unit, the relationship among planning units, and the overall regional situation. An aerial photo, soil maps, topographic map, or good knowledge of the regional landscape will be needed to address some of the questions.

The second worksheet is for compiling a species list for a restoration project on a specific planning unit. It contains four categories of information in addition to the species name. The categories are as follows:

- **Range of species**—The first category to complete is range information about the species. All species used in a restoration should naturally occur in the region. In the space provided, a "yes" or "no" can be written. The reference from which range information was obtained can be listed or other qualifiers can be used such as marginal, unknown, probable, or county record only.

- **Environmental tolerance**—The information for this category can be obtained from the matrices in Appendix C. If other information is available such as requirements for certain pH ranges, that can be noted in this column.

- **Wildlife**—The wildlife values also come from the matrices in Appendix C. These values relate to animal use of particular species of plants. This is not critical to restoration, but if certain animal species are targeted, then the plant species to support them must be present.

- **Nursery**—This category is to identify the nursery or nurseries selling this species. Many nurseries will also contract to grow particular species. Appendix D is a list of nurseries that carry native plants.

Site Planning Check List

Site name: _____

Natural region: _____

Size of the site: _____ acres. Sketch the site and identify the planning units.

Planning unit name or number and size: _____

Planning unit type and conditions:

    Upland   wet_____   mesic_____   dry _____

    Wetland   standing water __   depth _____   saturated soils _____

    Estuarine   depth_____   salinity_____

Desired vegetation type:

    closed forest   open forest   grassland   shrubland

Ecological community chosen:_____

Local ecological community survey completed?   yes   no

Location: _____

Results:

Proximity to natural communities in the region:

Proximity to natural communities on the site:

Do corridors currently exist between patches of natural communities?

Figure. 5.1. Site Planning Check List.

Preferred Species List for Restoration

Plant Name	Range	Env. Tol.	Wildlife	Nursery

Figure. 5.2. Preferred Species List For Restoration.

Analyzing the Presettlement/ Predisturbance Landscape

Although information about presettlement vegetation is primarily from sketchy historical accounts of early explorers and settlers, other sources can be used as additional evidence of the original landscape in a given area. Other sources include old deeds and property surveys, either of the site which is being restored or adjacent areas within the region. Many times these old deeds and survey records contain information about the flora and fauna of the area, especially notes on larger trees and the quantity and quality of timber. Early surveyors often referred to particular trees for the purpose of determining property boundaries. However, the colloquial names used and the level of botanical expertise of these early surveyors limit the usefulness of these records. If such documents are available for a particular restoration site or planning unit, they should be investigated. Often these documents are available at local or regional courthouses, land survey offices, historical societies, or college and university libraries in the special collections holdings.

Most colleges and universities have a special collections department that contains maps and historical records of the area. These can be useful if they contain references to the vegetation or physiognomy. Although many historical maps are not accurate by today's standards, some contain references to dominant or extensive vegetation types of a county, state, or region. These maps and historical records may also contain interesting notes on the natural or special features of an area, i.e., buffalo traces and wallows, springs, licks, geologic formations, caves and sinks, soils, mineral resources, and native peoples and their communities.

Some excellent general texts also exist which are worth consulting for particular geographical areas. Some of these sources emphasize the human perspective instead of an ecological or natural history perspective, but are useful especially if educational outreach is a component of a restoration project. Recent texts that contain historical, as well as ecological accounts, of the history of certain regions include Cronan (1983), Hart (1992), Kirby (1995), Larson (1995), Mills (1995), and Whitney (1994). It may be worthwhile to search theses and dissertations that have focused on presettlement conditions. Two examples from Kentucky are Campbell (1980) and Davidson (1950).

Aerial photographs are another good source of landscape-cover type information. The oldest aerial photographs for most areas are from the 1930s–1940s and are available from the National Cartographic Records office in Washington, D.C. The aerial photographs housed at the Natural Resources Conservation Service (formerly Soil Conservation Service), Farm Services Agency, and Property Valuation Administration offices in the county or parish of the restoration site can be useful sources to determine previous land use and cover information. These aerial photographs are generally ten to thirty years old; however, sometimes these offices have older sets of photographs that are no longer used. The Natural Resources Conservation Service and the Farm Services Agency are part of the United States Department of Agriculture and most states have one of these offices in many of the counties or parishes. The primary purpose of the Natural Resources Conservation Service is to provide technical assistance to landowners (primarily farmers) in planning and implementation of erosion control measures for farms and timberlands. They generally welcome researchers or land managers who want to use their photographs. The Property Valuation Administration offices located at county or parish courthouses also may have useful aerial photographs, although their main purpose is to keep current property records of lands within their jurisdiction.

State Natural Heritage Programs and Nature Conservancy field offices use these types of aerial photographs as a coarse filter in identifying natural areas and have developed methods for useful interpretation of aerial photographs. For the restorationist, these aerial photographs can provide useful information about a restoration site as well as nearby intact natural ecological communities which may be useful for seed and plant material sources or as comparison for restoration goals.

Restoration in Practice

Fernwood Prairie

In 1975, Don and Kay Harker, working with Stan Beikmann and Max Medley, conceived and planned a prairie restoration project at Fernwood (a botanic garden and nature center near Niles, Michigan). The early planning was inspired and guided by Ray Schulenberg from the Morton Arboretum. The Fernwood prairie project is now twenty years old and considered complete (Photos 5.1–5.4). Six botanists and scores of weeders and other workers have contributed to the prairie restoration over the years.

The original goals of the project were the following:

- Create a five-acre natural tallgrass prairie representative of southwestern Michigan. High diversity was desired.
- Use the restored prairie to help preserve the local gene pool by collecting seeds from local remnant prairies as close to Fernwood as possible.
- Use the prairie as a focus for educational and interpretive programs about this disappearing ecosystem.

The actual planting plan took several different directions over the years. Initially, seeds were collected in the late summer and early fall from the surrounding remnant prairies (Figure. 5.3). These prairie remnants include wet, mesic, and dry tallgrass prairies. Most of the Fernwood site was planted in mesic species. A small portion was planted in wet species and a small sandy knoll was created for dry species.

The original site was a fescue-covered field. Soil preparation began with a late summer plowing and then a disking. In the spring the soil was deep plowed (not recommended), then disked and raked prior to planting. Initially all weed control was through hand weeding. It was very labor intensive.

The seeds were scarified and stratified where appropriate, and planted in flats in the Fernwood greenhouse. Both greenhouse plants and the treated seeds were planted in the prepared soil. The prairie was hand weeded throughout the year.

The largest percentage of species planted were the warm season grasses, big bluestem (*Andropogon gerardii*), Indian grass (*Sorghastrum nutans*), and little bluestem (*Schizachyrium scoparium*). The forbs were planted randomly. Approximately one-half acre of prairie was added each year.

The final restored prairie has a very high diversity of species. The initial planting included 90 species and the final prairie has 150 native prairie species. The species are not all conservative prairie species, meaning some also grow places other than prairies.

Ecological restoration takes considerable time and effort. Educating the public about the project, especially in the early stages, is important. If the public shares the vision, they will support the long effort necessary for restoration. The Fernwood prairie is now a major attraction for visitors.

Photo. 5.1. Restored prairie and a new area being prepared for spring planting with a fall plowing.

Photo. 5.2. Newly planted section of prairie.

Photo. 5.3. Restored tallgrass prairie.

Photo. 5.4. Fire is used to maintain prairies.

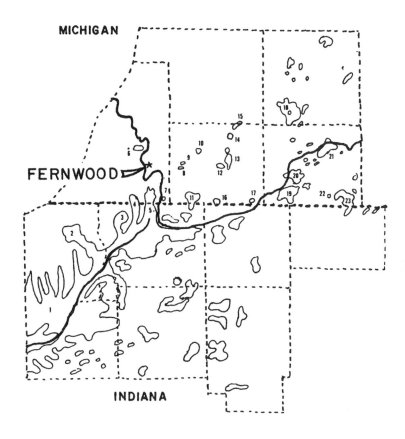

Figure. 5.3. Remnant natural prairies near Fernwood. Many remnants were used as seed sources (Medley 1976).

	Prairie	County	Town		Prairie	County	Town
1.	Grande Prairie	LaPorte & Starke		12.	Young's Prairie	Cass	Vandalia
2.	Door Prairie	LaPorte	Door Village	13.	LaGrange Prairie*	Cass	Cassopolis
3.	Rolling Prairie*	St. Joseph (IN)	Rolling Prairie	14.	Gard's Prairie	Cass	
4.	Terre Coupe Prairie*	St. Joseph (IN) & Berrien		15.	Little Prairie Ronde	Cass, VanBuren	Nicholasville
				16.	Shavehead's Prairie*	Cass	Union
5.	Portage Prairie*	St. Joseph (IN) & Berrien	South Bend	17.	Baldwin's Prairie	Cass	Union
				18.	Prairie Ronde	Kalamazoo	Schoolcraft
6.	Wolf's Prairie*	Berrien	Berrien Springs	19.	Indian Prairie	St. Joseph (MI)	White Pigeon
7.	Parc aux Vaches*	Berrien	Bertrand	20.	White Pigeon Prairie	St. Joseph (MI)	White Pigeon
8.	Sand Prairie*	Cass	Pokagon	21.	Goodrich Prairie	St. Joseph (MI)	Centerville, Mendon
9.	Pokagon's Prairie*	Cass	Pokagon				
10.	McKenney's Prairie*	Cass	Dowagiac	22.	Sturgis Prairie	St. Joseph (MI)	Sturgis
11.	Beardsley's Prairie*	Cass	Edwardsburg	23.	Dry Prairie	St. Joseph (MI)	Sturgis

*Prairies within 25 miles of Fernwood.

CHAPTER 6

Natural Regions of the United States and Their Dominant Ecological Communities

Introduction

Appendix A lists the dominant ecological communities within each ecological system and class for the thirty natural regions outlined for this book. This chapter explains how we developed the classification system for the natural regions, ecological systems, and classes. This classification was developed to make it possible for someone anywhere in the country to determine what general types of ecological communities could occur at a particular site.

Classification of the landscape and its natural vegetation has been approached from many different views and levels of detail, depending upon the purpose. It is fairly naive and simplistic to write down a series of categories and definitions and assume that nature has been neatly classified. It is uncommon in nature for regions or communities to be clearly delineated. Rather, they tend to grade into one another or form a mosaic, making it difficult to distinguish entities. However, classifying the landscape is the only way for us to study and comprehend an incredibly complex system. It is certainly necessary to provide a basis for what to restore and where.

Natural Regions

In the pocket of this book you will find a map showing thirty natural regions. This natural regions classification system was developed using numerous sources. Because the level of detail in any single source was not thought to be adequate for

our purposes, multiple sources of information were used to delineate the natural regions at the appropriate scale. The primary sources used to develop the natural regions were *Potential Natural Vegetation of the Conterminous United States* (Küchler 1964b, 1975) (see Appendix A and map in pocket), *Ecoregions of the United States* (Bailey 1978, 1994) (Figure. 6.1), *Ecoregions of the Conterminous United States* (Omernik 1986), and *North American Terrestrial Vegetation* (Barbour and Billings 1988).

Using the Küchler (1964) and Bailey (1978, 1994) maps combined as a base, the country was divided into thirty natural regions. These natural regions generally follow Bailey's regions and subregions as laid over Küchler's potential natural vegetation map. Bailey has updated and changed the names of his regions and subregions from vegetation descriptions to other descriptors. Bailey's vegetation descriptors are included in Figure 6.2 for reference purposes. The lines dividing regions generally follow a Küchler vegetation type. These natural regions are areas of the country that have certain natural features in common. These features include soils, geologic history, landforms, topography, vegetation types, plant and animal distributions, and climate. These natural regions are broad or coarse divisions which make it easier to subdivide and define the landscape and its vegetation. Appendix A provides descriptions of each natural region.

Figure. 6.1. Bailey's Ecoregions of the United States.

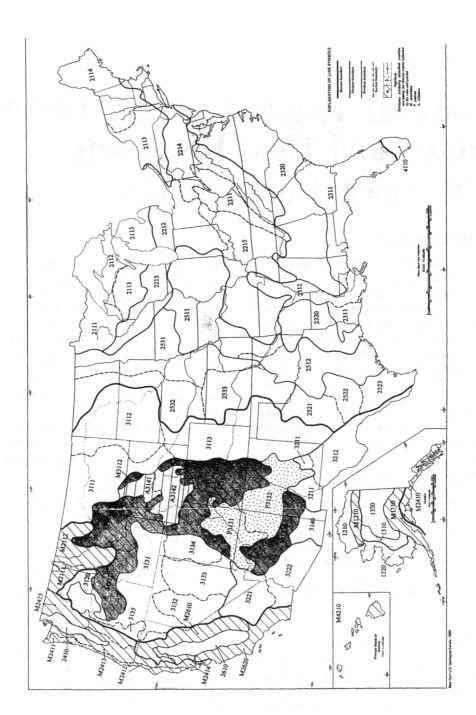

Lowland Ecoregions		Highland Ecoregions*	
Province	**Section**	**Province**	**Section**
2110 Laurentian Mixed Forest	2111 Spruce–fir Forest 2112 Northern Hardwoods–Fir Forest 2113 Northern Hardwoods Forest 2114 Northern Hardwoods–Spruce Forest	M2110 Columbia Forest (Dry Summer)	M2111 Douglas-fir Forest M2112 Cedar–Hemlock–Douglas-fir Forest
2210 Eastern Deciduous Forest	2211 Mixed Mesophytic Forest 2212 Beech–Maple Forest 2213 Maple–Basswood Forest + Oak Savanna 2214 Appalachian Oak Forest 2215 Oak–Hickory Forest		
2310 Outer Coastal Plain Forest	2311 Beech–Sweetgum–Magnolia–Pine–Oak Forest 2312 Southern Floodplain Forest		
2320 Southeastern Mixed Forest			
2410 Willamette-Puget Forest		M2410 Pacific Forest	M2411 Sitka Spruce–Cedar Hemlock Forest M2412 Redwood Forest M2413 Cedar–Hemlock–Douglas-fir Forest M2414 California Mixed Evergreen Forest M2415 Silver Fir–Douglas-fir Forest
2510 Prairie Parkland	2511 Oak–Hickory–Bluestem Parkland 2512 Oak + Bluestem Parkland		
2520 Prairie Brushland	2521 Mesquite–Buffalo Grass 2522 Juniper–Oak–Mesquite 2523 Mesquite–Acacia		
2530 Tall-grass Prairie	2531 Bluestem Prairie 2532 Wheatgrass–Bluestem–Needlegrass 2533 Bluestem–Grama Prairie		
2610 California Grassland		M2610 Sierran Forest M2620 California Chaparral	
3110 Great Plains Short-grass Prairie	3111 Grama–Needlegrass–Wheatgrass 3112 Wheatgrass–Needlegrass 3113 Grama–Buffalo Grass	M3110 Rocky Mountain Forest	M3111 Grand Fir–Douglas-fir Forest M3112 Douglas-fir Forest M3113 Ponderosa Pine–Douglas-fir Forest
3120 Palouse Grassland		M3120 Upper Gila Mountains Forest	
3130 Intermountain Sagebrush	3131 Sagebrush–Wheatgrass 3132 Lahontan Saltbush–Greasewood 3133 Great Basin Sagebrush 3134 Bonneville Saltbush–Greasewood 3135 Ponderosa Shrub Forest	P3130 Colorado Plateau	P3131 Juniper–Pinyon Woodland + Sagebrush–Saltbush Mosaic P3132 Grama–Galleta Steppe + Juniper–Pinyon Woodland Mosaic
3140 Mexican Highlands Shrub Steppe		A3140 Wyoming Basin	A3141 Wheatgrass–Needlegrass–Sagebrush
3210 Chihuahuan Desert	3211 Grama-Tobosa 3212 Tarbush-Creosote Bush		
3220 American Desert (Mojave-Colorado-Sonoran)	3221 Creosote Bush 3222 Creosote Bush-Bur Sage		*M–mountains, P–plateaus, A–altiplano
4110 Everglades			

Figure. 6.2. Codes for Bailey's Ecoregion Map

Ecological Systems and Classes

Ecological systems are categories which describe general ecological functions and processes. Upland (terrestrial), wetland (palustrine), and estuarine systems are included in this classification. The wetland definitions mainly follow the classification of Cowardin et al. (1979). True aquatic systems such as lakes, oceans, and rivers (called lacustrine, marine, and riverine systems, respectively) are not included in this book.

The upland, wetland, and estuarine systems are further subdivided into more specific classes. The classes are defined by the dominant life form of the vegetation which reflects the basic physiognomy or appearance of the habitat. Forest, woodland–barrens, shrub, and herbaceous are the classes used in this classification. The systems and classes are combined in the following definitions. These definitions are necessary because they help us distinguish between a wetland and an upland—or a forest and a woodland. The differences between systems or classes are often technical and difficult to recognize in the field. Again, nature is not so neatly classified. In order to analyze a site and determine the appropriate ecological community, it is necessary to be familiar with the systems and classes. The species lists for ecological communities in Appendix A are organized by regions and according to the following systems and classes.

Upland Systems

The upland systems include all areas that are typically found on dry land and have soil that is rarely saturated with water or is saturated only briefly, such as after prolonged rains. The upland systems are dominated by plants that cannot withstand prolonged flooding or saturated soils.

Upland forests occur in areas where sufficient moisture and soil is available to support relatively dense stands of trees. Upland forests are dominated by medium to tall trees which form a closed to mostly closed canopy which restricts most direct sunlight from reaching the forest floor during the growing season. These forests can be either evergreen or deciduous or a combination of both. Beneath the canopy, shade-tolerant species of small trees, shrubs, and forbs can occur. Because of their closed canopy, forests often modify the local environment creating cooler, more humid conditions.

Upland woodlands and barrens, sometimes called savannas, are intermediate between forested and nonforested habitats. Woodlands and barrens generally have a canopy cover ranging from 10 to 60 percent which allows variable amounts of sunlight to reach the ground. The understory may vary considerably from prairie grass dominated to interspersed forest and prairie herb dominated. Often this habitat is "park-like" with open-grown trees with wide spreading branches and an open grassy understory with few shrubs. Most woodland–barrens are fire maintained and require periodic fires to exist. Some woodland–barrens are edaphically maintained (i.e., have poor, thin, or doughy soil which retards dense tree growth) and exist without the benefit of fire.

Upland shrub, sometimes called scrub or shrub, occurs in areas that are dominated by dense to open stands of shrubs with few to no trees present. Generally shrubs are considered to be woody plants less than twenty feet tall. This can include true shrubs, small trees, and trees stunted due to environmental conditions. This type of community class is most common in the western United States where more arid conditions prevail. Most deserts are considered shrub communities. Herbaceous vegetation of grasses and forbs is often present but depends upon the density of the shrub cover and degree of available moisture. Some shrublands result from fire, a lack of fire, or overgrazing, depending upon the local situation and are thus considered successional or disturbance communities.

Upland herbaceous communities are those that are dominated by herbaceous vegetation with woody trees or shrubs absent, or restricted to scattered individuals or groves. Natural herbaceous dominated communities occur throughout the United States but are extensive only in the western and some mid-western states. Prairies and meadows, dominated by grasses, are the most widespread upland herbaceous communities. Other communities such as tundra can be dominated by herbaceous vegetation. Many upland herbaceous communities are maintained by fire and/or grazing. Fire also controls the invasion of woody species, except in the western states where climatic conditions (lack of rainfall) control.

Wetland Systems

Wetland systems include freshwater areas that are transitional between upland and aquatic systems. They typically have a high water table with saturated soil or shallow, standing water during all or a significant part of the growing season. They are dominated by emergent hydrophytic plants that can tolerate prolonged flooding or saturated soils. Wetland systems can grade into deep, open water (aquatic) systems.

Wetland forests are dominated by trees that form a closed to partly closed canopy. They occur in areas where water is not too deep to support trees and favorable environmental conditions exist, at least periodically, for the germination of tree seeds. Wetland forests range from deep water swamps to periodically flooded bottomland forests. They can be evergreen, deciduous, or a combination of both.

Wetland shrub communities occur in areas dominated by a shrub layer with few or no trees. Also called shrub swamps, these communities are often a successional stage of forested wetland. Some shrub swamps do appear to be relatively stable.

Wetland herbaceous communities are wetlands dominated by emergent herbaceous vegetation with little or no woody vegetation present. Wet prairie, marsh, sedge meadow, fen, and open bog are some of the different names applied to herbaceous wetlands.

Estuarine Systems

These systems include coastal wetlands that occupy intertidal zones where sea water is diluted by freshwater run-off from the land. Plants that grow in this community can tolerate salty or brackish water and fluctuating water levels caused by tidal activity or land run-off.

Estuarine forests occur primarily on shallow flats in the intertidal zone of coastal areas such as along tidal rivers and embayments. They are of limited occurrence in the United States, being primarily restricted to the more tropical climate of Florida. They are broadleaf, evergreen forests dominated exclusively by three species of mangrove trees, thus called mangrove forests or mangrove swamps.

Estuarine shrub communities occupy shallow flats in the intertidal zone of coastal areas. They are dominated by sometimes dense stands of shrubs or stunted trees usually less than twenty feet tall.

Estuarine herbaceous communities, often referred to as salt marshes, are the most common or typical estuarine communities in the United States. They are dominated by nonwoody, salt-tolerant plants and often cover large areas where the intertidal zone is extensive due to low relief and high tidal range.

Ecological Communities

The ecological communities used in this classification are interrelated assemblages of flora, fauna, and other biotic and abiotic features. A particular ecological community tends to occur repeatedly under a similar set of environmental conditions (i.e., soils, geologic substrate, hydrology, climate, microclimate, topography, and aspect). These communities are delineated based on the potential natural vegetation and the dominant or characteristic plant species that naturally occur (or occurred) in a given region of the country.

In today's landscape, very little undisturbed natural vegetation remains. Almost all areas of the country have either been converted to other uses (i.e., cropland, pasture, urbanization, and roads) or have been severely modified through logging, grazing, mining, or flooding. The term "potential natural vegetation" is used to indicate the vegetation which did or would naturally occur without the influence of modern civilization. Natural vegetation is considered from the time frame of the natural conditions which greeted the first Euro-American settlers. Because most of the landscape has been modified, determining the potential natural vegetation for a specific site may be difficult; however, it should be one of the first steps in choosing the appropriate ecological communities for your site.

In Appendix A, the dominant ecological communities in each natural region are briefly described and lists of dominant and characteristic plant species are provided. The information available from published literature on different communities varies considerably. Some community types are well documented while others have very little information available. The community descriptions reflect this availability of data. Ecological communities considered rare, or of very limited extent within a region, are not included in the community descriptions although they may be mentioned in the regional description. For example, glades are mentioned in the New England Region but a description and species list are not given because they are extremely rare and are of very limited extent. However, in the Ozark/Interior Plateaus Region, glades are described as a dominant ecological community because they are more common and occupy considerable acreage.

The ecological community descriptions and accompanying lists of species were also compiled from many sources. State and regional floras, state Natural Heritage Program data, and many other books and articles were used. Because of this variety of sources, all plant names, both common and scientific, have been standardized with Kartesz (1990 and 1991). Although locally important, subspecies and varieties have not been used because of the large areas covered by the dominant ecological communities in each natural region.

Many judgements were made in deciding which lesser or smaller communities to combine and which larger communities to divide in creating the ecological communities described in this book. Communities can be identified and described at many levels from the largest, the earth, to the smallest, a drop of pond water. This classification falls somewhere in between. We attempted to define the ecological communities at a useful level for national reference without losing too much of the detail of smaller community divisions. The references used in this effort can be found in the bibliography.

The ecological community descriptions in Appendix A are to be used as a guide to the species composition of the dominant vegetation types as they occur in a particular region. They cannot be used to define the specific species composition of a point on the ground. It must be realized that individual species, although a component of a community in a particular area, may not occur over the entire geographical range of the community. For that level of detail, one should consult local experts in the state Natural Heritage Programs, native plant societies, environmental groups, or universities (Appendix E).

The authors recognize that this list of communities and species is a first effort. Attempting to accurately delineate the ecological communities of the United States is a huge undertaking and was not the primary purpose of this book. Organizations, such as The Nature Conservancy through its nationwide network of Natural Heritage Programs, have been working to develop detailed community descriptions of the United States for many years. These lists of communities and species in this book are, in part, a way of both asking and partially answering the questions—what are the dominant ecological communities and what are their characteristic species?

The authors invite and encourage comments, suggestions, additions, and deletions. Please send any comments to Donald Harker, 433 Chestnut Street, Berea, Kentucky 40403.

CHAPTER 7

Ecological Restoration Types

Introduction

For the purposes of this handbook an Ecological Restoration Type (or Restoration Type) is a group of similar ecological communities (following the Dominant Ecological Communities classification in Appendix A). The ecological communities included in each Ecological Restoration Type have similar hydrological, biological, and landscape features. This allows for a presentation of methods, techniques, and ideas that are applicable to all or most ecological communities in an Ecological Restoration Type. Since many ecological communities have never been the focus of a restoration project, this is a way in which knowledge from similar communities can be applied to communities that have not been studied or for which no published references exist.

The Ecological Restoration Type descriptions in Appendix B contain information that should be both useful and practical to land managers (public and private), consultants, scientists, landscape architects, or anyone interested in ecological restoration. Descriptions consist of basic restoration techniques, considerations, ecological/biological aspects, and fundamental logistics that should help professionals design a restoration plan. Due to the fact that information in the Ecological Restoration Type descriptions was gleaned from many literature sources, some concepts, applications, or methods may seem contradictory. In these cases, attempts were made based on literature and other sources to present both sides and their associated advantages and disadvantages.

No detailed species lists accompany the Ecological Restoration Type descriptions for three reasons: (1) a list of characteristic species for each Dominant Ecological Community is included in Appendix A; (2) due to the broad scope of this handbook, it would take many volumes to adequately list all species

associated with the ecological communities discussed, especially considering the changes in species composition due to geographic variation of many communities; and (3) many excellent sources already exist for determining what species are found in particular communities and areas. We have listed sources in the bibliography as well as referenced them with each Ecological Restoration Type description. The best way to determine species needed in any restoration project is to not only use primary literature sources but to consult with experts in the state in which the project is being designed. These experts include Native Plant Societies, National Wildflower Research Center, Natural Heritage Programs, Department of Natural Resources, The Nature Conservancy, professors and instructors at colleges and universities, native plant nurseries, and consultants specializing in restoration (see Appendix E). Also, much can be learned from observing intact ecological communities (that are of the same kind) in nearby natural areas.

As with any scientific or technical publication, the ideas presented are not the final word but instead represent an analysis of available (widely distributed as well as some more obscure) literature (recent and historical). As concepts, applications, and methods are updated, improved, or discovered, the ideas presented here may become obsolete. Realizing this, any comments or suggestions are welcomed by the authors. Also Appendix E contains a variety of restoration resources.

Ecological Restoration Type Descriptions

The following list of topics and subtopics are common to most Ecological Restoration Type descriptions contained in Appendix B. The name of the restoration type will be listed first, followed by the Dominant Ecological Community types it includes and their page numbers from Appendix A. The description for each Restoration Type will follow the general headings and format outlined below.

Introduction

This section provides a general description of the restoration type, the factors affecting the ecological communities within the type, and the areal extent of these communities. A summary of impacts and general land use of these communities may be provided if such trends are known.

Site Conditions

Soils—This section covers relevant soil considerations for the Ecological Restoration Type. In addition, soil surveys available from the Natural Resources Conservation Service office in most county seats should also be consulted when planning a restoration project. Actual soil type determination and a soil analysis may be necessary if soil has been altered or if areas have not been previously typed as in most urban sites. On severely degraded sites, soil may be completely absent. But the remaining materials may not be, however, very different from the materials left by natural processes, such as by glaciation or volcanic activity (Bradshaw 1983). The decreased pH of a degraded site limits establishment of plants and may need to be corrected using lime. If the soil of a degraded site has been set aside, it can and should be replaced. Where it has been lost, it is unlikely to be economical to purchase topsoil from elsewhere, and the restorationist will have to work to improve the remaining material. The contribution of earthworms must not be underrated. They generate an impressive accumulation of fine soil on the surface of reclaimed land by an equivalent amount of tunneling (Bradshaw 1983).

Topography—This section contains general information about the topography of the communities within the Ecological Restoration Type. The United States Geological Survey (USGS) 7.5 minute quadrangle maps should be obtained for the planning and development of a restoration plan. The potential natural vegetation of a site is dependent upon particular topographic features (e.g., the pit and mound topography of red maple swamps in the

northeast). Minor topographic features such as the pit and mound topography of most forested wetlands, although very important to ecosystem structure and function, are not identifiable on USGS maps. A site specific map may need to be drawn to identify important topographic features not shown on USGS maps. Starting a restoration project with the proper topography is essential to the development and persistence of the desired ecological community. Slope, elevation, and aspect can determine the species composition of both herbaceous and forested communities. Through its influence on water availability, aspect can also directly influence important functions such as primary productivity. Considerations of these factors are important in selecting species for restoration efforts.

Moisture and Hydrology—This section contains general information about the relative moisture requirements and a description of the basic hydrological features (where the water comes from, how it is transported, and where it exits) of the ecological communities in an Ecological Restoration Type. The United States Geological Survey also publishes papers and reports on the hydrology and water resources of many areas. These can be found in the "Documents" section of most state colleges and universities. State government agencies (agencies regulating water and geologic surveys) may also be valuable sources of information on hydrology and water quantity and quality.

Establishment

Site Preparation—This section outlines various methods and considerations for site preparation that have been used for restoring communities within the Ecological Restoration Type. Some sites may not be appropriate for certain ecological communities. Sites are not appropriate for many reasons. The soil, hydrology, or microclimate may not be suitable for the development of a particular community. In these cases an alternative community should be selected. The restorationist should be on site during site preparation to supervise machinery operators.

Planting Considerations—This section contains information relevant to the planting of species in a given Ecological Restoration Type. Planting techniques for restoration projects include an array of options. These include transplanting roots, rhizomes, tubers, seedlings, or mature plants; broadcasting seeds obtained commercially or collected from other sites; importing soil and its seed bank from nearby sites (primarily used in wetland restoration); or relying completely on the seedbank of the original site. As a general rule, plants collected from the wild and transplanted to the restoration site have higher survival and longevity than those obtained from nursery stock. However, the former may not be an option depending upon the type of ecological community being restored, the availability of desired species in the desired quantities, and any regulations concerning collection of native plants. If collection of plants from the wild is permitted, the collectors should use good judgement and not overcollect a particular species in an area unless the area will be developed. Proper collecting techniques and numbers may be determined through consultation with someone familiar with the life history of certain species or found in the literature. As Bradshaw (1983) states, "It is relatively easy to reintroduce agricultural and forest tree species. Wild species are more difficult, yet if their reproductive biologies are properly understood they can be handled with ease and reliability." The National Wildflower Research Center has compiled useful information and bibliographies regarding the planting of native plants for the conterminous U.S. Other excellent sources concerning the collection and planting of native plants are all found in the literature (Aiken 1935, Bir 1992, Dirr and Heuser 1987, Hightsoe 1988, Jones and Foote 1990, MacDonald 1986, Martin 1990, Nellis 1994, Nokes 1986, Oldenwald and Turner 1987, Phillips 1985, Schmidt 1980, Taylor and Hamblin 1963, United States Department of Agriculture 1974, Vories, 1981).

There are many other books, manuals, reports, pamphlets, and periodicals that provide detailed procedures for propagating and cultivating native plants. An attempt was made to cite as many valuable sources as possible, however, some were surely overlooked. The National Wildflower Research Center publishes "Individual Propagation Factsheets" for many of the more common or showy native plants.

Forest trees are often established with little difficulty using traditional methods. Saving the surface soil of degraded land and replacing it afterward can be useful for herbaceous species, but is often limited for tree and shrub regeneration. In some areas hydraulic seeding, in which the seed is sprayed onto the site in a slurry containing an organic mulch, fertilizer, and usually a stabilizer, is used. This technique may have limited value because of toxic interactions between seed, fertilizer, and stabilizer (Sheldon and Bradshaw 1977, Roberts and Bradshaw 1985) which may limit germination of some species. With ecological understanding and more experience, our knowledge of the requirements of native species could equal that of cultivated ones.

Spacing Considerations—Recommendations for spacing or planting design are covered in this section. Often the spacing of plantings is somewhat experimental since mortality and vigor vary considerably depending upon the source of plant material and site conditions. As a general rule, the spacing should be similar to that found in the natural setting of a particular species. However, the restorationist will want to consider such factors as mortality (how many plantings will die?), competition from associate plants, herbivory (how many plants will be browsed or eaten by local wildlife?), and the potential for the species planted to spread or cover an area (e.g., under the right conditions, a single stem of cordgrass may cover a twenty-four to thirty foot diameter area in just two growing seasons). The desired goals of the project will also affect spacing requirements (e.g., if erosion control is the goal, plants may need to be more closely spaced from the beginning).

Temporal Spacing for Different Plants and Layers—This section describes various temporal spacing considerations for installation and establishment of vegetation in a Ecological Restoration Type. Ecological processes vary in their effects or importance at different spatial and temporal scales. One of the challenges in ecological restoration is contending with the large array of spatial and temporal scales of ecological processes and disturbance regimes (O'Neill et al. 1986). For example, in considering the mosaic of grassland community types, spatial scales range from square inches to hundreds of miles and time scales from minutes to millennia (Delcourt et al. 1983, Risser 1987). In the restoration of many communities, time is an important consideration. The restorationist cannot merely plant all species comprising a community at one time. The requirements of shade tolerant species or species dependent on other community processes must be considered during the planning process. For example, in forest restoration many understory herbs or shrubs cannot tolerate an open canopy and, therefore, must be planted after a canopy has been established.

Another example would be the importance of nitrogen-fixers on degraded soils. Palaniappan et al. (1979) found that tree lupine *(Lupinus arboreus)* facilitated the growth of grasses on china clay wastes (England). Without the lupine, there was no grass growth. Some later successional species will only begin to grow rapidly once soil development, in terms of nitrogen accumulation, has occurred. Restoration of forest on previously mined land in the Southern Appalachians is accelerated by the use of native black locust *(Robinia pseudoacacia)* as a nurse species.

Maintenance and Management

Weed Control—Methods for the control of problematic weeds within the Ecological Restoration Type are outlined in this section. These generally include biological, chemical (herbicide), and mechanical. Chemical control should be avoided if possible as herbicides are typically nonspecific (killing many species of plants) and may be harmful to animals as well. In areas dominated by extremely invasive exotics (e.g., leafy spurge [*Euphorbia esula*], common reed [*Phragmites australis*], or kudzu [*Pueraria lobata*]) herbicides may be the only way to restore the area. If herbicides are used, they should be the best available for the control of target plant species and should be applied carefully and at recommended or lower rates. Local Integrated Pest Management (IPM) experts should be consulted. Biological control is using organisms, usually nonnative, such as beetles, to eliminate weeds. Hokkanen and Lynch (1995) provide an excellent discussion of the benefits and risks of biological control. Mechanical control is using cutting, digging, or burning to control undesired plants. It involves less risk and no side-effects but may be costly, time-consuming, and may not completely eliminate the problem. A combination of

control methods properly timed and carried out is often most effective.

An Integrated Pest Management (IPM) program is recommended for all restoration projects which involve weed removal. Integrated Pest Management evolved in response to problems associated with an overreliance on chemical pesticides since the late 1940s (Evans and Heitlinger 1984). These problems included acquired resistance of many pests to pesticides, the destruction of natural pest enemies by broad-spectrum pesticides, the destruction and injury of nontarget species, and the increasing cost of petrochemicals. While IPM programs may involve pesticides, they are unlike traditional techniques in that they require a more sophisticated understanding of the ecology and pesticide sensitivity of pests. When pesticides are used in IPM programs, the emphasis on timing and accuracy generally results in reduced pesticide volumes (ibid.).

Sosebee (1983) found that genetic constitution, physiological and phenological stages, and environmental conditions influenced the effectiveness of herbicides in controlling weeds. In general, annuals are most effectively controlled if herbicides are applied during the active growth period before flowering. Herbaceous annuals are usually controllable by mowing or burning. Winter annuals and biennials are most susceptible in the rosette stage, either in fall or spring, but prior to bolting. Some herbaceous perennials are most vulnerable from the late vegetative stage through flowering, but before fruit set. Others are more vulnerable after fruiting. Control of a biennial or perennial species requires a method that will destroy a persistent tuber or extensive root system. Many suffrutescent shrubs (only slightly woody at the base) are most vulnerable during flowering and before fruiting. Woody perennials that do not sprout readily can usually be controlled during active vegetative growth, while freely sprouting woody perennials vary considerably in this respect.

As Evans and Heitlinger (1984) stated, "These variations make it clear why the effectiveness of a weed management treatment depends on its timing, whether the technique involves cutting, burning, or herbicide application." Clearly, any restoration projects that involve control of particular undesirable species can benefit from an Integrated

Pest Management program.

At some highly degraded sites, getting any sort of plants established, even weeds, will eventually improve chances for natural recovery by increasing soil organic matter and waterholding capacity. The control of some exotic plants may not be necessary for two reasons: (1) some species may not be a long-term problem; they may persist only as pioneer species and eventually be replaced by native plants as succession proceeds, and (2) some exotics that are not invasive possess significant values for wildlife habitat, erosion control, or aesthetic purposes (Purves 1995).

Nutrient Requirements—This section contains information concerning the nutrient requirements of the dominant plant species within an Ecological Restoration Type and the use of fertilizers and/or soil amendments in restoration projects. Nutrient requirements of many native plants are poorly understood compared to our knowledge of the requirements for agricultural plants. Basic nutrient requirements of native plants are often inferred from the nutrients present or available in the habitat where a particular plant most commonly occurs.

As a general rule fertilizer applications improve growth and survivorship of planted trees, shrubs, grasses, and herbaceous plants. However, they also encourage weed growth or even facilitate it in areas that would otherwise be relatively free of weeds. So, when we consider nutrient requirements, it is also helpful to think about nutrient limitations. Kimmerer (1985) suggested that some colonizers of extreme sites exhibit a stress-tolerant strategy of allocation of nutrients to internal storage rather than to new productivity. Thus, restorationists should not always be concerned with low biomass if the plants are healthy. In addition, if we start with low-nutrient soils and plant species which further reduce levels of available nutrients, we would be accomplishing what Luken (1990) refers to as the resources-depletion method for maintaining a plant community in an early successional stage. The starting level for many major nutrients in the successional process is often extremely low, especially nitrogen, because it is not a constituent of soil minerals (Bradshaw 1983). Biological nitrogen fixation is important in natural succession and for restoring degraded lands. Nitrogen-fixing plants (e.g., legumes, alders) are useful as a nurse crop on degraded sites. In the restoration of degraded lands nitrogen may need to be added over an extended period (Bradshaw 1983). Applications of calcium and phosphorus may also be necessary. Adjusting the pH may be the most important factor due to its affects upon nutrient availability.

Water Requirements and Hydrology—The water requirements of dominant plant species within the Ecological Restoration Type are discussed in this section, including the use of irrigation and flooding. Supplemental watering may be necessary for initial plant establishment on some sites. In certain situations, however, just as with fertilizers, watering may encourage weed growth. This section also summarizes the management considerations associated with maintaining the proper hydrology of ecological communities in a particular Ecological Restoration Type. Hydrology is especially important in wetland restorations. If the water level, flow, or period are not correct, the desired community will not develop.

Disturbance—Not all communities are dependent on disturbance, but some communities require it (e.g., flooding and fire) to maintain themselves or "reset" succession. Fire is the most common disturbance that is essential for the maintenance of certain communities. Considerations for using prescribed burning as a management tool are presented in this section. As Tom Bonnicksen stated (in Anonymous 1984), "To restore the forest, even incrementally, you may have to use fire the same way you use any other tool, recognizing that it is artificial and a way of working back to the natural condition. Once the natural condition is restored, fire will operate as a natural process, and future changes can be expected to mimic the normal development of the ecosystem."

Case Studies

This section includes summaries of, or references to, case studies (where available) of restoration projects of ecological communities within the Ecological Restoration Type. Since many restoration projects are not readily available or published in journals, the case studies that are presented in this section vary greatly in level of detail and types of information presented. The Society for Ecological Restoration publishes two journals, *Restoration and Management Notes* (since 1983) and *Restoration Ecology* (since 1993), that are the primary sources for case studies and current information about the field of ecological restoration.

An integral part of restoration ecology in the future must be the sharing of information and case studies. Real life examples, whether considered successful or not, often provide significant insights about the functioning of ecosystems. We hope that you find this format useful in your restoration endeavors and welcome any comments or suggestions.

Bibliography

Aber, J. D. 1990. Forest ecology and the forest ecosystem. In Introduction to forest science, eds., R. A. Young and R. L. Giese. New York: John Wiley & Sons.

Abrahamson, W. G. and D. C. Hartnett. 1990. Pine flatwoods and dry prairies. In Ecosystems of Florida, eds., R. L. Myers and J. J. Ewel. Orlando: University of Central Florida.

Abrams, P. A. 1988. How should resources be counted? Theoretical Population Biology 33: 226-242.

Adam, P. 1990. Saltmarsh ecology. New York: Cambridge University Press.

Adams, L. W. 1994. Urban wildlife habitats: a landscape perspective. Minneapolis: University of Minnesota Press.

Adams, L. W. and L. E. Dove. 1989. Wildlife reserves and corridors in the urban environment. Columbia: National Institute for Urban Wildlife.

Adkinson, S., ed. 1991. Garden pools, fountains, and waterfalls. Menlo Park: Sunset Publishing Company.

Agee, J. K. 1993. Fire ecology of Pacific northwest forests. Washington, D. C.: Island Press.

Ahlgren, C. E. 1966. Small mammals and reforestation following prescribed burning. Journal of Forestry (September): 614-618.

Ahrenhoerster, R. and T. Wilson. 1981. Prairie restoration for the beginner. Des Moines: Prairie Seed Source.

Aiken, G. D. 1935 (reprinted many times). Pioneering with Wildflowers. Brattleboro: Alan C. Hood and Company, Inc.

Ajilvsgi, G. 1979. Wild flowers of the big thicket. Number Four: The W. L. Moody, Jr., Natural History Series. College Station: Texas A and M University Press.

Aldon, E. F. and J. R. C. Doria. 1995. Growing and harvesting fourwing saltbush (*Atriplex canescens*) under saline conditions. In Proceedings: Wildland shrub and arid land restoration symposium, comps., B. A Roundy, E. D. McArthur, J. S. Haley, and D. K. Mann. Ogden: U.S. Forest Service General Technical Report INT-GTR-315.

Aldous, A. E. 1929. The eradication of brush and weeds from pasture lands. Agronomy Journal 21: 660-666.

Aldous, A. E. 1934. Effect of burning on Kansas bluestem prairie. Kansas Agricultural Experiment Station Technical Bulletin 38.

Alexander, C. E., M. A. Boutman, and D. W. Field. 1986. An inventory of coastal wetlands of the USA. Washington D.C.: U. S. Department of Commerce.

Allen, E. B. 1988. The reconstruction of disturbed arid lands: an ecological approach. Boulder: Westview Press.

Allen, E. B. and M. F. Allen. 1984. Competition between plants of different successional stages: mycorrhizae as regulators. Canadian Journal of Botany 62: 2625-2629.

Allen, M. F. 1988. Below ground structure: a key to reconstructing a productive arid ecosystem. In The reconstruction of disturbed arid lands: an ecological approach, ed., E. B. Allen. Boulder: Westview Press.

Allen, M. F. 1991. The ecology of mycorrhizae. New York: Cambrige University Press.

Allen, M. F., ed. 1992. Mycorrhizal functioning, an integrative plant-fungal process. New York: Chapman and Hall Press.

Allen, S. D., F. C. Golet, A. F. Davis, and T. E. Sokoloski. 1989. Soil vegetation correlations in transition zones of Rhode Island red maple swamps. Washington, D.C.: U.S. Fish and Wildlife Service Biological Report 89(8).

Ambuel, B. and S. A. Temple. 1983. Area dependent changes in the bird communities and vegetation of southern Wisconsin forests. Ecology 65(5): 1057-1068.

Amme, D. 1984. Prairie restoration and gorse control research underway (California). Restoration and Management Notes 2(1): 36.

Amon, J. P. and E. Briuer. 1993. Groundwater is hydrology source in Ohio fen creation. The Wetlands Research Program Bulletin 3(4): 5-8.

Amos, B. B. and F. R. Gehlback, eds. 1988. Edwards Plateau vegetation: plant ecological studies in central Texas. Waco: Baylor University Press.

Andersen, D. C. 1994. Demographics of small mammals using anthropogenic desert riparian habitat in Arizona. Journal of Wildlife Management 58(3): 445-453.

Anderson, A. N. 1993. Ants as indicators of restoration success at a uranium mine in tropical Australia. Restoration Ecology 1(3): 156-167.

Anderson, M. G. 1995. Interactions between *Lythrum salicaria* and native organisms: a critical review. Environmental Management 19(2): 225-231.

Anderson, M. K. 1993 Native Californians as ancient and contemporary cultivators. In Before the wilderness: native Californians as environmental managers, eds., T. C. Blackburn and M. K. Anderson. Menlo Park: Ballena Press.

Anderson, M. L. and V. Kurmis. 1981. Revegetation of mined peatlands: I. environmental properties of a mined area. St. Paul: Minnesota Peat Program, Minnesota Department of Natural Resources.

Anderson, R. C., E. B. Allen, M. R. Anderson, J. S. Fralish, R. M. Miller, and W. A. Niering. 1993. Science and restoration. Science 262: 14-15.

Anderson, R. C. and M. R. Anderson. 1995. North American conference on savannas and barrens. Restoration and Management Notes 13(1): 61-63.

Anderson, S. H., K. Mann, and H. H. Shugart, Jr. 1977. The effect of transmission-line corridors on bird populations. American Midland Naturalist 97: 216-222.

Anderson, S. H. and H. H. Shugart, Jr. 1974. Habitat selection of breeding birds in an east Tennessee deciduous forest. Ecology 55: 828-837.

Andreas, B. K. 1989. Vascular flora of the glaciated Allegheny Plateau region of Ohio. Columbus: Ohio Biological Survey.

Anonymous. No date. Guide to the natural communities of Florida. Tallahassee: Florida Natural Areas Inventory and Florida Department of Natural Resources.

Anonymous. 1984. Viewpoint: a call for accountability, an interview with Tom Bonnicksen. Restoration and Management Notes 2(1): 12-13.

Anonymous. 1987. We don't move until the baby hawks fly. Golf Course Management 55(2): 145-147.

Anonymous. 1991a. Natural landscapes of Maine: a classification of ecosystems and natural communities. Augusta: Maine Natural Heritage Program.

Anonymous. 1991b. Plant communities of Texas. Austin: Texas Natural Heritage Program.

Anonymous. 1991c. Wyoming plant community classification. Larami: Wyoming Natural Heritage Program.

Anonymous. 1995. Insect biological control of purple loosestrife. Fish and Wildlife Reference Service Newsletter 104 (Spring): 1.

Aplet, G. H., R. D. Laven, and P. L. Fiedler. 1992.The relevance of conservation biology to natural resource management. Conservation Biology 6(2): 298-300.

Arbuckle, N. and C. Crocker, eds. 1991. How to attract hummingbirds and butterflies. San Ramon: Ortho Books.

Archibold, O. W. 1995. Ecology of world vegetation. New York: Chapman and Hall.

Armentano, T. V. 1980. Drainage of organic soils as a factor in the world carbon cycle. Bioscience 30: 825-830.

Armitage, A. M. 1989. Herbaceous perennial plants. Athens: Varsity Press, Inc.

Arno, S. F. 1988. Fire ecology and its management implications in ponderosa pine forests. In Ponderosa pine: the species and its management; symposium proceedings, eds., D. M. Baumgartner and J. E. Lotan. Pullman: Washington State University Cooperative Extension publication.

Arno, S. F., M. G. Harrington, C. E. Fielder, and C. E. Carlson. 1995. Restoring fire-dependent ponderosa pine forests in western Montana. Restoration and Management Notes 13(1): 32-36.

Aronson, J., S. Dhillion, and E. Le Floc'h. 1995. On the need to select an ecosystem of reference, however imperfect: a reply to Pickett and Parker. Restoration Ecology 3(1): 1-3.

Aronson, J., C. Floret, E. Le Floc'h, C. Ovalle, and R. Pontanier. 1993a. Restoration and rehabilitation of degraded ecosystems in arid and semi-arid lands. I. A view from the south. Restoration Ecology 1(1): 8-17.

Aronson, J., C. Floret, E. Le Floc'h, C. Ovalle, and R. Pontanier. 1993b. Restoration and rehabilitation of degraded ecosystems in arid and semi-arid lands. II. Case studies in Chile, Tunisia and Cameroon. Restoration Ecology 1(2): 168-187.

Asher, S. C. and V. G. Thomas. 1985. Analysis of temporal variation in the diversity of a small mammal community. Canadian Journal of Zoology 63: 1106-1109.

Ashton, P. S. 1988. Conservation of biological diversity in botanical gardens. In Biodiversity, ed., E. O. Wilson. Washington: National Academy Press.

Askins, R. A. and M. J. Philbrick. 1987. Effects of changes in regional forest abundance on the decline and recovery of a forest bird community. Wilson Bulletin 99(1): 7-21.

Askins, R. A., M. J. Philbrick, and D. S. Sugeno. 1987 Relationship between the regional abundance of forest and the composition of forest bird communities. Biological Conservation 39: 129-152.

Atwater, W. G. 1954. Hair grass takes over. Everglades Natural History 2: 43.

Atwood, W. W. 1940. The physiographic provinces of North America. New York: Ginn and Company.

Aulbach-Smith, C. A. and S. J. de Kozlowski. 1990. Aquatic and wetland plants of South Carolina. Columbia: South Carolina Aquatic Plant Management Council and South Carolina Water Resources Commission.

Ault, E. B. 1983. Charting the course going "natural." Golf Course Management 51(8): 59-60.

Averitt, E., F. Steiner, R. A. Yabes, and D. Patten. 1994. An assessment of the Verde River Corridor Project in Arizona. Landscape and Urban Planning 28(2-3): 161-178.

Axelrod, D. I. 1967. Drought, diastrophism, and quantum evolution. Evolution 21: 201-209.

Baars, D. L. 1995. Navajo country, a geology and natural history of the four corners region. Albuquerque: University of New Mexico.

Bader, B. J. 1996. Midwest oak savanna and woodland ecosystems conference. Restoration and Management Notes 14(1): 43-45.

Bahre, C. J. 1991. A legacy of change: historic human impact on vegetation of the Arizona borderlands. Tucson: University of Arizona Press.

Bailey, C. A. 1985. Planting of sanctuary marsh (Ohio). Restoration and Management Notes 3(1): 53-54.

Bailey, R. G. 1978. Ecoregions of the United States. Ogden: Forest Service, U. S. Department of Agriculture.

Bailey, R. G. 1995. Descriptions of the ecoregions of the United States. 2nd ed. revised and expanded (1st ed. 1980). Miscellaneous Publication Number 1391. Washington, D.C.: United States Department of Agriculture Forest Service.

Bailey, R. G. 1996. Ecosystem geography. New York: Springer-Verlag.

Bainbridge, D. A. 1990. The restoration of agricultural lands and drylands. In Environmental restoration, ed., J. J. Berger. Washington: Island Press.

Bainbridge, D. A. 1994. Treeshelters improve establishment on dry sites. Tree Planters Notes 45(1): 13-16.

Bainbridge, D. A. 1996. Vertical mulch controls erosion, aids revegetation (California). Restoration and Management Notes 14(1): 82.

Bainbridge, D. A. and M. W. Fidelibus. 1994. Treeshelters improve woody transplant survival in arid lands (California). Restoration and Management Notes 12(1): 86.

Bainbridge, D. A., M. Fidelibus, and R. MacAller. 1995. Techniques for plant establishment in arid ecosystems. Restoration and Management Notes 13(2): 190-197.

Bainbridge, D. A. and R. A. Virginia. 1990. Restoration in the Sonoran Desert of California. Restoration and Management Notes 8(1): 3-14.

Baker, W. L. 1994. Restoration of landscape structure altered by fire suppression. Conservation Biology 8(3): 763-769.

Bakker, E. 1995. An island called California: an ecological introduction to its natural communities. 2nd ed. Berkeley: University of California.

Balogh, J. C.,and W. J. Walker, eds. 1992. Golf course management and construction. Chelsea: Lewis Publishers.

Balser, D., A. Bielak, G. De Boer, T. Tobias, G. Adindu, and R. S. Dorney. 1981. Nature reserve designation in a cultural landscape, incorporating island biogeography theory. Landscape Planning 8: 329-347.

Barber, M. 1994. Wiregrass research. Resource Management Notes 6(3): 12-13.

Barbour, M. G. 1968. Germination requirements for the desert shrub *Larrea divaricata*. Ecology 50: 679-685.

Barbour, M. G. 1988. Californian upland forests and woodlands. In North American terrestrial vegetation, eds., M. G. Barbour and W. D. Billings. Cambridge: Cambridge University Press.

Barbour, M. G. and W. D. Billings, eds. 1988. North American terrestrial vegetation. Cambridge: Cambridge University Press.

Barbour M. G., J. H. Burk, and W. D. Pitts. 1987. Terrestrial plant ecology. Menlo Park: The Benjamin Cummings Publishing Company, Inc.

Barbour, M. G. and J. Major, eds. 1977 (1988 expanded edition). Terrestrial vegetation of California. Sacramento: California Native Plant Society, Special publication No. 9.

Barbour, M. G. and J. Major, eds. 1988. Terrestrial vegetation of California. Sacramento: California Native Plant Society.

Barden, L. S. 1982. Effects of prescribed fire on honeysuckle and other ground flora (North Carolina). Restoration and Management Notes 1: 127.

Barden, L. S. and J. F. Matthews. 1980. Changes in abundance of honeysuckle (*Lonicera japonica*) and other ground flora after prescribed burning of a piedmont pine forest. Castanea 45: 257-260.

Bare, J. E. 1979. Wildflowers and weeds of Kansas. Lawrence: The Regents Press of Kansas.

Barnes, B. W. 1981. Michigan trees. Ann Arbor: University of Michigan.

Barrett, G. W., G. M. Van Dyne, and E. P. Odum. 1976. Stress ecology. Bioscience 26: 192-194.

Barrow, C. J. 1994. Land degradation, development and breakdown of terrestrial environments. Cambridge: Cambridge University Press.

Barrow, J. R. and K. M. Havstad. 1992. Recovery and germination of gelatin-encapsulated seeds fed to cattle. Journal of Arid Environments 22: 395-399.

Barry, J. M. 1980. Natural vegetation of South Carolina. Columbia: University of South Carolina Press.

Barry, W. J., A. S. Garlo, and C. A. Wood. 1996. Duplicating the mound-and-pool microtopography of forested wetlands. Restoration and Management Notes 14(1): 15-21.

Barth, F. G. 1985. Insects and flowers, the biology of a partnership. Princeton: Princeton University.

Beal, E. O. 1977. A manual of marsh and aquatic vascular plants of North Carolina with habitat data. Raleigh: North Carolina Agricultural Experiment Station.

Beal, E. O. and J. W. Thieret. 1986. Aquatic and wetland plants of Kentucky. Frankfort: Kentucky Nature Preserves Commission, Scientific and Technical Series No. 5.

Beatley, J. C. 1976. Vascular plants of the Nevada test site and central-southern Nevada: ecologic and geographic distributions. Springfield: Energy Research and Development Administration, NTIS, TID-26881.

Bedinger, M. S. 1981. Hydrology of bottomland hardwood forests of the Mississippi Embayment. In Wetlands of bottomland forests, eds., J. R. Clark, and J. Benforado. Amsterdam: Elsevier Scientific Publications.

Begon, M., J. L. Harper, and C. R. Townsend. 1986. Ecology: individuals, populations, and communities. Oxford: Blackwell Scientific Publications.

Begon, M., J. L. Harper, and C. Townsend. 1995. Ecology, individuals, populations, and communities. 3rd ed. Cambridge: Blackwell Science, Inc.

Beikmann, S. 1992. Personal communication during review.

Belaire, C. E. and D. Templet. 1995. Innovative use of dredged material in Texas benefits endangered whooping cranes. Restoration and Management Notes 13(1): 141.

Bender, G. L., ed. 1982. Reference handbook on the deserts of North American. Westport: Greenwood Press.

Bendix, J. 1994. Among-site variation in riparian vegetation of the Southern California Transverse Ranges. American Midland Naturalist 132(1): 136-151.

Bennett, G. W. 1970 (reprinted in 1983). Management of lakes and ponds. 2d ed. Melbourne: Kreiger Publishing Company.

Benning, T. L. and T. B. Bragg. 1993. Response of big bluestem (*Andropogon gerardii*) to timing of spring burning. American Midland Naturalist 130(1): 127-132.

Benson, L. 1969. The cacti of Arizona, 3rd ed. Tucson: The University of Arizona Press.

Benson, L. and R. A. Darrow. 1981. Trees and shrubs of the southwestern deserts, 3rd ed. Tucson: The University of Arizona Press.

Bentham, H., J. A. Harris, P. Birch, and K. C. Short. 1992. Habitat classification and soil restoration assessment using analysis of soil microbial and physico-chemical characteristics. Journal of Applied Ecology 29: 711-718.

Berendse, F., M. J. M. Oomes, H. J. Altena, and W. T. Elberse. 1992. Experiments on the restoration of species-rich meadows in The Netherlands. Biological Conservation 62: 59-65.

Bergen, A. and M. Levandowski. 1994. Salt marsh restoration: fertilizer and TPH study. New York: Natural Resources Group.

Berger, J. J. 1985. Restoring the earth. New York: Alfred A. Knopf.

Berger, J. J., ed. 1990. Environmental restoration. Washington, D.C.: Island Press.

Beson, L. and R. A. Darrow. 1981. Trees and shrubs of the southwestern deserts. 3rd ed. Tucson: The University of Arizona Press.

Betz, R. F. 1986. One decade of research in prairie restoration at Fermi National Accelerator Laboratory, Batavia, Illinois. In Proceedings of the Ninth North American Prairie Conference, eds., G. K. Clambey and R. H. Pemble. Fargo: Tri-College University.

Bir, R. E. 1992. Growing and propagating showy native woody plants. Chapel Hill: The University of North Carolina Press.

Bisset, N. 1995. Large scale harvest of wiregrass seed and use in reclamation on mined lands. Resource Management Notes 7(3): 10-11.

Blake, J. G. 1983. Trophic structure of bird communities in forest patches in east-central Illinois. Wilson Bulletin 95(3): 416-430.

Blake, J. G. 1987. Species-area relationships of winter residents in isolated woodlots. Wilson Bulletin 99(2): 243-252.

Blake, J. G. and J. R. Karr. 1984. Species composition of bird communities and the conservation benefit of large versus small forests. Biological Conservation 30: 173-187.

Blake, J. G. and J. R. Karr. 1987. Breeding birds of isolated woodlots: area and habitat relationships. Ecology 68(6): 1724-1734.

Blakeman, J. A. 1996. Placing stalks of common reed in treeshelters saves songbirds (Ohio). Restoration and Management Notes 14(1): 81-82.

Blouin, M. S. and E. F. Connor. 1985. Is there a best shape for nature reserves? Biological Conservation 32: 277-288.

Bock, C. E., V. A. Saab, T. D. Rich, and D. S. Dobkin. 1993. Effects of livestock grazing on neotropical landbirds in western North America. In Proceedings of the status and management of neotropical migratory birds symposium, eds., D. M. Finch, and P. W. Stangel. Fort Collins: U.S. Forest Service General Technical Report RM-229.

Bockheim, J. G. 1990. Forest soils. In Introduction to forest science, eds., R. A. Young and R. L Giese. New York: John Wiley & Sons.

Boecklen, W. J. and D. Simberloff. 1986. Area-based extinction models in conservation. In Dynamics of extinctions, ed., D. K. Elliott. New York: John Wiley & Sons.

Boettcher, J. F. and T. B. Bragg. 1988. Tallgrass prairie remnants of eastern Nebraska. In Proceedings of the Eleventh North American Prairie Conference, eds., T. B. Bragg and J. Stubbendieck. Lincoln: University of Nebraska Press.

Bond, W. J. 1995. Fire and plant population ecology. New York: Chapman and Hall.

Bookhout, T. A., ed. 1994. Research and management techniques for wildlife and habitats. 5th ed. Betheseda: The Wildlife Society.

Boon, W. and H. Groe. 1990. Nature's heartland, native plant communities of the Great Plains. Ames: Iowa State University Press.

Bornstein, C. J. 1985. Creating a California meadow. Restoration and Management Notes 3(1): 51-52.

Bourn, W. S. and C. Cottam. 1950. Some biological effects of ditching tidal marshes. Washington, D.C.: U.S. Fish and Wildlife Service Research Report 19.

Bowers, J. K. 1995. Innovations in tidal marsh restoration: the Kenilworth Marsh account. Restoration and Management Notes 13(2): 155-161.

Bowler, P. A. 1996. Hand-pulling controls invasive sea fig (California). Restoration and Management Notes 14(1): 78-79.

Bowles, M. L. and C. J. Whelan, eds. 1994. Restoration of endangered species: conceptual issues, planning, and implementation. New York: Cambridge University Press.

Bowman, C. W. 1994. Barrier island reconstruction. Land and Water (May/June): 34-36.

Boyd, R. 1986. Strategies of Indian burning in the Willamette Valley. Canadian Journal of Anthropology 5(1).

Bradshaw, A. D. 1983. The reconstruction of ecosystems. Journal of Applied Ecology 20: 1-17.

Bradshaw, A. D. 1987. Restoration: an acid test for ecology. In Restoration Ecology, eds., W. R. Jordan, III, M. E. Gilpin, and J. D. Aber. United Kingdom: Cambridge University Press.

Bradshaw, A. D. and M. J. Chadwick. 1980. The restoration of land. Berkeley: University of California Press.

Bragg, T. B. 1978. Allwine Prairie Preserve: a reestablished bluestem grassland research area. In Fifth Midwest Prairie Conference Proceedings, eds., D. C. Glenn-Lewin and R. Q. Landers. Ames: Iowa State University.

Bragg, T. B. and D. M. Sutherland. 1988. Establishing warm-season grasses and herbs using herbicides and mowing. In Proceedings of the Eleventh North American Prairie Conference, eds., T. B. Bragg and J. Stubbendieck. Lincoln: University of Nebraska Press.

Brandel, M. J. 1985. Human impact and restoration of a swamp conifer forest (Wisconsin). Restoration and Management Notes 3(1): 54-55.

Bratton, S. P. 1982. The effects of exotic plant and animal species on nature preserves. Natural Areas Journal 2(3): 3-13.

Bratton, S. P. 1986. Manager reflects on new environmental ethics program at University of Georgia. Restoration and Management Notes 4(1): 3-4.

Braun, L. E. [1950] 1967. Deciduous forests of eastern North America. Reprint. New York: Hafner Publishing Company.

Bren, L. J. 1993. Riparian zone, stream, and floodplain issues: a review. Journal of Hydrology 150: 277-299.

Brenholm, T. L. and A. G. van der Valk. 1994. Sedge establishment studies: soil amendments improve growth of *Carex stricta* seedlings. Chicago: Wetlands Research, Inc., Technical Paper No. 4.

Brenner, F. J. and N. L. Simon. 1984. Mast-producing natives direct seeded on surface-mined sites (Pennsylvania). Restoration and Management Notes 2(1): 38-39.

Briggs, M. K., B. A. Roundy, and W. W. Shaw. 1994. Trial and error: assessing the effectiveness of riparian revegetation in Arizona. Restoration and Management Notes 12(2): 160-167.

Brinson, M. M., B. L. Swift, R. C. Plantico, and J. S. Barclay. 1981. Riparian ecosystems: their ecology and status. Washington, D.C.: U.S. Fish and Wildlife Service, FWS OBS-81/17.

Brooks, R. P. and D. E. Samuel. 1985. Wetlands and water management on mined lands. University Park: Pennsylvania State University Press.

Broome, S. W. 1990. Creation and restoration of tidal wetlands of the southeastern United States. In Wetland creation and restoration: the status of the science, eds., K. A. Kusler and M. E. Kentula. Washington, D.C.: Island Press.

Broome, S. W. and E. D. Seneca. 1985. Elevation, moisture, and fertilization affect restoration of brackish-water marshes (North Carolina). Restoration and Management Notes 3(1): 36-37.

Broome, S. W., E. D. Seneca, and W. W. Woodhouse, Jr. 1981. Planting marsh grasses for erosion control. Raleigh: University of North Carolina Sea Grant College Publication UNC-SG-81-09.

Broome, S. W., E. D. Seneca, and W. W. Woodhouse, Jr. 1982. Establishing brackish marshes on graded upland sites in North Carolina. Wetlands 2: 152-178.

Broome, S. W., W. W. Woodhouse, Jr., and E. D. Seneca. 1975. The relationship of mineral nutrients to growth of *Spartina alterniflora* in North Carolina. I. Nutrient status of plants and soils in natural stands. Soil Science of America Proceedings 39(2): 295-301.

Brothers, T. S. 1992. Postsettlement plant migrations in northeastern North America. American Midland Naturalist 128: 72-82.

Brothers, T. S. and A. Spingarn. 1992. Forest fragmentation and alien plant invasion of central Indiana old-growth forests. Conservation Biology 6(1): 91-100.

Browder, J. A., P. J. Gleason, and D. R. Swift. 1994. Periphyton in the Everglades: spatial variation, environmental correlates, and ecological implications. In Everglades: the ecosystem and its restoration, eds., S. M. Davis, and J. C. Ogden. Delray Beach: St. Lucie Press.

Brown, C. A. [1945] 1972. Louisiana trees and shrubs. Reprint. Baton Rouge: Claitor's Publishing Division.

Brown, D. A. and K. D. Brown. 1996. Disturbance plays key role in distribution of plant species. Restoration and Management Notes 14(2): 140-147.

Brown, D. E., ed. 1982. Desert plants. In Biotic communities of the American southwest—United States and Mexico, ed., F. S. Crosswhite, Vol. 4 (Vols. 1-4). Superior: University of Arizona for the Boyce Thompson Southwestern Arboretum.

Brown, L. 1995. Grasslands. New York: Alfred A. Knopf.

Brown, M. T. and R. E. Tighe. 1991. Techniques and guidelines for reclamation of phosphate mined lands. Center for Wetlands, University of Florida. FIPR Pub. No. 03-044-095. Bartow: Florida Institute of Phosphate Research.

Bryan, H. D. 1987. Important habitats and quantitative environmental assessment. In Proceedings of the Fourth Symposium on Environmental Concerns in Rights-of-Way Management, eds., W. R. Byrnes and H. A. Holt. Indianapolis, Indiana. October 15-28.

Bryan, H. D. 1991. The distribution, habitat, and ecology of shrews (Soricidae: *Blarina*, *Sorex*, an *Cryptotis*) in Kentucky. Journal of the Tennessee Academy of Science 66(4): 187-189.

Bryan, H. 1992. Personal communication during review.

Bryant, L. D. and J. M. Skovlin. 1982. Effect of grazing strategies and rehabilitation on an eastern Oregon stream. In Habitat disturbance and recovery: Proceedings of a Symposium. San Francisco: California Trout, Inc.

Buckley, G. P., ed. 1989. Biological habitat reconstruction. New York: Belhaven Press.

Burgess, R. L. and D. M. Sharpe, eds. 1981. Forest island dynamics in man-dominated landscapes. New York: Springer-Verlag.

Burk, J. H. 1988. Sonoran desert vegetation. In Terrestrial vegetation of California, eds., M. G. Barbour and J. Major. Davis: California Native Plant Society.

Burns, R. M. and B. H. Honkala. 1990. Silvics of North America. Vols. 1–2, Hardwoods. Agriculture Handbook 654. Washington: Forest Service, United States Department of Agriculture.

Busciano, M. 1995. Lower rates of herbicide control oriental bittersweet (New York). Restoration and Management Notes 13(2): 225-226.

Busse, K. G. 1989. Ecology of *Salix* and *Populus* species of the Crooked River National Grassland. M.S. thesis. Oregon State University, Corvallis.

Butcher, G. S., W. A. Niering, W. J. Barry, and R. H. Goodwin. 1981. Equilibrium biogeography and the size of nature preserves: an avian case study. Oecologia 49: 29-37.

Caduto, M. J. 1990. Pond and brook, a guide to freshwater environments. Hanover: University Press of New England.

Cairns, J., Jr. 1981. Restoration and management: an ecologist's perspective. Restoration and Management Notes 1: 6-8.

Cairns, J. 1988a. Restoration and the alternative: a research strategy. Restoration and Management Notes 6(2): 65-67.

Cairns, J., Jr. 1988b. Restoration of damaged ecosystems as opportunities for increasing diversity. In Biodiversity, ed., E. O. Wilson. Washington, D.C.: National Academy Press.

Cairns, J., Jr. 1989. Restoring damaged ecosystems: is predisturbance condition a viable option? Environmental Professional 11: 152-159.

Cairns, J., Jr. 1991. The status of the theoretical and applied science of restoration ecology. The Environmental Professional 13: 186-194.

Cairns, J., Jr. 1993. Is restoration ecology practical? Restoration Ecology 1(1): 3-7.

Cairns, J., Jr., ed. 1988c. Rehabilitating damaged ecosystems. 2 Vols. Boca Raton: CRC Press.

Cairns, J., Jr., J. S. Crossman, K. L. Dickson, and E. E. Herricks. 1971. The recovery of damaged streams. Association of Southeastern Biologists Bulletin 18(3): 79-106.

Cairns, J., Jr., K. L. Dickson, and E. E. Herricks, eds. 1977. Recovery and restoration of damaged ecosystems. Charlottesville: University Press of Virginia.

Cameron, C. C. 1973. Peat, in United States Mineral Resources. In U.S. Geological Survey Professional Paper No. 820. Washington, D.C.: U.S. Geological Survey.

Cammen, L. M. 1976a. Abundance and production of macroinvertebrates from natural and artificially established salt marshes in North Carolina. American Midland Naturalist 96: 244-253.

Cammen, L. M. 1976b. Macroinvertebrate colonization of *Spartina* marshes artificially established on dredge spoil. Estuarine Coastal Marine Science 4: 357-372.

Campbell, J. J. N. 1980. Present and presettlement forest conditions in the Inner Bluegrass of Kentucky. Ph.D. dissertation. University of Kentucky, Lexington.

Carey, R. A. and D. J. Robertson. 1995. Treeshelters lead to unexpected problems (Pennsylvania). Restoration and Management Notes 13(1): 134.

Carlile, D. W., J. R. Skalski, J. E. Batker, J. M. Thomas, and V. I. Cullinan. 1989. Determination of ecological scale. Landscape Ecology 2: 203-213.

Carlton, J. M. 1974. Land-building and stabilization by mangroves. Environmental Conservation 1: 285-294.

Carothers, S. W. 1977. Importance, preservation, and management of riparian habitats: an overview. In Importance, preservation and management of riparian habitat: a symposium, tech. coords., R. R. Johnson and D. A. Jones, Jr. Fort Collins: U. S. Department of Agriculture Forest Service General Technical Report RM-43. Rocky Mountain Forest and Range Experiment Station.

Carothers, S. W., R. R. Johnson, and S. W. Aitchinson. 1974. Population structure and social organization of Southwestern riparian birds. American Zoologist 14: 77-108.

Carr, B. 1992. Personal communication during review.

Carter, M. R., L. A. Burus, T. R. Cavinder, K. R. Dugger, P. L. Fore, D. B. Hicks, H. L. Revells, and T. W. Schmidt. 1973. Ecosystem analysis of the Big Cypress Swamp and estuaries. Atlanta: U.S. Environmental Protection Agency Region IV.

Cathey, H. M. 1990. U. S. D. A. plant hardiness zone map. Agricultural Research Service Miscellaneous Publication Number 1475. Washington, D.C.: United States Department of Agriculture.

Case, R. L. 1995. Structure, biomass and recovery of riparian ecosystems of northeast Oregon. M.S. thesis. Oregon State University, Corvallis.

Chabot, B. F. and H. A. Mooney, eds. 1985. Physiological ecology of North American plant communities. New York: Chapman and Hall.

Chabreck, R. H. 1970. Marsh zones and vegetative types in the Louisiana coastal marshes. Ph.D. diss. Louisiana State University, Baton Rouge.

Chabreck, R. H. 1972. Vegetation, water and soil characteristics of the Louisiana coastal region. Baton Rouge: Louisiana Agricultural Experiment Station Bulletin 664.

Chabreck, R. H. 1988. Coastal marshes: ecology and wildlife management. Minneapolis: University of Minnesota Press.

Chabreck, R. H. and R. G. Linscombe. 1982. Changes in vegetative types in the Louisiana coastal marshes over a 10-year period. Proceedings of the Louisiana Academy of Science 45: 98-102.

Chalmers, A. 1982. Soil dynamics and the productivity of *Spartina alterniflora*. In Estuarine comparisons, ed., V. S. Kennedy. New York: Academic Press.

Chaney, E., W. Elmore, and W. S. Platts. 1990. Livestock grazing on western riparian areas. Denver: United States Environmental Protection Agency, Region 8.

Chapman, K. A. 1986. Draft descriptions of Michigan natural community types. Lansing: Michigan Natural Features Inventory.

Chapman, V. J. 1960. Salt marshes and salt deserts of the world. New York: Interscience Publishers.

Chapman, V. J. 1976a. Coastal vegetation. 2nd ed. Oxford: Pergamon Press.

Chapman, V. J. 1976b. Mangrove vegetation. Vaduz: J. Kramer.

Chapman, V. J. 1977. Wet coastal ecosystems. Amsterdam: Elsevier Scientific Publications.

Charlet, D. A. 1995. Atlas of Nevada conifers, a phytogeographic reference. University of Nevada Press.

Cheater, M. 1992. Alien invasion. Nature Conservancy: (September/October): 24-29.

Cheskey, E. D. 1993. Habitat restoration: a guide for proactive schools. Ontario Canada: The Waterloo County Board of Education, Curriculum and Program Development, Outdoor Education Department.

Choi, Y. D. and N. D. Pavlovic. 1994. Comparison of fire, herbicide, and sod removal to control exotic vegetation. Natural Areas Journal 14(3): 217-218.

Chow-Fraser, P. and L. Lukasik. 1995. Cootes Paradise Marsh: community participation in the restoration of a Great Lakes coastal wetland. Restoration and Management Notes 13(2): 183-189.

Christensen, N. L. 1988. Vegetation of the southeastern coastal plains. In North American terrestrial vegetation, eds., M. G. Barbour and W. D. Billings. Cambridge: Cambridge University Press.

Cintron, G., A. E. Lugo, D. J. Pool, and G. Morris. 1978. Mangroves of arid environments in Puerto Rico and adjacent islands. Biotropica 10: 110-121.

Clady, M. D. 1994. Conserving the Oregon silverspot butterfly on Siuslaw National Forest. Endangered Species Technical Bulletin 19(4): 12.

Clark, F. E. 1975. Viewing the invisible prairie. In Prairie: a multiple view, ed., M. K. Wali. Grand Forks: University of North Dakota Press.

Clark, J., ed. 1976. Barrier islands and beaches. Technical Proceedings of the 1976 Barrier Islands Workshop. Washington, D.C.: The Conservation Foundation.

Clark, J. R. and J. Benforado, eds. 1981. Wetlands of bottomland hardwood forests. Amsterdam: Elsevier Scientific Publications.

Clark, K., D. Euler, and E. Armstrong. 1983. Habitat associations of breeding birds in cottage and natural areas of central Ontario. Wilson Bulletin 95(1): 77-96.

Clements, F. E. 1905. Research methods in ecology. Lincoln: University Publishing Company.

Clewell, A. F. 1981. Vegetation restoration techniques on reclaimed phosphate strip mines in Florida. Wetlands 1: 158-170.

Clewell, A. F. 1985. Guide to the vascular plants of the Florida panhandle. Tallahassee: Florida State University Press.

Cobb, B. 1963. A field guide to the ferns and their related families. Boston: The Peterson Field Guide Series, Houghton Mifflin Company.

Cody, M. L. 1985a. Habitat selection in grassland and open-country birds. In Habitat selection in birds, ed., M. L. Cody. Orlando: Academic Press Inc.

Cody, M. L., ed. 1985b. Habitat selection in birds. Orlando: Academic Press Inc.

Cody, M. L. and J. M. Diamond, eds. 1975. Ecology and evolution of communities. Cambridge: The Belknap Press of Harvard University Press.

Cohen, A. D. 1984. Evidence of fires in the ancient Everglades and coastal swamps of southern Florida. In Environments of south Florida: present and past, memoir 2, ed., P. J. Gleason. Coral Gables: Miami Geological Society.

Cole, M. A. R. 1991. Vegetation management guideline: leafy spurge (*Euphorbia esula* L.). Natural Areas Journal 11(3): 271.

Cole, S. 1995. "Rescue" transplanting of wire grass at Oscar Scherer. Resource Management Notes 6(6): 9-11.

Collard III, S. B. 1993. Alien invaders: the continuing threat of exotic species. Danbury: Grolier Publishing.

Collins, B. J. 1976. Key to trees and shrubs of the deserts of southwestern Arizona. Thousand Oaks: California Lutheran College.

Collins, B. J. 1979. Key to wildflowers of the deserts of southern California. Thousand Oaks: California Lutheran College.

Collins, B. R. and K. H. Anderson. 1994. Plant communities of New Jersey, a study in landscape diversity. New Brunswick: Rutgers University Press.

Collins, E. R. 1985. Rare plants, insects are focus of prairie restoration (Illinois). Restoration and Management Notes 3(1): 51.

Collins, J. T., S. L. Collins, J. Horak, D. Mulhern, W. Busby, C. C. Freeman, and G. Wallace. 1996. Illustrated guide to endangered or threatened species in Kansas. University of Kansas.

Collins, S. L. and S. M. Glenn. 1991. Importance of spatial and temporal dynamics in species regional abundance and distribution. Ecology 72(2): 654-664.

Collins, S. L. and L. Wallace, eds. 1990. Fire in North American tallgrass prairies. University of Oklahoma.

Collins, T. M., A. H. Winward, D. Duff, H. Forsgren, T. Burton, N. Bar, G. Ketcheson, H. Hudak, W. Little, and W. Grow. 1992. Integrated riparian evaluation guide. Ogden: U.S. Forest Service Intermountain Regional Office.

Conner, W. H. and J. R. Tolliver. 1990. Long-term trends in the bald-cypress (*Taxodium distichum*) resource in Louisiana (U.S.A.). Forest Ecology Management 33/34: 385-403.

Conners, D. H., F. Riesenberg, IV, R. D. Charney, M. A. McEwen, R. B. Krone, and G. Tchobanoglous. 1990. Research needs: salt marsh restoration, rehabilitation, and creation techniques for Caltrans construction projects. Davis: Department of Civil Engineering, University of California.

Constant, P., E. M. C. Eybert, and R. Maheo. 1973. Recherches sur les oiseaux nicheurs dans les plantations des resineaux de la foret de Paimpont (Bertagne). Ardea 41: 371-384.

Conway, V. M. 1949. The bogs of central Minnesota. Ecological Monographs 19: 173-206.

Cooper, D. J. 1990. Ecology of wetlands in Big Meadows, Rocky Mountain Naitonal Park, Colorado. Washington, D.C.: U.S. Fish and Wildlife Service Biological Report 90(15).

Cooper, W. S. 1967. Coastal dunes of California. Geological Society of America Memoir 104: 1-131.

Copeland, B. J., R. G. Hodson, S. R. Riggs, and J. E. Easley. 1983. The ecology of Albermarle Sound, North Carolina: an estuarine profile. Washington, D.C.: U.S. Fish and Wildlife Service, FWS/OBS-83/01.

Core, E. L. 1966. Vegetation of West Virginia. Parsons: McClain Printing Company.

Corley, W. L. 1989. Wildflower establishment and cultivation for roadsides, meadows and beauty spots. Research News, Georgia Department of Transportation, Office of Materials and Research.

Correll, D. S. and H. B. Corell. 1972. Aquatic and wetland plants of the southwestern United States. Washington, D.C.: U.S. Environmental Protection Agency.

Correll, D. S. and M. C. Johnston. 1970. Manual of the vascular plants of Texas. Renner: Texas Research Foundation.

Cottam, W. P. and F. R. Evans. 1945. A comparative study of the vegetation of grazed and ungrazed canyons of the Wasatch Range, Utah. Ecology 26: 171-181.

Coultas, C. and Y. P. Hsieh. 1996. Ecology and management of tidal marshes, a model from the Gulf of Mexico. Delray Beach: St. Lucie Press.

Cousans, R. and M. Mortimer. 1996. Dynamics of weed populations. New York: Cambridge University Press.

Covin, J. D. and J. B. Zedler. 1988. Nitrogen effects on *Spartina foliosa* and *Salicornia virginica* in the salt marsh at Tijuana Estuary, California. Wetlands 8: 51-65.

Cowan, B. 1995. Coastal dune and bluff restoration. Fremontia 23(1): 29-31.

Cowardin, L. M., V. Carter, F. C. Golet, and E. T. LaRoe. 1979. Classification of wetlands and deepwater habitats of the United States. Washington, D.C.: U.S. Fish and Wildlife Service Biological Service Program, FMS/OBS.79/31.

Cox, J. 1991. Landscaping with nature. Emmaus: Rodale Press.

Cox, J. R., H. L. Morton, T. N. Johnson, G. L. Jordan, S. C. Martin, and L. C. Fierro. 1982. Vegetation restoration in the Chihuahuan and Sonoran Deserts of North America. Tucson: Agriculture Reviews and Manuals #28, U.S. Department of Agriculture, Agricultural Research Service.

Craft, C. B., S. W. Broome, and E. D. Seneca. 1986. Carbon, nitrogen and phosphorus accumulation in maniniated marsh soils. In Proceedings of the 29th annual meeting of the Soil Science Society of North Carolina, ed., A. Amoozegar. Raleigh: Soil Science Society of North Carolina.

Craft, C. B., S. W. Broome, and E. D. Seneca. 1988. Nitrogen, phosphorous and organic carbon pools in natural and transplanted marsh soil. Estuaries 11: 272-280.

Craig, R. M. 1984. Plants for coastal dunes of the gulf and south Atlantic coasts and Puerto Rico. Washington, D.C.: U.S. Government Printing Office.

Craighead, F. C., Sr. 1971. The trees of south Florida, Vol. 1: the natural environments and their succession. Coral Gables: University of Miami Press.

Craighead, J. J., F. C. Craighead, Jr., and R. J. Davis. 1963. A field guide to Rocky Mountain wildflowers from northern Arizona and New Mexico to British Columbia. Boston: Houghton Mifflin Company.

Crispin, D. J. and A. D. Randall. 1990. Techniques used in the Great Cedar Swamp restoration. Norwell: The BSC Group-Norwell Inc.

Crist, A. and D. C. Glenn-Lewin. 1978. The structure of community and environmental gradients in a northern Iowa prairie. In Fifth Midwest Prairie Conference Proceedings, eds., D. C. Glenn-Lewin and R. Q. Landers. Ames: Iowa State University.

Cronan, W. 1983. Changes in the land, Indians, colonists, and the ecology of New England. Hill and Wang.

Cronk, Q. C. B. and J. L. Fuller. 1995. Plant invaders, the threat to natural ecosystems. New York: Chapman and Hall Press.

Cronquist, A., A. H. Holmgren, N. H. Holmgren, J. L. Reveal, and P. K. Holmgren. 1972. Intermountain flora: vascular plants of the intermountain west, U.S.A. 4 Vols. New York: Hafner Publishing Company.

Crow, T. R., W. C. Johnson, and C. S. Adkisson. 1994. Fire and recruitment of *Quercus* in a postagricultural field. American Midland Naturalist 131(1): 84-97.

Crum, H. 1992. A focus on peatlands and peat mosses. Ann Arbor: University of Michigan.

Crum H. A. and L. E. Anderson. 1981. Mosses of eastern North America. 2 Vols. New York: Columbia University Press.

Currie, D. J. 1991. Energy and large-scale patterns of animal- and plant-species richness. American Naturalist 137: 27-49.

Curry, R. R. 1977. Reinhabiting the earth: life support and the future primitive. In Recovery and restoration of damaged ecosystems, eds., J. Cairns, Jr., K. L. Dickson, and E. E. Herricks. Charlottesville: University of Virginia Press.

Curtis, J. T. 1946. Use of mowing in management of white ladyslipper. Journal of Wildlife Management 10: 303-308.

Curtis, J. T. 1959 (second printing in 1971). The vegetation of Wisconsin, an ordination of plant communities. Madison: The University of Wisconsin Press.

Curtis, J. T. and H. C. Greene. 1949. A study of relic Wisconsin prairie by species-presence method. Ecology 30: 83-92.

Cusick, A. and G. M. Silberhorn. 1977. Vascular plants of unglaciated Ohio. Columbus: Ohio Biological Survey.

Dachnowski-Stokes, A. P. 1940. Structural characteristics of peats and mucks. Soil Science 50: 389-400.

Dahl, T. E. 1990. Wetland losses in the United States, 1780s to 1980s. Washington, D.C.: U.S. Department of the Interior, Fish and Wildlife Service.

Dahlem, E. A. 1979. The Mahogany Creek watershed—with and without grazing. In Proceedings of the Forum—grazing and riparian/stream ecosystems, ed., O. B. Cope. Denver: Trout Unlimited.

Dancer, W. S. 1985. Prairie soil restoration research summarized (Illinois). Restoration and Management Notes 3(1): 30-31.

Danielsen, C. W. 1996. Restoration of a native bunchgrass and wildflower grassland at Mount Diablo State Park (California). Restoration and Management Notes 14(1): 65.

Dansereau, P. and F. Segadas-Vianna. 1952. Ecological study of the peat bogs of eastern North America. Canadian Journal of Botany 30: 490-520.

D'Antonio, C. M. and P. M. Vitousek. 1992. Biological invasion by exotic grasses, the grass/fire cycle, and global change. Annual Review of Ecology and Systematics 23: 63-87.

Dasmann, R. F. 1976. Environmental conservation. 4th ed. New York: John Wiley & Sons.

Daubenmire, R. 1978. Plant geography. New York: Academic Press.

Davidson, U. M. 1950. The original vegetation of Lexington, Kentucky and vicinity. M.S. thesis. Lexington: University of Kentucky.

Davis, A. M. and T. F. Glick. 1978. Urban ecosystems and island biogeography. Environmental Conservation 5: 299-304.

Davis, G. A. 1977. Management alternatives for the riparian habitat in the Southwest. In Importance, preservation and management of riparian habitat: a symposium, tech. coords., R. R. Johnson and D. A. Jones, Jr. Fort Collins: U. S. Department of Agriculture Forest Service General Technical Report RM-43. Rocky Mountain Forest and Range Experiment Station.

Davis, J. H. 1940. The ecology and geologic role of mangroves in Florida. Washington, D.C.: Carnegie Institution, Publication No. 517.

Davis, J. H. 1943. The natural features of southern Florida, especially the vegetation, and the Everglades. Florida Geological Survey Bulletin No. 25.

Davis, R. J. 1952. Flora of Idaho. Dubuque: Wm. C. Brown Company.

Davis, S. M. and J. C. Ogden, eds. 1994. Everglades: the ecosystem and its restoration. Delray Beach: St. Lucie Press.

Day, J. H. 1981. Estuarine ecology, with particular reference to southern Africa. Rotterdam: A. A. Balkema.

De Bano, L. F. and L. J. Scmidt. 1990. Potential for enhancing riparian habitats in the southwestern United States with watershed practices. Forest Ecology Management 33/34: 385-403.

Dean, B. E., A. Mason, and J. L. Thomas. 1983. Wildflowers of Alabama and adjoining states. Birmingham: University of Alabama.

DeGraaf, R. M. and R. I. Miller, eds. 1995. Conservation of faunal diversity in forested landscapes. New York: Chapman and Hall.

Dehgan, B., T. J. Sheehan, D. M. Sylvia, M. Kane, B. C. Poole, and M. Niederhafer. 1987. Propagation and mycorrhizal inoculation of indigenous Florida plants for phosphate mine reclamation. Bartow: Florida Institute of Phosphate Research, Publication No. 03-053-076.

Deitz, K. B., J. A. O'Reilly, G. S. Podniesinski, and D. J. Leopold. 1996. Rebuilding microtopography and planting woody species restores abandoned agricultural land (New York). Restoration and Management Notes 14(2): 140-147.

Delcourt, H. R. and P. A. Delcourt. 1988. Quaternary landscape ecology: relevant scales in space and time. Landscape Ecology 2: 23-44.

Delcourt, H. R., P. A. Delcourt, and T. Webb, III. 1983. Dynamic plant ecology: the spectrum of vegetation change in space and time. Quaternary Science Review 1: 153-175.

Delcourt, P. A. and H. R. Delcourt. 1987. Long-term forest dynamics of the temperate zone. New York: Springer-Verlag.

de Waal, L. C., L. E. Child, P. M. Wade, and J. H. Brock, eds. 1994. Ecology and management of invasive riverside plants. New York: John Wiley & Sons.

Diamond, D. D., G. A. Rowell, and D. P. Keddy-Hector. 1995. Conservation of Ashe juniper (*Juniperus ashei*) woodlands of the central Texas Hill Country. Natural Areas Journal 15(2): 189-197.

Diamond, J. 1997. Reflections on goals and on the relationship between theory and practice. In Restoration ecology, eds., W. R. Jordan, III, M. E. Gilpin, and J. D. Aber. Cambridge, England: Cambridge University Press.

Diamond, J. and T. J. Case, eds. 1985. Community ecology. New York: Harper and Row.

Diamond, J. M. 1978. Critical areas for maintaining viable populations of species. In The breakdown and restoration of ecosystems, eds., M. W. Holdgate and M. J. Woodman. New York: Plenum Press.

Diamond, J. M. and M. E. Gilpin. 1982. Examination of the "null" model of Connor and Simberloff for species co-occurrences on islands. Oecologia 52: 64-74.

Diamond, J. M. and R. M. May. 1981. Island biogeography and the design of natural reserves. In Theoretical ecology: principles and applications, ed., R. M. May. Sunderland: Sinauor Associates, Inc.

di Castri, F. and A. Hansen, eds. Landscape boundaries: consequences for biotic diversity and ecological flows. SCOPE book series. New York: Springer-Verlag.

di Castri, F., A. J. Hansen, and M. M. Holland, eds. 1988. A new look at ecotones: emerging international projects on landscape boundaries. Biology International, Special Issue 17: 1-163.

Diboll, N. 1986. Mowing as an alternative to burning for control of cool season exotic grasses in prairie grass plantings. In Proceedings of the Ninth North American Prairie Conference, eds., G. K. Clambey and R. H. Pemble. Fargo: Tri-College University.

Dick-Peddie, W. A. 1992. New Mexico vegetation: past, present and future. Albuquerque: University of New Mexico Press.

Diekelmann, J. and R. Schuster. 1982. Natural landscaping: designing with native plant communities. New York: McGraw-Hill Book Company.

Dirr, M. A. 1990. Manual of woody landscape plants their identification, ornamental characteristics, culture, propagation and uses. Champaign: Stipes Publishing Company.

Dirr, M. A. and C. W. Heuser, Jr. 1987. The reference manual of woody plant propagation: from seed to tissue culture. Athens: Varsity Press, Inc.

Disney, H. J. S. and A. Stokes. 1976. Birds in pine and native forests. Emu 78: 133-138.

Dixon, K. R. 1974. A model for predicting the effects of sewage effluent on wetland ecosystems. Ph.D. diss. University of Michigan, Ann Arbor.

Dixon, R. M. 1988. Land imprinting for vegetative restoration. Restoration and Management Notes 6: 24-25.

Dixon, R. M. 1990. Land imprinting for dryland revegetation and restoration. In Environmental restoration, ed., J. J. Berger. Washington, D.C.: Island Press.

Donoghue, L. R. and V. J. Johnson. 1975. Prescribed burning in the north central states. St. Paul: North Central Forest Experiment Station, U.S. Department of Agriculture, Forest Service Research Paper NC-111.

Dorn, R. D. 1988. Vascular plants of Wyoming. Cheyenne: Mountain West Publishing.

Downs, J. C., W. H. Rickard, and L. L. Cadwell. 1995. Restoration of big sagebrush habitat in southeastern Washington. Restoration and Management Notes 13(1): 128.

Drake, L. and R. Langel. 1995. Hydraulic jetting effective for planting willow cuttings (Iowa). Restoration and Management Notes 13(2): 232.

Dramstad, W. E., J. D. Olson, and R. T. T. Forman. 1996. Landscape ecology principles in landscape architecture and land-use planning. Washington, D. C.: Harvard University Graduate School of Design, Island Press, and American Society of Landscape Architects.

Dregne, H. E. 1983. Desertification of arid lands. Chur: Hardwood Academic Publishers.

Dremann, C. C. 1996. Grasses and mulch control yellow-star thistle (California). Restoration and Management Notes 14(1): 79.

Drury, W. H. and I. C. T. Nisbet. 1973. Succession. Journal of the Arnold Arboretum 54: 331-368.

Duebendorfer, T. 1985. Habitat survey of *Erysimum menziesii* on the North Spit of Humbolt Bay. Eureka: County of Humbolt, Department of Public Works, Natural Resources Division.

Duffy, D. C. and A. J. Meier. 1992. Do Appalachian understories ever recover from clearcutting? Conservation Biology 6: 196-201.

Duft, J. F. and R. K. Moseley. 1989. Alpine wildflowers of the Rocky Mountains. Mountain Press.

Duncan, W. H. and M. B. Duncan. 1987. The Smithsonian guide to seaside plants of the Gulf and Atlantic coasts, from Louisiana to Massachusetts. Washington, D.C.: Smithsonian Institution Press.

Duncan, W. H. and M. B. Duncan. 1988. Trees of the southeastern United States. Athens: University of Georgia.

Duncan, W. H. and L. E. Foote. 1975. Wildflowers of the southeastern United States. Athens: The University of Georgia Press.

Dunn, C. D. and C. Loehle. 1988. Species-area parameter estimation: testing the null model of lack of relationship. Journal of Biogeography 15: 721-728.

Dyksterhuis, E. J. 1957. The savannah concept and its use. Ecology 38: 435-442.

Ebinger, J. E. 1983. Exotic shrubs (*Elaeagnus umbellata*, *Ligustrum obtusifolium*) a potential problem in natural area management in Illinois. Natural Areas Journal 3(1): 3-6.

Ebinger, J. E. 1991. Naturalized amur maple (*Acer ginnala* Maxim.) in Illinois. Natural Areas Journal 11(3): 170-171.

Ebinger, J. E., J. Newman, and R. Nyboer. 1984. Naturalized winged wahoo (*Euonymus alatus*) in Illinois. Natural Areas Journal 4(2): 26-29.

Edelman, C. H. and J. M. van Staveren. 1958. Marsh soils in the United States and in The Netherlands. Journal of Soil and Water Conservation 13: 5-17.

Edwards, P. J., R. M. May, and N. R. Webb, eds. 1994. Large scale ecology and conservation biology. Cambridge: Blackwell Scientific Publications, Inc.

Egler, F. E. 1952. Southeast saline everglades vegetation, Florida, and its management. Vegetatio 3: 213-265.

Ehleringer, J.R., L. A. Arnow, T. Arnow, I. R. McNulty, and N. C. Negus. 1992. Red Butte Canyon Research Natural Area: history, flora, geology, climate, and ecology. Great Basin Naturalist 52(2): 95-121.

Ehrenfield, D., ed. 1995a. Readings from conservation biology: to preserve biodiversity, an overview. Cambridge: Blackwell Science, Inc.

Ehrenfield, D., ed. 1995b. Readings from conservation biology: wildlife and forests. Cambridge: Blackwell Science, Inc.

Ehrenfield, D., ed. 1995c. Readings from conservation biology: plant conservation. Cambridge: Blackwell Science, Inc.

Ehrenfield, D., ed. 1995d. Readings from conservation biology: the landscape perspective. Cambridge: Blackwell Science, Inc.

Eilers, L. J. and D. M. Roosa. 1994. The vascular plants of Iowa, an annotated checklist and natural history. Iowa City: University of Iowa Press.

Eisenlohr, W. S., Jr., C. E. Sloan, and J. S. Shjeflo. 1972. Hydrologic investigations of prairie potholes in North Dakota, 1959-1968. Geological Survey Professional Paper 585-A.

Elder, J. F. 1985. Nitrogen and phosphorus speciation and flux in a large Florida river-wetland system. Water Resources Research 21: 724-732.

Elder, J. F. and H. C. Mattraw, Jr. 1982. Riverine transport of nutrient and detritus to the Appalachicola Bay estuary, Florida. Water Resources Research 18: 849-856.

Elmore, W. and R. L. Beschta. 1987. Riparian areas: perceptions in management. Rangelands 9(6): 260-265.

Environmental Laboratory. 1987. Corps of engineers wetland delineation manual. Vicksburg: U. S. Corps of Engineers, Waterways Experimental Station Technical Report 487-1.

Epple, A. O. 1995. A field guide to the plants of Arizona. Skyhouse Publishers.

Evans, J. E. 1983a. A literature review of management practices for multiflora rose (*Rosa multiflora*). Natural Areas Journal 3(1): 6-15.

Evans, J. E. 1983b. A literature review of management practices for smooth sumac (*Rhus glabra*), poison ivy (*Rhus radicans*) and other sumac species. Natural Areas Journal 3(1): 16-26.

Evans, J. E. and M. Heitlinger. 1984. IPM: a review for natural area managers. Restoration and Management Notes 11(1): 18-24.

Everett, R. L., comp. 1994. Restoration of stressed sites and processes. Portland: U.S. Department of Agriculture Forest Service, Pacific Northwest Research Station, General Technical Report PNW-GTR-330.

Ewel, J. J. 1990. Introduction. In Ecosystems of Florida, eds., R. L. Myers and J. J. Ewel. Orlando: University of Central Florida.

Ewel, K. C. 1990. Swamps. In Ecosystems of Florida, eds., R. L. Myers and J. J. Ewel. Orlando: University of Central Florida.

Ewel, K. C. and W. J. Mitsch. 1978. The effects of fire on species composition in cypress dome ecosystems. Florida Science 41: 25-31.

Ewel, K. C. and H. T. Odum, eds. 1984. Cypress swamps. Gainesville: University Press of Florida.

Eyre, F. H. ed. 1980. Forest cover types of the United States and Canada. Washington: Society of American Foresters.

Faber, P. 1982. Common wetland plants of coastal California, a field guide for the layman. Pickleweed.

Faber, P. M. and R. F. Holland. 1988. Common riparian plants of California, a field guide for the layman. Pickleweed.

Fahrig, L. and G. Merriam. 1990. Conservation of fragmented populations. In A landscape perspective: readings in conservation biology, ed., E. Ehrenfeld. Cambridge, Mass.: Society for Conservation Biology and Blackwell Science.

Fahrig, L. and G. Merriam. 1985. Habitat patch connectivity and population survival. Ecology 66: 1762-1768.

Falk, D. 1990. Discovering the future, creating the past: some reflections on restoration. Restoration and Management Notes 8(2): 71-72.

Falk, D. A., C. I. Millar, and M. Olwell, eds. 1995. Restoring diversity: reintroduction strategies for threatened species. Washington, D.C.: Island Press.

Falk, D. A. and O. Olwell. 1992. Scientific and policy consideration in restoration and reintroduction of endangered species. Rhodora 94: 287-315.

Fassett, N. C. 1957. A manual of aquatic plants. Madison: The University of Wisconsin Press.

Felker, P., Wiesman, and D. Smith. 1988. Comparison of seedling containers on growth and survival of *Prosopis alba* and *Leucaena leucocephala* in semi-arid conditions. Forest Ecology and Management 24: 177-182.

Fellows, D. P. 1995. Preliminary assessment of effect of prescribed fire upon establishment of *Aphthona nigriscutis* in leafy spurge. Leafy Spurge News 17(3): 6.

Fenton, T. E. 1983. Mollisols. In Pedogenesis and soil taxonomy, Vol. 2: the soil orders, eds., L. P. Wilding, N. E. Smeck, and G. F. Hall. New York: Elsevier Science Publishing Company.

Fernald, M. L. 1970. Gray's manual of botany. 8th ed., corrected printing. New York: Van Nostrand Reinhold Company.

Fidelibus, M. F. 1994. Jellyrolls reduce outplanting costs in arid land restoration. Restoration and Management Notes 12(1): 87.

Fidelibus, M. F. and D. A. Bainbridge. 1994. The effect of containerless transport on desert shrubs. Tree Planters Notes 45(3): 82-85.

Fielder, P. and S. Jain, eds. 1992. Conservation biology: the theory and practice of nature conservation, preservation, and management. New York: Chapman and Hall.

Fimbel, R. A. 1992. Restoring drastically disturbed sites within the pygmy pine forests of southern New Jersey's Pinelands National Reserve. Ph.D. diss. Rutgers University, New Brunswick.

Finch, D. M. 1991. Population ecology, habitat requirements, and conservation of neotropical migratory birds. Fort Collins: Forest Service, United States Department of Agriculture, General Technical Report RM-205.

Finch, D. M. and L. F. Ruggiero. 1993. Wildlife habitats and biological diversity in the Rocky Mountains and Northern Great Plains. Natural Areas Journal 13(3): 191-203.

Fleischner, T. L. 1994. Ecological costs of livestock grazing in western North America. Conservation Biology 8(3): 629-644.

Flora of North America Editorial Committee. 1993a. Flora of North America, Vol. 1: introduction. New York: Oxford University Press.

Flora of North America Editorial Committee. 1993b. Flora of North America, Vol. 2: pteridophytes and gymnosperms. New York: Oxford University Press.

Foote, L. E. and S. B. Jones, Jr. 1989. Native shrubs and woody vines of the southeast, landscape uses and identification. Portland: Timber Press.

Forman, R. T. T. 1986. Emerging directions in landscape ecology and applications in natural resource management. In Proceedings of the Conference on Science in National Parks, eds., R. Herrman and T. Bostedt-Draig. Vol. 1. Fort Collins: Colorado State University.

Forman, R. T. T. 1995. Land mosaics, the ecology of landscapes and regions. New York: Cambridge University Press.

Forman, R. T. T., ed. 1979. Pine barrens: ecosystem and landscape. New York: Academic Press.

Forman, R. T. T., A. Galli, and C. Leck. 1976. Forest size and avian diversity in New Jersey woodlots with some land use implications. Oecologia 26: 1-8.

Forman, R. T. T. and M. Godron. 1981. Patches and structural components for a landscape ecology. Bioscience 31: 733-740.

Forman, R. T. T. and M. Godron. 1986. Landscape ecology. New York: John Wiley & Sons.

Fournier, M. 1996. Sand dune restoration techniques utilized along New Jersey's coastline. Society for Ecological Restoration 1996 International Conference, June 17-22, 1996, New Brunswick, New Jersey.

Fralish, J. S., R. C. Anderson, J. E. Ebinger, and R. Szafoni, eds. 1994. Proceedings of the North American Conference on Barrens and Savannas. Normal: Illinois State University, October 15-16, 1994.

Frankel, O. H. and M. E. Soule. 1981. Conservation and evolution. New York: Cambridge University Press.

Frankenberg, D. 1995. The nature of the Outer Banks, environmental processes, field sites, and development issues, Corolla to Ocracoke. Chapel Hill: University of North Carolina.

Franklin, J. F. and C. T. Dyrness. 1973. Natural vegetation of Oregon and Washington. Salem: Oregon State University Press.

Franklin, J. F., W. H. Moir, M. A. Hemstrom, S. E. Greene, and B. G. Smith. 1988. The forest communities of Mount Rainier National Park. Washington, D.C.: U. S. Department of the Interior, Scientific Monograph Series No. 19.

Franzreb, K. E. 1989. Ecology and conservation of the endangered least Bell's vireo. Washington, D.C.: U.S. Fish and Wildlife Service Biological Report 89: 1-17.

Freemark, K. E. and H. G. Merriam. 1986. Importance of area and habitat heterogeneity to bird assemblages in temperate forest fragments. Biological Conservation 36: 115-141.

Fredrickson, E., J. Barrow, J. Herrick, K. Havstad, and B. Longland. 1996. Low-cost seeding practices for restoring desert environments (New Mexico). Restoration and Management Notes 14(1): 72-73.

Fulbright, T. E., G. L. Waggerman, and R. L. Bingham. 1992. Growth and survival of shrub seedlings planted for white-winged dove habitat restoration. Wildlife Society Bulletin 20: 286-289.

Furley, P. A., J. Procter, and J. Ratter, eds. 1992. The nature and dynamics of forest-savanna boundaries. New York: Chapman and Hall.

Galatowitsch, S. M. and A. G. van der Valk. 1995. Restoring prairie wetlands: an ecological approach. Ames: Iowa State University Press.

Gallagher, J. L. 1980. Salt marsh development. In Rehabilitation and creation of selected coastal habitats: proceedings of a workshop, eds., J. C. Lewis and E. W. Bunce. Washington, D.C.: U.S. Fish and Wildlife Service FWS/OBS-80/27.

Galli, A. E., C. F. Leck, and R. T. Forman. 1976. Avian distribution patterns in forest islands of different sizes in central New Jersey. Auk 93: 356-364.

Gameson, A. L. H. and A. Wheeler. 1977. Restoration and recovery of the Thames Estuary. In Recovery and restoration of damaged ecosystems, eds., J. Cairns, Jr., K. L. Dickson, and E. E. Herricks. Charlottesville: University Press of Virginia.

Garbisch, E. and J. Garbisch. 1994. The effects of forests along eroding shoreline banks of the Chesapeake Bay. Wetland Journal 6(1): 18-19.

Garbisch, E. W. 1995. The do's and don'ts of wetland planning. Wetland Journal 7(2): 12-14.

Garbisch, E. W. and S. M. McIninch. 1994. The establishment of *Scirpus tabernaemontani* (formerly *Scirpus validus*) from large and small rhizomes as a function of water depth. Wetland Journal 6(4): 17-21.

Gardner, J. L. 1951. Vegetation of the creosote bush area of the Rio Grande Valley in New Mexico. Ecological Monographs 21: 379-403.

Garfitt, J. E. 1995. Natural management of woods, continuous cover forestry. New York: John Wiley & Sons.

Gasaway, R. D. and T. F. Drda. 1977. Effects of grass carp introduction on water fowl habitat. Transactions of the North American Wildlife Natural Resources Conference 42: 73-85.

Gates, F. C. 1942. The bogs of northern lower Michigan. Ecological Monographs 12: 213-254.

Gates, J. E. and L. W. Gysel. 1978. Avian nest dispersion and fledgling success in field-forest ecotones. Ecology 59: 871-883.

Gelt, J. 1993. Abandoned farmland often is troubled land in need of restoration. Arroyo 7(2): 1-8.

Gilbert, J., D. L. Danielopol, and J. Stanford, eds. 1991. The ecology of urban habitats. New York: Chapman and Hall.

Gilbert, L. O. 1989. The ecology of urban habitats. London: Chapman and Hall, Ltd.

Gilkey, H. M. and L. J. Dennis. 1967. Handbook of northwestern plants. Corvallis: Oregon State University Bookstores, Inc.

Gill, J. D. and W. M. Healy, eds. 1974. Shrubs and vines for northeastern wildlife. Upper Darby: U. S. Department of Agriculture, Forest Service General Technical Report NE-9.

Giller, P. S. 1984. Community structure and the niche. London: Chapman and Hall.

Gilman, K. 1994. Hydrology and wetland conservation. New York: John Wiley & Sons.

Glaser, P. H. 1987. The ecology of patterned boreal peatlands of northern Minnesota: a community profile. Washington, D.C.: U.S. Fish and Wildlife Service Biological Report 85(7.14).

Glass, W. D. 1991. Vegetation management guideline: cut-leaved teasel (*Dipsacus laciniatus* L.) and common teasel (*D. sylvestris* Huds.). Natural Areas Journal 11(4): 213-214.

Glass, W. D. 1992. Vegetation management guideline: white poplar (*Populus alba* L.). Natural Areas Journal 12(1): 39-40.

Glattstein, J. 1994. Waterscaping: plants and ideas for natural and created water gardens. Pownal: Garden Way Publishing.

Gleason, H. A. and A. Cronquist. 1963. Manual of vascular plants of northeastern United States and adjacent Canada. New York: D. Van Nostrand Company.

Gleason, H. A. and A. Cronquist. 1964. The natural geography of plants. New York: Columbia University Press.

Gleason, H. A. and A. Cronquist. 1991. Manual of vascular plants of northeastern United States and adjacent Canada. 2nd ed. Bronx: New York Botanical Garden.

Gleason, P. J. and P. A. Stone. 1994. Age, origin, and landscape evolution of the Everglades peatland. In Everglades: the ecosystem and its restoration, eds., S. M. Davis and J. C. Ogden. Delray Beach: St. Lucie Press.

Glinski, R. L. 1977. Regeneration and distribution of sycamores and cottonwood trees along Sonoita Creek, Santa Cruz County, Arizona. In Importance, preservation and management of riparian habitat: a symposium, tech. coords., R. R. Johnson and D. A. Jones, Jr. Fort Collins: U. S. Department of Agriculture Forest Service General Technical Report RM-43. Rocky Mountain Forest and Range Experiment Station.

Godfrey, P. J. and M. M. Godfrey. 1976. Barrier island ecology of Cape Lookout National Seashore and Vicinity, North Carolina. Washington, D.C.: U.S. Government Printing Office, National Park Service Monograph Series, No. 9.

Godfrey, P. J., E. R. Kaynor, S. Pelczarski, and J. Benforado, eds. 1984. Ecological considerations in wetlands treatment of municipal wastewaters. New York: Van Nostrand Reinhold.

Godfrey, R. K. 1988. Trees, shrubs, and woody vines of northern Florida and adjacent Georgia and Alabama. Athens: The University of Georgia Press.

Godfrey, R. K. and J. W. Wooten. 1981. Aquatic and wetland plants of southeastern United States, dicotyledons. Athens: The University of Georgia Press.

Godfrey, R. K. and J. W. Wooten. 1986 (reissue of 1979 edition). Aquatic and wetland plants of southeastern United States, monocotyledons. Athens: The University of Georgia Press.

Goforth, H. W. and J. R. Thomas. 1979. Plantings of red mangroves (*Rhizophora mangle*) for stabilization of marl shorelines in the Florida Keys. In Proceedings of the 6th Annual Conference on the Restoration and Creation of Wetlands. Tampa: Hillsborough Community College.

Golet, F. C., A. J. K. Calhoun, W. R. DeRagon, D. J. Lowry, and A. J. Gold. 1993. Ecology of red maple swamps in the glaciated northeast: a community profile. Washington, D.C.: U.S. Fish and Wildlife Service Biological Report 12.

Good, R. E., D. F. Whigham, and R. L. Simpson, eds. 1978. Freshwater wetlands: ecological processes and management potential. New York: Academic Press.

Goodall, D. W. and R. A. Perry, eds. 1979. Aridland ecosystems: structure, functioning and management. Vol 1. Cambridge: Cambridge University Press.

Goodman, G. T. and D. F. Perkins. 1968. The role of mineral nutrients in *Eriophorum* communities. IV. Potassium supply as a limiting factor in an *E. vaginatum* community. Journal of Ecology 56: 685-696.

Goodwin, R. H. and W. A. Niering. 1975. Inland wetlands of the United States. New London: Connecticut College, National Park Service Natural History Theme Studies Number 2.

Gordon, R. A. and C. J. Scifries. 1977. Burning for improvement of Macartney rose-infested coastal prairie. Texas Agricultural Experiment Station Bulletin 1183.

Gordon, R. A., C. J. Scifres, and J. L. Mutz. 1982. Integration of burning and picloram pellets for Macartney rose control. Journal of Range Management 35: 427-430.

Gore, A. J. P., ed. 1983. Ecosystems of the world, vol. 4A, mires: swamp, bog, fen, and moor. Amsterdam: Elsevier Scientific Publications.

Gorham, E. 1967. Some chemical aspects of wetland ecology. Technical Mem. Committee on Geotechnical Research, National Research Council of Canada, No. 90.

Gosselink, J. G. 1980. Tidal marshes–the boundary between land and ocean. Washington, D.C.: U.S. Fish and Wildlife Service, FWS/OBS-80/15.

Gosselink, J. G. 1984. The ecology of delta marshes of coastal Louisiana: a community profile. Washington, D.C.: U.S. Fish and Wildlife Service, FWS/OBS-84/09.

Gosselink, J. G., C. L. Cordes, and J. W. Parsons. 1979. An ecological characterization study of the Chenier Plain coastal ecosystem of Louisiana and Texas. Washington, D.C.: U.S. Fish and Wildlife Service, FWS/OBS-78/9.

Gould, F. W. 1973 (reissue of 1951 edition). Grasses of the southwestern United States. Tucson: University of Arizona.

Graetz, K. E. 1973. Seacoast plants of the Carolinas. Raleigh: U.S. Department of Agriculture, Soil Conservation Service.

Granholm, S. L. 1989. A local government's mitigation guidebook for seasonal wetlands along San Francisco Bay. In Proceedings of the 1st Annual Conference of the Society for Ecological Restoration, Oakland, California.

Grant, J. A. and C. L. Grant. 1990. Trees and shrubs for Pacific northwest gardens. Portland: Timber Press.

Grant, K. 1994. Oregon river restoration: a sensitive management strategy boosts natural healing. Restoration and Management Notes 12(2): 152-159.

Gray, A. 1987 (reissue of the 8th and last edition). Gray's manual of botany. New York: Van Nostrand Reinhold Company.

Gray, A. J., M. J. Crawley, and P. J. Edwards, eds. 1987. Colonization, succession and stability. Oxford: Blackwell Scientific Publications.

Gray, D. H. and A. T. Leiser. 1982. Biotechnical slope protection and erosion control. New York: Van Nostrand Reinhold Company.

Gray, G. J. 1992. Health emergency imperils western forests. American Forests 98 (9 and 10): special insert.

Grayson, D. K. 1993. The desert's past, a natural prehistory of the Great Basin. Washington, D.C.: Smithsonian Institution Press.

Green, A. W. and R. C. Conner. 1989. Forests in Wyoming. Ogden: Forest Service, United States Department of Agriculture, Resource Bulletin INT-61.

Green, B. H. and I. C. Marshall. 1987. An assessment of the role of golf courses in Kent, England, in protecting wildlife and landscapes. Landscape and Urban Planning 14: 143-154.

Green, P. E. 1983. Natural vegetation of mined peatlands in northern Minnesota. M.S. thesis, University of Minnesota, St. Paul.

Gregory, S. V., F. J. Swanson, W. A. McKee, and K. W. Cummins. 1991. An ecosystem perspective of riparian zones. Bioscience 41: 540-551.

Griffin, J. R. 1988. Oak woodland. In Terrestrial vegetation of California, eds., M. G. Barbour and J. Major. Davis: California Native Plant Society.

Grigg, G. T. 1990. Seeking a fresh vision of environmental responsibility. Golf Course Management 58(9): 38-46.

Griggs, F. T. 1994. Adaptive management strategy helps assure cost-effective, large-scale riparian forest restoration (California). Restoration and Management Notes 12(1): 80-81.

Griggs, R. F. 1937. Timberlines as indicators of climatic trends. Science 85: 251-255.

Grilz, P.L. and J. T. Romo. 1995. Management considerations for controlling smooth brome in fescue prairie. Natural Areas Journal 15: 148-156.

Griswold, T. 1988. Physical factors and competitive interactions affecting salt marsh vegetation. M.S. thesis. San Diego State University, San Diego.

Groff, D. 1994. Revegetating disturbed mineral soils. Hortus Northwest 5(1): 12.

Grumbine, E. 1990. Protecting biological diversity through the greater ecosystem concept. Natural Areas Journal 10(3): 114-120.

Guinon, M. and D. Allen. 1990. Restoration of dune habitat at Spanish Bay. In Environmental restoration, ed., J. J. Berger. Washington, D.C.: Island Press.

Gunderson, L. H. 1990. Historical hydropatterns in vegetation communities of Everglades National Park. Freshwater Wetlands and Wildlife, Charlseston, South Carolina. Aiken: Savannah River Ecology Laboratory.

Gunderson, L. H. 1994. Vegetation of the Everglades: determinants of community composition. In Everglades: the ecosystem and its restoration, eds., S. M. Davis and J. C. Ogden. Delray Beach: St. Lucie Press.

Hackney, C. T., S. M. Adams, and W. H. Martin, eds. 1992. Biodiversity of the southeastern United States: aquatic communities. New York: John Wiley & Sons.

Hackney, C. T. and A. A. de la Cruz. 1981. Effects of fire on brackish marsh communities: management implications. Wetlands 1: 75-86.

Hall, D. W. 1993. Illustrated plants of Florida and the Coastal Plain, based on the collections of Leland and Lucy Blatzell. Maupin House.

Hammer, D. A. 1992. Creating freshwater wetlands. Chelsea: Lewis Publishers.

Handel, S. N., G. R. Robinson, and A. J. Beattie. 1994. Biodiversity resources for restoration ecology. Restoration Ecology 2(4): 230-241.

Hanes, T. L. 1988. Chaparral. In Terrestrial vegetation of California, eds., M. G. Barbour and J. Major. Davis: California Native Plant Society.

Haney, A. and S. Apfelbaum. 1994. Characterization of midwestern oak savannas. Wildflower 10(4): 39-43.

Hansen, A. J., F. Di Castri, and R. J. Naiman. 1988. Ecotones: what and why? Biology International, Special Issue 17: 9-46.

Hanson, L. and H. Lipke. 1995. Broadcasting prairie seed on snow. Habitat Restoration News 3(June): 2.

Hansson, L., ed. 1992. Ecological principles of nature conservation. New York: Elsevier Applied Science.

Hansson, L. and P. Angelstam. 1991. Landscape ecology as a theoretical basis for nature conservation. Landscape Ecology 5(4): 191-201.

Harker, D. F., S. Evans, M. Evans, and K. Harker. 1993. Landscape restoration handbook. Boca Raton: Lewis Publishers.

Harley, J. L. and S. E. Smith. 1983. Mycorrhizal symbiosis. Academic Press.

Harper, K. T., L. L. St. Clair, K. H. Thorne, and W. M. Hess, eds. 1994. Natural history of the Colorado Plateau and Great Basin. Boulder: University of Colorado.

Harper, J. L. 1982. After description. In The plant community as a working mechanism, ed., E. I. Newman. Oxford: Blackwell Scientific Publications.

Harper, J. L. 1987. The heuristic value of ecological restoration. In Restoration ecology, eds, W. R. Jordan, III, M. E. Gilpin, and J. D. Aber. United Kingdom: Cambridge University Press.

Harrington, J. and E. Howell. 1990. Pest plants in woodland restorations. In Environmental restoration, ed., J. J. Berger. Washington, D.C.: Island Press.

Harris, J. A. and T. C. J. Hill. 1995. Soil biotic communities and new woodland. In The ecology of woodland creation, ed., R. Ferris-Kaan. Chichester, England: John Wiley & Sons.

Harris, L. D. 1984. The fragmented forest. Chicago: University of Chicago Press.

Harris, P. 1994. Biological control of leafy spurge on the prairies. Leafy Spurge News 16(3): 2.

Harrison, R. L. 1992. Toward a theory of interrefuge corridor design. Conservation Biology 6(2): 293-295.

Harshberger, J. W. 1904. A phyto-geographic sketch of extreme southeastern Pennsylvania. Bulletin of the Torrey Botany Club 31: 125-159.

Harshberger, J. W. 1989 (reissue of 1916 edition). The vegetation of the New Jersey Pine Barrens. Mineola: Dover Publications, Inc.

Hart, J. 1992. Montana native plants and early people. Montana Historical Society Press.

Hart, M. G. R. 1962. Observations on the source of acid in empoldered mangrove soils. I. Formation of elemental sulphur. Plant Soil 17: 87-98.

Hart, M. G. R. 1963. Observations on the source of acid in empoldered mangrove soils. II. Oxidation of soil polysulphides. Plant Soil 19: 106-114.

Harty, F. M. 1991. How Illinois kicked the exotic habit. Conference on Biological Pollution: The Control and Impact of Invasive Exotic Species. Sponsored by the Indiana Academy of Science, October 25-26, Indianapolis, Indiana.

Hastings, J. R. and R. M. Turner. 1965. The changing mile: an ecological study of vegetation change with time in the lower mile of an arid and semiarid region. Tucson: University of Arizona Press.

Hayden, A. 1945. The selection of prairie areas in Iowa which should be preserved. Proceedings of the Iowa Academy of Science 52: 127-148.

Hayman, D. S. 1982. Influence of soils and fertility on activity and survival of vesicular-arbuscular mycorrhizal fungi. American Phytopathology Society 72(8): 1119-1125.

Hazen, W. E., ed. [1964] 1975. Readings in population and community ecology. 3rd ed. Philadelphia: W. B. Saunders Company.

Heady, H. F. 1988. Valley grassland. In Terrestrial vegetation of California, eds., M. G. Barbour and J. Major. Davis: California Native Plant Society.

Heckert, J. R. 1994. Breeding bird communities of midwestern prairie fragments: the effects of prescribed burning and habitat-area. Natural Areas Journal 14(2): 128-135.

Hedgpeth, J. W. and S. Obrebski. 1981. Willapa Bay: a historical perspective and a rationale for research. Washington, D.C.: U.S. Fish and Wildlife Service, FWS/OBA-81/03.

Hefner, J. M., B. O. Wilen, T. E. Dahl, and W. E. Frayer. 1994. Southeast wetlands: status and trends, mid-1970s to mid-1980s. Washington, D.C.: U.S. Department of the Interior, Fish and Wildlife Service.

Hefty, R. 1985. Late germinating prairie grasses survive first winter (Wisconsin). Restoration and Management Notes 3(1): 32-33.

Heidorn, R. 1991. Vegetation management guideline: exotic buckthorns, common buckthorn (*Rhamnus cathartica* L.), glossy buckthorn (*Rhamnus frangula* L.), dahurian buckthorn (*Rhamnus davurica* Pall.). Natural Areas Journal 11(4): 216-217.

Heilman, P. E. 1968. Relationship of availability of phosphorus and cations to forest succession and bog formation in interior Alaska. Ecology 49: 331-336.

Heimburg, K. 1984. Hydrology of north-central Florida cypress domes. In Cypress swamps, eds, K. C. Ewel and H. T. Odum. Gainesville: University Press of Florida.

Heinselman, M. L. 1963. Forest sites, bog processes, and peatland types in the glacial Lake Agassiz Region. Minnesota Ecological Monographs 33: 327-374.

Heinselman, M. L. 1970. Landscape evolution and peatland types, and the Lake Agassiz Natural Area. Minnesota Ecological Monographs 40: 235-261.

Hellmann, R. 1984. Forest restoration at Fancher Center. Restoration and Management Notes 2(1): 16-17.

Henderson, C. L. 1987. Landscaping for wildlife. St. Paul: Minnesota's Bookstore, State of Minnesota Department of Administration Print Communications Division.

Henderson, R. A. 1995. Plant species composition of Wisconsin prairies: an aid to selecting species for plantings and restorations based upon University of Wisconsin-Madison plant ecology laboratory data. Technical Bulletin No. 188. Madison: Department of Natural Resources.

Hesse, P. R. 1961. The decomposition of organic material in a mangrove swamp soil. Plant Soil 14: 249-263.

Hester, F. E. 1991. The U. S. National Park Service experience with exotic species. Natural Areas Journal 11(3): 127-128.

Heywood, V. H., ed. 1993 (updated edition). Flowering plants of the world. Oxford: Oxford Press.

Hickin, E. J., ed. 1995. River geomorphology. New York: John Wiley & Sons.

Hickman, J. C., ed. 1993. The Jepson manual, higher plants of California. Berkeley: University of California Press.

Hickman, K. R., D. C. Harnett, and R. C. Cochran. 1995. Plant community responses to bison and cattle grazing in tallgrass prairie: effects of grazing system and intensity. Restoration and Management Notes 14(1): 66.

Higgs, E. 1994. Expanding the scope of restoration ecology. Restoration Ecology 2(3): 137-146.

Hightsoe, G. L. 1988. Native, trees, shrubs, and vines for urban and rural America, a planting design manual for environmental designers. New York: Van Nostrand Reinhold Company.

Hill, G. R. and W. J. Platt. 1975. Some effects of fire upon a tallgrass prairie plant community in northwestern Iowa. In Proceedings of the Fourth Midwest Prairie Conference, ed., M. K. Wali. Grand Forks: University of North Dakota Press.

Hitchcock, A. S. [1950] 1971. Manual of the grasses of the United States. 2 Vols. 2nd Edition, ed. Agnes Chase. New York: Dover Publications, Inc.

Hitchcock, C. L. and A. Cronquist. 1973. Flora of the Pacific northwest. Seattle: University of Washington Press.

Hobbs, R. J. and L. F. Huenneke. 1992. Disturbance, diversity, and invasion: implications for Conservation. Conservation Biology 6: 324-337.

Hobbs, R. J. and D. A. Saunders, eds. 1993. Reintegrating fragmented landscapes, towards sustainable production and nature conservation. New York: Springer-Verlag.

Hoeger, S. 1995. Custom-grown, coconut-fiber sods prove effective in high-energy wetland restorations (New York). Restoration and Management Notes 13(1): 117-118.

Hogan, T. 1994. Low-elevation biodiversity in Colorado. Natural Areas Journal 14(1): 58-59.

Hokkanen, H. M. T. and J. M. Lynch, eds. 1995. Biological control, benefits and risks. New York: Cambridge University Press.

Holden, M. 1992. The greening of a desert. American Nurseryman 4(15): 22-29.

Holdgate, M. W. and M. J. Woodman, eds. 1978. The breakdown and restoration of ecosystems. New York: Plenum Press.

Holland, M. M. 1988. SCOPE/MAB technical consultations on landscape boundaries: report of a SCOPE/MAB workshop on ecotones. Biology International, Special Issue 17: 137-163.

Holland, M. M. and H. Décamps. 1989. A new international programme: research and management of landscape boundaries. In Proceedings of the Third International Wetlands Conference. Conservation and development: the sustainable use of wetland resources, ed., J. C. LeFeuvre. Paris: Museum of Natural History.

Holland, M. M., P. G. Risser, and R. J. Naiman, eds. 1991. Ecotones: the role of landscape boundaries in the management and restoration of changing environments. New York: Chapman and Hall.

Holland, R. F. 1978a. Biogeography and ecology of vernal pools in California. Ph.D. diss. University of California, Davis.

Holland, R. F. 1978b. The geographic and edaphic distribution of vernal pools in the Great Central Valley, California. Davis: California Native Plant Society, Special Publication No. 4.

Holland, R. F. 1986. Preliminary descriptions of the terrestrial natural communities of California. Sacramento: State of California, Department of Fish and Game.

Holland, R. F. and S. K. Jain. 1977. Vernal pools. In Terrestrial vegetation of California, eds., M. G. Barbour and J. Major. California Native Plant Society, Special Publication No. 9.

Holland, R. and S. Jain. 1988. Vernal pools. In Terrestrial vegetation of California, eds., M. G. Barbour and J. Major. Davis: California Native Plant Society.

Holling, C. F., ed. 1978. Adaptive environmental assessment and management. New York: John Wiley & Sons.

Holloway, M. 1994. Nurturing nature. Scientific American 270: 76-84.

Hoover, R. F. 1937. Endemism in the flora of the Great Valley of California. Ph.D. diss. University of California, Berkeley.

Hoover, R. F. 1970. The vascular plants of San Luis Obispo County, California. Berkeley: University of California Press.

Hotchkiss, N. 1967. Underwater and floating-leaved plants of the United States and Canada. Washington, D.C.: Bureau of Sport Fisheries and Wildlife, Resource Publication 44.

Hotchkiss, N. 1970. Common marsh plants of the United States and Canada. Washington, D.C.: Bureau of Sports Fisheries and Wildlife, Resource Publication 93.

Houghton, R. A. 1994. The worldwide extent of land-use change. Bioscience.

House, F. 1996. Restoring relationships: the vernacular approach to ecological restoration. Restoration and Management Notes 14(1): 57-61.

Howald, A. 1996. The California Exotic Pest Plant Council: Symposium '95. Restoration and Management Notes 14(1): 41-42.

Howe, H. F. 1995. Succession and fire season in experimental prairie plantings. Ecology 76(6): 1917-1925.

Howe, H. and L. Westley. 1988. Ecological relationships of plants and animals. New York: Oxford University Press.

Howe, R. H. 1984. Wings over the prairie. Iowa Conservationist 43(9): 5-7.

Howell, E. A. 1986. Woodland restoration: an overview. Restoration and Management Notes 4(1): 13-17.

Hubner, S. and M. K. Leach. 1995. Prairie parsley reappears following brush cutting and burning (Wisconsin). Restoration and Management Notes 13(2): 209-210.

Hudson, W. E., ed. 1991. Landscape linkages and biodiversity. Washington, D.C.: Island Press.

Huffman, R. T. and S. W. Forsythe. 1981. Bottomland hardwood forest communities and their relation to anaerobic soil conditions. In Wetlands of bottomland hardwood forests, eds., J. R. Clark and J. Benforado. Amsterdam: Elsevier Scientific Publications.

Hujik, P. and F. T. Griggs. 1995a. Cutting size, horticultural treatments affects growth of riparian species (California). Restoration and Management Notes 13(2): 219-220.

Hujik, P. and F. T. Griggs. 1995b. Field-seeded riparian trees and shrubs thrive in non-irrigated plots (California). Restoration and Management Notes 13(2): 220-221.

Hulbert, L. C. 1986. Fire effects on tallgrass prairie. In Proceedings of the Ninth North American Prairie Conference, eds., G. K. Clambey and R. H. Pemble. Fargo: Tri-College University.

Hull-Sieg, C. 1994. Herbicides and fire effects on leafy spurge density and seed germination. Leafy Spurge News 16(3): 10.

Hunter, M. L., Jr. 1990. Wildlife, forests, and forestry. Engelwood Cliffs: Prentice Hall.

Hunter, M. L., Jr. 1996. Fundamentals of conservation biology. Cambridge: Blackwell Science, Inc.

Hurley, L. M. 1990. Field guide to the submerged aquatic vegetation of Chesapeake Bay. Annapolis: U.S. Fish and Wildlife Service, Chesapeake Bay Estuary Program.

Huston, M. A. 1993. Biological diversity, the coexistence of species on changing landscapes. New York: Cambridge University Press.

Hutchison, M. 1992. Vegetation management guideline: reed canary grass (*Phalaris arundinacea* L.). Natural Areas Journal 12(3): 159-160.

Ibarra-Obando, S. E. and A. Escofet. 1987. Industrial development effects on the ecology of a Pacific Mexican estuary. Environmental Conservation 14: 135-141.

Ikeda, D. H. and R. A. Schisling, eds. 1990. Vernal pool plants: their habitat and biology. Studies from the Herbarium, No. 8, Chico: California State University.

Irwin, H. S. 1961. Roadside flowers of Texas. Austin: University of Texas Press.

Isaacson, D. 1995. Considerations in purchasing native seed. Hortus Northwest 6(1): 13-15.

Ischinger, L. S. and P. B. Shafroth. 1995. Induced root-suckering shows potential for reestablishing riparian trees (New Mexico). Restoration and Management Notes 13(1): 121.

Isley, D. 1990. Vascular flora of the southeastern United States, Vol. 3, Part 2: Leguminosae (Fabaceae). Chapel Hill: University of North Carolina Press.

Jackson, B., ed. 1990. The international forested wetland resource: identification and inventory. Forest Ecology Management 33/34: 1-648.

Jackson, L. L., N. Lopoukhine, and D. Hillyard. 1995. Ecological restoration: a definition and comments. Restoration Ecology 3(2): 71-75.

Jackson, L. L., J. R. McAuliffe, and B. A. Roundy. 1991. Desert restoration: revegetation trials on abandoned farmland in the Sonoran Desert lowlands. Restoration and Management Notes 9(2): 71-80.

Jain, S. K., ed. 1976. Vernal pools, their ecology and conservation. Davis: Institute of Ecology, University of California, Special Publication 9.

Jain, S. K. and P. Moyle. 1984. Vernal pools and intermittent streams. Davis: Institute of Ecology, University of California, Special Publication 28.

James, D. W. and J. J. Jurinak. 1978. Nitrogen fertilization of dominant plants in the northeastern Great Basin desert. In Nitrogen in desert ecosystems, eds., N. E. West and J. Skujins. Stroundsburg: Dowden, Hutchinson, and Ross, Inc.

Jastrow, J. D. 1987. Changes in soil aggregation associated with tallgrass prairie restoration. American Journal of Botany 74(11): 1656-1664.

Jastrow, J. D. 1994. Mechanisms of aggregate formation and stabilization in prairie soils. Ph.D. diss., University of Illinois, Chicago.

Jeffries, R. L. and A. J. Davies, eds. 1979. Ecological processes in coastal environments. Oxford: Blackwell Scientific Publications.

Jellinek, A. 1995. Restoring old-growth forests. College of Agricultural and Life Sciences (CALS) Quarterly (University of Wisconsin-Madison) 13(3 and 4): 1,3.

Jepson, W. L. 1951. A manual of the flowering plants of California. Berkeley: University of California Press.

Joern, A. and K. Keeler. 1994. The changing prairie, North American grasslands. New York: Oxford University Press.

Johnson, A. F. and M. G. Barbour. 1990. Dunes and maritime forests. In Ecosystems of Florida, eds. R. L. Myers and J. J. Ewel. Orlando: University of Central Florida.

Johnson, C. W. 1985. Bogs of the northeast. Hanover: University Press of New England.

Johnson, E. 1994a. What makes an exotic plant invasive?–the role of the proliferation threshold. Resource Management Notes 6(3): 6-7.

Johnson, E. 1994b. Hardwoods that belong in high pine. Resource Management Notes 6(4): 9-10.

Johnson, R. R. 1979. The lower Colorado River, a midwestern system. In Strategies for the protection and management of floodplain wetlands and other riparian ecosystems, eds., R. R. Johnson and J. F. McCormick. Washington, D.C.: United States Department of Agriculture Forest Service General Technical Report WO-12.

Johnson, R. R., ed. 1985. Riparian ecosystems and their management: reconciling conflicting uses. Fort Collins: United States Department of Agriculture Forest Service General Technical Report RM-120.

Johnson, R. R. and D. A. Jones, technical coordinators. 1977. Importance, preservation and management of riparian habitat: a symposium. Fort Collins: U.S. Department of Agriculture Forest Service General Technical Report RM-43. Rocky Mountain Forest and Range Experiment Station.

Johnson, R. R. and J. F. McCormick, eds. 1979. Strategies for the protection and management of floodplain wetlands and other riparian ecosystems. Washington, D.C.: United States Department of Agriculture Forest Service General Technical Report WO-12.

Johnson, W. C., R. L. Burgess, and W. R. Keamerer. 1976. Forest overstory vegetation on the Missouri River floodplain in North Dakota. Ecological Monographs 46: 59-84.

Johnston, C. A., N. E. Detenbeck, and G. J. Niemi. 1990. The cumulative effect of wetlands on stream water quality and quantity: a landscape approach. Biogeochemistry 10: 105-141.

Johnston, C. A. and R. J. Naiman. 1987. Boundary dynamics at the aquatic-terrestrial interface: the influence of beaver and geomorphology. Landscape Ecology 1: 47-57.

Johnston, D. W. and E. P. Odum. 1956. Breeding bird populations in relation to plant succession of the piedmont of Georgia. Ecology 37(1): 50-62.

Jones, F. B. 1977. Flora of the Texas coastal bend. Sinton: Rob and Bessie Welder Wildlife Foundation.

Jones, S. B., Jr. and L. E. Foote. 1990. Gardening with native wildflowers. Portland: Timber Press.

Jontos, R. J., Jr. and C. P. Allan. 1984. Test salt to control *Phragmites* in salt marsh restoration (Connecticut). Restoration and Management Notes 2(1): 32.

Jordan, P. W. and P. S. Nobel. 1979. Infrequent establishment of seedlings of *Agave deserti* (Agavaceae) in the northwestern Sonoran Desert. American Journal of Botany 66: 1079-1084.

Jordan, P. W. and P. S. Nobel. 1981. Seedling establishment of *Ferocactus acanthodes* in relation to drought. Ecology 62: 901-906.

Jordan, W. R., III. 1984. Working with the river. Restoration and Management Notes 2(1): 4-11.

Jordan, W. R., III. 1985. On the imitation of nature. Restoration and Management Notes 3(1): 2-3.

Jordan, W. R., III. 1995a. "Restoration" (the word). Restoration and Management Notes 13(2): 151-152.

Jordan, W. R., III. 1995b. Salmon Country: a sampler of restoration projects in the Pacific Northwest. Restoration and Management Notes 13(1): 5-6.

Jordan, W. R., III, M. E. Gilpin, and J. D. Aber, eds. 1987. Restoration ecology: a synthetic approach to ecological research. New York: Cambridge University Press.

Jorgensen, E. E. and L. E. Nauman. 1994. Disturbance in wetlands associated with commercial cranberry (*Vaccinium macrocarpon*) production. American Midland Naturalist 132: 152-158.

Josselyn, M. 1983. The ecology of San Francisco Bay tidal marshes: a community profile. Washington, D.C.: U.S. Fish and Wildlife Service, FWS/OBS-83/23.

Kadlec, R. H. and D. L. Tilton. 1979. The use of freshwater wetlands as a tertiary wastewater treatment alternative. Critical Review of Environmental Control 9: 185-212.

Kantrud, H., G. L. Krapu, and G. A. Swanson. 1989. Prairie basin wetlands of the Dakotas: a community profile. Washington, D.C.: U.S. Fish and Wildlife Service Biological Report 85.

Karpiscak, M. M. 1980. Secondary succession of abandoned field vegetation in southern Arizona. Ph.D. diss., University of Arizona, Tucson.

Karr, J. R. 1992. Defining and assessing ecological integrity beyond water quality. Environmental Toxicology and Chemistry 12:1521-1531.

Karr, J. R. and I. J. Sclosser. 1994. Landscape disturbance and stream ecosystems. New York: Chapman and Hall.

Kartesz, J. T. 1990. Common names for the North American flora. Portland: Timber Press.

Kartesz, J. T. 1991. Synonymized checklist of the vascular flora of the United States. Chapel Hill: North Carolina Botanical Garden, The University of North Carolina.

Kauffman, J. B., R. L. Case, D. Lytjen, N. Otting, and D. L. Cummings. 1995. Ecological approaches to riparian restoration in northeast Oregon. Restoration and Management Notes 13(1): 12-15.

Kauffman, J. B., W. C. Krueger, and M. Vavra. 1983. Effects of late season cattle grazing on riparian plant communities. Journal of Range Management 36: 685-691.

Kay, B. L. 1978. Mulches for erosion control and plant establishment on disturbed sites. Davis: Agronomy Progress Report No. 87, University of California Agronomy Department.

Keamerer, W. R., W. C. Johnson, and R. L. Burgess. 1975. Floristic analysis of the Missouri River bottomland forests in North Dakota. Canada Field Naturalist 89: 5-19.

Kearney, T. H. and R. H. Peebles, and collaborators. 1960. Arizona flora, 2nd ed. Supplement by J. T. Howell and E. McClintock and collaborators. Berkeley: University of California Press.

Kearns, S. K. 1985. Optimum times and transplanting methods for salvaging mature prairie plants (Wisconsin). Restoration and Management Notes 3(1): 31-32.

Keator, G. 1990. Complete garden guide to the native perennials of California. San Francisco: Chronicle Books.

Keeley, J. E., ed. 1993. Interface between ecology and land development in California. Los Angeles: Southern California Academy of Sciences.

Keeley J. E. and S. C. Keeley. 1988. Chaparral. In North American terrestrial vegetation, eds., M. G. Barbour and W. D. Billings. Cambridge: Cambridge University Press.

Keiter, R. E., ed. 1998. Reclaiming the native home of hope. Salt Lake City: The University of Utah Press.

Keller, C. R. and K. P. Burham. 1982. Riparian fencing, grazing, and trout habitat preference on Summit Creek, Idaho. North American Journal of Fisheries Management 2: 53-59.

Kellert, S. R. and E. O. Wilson, eds. 1993. The biophilia hypothesis. Washington, D.C.: Island Press.

Kelting, R. W. and W. T. Penfound. 1950. The vegetation of stock pond dams in Central Oklahoma. American Midland Naturalist 44: 69-75.

Kempton, W., J. S. Boster, and J. A. Hartley. 1995. Environmental values in America culture. Cambridge: Massachusetts Institute of Technology Press.

Kennay, J. and G. Fell. 1992. Vegetation management guideline: siberian elm (*Ulmus pumila* L.). Natural Areas Journal 12(1): 40-41.

Kent, D. M. 1994. Applied wetlands science and technology. Boca Raton: CRC Press, Inc./Lewis Publishers.

Kentula, M. E., R. P. Brooks, S. E. Gwin, C. C. Holland, A. D. Sherman, and J. C. Sifneos. 1992. Wetlands, an approach to improving decision making in wetland restoration and creation. Washington, D.C.: Island Press.

Kilgour, M. and B. Kilgour. 1995. Cutting Queen Anne's lace reduces seed supply (Wisconsin). Restoration and Management Notes 13(2): 227-228.

Kimmeins, J. P. 1996. The health and integrity of forest ecosystems: Are they threatened by forestry? Ecosystem Health 2: 5-18.

Kimmerer, R. W. 1985. Environmental stresses affect vegetation succession on lead-zinc mine waste (Wisconsin). Restoration and Management Notes 3(1): 43.

Kirby, J. T. 1995. Poquosin, a study of rural landscape and society. Chapel Hill: University of North Carolina Press.

Kirt, R. R. 1995. Prairie plants of the midwest, identification and ecology. Stipes Publishing.

Kituku, V. M., W. A. Laycock, J. Powell, and A. A. Beetle. 1995. Propagating bitterbush twigs for restoring shrublands. In Proceedings: Wildland shrub and arid land restoration symposium, comps., B. A. Roundy, E. D. McArthur, J. S. Haley, and D. K. Mann. Ogden: U.S. Forest Service General Technical Report INT-GTR-315.

Kleiner, E. F. and K. T. Harper. 1972. Environment and community organizations in grasslands of Canyonlands National Park. Ecology 53: 229-309.

Kleiner, E. F. and K. T. Harper. 1977. Soil properties in relation to crytogamic groundcover in Canyonlands National Park. Journal of Range Management 30: 202.

Kline, V. M. 1981. Control of honeysuckle and buckthorn in oak forests. Restoration and Management Notes 1: 31.

Kline, V. M. 1982. Control of sumac in a sand prairie by repeated cutting. Restoration and Management Notes 1: 133.

Knight, D. H. 1994. Mountains and plains, the ecology of Wyoming landscapes. New Haven:Yale Press.

Knight, D. H., R. J. Hill, and T. A. Harrison. 1976. Potential natural landmarks in the Wyoming Basin. Laramie: University of Wyoming, Department of Botany.

Knopf, F. L. and R. W. Cannon. 1982. Structural resilience of a willow riparian community to changes in grazing practices. In Proceedings of the wildlife - livestock relationship symposium, eds., L. Nelson, J. M. Peck, and P. D. Dalke. Moscow: Forest, Wildlife, and Range Experiment Station, University of Idaho.

Knopf, F. L. and F. B. Sampson. 1994. Scale perspectives on avian diversity in western riparian ecosystems. Conservation Biology 8(3): 669-676.

Knutson, P. L., J. C. Ford, M. R. Inskeep, and J. Oyler. 1981. National survey of planted salt marshes (vegetative stabilization and wave stress). Wetlands 1: 129-157.

Knutson, P. L. and M. R. Inskeep. 1982. Shore erosion control with salt marsh vegetation. U.S. Army Coastal Engineering Research Center, Technical Aid 8203.

Koebel, J. W., Jr. 1995. An historical perspective on the Kissimmee River restoration project. Restoration Ecology 3(3): 149-159.

Kondolf, G. M. 1993. Lag in stream channel adjustment to livestock exclosure, White Mountains, California. Restoration Ecology1(4): 226-230.

Kondolf, G. M. and E. R. Micheli. 1995. Forum: evaluating stream restoration projects. Environmental Management 19(1): 1-15.

Kopecko, K. J. P. and E. W. Lathrop. 1975. Vegetation zonation in a vernal marsh on the Santa Rosa Plateau of Riverside County, California. Aliso 8: 281-288.

Koske, R. E. and J. N. Gemma. 1992. Restoration of early and late successional dune communities at Province Lands, Cape Cod National Seashore. Kingston: University of Rhode Island.

Kozloff, E. N. and L. H. Beidleman. 1994. Plants of the San Francisco Bay region, Mendocino to Monterey. Sagen Press.

Krishnan, V. V. 1987. The effects of edge permeability and habitat geometry on emigration from patches of habitat. The American Naturalist 129(4): 533-552.

Kruckeberg, A. R. 1982. Gardening with native plants of the Pacific Northwest: an illustrated guide. Seattle: University of Washington Press.

Küchler, A. W. 1964a. Manual to accompany the map potential natural vegetation of the conterminous United States. American Geographical Society Special Publication No. 36. New York: American Geographical Society.

Küchler, A. W. 1964b. The potential natural vegetation of the conterminous United States. New York: American Geographical Society Special Publication Number 36.

Küchler, A. W. 1967. Vegetation mapping. New York: The Ronald Press Company.

Küchler, A. W. [1964] 1975. Potential natural vegetation of the conterminous United States. 2nd ed. New York: American Geographical Society.

Kurz, H. and R. K. Godfrey. 1962. Trees of northern Florida. Gainesville: University Press of Florida.

Kushlan, J. A. 1990. Freshwater marshes. In Ecosystems of Florida, eds., R. L. Myers and J. J. Ewel. Orlando: University of Central Florida.

Kusler, J.A. 1979. Strenthening state wetland regulations. Washington, D.C.: U. S. Fish and Wildlife Service FWS/OBS-79/78.

Kusler, J. A. and M. E. Kentula, eds. 1989. Wetland creation and restoration: the status of the science. 2 Vols. Washington, D.C.: U.S. Environmental Protection Agency, EPA/600/3-89/038a,b.

Kusler, J. A. and M. E. Kentula. 1990. Wetland creation and restoration: the status of the science. Washington, D.C.: Island Press.

Ladd, D. 1995. Tallgrass prairie wildflowers, a field guide. Falcon Press.

Langis, R., M. Zalejko, and J. B. Zedler. 1991.
Nitrogen assessments in a constructed and a natural
salt marsh of San Diego Bay, California. Ecological Applications 1: 40-51.

Larson, P. 1970. Deserts of America. Englewood
Cliffs: Prentice-Hall, Inc.

Larson, R. 1995. Swamp song, a natural history of
Florida's swamp. Gainesville: University of Florida
Press.

Lathrop, B. 1995. Demonstration project converts
poor-quality cropland to high-grade prairie
(Nebraska). Restoration and Management Notes
13(2): 214.

Lathrop, E. W. and R. F. Thorne. 1968. Flora of the
Santa Rosa Plateau of the Santa Ana Mountains,
California. Aliso 6: 17-40.

Latting, J., ed. 1976. Plant communities of southern
California. Symposium Proceedings. Special Publi-
cation Number 2. Davis: California Native Plant
Society.

Lauver, C. L. 1989. Preliminary classification of the
natural communities of Kansas. Kansas Biological
Survey Report No. 50. Lawrence: Kansas Natural
Heritage Program.

Leach, M. K. and L. Ross, eds. 1995. Midwest oak
ecosystems recovery plan: a call to action. Springfield: Midwest Oak Savanna and Woodland
Ecosystems Conference.

Lee, P. 1995. Trilliums from seed. Wildflower 11(1):
36-37.

Leisman, G. A. 1957. A vegetation and soil
chronosequence on the Mesabi Iron Range spoil in
Minnesota. Ecological Monographs 27: 221-245.

Lekwa, S. 1984. Prairie restoration and management.
Iowa Conservationist 43(9): 12-14.

Lellinger, D. B. 1985. A field manual of the ferns and
fern allies of the United States and Canada.
Washington, D.C.: Smithsonian Institution Press.

Leopold, A. 1949. A Sand County almanac and
sketches here and there. New York: Oxford
University Press.

Leopold, E. and V. C. Wright. 1995. Scotch broom
eradicated on Shaw Island Prairies (Washington).
Restoration and Management Notes 13(2). 228.

Leopold, L. B. 1951. Vegetation of southwestern
watersheds in the nineteenth century. Geographical
Review 41: 295-316.

Levin, S. A., D. Cohen, and A. Hastings. 1984.
Dispersal strategies in patchy environments.
Theoretical Population Biology 26: 165-191.

Lewis, R. R., III. 1979. Large scale mangrove
restoration on St. Croix, V.I. In Proceedings of the
6th Annual Conference on Restoration and
Creation of Wetlands. Tampa: Hillsborough
Community College.

Lewis, R. R., III, ed. 1982. Creation and restoration of
coastal plant communities. Boca Raton: CRC
Press, Inc.

Lewis, R. R., III. 1990a. Creation and restoration of
coastal plain wetlands in Florida. In Wetland
creation and restoration, eds., J. A. Kusler and M.
E. Kentula. Washington, D.C.: Island Press.

Lewis, R. R., III. 1990b. Creation and restoration of
coastal plain wetlands in Puerto Rico and the U.S.
Virgin Islands. In Wetland creation and restoration,
eds., J. A. Kusler and M. E. Kentula. Washington,
D.C.: Island Press.

Lewis, R. R., III, C. S. Lewis, W. K. Fehring, and J.
A. Rodgers. 1979. Coastal habitat mitigation in
Tampa Bay, Florida. In Proceedings of the Mitiga-
tion Symposium. Fort Collins: General Technical
Report RM-65.

Li, H. W., G. A. Lamberti, T. N. Pearsons, C. K. Tait,
J. L. Li, and J. C. Buckhouse. 1994. Cumulative
effects of riparian disturbances along High Desert
trout streams of the John Day Basin, Oregon.
Tranactions of the American Fisheries Society
123(4): 627-640.

Liegel, K. and L. Luthin. 1984. Spring and fall prairie
plantings compared (Wisconsin). Restoration and
Management Notes 2(1): 23-24.

Liegel, K. and J. Lyon. 1985. Change in grass to forb
ratio documented at ICF prairie restoration
(Wisconsin). Restoration and Management Notes
3(1): 30.

Lin, J. 1970. The floristic and plant succession in
vernal pools vegetation. M.A. thesis., San Francisco State College, San Francisco.

Lindau, C. W. and L. R. Hossner. 1981. Substrate
characterization of an experimental marsh and
three natural marshes. Soil Science Society of
America Journal 45: 1171-1176.

Linde, A. F. 1969. Techniques for wetlands management. Wisconsin Department of Natural Resources
Report 45.

Lippitt, L., M. W. Fidelibus, and D. A. Bainbridge.
1994. Native seed collection, processing, and
storage for revegetation projects in the western
United States. Restoration Ecology 2(2): 120-131.

Little, C. E. 1995. The dying of the trees: the pan-
demic in American forests. New York: Viking
Press, 1995.

Little, C. E. 1990. Greenways for America. Baltimore:
The Johns Hopkins University Press.

Littlehales, B. and W. A. Niering. 1991. Wetlands of
North America. Charlottesville: Thomasson-Grant.

Lodge, T. E. 1994. The Everglades handbook, understanding the ecosystem. Delray Beach: St. Lucie Press.

Loope, L. L. 1980. Phenology of flowering and fruiting in plant communities of Everglades National Park and Biscayne National Monument, Florida. Homestead: Everglades National Park, South Florida Research Center Report T-593.

Lorimer, C. G. 1990. Behavior and management of forest fires. In Introduction to forest science, eds., R. A. Young and R. L. Giese. New York: John Wiley & Sons.

Loughmiller, C. and L. Loughmiller. 1984. Texas wildflowers. Austin: University of Texas Press.

Lovejoy, T. 1988. Will unexpectedly the top blow off? Bioscience 38(10): 722-726.

Loveless, C. M. 1959. A study of the vegetation of the Florida Everglades. Ecology 40: 1-9.

Lowe, C. H., 1964. Arizona's natural environment. Tucson: University of Arizona Press.

Lowrance, R. R., R. L. Todd, J. Fail, Jr., O. Hendrickson, Jr., R. Leonard, and L. E. Asmussen. 1984. Riparian forests as nutrient filters in agricultural watersheds. Bioscience 34: 374-377.

Lugo, A. E. 1980. Mangrove ecosystems: successional or steady state? Biotropica (supplement) 12: 65-73.

Lugo, A. E. 1981. The island mangroves Inagua. Journal of Natural History 15: 845-852.

Lugo, A. E. and S. C. Snedaker. 1974. The ecology of mangroves. Annual Review of Ecological Systematics 5: 39-64.

Luken, J. O. 1990. Directing ecological succession. New York: Chapman and Hall.

Luken, J. O. and D. T. Mattimiro. 1991. Habitat-specific resilience of the invasive shrub amur honeysuckle (*Lonicera maackii*) during repeated clipping. Ecological Applications 1(1): 104-109.

Luttenberg, D., D. Lev, and M. Fedler. 1993. Native species planting guide for New York City and vicinity. New York: City of New York, Parks and Recreation.

Lym, R. G. 1994. Integration of herbicides with *Aphthona* spp. flea beetles for leafy spurge control and integration of herbicides with grazing for leafy spurge control. Leafy Spurge News 16(3): 7, 11.

Lynch, J. F. and D. F. Whigham. 1984. Effects of forest fragmentation on breeding bird communities in Maryland, U.S.A. Biological Conservation 28: 287-324.

Lytle, S. A. and B. N. Driskell. 1954. Physical and chemical characteristics of the peats, mucks, and clays of the coastal marsh area of St. Mary Parish, Louisiana. Louisiana Agricultural Experiment Station Bulletin 484.

MacArthur, J. W. 1975. Environmental fluctuations and species diversity. In Ecology and evolution of communities, eds., M. Cody and J. M. Diamond. Cambridge: The Belknap Press of Harvard University Press.

MacArthur, R. H. 1972. Geographical ecology. New York: Harper and Row.

MacArthur, R. H. and J. MacArthur. 1961. On bird species diversity. Ecology 42(3): 594-599.

MacArthur, R. H. and E. O. Wilson. 1967. The theory of island biogeography. Princeton: Princeton University Press.

McClain, W. E. 1986. Illinois prairie: past and future. A restoration guide. Springfield: Illinois Department of Conservation, Division of Natural Heritage.

McClaren, M. P. and T. R. Van Devender, eds. The desert grassland. Tucson: University of Arizona.

McClay, A. 1994. Overview of leafy spurge biological control in Alberta. Leafy Spurge News 16(3): 8.

McCoy, E. D. and H. R. Mushinsky. 1994. Effects of fragmentation on the richness of vertebrates in the Florida scrub habitat. Ecology 75(2): 446-457.

McCune, B. and G. Cottam. 1985. The successional status of a southern Wisconsin oak woods. Ecology 66: 1270-1278.

MacDonald, B. 1986 (4th printing 1993). Practical woody plant propagation for nursery growers. Portland: Timber Press.

Macdonald, K. B. 1977. Plant and animal communities of Pacific North American salt marshes. In Ecosystems of the World. I. Wet coastal ecosystems, ed., V. J. Chapman. New York: Elsevier Scientific Publications.

Macdonald, K. B. 1988. Coastal salt marsh. In Terrestrial vegetation of California, eds., M. G. Barbour and J. Major. Davis: California Native Plant Society.

McGinnies, W. G. and B. J. Goldman, eds. 1969. Arid lands in perspective. Tucson: University of Arizona Press.

McGregor R. L., T. M. Barkley, R. E. Brooks, and E. K. Schofield. 1986. Flora of the Great Plains. Lawrence: University Press of Kansas.

McIninch, S. M. and E. W. Garbisch. 1994. The establishment of *Sagittaria latifolia* large and small tubers as a function of water-depth. Wetland Journal 6(3): 19-21.

McIninch, S. M. and E. W. Garbisch. 1995. The establishment of *Peltandra virginica* from large and small bulbs as a function of water depth. Wetland Journal 7(1): 17-20.

McIninch, S. M., E. W. Garbisch, and D. Biggs. 1994. The benefits of wet-acclimating woody wetland plant species. Wetland Journal 6(2): 19-23.

McIntosh, R. P. 1980. The relationship between succession and the recovery process in ecosystems. In The recovery process in damaged ecosystems, ed., J. Cairns, Jr. Ann Arbor: Ann Arbor Science Publishers, Inc.

McIntyre, S. and G. W. Barrett. 1992. Habitat variegation, an alternative to fragmentation. Conservation Biology 6(1): 146-147.

Mack, R. N. 1981. Invasion of *Bromus tectorum* L. into western North America: an ecological chronicle. Agro-Ecosystems 7: 145-165.

McKell, C. M., J. P. Blaisdell, and J. R. Goodin, eds. 1972. Wildland shrubs–their biology and utilization. Ogden: U.S. Forest Service General Technical Report INT-1.

Mackintosh, G. 1989. Preserving communities and corridors. Washington: Defenders of Wildlife.

McLean, H. F. 1992. The Blue Mountains: forest out of control. American Forests 98(9 and 10): 32-35, 58, 61.

McMahan, C. A., R. G. Frye, and K. L. Brown. 1984. The vegetation types of Texas including cropland. Austin: Wildlife Division, Texas Parks and Wildlife Department.

MacMahon, J. A. 1988. Chapter: Warm deserts. In North American terrestrial vegetation, eds., M. G. Barbour and W. D. Billings. Cambridge: Cambridge University Press.

McMillan, C. 1971. Environmental factors affecting seedling establishment of the black mangrove on the central Texas coast. Ecology 52: 927-930.

McMillan, C. 1975. Interaction of soil texture with salinity tolerance of black mangrove (*Avicennia*) and white mangrove (*Laguncularia*) from North America. In Proceedings of the international symposium on the biology and management of mangroves, eds., G. Walsh, S. Snedaker, and H. Teas. Gainesville: University of Florida.

McMinn, H. E. 1939. An illustrated manual of California shrubs. Berkeley: University of California Press.

McMinn, H. E. and E. Maino. 1980. Illustrated manual of Pacific Coast trees. 2nd ed. Berkeley: University of California.

Macnae, W. 1963. Mangrove swamps in South Africa. Journal of Ecology 51: 1-25.

Magnuson, J. J., H. A. Regier, W. J. Christie, and W. C. Sonzogni. 1980. To rehabilitate and restore Great Lakes ecosystems. In The recovery process in damaged ecosystems, ed., J. Cairns, Jr. Ann Arbor: Science Publishers/The Butterworth Group.

Majer, J. and H. Recher. 1994. Restoration ecology: an international science? Restoration Ecology 2(4): 215-217.

Malanson, G. P. 1993. Riparian landscapes. New York: Cambridge University Press.

Malanson, G. P., ed. 1989. Natural areas facing climate change. The Hague: SPB Academic Publishing.

Malmer, N. 1975. Development of bog mires. In Coupling of land and water systems, Ecology Studies No. 10, ed., A. D. Hasler. New York: Springer-Verlag.

Marble, A. D. 1990. A guide to wetland functional design. McLean: U.S. Department of Transportation, Federal Highway Administration Report No. FHWA-IP-90-010.

Marble, A. D. 1992. A guide to wetland functional design. Boca Raton: Lewis Publishers.

Marinelli, J. 1996. Introduction: redefining the weed. In Invasive plants: weeds of the global garden, eds., J. M. Randall and J. Marinelli. New York: Brooklyn Botanic Garden.

Marks, M., B. Lapin, and J. Randall. 1994. *Phragmites australis (P. communis)*: threats, management and monitoring. Natural Areas Journal 14(4): 285-295.

Marmer, H. A. 1954. Tides and sea levels in the Gulf of Mexico. In Gulf of Mexico, its origin, waters, and marine life, ed., P. S. Galtsoff. Washington, D.C.: U.S. Fish and Wildlife Service Fishery Bulletin 89.

Martin, A. C., H. S. Zim, and A. L. Nelson. 1951. American wildlife and plants, a guide to wildlife food habits. New York: Dover Publications.

Martin, H. and G. Sick. 1995. American beautyberry for borrow pit reclamation in South Carolina. Restoration and Management Notes 13(1): 90-97.

Martin, L. C. 1990. The wildflower meadow book: a gardeners guide. 2nd ed. Charlotte: East Woods Press.

Martin, M. A. 1981. Control of smooth sumac by cutting in a mesic prairie. Restoration and Management Notes 1: 13.

Martin, W. H., S. G. Boyce, and A. S. Echternacht, eds. 1993. Biodiversity of the southeastern United States. 2 vols. New York: John Wiley & Sons.

Marx, D. H. 1991a. The practical significance of ectomycorrhizae in forest establishment. Symposium at the Marcus Wallenberg Prize Ceremonies. September 27, 1991, Stockholm, Sweden.

Marx, D. H. 1991b. Forest application of the ectomychorrhizal fungus *Pisolithus tinctorius*. The Marcus Wallenburg Prize lecture. September 26, 1991, Stockholm, Sweden.

Maser, C. 1988. The redesigned forest. San Pedro: R. and E. Miles.

Maser, C. and J. R. Sedell. 1994. From forest to the sea: the ecology of wood in streams, rivers, estuaries, and oceans. Delray Beach: St. Lucie Press.

Maser, C., J. W. Thomas, and R. G. Anderson. 1984. The relationship of terrestrial vertebrates to plant communities, part I. In Wildlife habitats in managed rangelands—the Great Basin of southeastern Oregon. Portland: U.S. Forest Service General Technical Report PNW-172.

Mason, H. L. 1957. A flora of the marshes of California. Berkeley: University of California Press.

Mason, H. L. 1980. Techniques for creating salt marshes along the California coast. In Rehabilitation and creation of selected coastal habitats: proceedings of a workshop, eds., J. C. Lewis and E. W. Bunce. Washington, D.C.: U.S. Fish and Wildlife Service, FWS/OBS-80/27.

Massey, M. 1983. Carrying capacity and the Greek dark ages. CoEvolution Quarterly 40: 29-35.

Mathiak, H. 1965. Pothole blasting for wildlife. Madison: Wisconsin Conservation Department Publication 352.

Matsil, M. A. and M. J. Feller. 1996. Natural areas restoration in New York City: a bit of the apple. Restoration and Management Notes 14(1): 5-14.

Matsui, T. 1996. Meiji Shrine: an early old-growth forest creation in Tokyo. Restoration and Management Notes 14(1): 46-52.

Mattoni, R. 1989. Unnatural acts: succession on the El Segundo sand dunes in California. In Restoration '89: the new management challenge, eds., H. G. Hughes, and T. M. Bonnicksen. Oakland: First Annual Meeting of the Society for Ecological Restoration, Jan. 16-20, 1989.

Mattoon, W. R. 1915. The southern cypress. Washington, D.C.: U.S. Department of Agriculture Bulletin 272.

Mauk, R. L. and J. A. Henderson. 1984. Coniferous forest habitat types of northern Utah. Ogden: United States Department of Agriculture, Forest Service General Technical Report INT-170.

Medley, M. 1976. The prairie, a historical perspective. Fernwood Notes: No. 111.

Meeks, G. and L. C. Runyon. 1990. Wetlands protection and the states. Denver: National Conference of State Legislatures.

Merchant, S. S. and B. J. Olson. 1995. Species composition of seed mechanically harvested from remnant native prairies in west-central Minnesota. Native Warm-Season Grass Newsletter 14(1): 7.

Merriam, G. 1984. Connectivity: a fundamental ecological characteristic of landscape pattern. In Methodology in landscape ecological research and planning, eds., J. Brandt and P. Agger. Denmark: Roskilde University Centre.

Messina, M. G. and W. H. Conner. 1997. Southern forested wetlands: ecology and management. Boca Raton: Lewis Publishers.

Miller, R. F. 1995. Pushing back juniper. Restoration and Management Notes 13(1): 51-52.

Miller, R. F. and P. E. Wigand. 1994. Holocene changes in semiarid pinyon-juniper woodlands: response to climate, fire, and human activities in the U.S. Great Basin. Bioscience 44(7): 465-474.

Miller, R. M. 1985. Mycorrhizae. Restoration and Management Notes 3(1): 14-20.

Mills, S. 1995. In service of the wild: restoring and reinhabiting damaged land. Boston: Beacon Press.

Minnich, R. A., M. G. Barbour, J. H. Burk, and R. F. Fernau. 1995. Sixty years of change in California conifer forests of the San Bernardino Mountains. Conservation Biology 9(4): 902-914.

Mitsch, W. J. and J. G. Gosselink. 1986. Wetlands. New York: Van Nostrand Reinhold Company.

Mitsch, W. J. and J. G. Gosselink. 1993. Wetlands. New York: Van Nostrand Reinhold Company.

Mohlenbrock, R. H. 1986. Guide to the vascular flora of Illinois. Urbana: University of Illinois.

Mohlenbrock, R. H. and D. M. Ladd. 1978. Distribution of Illinois vascular plants. Urbana: Southern Illinois University Press.

Mohlenbrock, R. H. and J. W. Voigt. 1959. A flora of southern Illinois. Urbana: Southern Illinois University Press.

Monk, C. D. 1966. An ecological significance of evergreenness. Ecology 47: 504-505.

Mooberry, F. M. 1984. Meadow response to April, June, July mowing (Pennsylvania). Restoration and Management Notes 2(1): 27-28.

Mooney, A. A. and J. A. Drake, eds. 1986. Ecology of biological invasions of North America and Hawaii. New York: Springer-Verlag.

Mooney, H. A. and M. Godron, eds. 1983. Disturbance and ecosystems. New York: Springer-Verlag.

Moore, P. D., ed. 1984. European mires. London: Academic Press.

Moore, P. D. and D. J. Bellamy. 1974. Peatlands. New York: Springer-Verlag.

Moreno-Casasola, P. and I. Espejel. 1986. Classification and ordination of coastal sand dune vegetation along the Gulf and Caribbean Sea of Mexico. Vegetatio 66: 147-182.

Morgan, J. 1994. Soil improvement: a little known technique holds potential for establishing prairie. Restoration and Management Notes 12(1): 55-56.

Morrison, M. L. 1994. Resource inventory and monitoring: concepts and applications for ecological restoration. Restoration and Management Notes 12: 179-183.

Morrison, M. L. 1995. Wildlife conservation and restoration ecology: toward a new synthesis. Restoration and Management Notes 13(2): 203-208.

Morrison, M. L., B. G. Marcot, and R. W. Mannan. 1992. Wildlife-habitat relationships. Madison: University of Wisconsin Press.

Morrison, M. L., T. Tennant, and T. A. Scott. 1994. Wildlife-habitat restoration in an urban park in southern California. Restoration Ecology 2: 17-30.

Montague, C. L. and R. G. Wiegert. 1990. Salt marshes. In Ecosystems of Florida, eds., R. L. Myers and J. J. Ewel. Orlando: University of Central Florida Press.

Moy, L. D. 1989. Are *Spartina* marshes renewable resources? A faunal comparison of a man-made marsh and two adjacent natural marshes. M.S. thesis, North Carolina State University, Raleigh.

Muenscher, W. C. 1964. Aquatic plants of the United States. Ithaca: Cornell University Press.

Mulyani, Y. A. and P. J. Dubowy. 1993. Avian use of wetlands in reclaimed minelands in southwestern Indiana. Restoration Ecology 1(3): 142-155.

Munz, P. A. 1974. Flora of southern California. Berkeley: University of California Press.

Murn, T. J. 1993. Wet meadow restoration: on soft ground. Wildflower 6: 31-32.

Myers, R. L. 1990. Scrub and high pine. In Ecosystems of Florida, eds., R. L. Myers and J. J. Ewel. Orlando: University of Central Florida Press.

Myers, R. L. and J. J. Ewel, eds. 1990. Ecosystems of Florida. Orlando: University of Central Florida Press.

Myers, R. S., G. P Shaffer, and D. W. Llewellyn. 1995. Baldcypress (*Taxodium distichum*) restoration in southeast Louisiana: the relative effects of herbivory, flooding competition, and macronutrients. Wetlands 15(2): 141-148.

Nabhan, G. P. 1995. The dangers of reductionism in biodiversity conservation. Conservation Biology 9(3): 479-481.

Nagel, T. 1996. Prescribed burns of savannas require different strategies and precautions (Missouri). Restoration and Management Notes 14(1): 72.

Naiman, R. J. and H. Déchamps, eds. 1990. The ecology and management of aquatic terrestrial ecotones. Man and the Biosphere Series. Canforth: The Parthenon Publishing Group.

Naiman, R. J., H. Déchamps, and F. Fournier, eds. 1989. The role of land/inland water ecotones in landscape management and restoration. Paris, France: Man and the Biosphere Digest 4, UNESCO.

Naiman, R. J., M. M. Holland, H. Déchamps, and P. G. Risser. 1988. A new UNESCO programme: research and management of land/inland water ecotones. Biology International, Special Issue 17: 107-136.

National Golf Foundation. 1997a. Golf facilities in the United States, 1997 ed. Jupiter: National Golf Foundation.

National Golf Foundation. 1997b. Golf participation in the United States, 1997 ed. Jupiter: National Golf Foundation.

National Ocean Service. 1986a. Tide tables 1987 high and low water predictions East Coast of North and South America. Washington, D.C.: U.S. Department of Commerce.

National Ocean Service. 1986b. Tide tables 1987 high and low water predictions West Coast of North and South America including the Hawaiian Islands. Washington, D.C.: U.S. Department of Commerce.

National Research Council. 1990. Managing coastal erosion. Washington, D.C.: National Academy Press.

National Research Council, Committee on Restoration of Aquatic Ecosystems. 1992. Restoration of aquatic ecosystems: Science, technology and public policy. Washington, D.C.: National Academy Press.

National Wetlands Policy Forum. 1998. Protecting America's wetlands: an action agenda. Washington, D.C.: Conservation Foundation.

National Wildflower Research Center. 1989. Wildflower handbook. Austin: Texas Monthly Press.

National Wildflower Research Center. 1992. Wildflower handbook. Stillwater: Voyageur Press.

Natural Resources Conservation Service (NRCS). 1996. Case summary report: examining a 1930s restoration of the Winooski River watershed, Vermont. Burlington: NRCS Watershed Science Institute.

Natural Resource Management Staff of the McHenry County (Illinois) Conservation District. 1996. Building and burning brush piles. Restoration and Management Notes 14(1): 22-25.

Naveh, Z. and A. S. Lieberman. 1990. Landscape ecology: theory and application. New York: Springer-Verlag

Neill, W. M. 1993. The tamerisk invasion of desert riparian areas. CalEPPC News 1(1): 6-7.

Neilson, R. P. and L. H. Wullstein. 1986. Microhabitat affinities of Gambel oak seedlings. Great Basin Naturalist 46: 294-298.

Nellis, D. W. 1994. Seashore plants of south Florida and the Caribbean: a guide to identification and propagation of xeriscape plants. Sarasota: Pineapple Press, Inc.

Nelson, J. B. 1986. The natural communities of South Carolina. No place. South Carolina Wildlife and Marine Resources Department.

Nelson, S. M. and D. C. Andersen. 1994. An assessment of riparian environmental quality by using butterflies and disturbance susceptibility scores. The Southwestern Naturalist 39(2): 137-142.

Neuenschwander, L., T. H. Thorsted, Jr., and R. Vogl. 1979. The salt marsh and transitional vegetation of Bahia de San Quentin. Bulletin of the Southern California Academy of Science 78: 163-182.

Neumann, V. 1973. Succession of soil fauna in afforested spoil banks of the brown-coal district of Columbia. In Ecology and Reclamation of devastated land, eds., R. J. Hutnik and G. Davis. New York: Gordon and Beard

Newcomb, L. 1988. Newcomb's wildflower guide.

Newling, C. J. and M. C. Landin. 1985. Long-term monitoring of habitat development at upland and wetland dredged material disposal sites, 1974-1982. U.S. Army Engineer Waterways Experiment Station Technical Report D-85-5.

Newman, E. I. 1993. Applied ecology. Cambridge: Blackwell Science, Inc.

Newman, E. I., ed. 1982. The plant community as a working mechanism. Oxford: Blackwell Scientific Publications.

Newsholme, C. 1992. Willows, the genus *Salix*. Portland: Timber Press.

Newsom-Brighton, M. 1987. Plant a garden for beauty and the butterflies. Plants and Gardens, Brooklyn Botanic Garden Record 43(3):44.

Newson, M. 1994. Hydrology and the river environment. Oxford: Oxford Press

Newton, G. A. 1989. A summary of three dune revegetation/stabilization projects. In Restoration '89: the new management challenge, eds., H. G. Hughes and T. M. Bonnicksen. Oakland: First Annual Meeting of the Society for Ecological Restoration.

Nickens, E. 1994. Muddy time for the bog turtle. Nature Conservancy 44(6): 8-9.

Nilsson, C., G. Grelsson, M. Johansson, and U. Sperens. 1989. Patterns of plant richness along riverbanks. Ecology 70: 77-84.

Nixon, S. W. 1982. The ecology of New England salt marshes: a community profile. Washington, D.C.: U.S. Fish and Wildlife Service, FWS/OBA-81/55.

Nixon, S. W., and C. A. Oviatt. 1973. Ecology of a New England salt marsh. Ecological Monographs 43: 463-498.

Nokes, J. 1986. How to grow native plants of Texas and the Southwest. Austin: Texas Monthly Press.

Nordby, C. S. and J. B. Zedler. 1991. Responses of fishes and benthos to hydrologic disturbances in Tijuana Estuary and Los Penasquitos Lagoon, California. Estuaries 14: 80-93.

Northcutt, G. 1994. Biotechnical erosion control: recent applications. Landscape Architecture 84(7): 30-33.

Norton, B., ed. 1986. The preservation of species. Princeton: Princeton University Press.

Noss, R. F. 1983. A regional landscape approach to maintain diversity. Bioscience 33(11): 700-706.

Noss, R. F. and A. Y. Cooperrider, and Defenders of Wildlife. 1994. Saving nature's legacy: protecting and restoring biodiversity.Washington, D.C.: Island Press.

Noss, R. F. and L. D. Harris. 1986. Nodes, networks, and MUMs: preserving diversity at all scales. Environmental Management 10: 399-409.

Noss, R. F., E. T. La Roe, and J. M. Scott. 1995. Endangered ecosystems of the United States: a preliminary assessment of loss and degradation. Washington, D.C.: U. S. Fish and Wildlife Service.

Noy-Meir, I. 1973. Desert ecosystems: environment and producers. Annual Review of Ecology and Systematics 4: 25-51.

Odum, E. P. 1971. Fundamentals of ecology. 3rd ed. Philadelphia: W. B. Saunders Publishing Company.

Odum, W. E. and C. C. McIvor. 1990. Mangroves. In Ecosystems of Florida, eds., R. L. Myers and J. J. Ewel. Orlando: University of Central Florida Press.

Odum, W. E., C. C. McIvor, and T. J. Smith, III. 1982. The ecology of the mangroves of south Florida: a community profile. Washington, D.C.: U.S. Fish and Wildlife Service, FWS/OBS-81/24.

Odum, W. E., T. J. Smith, III, J. K. Hoover, and C. C. McIvor. 1984. The ecology of tidal freshwater marshes of the United States east coast: a community profile. Washington, D.C.: U.S. Fish and Wildlife Service, FWS/OBS-83/17.

Odum, W. T. 1987. Predicting ecosystem development following creation and restoration of wetlands. In Increasing our wetland resources, eds., J. Zelanzny and J. S. Feierabend. Washington, D.C.: National Wildlife Federation Corporate Conservation Council Proceedings, Oct. 4-7, 1987.

Office of Technology Assessments. 1984. Wetlands: their use and regulation. Washington, D.C.: U.S. Government Printing Office.

Ohmart, R. D. and B. W. Anderson. 1982. North American desert riparian ecosystems. In Reference book on the deserts of North America, ed., G. Bender. Westport: Greenwood Press.

Ohmart, R. D., B. W. Anderson, and W. C. Hunter. 1988. The ecology of the lower Colorado River from Davis Dam to the Mexico-United States international boundary: a community profile. Washington, D.C.: U.S. Fish and Wildlife Service Biological Report 85(7.19).

O'Keefe, M. A. 1995a. Frequent mowing may increase quality of prairie restorations (Iowa). Restoration and Management Notes 13(1): 109-110.

O'Keefe, M. A. 1995b. Fitting in fire: a statistical approach to scheduling prescribed burns. Restoration and Management Notes 13(2): 198-202.

O'Keefe, M. A. 1996. Rainfall and prairie establishment. Restoration and Management Notes 14(1): 26-29.

Oldenwald, N. and J. Truner. 1987. Identification, selection, and use of southern plants for landscape design. Baton Rouge: Claiter's Publishing Division.

Olmsted, I. C. and L. L. Loope. 1984. Plant communities of Everglades National Park. In Environments of south Florida: present and past II, ed., P. J. Gleason. Coral Gables: Miami Geological Society.

Olmsted, I. C., L. L. Loope, and R. E. Rintz. 1980. A survey and baseline analysis of aspects of the vegetation of Taylor Slough, Everglades National Park. Homestead: South Florida Research Center, National Park Service, U.S. Department of the Interior, Report T-586.

Olmsted, I. C., L. L. Loope, and R. P. Russell. 1981. Vegetation of the southern coastal region of the Everglades National Park between Flamingo and Joe Bay. Homestead: Everglades National Park, South Florida Research Center, National Park Service, U.S. Department of the Interior, Report T-620.

Omernik, J. M. 1986. Ecoregions of the conterminous United States. Corvallis: Corvallis Environmental Research Laboratory, U. S. Environmental Protection Agency.

Omernik, J. M. 1987. Ecoregions of the conterminous United States. Annals of the Association of American Geographers 77: 118-125.

O'Neill, R. V., D. L. DeAngelis, J. B. Waide, and T. F. H. Allen. 1986. Hierarchical concept of ecosystems. Princeton: Princeton University Press.

Onuf, C. 1987. The ecology of Mugu Lagoon, California: an estuarine profile. Washington, D.C.: U.S. Fish and Wildlife Service Biological Report 85(7.15).

Opdam, P., G. Rijsdijk, and F. Hustings. 1985. Bird communities in small woods in an agricultural landscape: effects of area and isolation. Biological Conservation 34: 333-352.

Oosting, H. J. [1948] 1956. The study of plant communities. 2nd ed. San Francisco: W. H. Freeman and Company.

Ortega-Rubio, A., F. Salinas, A. Naranjo, C. Arguelles, J. L. Leon, A. Nieto, R. Aguilar, and H. Romero. 1995. Survival of transplanted xerophytic plants assessed (Baja California Sur, Mexico). Restoration and Management Notes 13(2): 223-225.

Osbourne, L. L. and D. A. Kovacic. 1993. Riparian vegetated buffer strips in water-quality restoration and stream management. Freshwater Biology 29(2): 243-258.

Outcalt, K. W. 1995. Re-establishing wiregrass by direct seeding. Poster presentation, Seattle: Society for Ecological Restoration Conference.

Pacific Estuarine Research Laboratory (PERL). 1990. A manual for assessing restored and natural coastal wetlands with examples from southern California. La Jolla: California Sea Grant Report T-CSGCP-021.

Packard, S. and C. F. Mutel, eds. 1997. The tallgrass restoration handbook, for prairies, savannas, and woodlands. Washington, D. C.: Island Press.

Packham, J. 1995. Ecology of dunes, salt marsh, and shingle. New York: Chapman and Hall.

Palaniappan, V. M., R. H. Marrs, and A. D. Bradshaw. 1979. The effect of *Lupinus arboreus* on the nitrogen status of china clay wastes. Journal of Applied Ecology 16: 825-831.

Palmisano, A. W. 1967. Ecology of *Scirpus olneyi* and *Scirpus robustus* in Louisiana coastal marshes. M.S. thesis, Louisiana State University, Baton Rouge.

Palmisano, A. W. 1970. Plant communities-soil relationships in Louisiana coastal marshes. Ph.D. diss., Louisiana State University, Baton Rouge.

Palmisano, A. W. 1971. The effects of salinity on the germination and growth of plants important to wildlife in the Gulf Coast marshes. Proceedings of the Annual Conference of the Southeastern Association of Game and Fish Commissions 25: 215-223.

Palmisano, A. W. and R. H. Chabreck. 1972. The relationship of plant communities and soils of the Louisiana coastal marshes. Proceedings of the Louisiana Association of Agronomists 13: 72-101.

Parker, K. F. 1972. An illustrated guide to Arizona weeds. Tucson: The University of Arizona Press.

Parker, P. E. and R. H. Kadlec. 1974. A dynamic ecosystem simulator. Salt Lake City: American Institute of Chemical Engineers 78th National Paper No. 8c.

Paton, P. W. C. 1994. The effect of edge on avian nest success: how strong is the evidence? Conservation Biology 8: 17-26.

Patrick, R. 1968. Natural and abnormal communities of aquatic life in streams. Via 1, Ecology in design (University of Pennsylvania, Graduate School of Fine Arts,): 36-41.

Patrick, R. 1994. Rivers of the United States, Vol. 1, estuaries. New York: John Wiley & Sons.

Patrick, R. 1995. Rivers of the United States, Vol. 2, chemical and physical characteristics. New York: John Wiley & Sons.

Patrick, R. 1996a. Rivers of the United States, Vol. 3, rivers of the east and southeast. New York: John Wiley & Sons.

Patrick, R. 1996b. Rivers of the United States, Vol. 4, Mississippi drainage. New York: John Wiley & Sons.

Patrick, R. 1996c. Rivers of the United States, Vol. 5, rivers of the west and southwest. New York: John Wiley & Sons.

Patrick, R. 1996d. Rivers of the United States, Vol. 6, pollution and environmental management. New York: John Wiley & Sons.

Patrick, W. H., Jr. 1981. Bottomland soils. In Wetlands of bottomland hardwood forests, eds., J. R. Clark and J. Benforado. Amsterdam: Elsevier Scientific Publications.

Patterson, J. and G. Stevenson. 1977. Native trees of the Bahamas. Hope Town: Jack Patterson.

Patton, D. R. 1992. Wildlife habitat relationships in forested ecosystems. Portland: Timber Press.

Pauly, W. R. 1984. No-till prairie establishment technique tested (Wisconsin). Restoration and Management Notes 2(1): 23.

Pauly, W. R. 1988. How to manage small prairie fires. Madison: Dane County Park Commission.

Payne, N. F. 1992. Techniques for wildlife habitat management of wetlands. New York: McGraw-Hill.

Payne, N. F. and F. C. Bryant. 1994. Techniques for wildlife management of uplands. New York: McGraw-Hill.

Pearson, D. 1994. Return of native maritime forest at Fort George Island. Resource Management Notes 6(4): 12.

Peattie, D. C. 1981. A natural history of western trees. Boston: Houghton Mifflin Company.

Peet, R. K. 1988. Forests of the Rocky Mountains. In North American terrestrial vegetation, eds., M. G. Barbour and W. D. Billings. Cambridge: Cambridge University Press.

Penfound, W. T. 1952. Southern swamps and marshes. Botanical Review 18: 413-446.

Penfound, W. T. and E. S. Hathaway. 1938. Plant communities in the marshlands of southeastern Louisiana. Ecological Monographs 8: 1-56.

Perkins, E. J. 1974. The biology of estuaries and coastal waters. New York: Academic Press.

Perrier, G. K., W. A. Williams, and S. R. Radosevich. 1981. Managing range and pasture to suppress tarweed. California Agriculture 35: 18-19.

Perry, D. A. 1994. Forest ecosystems. Baltimore: Johns Hopkins University Press.

Perry, D. A. and M. P. Amaranthus. 1990. The plant-soil bootstrap: microorganisms and reclamation of degraded ecosystems. In Environmental restoration, ed., J. J. Berger. Washington, D.C.: Island Press.

Peterjohn, W. T. and D. L. Correll. 1984. Nutrient dynamics in an agricultural watershed: observations on the role of a riparian forest. Ecology 65: 1466-1475.

Petersen, L. A. and H. M. Phipps. 1976. Water soaking pretreatment improves rooting and early survival of hardwood cuttings of some *Populus* clones. Tree Planters' Notes 27(1): 12-13.

Peterson, C. H. and N. M. Peterson. 1979. The ecology of intertidal flats of North Carolina: a community profile. Washington, D.C.: U.S. Fish and Wildlife Service, Office of Biological Services, FWS/OBS-79/39.

Petranka, J. W., M. E. Eldrige, and K. E. Haley. 1993. Effects of timber harvesting on southern Appalachian salamanders. Conservation Biology 7: 363-370.

Petrides, G. A. 1986. A field guide to trees and shrubs, northeastern and north-central United States and southeastern and south-central Canada. Peterson Field Guide. Boston: Houghton Mifflin Company.

Petrides, G. A. 1988. A field guide to eastern trees, eastern United States and Canada. Peterson Field Guide. Boston: Houghton Mifflin Company.

Petrides, G. A. 1992. A field guide to western trees. New York: Houghton Mifflin Company.

Pfister, R. D., B. L. Kovalchik, S. F. Arno, and R.C. Presby. 1977. Forest habitat types of Montana. Ogden: United States Department of Agriculture Forest Service.

Phillips, H. 1985. Mountain cranberry bog created (North Carolina). Restoration and Management Notes 3(1): 38-39.

Phillips, H. R. 1985. Growing and propagating wildflowers. Chapel Hill: The University of North Carolina Press.

Phleger, F. B. 1977. Soils of marine marshes. In Ecosystems of the World. I, wet coastal ecosystems, ed., V. J. Chapman. New York: Elsevier Scientific Publications.

Pickett, S. T. A. and V. T. Parker. 1994. Avoiding the old pitfalls: opportunities in a new discipline. Restoration Ecology 2: 75-79.

Pickett, S. T. A., V. T. Parker, and P. Fielder. 1992. The new paradigm in ecology: implications for conservation biology above the species level. In Conservation biology: the theory and practice of nature conservation, preservation and management, eds., P. Fielder and S. Jain. New York: Chapman and Hall.

Pickett, S. T. A. and P. S. White, eds. 1985. The ecology of natural disturbance and patch dynamics. New York: Academic Press.

Pielou, E. C. 1975. Ecological diversity. New York: John Wiley & Sons.

Pieterse, A. H. and K. J. Murphy, ed. 1990. Aquatic weeds, the ecology and management of nuisance aquatic vegetation. Oxford: Oxford University Press.

Pinkney, F. C. 1992. Revegetation and enhancement of riparian communities along the lower Colorado River. Unpublished report. Denver: Ecological Resources Branch, Resources Division, Denver Office, U.S. Bureau of Reclamation.

Pitschel, B. M. 1984. Grassland restoration in California surveyed: role for gardens, arboreta identified. Restoration and Management Notes 2(1): 26.

Platt, D. R. 1985. A prairie reconstruction at Kauffman Museum in south central Kansas. Restoration and Management Notes 3(1): 55-56.

Platt, W. J. 1975. Vertebrate fauna of the Cayler Prairie Reserve, Dickinson County, Iowa. Proceedings of the Iowa Academy of Science 82: 106-108.

Platt, W. J. and M. W. Schwartz. 1990. Temperate hardwood forests. In Ecosystems of Florida, eds., R. L. Myers and J. J. Ewel. Orlando: University of Central Florida Press.

Plochner, A. E. 1994. Population dynamics in response to fire in *Quercus laevis-Pinus palustris* barrens and related communities in southeast Virginia. Ph.D. diss., Old Dominion University, Norfolk.

Pojer, J. and A. MacKinnon, eds. 1994. Plants of the Pacific northwest coast, Washington, Oregon, British Columbia, and Alaska. Vancouver: British Columbia Ministry of Forests and Lone Pine Publishing.

Pomeroy, L. R. and J. J. Alberts, eds. 1988. Concepts of ecosystem ecology. New York: Springer-Verlag.

Pomeroy, L. R. and W. G. Wiegert, eds. 1981. The ecology of a salt marsh. New York: Springer-Verlag.

Pool, J. R. 1913. A study of the vegetation of the Sandhills of Nebraska. Ph.D. diss., University of Nebraska, Lincoln.

Por, F. D. and I. Dor, eds. 1984. Hydrobiology of the mangal. The Hague: Dr. W. Junk Publishers.

Porter, D. R. and D. A. Salvesen, eds. 1994. Collaborative planning for wetlands and wildlife, issues and examples. Washington, D.C.: Island Press.

Powell, C. L. and D. J. Bagyaraj, eds. 1984. Vesicular-arbuscular mycorrhiza. Boca Raton: CRC Press.

Preston, R. J., Jr. 1948. North American trees. Ames: Iowa University Press.

Primack, A., D. R. Modal, and M. D. Hilton. 1995. Restoration for clean water. Restoration and Management Notes 13(2): 176-178.

Prince, H. H. and F. M. D'Itri, eds. 1985. Coastal wetlands. Chelsea: Lewis Publishers.

Pritchett, D. A. and W. R. Ferren, Jr. 1990. Enhancement, restoration, and creation of vernal pools at Santa Barbara, California. Chicago: In the Program of the 2nd Annual Conference of the Society for Ecological Restoration, April 29-May 3, 1990.

Provost, M. W. 1974. Mean high water mark and use of tidelands in Florida. Florida Science 36: 50-66.

Pruka, B. 1995. Lists indicate recoverable oak savannas and open oak woodlands in southern Wisconsin. Restoration and Management Notes 13(1): 124-126.

Pulver, T. R. 1976. Transplant techniques for sapling mangrove trees, *Rhizophora mangle, Avicennia germinans,* and *Languncularia racemosa.* Florida Marine Research Publication 22.

Purer, E. A. 1939. Ecological study of vernal pools, San Diego County, California. Ecology 20: 217-229.

Purves, G. 1995. A wood for South Uist. Reforesting Scotland 12: 28-29.

Putnam, C. 1988. The Development and application of habitat standards for maintaining vertebrate species diversity on a national forest. Natural Areas Journal 8(4): 256-266.

Putnam, R. J. 1994. Community ecology. New York: Chapman and Hall.

Pyle, W. H. 1995. Riparian habitat restoration at Hart Mountain National Antelope Refuge. Restoration and Management Notes 13(1): 40-44.

Pyne, S. J. 1982. Fire in America: a cultural history of wildland and rural fire. Princeton: Princeton University Press.

Pyne, S. J. 1984. Introduction to wildland fire, fire management in the United States. New York: John Wiley & Sons.

Quammen, M. L. 1986. Measuring the success of wetland mitigation. National Wetlands Newsletter 8: 6-8.

Quigley, T. M. 1981. Estimating contribution of overstory vegetation to stream surface shade. Wildlife Society Bulletin 9(1): 22-27.

Quinn, J. F. and A. Hastings. 1987. Extinction in subdivided habitats. Conservation Biology 1: 198-208.

Rabinowitz, D. 1978a. Dispersal properties of mangrove propagules. Biotropica 10: 47-57.

Rabinowitz, D. 1978b. Early growth of mangrove seedlings in Panama, and a hypothesis concerning the relationship of dispersal and zonation. Journal of Biogeography 5: 113-133.

Radford, A. E., H. E. Ahles, and C. R. Bell. 1968. Manual of the vascular flora of the Carolinas. Chapel Hill: The University of North Carolina Press.

Rafaelli, D. and S. Hawkins. 1996. Intertidal ecology. New York: Chapman and Hall.

Randell, J. M. and J. Marinelli, eds. 1996. Invasive plants: weeds of the global garden. Brooklyn: Brooklyn Botanic Garden, Inc.

Ranwell, D. S. 1975. Ecology of salt marshes and sand dunes. London: Chapman and Hall.

Raven, P. H., H. J. Thompson, and B. A. Prigge. 1986. Flora of the Santa Monica Mountains, California. 2nd ed. Southern California Botanists.

Read, D. J., D. H. Lewis, A. Fitter, and I. J. Alexander, eds. 1992. Mycorrhizas in ecosystems. Wallingford: CAB International.

Reaka-Kudla, M. L., D. E. Wilson, and E. O. Wilson, eds. 1996. Biodiversity II, understanding and protecting our natural resources. Washington, D.C.: National Academy Press.

Reber, R. T. and P. E. Pope. 1995. Nitrogen and phosphorus fertilization affect mycorrhizal sybiosis and growth parameters of northern red oak seedlings. Restoration and Management Notes 13(1): 115.

Redfield, A. C. 1967. The ontogeny of a salt marsh. In Estuaries, ed., G. H. Lauff. Washington, D.C.: American Association for the Advancement of Science Publication 83.

Redfield, A. C. 1972. Development of a New England salt marsh. Ecological Monographs 42: 201-237.

Redington, C. B. 1994. Redington field guides to biological interactions: plants in wetlands. Kendall/ Hunt Publishers.

Reimhold, R. J. 1977. Mangals and salt marshes of the eastern United States. In Ecosystems of the World. I. Wet coastal ecosystems, ed., V. J. Chapman. New York: Elsevier Scientific Publications.

Reiner, R. and F. T. Griggs. 1989. TNC undertakes riparian forest restoration projects in California. Restoration and Management Notes 7(1): 3-8.

Reschke, C. 1990. Ecological communities of New York State. Latham: New York Natural Heritage Program.

Rhodes, A. F. and W. M. Klein, Jr. 1993. The vascular flora of Pennsylvania: annotated checklist and atlas. Philadelphia: American Philosophical Society.

Rice, E. 1972. Allelopathic effects of *Andropogon virginicus* and its persistence in old fields. American Journal of Botany 59(7): 752-755.

Rice, E. 1974. Allelopathy. New York: Academic Press.

Rich, A. C., D. S. Dobkin, and L. J. Niles. 1994. Defining forest fragmentation by corridor width: the influence of narrow forest-dividing corridors on forest-nesting birds in southern New Jersey. Conservation Biology 8: 1109-1121.

Richards, M. S. and R. Q. Landers. 1973. Responses of species in Kalsow Prairie, Iowa, to an April fire. Proceedings of the Iowa Academy of Science 80: 159-161.

Richardson, A. 1995. Plants of the Rio Grande delta. Austin: University of Texas.

Rieger, J. P. 1990. A quantitative basis for evaluating riparian habitat creation. Chicago: In the Program of the 2nd Annual Conference of the Society for Ecological Retsoration, April 29-May 3, 1990.

Riggs, G. B. 1925. Some sphagnum bogs of the North Pacific coast of America. Ecology 6: 260-278.

Ripp, M. 1985. Native vegetation reintroduction efforts using 2 and 3 year old rootstock. Restoration and Manangement Notes 3(1): 33.

Riskind, D. H. and D. O. Diamond. 1988. An introduction to environments and vegetation. In Edwards Plateau vegetation: plant ecological studies in Central Texas, eds., B. B. Amos and F. R. Gehlbach. Waco: Baylor University Press.

Risser, P. G. 1987. Landscape ecology: state of the art. In Landscape heterogeneity and disturbance, ed., M. G. Turner. New York: Springer-Verlag.

Risser, P. G., E. C. Birney, H. D. Blocker, S. W. May, W. J. Parton, and J. A. Wiens. 1981. The true prairie ecosystem. Stroudsburg: Hutchinson Ross Publishing Company.

Roberts, R. D. and A. D. Bradshaw. 1985. The development of a hydraulic seeding technique for unstable sand slopes, II: field evaluation. Journal of Applied Ecology 22: 979-994.

Robertson, D. J., ed. 1983. Reclamation and the phosphate industry: Proceedings of the Symposium. Bartow: Florida Institute of Phosphate Research.

Robertson, D. J. and M. C. Robertson. 1995. Eastern mixed mesophytic forest restoration. Restoration and Management Notes 13(1): 64-70.

Robinette, G. O. ed. [1977] 1983. Landscape planning for energy conservation. New York: Van Nostrand Reinhold Company.

Robinson, G. R. and S. N. Handel. 1993. Forest restoration on a closed landfill: rapid addition of new species by bird dispersal. Conservation Biology 7(2): 271-278.

Robinson. S. K., F. R. Thompson, III, T. M. Donovan, D. R. Whitehead, and J. Faaborg. 1995. Regional forest fragmentation and the nesting success of migratory birds. Science 267: 1987-1990.

Rock, H. W. 1981. Prairie propagation handbook. 6th ed. Milwaukee: Wehr Nature Center, Whitnall Park.

Rodiek, J. E. and E. G. Bolen, eds. 1991. Wildlife and habitats in managed landscapes. Washington, D.C.: Island Press.

Romme, W. H. 1982. Fire and landscape diversity in subalpine forests of Yellowstone National Park. Ecological Monographs 52: 199-221.

Romney, L. A., C. E. Walls, L. B. Gillham, and D. L. North. 1976. Progress report: Velpar weed killer for control of woody plants on industrial sites. Proceedings of the Southern Weed Science Society 29: 330-333.

Romo, J. T., P. L. Grilz, and L. Delanoy. 1994. Selective control of crested wheatgrass (*Agropyron cristatum* and *A. desertorum*) in the Northern Great Plains. Natural Areas Journal 14(4): 308-309.

Romo, J. T., P. L. Grilz, R. E. Redmann, and E. A. Driver. 1993. Standing crop biomass allocation patterns and soil-plant water relations in *Symphoricarpos occidentalis* following autumn or spring burning. American Midland Naturalist 130(2): 106-115.

Ross, J. E. 1985. NPI (Native Plants, Inc.). Restoration and Management Notes 3(1): 6-11.

Ross, L. M. and T. Vanderpoel. 1991. Mowing encourages establishment of prairie species. Restoration and Management Notes 9(1): 34-35.

Ross, W. M. and R. H. Chabreck. 1972. Factors affecting the growth and survival of natural and planted stands of *Scirpus olneyi*. Proceedings of the Southeastern Association of Game and Fish Commissions 26: 178-188.

Roth, S. A., ed. 1988. Garden pools and fountains. San Ramon: Ortho Books.

Rothermel, R. C., R. Wilson, Jr., G. A. Morris, and S. Sackett. 1986. Modeling moisture content of fine dead wildland fuels: input to the BEHAVE fire prediction system. Ogden: Intermountain Research Station Research Paper INT-359.

Rozsa, R. and R. A. Orson. 1995. Restoration of degraded salt marshes in Connecticut. Restoration and Management Notes 13(1): 123.

Rucks, J. A. 1978. Comparisons of riparian communities influenced by grazing. In Lowland river and stream habitat in Colorado: a symposium. Colorado Wildlife Society and Audubon Council.

Rundel, P. W. and A. C. Gibson. 1995. Ecological communities and processes in a Mojave desert ecosystem. New York: Cambridge University Press.

Rundel, P. W., D. J. Parsons, and D. T. Gordon. 1988. Montane and subalpine vegetation of the Sierra Nevada and Cascade Ranges. In Terrestrial vegetation of California, eds., M. G. Barbour and J. Major. Davis: California Native Plant Society.

Russell, R. J. 1967. Origins of estuaries. In Estuaries, ed., G. H. Lauff. Washington, D.C.: American Association for the Advancement of Science, Publication No. 83.

Ruttner, F. 1963. Fundamentals of limnology. 3rd ed. Canada: University of Toronto Press.

Rychert, R. C., J. Skujins, D. Sorensen, and D. Porcello. 1978. Nitrogen fixation by lichens and free-living microorganisms in deserts. In Nitrogen in desert ecosystems, eds., N. E. West and J. Skujins. Stroundburg: Dowden, Hutchinson, and Ross, Inc.

Rydberg, P. A. [1922] 1954. Flora of the Rocky Mountains and adjacent plains. 2nd ed. Reprint. New York: Hafner Publishing Company.

Rykiel, E. J. 1985. Towards a definition of ecological disturbance. Australian Journal of Ecology 10: 361-365.

Sacchi, C. F. and P. W. Price. 1992. The relative roles of abiotic and biotic factors in seedling demography of arroyo willow (*Salix lasiolepis*: Salicaeae). American Journal of Botany 79(4): 395-405.

Sacco, J. N. 1989. Infaunal community development of artificially established salt marshes in North Carolina. M.S. thesis., North Carolina State University, Raleigh.

Sacco, J. N., F. L. Booker, and E. D. Seneca. 1987. Comparison of the macrofaunal communities of a human-initiated salt marsh at two and fifteen years of age. In Increasing our wetland resources, eds., J. Zelazny and S. Feierabend. Washington, D.C.: National Wildlife Federation.

Safir, G., ed. 1985. The ecophysiology of vesicular-arbuscular mycorrhiza. Boca Raton: CRC Press.

St. Clair, L. L., B. L. Webb, J. R. Johansen, and G. T. Nebeker. 1984. Crytogramic soil crusts: enhancement of seedling establishment in disturbed and undisturbed areas. Reclamation and Revegetation Research 3: 129-136.

St. John, T. V. 1990. Mycorrhizal inoculation of container stock for restoration of self-sufficient vegetation. In Environmental restoration, ed., J. J. Berger. Washington, D.C.: Island Press.

St. John, T. 1996. Specially-modified land imprinter inoculates soil with mycorrhizal fungi (California). Restoration and Management Notes 14(1): 84-85.

Sala, O. E., W. P. Parton, L. A. Joyce, and W. K. Lauenroth. 1988. Primary production of the central grassland region of the United States. Ecology 69: 40-45.

Salmon, J., D. Henningsen, and T. McAlpin. 1982. Dune restoration and revegetation manual. Gainesville: Florida Sea Grant College Report Number 48.

Sampson, F. B. and F. L. Knopf. 1982. In search of a diversity ethic for wildlife management. Transactions of the North American Wildlife Natural Resources Conference 47: 421-431.

Santha, C. R. 1994. Coir, an abundant natural fiber resource. Land and Water: May/June 42.

Sather, J. H., ed. 1976. Proceedings of the national wetland classification and inventory workshop. Washington, D.C.: Fish and Wildlife Service, U.S. Department of the Interior.

Satterlund, D. R. and P. W. Adams. 1992. Wildland watershed management. 2nd ed. New York: John Wiley & Sons.

Sauer, L. J. and Andropogon Associates. 1998. The once and future forest, a guide to forest restoration strategies. Washington, D. C.: Island Press.

Saunders, D. A., G. W. Arnold, A. A. Burbidge, and A. J. M. Hopkins. 1987. Nature conservation: the role of remnants of native vegetation. Australia: Surrey Beatty and Sons Limited.

Saunders, D., R. Hobbs, and P. Ehrlich, eds. 1993. Reconstruction of fragmented ecosystems: global and regional perspectives. Australia: Surrey Beatty and Sons Limited.

Savage, T. 1972. Forida mangroves as shoreline stabilizers. Tallahassee: Florida Department of Natural Resources Professional Paper 19.

Sawyer, J., Jr. and T. Keeler-Wolf. 1996. A manual of California vegetation. Sacramento: California Native Plant Society Press,.

Sawyer, J. O., D. A. Thornburgh, and J. R. Griffin. 1988. Mixed evergreen forest. In Terrestrial vegetation of California, eds., M. G. Barbour and J. Major. Davis: California Native Plant Society.

Schaal, K. 1996. Mulch, fertilizer stimulate growth of white oak seedlings (Illinois). Restoration and Management Notes 14(1): 75.

Schafer, C. L. 1990. Nature Preserves. Washington, D.C.: Smithsonian Institution Press.

Schaid, T. A., D. W. Ureska, W. L. Tucker, and R. L. Linder. 1983. Effects of surface mining on the vesper sparrow in the northern Great Plains. Journal of Range Management 36: 500-503.

Schaller, F. W. and P. Sutton, eds. 1987. Reclamation of drastically disturbed lands. Madison: American Society of Agronomy.

Scheimer, F. and M. Zalewski. 1992. The importance of riparian ecotones for diversity and productivity of riverine fish communities. Netherlands Journal of Zoology 42(2-3): 323-335.

Schiechtl, H. 1980. Bioengineering for land reclamation and conservation. Winnipeg: University of Alberta Press.

Schilling, D. G. and J. F. Gaffney. 1995. Cogongrass control requires integrated approach (Florida). Restoration and Management Notes 13(2): 227.

Schmalzer, P. 1994. Fire management strategies related to scrub restoration. The Palmetto 14(2): 13.

Schmalzer, P. A. and C. R. Hinkle. 1987. Effects of fire on composition, biomass, and nutrients in oak scrub vegetation on John F. Kennedy Space Center, Florida. NASA Technical Memorandum: NASA TM-X-100305.

Schmidt, M. G. 1980. Growing California native plants. Berkeley: University of California Press, California Natural History Guides 45.

Schmitt, R. J. and C. W. Osenberg, eds. 1995. Detecting ecological impacts, concepts and applications in coastal habitats. Washington, D.C.: Academic Press.

Schneck, M. 1990. Butterflies: how to identify and attract them to your garden. Emmaus: Rodale Press.

Schneegas, E. R. 1975. National forest nongame bird management. Tucson: Symposium on Management of Forest and Range Habitats for Nongame Birds, May 6-9, 1975.

Schoenherr, A. A. 1990. Endangered plant communities of southern California: proceedings of the 15th annual symposium. Southern California Botanists.

Schoenherr, A. A. 1992. A natural history of California. Berkeley: University of California Press, California Natural History Guides 56.

Schramm, P. 1978. The "do's and dont's" of prairie restoration. In Fifth Midwest Prairie Conference Proceedings, eds., D. C. Glenn-Lewin and R. Q. Landers. Ames: Iowa State University.

Schramm, P. and R. L. Kalvin. 1978. The use of prairie in strip mine reclamation. In Fifth Midwest Prairie Conference Proceedings, eds., D. C. Glenn-Lewin and R. Q. Landers. Ames: Iowa State University.

Schreibner, J. A. and J. J. Dinsmore. 1995. Community structure of breeding birds on natural and restored wetlands in northwestern Iowa. Restoration and Management Notes 13(1): 140.

Schreiner, E. and R. Scott. 1994. Olympic revegetation efforts continue to evolve. Park Science 14(2): 7.

Schroeder, R. L. 1986. Habitat suitability index models: wildlife species richness in shelterbelts. Washington, D.C.: U.S. Fish Wildlife Service Biological Report 82(10.128).

Schroeder, W. W., S. P. Dinnel, and W. J. Wiseman, Jr. 1992. Salinity structure of a shallow, tributary estuary. In Dynamics and exchanges in estuaries and the coastal zone, ed., D. Prandle. Washington, D.C.: American Geophysical Union.

Schulz, T. T. and W. C. Leininger. 1991. Nongame wildlife communities in grazed and ungrazed riparian sites. Great Basin Naturalist 51(3): 286-292.

Schuster, M. A. and R. H. Zuck, eds. 1986. Proceedings: high altitude revegetation workshop no. 7. Information Series No. 58. Fort Collins: Colorado Water Resources Institute.

Schwartz, O. A. and P. D. Whitson. 1987. A 12-year study of vegetation and mammal succession on a reconstructed tallgrass prairie in Iowa. The American Midland Naturalist 117(2): 240-249.

Schwarzmeier, J. A. 1984. Sweet clover control in planted prairies: refined mow/burn prescription tested (Wisconsin). Restoration and Management Notes 2(1): 35-36.

Scott, J. 1992. Field and forest: a guide to native landscapes for gardeners and naturalists. New York: Walker and Company.

Scott, T., C. Fleming, and F. Campbell. 1994. Invasive alien control methods create balancing act. Bulletin of the Virginia Native Plant Society 13(3): 3.

Seastedt, T. R., J. Briggs, and D. Gibson. 1991. Controls of nitrogen limitation in tallgrass prairie. Oecologia 87: 72-79.

Seastedt, T. R., P. A. Duffy, and J. N. Knight. 1996. Reverse fertilization experiment produces mixed results (Colorado). Restoration and Management Notes 14(1): 64.

Sedenko, J. 1991. The butterfly garden: creating beautiful gardens to attract butterflies. New York: Villard Books.

Seliskar, D. M. and J. L. Gallagher. 1983. The ecology of tidal marshes of the Pacific northwest coast: a community profile. Washington, D.C.: U.S. Fish and Wildlife Service, FWS/OBS-82/32.

Seneca, E. D. 1980. Techniques for creating salt marshes along the east coast. In Rehabilitation and creation of selected coastal habitats: proceedings of a workshop, eds., J. C. Lewis and E. W. Bunce. Washington, D.C.: U.S. Fish and Wildlife Service, FWS/OBS-80/27.

Seneca, E. D. and S. W. Broome. 1992. Restoring tidal marshes in North Carolina and France. In Restoring the nation's marine environment, ed., G. W. Thayer. College Park: Maryland Sea Grant.

Seymour, F. C. 1993 (reissue of 1989 edition). Flora of New England. Hanover: University Press of New England.

Shafer, C. L. 1990. Nature preserves. Washington, D.C.: Smithsonian Institution Press.

Shaffer, M. L. and F. B. Sampson. 1985. Population size and extinction: a note on determining critical population size. American Naturalist 125: 144-152.

Sharitz, R. R. and J. W. Gibbons, eds. 1989. Freshwater wetlands and wildlife. Aiken: Savannah River Ecology Laboratory.

Shaw, S. P. and C. G. Fredine. 1954. Wetlands of the United States. Washington, D.C.: U.S. Fish and Wildlife Service, Circular 39.

Shealy, S. 1989. Environment by design: landscaping the highways. Drive: (Spring).

Shear, T., T. Lent, and K. Nunnery. 1992. Study of 50-year bottomland forest restorations concentrates on functional attributes (Tennessee, Kentucky). Restoration and Management Notes 10: 183-184.

Shear, T. H., T. J. Lent, and S. Fraver. 1996. Comparison of restored and mature bottomland hardwood forests of southwestern Kentucky. Restoration Ecology 4(2): 111-123.

Sheldon, J. C. and A. D. Bradshaw. 1977. The development of a hydraulic seeding technique for unstable sand slopes, I: effects of fertilizers, mulch, and stabilizers. Journal of Applied Ecology 14: 905-918.

Shelford, V. E. 1974. The ecology of North America. Chicago: University of Illinois Press.

Shimek, B. 1931. Relation between migrant and native flora of the prairie region. University of Iowa Studies in Natural History 14(2): 10-16.

Shirley, S. 1994. Restoring the tallgrass prairie, an illustrated manual for Iowa and the upper midwest. Iowa City: University of Iowa Press.

Shreve, F. and I. L. Wiggins. 1951. Vegetation and flora of the Sonoran Desert. Publication No. 591. Vol. 1. Washington, D.C.: Carnegie Museum.

Shreve, F., and I. L. Wiggins. 1964. Vegetation and flora of the Sonoran Desert, 2 Vol. Stanford: Stanford University Press.

Shugart, H. H. 1984. A theory of forest dynamics: the ecological implications of forest succession models. New York: Springer-Verlag.

Shulz, T. T. and W. C. Leininger. 1990. Differences in riparian vegetation structure between grazed areas and exclosures. Journal of Range Management 43: 295-299.

Shulz, T. T. and W. C. Leininger. 1991. Nongame wildlife communities in grazed and ungrazed montane riparian sites. Great Basin Naturalist 51(3): 286-292.

Sigg, J. 1994. Invasive exotics report. Bulletin of the California Native Plant Society 24(3): 6.

Silverberg, S. M. and M. S. Dennison. 1993. Wetlands and coastal zone regulations, strategies and procedures. New York: John Wiley & Sons.

Simberloff, D. 1990. Reconstituting the ambiguous—can islands be restored? In Ecological restoration of New Zealand islands, eds., D. Towns, C. H. Dougherty, and I. A. E. Atkinson. Wellington: New Zealand Department of Conservation.

Simberloff, D. S. and L. G. Abele. 1982. Refuge design and island biogeographic theory: effects of fragmentation. American Naturalist 120: 41-50.

Simberloff, D. S. and L. G. Abele. 1984. Conservation and obfuscation: subdivision of reserves. Oikos 42: 399-401.

Simberloff, D., J. A. Farr, J. Cox, and D. W. Mehlman. 1992. Movement corridors: conservation bargains or poor investments? Conservation Biology 6(4): 493-504.

Simberloff, D. and N. Gotelli. 1984. Effects of insularization on plant species richness in the prairie-forest ecotone. Biological Conservation 29: 27-46.

Simenstad, C. A. and R. M. Thom. 1995. *Spartina alterniflora* (smooth cordgrass) as an invasive halophyte in Pacific Northwest estuaries. Hortus Northwest 6(1): 9-12, 38-40.

Simpson, G. G. 1964. Species density of North American recent mammals. Systematic Zoology 13: 73-75.

Sims, P. L. 1988. Grasslands. In North American terrestrial vegetation, eds., M. G. Barbour and W. D. Billings. Cambridge: Cambridge University Press.

Skujins, J., ed. 1991. Semiarid lands and deserts: soil resources and reclaimation. New York: Marcel Dekker, Inc.

Slagle, K. 1992. Revegetation efforts accompany campsite rehabilitation in a Pacific Silver Fir plant community (Oregon). Restoration and Management Notes 10: 82-83.

Sloan, J. P. 1994. Root dippings of conifer seedlings shows little benefit in the Northern Rocky Mountains. Ogden: U.S. Forest Service, Intermountain Research Station Research Paper INT-RP-476.

Smith, C. G. and D. N. Ueckert. 1983. Prescribed fire/
herbicide systems for pricklypear control. Society
for Range Management 36th Annual Meeting.
Abstracts.

Smith, D. L. 1995. A case study of polyacrylimide use
for revegetation. Land and Water 39(Jan./Feb.): 9-
10.

Smith, D. S. 1993. An overview of greenways: their
history, ecological context, and specific functions. In
Ecology of greenways, eds., D. S. Smith and P. C.
Hellmond. Minneapolis: University of Minnesota
Press.

Smith, E. B. 1994. Keys to the flora of Arkansas.
Fayetteville: University of Arkansas Press.

Smith, G. W. 1990. The fight for a restored wetland in
Huntington Beach. Coast and Ocean 6(2): 42-45.

Smith, K. D. 1974. The utilization of gum trees by birds
in Africa. Ibis 116: 155-164.

Smith, K. D. 1990. Standards developed for white
oak-hickory forest restoration (Ohio). Restoration
and Management Notes 8: 108.

Smith, R. L. 1966. Ecology and field biology. New
York: Harper and Row, Publishers.

Smith, T. M. and H. H. Shugart. 1987. Territory size
variation in the ovenbird: the role of habitat structure.
Ecology 68(3): 695-704.

Snyder, J. M. and L. H. Wullstein. 1973. The role of
desert cryptograms in nitrogen fixation. American
Midland Naturalist 90: 257-265.

Snyder, J. R., A. Herndon, and W. B. Robertson, Jr.
1990. South Florida rockland. In Ecosystems of
Florida, eds., R. L. Myers and J. J. Ewel. Orlando:
University of Central Florida Press.

Snyder, W. D. and G. C. Miller. 1992. Changes in
riparian vegetation along the Colorado River and Rio
Grande. Great Basin Naturalist 52(4): 357-363.

Soil Conservation Service (SCS). 1984. Prescribed burning for wildlife in the coastal marshes. Alexandria:
United States Department of Agriculture.

Soil Conservation Service. 1992. Soil bioengineering
for upland slope protection and erosion reduction. In
Soil Conservation Service, Engineering Field Handbook 18: 1-50.

Sollenberger, D. S. 1984. Fen construction in restored
prairie will provide opportunity for pH studies
(Illinois). Restoration and Management Notes 2(1):
31.

Sorensen, N. 1993. Physiological ecology of the
desert shrub *Larrea divaricata*: implications for
arid land revegetation. M.S. thesis., San Diego
State University, San Diego.

Sosebee, R. E. 1983. Physiological, phenological, and
environmental considerations in brush and weed
control. Society for Range Management 36th
Annual Meeting.

Soule, M., ed. 1986. Conservation biology: science of
scarcity and diversity. Sunderland: Sinauer
Associates, Inc.

Soule, M. E., A. C. Alberts, and D. T. Bolger. 1992.
The effects of fragmentation on chaparral plants
and vertebrates. Oikos 63: 39-47.

Soule, M. E. and B. A. Wilcox, eds. 1980. Conservation biology an evolutionary-ecological perspective. Sunderland: Sinauer Associates, Inc.

Spellenberg, R. 1979. The Audubon Society field guide
to North American wildflowers: western region. New
York: Alfred A. Knopf.

Spencer, N. R. 1995. Leafy spurge research database
now available on CD-ROM. Restoration and
Management Notes 13(2): 228.

Sperka, M. 1973. Growing wildflowers. New York:
Charles Scribner's Sons.

Sperry, T. M. 1994. The Curtis Prairie Restoration,
using the single species planting method. Natural
Areas Journal 14(2): 124-127.

Spitsbergen, J. M. 1980. Seacoast life: an ecological
guide to natural seashore communities in North
Carolina. Chapel Hill: The University of North
Carolina Press.

Sprugel, D. G. 1991. Disturbance, equilibrium and
environmental variability: what is "natural"
vegetation in a changing environment? Biological
Conservation 58: 1-18.

Sprunt, A. 1975. Habitat management implications of
migration. Tucson: Symposium on Management of
Forest and Range Habitats for Nongame Birds, May
6-9, 1975.

Spurr, S. H. and B. V. Barnes. 1980. Forest ecology.
3rd ed. New York: John Wiley & Sons.

Statler, R. 1993a. Barrier island botany. Dubuque:
published by the author.

Statler, R. 1993b. Barrier island botany, the southeastern United States. Dubuque: published by the
author.

Stauffer, D. F. and L. Best. 1980. Habitat selection by
birds of riparian communities: evaluating effects of
habitat alterations. Journal of Wildlife Management
44(1): 1-15.

Stebbins, G. L., Jr. 1952. Aridity as a stimulus to plant
evolution. The American Naturalist 86: 33-44.

Steele, R., R. D. Pfister, R. A. Ryker, and J. A. Kittams.
1981. Forest habitat types of central Idaho. Ogden:
Forest Service, United States Department of Agriculture General Technical Report INT-114.

Stewart, R. E. and H. A. Katrund. 1972. Vegetation of
prairie potholes, North Dakota, in relation to
quality of water and other environmental factors.
United States Geological Survey Professional
Paper 585-C.

Steyermark, J. 1996 (reprint of 1963 edition). Flora of
Missouri. Iowa City: University of Iowa Press.

Stokes, D. and L. Stokes. 1989. The hummingbird book. Boston: Little, Brown and Company.

Stokes, D. and L. Stokes. 1991. The bluebird book. Boston: Little, Brown and Company.

Stokes, D., L. Stokes, and E. Williams. 1991. The butterfly book. Boston: Little, Brown and Company.

Stone, D. 1957. A unique balanced breeding system in the vernal pool mousetails. Evolution 13: 151-174.

Stout, J. P. 1984. The ecology of irregularly flooded salt marshes of the northeastern Gulf of Mexico: a community profile. Washington, D.C.: U.S. Fish and Wildlife Service Biological Report 85(7.1).

Straughsbaugh, P. D. and E. L. Core. 1978. Flora of West Virginia, 2nd ed. Morgantown: Seneca Books, Inc.

Stritch, L. R. 1990. Barrens restoration in the cretaceous hills of Pope and Massac Counties, Illinois. In Environmental restoration, ed., J. J. Berger. Washington, D.C.: Island Press.

Strohmeyer, D. L. and L. H. Fredrickson. 1967. An evaluation of dynamited potholes in Northwest Iowa. Journal of Wildlife Management 31: 525-532.

Stromberg, M. R. and P. Kephart. 1996. Restoring native grasses in California old fields. Restoration and Management Notes 14(2): 102-111.

Strong, Jr., D. R., D. Simberloff, L. G. Abele, and A. B. Thistle, eds. 1984. Ecological communities, conceptual issues and the evidence. Princeton: Princeton University Press.

Stubbindieck, J., S. L. Hatch, and C. H. Butterfield. 1992. North American range plants, 4th ed. Lincoln: University of Nebraska Press.

Sullivan, M. J. and F. C. Daiber. 1974. Response in production of cordgrass, Spartina alterniflora, to inorganic nitrogen and phosphorous fertilizer. Chesapeake Scientist 15(2): 121-123.

Sumner, R. and M. E. Kentula. 1994. Restoration project. Wetlands Research Update (Nov.): 3-4.

Swanson, F. J., S. V. Gregory, J. R. Sedell, and A. G. Campbell. 1982. Land-water interactions: the riparian zone. In Analysis of coniferous forest ecosystems in the western United States., ed., R. L. Edmonds. US/IBP Synthesis Series 14. Stroudsburg: Dowden, Hutchinson, and Ross Publishing Company.

Swanson, R. E. 1994. Field guide to the trees and shrubs of the southern Appalachians. Baltimore: John Hopkins.

Sweeney, B. W. 1993. Effects of streamside vegetation on macroinvertebrate communities of White Clay Creek in eastern North America. Proceedings of the National Academy of Science 144: 291-340.

Swift, K. L. 1988. Salt marsh restoration: assessing a southern California example. M.S. thesis., San Diego State University, San Diego.

Swindells, P. 1985. The water garden. London: Ward Lock Limited.

Swink, F. 1974. Plants of the Chicago region: a check list of the vascular flora of the Chicago region with notes on local distribution and ecology, 2nd ed. Lisle: The Morton Arboretum.

Szafoni, R. E. 1991. Vegetation management guideline: autumn olive, Elaeagnus umbellata Thunb. Natural Areas Journal 11(2): 121-122.

Szaro, R. C. 1989. Riparian forest and scrubland community types of Arizona and New Mexico. Desert Plants 9(3-4): 69-138.

Szaro, R. C. and C. P. Pase. 1983. Short-term changes in a cottonwood-ash-willow association on a grazed and an ungrazed portion of Little Ash Creek in central Arizona. Journal of Range Management 36: 382-384.

Tanner, G. W., J. M. Wood, and S. A. Jones. 1992. Cogongrass (Imperata cylindrica) control with glyphosate. Florida Scientist 55: 112-115.

Taylor, D. D. 1996. Management strategies in Kentucky's Daniel Boone National Forest, why burn the forest? Kentucky Native Plant Society Newsletter 11(1): 3-5.

Taylor, D. L. 1980. Fire history and man-induced fire problems in subtropical south Florida. Proceedings, fire history workshop. Rapid City: Rocky Mountain Forest and Range Experiment Station, Technical Report RM-81.

Taylor, D. L. 1981. Fire history and fire records for Everglades National Park 1948-1979. Homestead: Everglades National Park, South Florida Research Center Report T-619.

Taylor, K. L. and J. B. Grace. 1995. The effects of vertebrate herbivory on plant community structure in the coastal marshes of the Pearl River, Louisiana. Wetlands 15(1): 68-73.

Taylor, K. S. and S. F. Hamblin. 1963. Handbook of wildflower cultivation. New York: Macmillan.

Teal, J. M. 1962. Energy flow in the salt marsh ecosystem of Georgia. Ecology 43: 614-624.

Teal, J. M. 1986. The ecology of regularly flooded salt marshes of New England: a community profile. Washington, D.C.: U.S. Fish and Wildlife Report 85(7.4).

Teal, J. and J. Kanwisher. 1966. Gas transport in the marsh grass, Spartina alterniflora. Journal of Experimental Botany 17: 355-361.

Teal, J. and M. Teal. 1969. Life and death of the salt marsh. New York: Ballantine Books.

Teal, J. M., I. Valiela, and D. Berla. 1979. Nitrogen fixation by rhizosphere and free-living bacteria in salt marsh sediments. Limnology and Oceanography 24: 126-132.

Teas, H. 1977. Ecology and restoration of mangrove shorelines in Florida. Environmental Conservation 4: 51-57.

Teas, H. J. and R. J. McEwan. 1982. An epidemic dieback gall disease of *Rhizophora* mangroves in Gambia, West Africa. Plant Disease 66: 522-523.

Temple, S. A. 1986. Predicting impacts of habitat fragmentation on forest birds: a comparison of two models. In Wildlife 2000, eds., J. Verner, M.L. Morrison, and C. J. Ralph. Madison: The University of Wisconsin Press.

Temple, S. A. 1992. Conservation biologists and wildlife managers getting together. Conservation Biology 6(1): 4.

Terrene Institute. 1994. Symposium on ecological restoration. Proceedings of a conference March 2-4, 1993, Chicago, Illinois. Washington, D.C.: U. S. Environmental Protection Agency, Office of Water, EPA 841-B-84-003.

Terres, J. K. 1980. The Audubon encyclopedia of North American birds. New York: Alfred A. Knopf, Inc.

Tharp, B. C. 1926. Structure of Texas vegetation east of the 98th meridian. Austin: University of Texas Press, University of Texas Bulletin No. 2606.

Tharp, B. C. 1939. The vegetation of Texas. Houston: Anson Jones Press.

Thayer, G. W., ed. 1992. Restoring the nations marine environment. College Park: A Maryland Seagrant Book.

Thirgood, J. V. 1981. Man and the Mediterranean forest: a history of resource depletion. London: Academic Press.

Thom, R. M. 1992. Accretion rates of low intertidal salt marshes in the Pacific Northwest. Wetlands 12: 147-156.

Thomas, B. 1976. The swamp. New York: Norton.

Thomas, J. W., ed. 1979. Wildlife habitats in managed forests. Washington, D.C.: Forest Service, United States Department of Agriculture, Agriculture Handbook No. 553.

Thomas, J. W., C. Maser, and J. E. Rodiek. 1979. Wildlife habitats in managed rangelands—the Great Basin of southeastern Oregon, edges. United States Department of Agriculture, Forest Service General Technical Report PNW-85.

Thompson, J. R. 1992. Prairies, forests, and wetlands. Iowa City: University of Iowa Press.

Thornburg, A. A. 1982. Plant materials for use on surface-mined lands in arid and semiarid regions. Washington, D.C.: U.S. Department of Agriculture, Soil Conservation Service, SCS-TP-157, EPA-600/7-79-134.

Thorne, J. F. 1993. Landscape ecology: a foundation for greenway design. In Ecology of greenways, eds., D. S. Smith and P. C. Hellmond. Minneapolis: University of Minnesota Press.

Thorne, R. F. 1976. The vascular plant communities of southern California. In Plant communities of southern California, ed., J. Latting. California Native Plant Society Special Publication Number 2. Berkeley: California Native Plant Society.

Thorne, R. F. 1988. Montane and subalpine forests of the transverse and peninsular ranges. In Terrestrial vegetation of California, eds., M. J. Barbour and J. Major. Davis: California Native Plant Society.

Thorne, R. F. and E. W. Lathrop. 1969. A vernal marsh on the Santa Rosa Plateau of Riverside County, California. Aliso 7: 85-95.

Thorp, R. W. and J. M. Leong. 1995. Native bee pollinators of vernal pool plants. Fremontia 23(2): 3-7.

Thullen, J. S. and D. R. Eberts. 1995. Effects of temperature, stratification, and seed origin on the germination of *Scirpus acutus* seeds for use in constructed wetlands. Wetlands 15(3): 298-304.

Thunhorst, G. A. 1993. Wetland planting guide for the northeastern United States, plants for wetland creation, restoration, and enhancement. St. Michaels: Environmental Concern, Inc.

Tibbits, T. H. 1979. Humidity and plants. Bioscience 29: 358-363.

Tilghman, N. G. 1987. Characteristics of urban woodlands affecting breeding bird diversity and abundance. Landscape and Urban Planning 14: 481-495.

Tilman, D. 1986. Nitrogen-limited growth in plants from different successional stages. Ecology 67: 555-563.

Tilman, D. 1987. Secondary succession and the pattern of plant dominance along experimental nitrogen gradients. Ecological Monographs 57(3): 189-214.

Tilton, D. L. 1977. Seasonal growth and foliar nutrients of *Larix laricina* in three wetland ecosystems. Canadian Journal of Botany 55: 1291-1298.

Tiner, R. W., Jr. 1984. Wetlands of the United States: current status and recent trends. Washington, D.C.: U.S. Department of Interior, Fish and Wildlife Service.

Tiner, R. W., Jr. 1987. Field guide to coastal wetland plants of the northeastern United States. Amherst: The University of Massachusetts Press.

Tiner, R. W., Jr. 1993. Field guide to coastal wetland plants of the southeastern United States. Amherst: The University of Massuchusetts Press.

Tiszler, J., D. Bainbridge, M. Darby, M. Fidelibus, R. MacAller, and D. Waldecker. 1995. Technique for sand stabilization and mesquite-dune reconstruction tested in the Yuha Desert, California. Restoration and Management Notes 13(2): 222-223.

Tomback, D. F. and S. F. Arno. 1995. Fire and Clark's nutcracker aid whitebark pine regeneration. In Proceedings: Symposium on fire in wilderness and park management, tech. coordinators, J. K. Brown, R. W. Mutch, C. W. Spoon, and R. H. Wakimoto. Ogden: U.S. Forest Service General Technical Report INT-GTR-320.

Tomlinson, P. B. 1986. The botany of mangroves. London: Cambridge University Press.

Transeau, E. N. 1903. On the geographic distribution and ecological relationships of the bog plant societies of northern North America. Botanical Gazette 36: 401-420.

Trettin, C. C., ed. 1997. Northern forested wetlands: ecology and management. Boca Raton: Lewis Publishers.

Trimble, G. R. 1959. Logging roads in northeastern municipal watersheds. Journal of the American Water Works Assocociation 51: 407-410.

Trimble, G. R., Jr., and R. S. Sartz. 1957. How far from a stream should a logging road be located. Journal of Forestry 55: 339-341.

Trubble, J. T. and L. T. Kok. 1980a. Impact of 2,4-D on *Ceuthorhynchidius horridus* (Coleoptera: Curculionidae) and their compatibility for integrated control of *Carduus* thistles. Weed Research 20: 73-75.

Trubble, J. T. and L. T. Kok. 1980b. Integration of a thistle-headed weevil and herbicide for *Carduus* thistle control. Protection Ecology 2: 57-64.

Trulio, L. A. 1995a. Burrowing owls thrive in artificial habitat (California). Restoration and Management Notes 13(2): 238-239.

Trulio, L. A. 1995b. Passive relocation: a method to preserve burrowing owls on disturbed sites. Journal of Field Ornithology 66: 99-106.

Tucker, G. F., B. A. Schrader, and E. G. Horvath. 1992. Restoration ecology of coastal Oregon riparian areas—an applied approach. Abstract, presented at the 77th annual Ecological Society of America meeting.

Tuley, G. 1985. The growth of young oak trees in shelters. Forestry 58: 181-95.

Turner, F. 1987. The self-effacing art. restoration as imitation of nature. In Restoration ecology, eds., W. R. Jordan, III, M. E. Gilpin, and J. D. Aber. United Kingdom: Cambridge University Press.

Turner, M. G. 1989. Landscape ecology: the effect of pattern of process. Annual Review of Ecological Systems 20: 171-197.

Turner, M. G., ed. 1987. Landscape heterogeneity and disturbance. New York: Springer-Verlag.

Turner, M. G. and R. H. Gardner, eds. 1991. Quantitative methods in landscape ecology. New York: Springer-Verlag.

Turner, R. M., J. Bowers, and T. L. Burgess. 1995. Sonoran Desert plants, an ecological atlas.Tucson: University of Arizona Press.

Turner, R. M. and D. E. Brown. 1982. Tropical-subtropical desertlands. In Biotic communities of the American Southwest–United States and Mexico, ed., D. E. Brown. Desert Plants 4.

Tuttle, M. D. and D. A. R. Taylor. 1994. Bats and mines. Bat Conservation International, Inc. Resource Publication No. 3.

Twilley, R. R. 1985. The exchange of organic carbon in basin mangrove forests in a southwest Florida estuary. Estuarine, Coastal, and Shelf Science 20: 543-557.

Twilley, R. R., A. E. Lugo, and C. Patterson-Zucca. 1986. Litter production and turnover in basin mangrove forests in southwest Florida. Ecology 67: 670-683.

United States Department of Agriculture. 1974. Seeds of woody plants in the United States. Washington, D.C.: U.S. Government Printing Office, Agricultural Handbook No. 450.

United States Department of Agriculture. 1975. Soil taxonomy. Washington, D.C.: Government Printing Office, Agriculture Handbook 436.

United States Forest Service. 1978. A guide for prescribed fire in southern forests. Atlanta: U.S. Forest Service.

United States Forest Service. 1979. User guide to vegetation. Ogden: Intermountain Forest and Range Experimental Station, United States Department of Agriculture, Forest Service General Technical Report INT-64.

United States General Accounting Office. 1988. Public rangelands: some riparian areas restored but widespread improvement will be slow. Washington, D.C.: United States General Accounting Office, GAO/RCED-88-105.

United States Geological Survey. 1970. The national atlas of the United States of America. Washington, D.C.: Department of the Interior.

Urban, D. L., R. V. O'Neill, and H. H. Shugart. 1987. Landscape ecology. Bioscience 37: 119-127.

Uresk, D. W. and T. Yamamoto. 1994. Field study of plant survival as affected by amendments to bentonite spoil. Great Basin Naturalist 54(2): 156-161.

Valiela, I. and J. M. Teal. 1974. Nutrient limitation in salt marsh vegetation. In Ecology of halophytes, eds., R. J. Reimold and W. H. Queen. New York: Academic Press.

Valiela, I., J. M. Teal, C. Cogswell, J. Hartman, S. Allen, R. Van Etten, and D. Goehringer. 1984. Some long-term consequences of sewage contamination in salt marsh ecosystems. In Ecological considerations in wetlands treatment of municipal wastewaters, eds., P. J. Godfrey, E. R. Kaynor, S. Pelczarski, and J. Benforado. 1984. New York: Van Nostrand Reinhold.

van der Valk, A. G., ed. 1988. Northern prairie wetlands. Ames: Iowa State.

van der Valk, A. G., and C. B. Davis. 1978. The role of seed banks in the vegetation dynamics of prairie glacial marshes. Ecology 59: 322-335.

Van-Epps, G. and C. McKell. 1983. Effect of weedy annuals on the survival and growth of transplants under arid conditions. Journal of Range Management 36(3): 366-369.

Van Vleet, S. M. 1994. Establishment, increase and impact of *Aphthona* spp. on leafy spurge in Fremont County, Wyoming. Leafy Spurge News 16(3): 10.

Vasek, F. C. and M. G. Barbour. 1988. Mojave desert scrub vegetation. In Terrestrial vegetation of California, eds., M. G. Barbour and J. Major. Davis: California Native Plant Society.

Vasek, F. C. and R. F. Thorne. 1988. Transmontane coniferous vegetation. In Terrestrial vegetation of California, eds., M. G. Barbour and J. Major. Davis: California Native Plant Society.

Vatikus, M. R., T. G. Ciravolo, K. W. McLeod, E. M. Mavity, and K. L. Novak. 1993. Growth and photosynthesis of seedlings of five bottomland tree species following nutrient enrichment. American Midland Naturalist 129(1): 42-51.

Verner, J. 1975. Avian behavior and habitat management. Tucson: The Symposium on Management of Forest and Range Habitats for Nongame Birds. May 6-9, 1975.

Verner, J., M. L. Morrison, and C. J. Ralph. 1986. Wildlife 2000. Madison: The University of Wisconsin Press.

Verry, E. S. 1985. Selection of water impoundment sites in the Lake States. In Water impoundments for wildlife; a habitat management workshop, ed., M. D. Knighton. St. Paul: U.S. Forest Service, North Central Forest Experiment Station Technical Report, NC-100.

Vines, R. A. 1960. Trees, shrubs, and woody vines of the southwest. Austin: University of Texas Press.

Virginia, R. A. 1990. Desert restoration: the role of woody legumes. In Environmental restoration, ed., J. J. Berger. Washington, D.C.: Island Press.

Vivrette, N. J. and C. H. Muller. 1977. Mechanism of invasion and dominance of coastal grassland by *Mesembryanthemum crystallinum* L. Ecological Monographs 47: 302-318.

Vogl, R. J., W. P. Armstrong, K. L. White, and K. L. Cole. 1988. The closed-cone pines and cypress. In Terrestrial vegetation of California, eds., M.G. Barbour and J. Major. Davis: California Native Plant Society.

Vogl, R. M. and L. T. McHargue. 1966. Vegetation of California fan palm oases on the San Andreas fault. Ecology 47: 532-540.

Vories, K. C. 1981. Growing Colorado plants from seed: a state of the art. Volume 1: Shrubs. Springfield: National Technical Information Service, United States Department of Agriculture, Forest Service, General Technical Report INT-103.

Voss, E. G. 1972. Michigan flora, part 1: gymnosperms and monocots. Bloomfield Hills: Cranbrook Institute of Science.

Voss, E. G. 1985. Michigan flora, part 2: dicots, Saururaceae through Cornaceae. Bloomfield Hills: Cranbrook Institute of Science.

Waaland, M. 1990. Recovery of endangered plants in vernal pools of Santa Rosa, California. In the Program of the 2nd Annual Conference of the Society for Ecological Restoration, Chicago, April 29-May 3, 1990.

Waggoner, G. S. 1975. Eastern Deciduous Forest, 2 vols. Washington, D.C.: U.S. Government Printing Office.

Wagner, R. H. 1975. The American prairie inventory: a preliminary report. Proceedings of the Fourth Midwest Prairie Conference, ed., M. K. Wali. Grand Forks: University of North Dakota Press.

Waisel, Y. 1972. Biology of halophytes. New York: Academic Press.

Wali, M. K., ed. 1992. Ecosystem rehabilitation. The Hague: SPB Academic Publishing.

Walker, B. H. and C. F. Wehrhahn. 1971. Relationships between derived vegetation gradients and measured environmental variables in Saskatchewan wetlands. Ecology 52: 85-95.

Waller, D. M. 1982. Effects of cutting and herbicide treatment on smooth sumac. Restoration and Management Notes 1: 132.

Walter, H. 1973. Vegetation of the Earth. New York: Springer-Verlag.

Walters, C. 1986. Adaptive management of renewable resources. New York: Macmillan Company.

Walters, T. W., D. S. Decker-Walters, and D. R. Gordon. 1994. Restoration considerations for wiregrass (*Aristida stricta*): allozymic diversity of populations. Conservation Biology 8(2): 581-585.

Want, W. L. 1990. Law of wetlands regulation. New York: Clark-Boardman Company, Ltd.

Ward, G. B. and O. M. Ward. No date. Wild flowers of the southwest deserts in natural color. Palm Desert: Best-West Publications.

Warner, R. E. and K. Hendrix, eds. 1984. California riparian systems: ecology, conservation, and productive management. Berkeley: University of California Press.

Warren, P. L. and L. S. Anderson. 1987. Vegetation recovery following livestock removal near Quitobaquito Spring, Organ Pipe Cactus National Monument. Tucson: Cooperative National Park Resources Studies Unit, University of Arizona, Technical Report No. 10.

Wasowski, S. with A. Wasowski. 1991. Native Texas plants: landscaping region by region. Houston: Gulf Publishing Company.

Weathers, L. A. and M. Hunter. 1990. Flowers wild and wonderful. Southern Living: April, 90-95.

Weaver, H. 1943. Fire as an ecological and silvicultural factor in the ponderosa pine region of the Pacific slope. Journal of Forestry 41: 7-14.

Weaver, J. E. 1954. North American prairie. Lincoln: Johnsen Publishing Company.

Weaver, J. E. 1965. Native vegetation of Nebraska. Lincoln: University of Nebraska Press.

Weaver. J. E. 1968. Prairie plants and their environment. Lincoln: University of Nebraska Press.

Weaver, J. E. 1991 (reissue to 1968 edition). Prairie plants and their environment, a fifty-year study in the midwest. Lincoln: University of Nebraska Press.

Weaver, J. E. and F. W. Albertson. 1956. Grasslands of the Great Plains. Lincoln: Johnsen Publishing Company.

Weber, W. A. 1976. Rocky Mountain flora. Boulder: Colorado Associated University Press.

Weber, W. A. 1987. Colorado flora: western slope. Boulder: Colorado Associated University Press.

Weber, W. A. 1990. Colorado flora: eastern slope. Boulder: University Press of Colorado.

Weddell, B. J. 1991. Distribution and movements of Columbian ground squirrels (*Spermophilus columbianus* (Ord)): are habitat patches like islands? Journal of Biogeography 18: 285-294.

Weddle, A. E., ed. 1983. Landscape techniques. New York: Van Rostrand Reinhold Co.

Wedin, D. and D. Tilman. 1990. Species effects on nitrogen cycling: a test with perennial grasses. Oecologia 84: 433-441.

Welch, D. 1985. Studies in the grazing of heather moorland in north-east Scotland. IV. Seed dispersal and plant establishment in dung. Journal of Applied Ecology 22: 461-472.

Weller, M. W. 1989. Plant and water-level dynamics in an east Texas shrub/hardwood bottomland wetland. Wetlands 9: 73-88.

Weller, M. W. 1994. Freshwater marshes: ecology and wildlife management, 3rd ed. Minneapolis: University of Minnesota Press.

Weller, M. W. and C. E. Spatcher. 1965. Role of habitat in the distribution and abundance of marsh birds. Iowa State University Agriculture and Home Economics Experiment Station Special Report No. 43.

Welsh, S. L., N. D. Atwood, S. Goodrich, and L. C. Higgins, eds. 1987. A Utah flora. Great Basin Naturalist Memoirs Number 9. Provo: Brigham Young University.

Went, F. W. 1948. Ecology of desert plants. I. Observations on germination in the Joshua Tree National Monument, California. Ecology 29: 242-253.

West, K. A. 1984. Major pest species listed, control measures summarized at natural areas workshop. Restoration and Management Notes II(1): 34-35.

West, N. E. 1988. Intermountain deserts, shrub steppes, and woodlands. In North American terrestrial vegetation, eds., M. G. Barbour and W. D. Billings. Cambridge: Cambridge University Press.

West, N. E., K. H. Rhea, and R. O. Harniss. 1979. Plant demographic studies in sagebrush grass communities of southeastern Idaho. Ecology 60: 376-388.

West, N. E. and J. Skujins, eds. 1978. Nitrogen in desert ecosystems. Stroudsburg: Dowden, Hutchinson, and Ross, Inc.

Wharton, C. H. 1989. The natural environments of Georgia. Atlanta: Geologic and Water Resources Division and Resource Planning Section, Georgia Department of Natural Resources Bulletin 114.

Wharton, C. H., W. M. Kitchens, E. C. Pendleton, and T. W. Sipe. 1982. The ecology of bottomland hardwood swamps of the southeast: a community profile. Washington, D.C.: U.S. Fish and Wildlife Service, Biological Services Program FWS/OBS-81/37.

Wheeler, B. D. et al., eds. 1995. Resotration of temperate wetlands. New York: Wiley.

Whelan, R. J. 1995. The ecology of fire. Cambridge: Cambridge University.

Wherry, E. T. 1948. Wild flower guide: northeastern and midland United States. New York: Doubleday and Company, Inc.

Whitall, D. 1995. High alpine restoration work at McDonald Basin. Restoration and Management Notes 13(1): 29-31.

Whitcomb, R. F. 1977. Island biogeography and "habitat islands" of eastern forests. American Birds 31: 3-5.

Whitcomb, R. F., C. S. Robbins, J. F. Lynch, B. L. Klimkiewicz, B. L. Whitcomb, and D. Bystrak. 1981. Effects of forest fragmentation on avifauna of the eastern deciduous forest. In Forest island dynamics in man-dominated landscapes, eds., R. L. Burgess and D. M. Sharpe. New York: Springer-Verlag.

Whitcomb, B. L., R. Whitcomb, and D. Bystrak. 1977. Long-term turnover and effects of selective logging on the avifauna of forest fragments. American Birds 31(1): 17-23.

White, J. A. and D. C. Glenn-Lewin. 1984. Regional and local variation in tallgrass prairie remnants of Iowa and eastern Nebraska. Vegetatio 57: 65-78.

White, P. S. 1979. Pattern, process, and natural disturbance in vegetation. Botanical Review 45: 229-299.

Whitely, S. R., ed. 1988. Rehabilitating damaged ecosystems, Vol. 1. Boca Raton: CRC Press.

Whitfield, C. J., and H. L. Anderson. 1938. Secondary succession in the desert plains grassland. Ecology 19: 171-180.

Whitlatch, R. B. 1982. The ecology of New England tidal flats: a community profile. Washington, D.C.: U.S. Fish and Wildlife Service, FWS/OBS-81/01.

Whitney, G. G. 1994. From coastal wilderness to fruited plain: a history of environmental change in temperate North America from 1500 to the present. Cambridge: Cambridge University Press.

Whittaker, R. H. 1972. Evolution and measurement of species diversity. Taxon 21: 213-251.

Whittaker, R. H. 1975. Communities and ecosystems. New York: MacMillan Publishing.

Wiedemann, A. M. 1984. The ecology of Pacific Northwest coastal sand dunes: a community profile. Washington, D.C.: U.S. Fish and Wildlife Service. FWS/OBS-84/04.

Wieder, R. K. and G. E. Lang. 1983. Net primary production of the dominant bryophytes in a *Sphagnum*-dominated wetland in West Virginia. Bryologist 86: 280-286.

Wierenga, P. J., J. M. H. Hendricks, M. H. Nash, J. Ludwig, and L. A. Daugherty. 1987. Variation of soil and vegetation with distance along a transect in the Chihuahuan Desert. Journal of Arid Environments 13: 53-63.

Wilcox. B. A. 1980. Insular ecology and conservation. In Conservation biology: an evolutionary-ecological perspective, eds., M. E. Soule and B. A. Wilcox. Sunderland: Sinauer.

Willard, T. R. 1988. Biology, ecology, and management of cogongrass (*Imperata cylindrica* (L.) Beauv.). Ph.D. diss., University of Florida, Gainesville.

Williams, K. S. 1993. Use of terrestrial arthropods to evaluate restored riparian woodlands. Restoration Ecology 1(2): 107-116.

Williams, P. B. and J. L. Florsheim. 1994. Designing the Baylands Project. California Coast and Ocean 10(2): 19-27.

Willson, M. F. 1974. Avian community organization and habitat structure. Ecology 55: 1017-1029.

Wilson, E. O. 1978. On human nature. Cambridge: Harvard University Press.

Wilson, E. O. 1992. The diversity of life. Cambridge: Harvard University Press.

Wilson, E. O., ed. 1988. Biodiversity. Washington, D.C.: National Academy Press.

Wilson, H. 1987. Bright berries for the garden, food for the birds. Plants and Gardens, Brooklyn Botanic Garden Record 43(3): 44.

Wilson, M. G. 1995. Prairies in Portland? Restoration and Management Notes 13(1): 22-25.

Wilson, M. V., E. R. Alverson, D. L. Clark, R. H. Hayes, C. A. Ingersoll, and M. B. Naughton. 1995. The Willamette Valley Natural Areas Network: a partnership for the Oregon prairie. Restoration and Management Notes 13(1): 26-28.

Wilson, R. C., M. G. Narog, A. L. Koonce, and B. M. Corcoran. 1995. Postfire regeneration in Arizona's giant saguaro community. Restoration and Management Notes 13(1): 128.

Windell, K. 1991. Tree shelters for seedling protection. Washington, D.C.: United States Department of Agriculture Forest Service Report No. 9124-2834MTDC.

Windell, K. 1993. Tree shelters for seedling survival and growth. Washington, D.C.: U.S. Department of Agriculture Forest Service Technical Tips 2400.

Winegar, H. H. 1977. Camp Creek channel fencing—plant, wildlife, soil, and water responses. Rangeman's Journal 4: 10-12.

Winjum, J. K. and R. P. Neilson. 1989. The potential impact of rapid climatic change on forests in the United States. In The potential effects of global climate change on the United States, eds., J. B. Smith and D. A. Tirpak. Washington, D.C.: United States Environmental Protection Agency, EPA-230-05-89-050.

Wood, W. 1995. A restoration scenario for the Klamath River basin. Restoration and Management Notes 13(1): 58-59.

Woodhouse, W. W., Jr. 1979. Building salt marshes along the coasts of the continental United States. Vicksburg: U.S. Army Corps of Engineers, Coastal Engineering Research Center, Special Report No. 4.

Woodley, S., J. Kay, and G. Francis, eds. 1993. Ecological integrity and the management of ecosystems. Delray Beach: St. Lucie Press.

Woods, B. 1984. Ants disperse seed of herb species in a restored maple forest (Wisconsin). Restoration and Management Notes 2(1): 29-30.

Woodward, R. A. and D. D. Cornman. 1996. Drip irrigation aids revegetation of south Texas blackbrush community. Restoration and Management Notes 14(1): 73-74.

Woodwell, G. M. 1994. Ecology: the restoration. Restoration Ecology 2(1): 1-3.

World Wildlife Fund. 1992. State wetlands strategies, a guide to protecting and managing the resource. Washington, D.C.: Island Press.

Wright, H. 1995. The risks of summer burning. Native Warm-Season Grass Newsletter 14(1): 2-3.

Wright, H. A., and A. W. Bailey. 1982. Fire ecology, United States and southern Canada. New York: John Wiley & Sons.

Wunderlin, R. P. Guide to the vascular plants of central Florida. Tampa: University of South Florida.

Xerces Society. 1990. Butterfly gardening: creating summer magic in your garden. San Francisco: Sierra Club Books.

Yahner, R. H. 1995. Eastern deciduous forest: ecology and wildlife conservation. Minneapolis: University of Minnesota Press.

Yamashita, I. S. and S. J. Manning. 1995. Results of four revegetation treatments on barren farmland in the Owens Valley, California. In Proceedings: Wildland shrub and arid land restoration symposium, compilers, B. A. Roundy, E. D. McArthur, J. S. Haley, and D. K. Mann. Odgen: U.S. Forest Service General Technical Report INT-GTR-315.

Yazvenko, S. B. and D. J. Rapport. 1996. A framework for assessing forest ecosystem health. Ecosystem Health 2: 40-51.

Young, J. A. and C. G. Young. 1986. Collecting, processing and germinating seeds of wildland plants. Portland: Timber Press.

Young, J. A. and C. G. Young. 1992. Seeds of woody plants in North America. Portland: Dioscorides Press.

Young, R. A. and R. L. Giese, eds. 1990. Introduction to forest science. New York: John Wiley & Sons.

Younker, D. 1994. A proposal: native plant translocations for natural community restoration or artificial community reclamation on managed areas. Resource Management Notes 6(4): 26-29.

Zajicek-Traeger, J., M. L. Albrecht, and B. A. D. Hetrick. 1985. Vesicular-arbuscular mycorrhizae and greenhouse production of three native tallgrass prairie forbs (Kansas). Restoration and Management Notes 3(1): 39-40.

Zalejko, M. 1989. Nitrogen fixation in a natural and constructed salt marsh. M.S. thesis., San Diego State University, San Diego.

Zamora, B. A. 1994. Use of *Eriogonum* for reclamation. Hortus Northwest 5(1): 9.

Zasoski, R. J. and R. L. Edmonds. 1986. Water quality in relation to sludge and wastewater applications to forest land. In The forest alternative for treatment and utilization of municipal and industrial wastes, eds., D. W. Cole, C. L. Henry, and W. L. Nutter. Seattle: University of Washington Press.

Zedler, J. B. 1980. Algal mat productivity: comparisons in a salt marsh. Estuaries 3: 122-131.

Zedler, J. B. 1982a. Salt marsh algal mat composition: spatial and temporal comparisons. Bulletin of the Southern California Academy of Science 81: 41-51.

Zedler, J. B. 1982b. The ecology of southern California salt marshes: a community profile. Washington, D.C.: U.S. Fish and Wildlife Service, FWS/OBS-81/54.

Zedler, J. B. 1984. Salt marsh restoration: a guidebook for southern California. La Jolla: California Sea Grant Report, T-CSGCP-009.

Zedler, J. B. 1986. Catastrophic flooding and distributional patterns of Pacific cordgrass (*Spartina foliosa* Trin.). Bulletin of the Southern California Academy of Science 85: 74-86.

Zedler, J. B. 1988a. Salt marsh restoration: lessons from California. In Rehabilitating damaged ecosystems, ed., J. Cairns, Jr. Boca Raton: CRC Press.

Zedler, J. B. 1988b. Restoring diversity in salt marshes: can we do it? In Biodiversity, ed., E. O. Wilson. Washington, D.C.: National Academy Press.

Zedler, J. B. 1990. A manual for assessing restored and natural coastal wetlands with examples from southern California. La Jolla: California Sea Grant Report No. T-CSGCP-021.

Zedler, J. B. 1992. Restoring cordgrass marshes in southern California. In Restoring the nation's marine environment, a Maryland Sea Grant Book, ed., G. W. Thayer. College Park: Maryland.

Zedler, J. B. 1996. Tidal wetland restoration: a scientific perspective and southern California focus. La Jolla: California Sea Grant College System Publication No. T-038.

Zedler, J. B. and C. S. Nordby. 1986. The ecology of Tijuana Estuary: an estuarine profile. Washington, D.C.: U.S. Fish and Wildlife Service Biological Report 85(7.5).

Zedler, J. B., C. S. Nordby, and T. Griswold. 1990. Linkages: among estuarine habitats and the watershed. Technical Memoir. Washington, D.C.: National Oceanic and Atmospheric Administration, National Ocean Service, Office of Coastal Resource Management.

Zedler, P. H. 1987. The ecology of southern California vernal pools: a community profile. Washington, D.C.: U.S. Fish and Wildlife Service Biological Report 85(7.11).

Zedler, P. H. and C. Black. 1989. Creation of artificial habitat for a rare plant, *Pogogyne abramsii*, in San Diego County California. Oakland: In Proceedings of the 1st Annual Conference of the Society for Ecological Restoration, ed., H. G. Hughes and T. M. Bonnicksen. Jan. 16-20, 1989.

Zieman, J. C., Jr. 1972. Origin of circular beds of *Thalassia testudinum* in south Biscayne Bay, Florida, and their relationship to mangrove hammocks. Bulletin of Marine Science 22: 559-574.

Zink, T. A. and M. F. Allen. 1995. Soil amendments may help displace exotic plants invading reserve from pipeline corridor (California). Restoration and Management Notes 13(1): 132-133.

Zinke, P. J. 1988. The redwood forest and associated north coast forests. In Terrestrial vegetation of California, eds., M. G. Barbour and J. Major. Davis: California Native Plant Society.

Zinn, J. A. and C. Copeland. 1982. Wetland management. Washington, D. C.: Congressional Research Service, The Library of Congress.

Zonneveld, I. S. and R. T. T. Forman, eds. 1990. Changing landscapes: an ecological perspective. New York: Springer-Verlag.

Zuberer, D. A. and W. S. Silver. 1978. Biological nitrogen fixation (acetylene reduction) associated with Florida mangroves. Applied Environmental Microbiology 35: 567-575.

Natural Regions and Dominant Ecological Communities*

*See Chapter 6 for a description of ecological systems, dominant ecological communities, and background information on how they were developed.

Table of Contents*

* Regional maps are on the facing page to each region.

Natural Regions of the United States

A map of the Natural Regions of the Conterminous United States and a list of the codes for the Küchler natural vegetation types appears in Appendix G. A large color version of the Natural Regions of the Conterminous United States map can be found in the inside back cover pocket.

Ecological Restoration Types (ERT)

At the end of each Dominant Ecological Community description is a reference to the appropriate Ecological Restoration Type (ERT). The Ecological Restoration Types found in Appendix B will provide the reader with additional information about how to restore the particular Dominant Ecological Community.

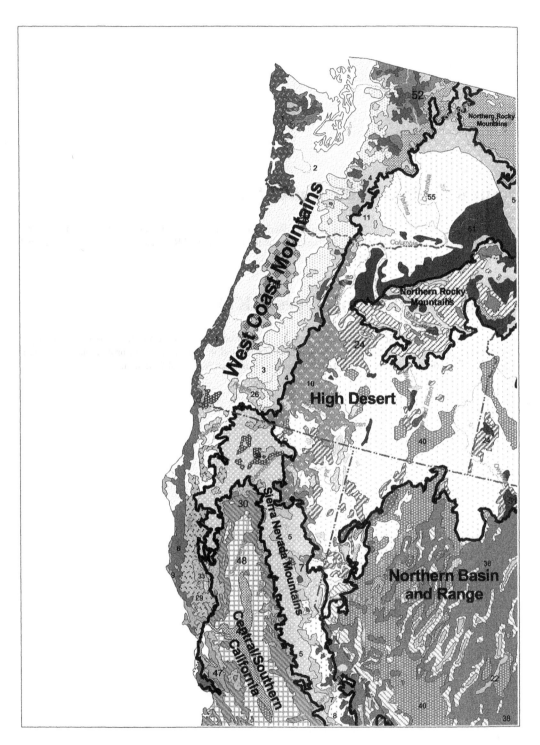

West Coast Mountains

West Coast Mountains

Introduction

The West Coast Mountain region includes the coastal mountains and the west slopes of the Cascade Mountains in Washington and Oregon and the north coastal region of California. The region also includes the Klamath Mountains of northwestern California and southwestern Oregon. This region is often referred to as the temperate rainforest region because rain or fog are part of the daily climate for most of the year. A number of climatic and geographic variables cause these forests to support some of the largest and tallest species in the world. Important factors in the success of the trees are the winter rains and the summer fog which provide moisture throughout the year. This prevents extensive moisture stress and temperature change and creates optimal growing conditions.

Many factors determine the composition of forest types; however, an elevational gradient is the most useful for describing changes in species composition. Douglas-fir is the most extensive species in this region occupying low- to mid-elevations in the Cascade and Olympic Mountains. At higher elevations Pacific silver fir and mountain hemlock become dominant. Sitka spruce and redwood forests are the most common along the coast. Deciduous forest, scrub, and grassland communities also occur in the region. The Klamath Mountains support a complex and unique assemblage of plant communities because of their location between the mesic coastal forests and the dry interior valleys. The floodplains of the Willamette and Columbia Rivers support Riparian Forests which are frequently dominated by deciduous trees. The interior valleys of the Umpqua, Rogue, and Willamette Rivers, which are in the rainshadow of the coast mountains, support communities such as oak woodlands, chaparral, and dry grasslands. Grasslands and shrub communities also occur along the coast.

Minor but unique communities to this region include: the prairies in the Willamette Valley and Puget Sound; alpine meadows and parklands occurring above treeline in the Cascade Mountains; and lava, mud flow, and serpentine areas associated with the geology of the Cascade Mountains.

The primary sources used to develop descriptions and species lists are Barbour and Billings 1988; Barbour and Major 1988; Franklin and Dyrness 1973; Holland 1986.

Dominant Ecological Communities

Upland Systems
Sitka Spruce–Hemlock Coastal Forest
Redwood Forest
Douglas-fir–Hemlock Forest
Silver Fir–Mountain Hemlock Forest
Mixed Evergreen Forest
Oregon Oak Woodland
Pine–Cypress Forest
Montane Chaparral
Coastal Mountain Grassland

Wetland Systems
Riparian Forest
Cedar–Alder Swamp

Estuarine Systems
Salt Marsh

UPLAND SYSTEMS
Sitka Spruce–Hemlock Coastal Forest

Sitka spruce characterizes the forest that occurs along the length of Washington's and Oregon's coasts. Sitka spruce forests are most often found below elevations of 495 feet but may reach elevations up to 1980 feet in the mountains closest to the coast. The climate is mild in the area near the ocean. Even in the characteristic dry northwest summers, frequent fog and fog drip protect the forest from extreme moisture stress. The understory of Sitka Spruce–Hemlock Coastal Forest is lush with dense growth of shrubs, herbs, and ferns. Unique in this community are the epiphyte-draped, big-leaf maples in the understory. This community corresponds to Küchler #1 and Coniferous Forest ERT.

Canopy
Characteristic Species
Picea sitchensis	sitka spruce
Thuja plicata	western arborvitae
Tsuga heterophylla	western hemlock

Associates
Abies amabilis	Pacific silver fir
Abies grandis	grand fir
Alnus rubra	red alder
Chamaecyparis lawsoniana	Port Orford cedar
Pinus contorta	lodgepole pine
Pseudotsuga menziesii	Douglas-fir
Sequoia sempervirens	redwood

Woody Understory
Acer circinatum	vine maple
Acer macrophyllum	big-leaf maple
Frangula purshiana	cacsara sagrada
Gaultheria shallon	salal
Menziesia ferruginea	fool's-huckleberry
Myrica californica	Pacific bayberry
Oplopanax horridum	devil's-club
Rhododendron macrophyllum	California rhododendron
Rubus spectabilis	salmon raspberry
Umbellularia californica	California-laurel
Vaccinium ovalifolium	oval-leaf blueberry
Vaccinium ovatum	evergreen blueberry
Vaccinium parvifolium	red blueberry

Herbaceous Understory
Athyrium filix-femina	subarctic lady fern
Blechnum spicant	deerfern
Claytonia sibirica	Siberian springbeauty
Disporum smithii	large-flower fairybells
Dryopteris campylpotera	mountain wood fern
Hylocomium splendens	feathermoss
Hypnum circinale	moss

Lysichiton americanus	yellow skunk-cabbage
Maianthemum dilatatum	two-leaf false Solomon's-seal
Oxalis oregana	redwood-sorrel
Polystichum munitum	pineland sword fern
Tiarella trifoliata	three-leaf foamflower
Viola glabella	pioneer violet
Viola sempervirens	redwood violet

Redwood Forest

Redwood Forest extends from southern Oregon into California along the coast and interior along streams or rivers. This community may occur just inland of sitka spruce and Douglas-fir-dominated forests. Although dependant on the summer fogs that are characteristic of the marine environment, redwoods do not tolerate salt spray as successfully as sitka spruce or Douglas-fir. This community corresponds to Küchler #6 and Coniferous Forest ERT.

Canopy
Characteristic Species
Pseudotsuga menziesii	Douglas-fir
Sequoia sempervirens	redwood

Associates
Abies grandis	grand fir
Lithocarpus densiflorus	tan-oak
Tsuga heterophylla	western hemlock

Woody Understory
Acer macrophyllum	big-leaf maple
Gaultheria shallon	salal
Myrica californica	Pacific bayberry
Rhododendron macrophyllum	California rhododendron
Torreya californica	California-nutmeg
Umbellularia californica	California-laurel
Vaccinium ovatum	evergreen blueberry

Herbaceous Understory
Oxalis oregona	redwood-sorrel
Polystichum munitum	pineland sword fern
Vancouveria hexandra	white inside-out-flower
Whipplea modesta	modesty

Douglas-fir–Hemlock Forest

Large areas from sea level up to 3300 feet in the West Coast Mountains region are dominated by Douglas-fir trees. Douglas-fir trees need an open canopy and exposure to the sun to become established. Periodic forest fires are important to create openings for regeneration. Western hemlock trees are also common within this community and may dominate on sites protected from windfall and fire. Unlike Douglas-fir, hemlock is successful in regenerating in the shade. Red cedar or western arborvitae, a common characteristic species throughout this region, is found along streams and in more mesic sites. This community corresponds to Küchler #2 and #12 and Coniferous Forest ERT.

Canopy
Characteristic Species
Pseudotsuga menziesii	Douglas-fir
Thuja plicata	western arborvitae
Tsuga heterophylla	western hemlock

Associates
Abies amabilis	Pacific silver fir
Abies concolor	white fir
Abies grandis	grand fir
Acer macrophyllum	big-leaf maple
Alnus rubra	red alder
Arbutus menziesii	Pacific madrone
Calocedrus decurrens	incense-cedar
Chamaecyparis lawsoniana	Port Orford-cedar
Larix occidentalis	western larch
Lithocarpus densiflorus	tan-oak
Picea sitchensis	sitka spruce
Pinus contorta	lodgepole pine
Pinus lambertiana	sugar pine
Pinus monticola	western white pine
Pinus ponderosa	ponderosa pine
Populus balsamifera	balsam poplar
Populus tremuloides	quaking aspen
Quercus garryana	Oregon white oak

Woody Understory
Acer circinatum	vine maple
Castanopsis chrysophylla	golden chinkapin
Cornus nuttallii	Pacific flowering dogwood
Corylus cornuta	western beaked hazelnut
Frangula purshiana	cassara sagrada
Gaultheria shallon	salal
Holodiscus discolor	hillside oceanspray
Linnaea borealis	American twinflower
Lonicera ciliosa	orange honeysuckle
Lonicera hispidula	pink honeysuckle
Mahonia nervosa	Cascate Oregon-grape
Oplopanax horridus	devil's-club
Prunus emarginata	bitter cherry
Rhododendron macrophyllum	California rhododendron
Rosa gymnocarpa	wood rose
Rubus spectabilis	salmon raspberry
Rubus ursinus	California dewberry
Salix scouleriana	Scouler's willow
Sambucus racemosa	European red elder
Symphoricarpos albus	common snowberry
Symphoricarpos mollis	creeping snowberry
Taxus brevifolia	Pacific yew
Umbellularia californica	California-laurel
Vaccinium ovatum	evergreen blueberry
Vaccinium parvifolium	red blueberry

Herbaceous Understory

Achlys triphylla	sweet-after-death
Adenocaulon bicolor	American trailplant
Asarum caudatum	wildginger long-tail
Athyrium filix-femina	subarctic lady fern
Blechnum spicant	deerfern
Bromus vulgaris	Columbian brome
Chimaphila menziesii	little prince's pine
Chimaphila umbellata	pipsissewa
Claytonia siberica	Siberian springbeauty
Collomia heterophylla	variable-leaf mountain-trumpet
Coptis laciniata	Oregon goldthread
Disporum hookeri	drops-of-gold
Dryopteris expansa	spreading wood fern
Festuca occidentalis	western fescue
Festuca subuliflora	crinkle-awn fescue
Galium aparine	sticky-willy
Galium triflorum	fragrant bedstraw
Goodyera oblongifolia	green-leaf rattlesnake-plantain
Hieracium albiflorum	white flower hawkweed
Hydrophyllum tenuipes	Pacific waterleaf
Iris tenax	tough-leaf iris
Lathyrus polyphyllus	leafy vetchling
Lysichitum americanum	yellow skunk-cabbage
Madia gracilis	grassy tarplant
Maiathemum stellatum	starry false Solomon's-seal
Melica subulata	Alaska melic grass
Moehringia macrophylla	large-leaf grove-sandwort
Osmorhiza chilensis	mountain sweet-cicely
Oxalis oregana	redwood-sorrel
Polystichum munitum	pineland sword fern
Pteridium aquilinum	northern bracken fern
Synthyris reniformis	snowqueen
Tiarella trifoliata	three-leaf foamflower
Trientalis borealis	American starflower
Trillium ovatum	western wakerobin
Trisetum cernuum	nodding false oat
Vancouveria hexandra	white inside-out-flower
Viola sempervirens	redwood violet
Whipplea modesta	modesty
Xerophyllum tenax	western turkeybeard

Silver Fir–Mountain Hemlock Forest

Silver Fir–Mountain Hemlock Forest occurs on the western slopes of the Cascade Range generally at elevations from 2600 to 6100 feet. Often referred to as "subalpine forests," silver fir is common between 2600 and 4200 feet and mountain hemlock common between 4200 and 6000 feet. Areas dominated by silver fir and mountain hemlock are wetter, cooler, and receive considerably more precipitation in the form of snow than areas dominated by Douglas-fir–Hemlock Forest. Much of the snow accumulates in snow packs of 3 to 10 feet in silver fir areas and to nearly 25 feet in higher mountain hemlock areas. In the highest elevations, the mountain hemlock canopy opens into alpine "parklands" and meadows. The parklands consist of patches of forest and tree groups interspersed with shrubby or herbaceous subalpine communities. This community corresponds to Küchler #3, #4, and #15 and Coniferous Forest ERT.

Canopy
Characteristic Species
Abies amabilis	Pacific silver fir
Abies lasiocarpa	subalpine fir
Abies procera	noble fir
Tsuga heterophylla	western hemlock
Tsuga mertensiana	mountain hemlock

Associates
Abies grandis	grand fir
Chamaecyparis nootkatensis	Alaska-cedar
Larix occidentalis	western larch
Picea engelmannii	Engelmann's spruce
Pinus contorta	lodgepole pine
Pinus monticola	western white pine
Pseudotsuga menziesii	Douglas-fir
Thuja plicata	western arborvitae

Woody Understory
Arctostaphylos nevadensis	pinemat manzanita
Cassiope mertensiana	western moss-heather
Chimaphila umbellata	pipsissewa
Gaultheria shallon	salal
Linnaea borealis	American twinflower
Mahonia nervosa	Cascade Oregon-grape
Menziesia ferruginea	fool's-huckleberry
Oplopanax horridus	devil's-club
Phyllodoce empetriformis	pink mountain-heath
Rhododendron albiflorum	cascade azalea
Rubus lasiococcus	hairy-fruit smooth dewberry
Rubus pedatus	strawberry-leaf raspberry
Rubus spectabilis	salmon raspberry
Sorbus sitchensis	Sitka mountain-ash
Vaccinium alaskense	Alaska blueberry
Vaccinium deliciosum	ranier blueberry
Vaccinium membranaceum	square-twig blueberry
Vaccinium ovalifolium	oval-leaf blueberry
Vaccinium parvifolium	red blueberry
Vaccinium scoparium	grouseberry

Herbaceous Understory
Achlys triphylla	sweet-after-death
Athyrium filix-femina	subarctic lady fern
Blechnum spicant	deerfern
Carex spectabilis	northwestern snowy sedge

Clintonia uniflora	bride's bonnet
Coptis asplenifolia	fern-leaf goldthread
Cornus canadensis	Canadian bunchberry
Erythronium montanum	white avalanche-lily
Gymnocarpium dryopteris	western oak fern
Linnea borealis	American twinflower
Listera caurina	northwestern twayblade
Lupinus latifolius	broad-leaf lupine
Lysichitum americanum	yellow skunk-cabbage
Maianthemum stellata	starry false Solomon's-seal
Maianthemum dilatatum	two-leaf false Solomon's-seal
Orthilia secunda	sidebells
Streptopus roseus	rosy twistedstalk
Streptopus streptopoides	small twistedstalk
Tiarella trifoliata	three-leaf foamflower
Trillium ovatum	western wakerobin
Valeriana stichensis	sitka valerian
Veratrum viride	American false hellebore
Viola orbiculata	evergreen yellow violet
Viola sempervirens	redwood violet
Xerophyllum tenax	western turkeybeard

Mixed Evergreen Forest

In the western Siskiyou and Klamath Mountains of southern Oregon and northern California, evergreen needle-leaved and sclerophyllous broad-leaved trees occur in Mixed Evergreen Forest. Mixed Evergreen Forest is geographically and biologically transitional between the dense coniferous forests of northwestern California and the open woodlands and savannas of the interior. Dominant species are Douglas-fir and tan-oak. This community corresponds to Küchler #29 and Western Deciduous and Mixed Forest ERT.

Canopy
Characteristic Species
Arbutus menziesii	Pacific madrone
Lithocarpus densiflorus	tan-oak
Pseudotsuga menziesii	Douglas-fir
Quercus chrysolepis	canyon live oak

Associates
Acer circinatum	vine maple
Acer macrophyllum	big-leaf maple
Aesculus californica	California buckeye
Calocedrus decurrens	incense-cedar
Castanopsis chrysophylla	golden chinkapin
Chamaecyparis lawsoniana	Port Orford-cedar
Pinus jeffreyi	Jeffrey pine
Pinus lambertiana	sugar pine
Pinus ponderosa	ponderosa pine
Quercus douglasii	blue oak
Quercus garryana	Oregon white oak
Quercus kelloggii	California black oak
Quercus wislizenii	interior live oak

Woody Understory

Arctosaphylos manzanita	big manzanita
Ceanothus parryi	ladybloom
Ceanothus thyrsiflorus	bluebrush
Cornus nuttallii	Pacific flowering dogwood
Corylus cornuta	beaked hazelnut
Gaultheria shallon	salal
Holodiscus discolor	hillside oceanspray
Lonicera hispidula	pink honeysuckle
Mahonia nervosa	cascade Oregon-grape
Philadelphus lewisii	Lewis' mock orange
Quercus sadleriana	deer oak
Quercus vaccinifolia	huckleberry oak
Rhododendron macrophyllum	California rhododendron
Rosa gymnocarpa	wood rose
Rubus ursinus	California dewberry
Taxus brevifolia	Pacific yew
Toxicodendron diversilobum	Pacific poison-oak
Umbellularia californica	California-laurel

Herbaceous Understory

Achlys triphylla	sweet-after-death
Apocynum androsaemifolium	spreading dogbane
Disporum hookeri	drops-of-gold
Festuca occidentalis	western fescue
Goodyera oblongifolia	green-leaf rattlesnake-plantain
Hieracium alboflorum	white-flower hawkweed
Linnaea borealis	American twinflower
Melica harfordii	Harford's melic grass
Polystichum munitum	pineland sword fern
Pteridium aquilinum	northern bracken fern
Trientalis borealis	American starflower
Wipplea mosesta	modesty
Xerophyllum tenax	western turkeybeard

Oregon Oak Woodland

Oregon Oak Woodland occupies very dry habitats and often occurs in the areas referred to as the "interior" valleys of Oregon and California. These areas include the valley bottoms and lowlands of the Umpqua, Rogue, and Willamette River valleys which occur between the Cascade mountain range to the east and the Coast or Siskiyou ranges to the west. The understory of the oak woodlands is often dominated by grasses and as a consequence, the areas are often subject to heavy grazing. Other plant communities associated with these dry valleys include grasslands and "chaparral." This community corresponds to Küchler #26 and Woodlands ERT.

Canopy
Characteristic Species

Pseudotsuga menziesii	Douglas-fir
Quercus garryana	Oregon white oak

Associates

Abies grandis	grand fir
Acer macrophyllum	big-leaf maple
Arbutus menziesii	Pacific madrone
Calocedrus decurrens	incense-cedar
Pinus ponderosa	ponderosa pine
Quercus chrysolepis	canyon live oak
Quercus kelloggii	California black oak

Woody Understory

Amelanchier alnifolia	Saskatoon service-berry
Corylus cornuta	beaked hazelnut
Crataegus douglassii	black hawthorn
Holodiscus discolor	hillside oceanspray
Lonicera ciliosa	orange honeysuckle
Mahonia aquifolium	holly-leaf Oregon-grape
Oemleria cerasiformis	oso-berry
Rosa gymnocarpa	wood rose
Rosa nutkana	Nootka rose
Rubus parviflorus	western thimble-berry
Rubus ursinus	California dewberry
Symphoricarpos albus	common snowberry
Toxicodendron diversilobum	Pacific poison-oak
Viburnum ellipticum	western blackhaw

Herbaceous Understory

Bromus laevipes	woodland brome
Camassia quamash	small camas
Carex inops	long-stolon sedge
Danthonia californica	California wild oat grass
Elymus glaucus	blue wild rye
Festuca californica	California fescue
Festuca idahoensis	bluebunch fescue
Festuca rubra	red fescue
Fragaria vesca	woodland strawberry
Fragaria virginiana	Virginia strawberry
Melica bulbosa	onion grass
Melica subulata	Alaska melic grass
Osmorhiza chilensis	mountain sweet cicely
Polystichum munitum	pineland sword fern
Pteridium aquilinum	northern bracken fern
Sanicula crassicaulis	Pacific black-snakeroot
Satureja douglasii	Oregon-tea
Tellima grandiflora	fragrant fringecup

Pine–Cypress Forest

Pine–Cypress Forest occurs on coastal headlands, bluffs, and islands along the length of the California coast. These forests are subject to nearly constant onshore winds. Areas where they occur have well-drained sandy soils and experience the typical coastal summer fog. The dominant species in these forests have closed or "serotinous" cones. The cones remain closed after maturity and accumulate on the tree until opened by fire. Periodic fires are essential to the reproduction of the dominant species which characterize this plant community. This community corresponds to Küchler #9 and Coniferous Forests ERT.

Canopy
Characteristic Species
Cupressus goveniana	Gowen cypress
Cupressus macrocarpa	Monterey cypress
Pinus contorta	lodgepole pine
Pinus muricata	Bishop pine
Pinus radiata	Monterey pine

Associates
Cupressus bakeri	Modoc cypress
Cupressus forbesii	Tecate cypress
Cupressus macnabiana	MacNab's cypress
Cupressus nevadensis	Paiute cypress
Cupressus pygmaea	Mendocino cypress
Cupressus sargentii	Sargent's cypress
Cupressus stephensonii	Cuyamaca cypress
Pinus attenuata	knob-cone pine
Pinus torreyana	Torrey pine
Quercus agrifolia	coastal live oak

Woody Understory
Arctostaphylos nummularia	Fort Braff manzanita
Arctostaphylos tomentosa	hairy manzanita
Artemisia californica	coastal sagebrush
Baccharis pilularis	coyotebrush
Ceanothus thyrsiflorus	bluebrush
Frangula californica	California coffee berry
Heteromeles arbutifolia	California-Christmas-berry
Ledum glandulosum	glandular Labrador tea
Rubus ursinus	California dewberry
Symphoricarpos mollis	creeping snowberry
Toxicodendron diversilobum	Pacific poison-oak
Vaccinium ovatum	evergreen blueberry

Herbaceous Understory
Agrostis dieogensis	leafy bent
Dudleya farinosa	powdery live-forever
Elymus glaucus	blue wild rye
Erigeron glaucus	seaside fleabane
Eriophyllum staechidifolium	seaside woolly-flower
Galium californicum	California bedstraw

Iris douglasiana	mountain iris
Pteridium aquilinum	northern bracken fern
Xerophyllum tenax	western turkeybeard

Montane Chaparral

Montane Chaparral is most widely distributed from the foothills of the Sierra Nevada Mountains of California to the Pacific Ocean. The northern limits of chaparral communities occur in the West Coast Mountain Region in the drier parts of the Rogue River watershed in Oregon. Chaparral consists of a continuous cover of closely spaced shrubs three to thirteen feet tall with intertwining branches. Herbaceous vegetation is sparse except immediately after fires, which can be frequent in the Chaparral. This community corresponds to Küchler #34 and Desert Lands ERT.

Characteristic Woody Species

Arctostaphylos canescens	hoary manzanita
Arctostaphylos viscida	white-leaf manzanita
Ceanothus cuneatus	sedge-leaf buckbrush

Associates

Adenocaulon fasciculatum	American trailplant
Aesculus californica	California buckeye
Amelanchier pallida	pale service-berry
Arctostaphylos glandulosa	Eastwood's manzanita
Arctostaphylos glauca	big-berry manzanita
Ceanothus cordulatus	mountain whitethorn
Ceanothus greggii	Mojave buckbrush
Ceanothus integerrimus	deerbrush
Ceanothus leucodermis	jackbrush
Ceanothus velutinus	tobacco-brush
Cercis canadensis	redbud
Cercocarpus montanus	alder-leaf mountain-mahogany
Chrysothamnus nauseosus	rubber rabbitbrush
Cornus glabrata	smooth-leaf dogwood
Eriodictyon californicum	California yerba-santa
Frangula californica	California coffee berry
Fraxinus dipetala	two-petal ash
Fremontodendron californicum	California flannelbush
Garrya fremontii	bearbush
Heteromeles arbutifolia	California-Christmas-berry
Lithocarpus densiflorus	tan-oak
Lonicera involucrata	four-line honeysuckle
Pickeringia montana	stingaree-bush
Prunus ilicifolia	holly-leaf cherry
Quercus chrysolepis	canyon live oak
Quercus dumosa	California scrub oak
Quercus wislizenii	interior live oak
Rhus ovata	sugar sumac
Rhus trilobata	ill-scented sumac
Toxicodendron diversilobum	Pacific poison-oak

Coastal Mountain Grassland

Coastal Mountain Grassland occurs along the coast on the peaks of coast range mountains. These areas, often called "grass balds," are incapable of supporting tree growth because of shallow soils. Grasslands also occur in the interior valleys. Examples of native grass communities are rare because of extensive grazing and the introduction of alien species into grassland areas. Forests have encroached on much of the natural grassland areas as a result of the suppression of natural fires. This community corresponds to Küchler #47 and Grasslands ERT.

Characteristic Species

Carex tumulicola	foothill sedge
Danthonia californica	California wild oat grass
Deschampsia cespitosa	tufted hair grass
Festuca idahoensis	bluebunch fescue

Associates

Agrostis hallii	Hall's bent
Armeria maritima	sea thrift
Calamagrostis nutkaensis	nootka reed grass
Calochortus luteus	yellow mariposa-lily
Danthonia intermedia	timber wild oat grass
Dichelostemma pulchellum	bluedicks
Elymus elymoides	western bottle-brush grass
Elymus glaucus	blue wild rye
Eriophyllum lanatum	common woolly-sunflower
Festuca rubra	red fescue
Fragaria chiloensis	beach strawberry
Grindelia hirsutula	hairy gumweed
Heterotheca bolanderi	Bolander's false golden-aster
Iris douglasiana	mountain iris
Koeleria macrantha	prairie Koeler's grass
Lupinus formosus	summer lupine
Lupinus versicolor	Lindley's varied lupine
Nassella lepida	tussock grass
Nassella pulchra	tussock grass
Pteridium aquilinum	northern bracken fern
Ranunculus californicus	California buttercup
Sanicula arctopoides	footsteps-of-spring
Sanicula bipinnatifida	purple black-snakeroot
Sisyrinchium bellum	California blue-eyed-grass
Stipa nelsonii	Nelson's needle grass
Veronica peregrina	neckweed
Vicia americana	American purple vetch

WETLAND SYSTEMS

Riparian Forest

Hardwood forests are not typical in the West Coast Mountain Region. However, hardwood forests do occur in this region along the floodplains of the Willamette and Columbia Rivers and other major rivers and on poorly drained sites subject to flooding. Balsam poplar (black cottonwood) is the most common species along the major rivers. Stands of balsam poplar occur on islands and along the shoreline. This community corresponds to Küchler #25 and Riparian Forests and Woodlands ERT.

Canopy
Characteristic Species

Alnus rubra	red alder
Fraxinus latifolia	Oregon ash
Populus balsamifera	balsam poplar

Associates

Acer macrophyllum	big-leaf maple
Alnus rhombifolia	white alder
Pinus ponderosa	ponderosa pine
Salix eriocephala	Missouri willow
Salix lucida	shining willow
Salix melanopsis	dusky willow
Salix scouleriana	Scouler's willow
Salix sessilifolia	sessile-leaf willow
Quercus garryana	Oregon white oak

Woody Understory

Oemleria cerasiformis	oso-berry
Physocarpus capitatus	Pacific ninebark
Rubus spectabilis	salmon raspberry
Symphoricarpos albus	common snowberry
Umbellularia californica	California-laurel

Herbaceous Understory

Aralia californica	California spikenard
Carex deweyana	round-fruit short-scale sedge
Carex obnupta	slough sedge
Deschampsia cespitosa	tufted hair grass
Equisetum hyemale	tall scouring-rush
Urtica dioica	stinging nettle

Cedar–Alder Swamp

Red alder and western red cedar (western arborvitae) are characteristic of swamps in and on the coastal plain of the West Coast Mountain Region. The best developed swamps can be found in the coastal plain of the Olympic Peninsula. Cedar–Alder Swamp often occurs along the mouths of rivers or around margins of lakes and springs. These swamps occur where there is a high water table, or even standing surface water, for all or a portion of the year. This community corresponds to Shrub Swamps ERT.

Canopy
Characteristic Species

Alnus rubra	red alder
Thuja plicata	western arborvitae
Tsuga heterophylla	western hemlock

Associates

Abies amabilis	Pacific silver fir
Picea sitchensis	sitka spruce
Pinus contorta	lodgepole pine
Pinus monticola	western white pine

Woody Understory

Gaultheria shallon	salal
Menziesia ferruginea	fool's-huckleberry
Rubus spectabilis	salmon raspberry
Salix hookeriana	coastal willow
Spiraea douglasii	Douglas' meadowsweet
Vaccinium ovalifolium	oval-leaf blueberry
Vaccinium ovatum	evergreen blueberry
Vaccinium parvifolium	red blueberry

Herbaceous Understory

Athyrium filix-femina	subarctic lady fern
Blechnum spicant	deerfern
Carex obnupta	slough sedge
Cornus canadensis	Canadian bunchberry
Lysichitum americanum	yellow skunk-cabbage
Maianthemum dilatatum	two-leaf false Solomon's-seal
Oenanthe sarmentosa	Pacific water-dropwort
Stachys mexicana	Mexican hedge-nettle
Tolmiea menziesii	piggyback-plant

ESTUARINE SYSTEMS

Salt Marsh

There are many tideland communities associated with estuaries along the Oregon, Washington, and California coasts. The predominant tideland communities are marshes on tidal flats. Salt Marsh is found along sheltered inland margins of bays, lagoons, and estuaries. The soils of these marshes are subject to regular tidal inundation for at least part of the year. Tidal influences are especially prevalent on islands in the lower Columbia basin. This community corresponds to Salt Marsh ERT.

Characteristic Species

Distichlis spicata	coastal salt grass
Jaumea carnosa	marsh Jaumea
Spartina foliosa	California cord grass

Associates

Argentina egedii	Pacific silverweed
Atriplex patula	halberd-leaf orache
Carex lyngbyei	Lyngbye's sedge
Cotula coronopifolia	common brassbuttons
Cressa truxillensis	spreading alkali-weed
Deschampsia cespitosa	tufted hair grass
Eleocharis parvula	little-head spike-rush
Frankenia salina	alkali sea-heath
Glaux maritima	sea-milkwort
Grindelia integrifolia	Pudget Sound gumweed
Grindelia paludosa	Suisun Marsh gumweed
Hainardia cylindrica	barb grass
Juncus balticus	Baltic rush
Juncus effusus	lamp rush
Juncus lesueurii	salt rush
Lasthenia minor	coastal goldfields
Limonium californicum	marsh-rosemary
Plantago maritima	goosetongue
Puccinellia kurilensis	dwarf alkali grass
Salicornia maritima	sea saltwort
Salicornia virginica	woody saltwort
Scirpus americanus	chairmaker's bulrush
Scirpus maritimus	saltmarsh bulrush
Spergularia salina	sandspurrey
Suaeda californica	broom seepweed
Triglochin concinnam	slender arrow-grass
Triglochin maritimum	seaside arrow-grass

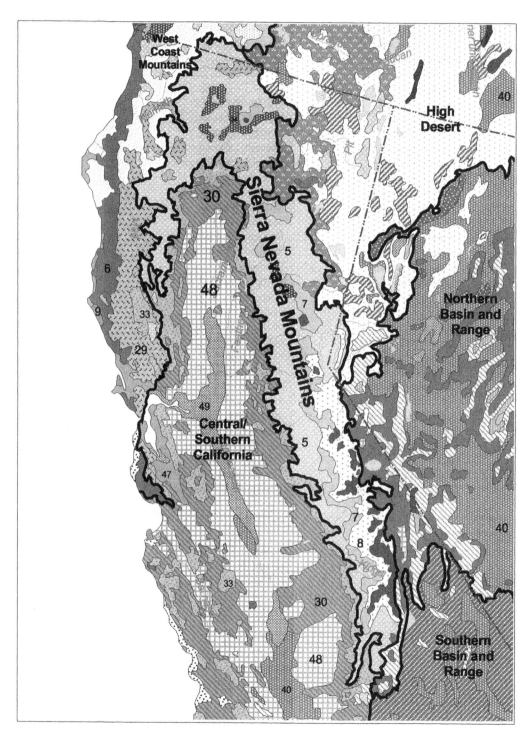

Sierra Nevada Mountains

Sierra Nevada Mountains

Introduction

The Sierra Nevada Mountains region is characterized by steep slopes and rough terrain. The region's vegetation is diverse, but conifer forests are the most prevalent vegetation type. This region includes both the east and west slopes of the Sierra Nevada Mountains. The southern portion of the Cascade Mountain Range and the Trinity Mountains, which extend along the northwestern rim of the Great Valley, are also included. The Klamath and Siskiyou Mountains of southern Oregon and northern California are not part of this region (see West Coast Mountains region for descriptions of the vegetation in these mountain areas).

The most extensive forest type is the Ponderosa Pine Forest. The forest composition changes on an altitudinal gradient. Altitude affects the length of the growing season, available moisture, and soil depth, therefore affecting species' dominance, abundance, and diversity. There are vegetation differences between the western slopes and the eastern desert-facing slopes of the Sierra Nevada Mountains. The eastern slopes are in a rain shadow and forest cover is less continuous with soils more "skeletal" than on western slopes. Although there are differences, the vegetation types described in this region are relevant to both sides of the mountains unless specifically noted.

Periodic fires were extremely common, helping maintain the structure of the vegetation. It is estimated that prior to 1875, fire frequency averaged about one fire every eight years in pine-dominated sites and about one per sixteen years in more mesic fir-dominated sites. Fire suppression and control have significantly changed many of the plant communities.

In the Sierra Nevada Mountains region, unique groves of the celebrated giant sequoia are found in areas that support Sierran white fir (see Sierran White Fir Forest description). Hardwood trees, primarily of oak and aspen species, are present in this region as successional phases or as isolated stands in particular sites such as canyons or steep slopes. Within pine communities, oak species tend to occur in openings in ponderosa pine areas and quaking aspens occur in areas with higher moisture availability. Although there is no riparian forest described in this region, aspen woodlands may occur along mountain streams. Forest types found in subalpine and treeline areas include mountain hemlock, western white pine, whitebark pine, foxtail pine, limber pine, and lodgepole pine.

The primary sources used to develop descriptions and species lists are Barbour and Billings 1988; Barbour and Major 1988; Thorne 1976.

Dominant Ecological Communities

Upland Systems
Ponderosa Pine Forest
Sierran White Fir Forest
Red Fir Forest
Lodgepole Pine Forest
Montane Chaparral

Wetland Systems
Mountain Meadow

UPLAND SYSTEMS
Ponderosa Pine Forest

Ponderosa Pine Forest is the most extensive forest type in California. The forest has been called "mid-montane conifer forest," "yellow pine forest," "white fir forest," "mixed conifer forest," and "big tree forest." Although each type listed may have dominants other than ponderosa pine, ponderosa pine occurs throughout each forest type. Generally, ponderosa pine occurs on dry sites at elevations between 1000 and 6000 feet in the north and 4000 and 7000 feet in the south. Where soil moisture increases, white fir may become dominant. Where soils are too poor for ponderosa pine, California black oak will become dominant. On the east side of the Sierras where conditions are generally drier and colder, Jeffrey pine is likely to share dominance with ponderosa pine. Ponderosa pine forests depend upon periodic fires for regeneration. Fires create openings where young trees can establish in full sunlight. This community corresponds to Küchler #5 and Coniferous Forests ERT.

Canopy
Characteristic Species
    Pinus ponderosa   ponderosa pine
Associates
Abies concolor	white fir
Calocedrus decurrens	incense-cedar
Pinus attenuata	knob-cone pine
Pinus coulteri	Coulter's pine
Pinus jeffreyi	Jeffrey pine
Pinus lambertiana	sugar pine
Pseudotsuga macrocarpa	big-cone Douglas-fir
Pseudotsuga menziesii	Douglas-fir
Quercus chrysolepis	canyon live oak
Quercus kelloggii	California black oak

Woody Understory
Arctostaphylos glandulosa	Eastwood's manzanita
Arctostaphylos patula	green-leaf manzanita
Ceanothus cordulatus	mountain whitethorn
Ceanothus integerrimus	deerbrush
Chamaebatia foliolosa	Sierran mountain-misery
Cornus nuttallii	Pacific flowering dogwood
Eriodictyon trichocalyx	hairy yerba-santa
Frangula californica	California coffee berry
Lithocarpus densiflorus	tan-oak
Ribes roezlii	Sierran gooseberry

Herbaceous Understory
Eriastrum densifolium	giant woolstar
Gilia splendens	splendid gily-flower
Iris hartwegii	rainbow iris
Lupinus excubitus	interior bush lupine
Lupinus formosus	summer lupine
Solanum xantii	chaparral nightshade
Streptanthus bernardinus	laguna mountain jewelflower

Sierran White Fir Forest

Sierran White Fir Forest occurs on relatively moist sites on the lower slopes of the Sierra Nevada Mountains. They generally occur from 4100 to 7300 feet. Giant sequoia (*Sequoiadendron giganteum*) groves occur within the Sierran White Fir Forest on west-side slopes. Giant sequoias are restricted to sites that have sufficient soil moisture throughout the summer drought months. Reproduction within sequoia groves depends upon periodic fire. Without regular fire, the accumulation of litter on the forest floor inhibits germination and establishment of sequoia seedlings and white fir will invade and eventually dominate sequoia groves. This community corresponds to Küchler #5 and Coniferous Forests ERT. Sequoia groves are marked on the Küchler map with an "S."

Canopy
Characteristic Species
Abies concolor	white fir
Abies magnifica	California red fir
Calocedrus decurrens	incense-cedar
Pinus lambertiana	sugar pine

Associates
Arbutus menziesii	Pacific madrone
Pseudotsuga menziesii	Douglas-fir

Woody Understory
Acer macrophyllum	big-leaf maple
Arctostaphylos patula	green-leaf manzanita
Castanopsis sempervirens	Sierran chinkapin
Ceanothus cordulatus	mountain whitethorn
Ceanothus integerrimus	deerbrush
Ceanothus parvifolius	cattlebush
Chamaebatia foliolosa	Sierran mountain-misery
Cornus nuttallii	Pacific flowering dogwood
Prunus emarginata	bitter cherry
Quercus chrysolepis	canyon live oak
Quercus kelloggii	California black oak
Ribes roezlii	Sierran gooseberry
Ribes viscosissimum	sticky currant
Rosa gymnocarpa	wood rose
Salix scouleriana	Scouler's willow
Symphoricarpos oreophilus	mountain snowberry
Taxus brevifolia	Pacific yew

Herbaceous Understory
Adenocaulon bicolor	American trailplant
Asarum hartwegii	Hartweg's wild ginger
Chimaphila menziesii	little prince's-pine
Clintonia uniflora	bride's-bonnet
Disporum hookeri	drops-of-gold
Fragaria vesca	woodland strawberry
Galium sparsiflorum	Sequoia bedstraw
Goodyera oblongifolia	green-leaf rattlesnake-plantain

Hieracium albiflorum	white-flower hawkweed
Iris hartwegii	rainbow iris
Lupinus latifolius	broad-leaf lupine
Osmorhiza chilensis	mountain sweet-cicely
Pteridium aquilinum	northern bracken fern
Pyrola picta	white-vein wintergreen
Viola glabella	pioneer violet
Viola lobata	moose-horn violet

Red Fir Forest

Red Fir Forest occurs between 6000 and 9000 feet in the Sierra Nevada Mountains. In these high elevations, snow is the major form of precipitation. Red Fir Forest is very dense, often limiting the understory growth. Red fir seedlings do, however, prefer shade to open areas for regeneration. Where fires or other disturbances cause breaks in the canopy, lodgepole pine is a local dominant at these altitudes. This community corresponds to Küchler #7 and Coniferous Forests ERT.

Canopy
Characteristic Species

Abies concolor	white fir
Abies magnifica	California red fir

Associates

Calocedrus decurrens	incense-cedar
Pinus contorta	lodgepole pine
Pinus jeffreyi	Jeffrey pine
Pinus lambertiana	sugar pine
Pinus monticola	western white pine
Tsuga mertensiana	mountain hemlock

Woody Understory

Acer glabrum	Rocky Mountain maple
Arctostaphylos patula	green-leaf manzanita
Castanopsis sempervirens	Sierran chinkapin
Ceanothus cordulatus	mountain white-thorn
Ceanothus velutinus	tobacco-brush
Lonicera conjugialis	purple-flower honeysuckle
Prunus emarginata	bitter cherry
Quercus vacciniifolia	huckleberry oak
Ribes roezlii	Sierran gooseberry
Ribes viscosissimum	sticky currant
Salix scouleriana	Scouler's willow
Symphoricarpos oreophilus	mountain snowberry

Herbaceous Understory

Arabis platysperma	pioneer rockcress
Aster breweri	Brewer's aster
Chimaphila umbellata	pipsissewa
Cistanthe umbellata	Mt. Hood pussypaws
Corallorrhiza maculata	summer coralroot
Elymus elymoides	western bottle-brush grass

Eriogonum nudum	naked wild buckwheat
Gayophytum humile	dwarf groundsmoke
Hieracium albiflorum	white-flower hawkweed
Monardella odoratissima	alpine mountainbalm
Orthilia secunda	sidebells
Pedicularis semibarbata	pinewoods lousewort
Poa bolanderi	Bolander's blue grass
Pterospora andromedea	pine drops
Pyrola picta	white-vein wintergreen
Sarcodes sanguinea	snowplant
Viola purpurea	goose-foot yellow violet
Wyethia mollis	woolly mule's-ears

Lodgepole Pine Forest

Extensive Lodgepole Pine Forest generally occurs above Red Fir Forest in the Sierra Nevada Mountains. Lodgepole pine is common between 6000 and 8000 feet in the north and between 7400 and 11000 feet in the south. The growing season in these high elevations is relatively short (two to three months) and the snow accumulation is heavy. Due to the short growing season and harsh environment, trees are shorter than those growing at lower elevations or in less severe environments. Limber pine is an associate at high elevations in the south. Suppression of naturally occurring fires in the forests over the years has resulted in increased density and cover in Lodgepole Pine Forest. Although fire may occur periodically in these pine forests and influence community structure, other significant environmental influences in Lodgepole Pine Forest and higher elevation communities are the lower mean annual temperature and precipitation in the form of persistent snow. These factors have been used to distinguish subalpine forests that are included here within the Lodgepole Pine Forest type. This community corresponds to Küchler #8 and Coniferous Forests ERT.

Canopy
Characteristic Species
Pinus contorta	lodgepole pine

Associates
Pinus albicaulis	white-bark pine
Pinus balfouriana	fox-tail pine
Pinus flexilis	limber pine
Populus tremuloides	quaking aspen
Tsuga mertensiana	mountain hemlock

Woody Understory
Arctostaphylos nevadensis	pinemat manzanita
Phyllodoce breweri	red mountain-heath
Ribes montigenum	western prickly gooseberry

Herbaceous Understory
Potentilla breweri	Sierran cinquefoil

Montane Chaparral

In the Sierra Nevada Mountains region, Montane Chaparral occurs on hills and lower mountain slopes that are dry. Although Montane Chaparral reaches maximum development in southern California, it is common along the eastern side of the Great Central Valley and higher regions of the Sierra Nevada Mountains. Several different types of chaparral, identified by associations and geographical location, have been identified in the literature. Montane Chaparral is the most common type in the Sierra Nevada Mountains. Montane Chaparral is lower and more "compact" than other chaparral types. The development of Montane Chaparral is slowed by cold temperatures, snow, and a short growing season. Many of the shrub species in the Montane Chaparral also occur as understory or gap colonizers in conifer forests. The interface of the Montane Chaparral and adjacent forests is strongly influenced by fire frequency. This community corresponds to Küchler #34 and Desert Lands ERT.

Characteristic Woody Species

Arctostaphylos nevadensis	pinemat manzanita
Arctostaphylos parryana	pineland manzanita
Arctostaphylos patula	green-leaf manzanita
Castanopsis sempervirens	Sierran chinkapin
Ceanothus cordulatus	mountain whitethorn
Ceanothus diversifolius	pinemat
Ceanothus integerrimus	deerbrush
Ceanothus sanguineus	Oregon teatree

Associates

Arctostaphylos canescens	hairy manzanita
Arctostaphylos viscida	white-leaf manzanita
Ceanothus fresnensis	fresnomat
Ceanothus parvifolius	cattlebush
Ceanothus prostratus	squawcarpet
Ceanothus tomentosus	ionebush
Ceanothus velutinus	tobacco-brush
Cercocarpus ledifolius	curl-leaf mountain-mahogany
Cercocarpus montanus	alder-leaf mountain-mahogany
Chrysothamnus nauseosus	rubber rabbitbush
Eriodictyon californicum	California yerba-santa
Frangula californica	California coffee berry
Garrya fremontii	bearbush
Heteromeles arbutifolia	California-Christmas-berry
Prunus emarginata	bitter cherry
Prunus virginiana	western choke cherry
Symphoricarpos oreophilus	mountain snowberry

WETLAND SYSTEMS
Mountain Meadow

Throughout the Sierra Nevada Mountains, there are many meadows. Meadows are common above the treeline. In lower areas meadows may be created by conditions too wet to support trees or may be the result of avalanche or fire activity. A shallow water table is the most important factor in explaining the occurrence and distribution of meadows. Although most of the species listed relate to wet site conditions, the woodland and upper elevation meadows also occur under drier conditions, the latter even in dry gravelly soils. This community corresponds to Meadowlands ERT.

Characteristic Species

Aster alpigenus	tundra aster
Calamagrostis breweri	short-hair reed grass
Camassia quamash	small camass
Cardamine breweri	Sierran bittercress
Carex bolanderi	Bolander's sedge
Carex exserta	short-hair sedge
Carex macloviana	Falkland Island sedge
Carex nebrascensis	Nebraska sedge
Carex rostrata	swollen beaked sedge
Carex scopulorum	Holm's Rocky Mountain sedge
Carex teneriformis	Sierran slender sedge
Cistanthe umbellata	Mt. Hood pussypaws
Deschampsia cespitosa	tufted hair grass
Dodecatheon jeffreyi	tall mountain shootingstar
Eleocharis bella	delicate spike-rush
Eleocharis quinqueflora	few-flower spike-rush
Eriogonum incanum	frosted wild buckwheat
Eriogonum ovalifolium	cushion wild buckwheat
Eriophorum cringerum	fringed cotton-grass
Festuca brachyphylla	short-leaf fescue
Gentiana newberryi	alpine gentian
Glyceria elata	tall manna grass
Heracleum maximum	cow-parsnip
Ivesia purpurascens	summit mousetail
Juncus nevadensis	Sierran rush
Juncus orthophyllus	straight-leaf rush
Lupinus breweri	matted lupine
Lupinus polyphyllus	blue-pod lupine
Mimulus primuloides	yellow creeping monkey-flower
Muhlenbergia filiformis	pullup muhly
Penstemon davidsonii	timberline beardtongue
Pteridium aquilinum	northern bracken fern
Ptilagrostis kingii	Sierran false needle grass
Scirpus congdonii	Congdon's bulrush
Senecio scorzonella	Sierran ragwort
Solidago multiradiata	Rocky Mountain goldenrod

Stipa occidentalis	western needle grass
Triglochin palustre	marsh arrow-grass
Trisetum spicatum	narrow false oat
Vaccinium cespitosum	dwarf blueberry
Veratrum californicum	California false hellebore
Veratrum fimbriatum	fringed false hellebore

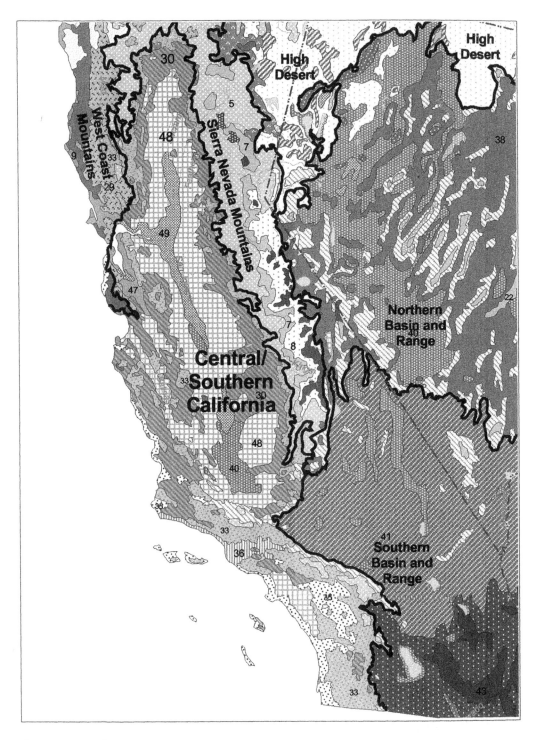

Central/Southern California

Central/Southern California

Introduction

The Central/Southern California region includes the Sacramento and San Joaquin Valleys (together known as the Central Valley), the Transverse and Peninsular Mountain Ranges, and the Central Coast Ranges. The islands off the coast of southern California are also included in the region. The region includes the salt marshes in the lowest areas near the coast to the subalpine meadows in the highest elevations of the San Bernardino and San Gabriel Mountains. A wide diversity of vegetation types is represented within the region.

Oak woodlands are the most extensive vegetation type in the region. Oak woodlands occur around the Central Valley and throughout the Coast Ranges. The woodlands occur on mountain slopes and foothills where sufficient moisture exists to support an open canopy of trees. Within the oak woodland, chaparral and prairie species are common understory components. Native prairies were originally quite extensive in this region, especially in the Sacramento and San Joaquin Valleys. Nearly all native prairie has been drained and converted to agriculture.

The Transverse Mountain Range, at the southwest edge of the state, supports several forest communities similar to Sierra Nevada Mountain communities. In fact, these mountains have been described as "islands of Sierran vegetation."

Unique to this region are the numerous freshwater wetlands. Unfortunately, much of the freshwater wetland has been drained and converted for agricultural purposes. The Vernal Pools are among the most unique features in this region.

The primary sources used to develop descriptions and species lists are Barbour and Billings 1988; Barbour and Major 1988; Holland 1986; Holland and Jain 1988; Thorne 1976.

Dominant Ecological Communities

Upland Systems
Yellow Pine Forest
White Fir–Sugar Pine Forest
Lodgepole Pine Forest
Coulter Pine Woodland
California Oak Woodland
Juniper–Pinyon Woodland
Chaparral
Coastal Sagebrush
San Joaquin Saltbush
California Prairie

Wetland Systems
Riparian Forest
Vernal Pools
Tule Marsh
Mountain Meadow

Estuarine Systems
Salt Marsh

UPLAND SYSTEMS

Yellow Pine Forest

Yellow Pine Forest is a mid-montane conifer forest type that occurs in the Transverse Mountains between 4500 and 9000 feet. Both dominant canopy species, Jeffrey and ponderosa pine, are often referred to as "yellow pine" and share dominance in this region. Ponderosa pine is more common on lower, more mesic slopes. Jeffrey pine is more common in higher areas. The two yellow pines are closely related and have many associated species in common. This community corresponds to Küchler #5 and Coniferous Forests ERT.

Canopy
Characteristic Species
Pinus jeffreyi	Jeffrey pine
Pinus ponderosa	ponderosa pine

Associates
Abies concolor	white fir
Calocedrus decurrens	incense-cedar
Pinus coulteri	Coulter's pine
Pinus lambertiana	sugar pine
Pseudotsuga macrocarpa	big-cone Douglas-fir
Quercus chrysolepis	canyon live oak
Quercus kelloggii	California black oak

Woody Understory
Arctostaphylos glandulosa	Eastwood's manzanita
Arctostaphylos patula	green-leaf manzanita
Arctostaphylos pringlei	pink-bract manzanita
Castanopsis sempervirens	Sierran chinkapin
Ceanothus integerrimus	deerbrush
Cornus nuttallii	Pacific flowering dogwood
Eriodictyon trichocalyx	hairy yerba-santa
Eriogonum wrightii	bastard-sage
Frangula californica	coffee berry
Garrya flavescens	ashy silktassel
Ribes roezlii	Sierran gooseberry
Symphoricarpos oreophilus	mountain snowberry

Herbaceous Understory
Arabis repanda	Yosemite rockcress
Asclepias eriocarpa	Indian milkweed
Bromus marginatus	large mountain brome
Bromus orcuttianus	Chinook brome
Calystegia occidentalis	chaparral false bindweed
Carex multicaulis	many-stem sedge
Castilleja martinii	Martin's Indian-paintbrush
Chaenactis santolinoides	Santolina pincushion
Chimaphila menziesii	little prince's-pine
Clarkia rhomboidea	diamond fairyfan

Collinsia childii	Child's blue-eyed Mary
Cordylanthus nevinii	Nevin's bird's-beak
Cordylanthus rigidus	stiff-branch bird's beak
Elymus elymoides	western bottle-brush grass
Eriastrum densifolium	giant woolstar
Eriogonum parishii	mountainmist
Fritillaria pinetorum	pinewoods missionbells
Galium johnstonii	Johnston's bedstraw
Gayophytum diffusum	spreading groundsmoke
Gayophytum heterozygum	zigzag groundsmoke
Gilia splendens	spendid gily-flower
Iris hartwegii	rainbow iris
Ivesia santolinoides	Sierran mousetail
Koeleria macrantha	prairie Koeler's grass
Linanthus ciliatus	whiskerbrush
Lotus nevadensis	Nevada bird's-foot-trefoil
Lupinus elatus	tall silky lupine
Lupinus excubitus	interior bush lupine
Lupinus formosus	summer lupine
Lupinus peirsonii	long lupine
Melica imperfecta	coast range melic grass
Melica stricta	rock melic grass
Penstemon caesius	San Bernardino beardtongue
Penstemon grinnellii	Grinnell's beardtongue
Penstemon labrosus	San Gabriel beardtongue
Penstemon rostriflorus	beaked beardtongue
Phacelia imbricata	imbricate scorpion-weed
Poa secunda	curly blue grass
Sarcodes sanguinea	snowplant
Silene lemmonii	Lemmon's catchfly
Solanum xantii	chaparral nightshade
Stipa coronata	giant needle grass
Streptanthus bernardinus	Laguna Mountain jewelflower
Viola purpurea	goose-foot yellow violet

White Fir–Sugar Pine Forest

White Fir–Sugar Pine Forest occurs at higher elevations (5500–8500 feet) than Yellow Pine Forest. White and sugar pine dominated areas are usually moist, steep, and on north and east facing slopes. This forest type grades into Lodgepole Pine Forest at higher elevations and Yellow Pine Forest at lower elevations. Cover contribution from each forest layer varies with site condition but four layers are usually present. This community corresponds to Coniferous Forests ERT.

Canopy
Characteristic Species

Abies concolor	white fir
Calocedrus decurrens	incense-cedar

Pinus lambertiana	sugar pine
Associates	
Arbutus menziesii	Pacific madrone
Castanopsis chrysophylla	golden chinkapin
Juniperus occidentalis	western juniper
Pinus contorta	lodgepole pine
Pinus jeffreyi	Jeffrey pine
Pinus monophylla	single-leaf pinyon
Quercus chrysolepis	canyon live oak
Quercus kelloggii	California black oak
Taxus brevifolia	Pacific yew

Woody Understory

Acer macrophyllum	big-leaf maple
Arctostaphylos patula	green-leaf manzanita
Castanopsis sempervirens	Sierran chinkapin
Ceanothus cordulatus	mountain whitethorn
Ceanothus greggii	Mojave buckbrush
Ceanothus integerrimus	deerbrush
Ceanothus parvifolius	cattlebush
Cerocarpus ledifolius	curl-leaf mountain-mahogany
Cornus nuttallii	Pacific flowering dogwood
Eriogonum umbellatum	sulphur-flower wild buckwheat
Eriogonum wrightii	bastard-sage
Fremontodendron californicum	California flannelbush
Holodiscus dumosus	glandular oceanspray
Leptodactylon pungens	granite prickly-phlox
Prunus emarginata	bitter cherry
Prunus virginiana	western choke cherry
Ribes viscosissimum	sticky currant
Rosa gymnocarpa	wood rose
Rubus parviflorus	western thimble-berry
Salix sitchensis	sitka willow
Sambucus cerulea	blue elder
Symphoricarpos oreophilus	mountain snowberry

Herbaceous Understory

Adenocaulon bicolor	American trailplant
Asarum hartwegii	Hartweg's wild ginger
Chimaphila menziesii	little prince's-pine
Clintonia uniflora	bride's bonnet
Disporum hookeri	drops-of-gold
Eriogonum kennedyi	Kennedy's wild buckwheat
Fragaria vesca	woodland strawberry
Galium sparsiflorum	Sequoia bedstraw
Goodyera oblongifolia	green-leaf rattlesnake-plantain
Hieracium albiflorum	white-flower hawkweed
Iris hartwegii	rainbow iris
Lupinus latifolius	broad-leaf lupine

Osmorhiza chilensis	mountain sweet-cicely
Pteridium aquilinum	northern bracken fern
Pyrola picta	white-vein wintergreen
Viola glabella	pioneer violet
Viola lobata	moose-horn violet

Lodgepole Pine Forest

Lodgepole Pine Forest, an upper montane forest type, occurs in the Transverse Mountain Ranges between 8000 and 10700 feet. Lodgepole Pine Forest is often a very dense forest, but in higher elevations the canopy may open and trees may take a "krummholz" form. Precipitation is primarily in the form of snow at this elevation. Soils are very shallow and rocky. This community corresponds to Coniferous Forests ERT.

Canopy
Characteristic Species

Pinus contorta	lodgepole pine

Associates

Pinus flexilis	limber pine

Woody Understory

Castanopsis sempervirens	Sierran chinkapin
Ceanothus cordulatus	mountain whitethorn
Eriogonum umbellatum	sulphur-flower wild buckwheat
Holodiscus dumosus	glandular oceanspray
Leptodactylon pungens	granite prickly-phlox
Ribes cereum	white squaw currant
Ribes montigenum	western prickly gooseberry

Herbaceous Understory

Antennaria rosea	rosy pussytoes
Arabis breweri	Brewer's rockcress
Arabis platysperma	pioneer rockcress
Calochortus invenustus	plain mariposa-lily
Carex macloviana	Falkland Island sedge
Carex rossii	Ross' sedge
Chimaphila umbellata	pipsissewa
Collinsia torreyi	Torrey's blue-eyed Mary
Corallorrhiza maculata	summer coralroot
Cryptogramma acrostichoides	American rockbrake
Draba corrugata	southern California whitlow-grass
Elymus elymoides	western bottle-brush grass
Erigeron breweri	Brewer's fleabane
Eriogonum saxatile	hoary wild buckwheat
Galium parishii	Parish's bedstraw
Heuchera abramsii	San Gabriel alumroot
Minuartia nuttallii	brittle stitchwort
Monardella cinerea	gray mountainbalm
Oreonana vestita	woolly mountain-parsley
Polypodium hesperium	western polypody

Pterospora andromedea	pine drops
Sarcodes sanguinea	snowplant
Sedum niveum	Davidson's stonecrop
Silene verecunda	San Francisco catchfly
Viola purpurea	goose-foot yellow violet
Woodsia oregana	Oregon cliff fern
Woodsia scopulina	Rocky Mountain cliff fern

Coulter Pine Woodland

This mixed evergreen forest community occurs in the southern and western Coast range and the Transverse and Peninsular ranges. Coulter Pine Woodland is best developed between 4000 and 6000 feet and may occur within the range of the Yellow Pine Forest and chaparral communities. Associates vary between sites, and particularly between different mountain ranges in the region. This community corresponds to Woodlands ERT.

Canopy
Characteristic Species

Pinus coulteri	coulter pine

Associates

Pinus ponderosa	ponderosa pine
Quercus chrysolepis	canyon live oak
Quercus kelloggii	California black oak

Woody Understory

Arctostaphylos canescens	hoary manzanita
Arctostaphylos viscida	white-leaf manzanita
Arctostaphylos glandulosa	Eastwood's manzanita
Rhamnus spp.	buckthorn
Ribes spp.	currant
Ceanothus integerrimus	deerbrush

Herbaceous Understory

Carex spp.	sedge
Galium spp.	bedstraw
Gayophytum spp.	groundsmoke
Lupinus spp.	lupine
Pyrola spp.	wintergreen

California Oak Woodland

California Oak Woodland occurs as a belt around the Central Valley of California generally at elevations between 250 to 3000 feet. Oak woodlands occur between grassland or scrub at lower elevations and montane forests at higher elevations. Several distinct oak "phases" are recognized such as "valley oak phase" and "blue oak phase." The phases can differ in canopy composition, but they all have a partial deciduous oak canopy and a grassy ground cover. Shrubs are usually present but contribute very little cover. The ground cover in the California Oak Woodland may include many grassland species listed in the California Prairie community. This community corresponds to Küchler #30 and Woodlands ERT.

Canopy
Characteristic Species

Pinus sabiniana	digger pine
Quercus douglasii	blue oak
Quercus lobata	valley oak

Associates

Aesculus californica	California buckeye
Juglans californica	southern California walnut
Quercus agrifolia	coastal live oak
Quercus engelmannii	Engelmann's oak
Quercus kelloggii	California black oak
Quercus wislizeni	interior live oak

Woody Understory

Arctostaphylos manzanita	big manzanita
Arctostaphylos viscida	white-leaf manzanita
Ceanothus cuneatus	sedge-leaf buckbrush
Frangula californica	coffee berry
Heteromeles arbutifolia	California-Christmas-berry
Prunus ilicifolia	holly-leaf cherry
Rhamnus crocea	holly-leaf buckthorn
Toxicodendron diversilobum	Pacific poison-oak

Juniper–Pinyon Woodland

Juniper–Pinyon Woodland occurs along the slopes of the Transverse Mountain ranges. This community is especially common on the north side of the San Gabriel and San Bernardino mountains. Junipers generally occur on the lower slopes and pinyon on the upper slopes in this area. Chaparral type shrubs may occur in the understory, especially on sites where these woodlands intergrade with chaparral. This community corresponds to Woodlands ERT.

Canopy
Characteristic Species

Juniperus californica	California juniper
Juniperus monosperma	one-seed juniper
Pinus monophylla	single-leaf pinyon

Associates

Juniperus osteosperma	Utah juniper
Pinus quadrifolia	four-leaf pinyon
Quercus dumosa	California scrub oak
Quercus turbinella	shrub live oak

Woody Understory

Arctostaphylos glauca	big-berry manzanita
Artemisia tridentata	big sagebrush
Cercocarpus montanus	alder-leaf mountain-mahogany
Chrysothamnus nauseosus	rubber rabbitbrush
Chrysothamnus teretifolius	needle-leaf rabbitbrush
Coleogyne ramosissima	blackbrush
Crossosoma bigelovii	ragged rockflower

Diplacus longiflorus	southern bush-monkey-flower
Echinocereus engelmannii	saints cactus
Encelia virginensis	Virgin River brittlebush
Ephedra nevadensis	Nevada joint-fir
Ericameria cuneata	cliff heath-goldenrod
Ericameria linearifolius	narrow-leaf heath-goldenrod
Eriogonum fasciculatum	eastern Mojave wild buckwheat
Krameria grayi	white ratany
Leptodactylon pungens	granite prickly-phlox
Nolina bigelovii	Bigelow's bear-grass
Opuntia basilaris	beaver-tail cactus
Oreonana vestita	woolly mountain-parsley
Prunus fasciculata	desert almond
Purshia glandulosa	antelope-brush
Salazaria mexicana	Mexican bladder-sage
Salvia dorrii	gray ball sage
Tetradymia axillaris	cottonhorn
Viguiera deltoidea	triangle goldeneye
Yucca brevifolia	Joshua-tree
Yucca schidigera	Mojave yucca

Herbaceous Understory

Anisocoma acaulis	scalebud
Caulanthus amplexicaulis	clasping-leaf wild cabbage
Erigeron parishii	Parish's fleabane
Eriogonum elongatum	long-stem wild buckwheat
Eriogonum ovalifolium	cushion wild buckwheat
Eriophyllum confertiflorum	yellow-yarrow
Gutierrezia sarothrae	kindlingweed
Hilaria rigida	big galleta
Layia glandulosa	white tidytips
Lotus rigidus	broom bird's-foot-trefoil
Mirabilis bigelovii	desert wishbonebush
Monardella linoides	flax-leaf mountainbalm
Pellaea mucronata	bird-foot cliffbrake
Penstemon grinnellii	Grinnell's beardtongue
Phacelia austromontana	southern Sierran scorpion-weed
Plagiobothrys kingii	Great Basin popcorn-flower
Poa secunda	curly blue grass
Silene parishii	Parish's catchfly
Stipa coronata	giant needle grass
Stipa speciosa	desert needle grass

Chaparral

In central and southern California, Chaparral occurs on most of the hills and lower mountain slopes from the Sierra Nevada Mountains to the coast. The characteristic species in chaparral communities are evergreen, woody shrubs. More than 100 evergreen shrub species are a part of chaparral communities. These species, however, do not occur together, but in the many different chaparral associations in the various climatic and geographic areas. The chaparral species are adapted to fire, drought, and nutrient-poor soils and grow best on rocky, steep slopes. Chaparral occurs in many areas but is best developed in southern California. Periodic fires occur frequently and are an important aspect of nutrient cycling in chaparral communities. This community corresponds to Küchler #33 and Desert Lands ERT.

Characteristic Species

Adenostoma fasciculatum	common chamise
Adenostoma sparsifolium	redshank
Arctostaphylos spp.	manzanita
Ceanothus thyrsiflorus	bluebrush

Associates

Arctostaphylos glandulosa	Eastwood's manzanita
Arctostaphylos glauca	big-berry manzanita
Arctostaphylos manzanita	big manzanita
Arctostaphylos myrtifolia	lone manzanita
Arctostaphylos parryana	pineland manzanita
Arctostaphylos viscida	white-leaf manzanita
Artemisia californica	coastal sagebrush
Ceanothus crassifolius	snowball
Ceanothus cuneatus	wedge-leaf buckbrush
Ceanothus dentatus	sandscrub
Ceanothus impressus	Santa Barbara buckbrush
Ceanothus jepsonii	muskbush
Ceanothus oliganthus	explorer's bush
Ceanothus palmeri	cuyamaca-bush
Ceanothus parryi	ladybloom
Ceanothus spinosus	redheart
Ceanothus tomentosus	ionebush
Ceanothus verrucosus	barranca-bush
Cercocarpus montanus	alder-leaf mountain-mahogany
Eriogonum fasciculatum	eastern Mojave wild buckwheat
Fraxinus dipetala	two-petal ash
Garrya flavescens	ashy silktassel
Garrya veatchii	canyon silktassel
Heteromeles arbutifolia	California-Christmas-berry
Leymus condensatus	giant lyme grass
Lonicera interrupta	chaparral honeysuckle
Lonicera subspicata	Santa Barbara honeysuckle
Malosma laurina	laurel-sumac
Prunus ilicifolia	holly-leaf cherry
Quercus dumosa	California scrub oak
Quercus durata	leather oak
Quercus wislizeni	interior live oak
Rhamnus crocea	holly-leaf buckthorn
Rhus integrifolia	lemonade sumac
Rhus ovata	sugar sumac
Ribes amarum	bitter gooseberry
Ribes californicum	California gooseberry
Ribes indecorum	white-flower currant
Ribes malvaceum	chaparral currant

Ribes viburnifolium	Santa Catalina currant
Salvia apiana	California white sage
Salvia mellifera	California black sage
Salvia spathacea	hummingbird sage
Yucca whipplei	Our-Lord's-candle

Coastal Sagebrush

Coastal Sagebrush occurs on the southern California coastal mountains on sites with lower moisture availability than sites which support chaparral. It generally occurs on the lower slopes of the California coastal mountains facing the ocean or in the interior coastal range in the rain shadow. This community corresponds to Küchler #35 and Desert Lands ERT.

Characteristic Species

Artemisia californica	coastal sagebrush
Eriogonum fasciculatum	eastern Mojave wild buckwheat
Salvia apiana	California white sage
Salvia mellifera	California black sage

Associates

Baccharis pilularis	coyote brush
Carpobrotus aequilateralus	baby sun-rose
Encelia californica	California brittlebush
Encelia farinosa	goldenhills
Eriophyllum confertiflorum	yellow-yarrow
Grindelia hirsutula	hairy gumweed
Hazardia squarrosa	saw-tooth bristleweed
Horkelia cuneata	wedge-leaf honeydew
Isocoma menziesii	jimmyweed
Malosma laurina	laurel-sumac
Opuntia littoralis	coastal prickly-pear
Rhus integrifolia	lemonade sumac
Salvia leucophylla	San Luis purple sage
Yucca whipplei	Our-Lord's-candle

San Joaquin Saltbush

Saltbush shrub communities may occur where the soil and available groundwater have high degrees of either salt or alkali. In the San Joaquin Valley, saltbush communities occur on wet sites near playas, sinks, and seeps that are fed with groundwater high in mineral content. Much of the area supporting saltbush communities has been lost to flood control, agricultural development, and groundwater pumping. This community corresponds to Desert Lands ERT.

Characteristic Species

Atriplex polycarpa	cattle-spinach

Associates

Allenrolfea occidentalis	iodinebush
Artemisia tridentata	big sagebrush
Arthrocnemum subterminale	Parish's glasswort
Atriplex fruticulosa	ball saltbush
Atriplex lentiformis	quailbush
Atriplex phyllostegia	arrow saltbush

Atriplex spinifera	spinescale
Distichlis spicata	coastal salt grass
Ephedra californica	California joint-fir
Kochia californica	California summer-cypress
Lycium cooperi	peachthorn
Nitrophila occidentalis	boraxweed
Pyrrocoma racemosa	clustered goldenweed
Sarcobatus vermiculatus	greasewood
Sporobolus airoides	alkali sacaton
Suaeda moquinii	shrubby seepweed

California Prairie

California Prairie once occurred throughout the Central Valley and in the surrounding mountains from sea level to 4000 feet. Today, grasslands occur as a ring around the Central Valley and species composition is dramatically different than the original prairie. The original prairie was a mixture of annuals and perennials. Since European settlement, the prairies changed by the invasion of exotic species, changes in grazing patterns, cultivation, and fire. This community corresponds to Küchler #48 and Grasslands ERT.

Characteristic Species

Nassella cernua	tussock grass
Nassella pulchra	tussock grass
Associates	
Aristida divaricata	poverty three-awn
Aristida oligantha	prairie three-awn
Aristida ternipes	spider grass
Deschampsia danthonioides	annual hair grass
Elymus glaucus	blue wild rye
Eschscholzia californica	California-poppy
Festuca idahoensis	bluebunch fescue
Gilia clivorum	purple-spot gily-flower
Gilia interior	inland gily-flower
Gilia minor	little gily-flower
Gilia tricolor	bird's-eyes
Koeleria macrantha	prairie Koeler's grass
Lasthenia californica	California goldfields
Leymus triticoides	beardless lyme grass
Lupinus bicolor	miniature annual lupine
Lupinus luteolus	butter lupine
Melica californica	California melic grass
Melica imperfecta	coast range melic grass
Nassella lepida	tussock grass
Orthocarpus attenuatus	valley-tassels
Orthocarpus campestris	field owl-clover
Orthocarpus erianthus	Johnnytuck
Orthocarpus linearilobus	pale owl-clover
Orthocarpus purpurascens	red owl-clover
Plagiobothrys nothofulvus	rusty popcorn-flower
Poa secunda	curly blue grass
Sisyrinchium bellum	Calfornia blue-eyed-grass

Stipa coronata	giant needle grass
Trifolium albopurpureum	rancheria clover
Trifolium depauperatum	balloon sack clover
Trifolium gracilentum	pin-point clover
Trifolium microdon	valparaiso clover
Trifolium olivaceum	olive clover
Vulpia microstachys	small six-weeks grass
Vulpia myuros	rat-tail six-weeks grass

WETLAND SYSTEMS

Riparian Forest

Riparian Forest in the Central Valley previously formed extensive stands along major streams. Due to flood control, water diversion, agricultural development, and urban expansion, these forests are now reduced and scattered or present as isolated young stands. Riparian Forest occurs on soils near streams that provide subsurface water even when the streams or rivers are dry. This community corresponds to Riparian Forests and Woodlands ERT.

Canopy
Characteristic Species
Populus fremontii	Fremont's cottonwood
Salix gooddingii	Goodding's willow

Associates
Acer negundo	ash-leaf maple
Alnus rhombifolia	white alder
Fraxinus latifolia	Oregon ash
Platanus racemosa	California sycamore
Quercus lobata	valley oak

Woody Understory
Adenostoma fasiculatum	common chamise
Baccharis salicifolia	mulefat
Cephalanthus occidentalis	common buttonbush
Clematis ligusticifolia	deciduous traveler's-joy
Eriodictyon crassifolium	thick-leaf yerba-santa
Rosa californica	California rose
Rubus vitifolius	Pacific dewberry
Salix bonplandiana	Bonpland's willow
Salix hindsiana	sandbar willow
Salix lasiolepis	arroyo willow
Salix lucida	shining willow
Vitis californica	California grape

Herbaceous Understory
Leymus triticoides	beardless lyme grass
Urtica dioica	stinging nettle

Vernal Pools

Vernal pools are small depressions typically in valley grasslands that fill with water during the winter and become dry during the spring and summer. As they dry, various annual plants flower in concentric rings. The pools most commonly border the east side of the Central Valley at the base of the Sierran Foothills, but are also found in southern and northern coastal areas. The vegetation of these pools is commonly described in terms of "zones." The zones are usually organized from pool bottom to mound top. Some pools support alkali or salt-associated plants. This community corresponds to Vernal Pools ERT.

Characteristic Species

Alopecurus saccatus	Pacific meadow-foxtail
Astragalus tener	alkali milk-vetch
Blennosperma nanum	common stickyseed
Boisduvalia glabella	smooth spike-primrose
Bromus hordeaceus	soft brome
Cressa truxillensis	spreading alkali-weed
Deschampsia danthonioides	annual hair grass
Distichlis spicata	coastal salt grass
Downingia bella	Hoover's calico-flower
Eleocharis palustris	pale spike-rush
Eryngium vaseyi	coyote-thistle
Evax caulescens	involucrate pygmy-cudweed
Gratiola ebracteata	bractless hedge-hyssop
Grindelia camporum	great valley gumweed
Isoetes howellii	Howell's quillwort
Juncus bufonius	toad rush
Lasthenia fremontii	Fremont's goldfields
Layia chrysanthemoides	smooth tidytips
Layia fremontii	Fremont's tidytips
Lepidium latipes	San Diego pepperwort
Lilaea scilloides	flowering-quillwort
Limnanthes douglasii	Douglas' meadowfoam
Machaerocarpus californicus	fringed-water-plantain
Marsilea vestita	hairy water-clover
Mimulus tricolor	tricolor monkey-flower
Minuartia californica	California stitchwort
Myosurus minimus	tiny mousetail
Navarretia intertexta	needle-leaf pincushion-plant
Navarretia leucocephala	white-flower pincushion-plant
Neostapfia colusana	colusa grass
Orthocarpus campestris	field owl-clover
Pilularia americana	American pillwort
Plagiobothrys acanthocarpus	adobe popcorn-flower
Plagiobothrys distantiflorus	Calfornia popcorn-flower
Plagiobothrys humistratus	dwarf popcorn-flower
Plagiobothrys hystriculus	bearded popcorn-flower
Plagiobothrys leptocladus	alkali popcorn-flower
Plagiobothrys stipitatus	stalked popcorn-flower
Pogogyne ziziphoroides	Sacramento mesa-mint

Psilocarphus tenellus	slender woollyheads
Trifolium barbigerum	bearded clover
Trifolium cyathiferum	bowl clover
Trifolium depauperatum	balloon sack clover
Trifolium fucatum	sour clover
Trifolium variegatum	white-tip clover
Veronica peregrina	neckweed
Vulpia myuros	rat-tail six-weeks grass

Tule Marsh

Tule Marsh is a community type consisting of "freshwater marshes" that occur along the coast in river deltas and in coastal valleys near lakes and springs. Freshwater marshes were common throughout the Sacramento and San Joaquin Valleys, but have been reduced due to development pressures. Although the marshes may occur in areas near the coast, they are not influenced by tidal currents and are permanently flooded with fresh water. This community corresponds to Nonestuarine Marshes ERT.

Characteristic Species

Scirpus acutus	hard-stem bulrush
Typha latifolia	broad-leaf cat-tail

Associates

Carex lanuginosa	woolly sedge
Carex obnupta	slough sedge
Carex senta	western rough sedge
Cyperus eragrostis	tall flat sedge
Cyperus esculentus	chufa
Eleocharis palustris	pale spike-rush
Hydrocotyle verticillata	whorled marsh-pennywort
Juncus balticus	Baltic rush
Juncus effusus	lamp rush
Juncus patens	spreading rush
Limosella aquatica	awl-leaf mudwort
Phragmites australis	common reed
Scirpus americanus	chairmaker's bulrush
Scirpus californicus	California bulrush
Scirpus robustus	seaside bulrush
Scirpus tabernaemontani	soft-stem bulrush
Sparganium eurycarpum	broad-fruit burr-reed
Typha angustifolia	narrow-leaf cat-tail
Typha domingensis	southern cat-tail
Verbena bonariensis	purple-top vervain

Mountain Meadow

Mountain Meadow occurs in the Transverse Mountain ranges where the water table is just below the surface of the soil. Meadows often occur along streams or snowmelt gullies. This community corresponds to Meadowlands ERT.

Characteristic Species

Agrostis idahoensis	Idaho bent
Barbarea orthoceras	American yellow-rocket

Carex hassei	Hasse's sedge
Carex heteroneura	different-nerve sedge
Carex jonesii	Jones' sedge
Carex schottii	Schott's sedge
Carex senta	western rough sedge
Castilleja miniata	great red Indian-paintbrush
Cystopteris fragilis	brittle bladder fern
Dodecatheon redolens	scented shootingstar
Draba albertina	slender whitlow-grass
Epilobium angustifolium	fireweed
Epilobium ciliatum	fringed willowherb
Epilobium glaberrimum	glaucous willowherb
Galium bifolium	twin-leaf bedstraw
Helenium bigelovii	Bigelow's sneezeweed
Heracleum sphondylium	eltrot
Hoita orbicularis	round-leaf leather-root
Hypericum anagalloides	tinker's penny
Iris missouriensis	Rocky Mountain iris
Juncus covillei	Coville's rush
Juncus nevadensis	Sierran rush
Lewisia nevadensis	Nevada bitter-root
Lilium parryi	lemon lily
Lithophragma tenellum	slender woodlandstar
Lotus oblongifolius	streambank bird's-foot-trefoil
Lupinus latifolius	broad-leaf lupine
Luzula congesta	heath wood-rush
Maianthemum stellatum	starry false Solomon's-seal
Mimulus moschatus	muskflower
Mimulus suksdorfii	minature monkey-flower
Oenothera elata	Hooker's evening-primrose
Perideridia parishii	Parish's yampah
Platanthera leucostachys	Sierra rein orchid
Platanthera sparsiflora	canyon bog orchid
Potentilla glandulosa	sticky cinquefoil
Pycnanthemum californicum	California mountain-mint
Ribes cereum	white squaw currant
Salix lasiolepis	arroyo willow
Senecio triangularis	arrow-leaf ragwort
Sisyrinchium bellum	California blue-eyed-grass
Solidago californica	northern California goldenrod
Sphenosciadium capitellatum	swamp whiteheads
Trifolium monanthum	mountain carpet clover
Trisetum spicatum	narrow false oat
Veratrum californicum	California false hellebore
Viola blanda	sweet white violet

ESTUARINE SYSTEMS

Salt Marsh

Salt Marsh occurs along the southern California coast in bays, lagoons, and estuaries. Salt marshes are not as extensive in southern California as they are in northern California. Salt marshes may be broken into two categories: deep-water sites and smaller, shallower sites. These two types have different tidal and flooding regimes which influence the community composition. This community corresponds to Estuarine Marshes ERT.

Characteristic species
Salicornia virginica	woody saltwort

Associates
Amblyopappus pusillus	dwarf coastweed
Anemopsis californica	yerba-mansa
Arthrocnemum subterminale	Parish's glasswort
Atriplex patula	halberd-leaf orache
Atriplex watsonii	Watson's saltbush
Batis maritima	turtleweed
Carpobrotus aequilateralus	baby sun-rose
Cotula coronopifolia	common brassbuttons
Cressa truxillensis	spreading alkali-weed
Distichlis spicata	coastal salt grass
Frankenia salina	alkali sea-heath
Heliotropium convolvulaceum	wide-flower heliotrope
Heliotropium curassavicum	seaside heliotrope
Jaumea carnosa	marsh jaumea
Juncus acutus	spiny rush
Juncus mexicanus	Mexican rush
Lasthenia glabrata	yellow-ray goldfields
Limonium californicum	marsh-rosemary
Monanthochloe littoralis	shore grass
Salicornia bigelovii	dwarf saltwort
Scirpus californicus	California bulrush
Spartina foliosa	California cord grass
Spergularia salina	sandspurrey
Suaeda californica	broom seepweed
Triglochin maritimum	seaside arrow-grass
Typha latifolia	broad-leaf cat-tail

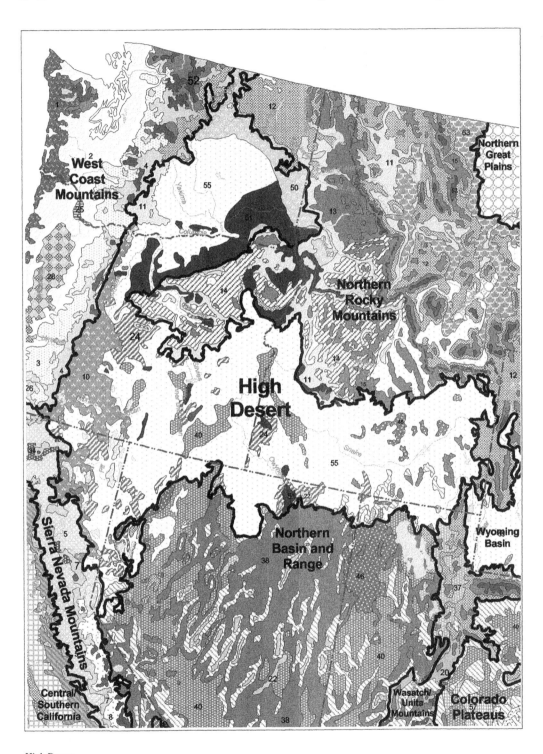

High Desert

High Desert

Introduction

The High Desert region occurs in the rain shadow of the Cascade Mountain Range. The region includes most of central and southeastern Washington, much of eastern Oregon (excluding the Blue Mountains), southern Idaho, and parts of northern California, Nevada, and Colorado. This region is arid to semiarid with warm-to-hot, dry summers and relatively cold winters. Sagebrush and dry grassland communities are common in this region, with woodland and forest communities in the mountains.

The vegetation of this region is related to the distribution of precipitation. Beginning in the dry Columbia Basin, extending eastward to the foothills of the Rocky Mountains and westward to the east side of the Cascades, the vegetation changes from shrub-steppe dominated by sagebrush to steppe dominated by bunchgrass. It then changes to meadow-steppe which supports grasses and forbs common to more mesic areas.

The High Desert Region has undergone many vegetational changes in the past century. Before European settlement, grazing animals were not an integral part of this region. Since the 1860s, however, cattle and sheep grazing have impacted much of the steppe and many exotic or nonnative species have been introduced into the ecosystem. One of the more aggressive weeds is cheatgrass (*Bromus tectorum*), which out competes the native grasses by its ability to grow through the fall and winter. Cheatgrass has also made this region more susceptible to earlier and more frequent fires.

Unique communities in this area include vegetation of the sand dunes near the Columbia River and marshes around the numerous ponds in the Columbia basin. Several other plant communities found in this region, but not included in the descriptions, are Juniper–Pinyon Woodland and Western Spruce–Fir Forest. These types are more widely represented in the adjacent regions and will be discussed at length in the Northern Basin and Range region or the Northern Rocky Mountain region.

The primary sources used to develop descriptions and species lists are Barbour and Billings 1988; Cronquist et al. 1972; Franklin and Dyrness 1973.

Dominant Ecological Communities

Upland Systems
Ponderosa Pine Forest
Juniper Woodland
Sagebrush Steppe
Saltbush–Greasewood Shrub
Palouse Prairie
Meadow Steppe

UPLAND SYSTEMS

Ponderosa Pine Forest

Within this region, Ponderosa Pine Forest occupies a narrow band on the eastern slopes of the Cascade Range. This area is very dry with a short growing season and very little summer precipitation. Much of the annual precipitation occurs in winter as snow. Ponderosa pine grows best on coarse-textured, sandy soils, often forming a mosaic with steppe or shrub-steppe communities. Ponderosa Pine Forest extends beyond the eastern slopes of the Cascades into the Ochoco, Blue, and Wallowa Mountains and into the Northern Rockies. In all of these areas, Idaho fescue (bluebunch fescue) and bitterbrush are understory species. This community corresponds to Küchler #10 and #11 and Coniferous Forests ERT.

Canopy
Characteristic Species
  Pinus ponderosa   ponderosa pine
Associates
  Juniperus occidentalis   western juniper
  Pinus contorta   lodgepole pine
  Populus tremuloides   quaking aspen
  Quercus garryana   Oregon white oak

Woody Understory
  Amelanchier alnifolia   Saskatoon service-berry
  Crataegus douglasii   black hawthorn
  Eriogonum niveum   snow erigonum
  Holodiscus discolor   hillside oceanspray
  Mahonia repens   creeping Oregon-grape
  Physocarpus malvaceus   mallow-leaf ninebark
  Prunus virginiana   choke cherry
  Purshia tridentata   bitterbrush
  Symphoricarpos albus   common snowberry

Herbaceous Understory
  Antennaria dimorpha   cushion pussytoes
  Apocynum androsaemifolium   spreading dogbane
  Arabis holboellii   Holboell's rockcress
  Balsamorhiza sagittata   arrow-leaf balsamroot
  Bromus vulgaris   Columbian brome
  Calamagrostis rubenscens   pinegrass
  Carex geyeri   Geyer's sedge
  Claytonia perfoliata   miner's-lettuce
  Collinsia parviflora   small-flower blue-eyed Mary
  Elymus glaucus   blue wild rye
  Epilobium brachycarpum   tall annual willowherb
  Erigeron compositus   dwarf mountain fleabane
  Eriogonum heracleoides   parsnip buckwheat
  Erythronium grandiflorum   yellow avalanche-lily
  Festuca idahoensis   bluebunch fescue
  Festuca occidentalis   western fescue
  Frasera albicaulis   white-stem elkweed
  Galium aparine   sticky-willy
  Galium boreale   northern bedstraw
  Koeleria macrantha   prairie Koeler's grass
  Lithophragma glabrum   bulbous woodlandstar
  Lithospermum ruderale   Columbian puccoon
  Lotus nevadensis   Nevada bird's-foot-trefoil
  Madia exigua   little tarplant
  Osmorhiza chilensis   mountain sweet-cicely
  Phlox gracilis   slender phlox
  Poa secunda   curly bluegrass
  Potentilla gracilis   graceful cinquefoil

Pseudoroegneria spicata	bluebunch-wheat grass
Ranunculus glaberrimus	sagebrush buttercup
Sisyrinchium douglasii	grass windows
Stellaria nitens	shining starwort
Stipa occidentalis	western needle grass
Vicia americana	American purple vetch
Vulpia microstachys	small six-weeks grass

Juniper Woodland

Juniper Woodland is the most xeric of the tree-dominated vegetation types in the Pacific Northwest. Juniper Woodland occupies habitats of intermediate moisture between the Ponderosa Pine Forest and the steppe or shrub-steppe vegetation. It is primarily in central Oregon around the Deschutes, Crooked, and John Day Rivers. Juniper is the dominant species, however, ponderosa pine may be found in canyon bottoms or on north slopes where soil moisture is greater. This community corresponds to Küchler #24 and Woodlands ERT.

Canopy
Characteristic Species
Juniperus occidentalis	western juniper

Associates
Pinus ponderosa	ponderosa pine

Woody Understory
Artemisia tridentata	big sagebrush
Chrysothamnus nauseosus	rubber rabbitbrush
Chrysothamnus viscidiflorus	green rabbitbrush
Eriogonum umbellatum	sulfur-flower wild buckwheat
Purshia tridentata	bitterbrush

Herbaceous Understory
Astragalus purshii	Pursh's milk-vetch
Claytonia perfoliata	miner's lettuce
Collinsia parviflora	small-flower blue-eyed Mary
Collomia grandiflora	large-flowered mountain-trumpet
Cryptantha ambigua	basin cat's-eye
Elymus elymoides	western bottle-brushgrass
Erigeron linearis	desert yellow fleabane
Eriogonum baileyi	Bailey's wild buckwheat
Eriophyllum lanatum	common woolly-sunflower
Festuca idahoensis	bluebunch fescue
Gayophytum humile	dwarf groundsmoke
Koeleria macrantha	prairie Koeler's grass
Lomatium triternatum	nineleaf desert-parsley
Mentzelia albicaulis	white-stem blazingstar
Penstemon humilis	low beardtongue
Phlox caespitosa	clustered phlox
Poa secunda	curly blue grass
Pseudoroegneria spicata	bluebunch-wheat grass
Stipa thurberiana	Thurber's needle grass

Sagebrush Steppe

Sagebrush Steppe is the driest and most widespread community in the High Desert. Sagebrush occupies the center of the Columbia river basin and extends west into the foothills of the Cascades and southeast into Oregon. Big sagebrush (*Artemisia tridentata*) is the most characteristic species. Fire within the Sagebrush Steppe community has increased since an exotic, cheatgrass (*Bromus tectorum*), has become dominant. Cheatgrass grows fast and then dries up, creating perfect tinder for fires ignited by lightening. This community corresponds to Küchler #55 and Desert Lands ERT.

Characteristic Woody Species
Artemisia tridentata	big sagebrush

Associates
Artemisia tripartita	three-tip sagebrush
Chrysothamnus nauseosus	rubber rabbitbrush
Chrysothamnus viscidiflorus	green rabbitbrush
Grayia spinosa	spiny hop-sage
Juniperus occidentalis	western juniper
Tetradymia canescens	spineless horsebrush

Herbaceous Understory
Elymus elymoides	western bottle-brush grass
Lappula occidentalis	flat-spine sheepburr
Poa fendleriana	mutton grass
Poa secunda	curly blue grass
Pseudoroegneria spicata	bluebunch-wheat grass
Stipa comata	needle-and-thread
Stipa thurberiana	Thurber's needle grass
Tortula brevipes	moss
Tortula princeps	moss
Tortula ruralis	moss

Saltbush–Greasewood Shrub

Saltbush–Greasewood Shrub is a desert-shrub community which occurs on saline soils and old lakebeds. Few plants can survive in this dry, salty habitat. A sizable representation of this community type occurs in southeast Oregon and southern Idaho. This community corresponds to Küchler #40 and Desert Lands ERT.

Characteristic Woody Species
Atriplex canescens	four-wing saltbush
Atriplex confertifolia	shadscale
Atriplex nuttallii	Nuttall's saltbush sage
Sarcobatus vermiculatus	greasewood

Associates
Allenrolfea occidentalis	iodinebush
Artemisia spinescens	bud sagebrush
Artemisia tridentata	big sagebrush
Grayia spinosa	spiny hop-sage
Krascheninnikovia lantata	winterfat

Lycium cooperi	peachthorn
Menodora spinescens	spiny menodora
Suaeda moquinii	shrubby seepweed

Herbaceous Understory

Distichlis spicata	coastal salt grass
Elymus elymoides	western bottle-brush grass
Kochia americana	greenmolly
Leymus cinereus	Great Basin lyme grass
Leymus triticoides	beardless lyme grass
Oryzopsis hymenoides	Indian mountain-rice grass
Poa secunda	curly blue grass
Salvia dorrii	gray ball sage

Palouse Prairie

Palouse Prairie depends upon a greater degree of moisture than Sagebrush Steppe and usually occurs at higher altitudes than sagebrush communities. Most native Palouse Prairie has been destroyed by excessive grazing and cultivation which has resulted in the replacement of native perennial bunchgrasses with sagebrush, cheatgrass, or agricultural crops. This community corresponds to Küchler #50 and Prairies ERT.

Characteristic Species

Festuca idahoensis	bluebunch fescue
Leymus condensatus	giant lyme grass
Poa secunda	curly blue grass
Pseudoroegneria spicata	bluebunch-wheat grass

Associates

Astragalus spaldingii	Spalding's milk-vetch
Chrysothamnus nauseosus	rubber rabbitbrush
Festuca campestris	prairie fescue
Koeleria macrantha	prairie Koeler's grass
Lappula occidentalis	flat-spine sheepburr
Montia linearis	linear-leaf candy-flower
Pascopyrum smithii	western-wheat grass
Phlox gracilis	slender phlox
Phlox longifolia	long-leaf phlox
Plantago patagonica	woolly plantain
Rosa woodsii	Woods' rose
Stellaria nitens	shiny starwort
Stipa comata	needle-and-thread

Meadow Steppe

Meadow Steppe is the most mesic of the steppe communities. This vegetation type encircles the Columbia Basin Province of Oregon and Washington. The dominant species are herbaceous; however, dwarfed shrubs may occur with the grasses forming shrubby islands or thickets. Where heavy grazing and fire have impacted community composition and structure, exotics such as Kentucky bluegrass (*Poa pratensis*) may dominate. This community corresponds to Meadowlands ERT.

Characteristic Species

Festuca idahoensis	bluebunch fescue
Koeleria macrantha	prairie Koeler's grass
Poa secunda	curly blue grass
Pseudoroegneria spicata	bluebunch-wheat grass
Symphoricarpos albus	common snowberry

Associates

Astragalus arrectus	palouse milk-vetch
Balsamorhiza sagittata	arrow-leaf balsamroot
Castilleja lutescens	stiff yellow Indian-paintbrush
Geranium viscosissimum	sticky purple crane's-bill
Geum triflorum	old-man's-whiskers
Helianthella uniflora	Rocky Mountain dwarf-sunflower
Hieracium cynoglossoides	hound-tongue hawkweed
Iris missouriensis	Rocky Mountain iris
Lupinus sericeus	Pursh's silky lupine
Potentilla gracilis	graceful cinquefoil
Rosa nutkana	Nootka rose
Rosa woodsii	Woods' rose

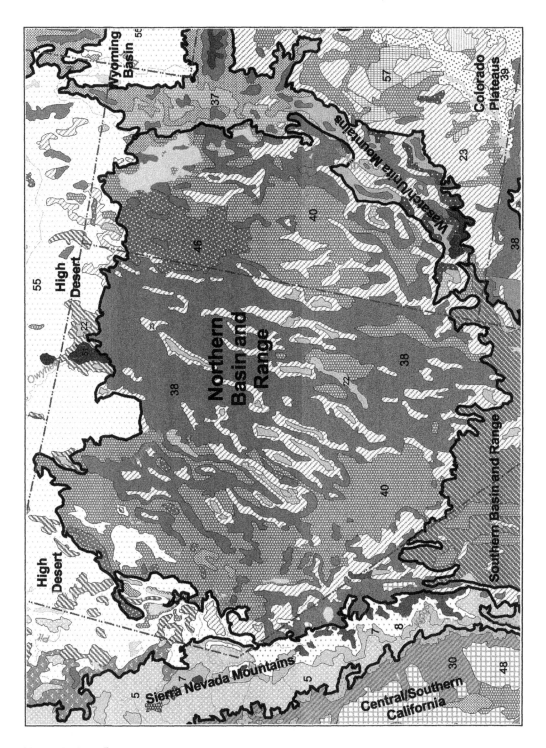

Northern Basin and Range

Northern Basin and Range

Introduction

The Northern Basin and Range region is also referred to as the intermountain region and the cold desert because of its cool climate and location between the Rocky Mountains to the east and the Sierra Nevada and Cascade Mountains to the west. The region encompasses most of Nevada and western Utah. The Snake River lava flows are the northern boundary with the Wasatch and Uinta Mountains to the west.

The Northern Basin and Range region, as its name implies, is characterized by numerous flat basins which occur between the many small mountain ranges. The soils are generally rocky, thin, low in organic matter, and high in minerals. The region is arid with the mountains capturing more moisture than the basins. The moisture that does make its way into the basins most often flows into low, flat playa lakes where the water evaporates leaving salt and mineral deposits. Because of this moisture regime, a large part of the vegetation is adapted to soils with a high mineral and salt content.

The primary sources used to develop descriptions and species lists are Barbour and Billings 1988; Cronquist et al. 1972.

Dominant Ecological Communities

Upland Systems
Great Basin Pine Forest
Juniper–Pinyon Woodland
Mountain-Mahogany–Oak Scrub
Great Basin Sagebrush
Saltbush Scrub
Greasewood Scrub
Wetland Systems
Inland Saline Marsh

UPLAND SYSTEMS

Great Basin Pine Forest

Great Basin Pine Forest occurs on several of the higher mountain ranges in the Northern Basin and Range region. The White Mountains, the Shoshone Mountains, and the Stillwater Range are among the ranges where the pine forest, dominated by limber pine and bristle-cone pine, can be found. These forests are best developed between 9500 feet and timberline (about 10500 feet). The trees are relatively short (between thirty and forty feet). Bristle-cone pine is dominant in the southern half of the region and limber pine is dominant in the north. This community corresponds to Küchler #22 and Coniferous Forests ERT.

Canopy
Characteristic Species
Pinus aristata   bristle-cone pine
Pinus flexilis   limber pine
Pinus longaeva   intermountain bristle-cone pine
Associates
Pinus ponderosa   ponderosa pine

Woody Understory

Artemisia tridentata	big sagebrush
Cercocarpus ledifolius	curl-leaf mountain-mahogany
Ericameria gilmanii	white-flower heath-goldenrod
Holodiscus dumosus	glandular oceanspray

Herbaceous Understory

Astragalus platytropis	broad-keel milk-vetch
Cymopterus cinerarius	gray spring-parsley
Cryptantha hoffmannii	Hoffman's cat's-eye
Cryptantha roosiorum	bristle-cone cat's-eye
Festuca idahoensis	bluebunch fescue
Poa secunda	curly blue grass
Pseudoroegneria spicata	bluebunch-wheat grass
Stipa pinetorum	pine-forest needle grass

Juniper–Pinyon Woodland

Juniper–Pinyon Woodland occurs on the slopes of the mountains throughout the Northern Basin and Range region. Pinyon and juniper are generally found at elevations between 5000 and 8000 feet. Pinyon tends to dominate at upper elevations and juniper at lower elevations. Much of the woodland area has been heavily grazed which has changed fire cycles. Historically, these communities were more likely open with grass-dominated understories. These changes have resulted in the loss of understory species and an increase in erosion. This community corresponds to Küchler #23 and Woodlands ERT.

Canopy

Characteristic Species

Juniperus monosperma	one-seed juniper
Juniperus osteosperma	Utah juniper
Pinus edulis	two-needle pinyon
Pinus monophylla	single-leaf pinyon

Associates

Acer glabrum	Rocky Mountain maple
Amelanchier alnifolia	Saskatoon service-berry
Amelanchier utahensis	Utah service-berry
Juniperus deppeana	alligator juniper
Juniperus occidentalis	western juniper
Quercus emoryi	Emory's oak
Quercus gumbelii	Gambel's oak
Quercus grisea	gray oak

Woody Understory

Artemisia arbuscula	dwarf sagebrush
Artemisia tridentata	big sagebrush
Ceanothus velutinus	tobacco-brush
Cercocarpus ledifolius	curl-leaf mountain-mahogany
Chrysothamnus nauseosus	rubber rabbitbrush
Chrysothamnus viscidiflorus	green rabbitbrush
Ephedra viridis	Mormon-tea

Eriogonum umbellatum	sulphur-flower wild buckwheat
Fallugia paradoxa	Apache-plume
Gutierrezia sarothrae	kindlingweed
Holodiscus dumosus	glandular oceanspray
Purshia mexicana	Mexican cliff-rose
Purshia tridentata	bitterbrush
Ribes cereum	white squaw currant
Ribes velutinum	desert gooseberry
Symphoricarpos oreophilus	mountain snowberry
Tetradymia canescens	spineless horsebrush

Herbaceous Understory

Balsamorhiza sagittata	arrow-leaf balsamroot
Bouteloua curtipendula	side-oats grama
Bouteloua gracilis	blue grama
Elymus elymoides	western bottle-brush grass
Eriogonum heracleoides	parsnip-flower wild buckwheat
Eriophyllum lanatum	common woolly-sunflower
Festuca idahoensis	bluebunch fescue
Festuca kingii	King's fescue
Frasera albomarginata	desert elkweed
Grindelia squarrosa	curly-cup gumweed
Heterotheca villosa	hairy false golden-aster
Hymneoxys richardsonii	Colorado rubberweed
Ipomopsis aggregata	scarlet skyrocket
Koeleria macrantha	prairie Koeler's grass
Lithospermum ruderale	Columbian puccoon
Lupinus sericeus	Pursh's silky lupine
Oryzopsis hymenoides	Indian mountain-rice grass
Pascopyrum smithii	western-wheat grass
Pentstemon speciosus	royal beardtongue
Pentstemon watsonii	Watson's beardtongue
Poa fendleriana	mutton grass
Poa secunda	curly blue grass
Pseudoroegneria spicatum	bluebunch-wheat grass
Sporbolus cryptandrus	sand dropseed
Stipa comata	needle-and-thread
Stipa nelsonii	Nelson's needle grass
Stipa thurberiana	Thurber's needle grass

Mountain-Mahogany–Scrub Oak

Mountain-Mahogany–Scrub Oak occurs in the Northern Basin and Range region in the foothills of several mountain ranges including the Toiyabee and Shoshone Mountains. This community is more widely represented in the Uinta Mountains at the northeast boundary of this region. Community dominants vary throughout the region and even locally with elevations and other factors. Within the Mountain-Mahogany–Oak Scrub community type there has been a decrease in herbaceous understory species due to extensive livestock grazing. This community corresponds to Küchler #37 and Woodlands ERT.

Canopy
Characteristic Species
Cercocarpus ledifolius	curl-leaf mountain-mahogany
Quercus gambelii	Gambel's oak

Associates
Acer glabrum	Rocky Mountain maple
Acer grandidentatum	canyon maple
Acer negundo	ash-leaf maple
Amelanchier alnifolia	Saskatoon service-berry
Amelanchier utahensis	Utah service-berry
Cercocarpus montanus	alder-leaf mountain-mahogany
Quercus havardii	Havard's oak
Quercus turbinella	shrub live oak

Woody Understory
Arctostaphylos spp.	manzanita
Artemisia arbuscula	dwarf sagebrush
Artemisia tridentata	big sagebrush
Betula occidentalis	water birch
Ceanothus fendleri	Fendler's buckbrush
Ceanothus velutinus	tobacco-brush
Chrysothamnus viscidiflorus	green rabbitbrush
Fallugia paradoxa	Apache-plume
Gutierrezia sarothrae	kindlingweed
Pachystima myrsinites	Oregon boxwood
Physocarpus malvaceus	mallow-leaf ninebark
Populus angustifolia	narrow-leaf cottonwood
Prunus virginiana	choke cherry
Purshia mexicana	Mexican cliff-rose
Purshia tridentata	bitterbrush
Rhamnus crocea	holly-leaf buckthorn
Rhus trilobata	ill-scented sumac
Ribes cereum	white squaw currant
Rosa woodsii	Arizona rose
Sambucus cerulea	blue elder
Symphoricarpos oreophilus	mountain snowberry

Herbaceous Understory
Claytonia lanceolata	lance-leaf springbeauty
Collinsia parviflora	small-flower blue-eyed Mary
Delphinium nuttallianum	two-lobe larkspur
Erigeron flagellaris	trailing fleabane
Erythronium grandiflorum	yellow avalanche-lily
Gayophytum ramosissimum	pinyon groundsmoke
Geranium caespitosum	purple cluster crane's-bill
Heliomeris multiflora	Nevada showy false goldeneye
Hydrophyllum capitatum	cat's-breeches
Leymus cinereus	great basin lyme grass
Lithophragma parviflorum	prairie woodlandstar

Nemophila breviflora	Great Basin baby-blue-eyes
Orthocarous luteus	golden-tongue owl-clover
Stipa comata	needle-and-thread
Stipa lettermanii	Letterman's needle grass
Wyethia amplexicaulis	northern mule's-ears

Great Basin Sagebrush

Sagebrush vegetation occurs in desert areas within the Northern Basin and Range region where annual precipitation is at least seven inches. The sagebrush communities are usually located in the foothills and mountainsides above 5000 feet; some sagebrush communities may extend up to 10000 feet. Great Basin Sagebrush communities are very arid and desert-like. The sagebrush seldom grows over one foot in height and the grasses are sparse. In the northern part of the region, the sagebrush communities are similar to the Sagebrush Steppe of the High Desert region. Much of the sagebrush area has been converted into farmland. This community corresponds to Küchler #38 and Desert Lands ERT.

Characteristic Woody Species

Artemisia tridentata	big sagebrush

Associates

Artemisia nova	black sagebrush
Atriplex confertifolia	shadscale
Chrysothamnus nauseosus	rubber rabbitbrush
Chrysothamnus viscidiflorus	green rabbitbrush
Coleogyne ramosissima	blackbrush
Ephedra torreyana	Torrey's joint-fir
Ephedra viridis	Mormon-tea
Grayia spinosa	spiny hop-sage
Purshia tridentata	bitterbrush
Ribes velutinum	desert gooseberry
Tetradymia glabrata	little-leaf horsebrush

Herbaceous Understory

Allium acuminatum	taper-tip onion
Aristida purpurea	purple three-awn
Balsamorhiza sagittata	arrow-leaf balsamroot
Calochortus nuttallii	sego-lily
Castilleja angustifolia	northwestern Indian-paintbrush
Crepis acuminata	long-leaf hawk's-beard
Delphinium andersonii	desert larkspur
Elymus elymoides	western bottle-brush grass
Elymus lanceolatus	streamside wild rye
Festuca idahoensis	bluebunch fescue
Heterotheca villosa	hairy false golden-aster
Hymenoxys richardsonii	Colorado rubberweed
Koeleria macrantha	prairie Koeler's grass
Leptodactylon pungens	granite prickly-phlox
Leymus cinereus	great basin lyme grass
Lupinus caudatus	Kellogg's spurred lupine
Lupinus sericeus	Pursh's silky lupine

Oryzopsis hymenoides	Indian mountain-rice grass
Pascopyrum smithii	western-wheat grass
Phlox hoodii	carpet phlox
Phlox longifolia	long-leaf phlox
Poa fendleriana	mutton grass
Poa secunda	curly blue grass
Pseudoroegneria spicata	bluebunch-wheat grass
Sporobolus airoides	alkali-sacaton
Stipa comata	needle-and-thread
Viola beckwithii	western pansy
Wyethia amplexicaulis	northern mule's-ears
Zigadenus paniculatus	sand-corn

Saltbush Scrub

Saltbush Scrub occurs in areas with little moisture and on saline valley soils. Structurally, this community is predominantly widely spaced shrubs. This community type is also often referred to as the "shadscale community." This community corresponds, in part, to Küchler #40 and Desert Lands ERT.

Characteristic Woody Species

Atriplex confertifolia	shadscale

Associates

Artemisia spinescens	bud sagebrush
Atriplex canescens	four-wing saltbush
Atriplex gardneri	Gardner's saltbush
Atriplex nuttallii	Nuttall's saltbush
Chrysothamnus viscidiflorus	green rabbitbrush
Ephedra nevadensis	Nevada joint-fir
Grayia spinosa	spiny hop-sage
Gutierrezia sarothrae	kindlingweed
Krascheninnikovia lanata	winterfat
Lycium cooperi	peachthorn
Sarcobatus vermiculatus	greasewood
Tetradymia glabrata	little-leaf horsebrush

Herbaceous Understory

Camissonia boothii	shredding suncup
Camissonia claviformis	browneyes
Camissonia scapoidea	Paiute suncup
Cardaria draba	heart-pod hoary cress
Cryptantha circumscissa	cushion cat's-eye
Eriogonum ovalifolium	cushion wild buckwheat
Halogeton glomeratus	saltlover
Iva nevadensis	Nevada marsh-elder
Kochia americana	greenmolly
Mirabilis alipes	winged four-o'clock
Sphaeralcea grossulariaefolia	currant-leaf globe-mallow
Vulpia octoflora	eight-flower six-weeks grass
Xylorhiza glabriuscula	smooth woody-aster

Greasewood Scrub

Greasewood Scrub is the most salt-tolerant community in the Northern Basin and Range region. This vegetation type occurs in the valley bottoms where the water table is relatively close to the surface. This community corresponds, in part, to Küchler #40 and Desert Lands ERT.

Characteristic Woody Species
Sarcobatus vermiculatus	greasewood

Associates
Allenrolfea occidentalis	iodinebush
Artemisia spinescens	bud sagebrush
Atriplex confertifolia	shadscale
Atriplex lentiformis	quailbush
Suaeda monquinii	shrubby seepweed
Thelypodium sagittatum	arrowhead thelypody

Herbaceous Understory
Cordylanthus maritimus	saltmarsh bird's-beak
Glaux maritima	sea-milkwort
Halogeton glomeratus	saltlover
Hutchinsia procumbens	ovalpurse
Iva axillaris	deer-root
Juncus balticus	Baltic rush
Kochia americana	greenmolly
Pyrrocoma lanceolata	lance-leaf goldenweed
Salicornia rubra	red saltwort
Sarcocornia pacifica	Pacific swampfire
Sporobolus airoides	alkali-sacaton

WETLAND SYSTEMS

Inland Saline Marsh

Inland Saline Marsh develops in areas with high salinity and poor drainage. This vegetation develops around shallow lake shores and across areas of standing shallow water. This community corresponds to Küchler #49 and Nonestuarine Marshes ERT.

Characteristic Species
Allenrolfea occidentalis	iodinebush
Berula erecta	cut-leaf-water-parsnip
Castilleja exilis	small-flower annual indian-paintbrush
Centaurium exaltatum	desert centaury
Ceratophyllum demersum	coontail
Distichlis spicata	coastal salt grass
Eleocharis rostellata	beaked spike-rush
Juncus balticus	Baltic rush
Najas marina	holly-leaf waternymph
Phragmites australis	common reed
Ruppia maritima	beaked ditch-grass

Sarcocornia pacifica	Pacific swampfire
Scirpus acutus	hard-stem bulrush
Scirpus americanus	chairmaker's bulrush
Scirpus maritimus	saltmarsh bulrush
Sporobolus airoides	alkali-sacaton
Triglochin maritimum	seaside arrow-grass
Typha latifolia	broad-leaf cat-tail

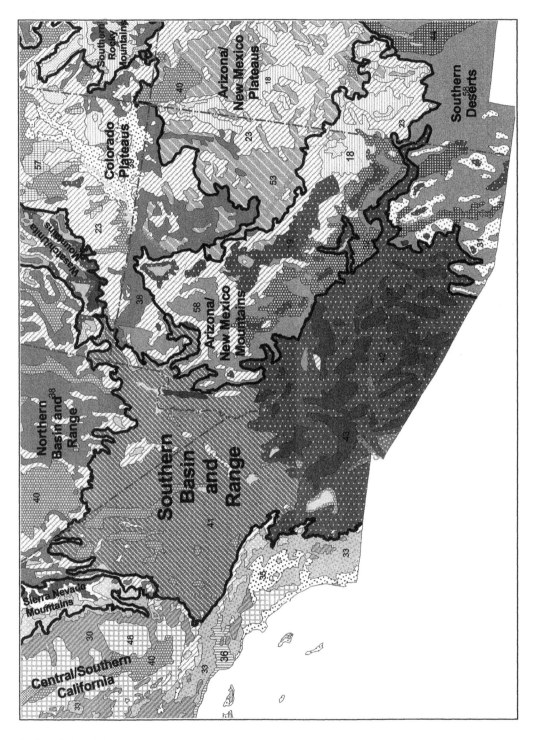

Southern Basin and Range

Southern Basin and Range

Introduction

The Southern Basin and Range region includes portions of southeast Arizona, southeast California, and southern Nevada. This is a very dry area with desert scrub the dominant vegetation in the basins and flats. The scattered mountain ranges support woodlands and coniferous forests at higher elevations.

The primary sources used to develop descriptions and species lists are Barbour and Billings 1977; Barbour and Major 1988; Brown 1982; Thorne 1976.

Dominant Ecological Communities

Upland Systems
Mojave Montane Forest
Great Basin Pine Woodland
Juniper–Pinyon Woodland
Mojave Desert Scrub
Sonoran Desert Scrub

Wetland Systems
Desert Oasis Woodland
Riparian Woodland

UPLAND SYSTEMS

Mojave Montane Forest

Mojave Montane Forest is found on Clark, Kingston, and New York Mountains in southeastern California and on the Charleston Mountains in southern Nevada at elevations from 5700 to 7100 feet. These open forests often contain white fir and are found on north-facing slopes or other mesic areas. This community corresponds to Western Deciduous and Mixed Forests ERT.

Canopy
Characteristic Species
Abies concolor	white fir

Associates
Acer glabrum	Rocky Mountain maple
Juniperus osteosperma	Utah juniper
Pinus monophylla	single-leaf pinyon
Quercus chrysolepis	canyon live oak
Quercus turbinella	shrub live oak

Woody Understory
Amelanchier utahensis	Utah service-berry
Fraxinus anomala	single-leaf ash
Holodiscus dumosus	glandular oceanspray
Leptodactylon pungens	granite prickly-phlox
Petrophyton caespitosum	Rocky Mountain rockmat

Philadelphus microphyllus	little-leaf mock orange
Ribes cereum	white squaw currant
Ribes velutinum	desert gooseberry
Sambucus cerulea	blue elder

Great Basin Pine Woodland

This woodland community is characterized by widely spaced limber pines and is the timberline woodland of eastern California. It is frequently found on granitic soils above 7,000 feet. The trees are short and widely spaced. The woodlands are similar to the pine forests of the Great Basin and may contain pure stands of bristle-cone pines. This community corresponds to Küchler # 22 and Woodlands ERT.

Canopy
Characteristic Species
Pinus flexilis	limber pine
Pinus longaeva	Intermountain bristle-cone pine

Associates
Acer glabrum	Rocky Mountain maple
Pinus jeffreyi	Jeffrey pine
Populus tremuloides	quaking aspen

Woody Understory
Artemisia tridentata	big sagebrush
Cerococarpus ledifolius	curl-leaf mountain-mahogany
Chamaebatiaria millefolium	fernbush
Chrysothamnus viscidiflorus	green rabbitbrush
Ericameria gilmanii	white-flower heath-goldenrod
Holodiscus dumosus	glandular oceanspray
Leptodactylon pungens	granite prickly-phlox
Ribes cereum	white squaw currant
Ribes montigenum	western prickly gooseberry
Symphoricarpos longiflorus	desert snowberry

Herbaceous Understory
Antennaria rosea	rosy pussytoes
Arenaria kingii	King's sandwort
Artemisia dracunculus	dragon wormwood
Astragalus kentrophyta	spiny milk-vetch
Astragalus platytropis	broad-keel milk-vetch
Cryptantha hoffmannii	Hoffmann's cat's-eye
Cryptantha roosiorum	bristle-cone cat's-eye
Cymopterus cinerarius	gray spring-parsley
Elymus elymoides	western bottle-brush grass
Erigeron spp.	fleabane
Festuca brachyphylla	short-leaf fescue
Festuca idahoensis	bluebunch fescue
Galium hypotrichium	alpine bedstraw
Heuchera rubescens	pink alumroot
Hymenoxys cooperi	Cooper's rubberweed

Koeleria macrantha	prairie Koeler's grass
Linanthus nuttallii	Nuttall's desert-trumpets
Mimulus bigelovii	yellow-throat monkey-flower
Muhlenbergia richardsonis	matted muhly
Oenothera cespitosa	tufted evening-primrose
Oxytropis parryi	Parry's locoweed
Pellaea breweri	Brewer's cliffbrake
Phlox covillei	Coville's phlox
Poa glauca	white blue grass
Poa secunda	curly blue grass
Pseudoroegneria spicata	bluebunch-wheat grass
Senecio spartioides	broom-like ragwort
Sphaeromeria cana	gray chicken-sage
Stenotus acaulis	stemless mock goldenweed
Stipa pinetorum	pine-forest needle grass

Juniper–Pinyon Woodland

Juniper–Pinyon Woodland of southern California, southern Nevada, and southwestern Arizona occurs on dry, rocky mountain slopes or other areas at elevations from 5000 to 8000 feet. These woodlands represent the westernmost extension of similar woodlands in the Northern Basin and Range and the Colorado Plateau regions. The pinyon and juniper components of this community may be somewhat distinct, forming a mosaic of pinyon or juniper areas. This community corresponds to Küchler #23 and Woodlands ERT.

Canopy
Characteristic Species

Juniperus californica	California juniper
Juniperus monosperma	one-seed juniper
Juniperus osteosperma	Utah juniper
Pinus edulis	two-needle pinyon
Pinus monophylla	single-leaf pinyon

Associates

Juniperus deppeana	alligator juniper
Juniperus occidentalis	western juniper
Pinus quadrifolia	four-leaf pinyon
Quercus chrysolepis	canyon live oak
Quercus emoryi	Emory's oak
Quercus gambelii	Gambel's oak
Quercus grisea	gray oak
Quercus turbinella	shrub live oak

Woody Understory

Arctostaphylos glauca	big-berry manzanita
Arctostaphylos pungens	Mexican manzanita
Artemisia arbuscula	dwarf sagebrush
Artemisia biglovii	flat sagebrush
Artemisia tridentata	big sagebrush
Baccharis sergiloides	squaw's false willow

Brickellia californica	California brickellbush
Ceanothus greggii	Mojave buckbrush
Cerocarpus intricatus	little-leaf mountain-mahogany
Cerocarpus ledifolius	cut-leaf mountain-mahogany
Cerocarpus montanus	alder-leaf mountain mahogany
Chrysothamnus depressus	long-flower rabbitbrush
Chrysothamnus nauseosus	rubber rabbitbrush
Chrysothamnus teretifolius	needle-leaf rabbitbrush
Chrysothamnus viscidiflorus	green rabbitbrush
Coleogyne ramosissima	blackbrush
Crossosoma bigelovii	ragged rockflower
Echinocereus engelmannii	saints cactus
Echinocereus triglochidiatus	king-cup cactus
Ephedra nevadensis	Nevada joint-fir
Ephedra viridis	Mormon-tea
Ericameria cuneata	cliff heath-goldenrod
Ericameria cooperi	Cooper's heath-goldenrod
Ericameria linearifolius	narrow-leaf heath-goldenrod
Eriodictyon angustifolia	narrow-leaf yerba-santa
Eriogonum fasciculatum	eastern Mojave wild buckwheat
Eriogonum heermannii	Heerman's wild buckwheat
Eriogonum umbellatum	sulphur-flower wild buckwheat
Eriogonum wrightii	bastard-sage
Fallugia paradoxa	Apache-plume
Forestieria pubescens	swamp-privet
Frangula californica	California coffee berry
Fraxinus anomala	single-leaf ash
Garrya flavescens	ashy silktassel
Glossopetalon spinescens	spiny greasebush
Gutierrezia sarothrae	Kindlingweed
Holodiscus dumosus	glandular oceanspray
Mahonia haematocarpa	red Oregon-grape
Menodora spinescens	spiny menodora
Opuntia erinacea	oldman cactus
Opuntia phaeacantha	tulip prickly-pear
Petradoria pumila	grassy rock-goldenrod
Petrophyton caespitosum	Rocky Mountain rockmat
Philadelphus microphyllus	little-leaf mock orange
Prunus fasciculata	desert almond
Purshia glandulosa	antelope-brush
Purshia mexicana	Mexican cliff-rose
Purshia tridentata	bitterbrush
Rhamnus crocea	holly-leaf buckthorn
Rhus trilobata	ill-scented sumac
Ribes cereum	white squaw currant
Ribes velutinum	desert gooseberry
Salazaria mexicana	Mexican bladder-sage
Symphoricarpos longiflorus	desert snowberry
Yucca baccata	banana yucca
Yucca brevifolia	Joshua-tree
Yucca schidigera	Mojave yucca

Herbaceous Understory

Anisocoma acaulis	scalebud
Arabis spp.	rockcress
Bouteloua curtipendula	side-oats grama
Bouteloua gracilis	blue grama
Caulanthus amplexicaulis	clasping-leaf wild cabbage
Elymus elymoides	western bottle-brush grass
Eriophyllum confertiflorum	yellow-yarrow
Hilaria rigida	big galleta
Layia glandulosa	white tidytips
Monardella linoides	flax-leaf mountainbalm
Oreonana vestita	woolly mountain-parsley
Oryzopsis hymenoides	Indian mountain-rice grass
Pascopyrum smithii	western-wheat grass
Pellaea mucronata	bird-foot cliffbrake
Penstemon grinnellii	Grinnell's beardtongue
Phacelia austromontana	southern Sierran scorpion-weed
Poa secunda	curly blue grass
Salvia dorii	gray ball sage
Salvia pachyphylla	rose sage
Sporobolus cryptandrus	sand dropseed
Stipa coronata	giant needle grass
Stipa speciosa	desert needle grass

Mojave Desert Scrub

Mojave Desert Scrub is found in southeastern California, southern Nevada, extreme southwestern Utah, and western and northwestern Arizona between the Great Basin Desert to the north and the Sonoran Desert to the south. Different shrubs form the dominant vegetation in different areas. The San Bernardino, Little San Bernardino, Cottonwood, and Eagle Mountains form a fairly definite southern boundary. At its northern limits, it gives way to Great Basin Scrub at elevations of 2700 to 3150 feet. The desert scrub occurs on elevations below the coniferous woodlands in the Mojave Desert and is composed mainly of low shrubs. Creosote bush frequently occurs alone or with a single associate, most commonly white burrobush. Saltbush, creosote bush, blackbrush, shadscale, and Joshua tree associations are typical vegetation types. Mojave Desert Scrub is unlike Sonoran Desert Scrub in that few trees occur even along arroyos and other drainage ways. Mojave Desert Scrub has lower diversity and contains fewer perennial plants than Sonoran Desert Scrub. Many ephemeral plants are endemic in this region. The Mojave Desert rarely has summer rain. Considerable variation in the vegetation occurs regionally, locally, seasonally, and annually. This community corresponds to Küchler #40 and #41 and Desert Lands ERT.

Characteristic Woody Species

Ambrosia dumosa	white burrobush
Atriplex canescens	four-wing saltbush
Atriplex confertifolia	shadscale
Baccharis sergiloides	squaw's false willow
Coleogyne ramosissima	blackbrush
Encelia farinosa	goldenhills
Grayia spinosa	spiny hop-sage

Larrea tridentata	creosote-bush
Lycium andersonii	red-berry desert-thorn
Salazaria mexicana	Mexican bladder-sage
Sarcobatus vermiculatus	greasewood
Sphaeralcea ambigua	apricot globe-mallow
Tetradymia axillaris	cottonthorn
Yucca brevifolia	Joshua-tree

Associates

Acacia greggii	long-flower catclaw
Acamptopappus shockleyi	Shockley's goldenhead
Agave utahensis	Utah century-plant
Allenrolfea occidentalis	iodinebush
Artemisia spinescens	bud sagebrush
Artemisia tridentata	big sagebrush
Atriplex hymenolytra	desert-holly
Atriplex polycarpa	cattle-spinach
Atriplex spinifera	spinescale
Chrysothamnus paniculatus	dotted rabbitbrush
Echinocactus polycephalus	cotton-top cactus
Echinocereus engelmannii	saints cactus
Ephedra california	California joint-fir
Ephedra nevadensis	Nevada joint-fir
Eriogonum fasciculatum	eastern Mojave wild buckwheat
Eriogonum wrighti	bastard-sage
Escobaria vivipara	desert pincushion
Ferocactus cylindraceus	barrel cactus
Fouquiera splendens	ocotillo
Hibiscus denudatus	paleface
Hymenoclea salsola	white cheesebush
Krameria grayi	white ratany
Krameria eracta	small-flower ratany
Krascheninnikovia lanata	winterfat
Lepidospartum latisquamum	Nevada scalebroom
Lycium andersonii	red-berry desert-thorn
Lycium cooperi	peachthorn
Lycium fremontii	Fremont's desert-thorn
Menadora spinescens	spiny menodora
Opuntia acanthocarpa	buck-horn cholla
Opuntia basilaris	beaver tail cactus
Opuntia bigelovii	teddy-bear cholla
Opuntia echinocarpa	golden cholla
Opuntia erinacea	oldman cactus
Opuntia parishii	matted cholla
Opuntia ramosissima	darning-needle cactus
Peucephyllum schottii	Scott's pygmy-cedar
Psorothamus arborescens	Mojave smokebush
Psorothamus emoryi	Emory's smokebush
Psorothamus fremontii	Fremont's smokebush
Psorothamus spinosus	smokethorn

Purshia gladulosa	antelope-brush
Salvia dorrii	gray ball sage
Salvia funerea	death valley sage
Salvia mohavensis	Mojave sage
Senna armata	desert wild sensitive-plant
Suaeda moquinii	shrubby seepweed
Thamnosma montana	turpentine-broom
Yucca baccata	banana yucca
Yucca schidigera	Mojave yucca

Herbaceous Understory

Asclepias subulata	rush milkweed
Atrichoseris platyphylla	parachute-plant
Bouteloua spp.	grama
Calycoseris parryi	yellow tackstem
Camissonia brevipes	golden suncup
Camissonia chamaenerioides	long-capsule suncup
Chaenactis carphoclinia	pebble pincushion
Chorizanthe brevicornu	brittle spineflower
Chorizanthe rigida	devil's spineflower
Cryptantha maritima	guadalupe cat's-eye
Distichlis spicata	coastal salt grass
Encelia farinosa	goldenhills
Eriogonum inflatum	Indian-pipeweed
Eriogonum trichopes	little desert trumpet
Gilia scopulorum	rock gily-flower
Hilaria mutica	tobosa grass
Hilaria rigida	big galleta
Kochia americana	greenmolly
Langloisia setosissima	bristly-calico
Lepidium lasiocarpum	hairy-pod pepperwort
Lotus salsuginosus	coastal bird's-foot-trefoil
Lupinus arizonicus	Arizona lupine
Mentzelia involucrata	white-bract blazingstar
Mirabilis bigelovii	desert wishbonebush
Nitrophila spp.	niterwort
Oenothera deltoides	devil's-lantern
Oryzopsis hymenoides	Indian mountain-rice grass
Perityle emoryi	Emory's rockdaisy
Phacelia crenulata	notch-leaf scorpion-weed
Phacelia distans	distant scorpion-weed
Plagiobothrys jonesii	Mojave popcorn-flower
Plantago ovata	blond plantain
Prenanthella exigua	brightwhite
Salicornia spp.	saltwort
Salvia columbariae	California sage
Stipa speciosa	desert needle grass
Stylocline micropoides	woolly-head neststraw
Trichoptilium incisum	yellowdome

Riparian trees only

Acacia greggii	long-flower catclaw
Chilopsis linearis	desert-willow
Prosopis glandulosa	honey mesquite

Sonoran Desert Scrub

Sonoran Desert Scrub covers most of southwestern Arizona and southeastern California extending into Sonora, Mexico. This region is commonly called the Colorado Desert in California. Trees and large succulents differentiate the Sonoran from the Mojave and Chihuahuan Deserts. The understory may be composed of one to several species of shrubs which form several layers. It is the hottest of the desert scrubs of the southwest. It is found from 100 to 4000 feet in elevation depending upon slope and exposure. Rainfall is distinctly biseasonal and occurs primarily during the winter. In Arizona, the desert scrub has two distinct communities, the creosote bush community and the palo-verde–saguaro community. The palo-verde–saguaro community contains small-leaved desert trees, cacti, and shrubs and is best developed on rocky bajadas or coarse-soiled slopes. This community corresponds to Küchler #42 and #43 and Desert Lands ERT.

Characteristic Woody Species

Ambrosia dumosa	white burrobush
Bursera microphylla	elephant-tree
Canotia holacantha	crucifixion-thorn
Carnegia gigantea	saguaro
Celtis pallida	shiny hackberry
Holocantha emoryi	thorn-of-Christ
Larrea tridentata	creosote-bush
Olneya tesota	desert-ironwood
Opuntia fulgida	jumping cholla
Pachycereus schottii	senita
Parkinsonia microphylla	yellow palo-verde
Stenocereus thurberi	organ-pipe cactus

Associates

Acacia constricta	mescat wattle
Acacia greggii	long-flower catclaw
Ambrosia deltoidea	triangle burr-ragweed
Atriplex polycarpa	cattle-spinach
Calliandra eriophylla	fairy-duster
Cleome isomeris	bladder-pod spider-flower
Condalia spathulata	squawbush
Echinocereus engelmanii	saints catcus
Encelia farinosa	goldenhills
Ephedra trifurca	long-leaf joint-fir
Eriogonum fasiculatum	eastern Mojave wild buckwheat
Ferocactus cylindraceus	California barrel cactus
Ferocactus wislezenii	candy barrel cactus
Fouguiera splendens	ocotillo
Jatropha cardiophylla	sangre-de-cristo
Krameria grayi	white ratany
Lycium richii	Santa Catalina desert-thorn
Mammillaria grahamii	Graham's nipple cactus
Mammillaria tetrancistra	corkseed catcus

Opuntia bigelovii	teddy-bear cholla
Opuntia echinocarpa	golden cholla
Opuntia engelmannii	catcus-apple
Opuntia kunzei	devil's cholla
Opuntia ramosissima	darning-needle cactus
Opuntia spinosior	walkingstick cactus
Opuntia versicolor	stag-horn cholla
Parkinsonia florida	blue palo-verde
Peucephyllum schottii	Schott's pygmy-cedar
Prosopis velutina	velvet mesquite
Psorothamnus arborescens	Mojave smokebush
Psorothamnus schottii	Schott's smokebush
Psorothamnus spinosus	smokethorn
Salvia vaseyi	bristle sage
Senna armata	desert wild sensitive-plant
Simmondsia chinensis	jojoba
Thamnosma montana	turpentine-broom

Herbaceous Understory

Chaenactis fremontii	morningbride
Emmenanthe penduliflora	yellow whispering-bells
Hilaria rigida	big galleta
Malacothrix glabrata	smooth desert-dandelion
Phacelia distans	distant scorpion-weed
Rafinesquia neomexicana	New Mexico plumseed
Zinnia acerosa	white zinnia

Riparian Trees of Dry Arroyos

Acacia greggi	long-flower catclaw
Baccharis sarothroides	rosinbush
Celtis laevigata	sugar-berry
Chilopsis linearis	desert-willow
Lycium andersonii	red-berry desert-thorn
Olneya tesota	desert-ironwood
Parkinsonia florida	blue palo-verde
Prosopis gladulosa	honey mesqutie
Psorothamnus spinosus	smokethorn
Sapium biloculare	Mexican jumping-bean
Ziziphus obtusifolia	lotebush

WETLAND SYSTEMS

Desert Oasis Woodland

These palm-dominated woodlands are found at the heads or in the bottoms of canyons and arroyos where permanent springs or seeps occur. Oases are also found in and around the Salton Basin in the United States and extend into Mexico. The palms are fire tolerant but the understory species are not. Periodic fire opens the understory allowing seedlings to establish. The reduced understory also removes competition for water. This community corresponds to Riparian Forests and Woodlands ERT.

Canopy
Characteristic Species

Washingtonia filifera	California fan palm

Associates

Fraxinus velutina	velvet ash
Platanus racemosa	California sycamore
Populus fremontii	Fremont's cottonwood

Woody Understory

Atriplex lentiformis	quailbush
Atriplex polycarpa	cattle-spinach
Baccharis sergiloides	squaw's false willow
Isocoma acradenia	alkali jimmyweed
Pluchea sericea	arrow-weed
Prosopis glandulosa	honey mesquite
Prosopis pubescens	American screw-bean
Salix bonplandiana	Bonpland's willow
Salix gooddingii	Goodding's willow
Suaeda moquinii	shrubby seepweed

Herbaceous Understory

Distichlis spicata	coastal salt grass
Juncus acutus	spiny rush
Juncus cooperi	Copper's rush
Phragmites australis	common reed
Sporobolus airoides	alkali-sacaton
Typha domingensis	southern cat-tail

Riparian Woodland

Riparian Woodland occurs along perennial streams throughout this region. It is composed of deciduous trees which grow in a relatively narrow band along riparian floodplains. Exotic tamarisk or salt cedar (*Tamarix* spp.) is dominant today in many of these woodlands. Some areas may be dominated by willows. This community corresponds to Riparian Forests and Woodlands ERT.

Canopy
Characteristic Species

Populus fremontii	Fremont's cottonwood

Associates

Acer glabrum	Rocky Mountain maple
Acer negundo	ash-leaf maple
Alnus rhombifolia	white alder
Betula occidentalis	water birch
Fraxinus velutina	velvet ash
Platanus racemosa	California sycamore
Populus balsamifera	balsam popular
Populus tremuloides	quaking aspen
Quercus chrysolepis	canyon live oak
Rubus vitifolius	Pacific dewberry

Salix bonplandiana	Bonpland's willow
Salix geyeriana	Geyer's willow
Salix lasiolepis	arroyo willow
Sambucus cerulea	blue elder

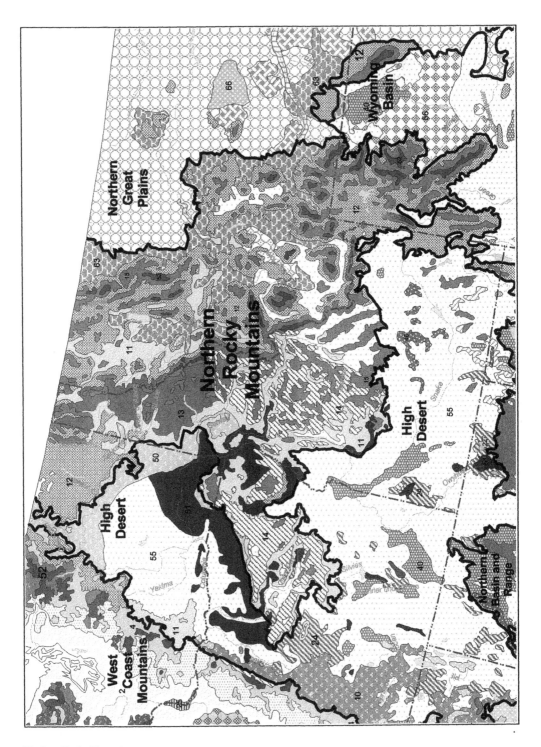

Northern Rocky Mountains

Northern Rocky Mountains

Introduction

The Northern Rocky Mountain region includes mountain ranges in eastern Washington, eastern Oregon, Idaho, Montana, and Wyoming. This region is bounded by the Northern Great Plains to the east, the High Desert to the west, and the Wyoming Basin to the south.

The characteristic vegetation of the Northern Rocky Mountain region is conifer forest. There are many conifer forest types in the region. Other vegetation types such as Sagebrush Steppe, Mountain Meadow, and Riparian Forest occur less commonly than the conifer forests.

In this very mountainous region, the forest composition changes along environmental gradients often related to altitude. Generally, as the elevation gets higher, the temperature drops, precipitation increases, solar and ultraviolet radiations increase, and snow depth and duration increase. Other considerations in mountainous areas include the topographical position. For example, south-facing slopes and ridge tops will generally be warmer and drier than north-facing slopes and sheltered valley bottoms. Soil texture is also an important factor in the type of vegetation that will grow on mountain slopes and in valleys.

The high elevation grassland communities are not included in the descriptions. Groves of aspen trees are also notable in the region but are more common in the Southern Rocky Mountain region (see that region for descriptions).

Intensive livestock grazing and fire suppression have altered the vegetation throughout the Rocky Mountains.

The primary sources used to develop descriptions and species lists are Barbour and Billings 1988; Franklin and Dyrness 1973; Green and Conner 1989.

Dominant Ecological Communities

Upland Systems
Western Ponderosa Forest
Douglas-fir Forest
Grand Fir Forest
Hemlock–Pine–Cedar Forest
Western Spruce–Fir Forest
Sagebrush Steppe
Foothills Prairie
Mountain Meadow
Wetland Systems
Riparian Forest

UPLAND SYSTEMS

Western Ponderosa Forest

Western Ponderosa Forest occurs on the lower mountain slopes. At its upper limits, Western Ponderosa Forest may grade into Douglas-fir or grand fir dominated forests and, at its lower limit, it may grade into Sagebrush Steppe. Ponderosa pine occupies relatively dry sites and is subject to periodic fires which regulate the density of the community. Fire also favors the reproduction of grasses in the understory. Intensive livestock grazing in some cases has slowed the growth and altered the composition of the herbaceous layer and encouraged a shrub layer to develop in the understory. This community corresponds to Küchler #11 and Western Deciduous and Mixed Forests ERT.

Canopy
Characteristic Species
    Pinus ponderosa   ponderosa pine
Associates
    Juniperus occidentalis   western juniper
    Juniperus scopulorum   Rocky Mountain juniper
    Pinus contorta   lodgepole pine
    Populus tremuloides   quaking aspen
    Pseudotsuga menziesii   Douglas-fir

Woody Understory
    Amelanchier alnifolia   Saskatoon service-berry
    Arctostaphylos uva-ursi   red bearberry
    Crataegus douglasii   black hawthorn
    Eriogonum heracleoides   parsnip-flower wild buckwheat
    Holodiscus discolor   hillside oceanspray
    Physocarpus malvaceus   mallow-leaf ninebark
    Prunus virginiana   choke cherry
    Purshia tridentata   bitterbrush
    Rosa nutkana   Nootka rose
    Rosa woodsii   Woods' rose
    Symphoricarpos albus   common snowberry

Herbaceous Understory
    Antennaria dimorpha   cushion pussytoes
    Arabis holboellii   Holboell's rockcress
    Balsamorhiza sagittata   arrow-leaf balsamroot
    Calamagrostis rubescens   pinegrass
    Carex geyeri   Geyer's sedge
    Carex rossii   Ross's sedge
    Claytonia perfoliata   miner's-lettuce
    Collinsia parviflora   small-flower blue-eyed Mary
    Draba cinerea   gray-leaf whitlow-grass
    Epilobium brachycarpum   tall annual willowherb
    Erigeron compositus   dwarf mountain fleabane
    Festuca idahoensis   bluebunch fescue
    Frasera albicaulis   white-stemmed elkwood
    Galium aparine   sticky-willy
    Koeleria macrantha   prairie Koeler's grass
    Lithophragma glabrum   bulbous woodlandstar
    Lithospermum ruderale   Columbian puccoon
    Lotus nevadensis   Nevada bird's-foot-trefoil
    Madia exigua   little tar plant
    Montia linearis   linear-leaf candy-flower
    Poa secunda   curly blue grass
    Pseudoroegneria spicata   bluebunch-wheat grass
    Ranunculus glaberrimus   sagebrush buttercup
    Sisyrinchium douglasii   grasswidows
    Stellaria nitens   shining chickweed

Stipa occidentalis	western needle grass
Triteleia grandiflora	large-flower triplet-lily
Vulpia microstachys	small six-weeks grass

Douglas-fir Forest

Douglas-fir Forest occurs at higher elevations than Western Ponderosa Forest. These higher areas are more mesic with cooler temperatures and higher annual precipitation. This community corresponds to Küchler #12 and Coniferous Forests ERT.

Canopy
Characteristic Species

Picea glauca	white spruce
Pinus ponderosa	ponderosa pine
Pseudotsuga menziesii	Douglas-fir

Associates

Abies concolor	white fir
Abies grandis	grand fir
Larix occidentalis	western larch
Picea engelmannii	Engelmann's spruce
Picea pungens	blue spruce
Pinus contorta	lodgepole pine
Populus tremuloides	quaking aspen

Woody Understory

Arctostaphylos uva-ursi	red bearberry
Holodiscus discolor	hillside oceanspray
Juniperus communis	common juniper
Physocarpus malvaceus	mallow-leaf ninebark
Rosa nutkana	Nootka rose
Rosa woodsii	Woods' rose
Symphoricarpos albus	common snowberry
Symphoricarpos oreophilus	mountain snowberry

Herbaceous Understory

Arnica cordifolia	heart-leaf leopard-bane
Arnica latifolia	daffodil leopard-bane
Aster conspicuus	eastern showy aster
Balsamorhiza sagittata	arrow-leaf balsamroot
Calamagrostis rubescens	pinegrass
Carex concinnoides	northwestern sedge
Carex geyeri	Geyer's sedge
Galium boreale	northern bedstraw

Grand Fir Forest

Grand Fir Forest typically occurs at higher elevations than Western Ponderosa and Douglas- fir Forest. They also occur at lower elevations than the subalpine Spruce–Fir Forest. The precipitation in Grand Fir Forest is higher and the temperature is lower than lower elevation forests, which often experience summer dryness and drought. These forests have higher temperatures and less accumulation of snow than subalpine forests. This community corresponds to Küchler #14 and Coniferous Forests ERT.

Canopy
Characteristic Species
Abies grandis	grand fir
Larix occidentalis	western larch
Pinus contorta	lodgepole pine
Pinus ponderosa	ponderosa pine
Pseudotsuga menziesii	Douglas-fir

Associates
Picea engelmannii	Engelmann's spruce
Pinus monticola	western white pine
Tsuga mertensiana	mountain hemlock

Woody Understory
Arctostaphylos uva-ursi	red bearberry
Ceanothus velutinus	tobacco-brush
Linnaea borealis	American twinflower
Pachystima myrsinites	myrtle boxleaf
Ribes lacustre	bristly black gooseberry
Rosa gymnocarpa	wood rose
Rubus lasiococcus	hairy-fruit smooth dewberry
Symphoricarpos albus	common snowberry
Taxus brevifolia	Pacific yew
Vaccinium membranaceum	square-twig blueberry

Herbaceous Understory
Adenocaulon bicolor	American trailplant
Anemone lyallii	little mountain thimbleweed
Anemone piperi	Piper's windflower
Apocynum androsaemifolium	spreading dogbane
Arnica cordifolia	heart-leaf leopard-bane
Asarum caudatum	long-tail wildginger
Bromus vulgaris	Columbian brome
Calamagrostis rubescens	pinegrass
Carex concinnoides	northwestern sedge
Carex geyeri	Geyer's sedge
Carex occidentalis	western sedge
Carex rossii	Ross' sedge
Chimaphila menziesii	little prince's-pine
Chimaphila umbellata	pipsissewa
Clintonia uniflora	bride's bonnet
Corallorhiza maculata	summer coralroot
Elymus elymoides	western bottle-brush grass
Epilobium angustifolium	fireweed
Fragaria chilosensis	beach strawberry
Fragaria virginiana	Virginia strawberry
Galium triflorum	fragrant bedstraw
Gayophytum humile	dwarf groundsmoke
Hieracium albiflorum	white-flower hawkweed
Listera convallariodes	broad-tip twayblade

Lupinus caudatus	Kellogg's spurred lupine
Lupinus latifolius	broad-leaf lupine
Maianthemum racemosum	feathery false Solomon's seal
Maianthemum stellatum	starry Solomon's seal
Mitella stauropetala	side-flower bishop's-cap
Moehringia macrophylla	big-leaf grove-sandwort
Monotropa hypopithys	many-flower Indian-pipe
Orthilia secunda	sidebells
Poa nervosa	Hooker's blue grass
Pyrola asariflia	pink wintergreen
Pyrola picta	white-vein wintergreen
Stipa occidentalis	western needle grass
Thalictrum occidentale	western meadow-rue
Trillium ovatum	western wakerobin
Viola glabella	pioneer violet

Hemlock–Pine–Cedar Forest

Hemlock–Pine–Cedar Forest occurs throughout the Northern Rocky Mountain region. This community corresponds to Küchler #13 and Coniferous Forests ERT.

Canopy
Characteristic Species

Pinus monticola	western white pine
Thuja plicata	western arborvitae (cedar)
Tsuga heterophylla	western hemlock

Associates

Abies grandis	grand fir
Larix occidentalis	western larch
Pinus ponderosa	ponderosa pine
Pseudotsuga menziesii	Douglas-fir

Woody Understory

Oplopanax horridus	devil's-club
Pachystima myrsinites	myrtle boxleaf
Taxus brevifolia	Pacific yew
Vaccinium membranaceum	square-twig blueberry

Herbaceous Understory

Asarum caudatum	long-tail wildginger
Athyrium filix-femina	subarctic lady fern
Clintonia uniflora	bride's bonnet
Dryopteris campyloptera	mountain wood fern
Gymnocarpium dryopteris	western oak-fern
Tiarella trifoliata	three-leaf foam flower

Western Spruce–Fir Forest

Western Spruce–Fir Forest may be bounded at its lower limits by hemlock, western red cedar, grand fir, or Douglas-fir. This Spruce–Fir Forest reaches its upper limit in a subalpine environment where the trees thin out and become a forest-meadow parkland. This community corresponds to Küchler #15 and Coniferous Forests ERT.

Canopy

Characteristic Species

Abies lasiocarpa	subalpine fir
Picea engelmannii	Engelmann's spruce
Pinus albicaulis	white-bark pine
Pinus contorta	lodgepole pine

Associates

Abies grandis	grand fir
Larix lyallii	subalpine larch
Larix occidentalis	western larch
Pinus monticola	western white pine
Populus tremuloides	quaking aspen
Pseudotsuga menziesii	Douglas-fir
Tsuga mertensiana	mountain hemlock

Woody Understory

Acer glabrum	Rocky Mountain maple
Amelanchier alnifolia	Saskatoon service-berry
Arctostaphylos uva-ursi	red bearberry
Juniperus communis	common juniper
Ledum glandulosum	glandular Labrador tea
Menziesia ferruginea	fool's-huckleberry
Rhododendron albiflorum	cascade azalea
Rubus parviflorus	western thimble-berry
Shepherdia canadensis	russet buffalo-berry
Symphoricarpos albus	common snowberry
Vaccinium cespitosum	dwarf blueberry
Vaccinium membranaceum	square-twig blueberry
Vaccinium scoparium	grouseberry

Herbaceous Understory

Actaea rubra	red baneberry
Adenocaulon bicolor	American trailplant
Arnica cordifolia	heart-leaf leopard-bane
Aster conspicuus	eastern showy aster
Calamagrostis canadensis	blue joint
Carex disperma	soft-leaf sedge
Carex geyeri	Geyer's sedge
Clintonia uniflora	bride's bonnet
Coptis occidentalis	Idaho goldthread
Galium triflorum	fragrant bedstraw
Hieracium albiflorum	white-flower hawkweed
Mitella pentandra	five-stamen bishop's-cap

Mitella stauropetala	side-flower bishop's-cap
Moehringia macrophylla	big-leaf grove-sandwort
Viola glabella	pioneer violet
Xerophyllum tenax	western turkeybeard

Sagebrush Steppe

Sagebrush Steppe is the driest community in this region. Big sagebrush is the most characteristic species. Changes in dominance are probably a result of grazing, fire, or cultivation. Fire within the Sagebrush Steppe community has increased since an exotic, cheatgrass (*Bromus tectorum*), has become dominant. This community corresponds to Küchler #55 and Desert Lands ERT.

Characteristic Woody Species

Artemisia tridentata	big sagebrush

Associates

Artemisia arbuscula	dwarf sagebrush
Artemisia nova	black sagebrush
Artemisia tripartita	three-tip sagebrush
Chrysothamnus nauseosus	rubber rabbitbrush
Chrysothamnus viscidiflorus	green rabbitbrush
Grayia spinosa	spiny hop-sage
Juniperus occidentalis	western juniper
Purshia tridentata	bitterbrush
Tetradymia canescens	spineless horsebrush

Herbaceous Understory

Balsamorrhiza sagittata	arrow-leaf balsamroot
Elymus elymoides	western bottle-brush grass
Festuca idahoensis	bluebunch fescue
Lappula occidentalis	flat-spine sheepburr
Lithospermum ruderale	Columbian puccoon
Lupinus sericeus	Pursh's silky lupine
Oryzopsis hymenoides	Indian mountain-rice grass
Poa fendleriana	mutton grass
Poa secunda	curly blue grass
Pseudoroegneria spicata	bluebunch-wheat grass
Stipa comata	needle-and-thread
Stipa thurberiana	Thurber's needle grass

Foothills Prairie

Foothills Prairie is characterized by perennial bunchgrasses. This community depends upon a greater degree of moisture than Sagebrush Steppe and usually occurs at higher altitudes than the sagebrush communities. Excessive grazing in the area has resulted in a replacement of the native perennial bunchgrasses with *Artemisia* and *Bromus* species. This community corresponds to Küchler #63 and Prairies ERT.

Characteristic Species

Festuca campestris	prairie fescue
Festuca idahoensis	bluebunch fescue
Leymus condensatus	giant lyme grass
Poa secunda	curly blue grass
Pseudoroegneria spicata	bluebunch-wheat grass
Stipa comata	needle-and-thread grass

Associates

Artemisa frigida	prairie sagebrush
Astragalus spaldingii	Spalding's milk-vetch
Bouteloua gracilis	blue grama
Carex filifolia	thread-leaf sedge
Chrysothamnus nauseosus	rubber rabbitbrush
Koeleria macrantha	prairie Koeler's grass
Lappula occidentalis	flat-spine sheepburr
Lithophragma glabrum	bulbous woodlandstar
Lupinus sericeus	Pursh's silky lupine
Montia linearis	linear-leaf candy-flower
Pascopyrum smithii	western-wheat grass
Phlox gracilis	slender phlox
Phlox longifolia	long-leaf phlox
Plantago patagonica	woolly plantain
Stellaria nitens	shiny starwort
Vulpia microstachys	small six-weeks grass

Mountain Meadow

Mountain Meadow is a permanent herbaceous community type found on relatively gentle topography along and near the heads of stream courses. In many areas, grazing pressure has deteriorated the meadow communities from a perennial grass into a weedy annual community or exotic perennials such as Kentucky bluegrass *(Poa pratensis)* and white clover *(Trifolium repens)*. This community corresponds to Meadowlands ERT.

Characteristic Species

Deschampsia cespitosa	tufted hair grass

Associates

Aster occidentalis	western mountain aster
Festuca rubra	red fescue
Juncus balticus	Baltic rush
Polygonum bistortoides	American bistort
Potentilla gracilis	graceful cinquefoil

WETLAND SYSTEMS

Riparian Forest

Riparian Forest occurs at low elevations in canyons and along streams in the Northern Rocky Mountains region. This is the least typical forest type of the region. This community corresponds to Riparian Forests and Woodlands ERT.

Canopy
Characteristic Species
    Populus angustifolia   narrow-leaf cottonwood
    Populus balsamifera   balsam poplar
Associates
    Cornus sericea   redosier
    Crataegus douglassii   black hawthorn

Woody Understory
    Rosa woodsii   *Woods' rose*
    Symphoricarpos albus   common snowberry

Herbaceous Understory
    Aralia nudicaulis   wild sarsaparilla
    Carex sprengelii   long-beak sedge
    Cicuta douglassii   western water-hemlock
    Heracleum lanatum   American cow-parsnip
    Ranunculus abortivus   kidney-leaf buttercup
    Ratibida pinnata   gray-head Mexican-hat

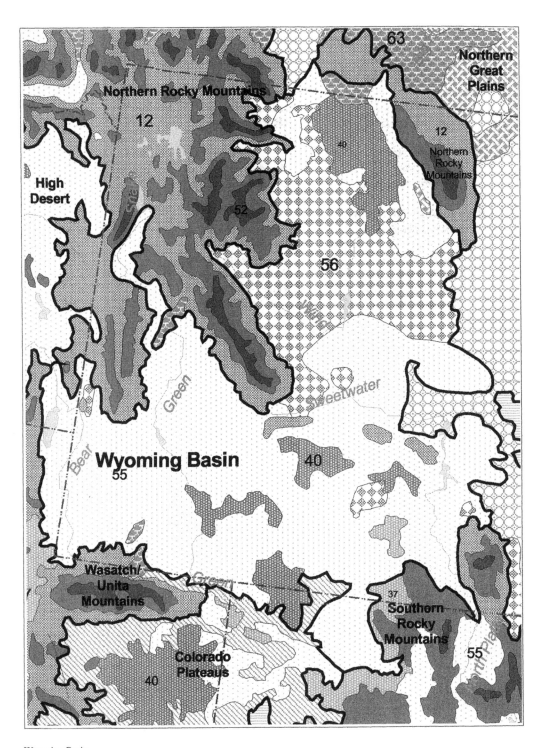

Wyoming Basin

Wyoming Basin

Introduction

The Wyoming Basin region occurs between the northern and southern Rocky Mountains. This region encompasses several major basins including the Big Horn, Wind River, Bridger, and Shirley Basins. The altitudes of the Wyoming Basin region are mostly between 6000 and 8000 feet above sea level. The basins are interrupted by mountains, buttes, river valleys, and badlands.

The characteristic vegetation is sagebrush and short grasses. Saltbush, greasewood, and mountain-mahogany are also common in specific sites. River floodplains support vegetation characterized by cottonwood, willow, and alder. Although shrub and grasslands are the most common vegetation types, there are also islands of forests in the mountains. A large portion of the region's floodplains has been converted to agriculture uses. Throughout the Wyoming Basin, natural areas have been altered by livestock grazing and the introduction of exotic species.

The primary sources used to develop descriptions and species lists are Barbour and Billings 1988; Knight et al. 1976; Wyoming Nature Conservancy 1991.

Dominant Ecological Communities

Upland Systems
   Sagebrush Steppe
   Greasewood Scrub
   Saltbush Scrub
   Mountain-Mahogany Shrub
   Shrub Grassland
Wetland Systems
   Floodplain Forest

UPLAND SYSTEMS

Sagebrush Steppe

Sagebrush Steppe is widespread throughout the Wyoming Basin. On drier sites, the sagebrush may be dwarfed, while on more mesic sites, it may be tall. Sagebrush Steppe may be separated based upon plant height. This community corresponds to Küchler #55 and Desert Lands ERT.

Characteristic Woody Species
   Artemisia arbuscula   dwarf sagebrush
   Artemisia tridentata   big sagebrush
Associates
   Amelanchier alnifolia   Saskatoon service-berry
   Artemisia cana   hoary sagebrush
   Artemisia frigida   prairie sagebrush
   Artemisia nova   black sagebrush
   Artemisia pedatifida   bird-foot sagebrush
   Artemisia tripartita   three-tip sagebrush
   Atriplex canescens   four-wing saltbush

Atriplex confertifolia	shadscale
Atriplex gardneri	Gardner's saltbush
Chrysothamnus nauseosus	rubber rabbitbrush
Chrysothamnus viscidiflorus	green rabbitbrush
Eriogonum microthecum	slender wild buckwheat
Grayia spinosa	spiny hop-sage
Gutierrezia sarothrae	kindlingweed
Krascheninnikovia lanata	winterfat
Opuntia polyacantha	hair-spine prickly-pear
Ribes cereum	white squaw currant
Tetradymia canescens	spineless horsebrush
Tetradymia spinosa	short-spine horsebrush

Herbaceous Understory

Agoseris glauca	pale goat-chicory
Allium acuminatum	taper-tip onion
Allium textile	wild white onion
Alyssum desertorum	desert madwort
Antennaria dimorpha	cushion pussytoes
Antennaria rosea	rosy pussytoes
Arabis holboellii	Holboell's rockcress
Arabis lignifera	Owens valley rockcress
Arabis sparsiflora	elegant rockcress
Arenaria hookeri	Hooker's sandwort
Artemisia ludoviciana	white sagebrush
Astragalus miser	timber milk-vetch
Astragalus pectinatus	narrow-leaf milk-vetch
Astragalus purshii	Pursh's milk-vetch
Astragalus spatulatus	tufted milk-vetch
Atriplex powellii	Powell's orache
Balsamorhiza incana	hoary balsamroot
Balsamorhiza sagittata	arrow-leaf balsamroot
Calochortus nuttallii	sego-lily
Carex duriuscula	spike-rush sedge
Carex filifolia	thread-leaf sedge
Carex obtusata	blunt sedge
Castilleja angustifolia	northwestern Indian-paintbrush
Chenopodium leptophyllum	narrow-leaf goosefoot
Comandra umbellata	bastard-toadflax
Cordylanthus ramosus	bushy bird's-beak
Crepis modocensis	siskiyou hawk's-beard
Cryptantha watsonii	Watson's cat's-eye
Delphinium bicolor	flat-head larkspur
Delphinium geyeri	Geyer's larkspur
Descurainia incana	mountain tansy-mustard
Distichilis spicta	coastal salt grass
Elymus elymoides	western bottle-brush grass
Elymus lanceolatus	streamside wild rye
Elymus trachycaulus	slender rye grass
Erigeron caespitosus	tufted fleabane

Erigeron engelmanni	Engelmann's fleabane
Eriogonum caespitosum	matted wild buckwheat
Eriogonum cernuum	nodding wild buckwheat
Eriogonum compositum	arrow-leaf wild buckwheat
Eriogonum ovalifolium	cushion wild buckwheat
Festuca idahoensis	bluebunch fescue
Gayophytum ramosissimum	pinyon groundsmoke
Halogeton glomeratus	saltlover
Hilaria jamesii	Jame's galleta
Hordeum jubatum	fox-tail barley
Hymenoxys richardsonii	Colorado rubberweed
Iva axillaris	deer-root
Juncus balticus	Baltic rush
Koeleria macrantha	prairie Koeler's grass
Lappula occidentalis	flat-spine sheepburr
Leptodactylon pungens	granite prickly-phlox
Lesquerella ludoviciana	Louisiana bladderpod
Leymus cinereus	Great Basin lyme grass
Lithospermum incisum	fringed gromwell
Lomatium foeniculaceum	carrot-leaf desert-parsley
Lupinus argenteus	silver-stem lupine
Machaeranthera canescens	hoary tansy-aster
Mertensia oblongifolia	languid-lady
Monolepis nuttalliana	Nuttall's poverty-weed
Musineon divaricatum	leafy wild parsley
Oryzopsis hymenoides	Indian mountain-rice grass
Pascopyrum smithii	western-wheat grass
Penstemon arenicola	red desert beardtongue
Penstemon fremontii	Fremont's beardtongue
Phlox hoodii	carpet phlox
Phlox longifolia	long-leaf phlox
Phlox multiflora	Rocky Mountain phlox
Poa fendleriana	mutton grass
Poa secunda	curly bluegrass
Polygonum douglasii	Douglas' knotweed
Pseudocymopterus montanus	alpine false mountain-parsley
Pseudoroegneria spicata	bluebunch-wheat grass
Sphaeralcea coccinea	scarlet globe-mallow
Stenotus acaulis	stemless mock goldenweed
Stipa comata	needle-and-thread
Stipa lettermani	Letterman's needlegrass
Stipa nelsonii	Nelson's needle grass
Thelypodiopsis elegans	westwater tumble-mustard
Thermopsis rhombifolia	prairie golden-banner
Trifolium gymnocarpon	holly-leaf clover
Vulpia octoflora	eight-flower six-weeks grass
Xylorhiza glabriuscula	smooth woody-aster
Zigadenus paniculatus	sand-corn
Zigadenus venenosus	meadow deathcamas

Greasewood Scrub

Greasewood Shrub is found throughout the Wyoming Basin in moist depressions frequently associated with floodplains and ponds. These flatlands are characterized by poor drainage and salt accumulations. Because Greasewood Scrub occurs near sources of water, many of these areas have been damaged by livestock as they find their way to the ponds associated with this vegetation type. This community corresponds in part to Küchler #40 and Desert Lands ERT.

Characteristic Woody Species

Atriplex confertifolia	shadscale
Sarcobatus vermiculatus	greasewood

Associates

Artemisia spinescens	bud sagebrush
Artemisia tridentata	big sagebrush
Atriplex gardneri	Gardner's saltbush
Chrysothamnus nauseosus	rubber rabbitbrush
Grayia spinosa	spiny hop-sage
Krascheninnikovia lanata	winterfat
Opuntia polyacantha	hair-spine prickly-pear
Suaeda moquinii	shrubby seepweed
Tetradymia spinosa	short-spine horsebrush

Herbaceous Understory

Allium textile	wild white onion
Atriplex patula	halberd-leaf orache
Atriplex argentea	silverscale
Bouteloua gracilis	blue grama
Chamaesyce serphyllifolia	thyme-leaf sandmat
Chenopodium dessicatum	arid-land goosefoot
Crepis occidentalis	large-flower hawk's-beard
Descurainia pinnata	western tansy mustard
Dodecatheon pulchellum	dark-throated shootingstar
Elymus elymoides	western bottle-brush grass
Erigeron pumilus	shaggy fleabane
Festuca brachyphylla	short-leaf fescue
Halogeton glomeratus	saltlover
Helianthus petilolaris	prairie sunflower
Hordeum jubatum	fox-tail barley
Hordeum pusillum	little barley
Hymenoxys richardsonii	Colorado rubberweed
Ipomopsis pumila	spike skyrocket
Iris missouriensis	Rocky Mountain iris
Iva axillaris	deer-root
Juncus balticus	Baltic rush
Lappula occidentalis	flat-spine sheepburr
Lepidium desiflorum	miner's pepperwort
Machaeranthera canescens	hoary tansy-aster
Machaeranthera tanacetifolia	takhoka-daisy
Monolepis nuttalliana	Nuttall's poverty-weed

Monroa squarrosa	false buffalo grass
Nassella viridula	green tussock
Oenothera pallida	white-pole evening-primrose
Pascopyrum smithii	western-wheat grass
Plantago eriopoda	red-woolly plantain
Plantago patagonica	woolly plantain
Poa secunda	curly blue grass
Puccinellia nuttalliana	Nuttall's alkali grass
Salicornia rubra	red saltwort
Schoenocrambe linifolium	Salmon River plains-mustard
Scirpus acutus	hard-stem bulrush
Spartina gracilis	alkali cord-grass
Sporobolus airoides	alkali-sacaton
Triglochin maritimum	seaside arrow-grass
Triglochin palustris	marsh arrow-grass
Vulpia octoflora	eight-flowered six-weeks grass

Saltbush Scrub

Saltbush Scrub occurs in the drier parts of the Wyoming Basin on old lake beds that are wet only in the spring. They are the most desert-like of all the plant communities in this region. Soils are characteristically dry and alkaline. This community corresponds in part to Küchler #40 and Desert Lands ERT.

Characteristic Woody Species

Atriplex confertifolia	shadscale
Atriplex gardneri	Gardner's saltbush

Associates

Artemisia spinescens	bud sagebrush
Opuntia polyacantha	hair-spine prickly-pear
Sarcobatus vermiculatus	greasewood

Herbaceous Understory

Chenopodium atrovirens	pinyon goosefoot
Cymopterus bulbosus	bulbous spring-parsley
Elymus elymoides	western bottle-brush grass
Eriogonum cernuum	nodding buckwheat
Halogeton glomeratus	saltlover
Hordeum jubatum	foxtail barley
Oxyzopsis hymenoides	Indian mountain-rice grass
Pascopyrum smithii	western-wheat grass
Poa secunda	curly blue grass
Thelypodiopsis elegans	westwater tumble-mustard

Mountain-Mahogany Shrub

Mountain-Mahogany Shrub occurs in the foothills of most mountain ranges in and adjacent to the Wyoming Basin. This community corresponds to Küchler #37 and Desert Lands ERT.

Characteristic Woody Species

Cercocarpus montanus	alder-leaf mountain-mahogany
Cercocarpus ledifolius	curl-leaf mountain-mahogany
Quercus gambelii	Gambel's oak

Associates

Acer grandidentatum	canyon maple
Amelanchier alnifolia	Saskatoon service-berry
Amelanchier utahensis	Utah service-berry
Artemisia frigida	prairie sagebrush
Artemisia tridentata	big sagebrush
Ceanothus velutinus	tobacco-brush
Chrysothamnus nauseosus	rubber rabbitbrush
Chrysothamnus parryi	Parry's rabbitbrush
Chrysothamnus viscidiflorus	green rabbitbrush
Eriogonum umbellatum	sulphur-flower wild buckwheat
Fallugia paradoxa	Apache-plume
Gutierrezia sarothrae	kindlingweed
Juniperus scopulorum	Rocky Mountain juniper
Juniperus osteosperma	Utah juniper
Krascheninnikovia lanata	winterfat
Pachystima myrsinites	myrtle boxleaf
Physocarpus malvaceus	mallow-leaf ninebark
Purshia mexicana	Mexican cliff-rose
Purshia tridentata	bitterbrush
Quercus havardii	Harvard's oak
Quercus turbinella	shrub live oak
Rhus trilobata	ill-scented sumac
Symphoricarpos albus	common snowberry

Herbaceous Understory

Antennaria rosea	rosy pussytoes
Besseya wyomingensis	Wyoming coraldrops
Carex duriuscula	spike-rush sedge
Cerastium arvense	field mouse-ear chickweed
Chenopodium leptophyllum	narrow-leaf goosefoot
Delphinium nuttallianum	two-love larkspur
Descurainia pinnata	western tansy-mustard
Elymus lanceolatus	streamside wild rye
Elymus trachycaulis	slender rye grass
Erigeron poliospermus	purple cushion fleabane
Festuca idahoensis	bluebunch fescue
Festuca kingii	King's fescue
Galium boreale	northern bedstraw
Grindelia squarrosa	curly-cup gumweed
Harbouria trachypleura	whiskybroom-parsley
Hedeoma drummondii	Drummond's false pennyroyal
Heuchera parvifolia	little-flower alumroot
Koeleria macrantha	prairie Koeler's grass
Lesquerella argentea	bladderpod
Linum lewisii	prairie flax

Mertensia lanceolata	prairie bluebells
Muhlenbergia filiculmis	slim-stem muhly
Orobanche ludoviciana	Louisiana broom-rape
Oryzopsis hymenoides	Indian mountain-rice grass
Pascopyrum smithii	western-wheat grass
Phlox multiflora	Rocky Mountain phlox
Pseudoroegneria spicatum	bluebunch-wheat grass
Stipa comata	needle-and-thread

Shrub Grassland

Shrub Grassland occurs in the foothills of the mountains above 7000 feet. Shrub Grassland requires a more moist environment than the shrublands in the lower areas. It is a community of short grasses and scattered small shrubs. This community corresponds to Küchler #56 and Desert Lands ERT.

Characteristic Woody Species

Artemisia frigida	prairie sagebrush

Associates

Atriplex gardneri	Gardner's saltbush
Chrysothamnus vaseyi	Vasey's rabbitbrush
Chrysothamnus viscidiflorus	green rabbitbrush
Gutierrezia sarothrae	kindlingweed
Krascheninnikovia lanata	winterfat
Opuntia polyacantha	hair-spine prickly-pear
Pediocactus simpsonii	snowball cactus
Tetradymia canescens	spineless horsebrush

Herbaceous Understory

Allium textile	white wild onion
Arabis holboellii	Holboell's rockcress
Arenaria hookeri	Hooker's sandwort
Astragalus adsurgens	standing milk-vetch
Astragalus purshii	Pursh's milk-vetch
Astragalus spatulatus	tufted milk-vetch
Bouteloua gracilis	blue grama
Calamovilfa longifolia	prairie sand-reed
Carex duriuscula	spike-rush sedge
Carex filifolia	thread-leaf sedge
Cymopterus montanus	mountain spring-parsley
Erigeron nematophyllus	needle-leaf fleabane
Eriogonum flavum	alpine golden wild buckwheat
Eriogonum ovalifolium	cushion wild buckwheat
Ipomopsis spicata	spiked skyrocket
Koeleria macrantha	prairie Koeler's grass
Lesquerella ludoviciana	Louisisana bladderpod
Lygodesmia juncea	rush skeleton-plant
Machaeranthera canescens	hoary tansy-aster
Orobanche fasciculata	clustered broom rape
Oryzopsis hymenoides	Indian mountain-rice grass

Oxytropis lagopus	hare-foot locoweed
Paronychia sessiliflora	low nailwort
Pascopyrum smithii	western-wheat grass
Penstemon angustifolius	broad-beard beardtongue
Poa secunda	curly blue grass
Polygonum douglasii	Douglas' knotweed
Pseudoroegneria spicata	bluebunch-wheat grass
Senecio canus	silver-woolly ratwort
Sphaeralcea coccinea	scarlet globe-mallow
Stenotus acaulis	stemless mock goldenweed
Trifolium gymnocarpon	holly-leaf clover

WETLAND SYSTEMS

Floodplain Forest

Floodplain Forest is commonly associated with rivers throughout the Wyoming Basin. Cottonwood is the most common tree species although blue spruce may be found at higher elevations. Alder thickets and meadows occur where the canopy is more open. This community corresponds to Floodplain Forests ERT.

Canopy
Characteristic Species
Betula occidentalis	water birch
Populus angustifolia	narrow-leaf cottonwood

Associates
Alnus incana	speckled alder
Juniperus osteosperma	Utah juniper
Picea pungens	blue spruce

Woody Understory
Cornus sericea	redosier
Elaeagnus commutata	American silver-berry
Prunus virginiana	choke cherry
Salix drummondiana	Drummond's willow
Salix exigua	sandbar willow
Salix geyeriana	Geyer's willow
Salix monticola	mountain willow

Herbaceous Understory
Agrostis scabra	rough bent
Artemisia ludoviciana	white sagebrush
Glycyrrhiza lepidota	American licorice
Hordeum jubatum	fox-tail barley
Juncus balticus	Baltic rush
Mentha arvensis	American wild mint
Oryzopsis hymenoides	Indian mountain-rice grass
Rumex maritimus	golden dock
Sporobolus airoides	alkali-sacaton

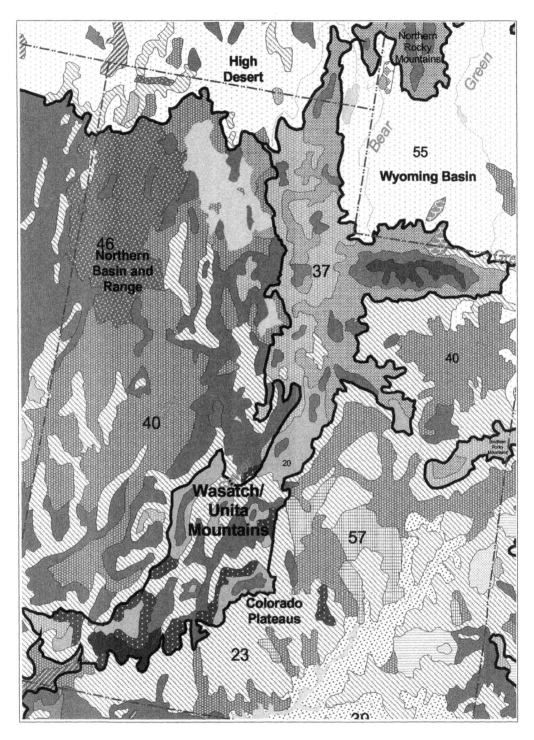

Wasatch/Uinta Mountains

Wasatch/Uinta Mountains

Introduction

The Wasatch and Uinta Mountains form a north/south ridge of high mountains between the Northern Basin and Range and the Colorado Plateau. The high elevation of these mountains creates a mesic environment supporting forests similar to the forests in the Northern Rocky Mountains. The southern part of the region supports communities that are closely related to the vegetation of the Colorado Plateaus and the Arizona and New Mexico mountains to the south.

The primary sources used to develop descriptions and species lists are Barbour and Billings 1988; Cronquist et al. 1972.

Dominant Ecological Communities

Upland Systems
Douglas-fir Forest
Western Spruce–Fir Forest
Spruce–Fir–Douglas-fir Forest
Ponderosa Pine Forest
Aspen–Lodgepole Pine Forest
Mountain-Mahogany–Oak Scrub
Juniper–Pinyon Woodland
Mountain Sagebrush
Dry Mountain Meadow
Wheat grass–Blue grass Prairie
Wetland Systems
Wet Mountain Meadow

UPLAND SYSTEMS

Douglas-fir Forest

Douglas-fir Forest occurs in the Wasatch and Uinta Mountains in high areas on mesic sites. This forest type is very similar to Douglas-fir Forest in the Northern Rocky Mountains and in the Front Range of the Southern Rocky Mountains. This community corresponds to Küchler #12 and Coniferous Forests ERT.

Canopy
Characteristic Species
   Pseudotsuga menziesii   Douglas-fir
Associates
   Abies concolor   white fir
   Larix occidentalis   western larch
   Picea glauca   white spruce
   Picea pungens   blue spruce
   Pinus contorta   lodgepole pine
   Pinus ponderosa   ponderosa pine
   Populus tremuloides   quaking aspen

Woody Understory

Arctostaphylos uva-ursi	red bearberry
Physocarpus malvaceus	mallow-leaf ninebark
Rosa nutkana	Nootka rose
Rosa woodsii	Woods' rose
Symphoricarpos albus	common snowberry

Herbaceous Understory

Arnica latifolia	daffodil leopard-bane
Calamagrostis rubescens	pinegrass
Carex concinnoides	northwestern sedge
Carex geyeri	Geyer's sedge

Western Spruce–Fir Forest

Western Spruce–Fir Forest, dominated by subalpine fir and Englemann spruce, occurs throughout the Wasatch and Uinta Mountains and throughout the entire Rocky Mountain area at high altitudes. Aspen is also abundant within the areas supporting spruce–fir forests. This community corresponds to Küchler #15 and Coniferous Forests ERT.

Canopy
Characteristic Species

Abies lasiocarpa	subalpine fir
Picea engelmannii	Engelmann's spruce

Associates

Pinus albicaulis	white-bark pine
Pinus contorta	lodgepole pine
Pinus flexilis	limber pine
Pinus longaeva	Intermountain bristle-cone pine
Populus tremuloides	quaking aspen
Pseudotsuga menziesii	Douglas-fir

Woody Understory

Arctostaphylos uva-ursi	red bearberry
Shepherdia canadensis	russet buffalo-berry
Symphoricarpos albus	common snowberry
Vaccinium myrtillus	whortle-berry
Vaccinium scoparium	grouseberry
Viburnum edule	squashberry

Herbaceous Understory

Arnica cordifolia	heart-leaf leopard-bane
Calamagrostis canadensis	blue joint
Orthilia secunda	sidebells
Polemonium pulcherrimum	beautiful Jacob's ladder
Xerophyllum tenax	western turkeybeard

Spruce–Fir–Douglas-fir Forest

In the southern part of the region, Douglas-fir becomes mixed and shares dominance with white fir and blue spruce. This forest type occurs at elevations between 8000 and 10000 feet. Within these communities, there may be areas dominated by aspen. This community corresponds to Küchler #20 and Coniferous Forests ERT.

Canopy

Characteristic Species

Abies concolor	white fir
Picea pungens	blue spruce
Pseudotsuga menziesii	Douglas-fir

Associates

Populus angustifolia	narrow-leaf cottonwood
Populus tremuloides	quaking aspen

Woody Understory

Acer glabrum	Rocky Mountain maple
Amelanchier alnifolia	Saskatoon service-berry
Betula occidentalis	water birch
Ceanothus fendleri	Fendler's buckbrush
Cornus sericea	redosier
Holodiscus dumosa	glandular oceanspray
Juniperus communis	common juniper
Mahonia repens	creeping Oregon-grape
Pachystima myrsinites	myrtle boxleaf
Physocarpus malvaceus	mallow-leaf ninebark
Prunus virginiana	choke cherry
Ribes cereum	white squaw currant
Ribes montigenum	western prickly gooseberry
Salix lutea	yellow willow
Salix scouleriana	Scouler's willow
Sambucus cerulea	blue elder
Shepherdia canadensis	russet buffalo-berry
Sorbus scopulina	Cascade Mountain-ash
Symphoricarpos oreophilus	mountain snowberry

Herbaceous Understory

Aconitum columbianum	Columbian monkshood
Angelica pinnata	small-leaf angelica
Aquilegia coerulea	Colorado blue columbine
Aquilegia formosa	crimson columbine
Arnica cordifolia	heart-leaf leopard-bane
Bromus marginatus	large mountain brome
Cardamine cordifolia	large mountain bittercress
Castilleja miniata	great red Indian-paintbrush
Corallorrhiza maculata	summer coralroot
Delphinium occidentale	dunce-cap larkspur
Dracocephalum parvifolium	American dragonhead
Elymus glaucus	blue wild rye

Fragaria vesca	woodland strawberry
Ipomopsis aggregata	scarlet skyrocket
Mimulus guttatus	seep monkey-flower
Mitella stauropetala	side-flower bishop's cap
Monardella odoratissima	alpine mountainbalm
Osmorhiza depauperata	blunt-fruit sweet-cicely
Polemonium foliosissimum	towering Jacob's ladder
Pseudoroegneria spicata	bluebunch-wheat grass
Pseudostellaria jamesiana	sticky-starwort
Rudbeckia occidentalis	western coneflower
Scrophularia lanceolata	lance-leaf figwort
Silene douglasii	seabluff catchfly
Stipa lettermanii	Letterman's needle grass
Stipa nelsonii	Nelson's needle grass
Thalictrum fendleri	Fendler's meadow-rue
Veronica americana	American-brooklime

Ponderosa Pine Forest

Ponderosa Pine Forest occurs at low elevations in the Wasatch and Uinta Mountains. Ponderosa Pine Forest is often similar to woodlands, with an open canopy and a grassy understory. In higher areas, ponderosa pine may form dense stands. The characteristic grassy understory of Ponderosa Pine Forest is highly flammable during dry summers. Suppression of wildfires and the introduction of cattle onto Ponderosa Pine Forest changed the species composition and the forest structure by increasing the shrub and understory species. This community corresponds to Coniferous Forests ERT.

Canopy
Characteristic Species

Pinus ponderosa	ponderosa pine

Associates

Pinus cembroides	Mexican pinyon
Pinus flexilis	limber pine
Pinus leiophylla	Chihuahuan pine
Populus tremuloides	quaking aspen
Pseudotsuga menziesii	Douglas-fir
Quercus gambelii	Gambel's oak

Woody Understory

Arctostaphylos patula	green-leaf manzanita
Artemisia nova	black sagebrush
Artemisia tridentata	big sagebrush
Ceanothus fendleri	Fendler's buckbrush
Chamaebatiaria millefolium	fernbush
Chrysothamnus parryi	Parry's rabbitbrush
Fallugia paradoxa	Apache-plume
Fendlerella utahensis	Utah-fendlerbush
Holodiscus boursieri	Boursier's oceanspray
Holodiscus dumosus	glandular oceanspray
Jamesia americana	five-petal cliffbush

Juniperus communis	common juniper
Mahonia repens	creeping Oregon-grape
Pachystima myrsinites	myrtle boxleaf
Philadelphus microphyllus	little-leaf mock orange
Physocarpus monogynus	mountain ninebark
Purshia mexicana	Mexican cliff-rose
Ribes cereum	white squaw currant
Robinia neomexicana	New Mexico locust
Rubus idaeus	common red raspberry

Herbaceous Understory

Artemisia ludoviciana	white sagebrush
Blepharoneuron tricholepis	pine-dropseed
Bothriochloa barbinodis	cane beard grass
Bouteloua gracilis	blue grama
Bromus spp.	brome
Comandra umbellata	bastard toadflax
Elymus elymoides	western bottle-brush grass
Festuca arizonica	Arizona fescue
Hedeoma dentatum	Arizona false pennyroyal
Iris missouriensis	Rocky Mountain iris
Koeleria macrantha	prairie Koeler's grass
Lupinus palmeri	Palmer's lupine
Monarda fistulosa	Oswego-tea
Muhlenbergia montana	mountain muhly
Muhlenbergia wrightii	Wright's muhly
Oxytropis lamberti	stemless locoweed
Pteridium aquilinum	northern bracken fern
Sphaeralcea fendleri	thicket globe-mallow
Thalictrum fendleri	Fendler's meadow-rue
Thermopsis rhombifolia	prairie golden-banner
Vicia americana	American purple vetch

Aspen–Lodgepole Pine Forest

Aspen–Lodgepole Pine Forest occurs on mountain slopes, generally between 8000 and 10000 feet. Aspen is common in both the Wasatch and Uinta Mountains, but only small areas of lodgepole pine are represented in the Wasatch Range. Aspen and lodgepole pine establish and grow well in areas that have burned although they are not exclusive to burned areas. This community corresponds to Western Deciduous and Mixed Forests ERT.

Canopy
Characteristic Species

Pinus contorta	lodgepole pine
Populus tremuloides	quaking aspen

Associates

Pinus flexilis	limber pine
Pseudotsuga menziesii	Douglas-fir

Woody Understory

Acer grandidentatum	canyon maple
Arctostaphylos uva-ursi	red bearberry
Arnica cordifolia	heart-leaf leopard-bane
Ceanothus velutinus	tobacco-brush
Mahonia repens	creeping Oregon grape
Pachystima myrsinites	myrtle boxleaf
Rubus parviflorus	western thimble-berry
Sambucus cerulea	blue elder
Shepherdia canadensis	russet buffalo-berry

Herbaceous Understory

Hydrophyllum capitatum	cat's breeches
Nemophila breviflora	Great Basin baby-blue-eyes
Penstemon subglaber	Utah smooth beardtongue
Phacelia sericea	purplefringe
Poa fendleriana	mutton grass
Pyrola asarifolia	pink wintergreen
Scrophularia lanceolata	lance-leaf figwort
Solidago spathulata	coastal-dune goldenrod
Thalictrum fendleri	Fendler's meadow rue
Wyethia amplexicaulis	northern mule's-ears

Mountain-Mahogany–Oak Scrub

Mountain-Mahogany–Scrub Oak occurs in the Wasatch and Uinta Mountains at elevations between 5000 and 9000 feet. Mountain-mahogany, Gambel's oak, and canyon maple are the dominant species. Unlike most upland forest types in this region, the dominants are deciduous. On some sites, understory shrubs form dense thickets. This community corresponds to Küchler #37 and Woodlands ERT.

Characteristic Woody Species

Acer grandidentatum	canyon maple
Cercocarpus ledifolius	curl-leaf mountain-mahogany
Quercus gambelii	Gambel's oak

Associates

Acer glabrum	Rocky Mountain maple
Acer negundo	ash-leaf maple
Amelanchier alnifolia	Saskatoon service-berry
Amelanchier utahensis	Utah service-berry
Arctostaphylos spp.	manzanita
Artemisia nova	black sagebrush
Artemisia tridentata	big sagebrush
Betula occidentalis	water birch
Ceanothus fendleri	Fendler's buckbrush
Ceanothus velutinus	tobacco-brush
Cercocarpus montanus	alder-leaf mountain-mahogany
Chrysothamnus viscidiflorus	green rabbitbrush
Fallugia paradoxa	Apache-plume
Gutierrezia sarothrae	kindlingweed

Physocarpus malvaceus	mallow-leaf ninebark
Populus angustifolia	narrow-leaf cottonwood
Prunus virginiana	choke cherry
Purshia mexicana	Mexican cliff-rose
Purshia tridentata	bitterbrush
Quercus havardii	Havard's oak
Quercus turbinella	shrub live oak
Rhamnus crocea	holly-leaf buckthorn
Rhus trilobata	ill-scented sumac
Ribes cereum	white squaw currant
Rosa woodsii	Arizona rose
Sambucus cerulea	blue elder
Symphoricarpos oreophilis	mountain snowberry

Herbaceous Understory

Claytonia lanceolata	lance-leaf springbeauty
Collinsia parviflora	small-flower blue-eyed Mary
Delphinium nuttallianum	two-lobe larkspur
Erigeron flagellaris	trailing fleabane
Erythronium grandiflorum	yellow avalanche-lily
Gayophytum ramosissimum	pinyon groundsmoke
Geranium caespitosum	purple cluster crane's-bill
Heliomeris multiflora	Nevada showy false goldeneye
Hydrophyllum capitatum	cat's-breeches
Leymus cinereus	Great Basin lyme grass
Lithophragma parviflora	prairie woodlandstar
Nemophila breviflora	Great Basin baby-blue-eyes
Orthocarpus luteus	golden-tongue owl-clover
Stipa comata	needle-and-thread
Stipa lettermanii	Letterman's needle grass
Wyethia amplexicaulis	northern mule's-ears

Juniper–Pinyon Woodland

Juniper–Pinyon Woodland occurs on the slopes of the Wasatch and Uinta Mountains at elevations between 5000 and 8000 feet. In lower areas, there is usually not enough precipitation to support these trees. The trees in the Juniper–Pinyon Woodland rarely grow over 20 feet in height. Much of the area has been heavily grazed resulting in the loss of understory species. There has also been an increase in erosion on the mountain slopes. This community corresponds to Küchler #23 and Woodlands ERT.

Canopy
Characteristic Species

Juniperus monosperma	one-seed juniper
Juniperus osteosperma	Utah juniper
Pinus edulis	two-needle pinyon
Pinus monophylla	single-leaf pinyon

Associates

Juniperus occidentalis	western juniper
Juniperus deppeana	alligator juniper

Quercus emoryi	Emory's oak
Quercus gambelii	Gambel's oak
Quercus grisea	gray oak

Woody Understory

Acer glabrum	Rocky Mountain maple
Amelanchier alnifolia	Saskatoon service-berry
Artemisia arbuscula	dwarf sagebrush
Artemisia tridentata	big sagebrush
Ceanothus velutinus	tobacco-brush
Cercocarpus ledifolius	curl-leaf mountain-mahogany
Chrysothamnus nauseosus	rubber rabbitbrush
Chrysothamnus viscidiflorus	green rabbitbrush
Ephedra viridis	Mormon-tea
Eriogonum microthecum	slender wild buckwheat
Eriogonum umbellatum	sulphur-flower wild buckwheat
Fallugia paradoxa	Apache-plume
Gutierrezia sarothrae	kindlingweed
Holodiscus dumosus	glandular oceanspray
Purshia mexicana	Mexican cliff-rose
Purshia tridentata	bitterbrush
Ribes cereum	white squaw currant
Ribes velutinum	desert gooseberry
Symphoricarpos oreophilus	mountain snowberry
Tetradymia canescens	spineless horsebrush

Herbaceous Understory

Balsamorhiza sagittata	arrow-leaf balsamroot
Bouteloua curtipendula	side-oats grama
Bouteloua gracilis	blue grama
Elymus elymoides	western bottle-brush grass
Eriogonum heracleoides	parsnip-flower wild buckwheat
Eriophyllum lanatum	common woolly-sunflower
Festuca idahoensis	bluebunch fescue
Festuca kingii	King's fescue
Frasera albomarginata	desert elkweed
Grindelia squarrosa	curly-cup gumweed
Heterotheca villosa	hairy false golden-aster
Hymenoxys richardsonii	Colorado rubberweed
Ipomopsis aggregata	scarlet skyrocket
Koeleria macrantha	prairie Koeler's grass
Lithospermum ruderale	Columbian puccoon
Lupinus sericeus	Pursh's silky lupine
Oryzopsis hymenoides	Indian mountain-rice grass
Pascopyrum smithii	western-wheat grass
Penstemon speciosus	royal beardtongue
Penstemon watsonii	Watson's beardtongue
Poa fendleriana	mutton grass
Poa secunda	curly blue grass
Pseudoroegneria spicata	bluebunch-wheat grass

Sporobolus cryptandrus	sand dropseed
Stipa comata	needle-and-thread
Stipa nelsonii	Nelson's needle grass
Stipa thurberiana	Thurber's needle grass

Mountain Sagebrush

Mountain Sagebrush occurs on mountain slopes between 7000 and 8000 feet. Big sagebrush is the most common species, although in more moist areas, hoary sagebrush may dominate. Where the soil is rocky, black sagebrush may be dominant. This community corresponds to Küchler #38 and Desert Lands ERT.

Characteristic Woody Species

Artemisia cana	hoary sagebrush
Artemisia nova	black sagebrush
Artemisia tridentata	big sagebrush

Associates

Amelanchier alnifolia	Saskatoon service-berry
Amelanchier utahensis	Utah service-berry
Cercocarpus montanus	alder-leaf mountain-mahogany
Eriogonum umbellatum	sulphur-flower wild buckwheat
Juniperus scopulorum	Rocky Mountain juniper
Purshia tridentata	bitterbrush
Symphoricarpos oreophilus	mountain snowberry

Herbaceous Understory

Arabis drummondii	Canadian rockcress
Balsamorhiza sagittata	arrow-leaf balsamroot
Castilleja flava	lemon-yellow Indian-paintbrush
Castilleja linariaefolia	Wyoming Indian-paintbrush
Collomia linearis	narrow-leaf mountain-trumpet
Cymopterus longipes	long-stalk spring-parsley
Elymus elymoides	western bottle-brush grass
Eriogonum heracleoides	parsnip-flower wild buckwheat
Erysimum asperum	plains wallflower
Frasera speciosa	monument plant
Hackelia patens	spotted stickseed
Heterotheca villosa	hairy false golden-aster
Hymenoxys richardsonii	Colorado rubberweed
Koeleria macrantha	prairie Koeler's grass
Leymus salinus	salinas lyme grass
Ligusticum porteri	Porter's wild lovage
Linum lewisii	prairie flax
Oenothera cespitosa	evening-primrose
Orthocarpus tolmiei	Tolmie's owl-clover
Phacelia sericea	purplefringe
Phlox longifolia	long-leaf phlox
Poa fendleriana	mutton grass
Sedum stenopetalum	worm-leaf stonecrop
Senecio multilobatus	lobe-leaf ragwort
Stipa lettermanii	Letterman's needle grass

Dry Mountain Meadow

Within the higher elevations, there are many dry meadows. These meadows may be the result of the slow filling-in of former glacial lakes or the results of past burns. These areas are usually dominated by grasses and forbs, but many rocky areas and forest interfaces support shrubs. This community corresponds to Meadowlands ERT.

Characteristic Species

Agastache urticifolia	nettle-leaf giant-hyssop
Aquilegia coerulea	Colorado blue columbine
Artemisia frigida	prairie sagebrush
Artemisia michauxiana	Michaux's wormwood
Calamagrostis canadensis	bluejoint
Carex microptera	small-wing sedge
Castilleja applegatei	wavy-leaf Indian-paintbrush
Castilleja rhexiifolia	rosy Indian-paintbrush
Castilleja sulphurea	sulphur Indian-paintbrush
Cerastium beeringianum	Bering Sea mouse-ear chickweed
Clematis columbiana	Columbian virgin's-bower
Danthonia intermedia	timber wild oat grass
Delphinium barbeyi	subalpine larkspur
Delphinium occidentale	dunce-cap larkspur
Deschampsia cespitosa	tufted hair grass
Elymus trachycaulum	slender wild rye
Festuca kingii	King's fescue
Frasera speciosa	monument plant
Geranium richardsonii	white crane's-bill
Heliomeris multiflora	Nevada showy false goldeneye
Heuchera parvifolia	little-flower alumroot
Ligusticum filicinum	fern-leaf wild lovage
Mertensia arizonica	aspen bluebells
Osmorhiza occidentalis	Sierran sweet-cicely
Penstemon rydbergii	meadow beardtongue
Penstemon whippleanus	dark beardtongue
Polemonium foliosissimum	towering Jacob's ladder
Sedum stenopetalum	worm-leaf stonecrop
Solidago multiradiata	Rocky Mountain goldenrod
Stipa lettermanii	Letterman's needle grass
Stipa nelsonii	Nelson's needle grass
Thalictrum fendleri	Fendler's meadow-rue
Trisetum spicatum	narrow false oat
Valeriana occidentalis	small-flower valerian

Wheat grass–Blue grass Prairie

There is only a very small representation of this community in the Wasatch and Uinta Mountain region. This community corresponds to Küchler #51 and Grasslands ERT.

Characteristic Species

Festuca idahoensis	bluebunch fescue
Poa secunda	curly blue grass
Pseudoroegneria spicata	bluebunch-wheat grass

Associates

Chrysothamnus nauseosus	rubber rabbitbrush
Lithophragma glabrum	bulbous woodlandstar
Lupinus sericeus	Pursh's silky lupine
Plantago patagonica	woolly plantain
Stellaria nitens	shiny starwort
Vulpia microstachys	small six-weeks grass

WETLAND SYSTEMS

Wet Mountain Meadow

In the Wasatch and Uinta mountains, there are many formerly glaciated areas with small lakes and wet meadows. The wet meadows develop in depressional areas of former small lakes. Grasses, sedges, and rushes predominate in these meadows, but a mix of emergent plants occur in deeper water, including both shrubs and forbs. This community corresponds to Meadowlands ERT.

Characteristic Species

Aconitum columbianum	Columbian monkshood
Alopecurus borealis	meadow-foxtail
Arnica chamissonis	leafy leopard-bane
Calamagrostis canadensis	bluejoint
Caltha leptosepala	white marsh-marigold
Cardamine cordifolia	large mountain bittercress
Carex aurea	golden-fruit sedge
Carex microptera	small-wing sedge
Castilleja rhexiifolia	rosy Indian-paintbrush
Deschampsia cespitosa	tufted hair grass
Eleocharis quinqueflora	few-flower spike-rush
Erigeron ursinus	Bear River fleabane
Hordeum brachyantherum	meadow barley
Mertensia ciliata	tall fringe bluebells
Muhlenbergia richardsonis	matted muhly
Pedicularis groenlandica	bull elephant's-head
Phleum alpinum	mountain timothy
Poa alpina	alpine blue grass
Poa reflexa	nodding blue grass
Polygonum bistortoides	American bistort
Primula parryi	brook primrose
Rorippa teres	southern marsh yellowcress
Saxifraga odontoloma	streambank saxifrage
Sedum rhodanthum	queen's-crown
Trisetum spicatum	narrow false oat
Veratrum californicum	California false hellebore
Zigadenus elegans	mountain deathcamas

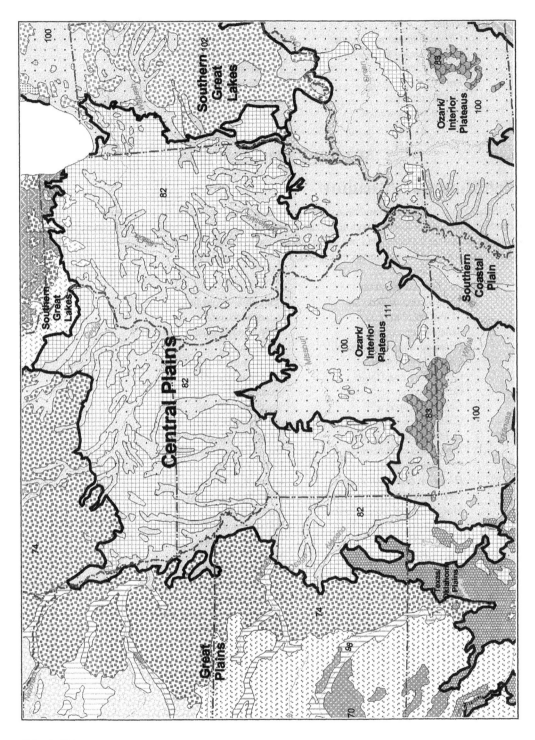

Colorado Plateaus

Colorado Plateaus

Introduction

The Colorado Plateaus region consists of plateaus, valleys, and canyons bordered by high mountains. The area is relatively high in elevation, between 5000 and 12000 feet, and arid. Some of the plateau areas are higher than the surrounding mountains. The region is in the rain shadow of the Rocky Mountains to east, the Wasatch and Uinta mountains to the north and west, and the Arizona and New Mexico Mountains to the south. There are only two major gaps in the mountainous rim that surrounds the region. One gap is a low divide between the San Juan and Rio Grande Rivers. The other gap is called the "Dixie Corridor" which connects this region to the Mojave desert to the southwest.

Canyons are a characteristic feature throughout the region. The Colorado, Green, and San Juan are the major rivers. These rivers and their tributaries dissect this region and have created numerous canyons. Beyond the spectacular geological formations, this region is of interest to botanists because it supports a large number of endemic species.

The vegetation in the region is predominantly Juniper–Pinyon Woodland and scrub. Forests are found in only a few locations where there is sufficient moisture. Two vegetation types which do not develop extensively elsewhere are the Blackbrush Shrub and the Galleta–Three–Awn Shrubsteppe.

The primary sources used to develop descriptions and species lists are Barbour and Billings 1988; Cronquist et al. 1972.

Dominant Ecological Communities

Upland Systems
Southwestern Spruce–Fir Forest
Pine–Douglas-fir Forest
Juniper–Pinyon Woodland
Blackbrush Scrub
Great Basin Sagebrush
Saltbush Scrub
Greasewood Scrub
Galleta–Three-Awn Shrubsteppe
Grama–Tobosa Shrubsteppe
Grama–Galleta Steppe

UPLAND SYSTEMS

Southwestern Spruce–Fir Forest

Southwestern Spruce–Fir Forest is most common in the Southern Rocky Mountain region, but extends into this region. This community corresponds to Küchler #21 and Coniferous Forests ERT.

Canopy
Characteristic Species

Abies lasiocarpa	subalpine fir
Picea engelmannii	Engelmann's spruce

Associates

Picea glauca	white spruce
Pinus aristata	bristle-cone pine
Pinus flexilis	limber pine
Populus tremuloides	quaking aspen

Woody Understory

Acer glabrum	Rocky Mountain maple
Alnus incana	speckled alder
Juniperus communis	common juniper
Lonicera involucrata	four-line honeysuckle
Mahonia repens	creeping Oregon-grape
Pachystima myrsinites	myrtle boxleaf
Pentaphylloides floribunda	golden-hardhack
Prunus emarginata	bitter cherry
Salix bebbiana	long-beak willow
Salix scouleriana	Scouler's willow
Symphoricarpos oreophilus	mountain snowberry
Vaccinium myrtillus	whortle-berry

Herbaceous Understory

Actaea rubra	red baneberry
Aquilegia chrysantha	golden columbine
Carex spp.	sedge
Dugaldia hoopesii	owl's-claws
Festuca rubra	red fescue
Gentiana spp.	gentian
Juncus spp.	rush
Pedicularis procera	giant lousewort
Pedicularis racemosa	parrot's-beak
Phleum alpinum	mountain timothy
Primula spp.	primrose
Trisetum spicatum	narrow false oat
Veratum californicum	California false hellebore
Viola canadensis	Canadian white violet
Viola nephrophylla	northern bog violet

Pine–Douglas-fir Forest

Pine–Douglas-fir Forest is found on high plateaus and mountains, extending southward from the Rocky Mountains to the Colorado Plateau. This forest type is also found in parts of Utah, New Mexico, and Arizona. This community responds to Küchler #18 and Coniferous Forests ERT.

Canopy
Characteristic Species

Pinus ponderosa	ponderosa pine
Pseudotsuga menziesii	Douglas-fir

Associates

Picea pungens	blue spruce
Pinus flexilis	limber pine

Woody Understory

Acer glabrum	Rocky Mountain maple
Alnus incana	speckled alder
Ceanothus fendleri	Fendler's buckbrush
Chamaebatiaria millifolium	fernbush
Holodiscus dumosus	glandular oceanspray
Jamesia americana	five-petal cliffbush
Juniperus communis	common juniper
Prunus emarginata	bitter cherry

Herbaceous Understory

Blepharoneuron tricholepis	pine-dropseed
Festuca arizonica	Arizona fescue

Juniper–Pinyon Woodland

Juniper–Pinyon Woodland is best developed at elevations of 5000 to 8000 feet under conditions of annual precipitation of 12 inches or more. The canopy of this community type is widely spaced and the woody species are low, evergreen trees. This community corresponds to Küchler #23 and Woodlands ERT.

Canopy
Characteristic Species

Juniperus monosperma	one-seed juniper
Juniperus osteosperma	Utah juniper
Pinus edulis	two-needle pinyon

Associates

Juniperus deppeana	alligator juniper
Juniperus occidentalis	western juniper
Pinus monophylla	single-leaf pinyon
Quercus emoryi	Emory's oak
Quercus gambelii	Gambel's oak
Quercus grisea	gray oak
Quercus turbinella	shrub live oak

Woody Understory

Amelanchier alnifolia	Saskatoon service-berry
Atriplex confertifolia	shadscale
Artemisia nova	black sagebrush
Artemisia tridentata	big sagebrush
Ceanothus fendleri	Fendler's buckbrush
Ceanothus integerrimus	deerbrush
Cercocarpus intricatus	little-leaf mountain-mahogany
Cercocarpus ledifolius	curl-leaf mountain-mahogany
Cercocarpus montanus	alder-leaf mountain-mahogany
Chamaebatiaria millifolium	fernbush
Chrysothamnus depressus	long-flower rabbitbrush
Chrysothamnus nauseosus	rubber rabbitbrush
Coleogyne ramosissima	blackbrush
Echinocereus coccineus	scarlet hedgehog cactus

Ephedra viridis	Mormon-tea
Fallugia paradoxa	Apache-plume
Garrya wrightii	Wright's silktassel
Krashcheninnikovia lanata	winterfat
Mahonia fremonti	desert Oregon-grape
Opuntia basilaris	beaver-tail cactus
Opuntia erinacea	oldman cactus
Opuntia fragilis	pygmy prickly-pear
Opuntia polyacantha	hair-spine prickly-pear
Opuntia whipplei	rat-tail cholla
Purshia mexicana	Mexican cliff-rose
Purshia tridentata	bitterbrush
Yucca baccata	banana yucca

Herbaceous Understory

Bouteloua curtipendula	side-oats grama
Bouteloua eriopoda	black grama
Bouteloua gracilis	blue grama
Calochortus ambiguus	doubting mariposa-lily
Castelleja integra	squawfeather
Elymus elymoides	western bottle-brush grass
Festuca arizonica	Arizona fescue
Keckiella antirrhinoides	chaparral bush-beardtongue
Koeleria macrantha	prairie Koeler's grass
Muhlenbergia torreyi	ringed muhly
Oryzopsis hymenoides	Indian mountain-rice grass
Pascopyrum smithii	western-wheat grass
Piptochaetium fimbriatum	pinyon spear grass
Sphaeralcea emoryi	Emory's globe-mallow
Sporobolus cryptandrus	sand dropseed
Stipa nelsonii	Nelson's needle grass

Blackbrush Scrub

Blackbrush Scrub occurs in lower parts of the Colorado Plateaus in areas with low rainfall and non-saline, sandy soil. Blackbrush occurs in an environment that is too harsh for many plants. Blackbrush occurs widely in an area between the cold deserts to the north and the warm deserts to the south. Other shrubs are rare in the area and very few herbaceous species can survive the climate. This community corresponds to Küchler #39 and Desert Lands ERT.

Characteristic Woody Species

Coleogyne ramosissima	blackbrush

Associates

Artemisia filifolia	silver sagebrush
Artemisia tridentata	big sagebrush
Atriplex confertifolia	shadscale
Ephedra nevadensis	Nevada joint-fir
Ephedra torreyana	Torrey's joint-fir
Ericameria linearifolius	narrow-leaf heath-goldenrod

Erigonum fasciculatum	eastern Mojave wild buckwheat
Gutierrezia sarothrae	kindlingweed
Opuntia ramosissima	darning-needle cholla
Prunus fasciculata	desert almond
Psorothamnus fremontii	Fremont's smokebush
Salazaria mexicana	Mexican bladder-sage
Thamnosa montana	turpentine-broom
Yucca baccata	banana yucca
Yucca brevifolia	Joshua-tree

Herbaceous Understory

Bouteloua eriopoda	black grama
Brickellia oblongifolia	narrow-leaf brickellbush
Hilaria jamesii	James' galleta
Hilaria rigida	big galleta
Muhlenbergia porteri	bush muhly
Oryzopsis hymenoides	Indian mountain-rice grass

Great Basin Sagebrush

Great Basin Sagebrush is a desert-like mix of shrubs, primarily sagebrush and perennial grasses. Great Basin Sagebrush occurs above 5000 feet and may extend up to 10000 feet. This community corresponds to Küchler #38 and #55 and Desert Lands ERT.

Characteristic Woody Species

Artemisia tridentata	big sagebrush

Associates

Artemisia arbuscula	dwarf sagebrush
Artemisia nova	black sagebrush
Atriplex confertifolia	shadscale
Chrysothamnus nauseosus	rubber rabbitbrush
Chrysothamnus viscidiflorus	green rabbitbrush
Coleogyne ramosissima	blackbrush
Ephedra torreyana	Torrey's joint-fir
Ephedra viridis	Mormon-tea
Grayia spinosa	spiny hop-sage
Leptodactylon pungens	granite prickly-pear
Purshia tridentata	bitterbrush
Ribes velutinum	desert gooseberry
Tetradymia glabrata	little-leaf horsebrush

Herbaceous Understory

Allium acuminatum	taper-tip onion
Aristida pupurea	purple three-awn
Balsamorhiza sagittata	arrow-leaf balsamroot
Calochortus nuttallii	sego-lily
Castilleja angustifolia	northwestern Indian-paintbrush
Crepis acuminata	long-leaf hawk's-beard
Delphinium andersonii	desert larkspur

Elymus elymoides	western bottle-brush grass
Elymus lanceolatus	streamside wild rye
Festuca idahoensis	bluebunch fescue
Heterotheca villosa	hairy false golden-aster
Hymenoxys richardsonii	Colorado rubberweed
Koeleria macrantha	prairie Koeler's grass
Leymus cinereus	Great basin lyme grass
Lupinus caudatus	Kellogg's spurred lupine
Lupinus sericeus	Pursh's silky lupine
Oryzopsis hymnoides	Indian mountain-rice grass
Pascopyrum smithii	western-wheat grass
Phlox hoodii	carpet phlox
Phlox longifolia	long-leaf phlox
Poa fendleriana	mutton grass
Poa secunda	curly blue grass
Pseudoroegneria spicata	bluebunch-wheat grass
Sporobolus airoides	alkali-sacaton
Stipa comata	needle-and-thread
Viola beckwithii	western pansy
Wyethia amplexicaulis	northern mule's-ears
Zigadenus paniculatus	sand-corn

Saltbush Scrub

Saltbush Scrub occurs in areas with little moisture and on saline valley soils. This community type is often referred to as the "shadscale community." Structurally, this desert community is made up of low, widely spaced shrubs, typically covering about 10 percent of the ground. Vegetation is characteristically grayish, spiny, and very small leaved. This community corresponds to Küchler #40 and Desert Lands ERT.

Characteristic Woody Species

Atriplex confertifolia	shadscale

Associates

Artemisia spinescens	bud sagebrush
Atriplex canescens	four-wing saltbush
Atriplex gardneri	Gardner's saltbush
Atriplex nuttallii	Nuttall's saltbush
Chrysothamnus viscidiflorus	green rabbitbrush
Ephedra nevadensis	Nevada joint-fir
Grayia spinosa	spiny hop-sage
Gutierrezia sarothrae	kindlingweed
Krascheninnikovia lanata	winterfat
Lycium cooperi	peachthorn
Sacrobatus vermiculatus	greasewood
Tetradymia glabrata	little-leaf horsebrush

Herbaceous Understory

Camissonia boothii	shredding suncup
Camissonia claviformis	browneyes
Camissonia scapoidea	Paiute suncup

Cryptantha circumscissa	cushion cat's-eye
Eriogonum ovalifolium	cushion wild buckwheat
Halogeton glomeratus	saltlover
Iva nevadensis	Nevada marsh-elder
Kochia americana	greenmolly
Mirabilis alipes	winged four-o'clock
Sphaeraliea grossulariifolia	currant-leaf globe-mallow
Vulpia octoflora	eight-flower six-weeks grass
Xylorhiza glabriuscula	smooth woody-aster

Greasewood Scrub

Greasewood Scrub is a salt-tolerant community that occurs in valley bottoms where the water table is relatively close to the surface. This community corresponds to Küchler #40 and Desert Lands ERT.

Characteristic Woody Species
Sarcobatus vermiculatus	greasewood

Associates
Allenrolfea occidentalis	iodinebush
Artemisia spinescens	bud sagebrush
Atriplex confertifolia	shadscale
Atriplex lentiformis	quailbush
Suaeda moquinii	shrubby seepweed

Herbaceous Understory
Cordylanthus maritimus	saltmarsh bird's-beak
Distichlis spicata	coastal salt grass
Glaux maritima	sea-milkwort
Halogeton glomeratus	saltlover
Hutchinsia procumbens	ovalpurse
Iva axillaris	deer-root
Juncus balticus	Baltic rush
Kochia americana	greenmolly
Pyrrocoma lanceolatus	lance-leaf goldenweed
Salicornia rubra	red saltwort
Sarcocornia pacifica	Pacific swampfire
Sporobolus airoides	alkali-sacaton
Thelypodium sagittatum	arrowhead thelypody

Galleta–Three-Awn Shrubsteppe

Galleta–Three-Awn Shrubsteppe is an arid grassland interspersed with shrubs. One of the community dominants, galleta, is a sod-forming grass and three-awn is a vigorous bunchgrass. The sandy soils in these areas are covered by a microphytic crust. Grazing breaks up the crust causing dunes to form, allowing shrubs to invade and increase. This community corresponds to Küchler #57 and Grasslands ERT.

Characteristic Species
Aristida purpurea	purple three-awn
Artemisia filifolia	silver sagebrush
Ephedra viridis	Mormon tea
Hilaria jamesii	James' galleta

Associates

Ambrosia acanthicarpa	flat-spine burr-ragweed
Bouteloua gracilis	blue grama
Chrysothamnus nauseosus	rubber rabbitbrush
Chrysothamnus viscidiflorus	green rabbitbrush
Encelia farinosa	goldenhills
Ephedra torryana	Torrey's joint-fir
Helianthus anomalus	western sunflower
Heliotropium convolvulaceum	wide-flower heliotrope
Heterotheca villosa	hairy false golden-aster
Lepidium fremontii	bush pepperwort
Machaeranthera canescens	hoary tansy-aster
Mahonia fremontii	desert Oregon-grape
Mentzelia pumila	golden blazingstar
Monroa squarrosa	false buffalo grass
Muhlenbergia pungens	sandhill muhly
Oenothera albicaulis	white-stem evening-primrose
Oryzopsis hymenoides	Indian mountain-rice grass
Poliomintha incana	hoary rosemary-mint
Sphaeralcea grossulariifolia	currant-leaf globe-mallow
Sphaeralcea leptophylla	scaly globe-mallow
Sporobolus cryptandrus	sand dropseed
Stephanomeria pauciflora	brown-plume wire-lettuce

Grama–Tobosa Shrubsteppe

Grama–Tobosa Shrubsteppe is a mosaic of short grasses and shrubs that vary from open to dense. This community corresponds to Küchler #58 and Grasslands ERT.

Characteristic Species

Bouteloua eriopoda	black grama
Hilaria mutica	tobosa grass
Larrea tridentata	creosote-bush

Associates

Acacia constricta	mescat wattle
Aristida californica	California three-awn
Aristida divaricata	poverty three-awn
Aristida purpurea	purple three-awn
Aristida ternipes	spider grass
Baileya multiradiata	showy desert-marigold
Bothriocholoa barbinodis	cane beard grass
Bouteloua curtipendula	side-oats grama
Bouteloua gracilis	blue grama
Gutierrezia sarothrae	kindlingweed
Hilaria belangeri	curly-mesquite
Hilaria jamessii	James' galleta
Muhlenbergia porteri	bush muhly
Prosopis glandulosa	honey mesquite
Sporobolus airoides	alkali-sacaton

Sporobolus cryptandrus	sand dropseed
Sporobolus flexuosus	mesa dropseed
Yucca baccata	banana yucca
Yucca elata	soaptree yucca
Zinnia acerosa	white zinnia
Zinnia grandiflora	little golden zinnia

Grama–Galleta Steppe

Grama–Galleta Steppe is characterized by low to medium tall grasslands with few woody plants. This community corresponds to Küchler #53 and Grasslands ERT.

Characteristic Species

Bouteloua gracilis	blue grama
Hilaria jamesii	James' galleta

Associates

Andropogon hallii	sand bluestem
Artemisia tridentata	big sagebrush
Atriplex canescens	four-wing saltbush
Bouteloua curtipendula	side-oats grama
Bouteloua hirsuta	hairy grama
Ephedra viridis	Mormon-tea
Optuntia whipplei	rat-tail cholla
Oryzopsis hymenoides	Indian mountain-rice grass
Schizachyrium scoparium	little false bluestem
Yucca glauca	soapweed yucca

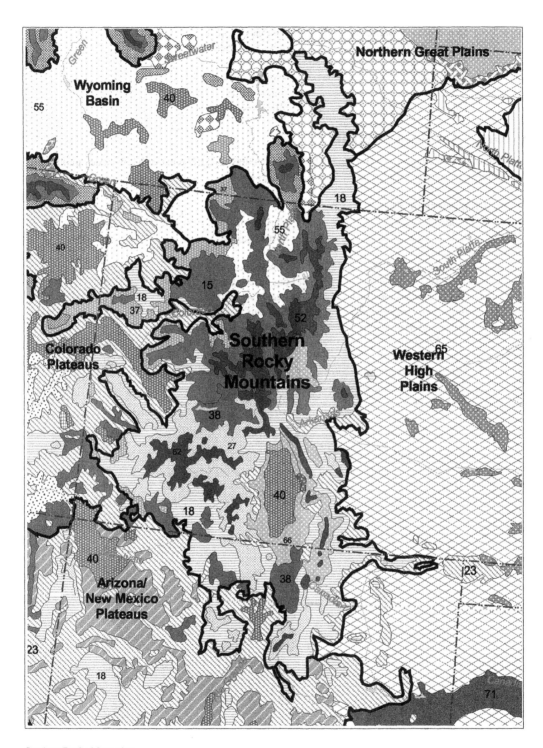

Southern Rocky Mountains

Southern Rocky Mountains

Introduction

The Southern Rocky Mountains region occupies most of central and western Colorado. The mountain ranges of this region also extend north into Wyoming and south into New Mexico. The mountains begin at approximately 4950 feet and rise to 14520 feet. Major ranges within the Southern Rockies include the San Juan Mountains, the Sangre de Cristo Mountains, the Front Ranges, and the Medicine Bow Mountains.

The predominant vegetation type of the region is conifer forest. Environmental gradients associated with elevation influence the vegetation types occurring in an area. The lowest elevation forests are dominated by ponderosa pine. As elevation increases, Douglas-fir and then subalpine fir and Engelmann's spruce become common species. The lower mountain slopes and basins support shrub communities of Sagebrush Steppe, Juniper-Pinyon Woodland, and Mountain Mahogany-Oak Scrub.

Frequent fire is characteristic of the Southern Rocky Mountains, as it is in the Northern Rockies. Other disturbances caused by wind, insects, avalanches, and browsing are also natural factors which periodically change the vegetation. Humans, however, have caused major changes in the region's natural vegetation by suppressing natural fires and introducing intensive domestic livestock grazing.

A unique feature of the Southern Rocky Mountain region is the "park." Parks are low elevation valley bottoms on fine-textured soils, dominated by grasses, sedges, and forbs. Although not covered by this book, many interesting alpine communities exist in extreme conditions in the highest altitudes of the Rocky Mountains.

The primary sources used to develop descriptions and species lists are Barbour and Billings 1988; Brown 1982.

Dominant Ecological Communities

Upland Systems
    Ponderosa Pine Forest
    Douglas-fir Forest
    Spruce–Fir Forest
    Juniper–Pinyon Woodland
    Mountain-Mahogany–Oak Scrub
    Sagebrush Steppe
    Saltbush–Greasewood Shrub
    Shortgrass Prairie
Wetland Systems
    Riparian Forest

UPLAND SYSTEMS

Ponderosa Pine Forest

Ponderosa Pine Forest is more open at lower elevations. It closes, becomes mixed with Douglas-fir, and grades to Douglas-fir Forest at higher elevations. This community corresponds, in part, to Küchler #18 and Coniferous Forests ERT.

Canopy
Characteristic Species
Pinus ponderosa	ponderosa pine
Pseudotsuga menziesii	Douglas-fir

Associates
Picea pungens	blue spruce
Pinus flexilis	limber pine
Pinus strobliformis	southwestern white pine
Quercus gambelii	Gambel's oak

Woody Understory
Acer glabrum	Rocky Mountain maple
Alnus incana	speckled alder
Ceanothus fendleri	Fendler's buckbrush
Chamaebatiaria millifolium	fernbush
Holodiscus dumosus	glandular oceanspray
Jamesia americana	five-petal cliffbush
Juniperus communis	common juniper
Mahonia repens	creeping Oregon-grape
Physocarpus monogynus	mountain ninebark
Prunus emarginata	bitter cherry
Rhus glabra	smooth sumac
Ribes aureum	golden currant
Ribes pinetorum	orange gooseberry
Ribes viscosissimum	sticky currant
Robinia neomexicana	New Mexico locust
Rosa woodsii	Arizona rose
Sambucus cerulea	blue elder
Symphoricarpos longiflorus	desert snowberry
Symphoricarpos oreophilus	mountain snowberry
Symphoricarpos rotundifolius	round-leaf snowberry

Herbaceous Understory
Blepharoneuron tricholepis	pine-dropseed
Bromus anomalus	nodding brome
Bromus ciliatus	fringed brome
Carex geophila	white mountain sedge
Cyperus fendlerianus	Fendler's flat sedge
Elymus elymoides	western botttle-brush grass
Festuca arizonica	Arizona fescue
Koeleria macrantha	prairie Koeler's grass
Muhlenbergia minutissima	least muhly
Muhlenbergia montana	mountain muhly
Muhlenbergia virescens	screw leaf muhly
Panicum bulbosum	bulb panic grass
Piptochaetium pringlei	Pringle's spear grass
Poa fendleriana	mutton grass

Douglas-fir Forest

Douglas-fir Forest occurs between Ponderosa Pine Forest and the colder Spruce–Fir Forest of higher elevations and is predominant on north-facing slopes and ravines. Douglas-fir and associated canopy species, *Pinus contorta* and *Pinus ponderosa,* are tolerant of fire. Frequent low-intensity fires favor regeneration of this forest type. This community corresponds to Küchler #12 and #18 and Coniferous Forests ERT.

Canopy
Characteristic Species
Pseudotsuga menziesii	Douglas-fir

Associates
Abies concolor	white fir
Larix occidentalis	western larch
Picea pungens	blue spruce
Pinus contorta	lodgepole pine
Pinus ponderosa	ponderosa pine
Populus tremuloides	quaking aspen

Woody Understory
Acer glabrum	Rocky Mountain maple
Alnus incana	speckled alder
Ceanothus fendleri	Fendler's buckbrush
Chamaebatiaria millefolium	fernbush
Holodiscus dumosus	glandular oceanspray
Jamesia americana	five-petal cliffbush
Juniperus communis	common juniper
Physocarpus malvaceus	mallow-leaf ninebark
Prunus emarginata	bitter cherry

Herbaceous Understory
Arnica latifolia	daffodil leopard-bane
Blepharoneuron tricholepis	pine-dropseed
Calamagrostis rubescens	pinegrass
Carex concinnoides	northwestern sedge
Carex geyeri	Geyer's sedge
Festuca arizonica	Arizona fescue

Spruce–Fir Forest

Spruce–Fir Forest occurs at elevations higher than Douglas-fir and Ponderosa Pine Forest, and is predominant in the subalpine portion of the Rocky Mountains. Spruce–Fir Forest occurs at high elevations throughout the Rocky Mountains, however, changes in species will occur along a north-south gradient. Shrubs and herbaceous species are sparse in this forest where the canopy is closed. In areas that are open due to past disturbances, species diversity and abundance increase in the herbaceous and shrub understory. This community corresponds to Küchler #15 and #21 and Coniferous Forests ERT.

Canopy
Characteristic Species
Abies lasiocarpa	subalpine fir
Picea engelmannii	Engelmann's spruce

Associates

Picea pungens	blue spruce
Pinus aristata	bristle-cone pine
Pinus flexilis	limber pine
Populus tremuloides	quaking aspen

Woody Understory

Acer glabrum	Rocky Mountain maple
Alnus incana	speckled alder
Arctostaphylos uva-ursi	red bearberry
Juniperus communis	common juniper
Lonicera involucrata	four-line honeysuckle
Mahonia repens	creeping Oregon-grape
Pachystima myrsinites	myrtle boxleaf
Pentaphylloides floribunda	golden-hardhack
Prunus emarginata	bitter cherry
Salix bebbiana	long-beak willow
Salix scouleriana	Scouler's willow
Sambucus cerulea	blue elder
Symphoricarpos oreophilus	mountain snowberry
Vaccinium myrtillus	whortle-berry

Herbaceous Understory

Arnica cordifolia	heart-leaf leopard-bane
Epilobium spp.	willowherb
Erigeron formosissimus	beautiful fleabane
Fragaria virginiana	Virginia strawberry
Lathyrus lanszwertii	Nevada vetchling
Pseudocymopterus montanus	alpine false mountain-parsley
Trisetum spicatum	narrow false oat
Veratrum californicum	California false hellebore
Vicia spp.	vetch

Juniper–Pinyon Woodland

Juniper–Pinyon Woodland is predominant in the foothills of eastern slopes in this region. This community corresponds to Küchler #23 and Woodlands ERT.

Canopy

Characteristic Species

Juniperus monosperma	one-seed juniper
Pinus edulis	two-needle pinyon

Associates

Juniperus occidentalis	western juniper
Quercus emoryi	Emory's oak
Quercus gambelii	Gambel's oak
Quercus grisea	gray oak

Woody Understory

Artemisia tridentata	big sagebrush
Fallugia paradoxa	Apache-plume
Purshia mexicana	Mexican cliff-rose
Purshia tridentata	bitterbrush

Herbaceous Understory

Bouteloua curtipendula	side-oats grama
Bouteloua gracilis	blue grama
Oryzopsis hymenoides	Indian mountain-rice grass
Pascopyrum smithii	western-wheat grass
Sporobolus cryptandrus	sand dropseed

Mountain-Mahogany–Scrub Oak

Mountain-Mahogany–Scrub Oak occupies a transition zone between coniferous forests and treeless plains and plateaus at margins of the Rocky Mountains. This community corresponds to Küchler #37 and Woodlands ERT.

Characteristic Woody Species

Cercocarpus ledifolius	curl-leaf mountain-mahogany
Quercus gambelii	Gambel's oak

Associates

Acer grandidentatum	canyon maple
Amelanchier utahensis	Utah service-berry
Arctostaphylos pringlei	pink-bract manzanita
Arctostaphylos pungens	Mexican manzanita
Artemisia tridentata	big sagebrush
Ceanothus greggii	Mojave buckbrush
Ceanothus velutinus	tobacco-brush
Cercocarpus montanus	alder-leaf mountain-mahogany
Eriodictyon angustifolium	narrow-leaf yerba-santa
Fallugia paradoxa	Apache-plume
Garrya flavescens	ashy silktassel
Garrya wrightii	Wright's silktassel
Pachystima myrsinites	myrtle boxleaf
Physocarpus malvaceus	mallow-leaf ninebark
Purshia mexicana	Mexican cliff-rose
Purshia tridentata	bitterbrush
Quercus chrysolepis	canyon live oak
Quercus dumosa	California scrub oak
Quercus emoryi	Emory's oak
Quercus havardii	Havard's oak
Quercus turbinella	shrub live oak
Rhamnus crocea	holly-leaf buckthorn
Rhus trilobata	ill-scented sumac

Sagebrush Steppe

Sagebrush Steppe is a mix of scattered shrubs, primarily sagebrush, and perennial grasses. However, the structure and floristic diversity of the sagebrush community is variable depending upon moisture availability, grazing, fire frequency, and other factors. Grazing pressure in the sagebrush community has resulted in an increase in shrubs over perennial grasses. This community corresponds to Küchler #55 and #38 and Desert Lands ERT.

Characteristic Woody Species

Artemisia tridentata	big sagebrush

Associates

Artemisia nova	black sagebrush
Atriplex confertifolia	shadscale
Balsamorrhiza sagittata	arrow-leaf balsamroot
Purshia tridentata	bitterbrush

Herbaceous Understory

Festuca idahoensis	bluebunch fescue
Lithospermum ruderale	Columbian puccoon
Lupinus sericeus	Pursh's silky lupine
Oryzopsis hymenoides	Indian mountain-rice grass
Pascopyrum smithii	western-wheat grass
Poa secunda	curly blue grass
Pseudoroegneria spicata	bluebunch-wheat grass

Saltbush–Greasewood Shrub

Saltbush–Greasewood Shrub occurs primarily on highly saline soils, although it can be found on other dry, hard desert soils. In some areas, vegetation is entirely lacking due to extreme conditions. Water availability can also vary from standing water much of the year to xeric. This community corresponds to Küchler #40 and Desert Lands ERT.

Characteristic Species

Atriplex confertifolia	shadscale
Sacrobatus vermiculatus	greasewood

Associates

Allenrolfea occidentalis	iodinebush
Artemisia spinescens	bud sagebrush
Distichlis spicata	coastal salt grass
Grayia spinosa	spiny hop-sage
Kochia americana	greenmolly
Krascheninnikovia lanata	winterfat
Lycium cooperi	peachthorn
Mendora spinescens	spiny mendora
Suaeda moquinii	shrubby seepweed

Shortgrass Prairie

This grassland community is found in dry, arid conditions within the Southern Rocky Mountains region. Although some ecologists believe this community is a result of grazing, primarily by buffalo, most believe it is the unique environmental conditions that have influenced the development of these grasslands. This community corresponds to Küchler #66 and Prairies ERT.

Characteristic Species

Bouteloua gracilis	blue grama
Pascopyrum smithii	western-wheat grass
Stipa comata	needle-and-thread

Associates

Artemisia frigida	prairie sagebrush
Chrysothamnus viscidiflorus	green rabbitbrush
Elymus trachycaulus	slender wild rye
Festuca arizonica	Arizona fescue
Koeleria macrantha	prairie Koeler's grass
Muhlenbergia montana	mountain muhly
Oryzopsis hymenoides	Indian mountain-rice grass
Tetradymia canescens	spineless horsebrush

WETLAND SYSTEMS

Riparian Forest

Riparian Forest, dominated by the deciduous cottonwood tree, *Populus angustifolia*, is the least common forest type in the Southern Rocky Mountain Region. The riparian forest, however, occurs throughout the region in canyons and along streams. Riparian Forest occurs from elevations as low as 4450 to 7600 feet. On some sites within the riparian zone, the forest canopy is absent and willow thickets are well developed. This community corresponds to Riparian Forests and Woodlands ERT.

Canopy
Characteristic Species

Populus angustifolia	narrow-leaf cottonwood

Associates

Abies concolor	white fir
Acer glabrum	Rocky Mountain maple
Acer grandidentatum	canyon maple
Alnus incana	speckled alder
Alnus oblongifolia	Arizona alder
Betula occidentalis	water birch
Pinus ponderosa	ponderosa pine
Populus deltoides	eastern cottonwood
Populus fremontii	Fremont's cottonwood
Populus tremuloides	quaking aspen
Prunus americanus	American plum
Quercus gambelii	Gambel's oak
Robinia neomexicana	New Mexico locust

Woody Understory

Acer negundo	ash-leaf maple
Cornus stolonlifera	redosier
Parthenocissus vitacea	thicket-creeper
Prunus emarginata	bitter cherry
Rhus glabra	smooth sumac
Salix spp.	willow
Sambucus cerulea	blue elder

Herbaceous Understory

Aralia nudicaulis	wild sarsaparilla
Blepharoneuron tricholepsis	pine-dropseed
Carex sprengelii	long-beak sedge
Ranunculus abortivus	kidney-leaf buttercup
Ratibida pinnata	gray-head Mexican-hat

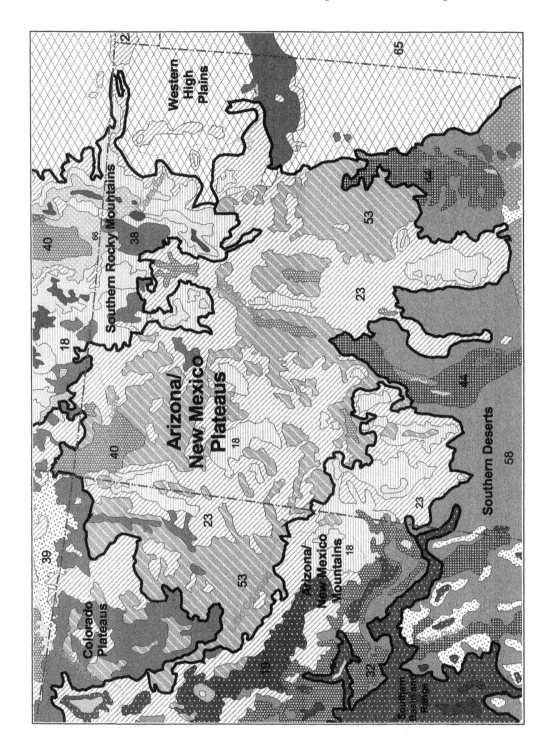

Arizona/New Mexico Plateaus

Arizona/New Mexico Plateaus

Introduction

The Arizona/New Mexico Plateau Region is located in northeastern Arizona and northwestern and central New Mexico. This region encompasses several large plateaus and scattered mountain ranges. Evergreen forests of pines and areas of fir with spruce dominate the higher elevations. Evergreen woodland generally occupies areas below the forests while grasslands and desert scrub occur in the driest areas.

The primary sources used to develop descriptions and species lists are Barbour and Billings 1988; Dick-Peddie 1992; Lowe 1964.

Dominant Ecological Communities

Upland Systems
Pine–Douglas-fir Forest
Ponderosa Pine Forest
Juniper–Pinyon Woodland
High Desert Scrub
Desert Grassland
Wetland Systems
Riparian Woodland

UPLAND SYSTEMS

Pine–Douglas-fir Forest

Pine–Douglas-fir Forest is found on high plateaus and mountains and extends southward from the Rocky Mountains to the southwest in Colorado, Utah, New Mexico, and Arizona. The forests range from open to dense with tall evergreen trees. In the southern mountains, there is often much undergrowth. This community corresponds to Kuchler #18 and Coniferous Forests ERT.

Canopy
Characteristic Species
Pinus ponderosa	ponderosa pine
Pseudotsuga menziesii	Douglas-fir

Associates
Picea pungens	blue spruce
Pinus flexilis	limber pine

Woody Understory
Acer glabrum	Rocky Mountain maple
Alnus incana	speckled alder
Ceanothus fendleri	Fendler's buckbrush
Chamaebatiaria millefolium	fernbush
Holodiscus dumosus	glandular oceanspray
Jamesia americana	five-petal cliffbush
Juniperus communis	common juniper
Prunus emarginata	bitter cherry

Herbaceous Understory

Blepharoneuron tricholepis	pine-dropseed
Festuca arizonica	Arizona fescue

Ponderosa Pine Forest

Ponderosa Pine Forest is similar to and shares many species with the Ponderosa Pine Forest of the Arizona/New Mexico Mountains and the Southern Deserts regions. The Ponderosa Pine Forest of this region, however, occurs on large flat mesas and plateaus. The forests are, as in the Southern Deserts region, found below the fir forests. The trees form an open to dense canopy. The shrubs are few, rarely dense, and not especially common. In the open canopy Ponderosa Pine Forest, the principle groundcovers are often grasses. There are fewer tree species than in the southern Arizona Ponderosa Pine Forest. This community corresponds to Küchler #19 and Coniferous Forests ERT.

Canopy
Characteristic Species

Pinus ponderosa	ponderosa pine

Associates

Pinus cembroides	Mexican pinyon
Pinus flexilis	limber pine
Pinus leiophylla	Chihuahuan pine
Populus tremuloides	quaking aspen
Pseudotsuga menziesii	Douglas-fir
Quercus gambelii	Gambel's oak

Woody Understory

Arctostaphylos patula	green-leaf manzanita
Artemisia nova	black sagebrush
Artemisia tridentata	big sagebrush
Ceanothus fendleri	Fendler's buckbrush
Chamaebatiaria millefolium	fernbush
Chrysothamnus parryi	Parry's rabbitbrush
Fallugia paradoxa	Apache-plume
Fendlerella utahensis	Utah-fendlerbush
Holodiscus boursieri	Boursier's oceanspray
Holodiscus dumosus	glandular oceanspray
Jamesia americana	five-petal cliffbush
Juniperus communis	common juniper
Mahonia repens	creeping Oregon-grape
Pachystima myrsinites	myrtle boxleaf
Philadelphus microphyllus	little-leaf mock orange
Physocarpus monogynus	mountain ninebark
Purshia mexicana	Mexican cliff-rose
Ribes cereum	white squaw currant
Robinia neomexicana	New Mexico locust
Rubus idaeus	common red raspberry

Herbaceous Understory

Artemisia ludoviciana	white sagebrush
Blepharoneuron tricholepis	pine-dropseed
Bothriochloa barbinodis	cane beard grass
Bouteloua gracilis	blue grama
Comandra umbellata	bastard-toadflax
Elymus elymoides	western bottle-brush grass
Festuca arizonica	Arizona fescue
Hedeoma dentatum	Arizona false pennyroyal
Iris missouriensis	Rocky Mountain iris
Koeleria macrantha	prairie Koeler's grass
Lupinus palmeri	Palmer's lupine
Monarda fistulosa	Oswego-tea
Muhlenbergia montana	mountain muhly
Muhlenbergia wrightii	Wright's muhly
Oxytropis lamberti	stemless locoweed
Pteridium aquilinum	northern bracken fern
Sphaeralcea fendleri	thicket globe-mallow
Thalictrum fendleri	Fendler's meadow-rue
Thermopsis rhombifolia	prairie golden-banner
Vicia americana	American purple vetch

Juniper–Pinyon Woodland

Juniper–Pinyon Woodland covers large areas below Ponderosa Pine Forest on the Mogollon, Coconino, and Kaibab plateaus. It usually occurs between 5500 and 7000 feet on flat-topped mesas and plateaus. The woodlands are an open mixture of shrubs with herbaceous plants forming the understory. The oaks may be in tree or shrub form. This community corresponds to Küchler #23 and Woodlands ERT.

Canopy

Characteristic Species

Juniperus monosperma	one-seed juniper
Juniperus osteosperma	Utah juniper
Pinus edulis	two-needle pinyon
Pinus monophylla	single-leaf pinyon

Associates

Juniperus occidentalis	western juniper
Juniperus scopulorum	Rocky Mountain juniper
Quercus emoryi	Emory's oak
Quercus gambelii	Gambel's oak
Quercus grisea	gray oak
Quercus turbinella	shrub live oak

Woody Understory

Amelanchier alnifolia	Saskatoon service-berry
Artemisia nova	black sagebrush
Artemisia tridentata	big sagebrush
Ceanothus fendleri	Fendler's buckbrush
Ceanothus integerrimus	deerbrush
Cercocarpus ledifolius	curl-leaf mountain-mahogany

Cerocarpus intricatus	little-leaf mountain-mahogany
Cerocarpus montanus	alder-leaf mountain-mahogany
Chamaebatiaria millifolium	fernbush
Chrysothamnus depressus	long-flower rabbitbrush
Chrysothamnus nauseosus	rubber rabbitbrush
Coleogyne ramosissima	blackbrush
Echinocereus coccineus	scarlet hedgehog cactus
Ephedra viridis	Mormon-tea
Fallugia paradoxa	Apache-plume
Garrya wrightii	Wright's silktassel
Krascheninnikovia lanata	winterfat
Mahonia fremonti	desert Oregon-grape
Opuntia basilaris	beaver-tail cactus
Opuntia erinacea	oldman cactus
Opuntia fragilis	pygmy prickly-pear
Opuntia polyacantha	hair-spine prickly-pear
Opuntia whipplei	rat-tail cholla
Purshia mexicana	Mexican cliff-rose
Purshia tridentata	bitterbrush
Yucca baccata	banana yucca

Herbaceous Understory

Bouteloua curtipendula	side-oats grama
Bouteloua eriopoda	black grama
Bouteloua gracilis	blue grama
Calochortus ambiguus	doubting mariposa-lily
Castilleja integra	squawfeather
Elymus elymoides	western bottle-brush grass
Festuca arizonica	Arizona fescue
Keckiella antirrhinoides	chapparral bush-beardtongue
Koeleria macrantha	prairie Koeler's grass
Muhlenbergia torreyi	ringed muhly
Oryzopsis hymenoides	Indian mountain-rice grass
Pascoyprum smithii	western-wheat grass
Piptochaetium fimbriatum	pinyon spear grass
Sphaeralcea emoryi	Emory's globe-mallow
Sporobolus cryptandrus	sand dropseed
Stipa nelsonii	Nelson's needle grass

High Desert Scrub

High Desert Scrub occurs in areas ranging in elevation from 3000 to 6500 feet and is sometimes referred to as cool or cold desert. It is the southeastern limit of the Great Basin desert system. Rainfall averages from seven to twelve inches per year and is more evenly distributed throughout the year than in the desert types to the south. Dominance within this community may vary from area to area with one or two species of shrub often forming pure stands. This community corresponds, in part, to Küchler #38 and #40 and Desert Lands ERT.

Characteristic Woody Species

Artemisia tridentata	big sagebrush
Atriplex canescens	four-wing saltbush
Atriplex confertifolia	shadscale
Coleogyne ramosissima	blackbrush
Sarcobatus vermiculatus	greasewood

Associates

Allenrolfea occidentalis	iodinebush
Artemisia bigelovii	flat sagebrush
Artemisia filifolia	silver sagebrush
Artemisia nova	black sagebrush
Artemisia spinescens	bud sagebrush
Chrysothamnus nauseosus	rubber rabbitbrush
Ephedra nevadensis	Nevada joint-fir
Ephedra viridis	Mormon-tea
Grayia spinosa	spiny hop-sage
Gutierrezia sarothrae	kindlingweed
Krascheninnikovia lanata	winterfat
Lycium cooperi	peachthorn
Lycium pallidum	pale desert-thorn
Menodora spinescens	spiny menodora
Opuntia erinacea	oldman cactus
Opuntia polyacantha	hair-spine prickly-pear
Opuntia whipplei	rat-tail cholla
Purshia tridentata	bitterbrush
Yucca angustissima	fine-leaf yucca

Herbaceous Understory

Balsamorhiza sagittata	arrow-leaf balsamroot
Bouteloua gracilis	blue grama
Camissonia boothii	shredding suncup
Camissonia claviformis	browneyes
Camissonia scapoidea	Paiute suncup
Cardaria draba	heart-pod hoarycress
Cryptantha circumscissa	cushion cat's-eye
Distichlis spicata	coastal salt grass
Eriogonum ovalifolium	cushion wild buckwheat
Festuca idahoensis	bluebunch grass
Halogeton glomeratus	saltlover
Iva nevadensis	Nevada marsh-elder
Kochia americana	greenmolly
Lupinus sericeus	Pursh's silky lupine
Mirabilis alipes	winged four-o'clock
Oryzopsis hymenoides	Indian mountain-rice grass
Pascopyrum smithii	western-wheat grass
Poa secunda	curly blue grass
Sphaeralcea grossulariifolia	currant-leaf globe-mallow
Sporobolus airoides	alkali-sacaton
Vulpia octoflora	eight-flower six-weeks grass
Xylorhiza glabriuscula	smooth woody-aster

Desert Grassland

The short grasses vary from dense to open cover and are very similar to the grasslands of the Southern Deserts region. See the Southern Deserts region for a species list of this community.

WETLAND SYSTEMS

Riparian Woodland

Riparian Woodland occurs along permanent or semipermanent streams. See the Southern Deserts region for a species list of this community.

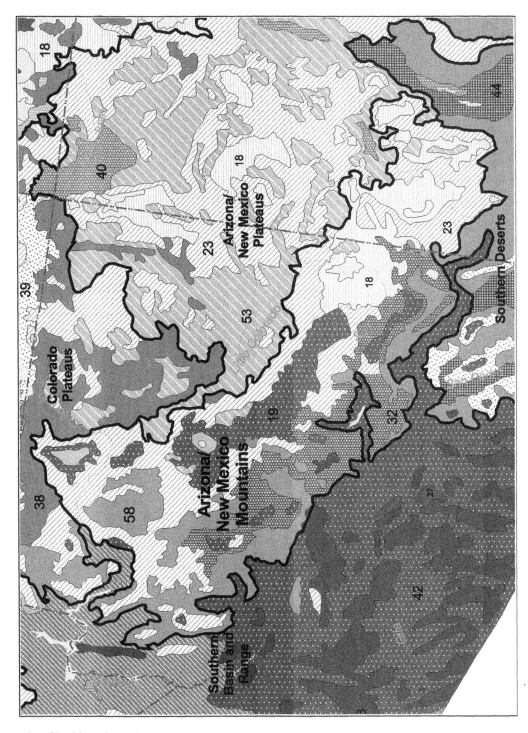

Arizona/New Mexico Mountains

Arizona/New Mexico Mountains

Introduction

The ecological communities of the Arizona and New Mexico Mountains region are similar to communities of the Southern Deserts region. Many of the plant species are shared by the two regions. The ecological communities change in response to the temperature and precipitation changes of increased elevation. The biseasonal rain pattern of widespread, gentle rains in the winter and localized, heavy rains accompanying thunderstorms in the summer is found in both regions and significantly impacts plant communities.

Aspen occurs throughout the region in burned or disturbed areas. Riparian Woodland occurs along streams that flow for at least several months during winter and spring each year.

The primary sources used to develop descriptions and species lists are Barbour and Billings 1988; Dick-Peddie 1992; Lowe 1964.

Dominant Ecological Communities

Upland Systems
Southwestern Spruce–Fir Forest
Spruce–Fir–Douglas-fir Forest
Ponderosa Pine Forest
Pine–Douglas-fir Forest
Juniper–Pinyon Woodland
Montane Scrub
Desert Grassland
Wetland Systems
Riparian Woodland

UPLAND SYSTEMS

Southwestern Spruce–Fir Forest

Southwestern Spruce–Fir Forest contains dense to open stands of low to medium tall needleleaf evergreen trees. Spruce–fir forests are found throughout the Southern Rocky Mountains with the southernmost extensions found on mountain peaks in the Arizona/New Mexico Mountains region between 8500 and 9000 feet. Because of the accumulation of litter and the often dense canopy, the herbaceous layer is not well developed in these forests. This community corresponds to Küchler #21 and Coniferous Forests ERT.

Canopy
Characteristic Species

Abies lasiocarpa	subalpine fir
Picea engelmannii	Engelmann's spruce

Associates

Abies concolor	white fir
Picea glauca	white spruce
Pinus aristata	bristle-cone pine
Pinus flexilis	limber pine
Populus tremuloides	quaking aspen

Woody Understory

Acer glabrum	Rocky Mountain maple
Alnus incana	speckled alder
Juniperus communis	common juniper
Lonicera involucrata	four-line honeysuckle
Mahonia repens	creeping Oregon-grape
Pachystima myrsinites	myrtle boxleaf
Pentaphylloides floribunda	golden-hardhack
Prunus emarginata	bitter cherry
Salix bebbiana	long-beak willow
Salix scouleriana	Scouler's willow
Symphoricarpos oreophilus	mountain snowberry
Vaccinium myrtillus	whortle-berry

Herbaceous Understory

Actaea rubra	red baneberry
Aquilegia chrysantha	golden columbine
Carex spp.	sedge
Dugaldia hoopesii	owl's-claws
Festuca rubra	red fescue
Gentiana spp.	gentian
Juncus spp.	rush
Pedicularis procera	giant lousewort
Pedicularis racemosa	parrot's-beak
Phleum alpinum	mountain timothy
Primula spp.	primrose
Trisetum spicatum	narrow false oat
Veratum californicum	California false hellebore
Viola canadensis	Canadian white violet
Viola nephrophylla	northern bog violet

Spruce–Fir–Douglas-fir Forest

Spruce Fir–Douglas-fir Forest is found at elevations lower than Southwestern Spruce–Fir Forest and is composed of large to medium trees. The forest has an open to dense canopy. If the canopy is open, broadleaf deciduous trees and shrubs are found in the understory. This community corresponds to Küchler #20 and Coniferous Forests ERT.

Canopy
Characteristic Species

Abies concolor	white fir
Picea pungens	blue spruce
Pseudotsuga menziesii	Douglas-fir

Associates

Pinus ponderosa	ponderosa pine
Populus tremuloides	quaking aspen

Woody Understory

Acer glabrum	Rocky Mountain maple
Amelanchier alnifolia	Saskatoon service-berry
Chamaebatiaria millefolium	fernbush
Pachystima myrsinites	myrtle boxleaf
Physocarpus malvaceus	mallow-leaf ninebark
Prunus virginiana	choke cherry
Sambucus cerulea	blue elder
Symphoricarpos oreophilus	mountain snowberry
Vaccinium myrtillus	whortle-berry

Herbaceous Understory

Adiantum pedatum	northern maidenhair
Dugaldia hoopesii	owl's-claws
Erigeron eximius	spruce-fir fleabane
Festuca arizonica	Arizona fescue
Leymus triticoides	beardless lyme grass
Muhlenbergia virescens	screw leaf muhly

Ponderosa Pine Forest

Ponderosa Pine Forest of this region is similar to and shares many species with Ponderosa Pine Forest of the Arizona/New Mexico Plateaus and Southern Deserts regions. The Ponderosa Pine Forest of this region, however, occurs on large flat mesas and plateaus. The forests are found below the spruce-fir forests. The trees form an open to dense canopy; however, the open canopy with grass understories are not found on the steep slopes in southern Arizona. The shrubs are few, rarely dense, and not especially common. In the open canopy Ponderosa Pine Forest, the principle groundcovers are often grasses. There are fewer tree species than in the southern Arizona Ponderosa Pine Forest. This community corresponds to Küchler #19 and Coniferous Forests ERT.

Canopy
Characteristic Species

Pinus ponderosa	ponderosa pine

Associates

Pinus cembroides	Mexican pinyon
Pinus flexilis	limber pine
Pinus leiophylla	Chihuahuan pine
Populus tremuloides	quaking aspen
Pseudotsuga menziesii	Douglas-fir
Quercus gambelii	Gambel's oak

Woody Understory

Arctostaphylos patula	green-leaf manzanita
Artemisia nova	black sagebrush
Artemisia tridentata	big sagebrush
Ceanothus fendleri	Fendler's buckbrush
Chamaebatiaria millefolium	fernbush
Chrysothamnus parryi	Parry's rabbitbrush
Fallugia paradoxa	Apache-plume

Fendlerella utahensis	Utah-fendlerbush
Holodiscus boursieri	Boursier's oceanspray
Holodiscus dumosus	glandular oceanspray
Jamesia americana	five-petal cliffbush
Juniperus communis	common juniper
Mahonia repens	creeping Oregon-grape
Pachystima myrsinites	myrtle boxleaf
Philadelphus microphyllus	little-leaf mock orange
Physocarpus monogynus	mountain ninebark
Purshia mexicana	Mexican cliff-rose
Ribes cereum	white squaw currant
Robinia neomexicana	New Mexico locust
Rubus idaeus	common red raspberry

Herbaceous Understory

Artemisia ludoviciana	white sagebrush
Blepharoneuron tricholepis	pine-dropseed
Bothriochloa barbinodis	cane beard grass
Bouteloua gracilis	blue grama
Comandra umbellata	bastard-toadflax
Elymus elymoides	western bottle-brush grass
Festuca arizonica	Arizona fescue
Hedeoma dentatum	Arizona false pennyroyal
Iris missouriensis	Rocky Mountain iris
Koeleria macrantha	prairie Koeler's grass
Lupinus palmeri	Palmer's lupine
Monarda fistulosa	Oswego-tea
Muhlenbergia montana	mountain muhly
Muhlenbergia wrightii	Wright's muhly
Oxytropis lamberti	stemless locoweed
Pteridium aquilinum	northern bracken fern
Sphaeralcea fendleri	thicket globe-mallow
Thalictrum fendleri	Fendler's meadow-rue
Thermopsis rhombifolia	prairie golden-banner
Vicia americana	American purple vetch

Pine–Douglas-fir Forest

Pine–Douglas-fir Forest is found on high plateaus and mountains and extends southward from the Rocky Mountains southwest to Colorado, Utah, New Mexico, and Arizona. The forests range from open to dense with tall, evergreen trees. In the southern mountains, there is often much undergrowth. This community type corresponds to Küchler #18 and Coniferous Forests ERT.

Canopy
Characteristic Species

Pinus ponderosa	ponderosa pine
Pseudotsuga menziesii	Douglas-fir

Associates

Picea pungens	blue spruce
Pinus flexilis	limber pine

Woody Understory

Acer glabrum	Rocky Mountain maple
Alnus incana	speckled alder
Ceanothus fendleri	Fendler's buckbrush
Chamaebatiaria millefolium	fernbush
Holodiscus dumosus	glandular oceanspray
Jamesia americana	five-petal cliffbush
Juniperus communis	common juniper
Prunus emarginata	bitter cherry

Herbaceous Understory

Blepharoneuron tricholepis	pine-dropseed
Festuca arizonica	Arizona fescue

Juniper–Pinyon Woodland

Juniper–Pinyon Woodland covers large areas below Ponderosa Pine Forest on the Mogollon, Coconino, and Kaibab plateaus. It is usually between 5500 and 7000 feet on flat-topped mesas and plateaus. The woodlands are open with a mixture of shrubs and herbaceous plants forming the understory. The oaks may be in tree or shrub form. This community corresponds to Küchler #23 and, in part, #32 and Woodlands ERT.

Canopy
Characteristic Species

Juniperus monosperma	one-seed juniper
Juniperus osteosperma	Utah juniper
Pinus edulis	two-needle pinyon

Associates

Juniperus deppeana	alligator juniper
Juniperus occidentalis	western juniper
Pinus monophylla	single-leaf pinyon
Quercus emoryi	Emory's oak
Quercus gambelii	Gambel's oak
Quercus grisea	gray oak
Quercus turbinella	shrub live oak

Woody Understory

Amelanchier alnifolia	Saskatoon service-berry
Amelanchier utahensis	Utah service-berry
Artemisia nova	black sagebrush
Artemisia tridentata	big sagebrush
Atriplex canescens	four-wing saltbush
Ceanothus fendleri	Fendler's buckbrush
Ceanothus integerrimus	deerbrush
Cercocarpus intricatus	little-leaf mountain-mahogany
Cercocarpus ledifolius	curl-leaf mountain-mahogany
Cerocarpus montanus	alder-leaf mountain-mahogany
Chamaebatiaria millifolium	fernbush

Chrysothamnus depressus	long-flower rabbitbrush
Chrysothamnus nauseosus	rubber rabbitbrush
Coleogyne ramosissima	blackbrush
Echinocereus coccineus	scarlet hedgehog cactus
Ephedra viridis	Mormon-tea
Fallugia paradoxa	Apache-plume
Garrya wrightii	Wright's silktassel
Gutierrezia sarothrae	kindlingweed
Krascheninnikovia lanata	winterfat
Mahonia fremonti	desert Oregon-grape
Opuntia basilaris	beaver-tail cactus
Opuntia erinacea	oldman cactus
Opuntia fragilis	pygmy prickly-pear
Opuntia polyacantha	hair-spine prickly-pear
Opuntia whipplei	rat-tail cholla
Purshia mexicana	Mexican cliff-rose
Purshia tridentata	bitterbrush
Robinia neomexicana	New Mexico locust
Symphoricarpos oreophilus	mountain snowberry
Yucca baccata	banana yucca

Herbaceous Understory

Bouteloua curtipendula	side-oats grama
Bouteloua eriopoda	black grama
Bouteloua gracilis	blue grama
Calochortus ambiguus	doubting mariposa-lily
Castelleja integra	squawfeather
Elymus elymoides	western bottle-brush grass
Festuca arizonica	Arizona fescue
Hilaria jamesii	James' galleta
Keckiella antirrhinoides	chaparral bush-beardtongue
Koeleria macrantha	prairie Koeler's grass
Muhlenbergia torreyi	ringed muhly
Oryzopsis hymenoides	Indian Mountain-rice grass
Pascopyrum smithii	western-wheat grass
Piptochaetium fimbriatum	pinyon spear grass
Schizachyrium scoparium	little false bluestem
Sphaeralcea emoryi	Emory's globe-mallow
Sporobolus cryptandrus	sand dropseed
Stipa nelsonii	Nelson's needle grass
Stipa neomexicana	New Mexico needle grass

Montane Scrub

Montane Scrub is found at elevations of 3500 to 7000 feet in Arizona from the foothills of the Mogollon Rim to south of the Gila River. The precipitation varies from thirteen to twenty-three inches per year and occurs during two periods of the year. The shrubs have small, evergreen leaves and regenerate rapidly from root crowns after fires. Natural fires occur every 50 to 100 years. The surface vegetation dates from the last fire; the root systems are much older. The shrub cover varies from dense to somewhat open and exhibits a uniform height varying from three to seven feet. Because of grazing, the native grasses are limited to rocky areas. This community corresponds, in part, to Küchler #58 and Desert Lands ERT.

Characteristic Woody Species

Quercus turbinella	shrub live oak

Associates

Arctostaphylos pringlei	pink-bract manzanita
Arctostaphylos pungens	Mexican manzanita
Brickellia californica	California brickellbrush
Ceanothus greggi	Mojave buckbrush
Ceanothus intergerrimus	deerbrush
Cercocarpus ledifolius	curl-leaf mountain-mahogany
Cercocarpus montanus	alder-leaf mountain-mahogany
Eriodictyon angustifolium	narrow-leaf yerba-santa
Fallugia paradoxa	Apache-plume
Forestiera pubescens	stretchberry
Frangula betulifolia	birch-leaf buckthorn
Frangula californica	California coffee berry
Fraxinus anomala	single-leaf ash
Garrya flavescens	ashy silktassel
Garrya wrightii	Wright's silktassel
Juniperus monosperma	one-seed juniper
Mahonia haematocarpa	red Oregon-grape
Mimosa biuncifera	cat-claw mimosa
Prunus virginiana	choke cherry
Purshia mexicana	Mexican cliff-rose
Quercus chrysolepis	canyon live oak
Quercus emoryi	Emory's oak
Rhamnus crocea	holly-leaf buckthorn
Rhus glabra	smooth sumac
Rhus ovata	sugar sumac
Rhus trilobata	ill-scented sumac
Toxicodendron diversilobum	Pacific poison-oak
Vauquelinia californica	Arizona-rosewood

Herbaceous Understory

Aristida orcuttiana	single-awn three-awn
Aristida purpurea	purple three-awn
Aristida ternipes	spider grass
Bothriochloa barbinodis	cane beard grass
Bouteloua curtipendula	side-oats grama
Bouteloua eriopoda	black grama
Bouteloua hirsuta	hairy grama
Dalea albiflora	white-flower prairie-clover
Eragrostis intermedia	plains love grass
Glandularia wrightii	Davis Mountain mock vervain
Lycurus phleoides	common wolf's-tail
Penstemon eatoni	Eaton's beardtongue
Penstemon linarioides	toadflax beardtongue
Penstemon palmeri	scented beardtongue
Solanum xantii	chaparral nightshade
Solidago velutina	three-nerve goldenrod

Desert Grassland

The short grasses of this community type vary from dense to open cover and are very similar to the grasslands of the Southern Deserts region. See Southern Deserts region for a species list of this community.

WETLAND SYSTEMS

Riparian Woodland

Riparian Woodland occurs along permanent or semipermanent streams. See the Southern Deserts region for a species list of this community.

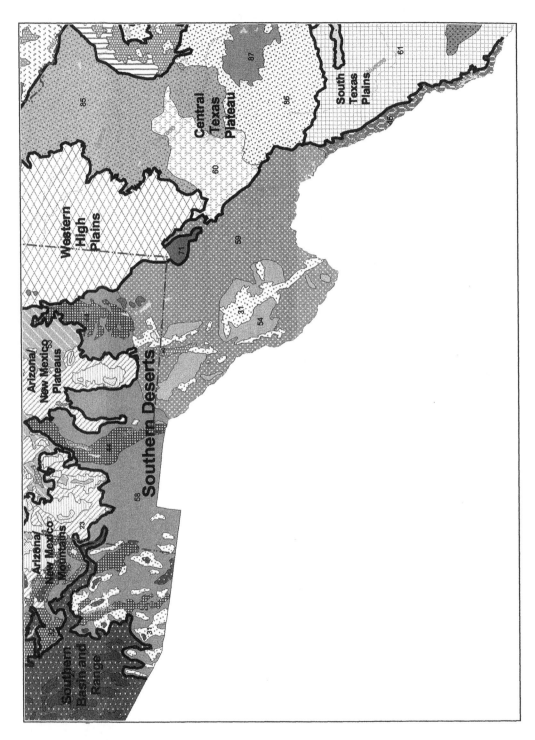

Southern Deserts

Southern Deserts

Introduction

The Southern Deserts region encompasses southeastern Arizona, Southern New Mexico, and Transpecos Texas extending in a narrow band along the Rio Grande River. The vegetation varies from Desert Scrub at low elevations to Spruce–Fir Forests in the mountains. Desert Grassland, Evergreen Woodland, and Ponderosa Pine Forest are found at middle elevations. Broad ecotones exist where slope, soil, and topographic changes create a mosaic of vegetation types. The forests and woodlands vary from sparsely spaced trees to dense forests. The vegetation of the Southern Deserts region contains plants associated with the vegetation of Mexico as well as the Rocky Mountains.

The climate is characterized by biseasonal rains. Widespread, gentle rains occur in winter. The summer precipitation, beginning in July and continuing through October, occurs as intense local thunderstorms. As elevation increases, so does precipitation and thus changes in vegetation. Additionally, exposure and soils influence the distribution of vegetation types.

The riparian communities that occur along perennial streams vary with altitude. The deciduous trees of the riparian community often extend far into the Desert Scrub along streams. Trees found along dry washes and arroyos are distinct and characterized by small deciduous leaves.

The primary sources used to develop descriptions and species lists are Barbour and Billings 1988; Dick-Peddie 1992; Lowe 1964.

Dominant Ecological Communities

Upland Systems
Southwestern Spruce–Fir Forest
Ponderosa Pine Forest
Evergreen Woodland
Desert Scrub
Desert Grassland
Wetland Systems
Riparian Woodland

UPLAND SYSTEMS

Southwestern Spruce–Fir Forest

Southwestern Spruce–Fir Forest is found at elevations of 7500 to 11500 feet on the highest peaks in the Southern Deserts region. Precipitation, mostly as snow, averages from twenty-five to thirty inches per year. Frequent afternoon thunderstorms occur during the summer. Where the conifers do not form a closed canopy, a few broadleaf, deciduous trees occur. Aspen is the most common tree in areas of fire or disturbances. Shrubs are more common on sites where the canopy is open. Sedges, mosses, lichens, and liverworts occur under dense canopies. In some areas, stands of fir dominate exclusively. Where fir is dominant, there is often little understory due to the considerable duff which accumulates on the forest floor. This community corresponds to Küchler #21 and Coniferous Forests ERT.

Canopy

Characteristic Species

Abies concolor	white fir
Abies lasiocarpa	subalpine fir
Picea engelmannii	Engelmann's spruce
Picea pungens	blue spruce
Pinus aristata	bristle-cone pine
Pinus flexilis	limber pine
Pseudotsuga menziesii	Douglas-fir

Associates

Acer negundo	ash-leaf maple
Pinus ponderosa	ponderosa pine
Pinus strobiformis	southwestern white pine
Populus tremuloides	quaking aspen
Quercus gambelii	Gambel's oak

Woody Understory

Acer glabrum	Rocky Mountain maple
Alnus incana	speckled alder
Arctostaphylos uva-ursi	red bearberry
Betula occidentalis	water birch
Chimaphila umbellata	pipsissewa
Juniperus communis	common juniper
Lonicera involucrata	four-line honeysuckle
Mahonia repens	creeping Oregon-grape
Pachystima myrsinites	myrtle boxleaf
Pentaphylloides floribunda	golden-hardhack
Prunus emarginata	bitter cherry
Ribes wolfii	winaha currant
Robinia neomexicana	New Mexican locust
Rubus spp.	raspberry
Salix bebbiana	long-beak willow
Salix scouleriana	Scouler's willow
Sambucus cerulea	blue elder
Vaccinium myrtillus	whortle-berry

Herbaceous Understory

Actaea spp.	baneberry
Adiantum pedatum	Northern maidenhair
Agrostis scabra	rough bent
Aquilegia spp.	columbine
Balsamorhiza spp.	balsamroot
Bromus anomalus	nodding brome
Bromus ciliatus	fringed brome
Bromus marginatus	large mountain brome
Carex spp.	sedge
Danthonia spp.	wild oat grass
Deschampsia cespitosa	tufted hair grass
Dugaldia hoopesii	owl's-claws

Epilobium spp.	willowherb
Erigeron formosissimus	beautiful fleabane
Festuca rubra	red fescue
Fragaria virginiana	Virginia strawberry
Frasera speciosa	monument plant
Gentiana spp.	gentian
Goodyera oblongifolia	green-leaf rattlesnake-plantain
Helenium spp.	sneezeweed
Heuchera rubescens	pink alumroot
Juncus spp.	rush
Lathyrus lanszwertii	Nevada vetchling
Maianthemum racemosum	feathery false Solomon's-seal
Oxalis spp.	wood-sorrel
Pedicularis spp.	lousewort
Phleum alpinum	mountain timothy
Poa spp.	blue grass
Primula spp.	primrose
Pseudocymopterus montanus	alpine false mountain-parsley
Pyrola chlorantha	green-flower wintergreen
Scrophularia parviflora	pineland figwort
Senecio spp.	ragwort
Trifolium spp.	clover
Trisetum spicatum	narrow false oat
Valeriana arizonica	Arizona valerian
Veratrum californicum	California false hellebore
Vicia americana	American purple vetch
Viola adunca	hook-spur violet
Viola canadensis	Canadian white violet

Ponderosa Pine Forest

Ponderosa Pine Forest is found between 6000 and 9000 feet. The development of ponderosa pine is controlled by precipitation which must be between eighteen and twenty-six inches annually. These pines, which may occur in pure stands, are isolated from the Ponderosa Pine Forest that occurs further north and have some different understory species. Because of the steep topography, with few level areas, grasslands and parks are uncommon in the southern Arizona pine forests. Such parks are more common in northern Arizona. Few well-developed riparian areas exist because of the steep topography. The importance and distribution of the associated trees, shrubs, and herbaceous vegetation vary across the Southern Deserts region. This community corresponds to Küchler #19 and Coniferous Forests ERT.

Canopy
Characteristic Species

Pinus ponderosa	ponderosa pine

Associates

Acer grandidentatum	canyon maple
Alnus oblongifolia	Arizona alder
Arbutus arizonica	Arizona madrone
Cupressus arizonica	Arizona cypress
Juniperus deppeana	alligator juniper
Juniperus flaccida	drooping juniper

Pinus leiophylla	Chihuahuan pine
Pinus strobiformis	southwestern white pine
Populus tremuloides	quaking aspen
Pseudotsuga menziesii	Douglas-fir
Quercus emoryi	Emory's oak
Quercus gambelii	Gambel's oak
Quercus gravesii	Graves' oak
Quercus grisea	gray oak
Quercus hypoleucoides	silver-leaf oak
Quercus muhlenbergii	chinkapin oak

Woody Understory

Arctostaphylos pringlei	pink-bract manzanita
Arctostaphylos pungens	Mexican manzanita
Ceanothus fendleri	Fendler's buckbrush
Ceanothus integerrimus	deerbrush
Echinocereus triglochidiatus	king-cup cactus
Frangula betulifolia	birch-leaf buckthorn
Frangula californica	California coffee berry
Garrya wrightii	Wright's silktassel
Holodiscus dumosus	glandular oceanspray
Mahonia repens	creeping Oregon-grape
Morus microphylla	Texas mulberry
Pachystima myrsinites	myrtle boxleaf
Prunus serotina	black cherry
Quercus rugosa	net-leaf oak
Rhamnus crocea	holly-leaf buckthorn
Rhus glabra	smooth sumac
Rhus trilobata	ill-scented sumac
Ribes aureum	golden currant
Ribes pinetorum	orange gooseberry
Ribes viscosissimum	sticky currant
Robinia neomexicana	New Mexico locust
Rosa woodsii	Wood's rose
Sambucus cerulea	blue elder
Symphoricarpos longiflorus	desert snowberry
Symphoricarpos oreophilus	mountain snowberry
Yucca schottii	hoary yucca

Herbaceous Understory

Astragalus amphioxys	Aladdin's-slippers
Astragalus castaneiformis	chestnut milk-vetch
Astragalus tephrodes	ashen milk-vetch
Astragalus troglodytus	creeping milk-vetch
Blepharoneuron tricholepis	pine-dropseed
Bothriochloa barbinodis	cane beard grass
Bouteloua curtipendula	side-oats grama
Bouteloua gracilis	blue grama
Castilleja integra	squawfeather

Castilleja linarifolia	Wyoming Indian-paintbrush
Castilleja minor	alkali Indian-paint
Comandra umbellata	bastard-toadflax
Elyonurus barbiculmis	wool-spike grass
Erigeron arizonicus	Arizona fleabane
Erigeron concinnus	Navajo fleabane
Erigeron divergens	spreading fleabane
Erigeron lemmonii	Lemmon's fleabane
Erigeron neomexicanus	New Mexico fleabane
Erigeron oreophilus	chaparral fleabane
Erigeron platyphyllus	broad-leaf fleabane
Festuca arizonica	Arizona fescue
Hedeoma costatum	ribbed false pennyroyal
Hedeoma dentata	Arizona false pennyroyal
Hedeoma drummondii	Drummond's false pennyroyal
Hedeoma hyssopifolium	aromatic false pennyroyal
Hedeoma oblongifolium	oblong-leaf false pennyroyal
Hymenoxys bigelovii	Bigelow's rubberweed
Hymenoxys quinquesquamata	Rincon rubberweed
Iris missouriensis	Rocky Mountain iris
Laennecia schiedeana	pineland marshtail
Lathyrus graminifolius	grass-leaf vetchling
Lathyrus lanszwertii	Nevada vetchling
Lotus pleblus	long-bract bird's-foot-trefoil
Lotus unifoliolatus	American bird's-foot-trefoil
Lotus wrightii	scrub bird's-foot-trefoil
Lupinus brevicaulis	short-stem lupine
Lupinus huachucanus	Huachuca Mountain lupine
Lupinus neomexicanus	New Mexico lupine
Lupinus palmeri	Palmer's lupine
Monarda fistulosa	Oswego-tea
Muhlenbergia emersleyi	bull grass
Muhlenbergia montana	mountain muhly
Muhlenbergia virescens	screw-leaf muhly
Panicum bulbosum	bulb panic grass
Penstemon barbatus	beard-lip beardtongue
Penstemon virgatus	upright blue beardtongue
Piptochaetium fimbriatum	pinyon spear grass
Piptochaetium pringlei	Pringle's spear grass
Potentilla crinita	bearded cinquefoil
Potentilla hippiana	woolly cinquefoil
Potentilla thurberi	scarlet cinquefoil
Pteridium aquilinum	northern bracken fern
Schizachyrium cirratum	Texas false bluestem
Schizachyrium scoparium	little false bluestem
Senecio cynthioides	white mountain ragwort
Senecio eremophilus	desert ragwort
Senecio parryi	mountain ragwort
Solidago missouriensis	Missouri goldenrod

Solidago velutina	three-nerve goldenrod
Solidago wrightii	Wright's goldenrod
Sphaeralcea fendleri	thicket globe-mallow
Tetraneuris acaulis	stemless four-nerve-daisy
Thermopsis rhombifolia	pairie golden-banner
Viola canadensis	Canadian white violet
Viola nephrophylla	northern bog violet
Viola umbraticola	ponderosa violet

Evergreen Woodland

Evergreen Woodland is dominated by evergreen oaks on hill and mountain slopes between 4000 and 6500 feet. Evergreen Woodland varies from open to very dense with numerous associated species of grasses, dry tropic shrubs, succulents, and some cacti. The center of the Evergreen Woodland is the Sierra Madre of Mexico but it extends northward to the mountains of southeastern Arizona, southwestern New Mexico, and Transpecos Texas. Evergreen Woodland is typically very open. On the upper elevations, the evergreen oaks are mixed with pine and juniper and on the lower elevations, grassland species are frequent in the open areas. This community corresponds, in part, to Küchler #31 and #71 and Woodlands ERT.

Canopy
Characteristic Species

Pinus engelmannii	Apache pine
Pinus leiophylla	Chihuahuan pine
Quercus arizonica	Arizona white oak
Quercus emoryi	Emory's oak
Quercus grisea	gray oak
Quercus hypoleucoides	silver-leaf oak
Quercus oblongifolia	Mexican blue oak
Quercus rugosa	net-leaf oak

Associates

Cupressus arizonica	Arizona cypress
Juniperus deppeana	alligator juniper
Juniperus erythrocarpa	red-berry juniper
Juniperus monosperma	one-seed juniper
Juniperus pinchotii	Pinchot's juniper
Pinus cembroides	Mexican pinyon
Pinus edulis	two-needle pinyon
Pinus remota	paper-shell pinyon

Woody Understory

Agave palmeri	Palmer's century-plant
Agave parryi	Parry's century-plant
Aplopappus laricifolius	turpentine bush
Arctostaphylos pungens	Mexican manzanita
Ceanothus fendleri	Fendler's buckbrush
Cercocarpus montanus	alder-leaf mountain-mahogany
Coryphantha recurvata	Santa Cruz beehive cactus
Dalea formosa	featherplume
Dasylirion wheeleri	common sotol
Echinocereus fendleri	pink-flower hedgehog cactus

Echinocereus rigidissimus	rainbow hedgehog cactus
Echinocereus triglochidiatus	king-cup cactus
Erythrina flabelliformis	coral-bean
Escobaria vivipara	spinystar
Eysenhardtia orthocarpa	Tahitian kidneywood
Fouquieria splendens	ocotillo
Mahonia haematocarpa	red Oregon-grape
Mammillaria heyderi	little nipple cactus
Mammillaria viridiflora	green-flower nipple cactus
Mimosa biuncifera	cat-claw mimosa
Mimosa dysocarpa	velvet-pod mimosa
Nolina microcarpa	sacahuista bear-grass
Opuntia phaeacantha	tulip prickly-pear
Opuntia santa-rita	Santa Rita prickly-pear
Opuntia spinosior	walkingstick cactus
Opuntia versicolor	stag-horn cholla
Prosopis glandulosa	honey mesquite
Purshia mexicana	Mexican cliff-rose
Quercus mohriana	Mohr's oak
Quercus toumeyi	Toumey's oak
Rhus virens	evergreen sumac
Robinia neomexicana	New Mexico locust
Vauquelinia californica	Arizona-rosewood
Yucca baccata	banana yucca
Yucca schottii	hoary yucca

Herbaceous Understory

Artemisia ludoviciana	white sagebrush
Bothriochloa barbinodis	cane beard grass
Bouteloua curtipendula	side-oats grama
Bouteloua eriopoda	black grama
Bouteloua gracilis	blue grama
Bouteloua hirsuta	hairy grama
Brickellia spp.	brickellbush
Calochortus spp.	mariposa lily
Cyperus spp.	flatsedge
Dalea versicolor	oakwoods prairie-clover
Elyonurus barbiculmis	wool-spike grass
Eragrostis intermedia	plains love grass
Eriogonum spp.	wild buckwheat
Heteropogon contortus	twisted tanglehead
Hibiscus spp.	rose-mallow
Leptochloa dubia	green sprangletop
Lupinus spp.	lupine
Lycurus phleoides	common wolf's-tail
Muhlenbergia emersleyi	bull grass
Muhlenbergia porteri	bush muhly
Muhlenbergia torreyi	ringed muhly
Oxalis spp.	wood sorrel

Penstemon spp.	beardtongue
Phaseolus spp.	bean
Salvia spp.	sage
Schizachyrium scoparium	little false bluestem
Senna hirsuta	woolly wild sensitive-plant
Sphaeralcea spp.	globe mallow
Verbena spp.	vervain

Desert Scrub

Desert Scrub is part of the Chihuahuan Desert and lies between 1800 feet in Transpecos Texas to above 3500 feet in Arizona. It covers southwest Texas, southern New Mexico, smaller portions of southeast Arizona, and extends into Mexico. Many shrub complexes occur in this complex transition community between the grassland plains and Sonoran Desert. The dominant shrubs can change over short distances in response to changes in soil conditions. Some ecologists believe that this was once desert grassland, but as it was degraded by grazing and fire control, shrubs were favored and the community structure changed. This community corresponds to Küchler #44, #45, and #59 and Desert Lands ERT.

Characteristic Woody Species

Flourensia cernua	American tarwort
Larrea tridentata	creosote-bush

Associates

Acacia berlandieri	guajillo
Acacia constricta	mescat wattle
Acacia greggii	long-flower catclaw
Acacia neovernicosa	trans-pecos wattle
Agave lechuguilla	lechuguilla
Agave neomexicana	New Mexico century-plant
Agave parryi	Parry's century-plant
Agave scabra	rough century-plant
Aloysia gratissima	whitebrush
Aloysia wrightii	Wright's beebrush
Artemisia filifolia	silver sagebrush
Atriplex canescens	four-wing saltbush
Castela erecta	goatbush
Condalia ericoides	javelin-bush
Coryphantha cornifera	rhinoceros cactus
Coryphantha macromeris	nipple beehive cactus
Coryphantha ramillosa	whiskerbush
Coryphantha scheeri	long-tubercle beehive cactus
Dasylirion leiophyllum	green sotol
Dasylirion wheeleri	common sotol
Echinocactus horizonthalonius	devil's-head
Echinocactus texensis	horse-crippler
Echinocereus chloranthus	brown-spine hedgehog cactus
Echinocereus enneacanthus	pitaya
Echinocereus pectinatus	Texas rainbow cactus
Echinocereus rigidissimus	rainbow hedgehog cactus
Echinocereus stramineus	strawberry hedgehog cactus
Echinocereus triglochidiatus	king-cup cactus

Ephedra trifurca	long-leaf joint-fir
Epithelantha micromeris	ping-pong-ball cactus
Escobaria tuberculosa	white-column fox-tail cactus
Escobaria vivipara	spinystar
Euphorbia antisyphilitica	candelilla
Eysenhardtia texana	Texas kidneywood
Ferocactus hamatacanthus	turk's head
Forestiera pubescens	stretchberry
Fouquieria splendens	ocotillo
Garrya ovata	lindheimer silktassel
Guajacum angustifolium	Texas lignumvitae
Gutierrezia microcephala	small-head snakeweed
Gutierrezia sarothrae	kindlingweed
Haploesthes greggii	false broomweed
Hechtia texensis	Texas false agave
Isocoma pluriflora	southern jimmyweed
Jatropha dioica	leatherstem
Koeberlinia spinosa	crown-of-thorns
Leucophyllum frutescens	Texas barometer-bush
Leucophyllum minus	Big Bend barometer-bush
Mahonia trifoliata	Laredo Oregon-grape
Mammillaria heyderi	little nipple cactus
Mammillaria pottsii	rat-tail nipple cactus
Menodora scabra	rough menadora
Mortonia sempervirens	Rio Grande saddlebush
Nolina erumpens	foothill bear-grass
Nolina microcarpa	sacahuista bear-grass
Nolina texana	Texas bear-grass
Opuntia engelmannii	cactus-apple
Opuntia imbricata	tree cholla
Opuntia kleiniae	candle cholla
Opuntia leptocaulis	Christmas cholla
Opuntia macrocentra	purple prickly-pear
Opuntia phaeacantha	tulip prickly-pear
Opuntia polyacantha	hair-spine prickly-pear
Opuntia schottii	dog cholla
Opuntia tunicata	thistle cholla
Parkinsonia texana	Texas palo-verde
Parthenium argentatum	guayule
Parthenium incanum	mariola
Peniocereus greggii	night-blooming-cereus
Prosopis glandulosa	honey mesquite
Quercus sinuata	bastard oak
Rhus aromatica	fragrant sumac
Rhus microphylla	little-leaf sumac
Schaefferia cuneifolia	desert-yaupon
Sclerocactus intertextus	white fish-hook cactus
Sclerocactus scheeri	Scheer's fish-hook cactus
Sclerocactus uncinatus	Chihuahuan fish-hook cactus

Senna wislizeni	Wislizenus' wild sensitive-plant
Thelocactus bicolor	glory-of-Texas
Tiquilia canescens	woody crinklemat
Tiquilia greggii	plumed crinklemat
Viguiera stenoloba	resinbush
Yucca baccata	banana yucca
Yucca elata	soaptree yucca
Yucca faxoniana	Eve's-needle
Yucca thompsoniana	Thompson's yucca
Yucca torreyi	Torrey's yucca
Ziziphus obtusifolia	lotebush

Herbaceous Understory

Aristida californica	California three-awn
Aristida divaricata	poverty three-awn
Aristida purpurea	purple three-awn
Aristida ternipes	spider grass
Baileya multiradiata	showy desert-marigold
Bothriochloa barbinodis	cane beard grass
Bothriochloa saccharoides	plumed beard grass
Bouteloua breviseta	gypsum grass
Bouteloua curtipendula	side-oats grama
Bouteloua eriopoda	black grama
Bouteloua gracilis	blue grama
Bouteloua hirsuta	hairy grama
Bouteloua ramosa	chino grama
Bouteloua rigidiseta	Texas grama
Bouteloua trifida	red grama
Buchloë dactyloides	buffalo grass
Chamaesaracha sordida	hairy five-eyes
Croton dioicus	grassland croton
Distichlis spicata	coastal salt grass
Engelmannia pinnatifida	Englemann's daisy
Erioneuron pulchellum	low woolly grass
Helianthus spp.	sunflower
Hilaria belangeri	curly-mesquite
Hilaria jamesii	James' galleta
Hilaria mutica	tobosa grass
Hymenoxys spp.	rubberweed
Mentzelia spp.	blazingstar
Muhlenbergia porteri	bush muhly
Nassella leucotricha	Texas wintergrass
Oryzopsis hymenoides	Indian mountain-rice grass
Pappophorum bicolor	pink pappus grass
Proboscidea louisianica	ram's horn
Schizachyrium scoparium	little false bluestem
Scleropogon brevifolius	burro grass
Senna roemeriana	two-leaved wild sensitive-plant
Setaria macrostachya	plains bristle grass

Sporobolus airoides	alkali-sacaton
Sporobolus contractus	narrow-spike dropseed
Sporobolus cryptandrus	sand dropseed
Sporobolus flexuosus	mesa dropseed
Sporobolus giganteus	giant dropseed
Sporobolus wrightii	Wright's dropseed
Thymophylla acerosa	American prickyleaf
Thymophylla pentachaeta	five-needle prickleleaf
Tidestromia spp.	honeysweet
Tridens muticus	awnless fluff grass
Xanthocephalum spp.	matchweed
Zinnia acerosa	white zinnia
Zinnia grandiflora	little golden zinnia

Desert Grassland

Desert Grassland is found throughout this region from Transpecos Texas through southern New Mexico to southeast Arizona and south into Mexico. These grasslands form extensive mosaics over large ecotones between the Desert Scrub and Evergreen Woodland. The two characteristic species are tobosa grass (on heavy soils subject to flooding) and black grama (gravelly uplands). The desert grasslands occur from 3500 to 7000 feet and receive approximately ten to fifteen inches of rain annually. Originally the grasses were perennial bunch grasses. In many areas, these grasses have been replaced with low growing sod grasses or desert scrublands as a result of grazing and the suppression of natural fire. Trees and shrubs occur but vary in size, composition, and density from site to site. The compositional mix of the most prevalent grasses varies across the grassland depending upon site specific characteristics. Desert grasslands vary from very open with bare ground between the bunches of grass to areas which have continuous to nearly uninterrupted cover. Tobosa grass may form dense stands in sites that receive excess run-off from the surrounding landscape and hence represent small, internally drained basins. Alkali-sacaton dominates in saline areas. This community corresponds, in part, to Küchler #53, #54, and #58 and Grasslands ERT.

Characteristic Herbaceous Species

Bouteloua eriopoda	black grama
Bouteloua gracilis	blue grama
Hilaria jamesii	James' galleta
Hilaria mutica	tobosa grass

Associates

Ageratina wrightii	Wright's snakeroot
Allionia incarnata	trailing windmills
Amaranthus palmeri	Palmer's amaranth
Andropogon hallii	sand bluestem
Aristida californica	California three-awn
Aristida divaricata	poverty three-awn
Aristida purpurea	purple three-awn
Aristida ternipes	spider grass
Artemisia ludoviciana	white sagebrush
Astragalus mollissimus	woolly milk-vetch
Astragalus wootonii	Wooton's milk-vetch
Baileya multiradiata	showy desert-marigold
Bothriochloa barbinodis	cane beard grass
Bouteloua breviseta	gypsum grama
Bouteloua chondrosioides	spruce-top grama

Bouteloua curtipendula	side-oats grama
Bouteloua hirsuta	hairy grama
Bouteloua repens	slender grama
Buchloë dactyloides	buffalo grass
Calcohortus ambiguus	doubting mariposa-lily
Cryptantha micrantha	red-root cat's-eye
Dalea candida	white prairie-clover
Dalea purpurea	violet prairie-clover
Digitaria californica	California crab grass
Eragrostis intermedia	plains love grass
Erioneuron pulchellum	low woolly grass
Froelichia gracilis	slender snakecotton
Gaillardia pulchella	firewheel
Gaura coccinea	scarlet beeblossom
Glandularia bipinnatifida	Dakota mock vervain
Gossypium thurberi	Thurber's cotton
Grindelia squarrosa	curly-cup gumweed
Gutierrezia sarothrae	kindlingweed
Heteropogon contortus	twisted tanglehead
Hilaria belangeri	curly-mesquite
Leptochloa spp.	sprangletop
Lesquerella gordonii	Gordon's bladderpod
Lycurus phleoides	common wolf's-tail
Mentzelia albicaulis	white-stem blazingstar
Mentzelia multiflora	Adonia blazingstar
Muhlenbergia porteri	bush muhly
Muhlenbergia torreyi	ringed muhly
Oryzopsis hymenoides	Indian mountain-rice grass
Pacopyrum smithii	western-wheat grass
Panicum obtusum	blunt panic grass
Pappophorum vaginatum	whiplash pappus grass
Schizachyrium scoparium	little false bluestem
Scleropogon brevifolius	burro grass
Senecio flaccidus	thread-leaf ragwort
Senna bauhinioides	shrubby wild sensitive-plant
Setaria macrostachya	plains bristle grass
Sphaeralcea angustifolia	copper globe-mallow
Sphaeralcea coccinea	scalet globe-mallow
Sphaeralcea incana	soft globe-mallow
Sporobolus airoides	alkali-sacaton
Sporobolus cryptandrus	sand dropseed
Sporobolus flexuosus	mesa dropseed
Sporobolus wrightii	Wright's dropseed
Stipa comata	needle-and-thread
Tridens muticus	awnless fluff grass
Verbesina encelioides	golden crownbeard
Xanthocephalum spp.	matchweeds
Zinnia acerosa	white zinnia
Zinnia grandiflora	little golden zinnia

Woody Associates

Acacia constricta	mescat wattle
Acacia greggii	long-flower catclaw
Acacia neovernicosa	trans-pecos wattle
Agave lechuguilla	lechuguilla
Agave palmeri	Palmer's century-plant
Agave parryi	Parry's century-plant
Agave parviflora	small-flower century-plant
Agave scabra	rough century-plant
Agave schottii	Schott's century-plant
Aloysia wrightii	Wright's beebrush
Anisacanthus thurberi	Thurber's desert-honeysuckle
Artemisia bigelovii	flat sagebrush
Artemisia frigida	prairie sagebrush
Artemisia tridentata	big sagebrush
Atriplex canescens	four-wing saltbush
Buddleja scordioides	escobilla butterfly-bush
Calliandra eriophylla	fairy-duster
Celtis laevigata	sugar-berry
Celtis pallida	shiny hackberry
Condalia ericoides	javelin-bush
Condalia spathulata	squawbush
Coryphantha recurvata	Santa Cruz beehive cactus
Coryphantha scheeri	long-tubercle beehive cactus
Coursetia glandulosa	rosary baby-bonnets
Dalea formosa	featherplume
Dasylirion leiophyllum	green sotol
Dasylirion wheeleri	common sotol
Dodonaea viscosa	Florida hopbush
Echinocactus horizonthalonius	devil's-head
Echinocereus fendleri	pink-flower hedgehog cactus
Echinocereus rigidissimus	rainbow hedgehog cactus
Ephedra antisyphilitica	clapweed
Ephedra torreyana	Torrey's joint-fir
Ephedra trifurca	long-leaf joint-fir
Ephedra viridis	Mormon-tea
Ericameria laricifolia	turpentine-bush
Eriogonum abertianum	Abert's wild buckwheat
Eriogonum rotundifolium	round-leaf wild buckwheat
Eriogonum wrightii	bastard-sage
Erythrina flabelliformis	coral-bean
Escobaria orcuttii	Orcutt's fox-tail cactus
Escobaria tuberculosa	white-column fox-tail cactus
Escobaria vivipara	spinystar
Eysenhardtia orthocarpa	Tahitian kidneywood
Ferocactus wislizeni	candy barrel cactus
Flourensia cernua	American tarwort
Fouquieria splendens	ocotillo
Isocoma pluriflora	southern jimmyweed

Isocoma tenuisecta	shrine jimmyweed
Juniperus monosperma	one-seed juniper
Koeberlinia spinosa	crown-of-thorns
Larrea tridentata	creosote-bush
Lysiloma watsonii	little-leaf false tamarind
Mahonia trifoliata	Laredo Oregon-grape
Mammillaria grahamii	Graham's nipple cactus
Mammillaria heyderi	little nipple cactus
Mammillaria mainiae	counter-clockwise nipple cactus
Mammillaria wrightii	Wright's nipple cactus
Menodora scabra	rough menodora
Mimosa biuncifera	cat-claw mimosa
Mimosa dysocarpa	velvet-pod mimosa
Nolina erumpens	foothill bear-grass
Nolina microcarpa	sacahuista bear-grass
Nolina texana	Texas bear-grass
Opuntia chlorotica	clock-face prickly-pear
Opuntia engelmannii	cactus-apple
Opuntia gosseliniana	violet prickly-pear
Opuntia imbricata	tree cholla
Opuntia kunzei	devil's cholla
Opuntia leptocaulis	Christmas cholla
Opuntia macrocentra	purple prickly-pear
Opuntia phaeacantha	tulip prickly-pear
Opuntia polyacantha	hair-spine prickly-pear
Opuntia santa-rita	Santa Rita prickly-pear
Opuntia spinosior	walkingstick cactus
Opuntia tetracantha	Tucson prickly-pear
Parthenium incanum	mariola
Prosopis glandulosa	honey mesquite
Prosopis velutina	velvet mesquite
Rhus microphylla	little-leaf sumac
Rhus virens	evergreen sumac
Sageretia wrightii	Wright's mock buckthorn
Sclerocactus erectocentrus	red-spine fish-hook cactus
Sclerocactus intertextus	white fish-hook cactus
Senecio flaccidus	thread-leaf ragwort
Sideroxylon lanuginosum	gum bully
Yucca baccata	banana yucca
Yucca elata	soaptree yucca
Yucca faxoniana	Eve's-needle
Yucca glauca	soapweed yucca
Yucca thompsoniana	Thompson's yucca
Yucca torreyi	Torrey's yucca
Yucca treculeana	Don Quixote's-lace
Ziziphus obtusifolia	lotebush

WETLAND SYSTEMS
Riparian Woodland

Very little riparian vegetation remains in the deserts of the southwest. The riparian vegetation of the southwest varies from large deciduous trees forming a full canopy to riparian scrublands and changes with altitude. Cottonwood and willows can still be found in some areas forming a canopy immediately adjacent to perennial streams. Large thickets of mesquite called bosques were once found on floodplains, but few representatives remain. The understory of mature bosques historically contained annual and perennial grasses and young trees. Mesquite also forms the understory on stream banks. Some trees and shrubs of the riparian habitat are found along dry washes far into Desert Scrub regions. This community corresponds to Riparian Forests and Woodlands ERT.

Canopy
Characteristic Species
Fraxinus velutina	velvet ash
Juglans major	Arizona walnut
Platanus wrightii	Arizona sycamore
Populus deltoides	eastern cottonwood
Populus fremontii	Fremont's cottonwood
Salix bonplandiana	Bonpland's willow
Salix gooddingii	Goodding's willow

Associates
Acer grandidentatum	canyon maple
Celtis laevigata	sugar-berry
Populus angustifolia	narrow-leaf cottonwood
Prunus serotina	black cherry

Woody Understory
Acacia constricta	mescat wattle
Acacia farnesiana	mealy wattle
Acacia greggii	long-flower catclaw
Acacia rigidula	chaparro-prieto
Acer glabrum	Rocky Mountain maple
Acer negundo	ash-leaf maple
Alnus oblongifolia	Arizona alder
Atriplex lentiformis	quailbush
Atriplex polycarpa	cattle-spinach
Baccharis sarothroides	rosinbush
Brickellia laciniata	split-leaf brickellbush
Cephalanthus occidentalis	common buttonbush
Chilopsis linearis	desert-willow
Clematis ligusticifolia	deciduous traveler's-joy
Fallugia paradoxa	Apache-plume
Fraxinus berlandieriana	Mexican ash
Hymenoclea monogyra	single-whorl cheesebush
Lycium andersonii	red-berry desert-thorn
Lycium berlandieri	silver desert-thorn
Lycium fremontii	Freemont's desert-thorn
Morus microphylla	Texas mulberry

Olneya tesota	desert-ironwood
Parkinsonia florida	blue palo-verde
Parkinsonia texana	Texas palo-verde
Pluchea camphorata	plowman's wort
Pluchea sericea	arrow-weed
Prosopis glandulosa	honey mesquite
Prosopis pubescens	American screw-bean
Rhus microphylla	little-leaf sumac
Salix amygdaloides	peach-leaf willow
Salix interior	sandbar willow
Salix lasiolepis	arroyo willow
Salix lucida	shining willow
Salix nigra	black willow
Salix scouleriana	Scouler's willow
Sambucus cerulea	blue elder
Sapindus saponaria	wing-leaf soapberry
Suaeda moquinii	shrubby seepweed
Vitis arizonica	canyon grape
Ziziphus obtusifolia	lotebush

Herbaceous Understory

Amaranthus palmeri	Palmer's amaranth
Chloracantha spinosa	Mexican devilwood
Cucurbita spp.	gourd
Heliotropium curassavicum	seaside heliotrope
Panicum obtusum	blunt panic grass
Phragmites australis	common reed
Pteridium aquilinum	northern bracken fern
Sarcostemma spp.	twinevine
Verbesina encelioides	golden crownbeard

Riparian Woody Species of Dry Arroyos and Washes
Arizona

Acacia greggii	long-flower catclaw
Celtis laevigata	sugar-berry
Chilopsis linearis	desert willow
Juniperus monosperma	one-seed juniper
Parkinsonia florida	blue paloverde
Prosopis glandulosa	honey mesquite
Psorothamnus spinosus	smoketree
Sapindus saponaria	wing-leaf soapberry
Sapium biloculare	Mexican jumping bean

Transpecos Texas

Acacia farnesiana	mealy wattle
Acacia greggii	long-flower catclaw
Brickellia laciniata	split-leaf brickellbush
Celtis laevigata	sugar-berry
Celtis pallida	shiny hackberry

Chilopsis linearis	desert willow
Fraxinus velutina	velvet ash
Juglans microcarpa	little walnut
Populus fremontii	Fremont's cottonwood
Prosopis glandulosa	honey mesquite
Quercus pungens	scrub oak
Rhus microphylla	little-leaf sumac
Rhus virens	evergreen sumac
Sapindus saponaria	wing-leaf soapberry
Ungnadia speciosa	Mexican-buckeye

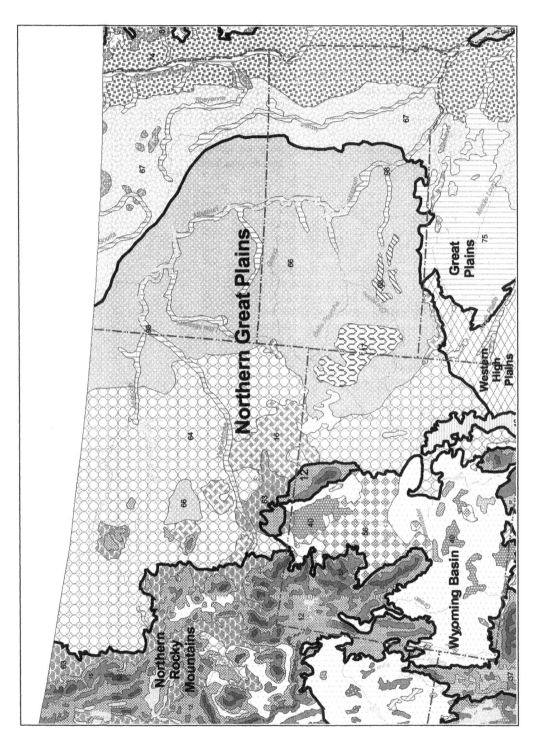

Northern Great Plains

Northern Great Plains

Introduction

The Northern Great Plains region includes western North and South Dakota, central and eastern Montana, and eastern Wyoming. The region is characterized by mixed and shortgrass prairies with forest communities more common at higher elevations. The region's boundaries are the tallgrass prairie to the east, the Rocky Mountains with conifer forests to the west, the blue grama and buffalo grass dominated shortgrass prairie to the south and to the north, grasslands extend into Canada.

Mixed Grass Prairie is characteristic of the eastern half of this region. In areas with sufficient moisture, the Mixed Grass Prairie community type may be dominated by tallgrass prairie species; shortgrass prairie species may dominate in drier situations. This area is an ecotone between the tallgrass and shortgrass prairies. Moving west into Wyoming and Montana, the short grasses become more dominant.

The Blackhills of South Dakota and Wyoming support ponderosa pine forests. The Bighorn, Little Belt, Bear Paw, and Little Rocky Mountains also occur within the region. These mountains support vegetation similar to that found in the Rocky Mountains to the West, including Douglas-fir Forest which is described in the Rocky Mountain region.

The primary sources used to develop descriptions and species lists are Barbour and Billings 1988; Boon and Groe 1990; Pfister et al. 1977.

Dominant Ecological Communities

Upland Systems
Blackhills Pine Forest
Eastern Ponderosa Forest
Sagebrush Steppe
Foothills Prairie
Mixed Grass Prairie
Northern Shortgrass Prairie
Wetland Systems
Northern Floodplain Forest
Wet Meadow
Pothole Marsh

UPLAND SYSTEMS

Blackhills Pine Forest

The Blackhills of South Dakota and eastern Montana support ponderosa pine forest communities. On the east side of the Blackhills, ponderosa pine forests may have broadleaf deciduous trees as a component of the canopy. On the west side of the Blackhills, shrubs are more likely to become important components of the community structure. Along streams within the Blackhills Pine Forest, hackberry, burr oak, and American elm are important community components. This community corresponds to Küchler #17 and Western Deciduous and Mixed Forests ERT.

Canopy
Characteristic Species
Pinus ponderosa　　　　　　　　　　ponderosa pine

Associates

Acer negundo	ash-leaf maple
Betula papyrifera	paper birch
Picea glauca	white spruce
Prunus virginiana	choke cherry
Quercus macrocarpa	burr oak
Ulmus americana	American elm

Woody Understory

Amelanchier alnifolia	Saskatoon service-berry
Arctostaphylos uva-ursi	red bearberry
Artemisia tridentata	big sagebrush
Celtis occidentalis	common hackberry
Chrysothamnus nauseosus	rubber rabbitbrush
Juniperus communis	common juniper
Mahonia repens	creeping Oregon-grape
Opuntia polyacantha	hair-spine prickly-pear
Ostrya virginiana	eastern hop-hornbeam
Physocarpus opulifolius	Atlantic ninebark
Rosa woodsii	Woods' rose
Symphoricarpos albus	common snowberry
Symphoricarpos occidentalis	western snowberry

Herbaceous Understory

Antennaria plantaginifolia	woman's-tobacco
Aristida purpurea	purple three-awn
Bouteloua gracilis	blue grama
Buchloë dactyloides	buffalo grass
Carex concinna	low northern sedge
Carex duriuscula	spike-rush sedge
Carex filifolia	thread-leaf sedge
Carex foenea	dry-spike sedge
Carex inops	long-stolon sedge
Danthonia spicata	poverty wild oat grass
Fragaria virginiana	Virginia strawberry
Galium boreale	northern bedstraw
Gutierrezia sarothrae	kindlingweed
Koeleria macrantha	prairie Koeler's grass
Lathyrus ochroleucus	cream vetchling
Lupinus argenteus	silver-stem lupine
Oryzopsis asperifolia	white-grain mountain-rice grass
Pascopyrum smithii	western-wheat grass
Plantago patagonica	woolly plantain
Pseudoroegneria spicatum	bluebunch-wheat grass
Schizachyrium scoparium	little false bluestem
Stipa comata	needle-and-thread

Eastern Ponderosa Forest

Eastern Ponderosa Forest occurs in central Montana in the Bighorn, Snowy, Little Belt, Bear Paw, and Little Rocky Mountains. This community corresponds to Küchler #16 and Western Deciduous and Mixed Forests ERT.

Canopy
Characteristic Species
Pinus ponderosa	ponderosa pine

Associates
Juniperus scopulorum	Rocky Mountain juniper
Prunus virginiana	choke cherry

Woody Understory
Amelanchier alnifolia	Saskatoon service-berry
Mahonia repens	creeping Oregon-grape
Opuntia spp.	prickly-pear
Purshia tridentata	bitterbrush
Rhus trilobata	ill-scented sumac
Symphoricarpos albus	common snowberry
Toxicodendron radicans	eastern poison-ivy
Yucca glauca	soapweed yucca

Herbaceous Understory
Andropogon gerardii	big bluestem
Antennaria rosea	rosy pussytoes
Arnica cordifolia	heart-leaf leopard-bane
Balsamorhiza sagittata	arrow-leaf balsamroot
Bouteloua gracilis	blue grama
Carex pensylvanica	Pennsylvania sedge
Cystopteris fragilis	brittle bladder fern
Danthonia unispicata	few-flower wild oat grass
Festuca campestris	prairie fescue
Festuca idahoensis	bluebunch fescue
Galium boreale	northern bedstraw
Lithospermum ruderale	Columbian puccoon
Pascopyrum smithii	western-wheat grass
Pseudoroegneria spicata	bluebunch-wheat grass
Schizachne purpurascens	false melic grass
Schizachyrium scoparium	little false bluestem
Stipa comata	needle-and-thread

Sagebrush Steppe

Sagebrush Steppe occurs in the Northern Great Plains region, in the northeastern corner of Wyoming. This community corresponds to Küchler #55 and Desert Lands ERT.

Characteristic Species
Artemisia tridentata	big sagebrush
Pseudoroegneria spicatum	bluebunch-wheat grass

Associates

Agoseris glauca	pale goat-chicory
Allium acuminatum	taper-tip onion
Allium textile	wild white onion
Alyssum desertorum	desert madwort
Amelanchier alnifolia	Saskatoon service-berry
Antennaria dimorpha	cushion pussytoes
Antennaria rosea	rosy pussytoes
Arabis holboellii	Holboell's rockcress
Arabis lignifera	Owen's valley rockcress
Arabis sparsiflora	elegant rockcress
Arenaria hookeri	Hooker's sandwort
Artemisia cana	hoary sagebrush
Artemisia ludoviciana	white sagebrush
Artemisia nova	black sagebrush
Artemisia pedatifida	bird-foot sagebrush
Artemisia tripartita	three-tip sagebrush
Artemisia frigida	prairie sagebrush
Astragalus miser	timber milk-vetch
Astragalus pectinatus	narrow-leaf milk-vetch
Astragalus purshii	Pursh's milk-vetch
Astragalus spatulatus	tufted milk-vetch
Atriplex canescens	four-wing saltbush
Atriplex confertifolia	shadscale
Atriplex gardneri	Gardner's saltbush
Balsamorhiza incana	hoary balsamroot
Balsamorhiza sagittata	arrow-leaf balsamroot
Calochortus nuttallii	sego-lily
Carex duriuscula	spike-rush sedge
Carex filifolia	threadleaf sedge
Carex obtusata	blunt sedge
Castilleja angustifolia	northwestern Indian-paintbrush
Chenopodium leptophyllum	narrow-leaf goosefoot
Chrysothamnus nauseosus	rubber rabbitbrush
Chrysothamnus viscidiflorus	green rabbitbrush
Comandra umbellata	bastard-toadflax
Cordylanthus ramosus	bushy bird's-beak
Crepis modocensis	siskiyou hawk's-beard
Cryptantha watsonii	Watson's cat's-eye
Delphinium bicolor	flat-head larkspur
Delphinium geyeri	Geyer's larkspur
Descurainia incana	mountain tansy-mustard
Distichilis spicata	coastal salt grass
Elymus elymoides	western bottle-brush grass
Elymus macrourus	thick-spike wild rye
Elymus trachycaulus	slender rye grass
Erigeron caespitosus	tufted fleabane
Erigeron engelmannii	Engelmann's fleabane
Eriogonum caespitosum	matted wild buckwheat

Eriogonum cernuum	nodding wild buckwheat
Eriogonum compositum	arrow-leaf wild buckwheat
Eriogonum microthecum	slender wild buckwheat
Eriogonum ovalifolium	cushion wild buckwheat
Festuca idahoensis	bluebunch fescue
Gayophytum ramosissimum	pinyon groundsmoke
Grayia spinosa	spiny hop-sage
Gutierrezia sarothrae	kindlingweed
Hilaria jamesii	James' galleta
Hordeum jubatum	fox-tail barley
Hymenoxys richardsonii	Colorado rubberweed
Iva axillaris	deer-root
Juncus balticus	Baltic rush
Koeleria macrantha	prairie Koeler's grass
Krascheninnikovia lanata	winterfat
Lappula occidentalis	flat-spine sheepburr
Leptodactylon pungens	granite prickly-phlox
Lesquerella ludoviciana	Louisiana bladderpod
Leymus cinereus	Great Basin lyme grass
Lithospermum incisum	fringed gromwell
Lithospermum ruderale	Columbian puccoon
Lomatium foeniculaceum	carrot-leaf desert-parsley
Lupinus argenteus	silver-stem lupine
Lupinus sericeus	Pursh's silky lupine
Machaeranthera canescens	hoary tansy-aster
Mertensia oblongifolia	languid-lady
Monolepis nuttalliana	Nuttall's poverty-weed
Musineon divaricatum	leafy wild parsley
Opuntia polyacantha	hair-spine prickly-pear
Oryzopsis hymenoides	Indian mountain-rice grass
Pascopyrum smithii	western-wheat grass
Penstemon arenicola	red desert beardtongue
Penstemon fremontii	Fremont's beardtongue
Phlox hoodii	carpet phlox
Phlox longifolia	long-leaf phlox
Phlox multiflora	Rocky Mountain phlox
Poa fendleriana	mutton grass
Poa secunda	curly blue grass
Polygonum douglasii	Douglas' knotweed
Pseudocymopterus montanus	alpine false mountain-parsley
Pseudoroegneria spicatum	bluebunch-wheat grass
Purshia tridentata	bitterbrush
Ribes cereum	white squaw currant
Sphaeralcea coccinea	scarlet globe-mallow
Stenotus acaulis	stemless mock goldenweed
Stipa comata	needle-and-thread
Stipa lettermani	Letterman's needle grass
Stipa nelsonii	Nelson's needle grass
Tetradymia canescens	spineless horsebrush

Tetradymia spinosa	short-spine horsebrush
Thermopsis rhombifolia	prairie golden-banner
Trifolium gymnocarpon	holly-leaf clover
Vulpia octoflora	eight-flower six-weeks grass
Xylorhiza glabriuscula	smooth woody-aster
Ziadenus venosus	meadow deathcamas
Zigadenus paniculatus	sand-corn

Foothills Prairie

Foothills Prairie occurs in the foothills of the few mountain ranges within the region and along the western boundary of the region in the foothills of the Rocky Mountains. This prairie type is dominated by short and medium grasses. This community corresponds to Küchler #63 and Prairies ERT.

Characteristic Species

Festuca campestris	prairie fescue
Festuca idahoensis	bluebunch fescue
Pseudoroegneria spicatum	bluebunch-wheat grass
Stipa comata	needle-and-thread

Associates

Artemisia frigida	prairie sagebrush
Bouteloua gracilis	blue grama
Carex filifolia	threadleaf sedge
Koeleria macrantha	prairie Koeler's grass
Pascopyrum smithii	western-wheat grass
Poa secunda	curly blue grass

Mixed Grass Prairie

Mixed Grass Prairie occurs in the eastern half of the Northern Great Plains region. Along with a mixture of short and tall grasses, this community supports scattered needle-leaf evergreen shrubs and small trees. This prairie community is found in western North and South Dakota. This community corresponds to Küchler #66 and #68 and Prairies ERT.

Characteristic Species

Bouteloua gracilis	blue grama
Buchloë dactyloides	buffalo grass
Nassella viridula	green tussock
Pascopyrum smithii	western-wheat grass
Stipa comata	needle-and-thread

Associates

Artemisia drancunculus	dragon wormwood
Artemisia frigida	prairie sagebrush
Artemisia ludoviciana	white sagebrush
Aster ericoides	white heath aster
Bouteloua curtipendula	side-oats grama
Carex duriuscula	spike rush sedge
Carex filifolia	threadleaf sedge
Carex inops	long-stolon sedge
Echinacea angustifolia	blacksamson

Elymus trachycaulus	slender rye grass
Juniperus scopulorum	Rocky Mountain juniper
Koeleria macrantha	prairie Koeler's grass
Liatris punctata	dotted gay feather
Oryzopsis hymenoides	Indian mountain-rice grass
Pediomelum argophyllum	silver-leaf Indian-breadroot
Schizachyrium scoparium	little false bluestem
Stipa spartea	porcupine grass

Northern Shortgrass Prairie

Northern Shortgrass Prairie occurs in western North and South Dakota, eastern Montana, Wyoming, and Colorado. This prairie, a short-medium tall grassland, is the dominant vegetation type in this region. This community correspond to Küchler #64 and Prairies ERT.

Characteristic Species
Bouteloua gracilis	blue grama
Pascopyrum smithii	western-wheat grass
Stipa comata	needle-and-thread

Associates
Artemisia frigida	prairie sagebrush
Carex filifolia	threadleaf sedge
Gutierrezia sarothrae	kindlingwood
Heterotheca villosa	hairy false golden-aster
Koeleria macrantha	prairie Koeler's grass
Liatris punctata	dotted gay feather
Muhlenbergia cuspidata	stony-hills muhly
Nassella viridula	green tussock
Poa secunda	curly blue grass
Pseudoroegneria spicatum	bluebunch-wheat grass
Schizachyrium scoparium	little false bluestem
Sporobolus cryptandrus	sand dropseed
Stipa spartea	porcupine grass

WETLAND SYSTEMS

Northern Floodplain Forest

Northern Floodplain Forest occurs along major rivers and streams from North Dakota south to Oklahoma. This community corresponds to Küchler #98 and Floodplain Forests ERT.

Canopy
Characteristic Species
Populus deltoides	eastern cottonwood
Salix nigra	black willow
Ulmus americana	American elm

Associates

Acer negundo	ash-leaf maple
Acer rubrum	red maple
Acer saccharinum	silver maple
Betula nigra	river birch
Celtis occidentalis	common hackberry
Fraxinus americana	white ash
Fraxinus pennsylvanica	green ash
Gleditsia triacanthos	honey-locust
Juglans nigra	black walnut
Platanus occidentalis	American sycamore
Prunus virginiana	choke cherry
Quercus macrocarpa	burr oak
Ulmus rubra	slippery elm

Woody Understory

Amorpha fruticosa	false Indigo-bush
Celastrus scandens	American bittersweet
Clematis virginiana	devil's-darning-needles
Parthenocissus quinquefolia	Virgina creeper
Rhus glabra	smooth sumac
Ribes missouriense	Missouri gooseberry
Salix amygdaloides	peach-leaf willow
Salix interior	sandbar willow
Smilax tamnoides	chinaroot
Symphoricarpos occidentalis	western snowberry
Symphoricarpos orbiculatus	coral-berry
Vitis riparia	river-bank grape
Vitis vulpina	frost grape

Herbaceous Understory

Elymus virginicus	Virginia wild rye
Galium aparine	sticky-willy
Laportea canadensis	Canadian wood-nettle

Wet Meadow

Wet Meadow occurs scattered throughout the eastern Great Plains region. Wet Meadow occupies areas slightly higher than the Pothole Marshes. The soils of the Wet Meadow are usually wet and soggy for most of the year. This community corresponds to Meadowlands ERT.

Characteristic Species

Carex trichocarpa	hairyfruit sedge
Eleocharis acicularis	needle spike-rush
Eleocharis compressa	flat-stem spike-rush
Eleocharis palustris	pale spike-rush
Juncus balticus	Baltic rush
Juncus marginatus	grass-leaf rush
Juncus nodosus	knotted rush

Juncus tenuis	poverty rush
Juncus torreyi	Torrey's rush
Scirpus americanus	chairmaker's bulrush

Associates

Agalinis tenuifolia	slender-leaf false foxglove
Agrostis gigantea	black bent
Agrostis hyemalis	winter bent
Alopecurus aequalis	short-awn meadow foxtail
Aureolaria virginica	downy yellow false-foxglove
Calamagrostis stricta	slim-stem reedgrass
Caltha palustris	yellow marsh-marigold
Campanula aparinoides	marsh bellflower
Carex festucacea	fescue sedge
Carex gravida	heavy sedge
Carex hystericina	porcupine sedge
Carex lanuginosa	woolly sedge
Carex nebrascensis	Nebraska sedge
Carex scoparia	pointed broom sedge
Carex stricta	uptight sedge
Catabrosa aquatica	water whorl grass
Cicuta maculata	spotted water-hemlock
Crepis runcinata	fiddle-leaf hawk's-beard
Cyperus squarrosus	awned flat sedge
Distichlis spicata	coastal salt grass
Dodecatheon meadia	eastern shootingstar
Epilobium palustre	marsh willowherb
Eriophorum gracile	slender cotton-grass
Galium trifidum	three-petal bedstraw
Galium triflorum	fragrant bedstraw
Gentiana andrewsii	closed bottle gentian
Gentiana saponaria	harvestbells
Glyceria striata	fowl manna grass
Helianthus tuberosus	Jerusalem-artichoke
Hordeum jubatum	fox-tail barley
Hypericum majus	greater Canadian St. John's-wort
Hypoxis hirsuta	eastern yellow star-grass
Juncus longistylis	long-style rush
Lilium philadelphicum	wood lily
Liparis loeselii	yellow wide-lip orchid
Lobelia siphilitica	great blue lobelia
Lotus unifoliolatus	American bird's-foot-trefoil
Lycopus asper	rough water-horehound
Lysimachia ciliata	fringed yellow-loosestrife
Lysimachia thyrsiflora	tufted yellow-loosestrife
Lythrum alatum	wide-angle loosestrife
Mentha arvensis	American wild mint
Menyanthes trifoliata	buck-bean
Plantago eriopoda	red-woolly plantain
Platanthera leucophaea	prairie white fringed orchid

Potentilla paradoxa	bushy cinquefoil
Ranunculus cymbalaria	alkali buttercup
Ranunculus sceleratus	cursed crowfoot
Scirpus pallidus	pale bulrush
Scutellaria galericulata	hooded skullcap
Spartina cynosuroides	big cord grass
Spiranthes cernua	white nodding ladies'-tresses
Spiranthes romanzoffiana	hooded ladies'-tresses
Stachys palustris	woundwort
Strophostyles leiosperma	slick-seed fuzzy-bean
Teucrium canadense	American germander
Triadenum virginicum	Virginia marsh-St. John's-wort
Triglochin maritimum	seaside arrow-grass

Pothole Marsh

Pothole Marsh occurs around the Minnesota-South Dakota border and northward into Canada. Pothole Marsh is a very important nesting habitat for ducks in North America. It is found on poorly drained soils of lowland and backwater bays or pothole depressions of mesic prairie. The outer margins of the pothole depressions are dominated by cord grass mixed with other species such as swamp milkweed and ironweed. The centers are typically dominated by smartweed which may be mixed with duck-potato and cat-tail. Draining activities have greatly diminished the extent of these valuable wetlands This community corresponds to Nonestuarine Marshes ERT.

Characteristic Species

Phragmites australis	common reed
Scirpus americanus	chairmaker's bulrush
Scirpus tabernaemontani	soft-stem bulrush
Spartina cynosuroides	big cord grass
Spartina pectinata	freshwater cord grass
Typha angustifolia	narrow-leaf cat-tail
Typha latifolia	broad-leaf cat-tail
Zizania aquatica	Indian wild rice

Associates

Alisma subcordatum	American water-plantain
Asclepias incarnata	swamp milkweed
Berula erecta	cut-leaf water-parsnip
Bidens frondosa	devil's pitchfork
Carex ssp.	sedge
Eleocharis acicularis	needle spike-rush
Eleocharis palustris	pale spike-rush
Equisetum hyemale	tall scouring-rush
Lobelia siphilitica	great blue lobelia
Polygonum amphibium	water smartweed
Sagitaria latfolia	duck-potato
Scirpus acutus	hard-stem bulrush
Scirpus fluviatilis	river bulrush
Scirpus pallidus	pale bulrush
Sparganium eurycarpum	broad-fruit burr-reed
Verbena hastata	simpler's-joy
Vernonia fasciculata	prairie ironweed

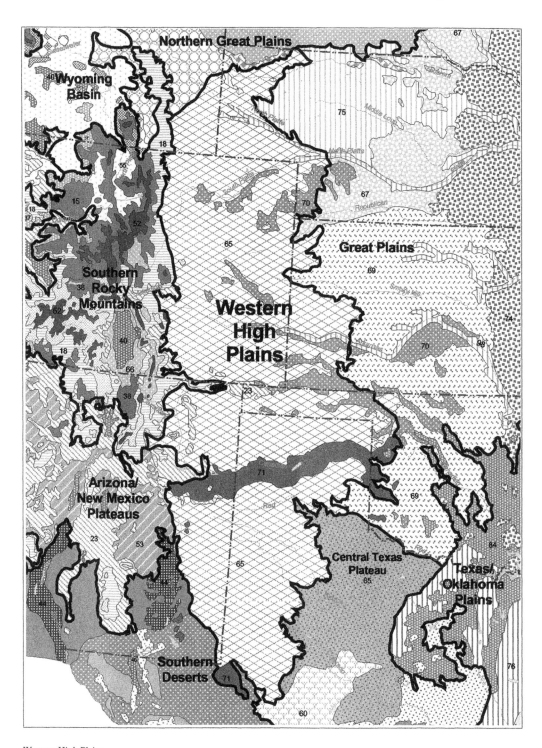

Western High Plains

Western High Plains

Introduction

The Western High Plains region extends from the eastern boundary of the southern Rocky Mountains in Wyoming through the panhandle of Nebraska into western Kansas and south into the panhandle of Oklahoma, Texas, and eastern New Mexico. The characteristic vegetation of the region is Shortgrass Prairie. The most common Shortgrass Prairie species include blue grama and buffalo grass. Unsuccessful attempts to farm large areas of shortgrass prairie have resulted in an increase of annual and perennial weeds to the detriment of the native grasses.

Although the region is mainly Shortgrass Prairie, mixedgrass and tallgrass communities may be found in moist areas and in areas along the eastern boundary where there is greater precipitation. The Nebraska panhandle contains ponderosa pine forests. Since this community is more typical of the area north and west of the high plains, it is described in the Northern Great Plains region. The Canadian River in the panhandle of Texas provides habitat for several uncommon species of oaks and tall grasses.

The primary sources used to develop descriptions and species lists are Barbour and Billings 1988; Dick-Peddie 1992; McMahan et al. 1984.

Dominant Ecological Communities

Upland Systems
> Juniper–Pinyon Woodland
> Shinnery
> Sagebrush–Bluestem Prairie
> Shortgrass Prairie
> Sandhill Prairie

Wetland Systems
> Floodplain Forest

UPLAND SYSTEMS

Juniper–Pinyon Woodland

Juniper–Pinyon Woodland is the southwestern-most community in this region occurring in the foothills of the Rocky Mountains. This community corresponds to Küchler #23 and Woodlands ERT.

Canopy
Characteristic Species

Juniperus monosperma	one-seed juniper
Juniperus osteosperma	Utah juniper
Juniperus scopulorum	Rocky Mountain juniper
Pinus edulis	two-needle pinyon

Associates

Juniperus occidentalis	western juniper
Quercus emoryi	Emory's oak
Quercus gambelii	Gambel's oak
Quercus grisea	gray oak
Quercus mohriana	Mohr's oak

Woody Understory

Artemisia filifolia	silver sagebrush
Artemisia tridentata	big sagebrush
Cerocarpus montanus	alder-leaf mountain-mahogany
Fallugia paradoxa	Apache-plume
Opuntia imbricata	tree cholla
Physocarpus monogynus	mountain ninebark
Purshia mexicana	Mexican cliff-rose
Purshia tridentata	bitterbrush
Rhus aromatica	fragrant sumac
Yucca glauca	soapweed yucca

Herbaceous Understory

Asragalus crassicarpus	ground-plum
Bouteloua curtipendula	side-oats grama
Bouteloua gracilis	blue grama
Bouteloua hirsuta	hairy grama
Buchloë dactyloides	buffalo grass
Carex gravida	heavy sedge
Cenchrus longispinus	innocent-weed
Chloris verticillata	tumble windmill grass
Cryptantha cinerea	James' cat's-eye
Dalea enneandra	nine-anther prairie-clover
Dalea purpurea	violet prairie-clover
Elymus elymoides	western bottle-brush grass
Eriogonum tenellum	tall wild buckwheat
Hilaria jamesii	James' galleta
Hordeum pusillum	little barley
Krascheninnikovia lanata	winterfat
Lithospermum incisum	fringed gromwell
Mentzelia decapetala	gumbo-lily
Mentzelia nuda	goodmother
Mirabilis linearis	narrow-leaf four-o'-clock
Oryzopsis hymenoides	Indian mountain-rice grass
Panicum obtusum	blunt panic grass
Pascopyrum smithii	western-wheat grass
Penstemon albidus	red-line beardtongue
Ribes cereum	white squaw currant
Schizachyrium scoparium	little false bluestem
Solanum elaeagnifolium	silver-leaf nightshade
Sporobolus cryptandrus	sand dropseed
Verbena bracteata	carpet vervain

Shinnery

Shinnery is a savanna community of midgrass prairie species with evergreen trees and shrubs. This community occurs on sandy soils and dunes in the panhandle of Texas and eastern New Mexico in close proximity to the Canadian River which runs through the Western High Plains region of these states. Areas along the Canadian River may also support tallgrass prairie species where there is sufficient water. This community corresponds, in part, to Küchler #65 and Woodlands ERT.

Characteristic Species

Quercus havardii	Havard's oak
Quercus mohriana	Mohr's oak
Schizachyrium scoparium	little false bluestem

Associates

Andropogon gerardii	big bluestem
Andropogon hallii	sand bluestem
Artemisia filifolia	silver sagebrush
Bouteloua curtipendula	side-oats grama
Bouteloua gracilis	blue grama
Buchloë dactyloides	buffalo grass
Calamovilfa gigantea	giant sand-reed
Celtis laevigata	sugar-berry
Eriogonum annuum	annual wild buckwheat
Juniperus pinchotii	Pinchot's juniper
Panicum virgatum	wand panic grass
Prosopis glandulosa	honey mesquite
Prunus angustifolia	chickasaw plum
Quercus pungens	sandpaper oak
Rhus trilobata	ill-scented sumac
Sorghastrum nutans	yellow Indian grass
Sporobolus cryptandrus	sand dropseed
Sporobolus giganteus	giant dropseed
Yucca campestris	plains yucca
Yucca glauca	soapweed yucca

Sagebrush–Bluestem Prairie

Sagebrush–Bluestem Prairie occurs throughout eastern Colorado, in the panhandle of Oklahoma, and in western Kansas. The community is characterized by medium-tall grasses and the dwarf shrub, silver sagebrush (sandsage). Sagebrush-Bluestem Prairie is considered a mixed prairie of both tall and short grasses. This community corresponds to Küchler #70 and Prairies ERT.

Characteristic Species

Andropogon hallii	sand bluestem
Artemisia filifolia	silver sagebrush
Bouteloua hirsuta	hairy grama
Schizachyrium scoparium	little false bluestem

Associates

Bouteloua gracilis	blue grama
Buchloë dactyloides	buffalo grass
Calamovilfa longifolia	prairie sand-reed
Eragrostis trichodes	sand love grass
Helianthus petiolaris	prairie sunflower
Hordeum jubatum	fox-tail barley
Panicum virgatum	wand panic grass
Redfieldia flexuosa	blowout grass
Sporobolus cryptandrus	sand dropseed
Stipa comata	needle-and-thread
Yucca glauca	soapweed yucca

Shortgrass Prairie

Shortgrass Prairie is the most common community in the High Plains region. Blue grama is the most abundant grass and shares dominance with buffalo grass. The extensive shortgrass prairie was once home to herds of buffalo and pronghorn antelope. Much of the area has been converted to agricultural and livestock purposes which has caused irreparable changes in the community structure. This community corresponds in part to Küchler #65 and Prairies ERT.

Characteristic Species

Bouteloua gracilis	blue grama
Buchloë dactyloides	buffalo grass

Associates

Aristida purpurea	purple three-awn
Bouteloua curtipendula	side-oats grama
Bouteloua hirsuta	hairy grama
Elymus elymoides	western bottle-brush grass
Gaura coccinea	scarlet beeblossom
Grindelia squarrosa	curly-cup gumweed
Lycurus phleoides	common wolf's-tail
Machaeranthera pinnatifida	lacy tansy-aster
Muhlenbergia torreyi	ringed muhly
Pascopyrum smithii	western-wheat grass
Plantago patagonica	woolly plantain
Psoralidium tenuiflorum	slender-flower lemonweed
Ratibida columnifera	red-spike Mexican-hat
Sphaeralcea coccinea	scarlet globe-mallow
Sporobolus cryptandrus	sand dropseed
Yucca glauca	soapweed yucca
Zinnia grandiflora	little golden zinnia

Sandhill Prairie

Sandhill Prairie is more characteristic of the Nebraska Sandhills in the Great Plains region. It is of limited extent in this region, confined to several isolated locations near the Platte River valley. This community corresponds to Küchler #75 and Prairies ERT.

Characteristic Species

Andropogon gerardii	big bluestem
Andropogon hallii	sand bluestem
Calamovilfa longifolia	prairic sand-reed
Schizachyrium scoparium	little false bluestem
Stipa comata	needle-and-thread

Associates

Androsace occidentalis	western rock-jasmine
Anemone caroliniana	Carolina thimbleweed
Aristida longespica	red three-awn
Artemisia campestris	pacific wormwood
Asclepias arenaria	sand milkweed
Asclepias lanuginosa	side-cluster milkweed
Asclepias latifolia	broad-leaf milkweed

Asclepias stenophylla	slim-leaved milkweed
Asclepias viridiflora	green comet milkweed
Aster ericoides	white heath aster
Astragalus crassicarpus	ground-plum
Bouteloua curtipendula	side-oats grama
Bouteloua hirsuta	hairy grama
Calylophus serrulatus	yellow sundrops
Carex duriuscula	spike-rush sedge
Carex filifolia	thread-leaf sedge
Carex inops	long-stolon sedge
Chamaesyce geyeri	Geyer's sandmat
Chamaesyce missurica	prairie sandmat
Cirsium canescens	prairie thistle
Collomia linearis	narrow-leaf mountain-trumpet
Commelina virginica	Virginia dayflower
Cymopteris montanus	mountain spring-parsley
Cyperus schweinitzii	sand flat sedge
Dalea candida	white prairie-clover
Dalea purpurea	violet prairie-clover
Dalea villosa	silky prairie-clover
Eragrostis trichodes	sand love grass
Erigeron bellidiastrum	western daisy fleabane
Eriogonum annuum	annual wild buckwheat
Erysimum asperum	plains wallflower
Euphorbia hexagona	six-angle spurge
Froelichia floridana	plains snake-cotton
Heterotheca villosa	hairy false golden-aster
Hymenopappus filifolius	fine-leaf woollywhite
Ipomopsis longiflora	white-flower skyrocket
Koeleria macrantha	prairie Koeler's grass
Lespedeza capitata	round-head bush-clover
Leucocrinum montanum	star-lily
Liatris squarrosa	scaly gayfeather
Linum rigidum	large-flower yellow flax
Lithospermum caroliniense	hairy puccoon
Lithospermum incisum	fringed gromwell
Machaeranthera canescens	hoary tansy-aster
Machaeranthera pinnatifida	lacy tansy-aster
Mirabilis hirsuta	hairy four-o'clock
Mirabilis linearis	narrow-leaf four-o'clock
Oenothera pallida	white-pole evening-primrose
Oenothera rhombipetala	greater four-point evening-primrose
Opuntia humifusa	eastern prickly-pear
Orobanche fasciculata	clustered broom-rape
Oryzopsis hymenoides	Indian mountain-rice grass
Oxytropis lamberti	stemless locoweed
Panicum virgatum	wand panic grass
Paspalum setaceum	slender crown grass
Pediomelum cuspidatum	large-bract Indian-breadroot

Pediomelum esculentum	large Indian-breadroot
Penstemon albidus	red-line beardtongue
Penstemon angustifolius	broad-beard beardtongue
Penstemon gracilis	lilac beardtongue
Phlox caespitosa	clustered phlox
Polygala alba	white milkwort
Polygala verticillata	whorled milkwort
Senecio plattensis	prairie ragwort
Sisyrinchium angustifolium	narrow-leaf blue-eyed-grass
Sorghastrum nutans	yellow Indian grass
Sporobolus asper	tall dropseed
Sporobolus cryptandrus	sand dropseed
Talinum teretifolium	quill fameflower
Thelesperma megapotamicum	hopi-tea
Tradescantia occidentalis	prairie spiderwort

WETLAND SYSTEMS

Floodplain Forest

Floodplain Forest occurs along major rivers including the Missouri, Little Missouri, Yellowstone, and White Rivers. This community corresponds to Floodplain Forests ERT.

Canopy
Characteristic Species
Populus deltoides	eastern cottonwood
Salix nigra	black willow
Ulmus americana	American elm

Associates
Acer negundo	ash-leaf maple
Acer rubrum	red maple
Acer saccharinum	silver maple
Betula nigra	river birch
Celtis occidentalis	common hackberry
Fraxinus americana	white ash
Fraxinus pennsylvanica	green ash
Gleditsia triacanthos	honey-locust
Juglans nigra	black walnut
Platanus occidentalis	American sycamore
Prunus virginiana	choke cherry
Quercus macrocarpa	burr oak
Ulmus rubra	slippery elm

Woody Understory
Amorpha fruticosa	false Indigo-bush
Celastrus scandens	American bittersweet
Clematis virginiana	devil's-darning-needles
Parthenocissus quinquefolia	Virginia-creeper
Rhus glabra	smooth sumac

Ribes missouriense	Missouri gooseberry
Salix amygdaloides	peach-leaf willow
Salix interior	sandbar willow
Smilax tamnoides	chinaroot
Symphoricarpos occidentalis	western snowberry
Symphoricarpos orbiculatus	coral-berry
Toxicodendron radicans	eastern poison-ivy
Vitis riparia	river-bank grape
Vitis vulpina	frost grape

Herbaceous Understory

Elymus virginicus	Virginia wild rye
Galium aparine	sticky-willy
Laportea canadensis	Canada wood-nettle

Natural Regions and Dominant Ecological Communities

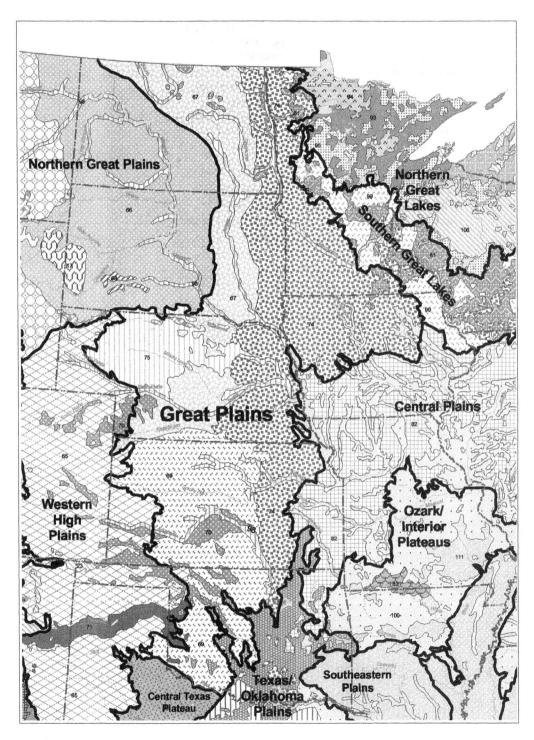

Great Plains

Great Plains

Introduction

The Great Plains extend from southern Manitoba, Canada, through the eastern Dakotas, western Minnesota, northern Iowa, Nebraska, Kansas, and western Oklahoma. This region is characterized by tallgrass and mixed grass prairie communities on flat and rolling plains. The boundaries of the region are the forested areas to the east and the shortgrass plains to the west.

The Great Plains support several kinds of prairie communities. Big bluestem, Indian grass, and yellow Indian grass are characteristic throughout the region in relatively moist areas. In the areas of the tall grasses, big bluestem sometimes reaches heights of eight feet above the ground with roots extending equally deep into the soil. Precipitation tends to decrease, evaporation tends to increase, and the grasses become shorter and less dense in a gradient from east to west across the grasslands. Where moisture is less abundant and evaporation greater, little false bluestem, dropseed, and side-oats grama are characteristic. These mid-height grasses generally grow in bunches or clumps two to four feet tall. The prairie soils in this region are very dark and rich with humus topsoil.

Other communities in the region include Northern Floodplain Forest along major rivers and streams, Oak–Hickory Forest and Oak Barrens along the eastern prairie border, and the Cross Timbers Woodland to the south. Oak–Hickory Forest and Oak Barrens are described in the Central Plains and Southern Great Lakes regions. Throughout the area, marsh and wet meadow communities occur in depressions. The area around the Minnesota-South Dakota border is notable because of the numerous small marshes called "marsh or prairie pot-holes."

Prairies were once widespread throughout the region, but, today, only small patches of prairie remain. Wild grazing animals such as the bison and pronghorn antelope were an integral component of the native prairie ecosystem. Natural fires were also very important in the maintenance of the prairie. Today, most of the region has been converted to agriculture.

The primary sources used to develop descriptions and species lists are Barbour and Billings 1988; Boon and Groe 1990; Pool 1913; Weaver 1954.

Dominant Ecological Communities

Upland Systems
Bluestem Prairie
Western Bluestem Prairie
Sandhill Prairie
Mixed Grass Prairie
Mixed Grass–Shrub Prairie

Wetland Systems
Northern Floodplain Forest
Wet Meadow
Pothole Marsh

UPLAND SYSTEMS

Bluestem Prairie

Bluestem Prairie occurs from North Dakota and Minnesota south to Iowa, Nebraska, Kansas, and Oklahoma. It is often referred to as the "True Prairie" and is one of three recognized "tallgrass" prairie types. Bluestem Prairie is dominated by big bluestem, wand panic grass, and yellow Indian grass. Bluestem Prairie has largely been converted to agricultural purposes. Suppression of fire has also encouraged the invasion of shrubs and trees into the prairie. This community corresponds to Küchler #74 and Prairies ERT.

Characteristic Species

Andropogon gerardii	big bluestem
Sorghastrum nutans	yellow Indian grass

Associates

Amorpha canescens	leadplant
Antennaria neglecta	field pussytoes
Artemisia ludoviciana	white sagebrush
Asclepias ovalifolia	oval-leaf milkweed
Asclepias speciosa	showy milkweed
Aster ericoides	white heath aster
Aster laevis	smooth blue aster
Astragalus crassicarpus	ground-plum
Baptisia alba	white wild indigo
Baptisia bracteata	long-bract wild indigo
Bouteloua curtipendula	side-oats grama
Cirsium flodmanii	Flodman's thistle
Dalea candida	white prairie-clover
Dalea purpurea	violet prairie-clover
Delphinium carolinianum	Carolina larkspur
Dichanthelium leibergii	Leiberg's rosette grass
Dichanthelium oligosanthes	Heller's rosette grass
Echinacea angustifolia	blacksamson
Erigeron strigosus	prairie fleabane
Galium boreale	northern bedstraw
Galium tinctorium	stiff marsh bedstraw
Helianthus grosseserratus	saw-tooth sunflower
Helianthus maximiliani	Michaelmas-daisy
Helianthus pauciflorus	stiff sunflower
Heuchera richardsonii	Richardson's alumroot
Koeleria macrantha	prairie Koeler's grass
Liatris aspera	tall gayfeather
Liatris punctata	dotted gayfeather
Liatris scariosa	devil's-bite
Lilium philadelphicum	wood lily
Lithospermum canescens	hoary puccoon
Panicum virgatum	wand panic grass
Pedicularis canadensis	Canadian lousewort
Pediomelum argophyllum	silver-leaf Indian-breadroot
Pediomelum esculentum	large Indian-breadroot

Phlox pilosa	downy phlox
Potentilla arguta	tall cinquefoil
Prenanthes racemosa	purple rattlesnake-root
Psoralidium tenuiflorum	slender-flower lemonweed
Ratibida columnifera	red-spike Mexican-hat
Ratibida pinnata	gray-head Mexican-hat
Rosa arkansana	prairie rose
Schizachyrium scoparium	little false bluestem
Silphium laciniatum	compassplant
Solidago canadensis	Canadian goldenrod
Solidago missouriensis	Missouri goldenrod
Solidago rigida	hard-leaf goldenrod
Spartina pectinata	freshwater cord grass
Sporobolus heterolepis	prairie dropseed
Stipa spartea	porcupine grass
Viola pedatifida	crow-foot violet
Zigadenus elegans	mountain deathcamas
Zizia aptera	heart-leaf alexanders

Western Bluestem Prairie

The western edge of the "true prairie" is indicated by the abundance of western-wheat grass and porcupine grass (needle grass). This tallgrass prairie community originally covered an area approximately thirty-five to forty-five miles wide extending from North Dakota into South Dakota and Nebraska. This community corresponds to Küchler #67 and Prairies ERT.

Characteristic Species

Pascopyrum smithii	western-wheat grass
Schizachyrium scoparium	little false bluestem
Stipa spartea	porcupine grass

Associates

Andropogon gerardii	big bluestem
Artemisia frigida	prairie sagebrush
Artemisia ludoviciana	white sagebrush
Aster ericoides	white heath aster
Bouteloua curtipendula	side-oats grama
Bouteloua gracilis	blue grama
Echinacea angustifolia	blacksamson
Elymus trachycaulus	slender wild rye
Koeleria macrantha	prairie Koeler's grass
Liatris punctata	dotted gayfeather
Nassella viridula	green tussock grass
Pediomelum argophyllum	silver-leaf Indian-breadroot
Rosa arkansana	prairie rose
Solidago missouriensis	Missouri goldenrod
Solidago mollis	velvet goldenrod
Stipa comata	needle-and-thread

Sandhill Prairie

Sandhill Prairie occurs in areas in central and south-central Kansas and widely in the Sandhills region in Nebraska. Sandhill Prairie is the third recognized tallgrass prairie type. Unique in the Nebraska Sandhills are formations on the hillsides referred to as "blow-outs." Plants associated with dune stabilization can be found in "blow-out" areas. The community corresponds to Prairies ERT.

Characteristic Species

Andropogon gerardii	big bluestem
Andropogon hallii	sand bluestem
Calamovilfa longifolia	prairie sand-reed
Schizachyrium scoparium	little false bluestem
Stipa comata	needle-and-thread

Associates

Androsace occidentalis	western rock-jasmine
Anemone caroliniana	Carolina thimbleweed
Aristida longespica	red three-awn
Artemisia campestris	Pacific wormwood
Asclepias amplexicaulis	clasping milkweed
Asclepias arenaria	sand milkweed
Asclepias lanuginosa	side-cluster milkweed
Asclepias latifolia	broad-leaf milkweed
Asclepias stenophylla	slim-leaf milkweed
Asclepias viridiflora	green comet milkweed
Aster ericoides	white heath aster
Astragalus crassicarpus	ground-plum
Bouteloua curtipendula	side-oats grama
Bouteloua hirsuta	hairy grama
Calylophus serrulatus	yellow sundrops
Carex duriuscula	spike-rush sedge
Carex filifolia	thread-leaf sedge
Carex inops	long-stolon sedge
Chamaesyce geyeri	Geyer's sandmat
Chamaesyce missurica	prairie sandmat
Cirsium canescens	prairie thistle
Collomia linearis	narrow-leaf mountain-trumpet
Commelina virginica	Virginia dayflower
Cymopterus montanus	mountain spring-parsley
Cyperus schweinitzii	sand flat sedge
Dalea candida	white prairie-clover
Dalea purpurea	violet prairie-clover
Dalea villosa	silky prairie-clover
Eragrostis trichodes	sand lover grass
Erigeron bellidiastrum	western daisy fleabane
Eriogonum annuum	annual wild buckwheat
Erysimum asperum	plains wallflower
Euphorbia hexagona	six-angle spurge
Froelichia floridana	plains snake-cotton
Heterotheca villosa	hairy false golden-aster
Hymenopappus filifolius	fine-leaf woollywhite

Ipomopsis longiflora	white-flower skyrocket
Koeleria macrantha	prairie Koeler's grass
Lespedeza capitata	round-head bush-clover
Leucocrinum montanum	star-lily
Liatris squarrosa	scaly gayfeather
Linum rigidum	large-flower yellow flax
Lithospermum caroliniense	hairy puccoon
Lithospermum incisum	fringed gromwell
Machaeranthera canescens	hoary tansy-aster
Machaeranthera pinnatifida	lacy tansy-aster
Mirabilis hirsuta	hairy four-o'clock
Mirabilis linearis	narrow-leaf four-o'clock
Oenothera pallida	white-pole evening-primrose
Oenothera rhombipetala	greater four-point evening-primrose
Opuntia humifusa	eastern prickly-pear
Orobanche fasciculata	clustered broom-rape
Oryzopsis hymenoides	Indian mountain-rice grass
Oxytropis lamberti	stemless locoweed
Panicum virgatum	wand panic grass
Paspalum setaceum	slender crown grass
Pediomelum cuspidatum	large-bract Indian-breadroot
Pediomelum esculentum	large Indian-breadroot
Penstemon albidus	red-line beardtongue
Penstemon angustifolius	broad-beard beardtongue
Penstemon gracilis	lilac beardtongue
Phlox caespitosa	clustered phlox
Polygala alba	white milkwort
Polygala verticillata	whorled milkwort
Senecio plattensis	prairie ragwort
Sisyrinchium angustifolium	narrow-leaf blue-eyed-grass
Sorghastrum nutans	yellow Indian grass
Sporobolus asper	tall dropseed
Sporobolus cryptandrus	sand dropseed
Talinum teretifolium	quill fameflower
Thelesperma megapotamicum	Hopi-tea
Tradescantia occidentalis	prairie spiderwort

Mixed Grass Prairie

Mixed Grass Prairie occurs in Kansas and Oklahoma west of the more mesic Bluestem and Western Bluestem Prairies. Mixed Grass Prairie is considered a "mixed prairie" due to the occurrence of both tall and short grasses. Because mixed grass prairies occur as a transition zone between the short and tall grass types, they support a very diverse mix of species. The transitional nature of mixed prairies makes them the most floristically rich of all the grasslands. Grazing pressure and drought have caused many changes in the original vegetation. This community corresponds to Küchler #69 and Prairies ERT.

Characteristic Species

Bouteloua curtipendula	side-oats grama
Bouteloua gracilis	blue grama
Schizachyrium scoparium	little false bluestem

Associates

Ambrosia psilostachya	western ragwort
Amorpha canescens	leadplant
Andropogon gerardii	big bluestem
Buchloë dactyloides	buffalo grass
Calylophus serrulatus	yellow sundrops
Clematis fremontii	Fremont's leather-flower
Dalea enneandra	nine-anther prairie-clover
Echinacea angustifolia	blacksamson
Erysimum asperum	plains wallflower
Hedeoma hispidum	rough false pennyroyal
Liatris punctata	dotted gayfeather
Panicum virgatum	wand panic grass
Paronychia jamesii	James' nailwort
Pascopyrum smithii	western-wheat grass
Psoralidium tenuiflorum	slender-flower lemonweed
Scutellaria resinosa	resin-dot skullcap
Sorghastrum nutans	yellow Indian grass
Sporobolus asper	tall dropseed
Stenosiphon linifolius	false gaura

Mixed Grass–Shrub Prairie

Mixed Grass–Shrub Prairie occurs in southcentral Nebraska, Kansas, and northwestern Oklahoma. The community is characterized by medium-tall grasses and the dwarf shrub, silver sagebrush (sandsage). This prairie community is considered a mixed prairie, having both tall and short grasses. This community corresponds to Küchler #70 and Prairies ERT.

Characteristic Species

Andropogon hallii	sand bluestem
Artemisia filifolia	silver sagebrush
Bouteloua hirsuta	hairy grama
Schizachyrium scoparium	little false bluestem

Associates

Bouteloua gracilis	blue grama
Buchloë dactyloides	buffalo grass
Calamovilfa longifolia	prairie sand-reed
Eragrostis trichodes	sand love grass
Helianthus petiolaris	prairie sunflower
Hordeum jubatum	fox-tail barley
Panicum virgatum	wand panic grass
Redfieldia flexuosa	blowout grass
Sporobolus cryptandrus	sand dropseed
Stipa comata	needle-and-thread
Yucca glauca	soapweed yucca

WETLAND SYSTEMS
Northern Floodplain Forest

Many rivers dissect the eastern Great Plains region. The Red, James, Platt, Kansas, and Arkansas Rivers are among the largest streams that support Northern Floodplain Forest. Floodplain species tolerate both flooding and drought. This community corresponds to Küchler #98 and Floodplain Forests ERT.

Canopy
Characteristic Species

Populus deltoides	eastern cottonwood
Salix nigra	black willow
Ulmus americana	American elm

Associates

Acer negundo	ash-leaf maple
Acer rubrum	red maple
Acer saccharinum	silver maple
Betula nigra	river birch
Celtis occidentalis	common hackberry
Fraxinus americana	white ash
Fraxinus pennsylvanica	green ash
Gleditsia triacanthos	honey-locust
Juglans nigra	black walnut
Platanus occidentalis	American sycamore
Prunus virginiana	choke cherry
Quercus macrocarpa	burr oak
Ulmus rubra	slippery elm

Woody Understory

Amorpha fruticosa	false indigo-bush
Celastrus scandens	American bittersweet
Clematis virginiana	devil's-darning-needles
Parthenocissus quinquefolia	Virgina-creeper
Rhus glabra	smooth sumac
Ribes missouriense	Missouri gooseberry
Salix amygdaloides	peach-leaf willow
Salix interior	sandbar willow
Smilax tamnoides	chinaroot
Symphoricarpos occidentalis	western snowberry
Symphoricarpos orbiculatus	coral-berry
Toxicodendron radicans	eastern poison-ivy
Vitis riparia	river-bank grape
Vitis vulpina	frost grape

Herbaceous Understory

Elymus virginicus	Virginia wild rye
Galium aparine	sticky-willy
Laportea canadensis	Canadian wood-nettle

Wet Meadow

Wet Meadow occurs scattered throughout the eastern Great Plains region. Wet Meadow occupies areas slightly higher than Pothole Marsh. The soils of Wet Meadow are usually wet and soggy for most of the year. This community corresponds to Meadowlands ERT.

Characteristic Species

Carex trichocarpa	hairy-fruit sedge
Eleocharis acicularis	needle spike-rush
Eleocharis compressa	flat-stem spike-rush
Eleocharis palustris	pale spike-rush
Juncus balticus	Baltic rush
Juncus marginatus	grass-leaf rush
Juncus nodosus	knotted rush
Juncus tenuis	poverty rush
Juncus torreyi	Torrey's rush
Scirpus americanus	chairmaker's bulrush

Associates

Agalinis tenuifolia	slender-leaf false foxglove
Agrostis gigantea	black bent
Agrostis hyemalis	winter bent
Alopecurus aequalis	short-awn meadow foxtail
Aureolaria virginica	downy yellow false-foxglove
Calamagrostis canadensis	bluejoint
Calamagrostis stricta	slim-stem reedgrass
Caltha palustris	yellow marsh-marigold
Campanula aparinoides	marsh bellflower
Carex festucacea	fescue sedge
Carex gravida	heavy sedge
Carex hystericina	porcupine sedge
Carex lanuginosa	woolly sedge
Carex nebrascensis	Nebraska sedge
Carex praegracilis	clustered field sedge
Carex scoparia	pointed broom sedge
Carex stricta	uptight sedge
Catabrosa aquatica	water whorl grass
Cicuta maculata	spotted water-hemlock
Crepis runcinata	fiddle-leaf hawk's-beard
Cyperus squarrosus	awned flat sedge
Distichlis spicata	coastal salt grass
Dodecatheon meadia	eastern shootingstar
Epilobium palustre	marsh willowherb
Eriophorum gracile	slender cotton-grass
Galium trifidum	three-petal bedstraw
Galium triflorum	fragrant bedstraw
Gentiana andrewsii	closed bottle gentian
Gentiana saponaria	harvestbells
Glyceria striata	fowl manna grass
Helianthus tuberosus	Jerusalem-artichoke

Hordeum jubatum	fox-tail barley
Hypericum majus	greater Canadian St. John's-wort
Hypoxis hirsuta	eastern yellow star-grass
Juncus longistylis	long-style rush
Lilium philadelphicum	wood lily
Liparis loeselii	yellow wide-lip orchid
Lobelia siphilitica	great blue lobelia
Lotus unifoliolatus	American bird's-foot-trefoil
Lycopus asper	rough water-horehound
Lysimachia ciliata	fringed yellow-loosestrife
Lysimachia thyrsiflora	tufted yellow-loosestrife
Lythrum alatum	wing-angle loosestrife
Mentha arvensis	American wild mint
Menyanthes trifoliata	buck-bean
Plantago eriopoda	red-woolly plantain
Platanthera leucophaea	prairie white fringed orchid
Platanthera praeclara	Great Plains white fringed orchid
Potentilla paradoxa	bushy cinquefoil
Ranunculus cymbalaria	alkali buttercup
Ranunculus sceleratus	cursed crowfoot
Scirpus pallidus	pale bulrush
Scutellaria galericulata	hooded skullcap
Spartina cynosuroides	big cord grass
Spiranthes cernua	white nodding ladies'-tresses
Spiranthes romanzoffiana	hooded ladies'-tresses
Stachys palustris	woundwort
Strophostyles leiosperma	slick-seed fuzzy-bean
Teucrium canadense	American germander
Triadenum virginicum	Virginia marsh-St. John's-wort
Triglochin maritimum	seaside arrow-grass

Pothole Marsh

Pothole Marsh extends from northern Iowa through western Minnesota and the eastern half of South Dakota into the eastern and central portion of North Dakota and north into Canada. Pothole Marsh is a very important nesting habitat for ducks in North America. It is found on poorly drained soils of lowland mesic prairie. The outer margins of the pothole depressions are dominated by cord grass mixed with other species such as swamp milkweed and ironweed. The centers are typically dominated by smartweed which may be mixed with duck-potato and cat-tail. Some potholes may be temporary, representing a stage of succession. Draining activities have greatly diminished the extent of these valuable wetlands. This community corresponds to Nonestuarine Marshes ERT.

Characteristic Species

Phragmites australis	common reed
Scirpus acutus	hard-stem bulrush
Scirpus americanus	chairmaker's bulrush
Scirpus tabernaemontani	soft-stem bulrush
Typha latifolia	broad-leaf cat-tail
Zizania aquatica	Indian wild rice

Associates

Alisma subcordatum	American water-plantain
Asclepias incarnata	swamp milkweed
Berula erecta	cut-leaf water-parsnip
Bidens frondosa	devil's pitchfork
Eleocharis acicularis	needle spike-rush
Eleocharis palustris	pale spike-rush
Equisetum hyemale	tall scouring-rush
Lobelia siphilitica	great blue lobelia
Polygonum amphibium	water smartweed
Sagittaria latifolia	duck-potato
Scirpus fluviatilis	river bulrush
Scirpus pallidus	pale bulrush
Spartina cynosuroides	big cord grass
Spartina pectinata	freshwater cord grass
Verbena hastata	simpler's-joy
Vernonia fasciculata	prairie ironweed

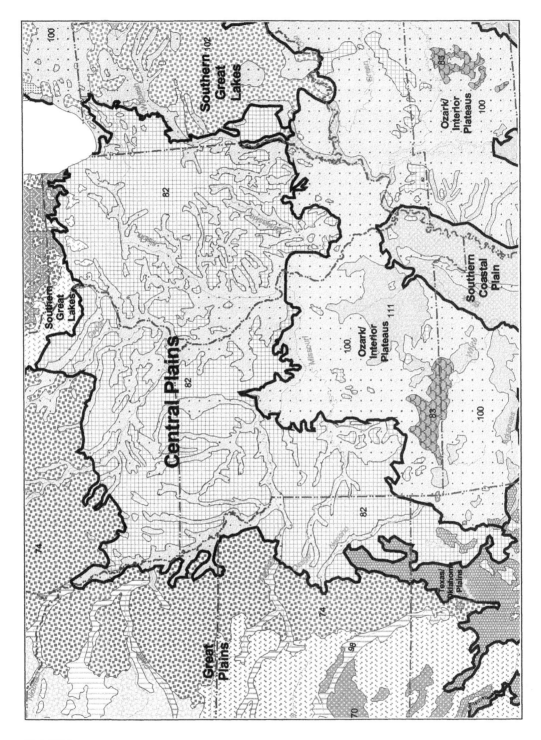

Central Plains

Central Plains

Introduction

The Central Plains region extends from extreme western Indiana west to southwestern Iowa and eastern Kansas. The region encompasses most of Illinois, southern Iowa, northern and western Missouri, parts of eastern Kansas, and northern Oklahoma.

At the time of settlement, most of this region was covered in a mosaic of Bluestem Prairie, Oak Barrens (savanna), and Oak–Hickory Forest. Floodplain Forest, Wet Prairie, and Marsh occupied the broad river floodplains. Periodic fires were important in maintaining the prairie and barrens.

Today, very little of the natural vegetation remains. Almost all of the prairie and forest have been destroyed, primarily for agricultural purposes.

The primary sources used to develop descriptions and species lists are Barbour and Billings 1988; Braun 1950.

Dominant Ecological Communities

Upland Systems
Bluestem Prairie
Oak–Hickory Forest
Oak Barrens
Wetland Systems
Floodplain Forest
Marsh
Wet Prairie

UPLAND SYSTEMS

Bluestem Prairie

Bluestem Prairie dominated this region at the time of settlement. Vast tracts of grassland covered much of the uplands between drainages. Many of these prairies were rich and mesic, but drier prairies occurred on sandy, rocky, or gravelly soil. The dominant grasses were big bluestem, yellow Indian grass, and wand panic grass (switchgrass). This community corresponds to Küchler #74 and Prairies ERT.

Characteristic Species
Andropogon gerardii	big bluestem
Panicum virgatum	wand panic grass
Schizachyrium scoparium	little false bluestem
Sorghastrum nutans	yellow Indian grass

Associates
Allium cernuum	nodding onion
Amorpha canescens	leadplant
Asclepias tuberosa	butterfly milkweed
Asclepias viridiflora	green comet milkweed
Aster ericoides	white heath aster
Aster laevis	smooth blue aster

Aster sericeus	western silver aster
Baptisia alba	white wild indigo
Baptisia leucophaea	long-bract wild indigo
Bouteloua curtipendula	side-oats grama
Ceanothus americanus	New Jersey-tea
Dalea candida	white prairie-clover
Dalea purpurea	violet prairie-clover
Desmanthus illinoensis	prairie bundle-flower
Dodecatheon meadia	eastern shootingstar
Echinacea pallida	pale purple-coneflower
Elymus canadensis	nodding wild rye
Eryngium yuccifolium	button eryngo
Gaura parviflora	velvetweed
Gentiana puberulenta	downy gentian
Helianthus grosseserratus	saw-tooth sunflower
Helianthus mollis	neglected sunflower
Heliopsis helianthoides	smooth oxeye
Lespedeza capitata	round-head bush-clover
Liatris aspera	tall gayfeather
Liatris pycnostachya	cat-tail gayfeather
Lobelia spicata	pale-spike lobelia
Parthenium integrifolium	wild quinine
Phlox pilosa	downy phlox
Potentilla arguta	tall cinquefoil
Ratibida pinnata	gray-head Mexican-hat
Rudbeckia hirta	black-eyed-susan
Silphium laciniatum	compass plant
Silphium terebinthinaceum	prairie rosinweed
Solidago missouriensis	Missouri goldenrod
Solidago rigida	hard-leaf goldenrod
Sporobolus asper	tall dropseed
Sporobolus heterolepis	prairie dropseed
Stipa spartea	porcupine grass
Veronicastrum virginicum	culver's-root
Zizia aptera	heart-leaf alexanders

Oak–Hickory Forest

Oak–Hickory Forest occurs scattered throughout the region on dry to dry-mesic uplands. Dominance and species composition varies in response to topography, soils, geology, and disturbance history. This community corresponds to Küchler #100 and Western Deciduous and Mixed Forests ERT.

Canopy
Characteristic Species

Carya glabra	pignut hickory
Carya ovata	shag-bark hickory
Carya tomentosa	mockernut hickory
Quercus alba	northern white oak
Quercus rubra	northern red oak
Quercus velutina	black oak

Associates

Acer rubrum	red maple
Acer saccharum	sugar maple
Carya ovalis	red hickory
Carya texana	black hickory
Diospyros virginiana	common persimmon
Fraxinus americana	white ash
Fraxinus quadrangulata	blue ash
Nyssa sylvatica	black tupelo
Pinus echinata	short-leaf pine
Quercus coccinea	scarlet oak
Quercus falcata	southern red oak
Quercus marlandica	blackjack oak
Quercus muhlenbergii	chinkapin oak
Quercus prinoides	dwarf chinkapin oak
Quercus shumardii	Shumard's oak
Quercus stellata	post oak
Ulmus rubra	slippery elm

Woody Understory

Amelanchier arborea	downy service-berry
Ceanothus americanus	New Jersey tea
Cercis canadensis	redbud
Cornus drummondii	rough-leaf dogwood
Cornus florida	flowering dogwood
Cornus racemosa	gray dogwood
Parthenocissus quinquefolia	Virginia-creeper
Rhus aromatica	fragrant sumac
Ulmus alata	winged elm
Vaccinium arboreum	tree sparkle-berry
Vaccinium pallidum	early lowbush blueberry
Vitis aestivalis	summer grape

Herbaceous Understory

Brachyelytrum erectum	bearded shorthusk
Carex pensylvanica	Pennsylvania sedge
Cunila marina	common dittany
Danthonia spicata	poverty wild oat grass
Desmodium nudiflorum	naked-flower tick-trefoil
Frasera caroliniensis	American-columbo
Galium concinnum	shining bedstraw
Galium pilosum	hairy bedstraw
Hieracium gronovii	queendevil
Podophyllum peltatum	may-apple
Prenanthes alba	white rattlesnake-root
Prenanthes altissima	tall rattlesnake-root
Solidago hispida	hairy goldenrod
Tradescantia virginiana	Virginia spiderwort

Oak Barrens

Oak Barrens is a widespread woodland community in this region. This community type is often considered a transition zone between prairie and forest. The herbaceous understory is made up primarily of prairie grasses and forbs. Periodic fires are important in maintaining the savanna-like conditions of this community. This community corresponds to Woodlands ERT.

Canopy
Characteristic Species
Quercus alba	northern white oak
Quercus macrocarpa	burr oak
Quercus marilandica	blackjack oak
Quercus stellata	post oak

Associates
Carya laciniosa	shell-bark hickory
Carya ovata	shag-bark hickory
Carya texana	black hickory
Juniperus virginiana	eastern red cedar
Quercus imbricaria	shingle oak
Quercus velutina	black oak

Woody Understory
Amorpha canescens	leadplant
Ceanothus americanus	New Jersey-tea
Ceanothus herbaceus	prairie redroot
Cornus drummondii	rough-leaf dogwood
Cornus racemosa	gray dogwood
Corylus americana	American hazelnut
Crataegus spp.	hawthorn
Malus ioensis	prairie crabapple
Rhus aromatica	fragrant sumac
Rosa carolina	Carolina rose
Salix humilis	small pussy willow

Herbaceous Understory
Andropogon gerardii	big bluestem
Antennaria neglecta	field pussytoes
Asclepias verticillata	whorled milkweed
Aster drummondii	Drummond's aster
Aster ericoides	white heath aster
Aster laevis	smooth blue aster
Baptisia alba	white wild indigo
Cacalia atriplicifolia	pale Indian-plantain
Camassia scilloides	Atlantic camas
Carex bicknellii	Bicknell's sedge
Claytonia virginica	Virginia springbeauty
Comandra umbellata	bastard-toadflax
Coreopsis tripteris	tall tickseed
Cyperus filiformis	wiry flat sedge

Desmodium illinoense	Illinois tick-trefoil
Dichanthelium oligosanthes	Heller's rosette grass
Elymus canadensis	nodding wild rye
Eragrostis spectabilis	petticoat-climber
Eryngium yuccifolium	button eryngo
Euphorbia corollata	flowering spurge
Fragaria virginiana	Virginia strawberry
Gentiana puberulenta	downy gentian
Helianthus occidentalis	few-leaf sunflower
Lespedeza capitata	round-head bush-clover
Lespedeza violacea	violet bush-clover
Lobelia spicata	pale-spike lobelia
Monarda fistulosa	Oswego-tea
Parthenium integrifolium	wild quinine
Pycnanthemum virginianum	Virginia mountain-mint
Ranunculus fascicularis	early buttercup
Ratibida pinnata	gray-head Mexican-hat
Rudbeckia hirta	black-eyed-Susan
Schizachyrium scoparium	little false bluestem
Silene regia	royal catchfly
Silene stellata	widow's-frill
Silphium integrifolium	entire-leaf rosinweed
Sisyrinchium albidum	white blue-eyed-grass
Solidago nemoralis	gray goldenrod
Solidago rigida	hard-leaf goldenrod
Sorghastrum nutans	Yellow Indian grass
Stipa spartea	porcupine grass
Tephrosia virginiana	goat's-rue
Tradescantia ohiensis	bluejacket
Verbena stricta	hoary vervain
Viola sororia	hooded blue violet

WETLAND SYSTEMS

Floodplain Forest

Floodplain Forest occurs throughout the region along large streams and rivers where floodplains have developed. The canopy is composed of a variety of large deciduous trees and the understory and herbaceous layer ranges from well developed to poorly developed depending upon the duration of annual flooding. This community corresponds to Küchler #98 and Floodplain Forests ERT.

Canopy
Characteristic Species

Acer saccharinum	silver maple
Betula nigra	river birch
Carya laciniosa	shell-bark hickory
Celtis occidentalis	common hackberry
Liquidambar styraciflua	sweet-gum

Populus deltoides	eastern cottonwood
Quercus bicolor	swamp white oak
Quercus lyrata	overcup oak
Quercus macrocarpa	burr oak
Quercus pagoda	cherry-bark oak
Quercus palustris	pin oak
Ulmus americana	American elm
Ulmus rubra	slippery elm

Associates

Acer negundo	ash-leaf maple
Acer rubrum	red maple
Carya cordiformis	bitter-nut hickory
Fraxinus pennsylvanica	green ash
Gleditsia triacanthos	honey locust
Juglans nigra	black walnut
Nyssa sylvatica	black tupelo
Platanus occidentalis	American sycamore
Quercus michauxii	swamp chestnut oak
Salix nigra	black willow

Woody Understory

Arundinaria gigantea	giant cane
Asimina triloba	common pawpaw
Campsis radicans	trumpet-creeper
Carpinus caroliniana	American hornbeam
Cornus foemina	stiff dogwood
Ilex decidua	deciduous holly
Lindera benzoin	northern spicebush
Toxicodendron radicans	eastern poison-ivy
Vitis spp.	grape

Herbaceous Understory

Boehmeria cylindrica	small-spike false nettle
Carex grayi	Gray's sedge
Carex tribuloides	blunt broom sedge
Cinna arundinacea	sweet wood-reed
Impatiens capensis	spotted touch-me-not
Laportea canadensis	Canadian wood-nettle
Mertensia virginica	Virginia bluebells
Onoclea sensibilis	sensitive fern
Phacelia purshii	Miami-mist
Sanicula canadensis	Canadian black-snakeroot
Viola cucullata	marsh blue violet

Marsh

Marsh is primarily associated with floodplains of larger streams and rivers although it also occurs in upland depressions. This community corresponds to Nonestuarine Marshes ERT.

Characteristic Species

Alisma subcordatum	American water-plantain
Boltonia asteroides	white doll's-daisy
Carex hyalinolepis	shoreline sedge
Carex lupulina	hop sedge
Carex tribuloides	blunt broom sedge
Carex vulpinoidea	common fox sedge
Eleocharis acicularis	needle spike-rush
Eleocharis obtusa	blunt spike-rush
Hibiscus laevis	halberd-leaf rose-mallow
Iris virginica	Virginia blueflag
Juncus effusus	lamp rush
Leersia oryzoides	rice cut grass
Leersia virginica	white grass
Nelumbo lutea	American lotus
Nuphar lutea	yellow pond-lily
Polygonum amphibium	water smartweed
Polygonum hydropiperoides	swamp smartweed
Polygonum pensylvanicum	pinkweed
Rumex verticillatus	swamp dock
Sagittaria latifolia	duck-potato
Scirpus americanus	chairmaker's bulrush
Scirpus atrovirens	dark-green bulrush
Scirpus cyperinus	cottongrass bulrush
Scirpus tabernaemontani	soft-stem bulrush
Sparganium eurycarpum	broad-fruit burr-reed
Typha latifolia	broad-leaf cat-tail

Wet Prairie

Wet prairie was often associated with marsh communities in floodplains of larger streams and in poorly drained depressions in the uplands. These grass-dominated wetlands are essentially eradicated with most areas converted to cropland. The community corresponds to Wet Prairies and Grasslands ERT.

Characteristic Species

Amsonia illustris	Ozark bluestar
Asclepias incarnata	swamp milkweed
Aster umbellatus	parasol flat-top white aster
Calamogrostis canadensis	bluejoint
Caltha palustris	yellow marsh-marigold
Carex arkansana	Arkansas sedge
Carex bicknellii	Bicknell's sedge
Carex lacustris	lakebank sedge
Carex vulpinoidea	common fox sedge
Cephalanthus occidentalis	common buttonbush

Cicuta maculata	spotted water-hemlock
Eupatorium maculatum	spotted joe-pye-weed
Helianthus angustifolius	swamp sunflower
Helianthus grosseserratus	saw-tooth sunflower
Juncus interior	inland rush
Leersia oryzoides	rice cut grass
Lysimachia quadrifolia	whorled yellow-loosestrife
Panicum virgatum	wand panic grass
Physostegia virginiana	obedient plant
Potentilla simplex	oldfield cinquefoil
Rhexia mariana	Maryland meadow-beauty
Sium suave	hemlock water-parsnip
Spartina pectinata	freshwater cord grass
Spiraea alba	white meadowsweet
Stachys palustris	woundwort
Tripsacum dactyloides	eastern mock grama

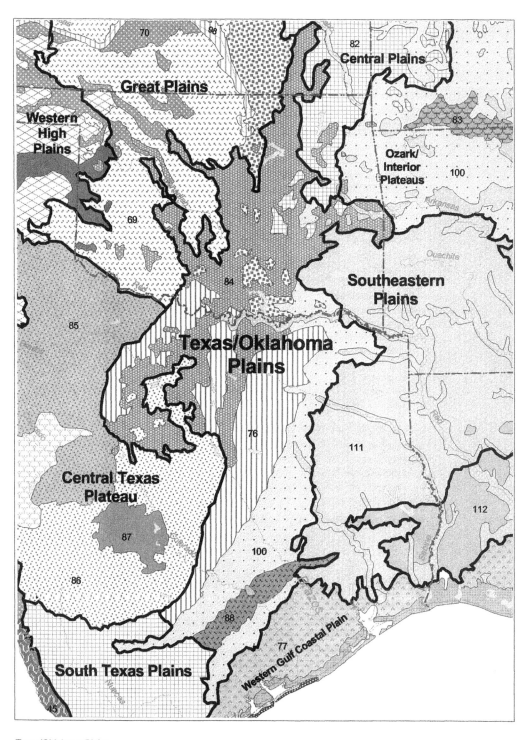

Texas/Oklahoma Plains

Texas/Oklahoma Plains

Introduction

The Texas/Oklahoma Plains region extends from southeastern Kansas into central Oklahoma and central Texas. This region is characterized by a mixture of both prairie and woodland vegetation. Major vegetation types include Blackland Prairie and Cross Timbers Woodland or Post Oak Savanna, as it is also known. Most species found in this area have ranges that extend northward into the Great Plains or eastward into the eastern deciduous forests.

The Blackland Prairie is a tallgrass prairie which extends from the Red River to the chaparral communities of south central Texas. Very few examples of the native vegetation of the Blackland Prairie remain because the rich soil in the area has either been converted to croplands or the prairie has been heavily grazed. In general, little false bluestem, yellow Indian grass, and big bluestem were dominant grasses in the presettlement prairie. Much of this area has been invaded by silver bluestem, Texas wintergrass, and buffalo grass.

The Cross Timbers Woodland is dominated by post oak and blackjack oak in the canopy and prairie species in the understory. It is presumed that early Texas settlers named the region "Cross Timbers" because they found belts of oak forests crossing strips of prairie grasslands.

The primary sources used to develop descriptions and species lists are Barbour and Billings 1988; Bill Carr (pers. com.); McMahan et al. 1984; Risser et al. 1981; Sims 1988; Texas Natural Heritage Program 1991.

Dominant Ecological Communities

Upland Systems
Oak–Hickory Forest
Cross Timbers Woodland
Bluestem Prairie
Fayette Prairie
Blackland Prairie
Wetland Systems
Floodplain Forest

UPLAND SYSTEMS

Oak–Hickory Forest

Oak–Hickory Forest extends along the eastern edge of the Texas/Oklahoma Plains region. This is the western-most range of the Oak–Hickory Forest and is generally much drier and more open than oak–hickory forests further east. This is reflected in the lower diversity of trees and greater diversity of understory prairie plants. This community corresponds to Küchler #100 and Western Deciduous and Mixed Forests ERT.

Canopy
Characteristic Species
Carya ovalis   red hickory
Carya texana   black hickory

Quercus incana	bluejack oak
Quercus marilandica	blackjack oak
Quercus stellata	post oak

Associates

Juniperus virginiana	eastern red-cedar
Prosopis glandulosa	honey mesquite
Quercus virginiana	live oak
Ulmus crassifolia	cedar elm

Woody Understory

Berchemia scandens	Alabama supplejack
Callicarpa americana	American beauty-berry
Campsis radicans	trumpet-creeper
Crataegus crus-galli	cock-spur hawthorn
Crataegus viridis	green hawthorn
Ilex vomitoria	yaupon
Prunus mexicana	bigtree plum
Prunus texana	peachbush
Rhus aromatica	fragrant sumac
Rhus copallina	winged sumac
Rubus trivialis	southern dewberry
Sideroxylon lanuginosa	gum bully
Smilax bona-nox	fringed greenbriar
Symphoricarpas orbiculatus	coral-berry
Toxicodendron pubescens	Atlantic poison-oak
Vaccinium arboreum	tree sparkle-berry
Viburnum rufidulum	rusty blackhaw
Vitis aestivalis	summer grape
Vitis mustangensis	mustang grape
Vitis rotundifolia	muscadine
Yucca arkansana	Arkansas yucca
Zanthoxylum clava-herculis	Hercules'-club

Herbaceous Understory

Alophia drummondii	propeller-flower
Antennaria plantaginifolia	woman's-tobacco
Ascyrum hypericoides	St. Andrew's cross
Berlandiera texana	Texas greeneyes
Bothriochloa saccaroides	plumed beard grass
Bouteloua hirsuta	hairy grama
Brachiaria ciliatissima	fringed signal grass
Carex caroliniana	Carolina sedge
Carex festucacea	fescue sedge
Carex laxiflora	broad loose-flower sedge
Carex muhlenbergii	Muhlenberg's sedge
Cenchrus carolinianus	sandburr
Centrosema virginianum	spurred butterfly-pea
Chaetopappa asteroides	Arkansas leastdaisy
Chamaecrista fasciculata	sleepingplant

Chamaesyce cordifolia	heart-leaf sand-mat
Chasmanthium laxum	slender wood-oats
Chloris cucullata	hooded windmill grass
Cnidoscolus texanus	Texas bullnettle
Coreopsis basalis	golden-mane tickseed
Coreopsis grandiflora	large-flower tickseed
Croton capitatus	hogwort
Croton glandulosus	vente-conmigo
Desmodium rotundifolium	prostrate tick-trefoil
Dichanthelium laxiflorum	open-flower rosette grass
Dichanthelium linearifolium	slim-leaf rosette grass
Dichanthelium malacophyllum	soft-leaf rosette grass
Dichanthelium oligosanthes	Heller's rosette grass
Dichanthelium ravenelii	Ravenel's rosette grass
Elymus virginicus	Virginia wild rye
Eragrostis secundiflora	red love grass
Eragrostis trichodes	sand love grass
Gaillardia aestivalis	lance-leaf blanket-flower
Galactia canescens	hoary milk-pea
Helianthemum georgianum	Georgia frostweed
Helianthus debilis	cucumber-leaf sunflower
Helianthus hirsutus	whiskered sunflower
Hymenopappus artemisiifolius	old-plains-man
Lechea tenuifolia	narrow-leaf pinweed
Lespedeza procumbens	trailing bush-clover
Lespedeza repens	creeping bush-clover
Lespedeza virginica	slender bush-clover
Liatris elegans	pink-scale gayfeather
Liatris pycnostachya	cat-tail gayfeather
Lupinus subcarnosus	Texas bluebonnet
Mirabilis albida	white four-o'clock
Mirabilis linearis	narrow-leaf four-o'clock
Monarda fistulosa	Oswego-tea
Monarda punctata	spotted beebalm
Nuttalanthus texanus	Texas-toadflax
Palafoxia rosea	rosy palafox
Panicum anceps	beaked panic grass
Panicum virgatum	wand panic grass
Paronychia drummondii	Drummond's nailwort
Phlox drummondii	annual phlox
Phyllanthus abnormix	Drummond's leaf-flower
Polypremum procumbens	juniper-leaf
Rhynchosia americana	American snout-bean
Rhynchospora harveyi	Harvey's beak sedge
Rudbeckia hirta	black-eyed-Susan
Schizachyrium scoparium	little false bluestem
Scleria triglomerata	whip nut-rush
Sorghastrum nutans	yellow Indian grass
Stillingia sylvatica	queen's-delight

Stylisma pickeringii	Pickering's dawnflower
Stylosanthes biflora	side-beak pencil-flower
Tephrosia virginiana	goat's-rue
Tetragonotheca ludoviciana	Louisiana nerveray

Cross Timbers Woodland

Cross Timbers Woodland is the most common woody vegetation type in the region. These woodlands are a combination of deciduous forest species and grassland species. The dominant species include post oak and blackjack oak, which make up about 92 percent of the mostly open canopy, and little false bluestem in the understory. This community corresponds to Küchler #84 and Woodlands ERT.

Canopy
Characteristic Species

Quercus marilandica	blackjack oak
Quercus stellata	post oak

Associates

Carya cordiformis	bitter-nut hickory
Carya ovalis	red hickory
Carya texana	black hickory
Celtis laevigata	sugar-berry
Fraxinus texensis	Texas ash
Juniperus virginiana	eastern red cedar
Prosopis glandulosa	honey mesquite
Quercus falcata	southern red oak
Quercus fusiformis	plateau oak
Quercus incana	bluejack oak
Quercus shumardii	Shumard's oak
Quercus sinuata	bastard oak
Quercus virginiana	live oak
Ulmus alata	winged elm
Ulmus crassifolia	cedar elm

Woody Understory

Ascyrum hypericoides	St. Andrew's cross
Berchemia scandens	Alabama supplejack
Callicarpa americana	American beauty-berry
Campsis radicans	trumpet-creeper
Cercis canadensis	redbud
Cornus drummondii	rough-leaf dogwood
Coryphantha sulcata	pineapple cactus
Crataegus crus-galli	cock-spur hawthorn
Echinocereus reichenbachii	lace hedgehog cactus
Forestiera pubescens	stretchberry
Ilex decidua	deciduous holly
Ilex vomitoria	yaupon
Lonicera albiflora	white honeysuckle
Opuntia engelmanii	cactus-apple
Opuntia leptocaulis	Christmas cholla

Prunus mexicana	bigtree plum
Rhus aromatica	fragrant sumac
Rhus glabra	smooth sumac
Sideroxylon lanuginosum	gum bully
Smilax bona-nox	fringed greenbriar
Symphoricarpus orbiculatus	coral-berry
Toxicodendron pubescens	Atlantic poison-oak
Viburnum rudifulum	rusty blackhaw
Vitis mustangensis	mustang grape
Yucca constricta	Buckley's yucca
Zizyphus obtusifolia	lotebush

Herbaceous Understory

Acalypha gracilens	three-seed mercury
Alcalypha virginica	Virginia three-seed mercury
Allium canadense	meadow garlic
Ambrosia psilostachya	western ragweed
Ambrosia trifida	great ragweed
Andropogon gerardii	big bluestem
Andropogon ternarius	split-beard bluestem
Antennaria parlinii	Parlin's pussytoes
Asclepias asperula	spider antelope horns
Bothriochloa saccharoides	plumed beard grass
Bouteloua curtipendula	side-oats grama
Bouteloua hirsuta	hairy grama
Carex muhlenbergii	Muhlenberg's sedge
Carex planostachys	cedar sedge
Chaetopappa asteroides	Arkansas leastdaisy
Desmodium sessilifolium	sessil-leaf tick-trefoil
Desmodium rotundifolium	prostrate tick-trefoil
Dichanthelium acuminatum	tapered rosette grass
Dichanthelium linearifolium	slim-leaf rosette grass
Dichanthelium oligosanthes	Heller's rosette grass
Dyschoriste linearis	polkadots
Elymus canadensis	nodding wild rye
Elymus virginicus	Virginia wild rye
Eragrositis trichodes	sand love grass
Eragrostis spectabilis	petticoat-climber
Euphorbia dentata	toothed spurge
Euphorbia heterophylla	Mexican-fireplant
Euphorbia tetrapora	weak spurge
Froelichia floridana	plains snake-cotton
Gaillardia pulchella	firewheel
Gymnopogon ambiguus	bearded skeleton grass
Houstonia pusilla	tiny bluet
Hymenopappus artemisiifolius	old-plains-man
Ipomopsis rubra	standing-cypress
Lespedeza virginica	slender bush-clover
Lotus unifoliolatus	American bird's-foot-trefoil

Matelea reticulata	netted milkvine
Monarda citriodora	lemon beebalm
Monarda clinopodioides	basil beebalm
Monarda fistulosa	Oswego-tea
Myosotis verna	spring forget-me-not
Neptunia lutea	yellow puff
Nothoscordum bivalve	crow poison
Panicum anceps	beaked panic grass
Panicum virgatum	wand panic grass
Penstemon australis	Eustis Lake beardtongue
Penstemon digitalis	foxglove beardtongue
Penstemon tubiflorus	white wand beardtongue
Psoralidium tenuiflorum	slender-flower lemonweed
Schizachyrium scoparium	little false bluestem
Sorghastrum nutans	yellow Indian grass
Sporobolus asper	tall dropseed
Sporobolus cryptandrus	sand dropseed
Tephrosia lindheimeri	Lindheimer's hoary-pea
Tephrosia virginiana	goats's-rue
Tradescantia ohiensis	bluejacket
Verbena halei	Texas vervain
Verbena bracteata	carpet vervain

Bluestem Prairie

Patches of Bluestem Prairie occur within the Cross Timbers Woodland in the northern part of this region. The Bluestem Prairie community in this region is very similar to the Bluestem Prairie in the Great Plains region. It is considered distinct, however, because it occurs on a different soil type. This community corresponds to Küchler #74 and Prairies ERT.

Characteristic Species

Andropogon gerardii	big bluestem
Panicum virgatum	wand panic grass
Schizachyrium scoparium	little false bluestem
Sorghastrum nutans	yellow Indian grass

Associates

Acalypha gracilens	slender three-seed mercury
Achillea millefolium	common yarrow
Aegilops cylindrica	jointed oat grass
Amorpha canescens	leadplant
Antennaria neglecta	field pussytoes
Asclepias syriaca	common milkweed
Asclepias tuberosa	butterfly milkweed
Aster ericoides	white heath aster
Aster laevis	smooth blue aster
Baptisia alba	white wild indigo
Baptisia bracteata	long-bract wild indigo
Bouteloua curtipendula	side-oats grama

Cacalia plantaginea	groove-stem Indian-plantain
Chamaecrista fasiculata	sleepingplant
Dalea candida	white prairie-clover
Dalea purpurea	violet prairie-clover
Dichanthelium leibergii	Leiberg's rosette grass
Dichanthelium oligosathes	Heller's rosette grass
Erigeron strigosus	prairie fleabane
Galium tinctorum	stiff marsh bedstraw
Helianthus grosseserratus	saw-tooth sunflower
Helianthus maximiliani	Michaelmas-daisy
Helianthus mollis	ashy sunflower
Koeleria macrantha	prairie Koeler's grass
Liatrus aspera	tall gayfeather
Liatrus punctata	dotted gayfeather
Liatrus scariosa	devil's-bite
Mentzelia oligosperma	chickenthief
Pediomelum argophyllum	silver-leaf Indian-breadroot
Phlox pilosa	downy phlox
Polytaenia nuttallii	Nuttall's prairie-parsley
Psoralidium tenuiflorum	slender-flower lemonweed
Ratibida columnifera	red-spike Mexican-hat
Ratibida pinnata	gray-head Mexican-hat
Rosa arkansana	wild prairie rose
Schrankia microphylla	little-leaf sensitive-briar
Silphium integrifolium	entire-leaf rosinweed
Silphium laciniatum	compassplant
Solidago canadensis	Canadian goldenrod
Solidago missouriensis	Missouri goldenrod
Solidago rigida	hard-leaf goldenrod
Sporobolus asper	tall dropseed
Sporobolus heterolepis	prairie dropseed
Stipa spartea	porcupine grass
Tripsacum dactyloides	eastern mock grama

Fayette Prairie

Fayette Prairie is a mixed grass prairie which occurs in the southern part of the region. Scattered oak and hickory trees occur throughout this prairie type. This community corresponds to Küchler #88 and Prairies ERT.

Characteristic Species

Andropogon gerardii	big bluestem
Buchloë dactyloides	buffalo grass
Schizachyrium scopariuum	little false bluestem
Sorghastrum nutans	yellow Indian grass

Associates

Acacia angustissima	prairie wattle
Andropogon ternarius	split-beard bluestem
Aristida purpurascens	arrow-feather three-awn

Aster ericoides	white heath aster
Aster patens	late purple aster
Bifora americana	prairie bishop
Bothriochloa saccharoides	plumed beard grass
Bouteloua curtipendula	side-oats grama
Bouteloua rigidiseta	Texas grama
Brickellia eupatorioides	false boneset
Camassia scilloides	Atlantic camas
Carex microdonta	little-tooth sedge
Castilleja indivisa	entire-leaf Indian-paintbrush
Coelorachis cylindrica	Carolina joint-tail grass
Dalea compacta	compact prairie-clover
Dalea multiflora	round-head prairie-clover
Echinacea angustifolia	blacksamson
Eragrostis intermedia	plains love grass
Helianthus maximilliani	Michaelmas-daisy
Houstonia pusilla	tiny bluet
Liatris squarrosa	scaly gayfeather
Lupinus subcarnosus	Texas bluebonnet
Marshallia caespitosa	puffballs
Monarda citriodora	lemon beebalm
Muhlenbergia capillaris	hair-awn muhly
Nemastylis geminiflora	prairie pleatleaf
Panicum obtusum	blunt panic grass
Panicum virgatum	wand panic grass
Paspalum floridanum	Florida crown grass
Paspalum plicatulum	brown-seed crown grass
Paspalum setaceum	slender crown grass
Penstemon cobaea	cobaea breadtongue
Penstemon tubiflorus	white wand beardtongue
Polytaenia nuttallii	Nuttall's prairie-parsley
Polytaenia texana	Texas false-parsley
Rudbeckia hirta	black-eyed-Susan
Ruellia humilis	finge-leaf wild petunia
Sabatia campestris	Texas-star
Salvia azurea	azure-blue sage
Sporobolus asper	tall dropseed
Tripsacum dactyloides	eastern mock grama
Vernonia texana	Texas ironweed

Blackland Prairie

Blackland Prairie is also referred to as the "Grand Prairie." Most of the Blackland Prairie has been converted to agriculture or has been heavily grazed. The dominant grass of the Blackland Prairie was little false bluestem, however, heavy grazing and disturbance have caused buffalo grass and Texas grama to invade and increase. This community corresponds to Küchler #76 and Prairies ERT.

Characteristic Species

Andropogon gerardii	big bluestem
Schizachyrium scoparium	little false bluestem

Associates

Acacia angustissima	prairie wattle
Ambrosia psilostachya	western ragweed
Aristida purpurea	purple three-awn
Aster ericoides	white heath aster
Bifora americana	prairie bishop
Bothriochloa saccharoides	plumed beard grass
Bouteloua curtipendula	side-oats grama
Bouteloua hirsuta	hairy grama
Bouteloua rigidiseta	Texas grama
Brickellia eupatorioides	false boneset
Buchloë dactyloides	buffalo grass
Carex meadii	Mead's sedge
Carex microdonta	little-tooth sedge
Castilleja indivisa	entire-leaf Indian-paintbrush
Dalea candida	white priarie-clover
Dalea purpurea	violet prairie-clover
Desmanthus illinoensis	prairie bundle flower
Echinacea angustifolia	blacksamson
Elymus canadensis	nodding wild rye
Erioneuron pilosum	hairy woolly grass
Eryngium leavenworthii	Leavenworth's eryngo
Eryngium yuccifolium	button eryngo
Fimbristylis puberula	hairy fimbry
Hedyotis nigricans	diamond-flowers
Helianthus maximiliani	Michaelmas-daisy
Horedeum pusillum	little barley
Liatris punctata	dotted gayfeather
Limnodea arkansana	Ozark grass
Monarda citriodora	lemon beebalm
Nothoscordum bivalve	cowpoison
Oenothera speciosa	pinkladies
Panicum virgatum	wand panic grass
Paspalum floridanum	Florida crown grass
Penstemon cobaea	cobaea beardtongue
Polytaenia texana	Texas false-parsley
Ratibida columnifera	red-spike Mexican-hat
Rudbeckia hirta	black-eyed-Susan
Ruellia humilis	fringe-leaf wild petunia
Salvia azurea	azure-blue sage
Silphium laciniatum	compassplant
Silphium radula	rough stem rosinweed
Solidago missouriensis	Missouri goldenrod
Sorgastrum nutans	yellow Indian grass
Sporobolus asper	tall dropseed
Sporobolus silveanus	Silveus' dropseed
Tragia ramosa	branched noseburn
Tridens strictus	long-spike fluff grass
Tripsacum dactyloides	eastern mock grama
Vernonia texana	Texas ironweed
Vulpia octoflora	eight-flower six-weeks grass

WETLAND SYSTEMS

Floodplain Forest

Floodplain Forest in the Texas/Oklahoma Plains region occurs along the major rivers. The floodplains may be dominated by pecan, sugar-berry, and cedar elm. These forests have many similarities to the Southern Floodplain Forest of the Southeastern Coastal Plain. This community corresponds to Floodplain Forests ERT.

Canopy

Characteristic Species

Carya illinoinensis	pecan
Celtis laevigata	sugar-berry
Ulmus americana	American elm
Ulmus crassifolia	cedar elm

Associates

Acer rubrum	red maple
Fraxinus pennslyvanica	green ash
Gleditsia triacanthos	honey-locust
Juglans nigra	black walnut
Liquidambar styraciflua	sweet-gum
Morus rubra	red mulberry
Plantanus occidentalis	American sycamore
Populus deltoides	eastern cottonwood
Quercus buckleyi	Buckley's oak
Quercus fusiformis	plateau oak
Quercus lyrata	overcup oak
Quercus macrocarpa	burr oak
Quercus nigra	water oak
Quercus pagoda	cherry-bark oak
Quercus virginiana	live oak
Sapindus saponaria	wing-leaf soapberry
Salix nigra	black willow
Taxodium distichum	southern bald-cypress

Woody Understory

Amorpha fruticosa	false indigo-bush
Ampelopsis cordata	heart-leaf peppervine
Cornus drummondii	rough-leaf dogwood
Forestiera pubescens	stretchberry
Fraxinus texensis	Texas ash
Ilex decidua	deciduous holly
Prosopis glandulosa	honey mesquite
Sideroxylon lanuginosa	gum bully
Smilax bona-nox	fringed greenbriar
Toxicodendron pubescens	Atlantic poison-oak
Vitis mustangensis	mustang grape

Herbaceous Understory

Ambrosia psilostachya	western ragweed
Aster drummondii	Drummond's aster
Carex amphibola	eastern narrow-leaf
Chasmanthium latifolium	Indian wood-oats
Clematis pitcheri	bluebill
Elephantopus carolinianus	Carolina elephant's foot
Elymus canadensis	nodding wild rye
Elymus virginicus	Virginia wild rye
Geum canadense	white avens
Paspalum pubiflorum	hairy-seed crown grass
Passiflora lutea	yellow passion-flower
Rivina humilis	rougeplant
Solidago canadensis	Canadian goldenrod
Teucrium canadense	American germander
Verbesina virginica	white crownbeard

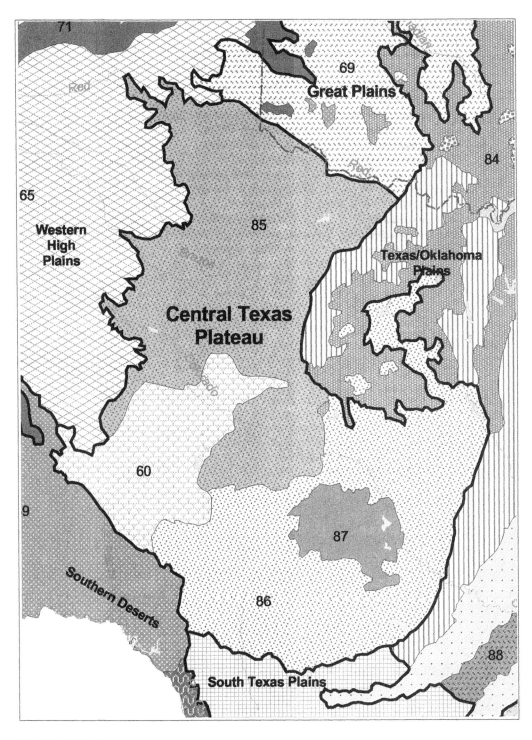

Central Texas Plateau

Central Texas Plateau

Introduction

The Central Texas Plateau region includes what is commonly known as the "Edwards Plateau" and "Rolling Plains." In its natural state, the area is predominantly shortgrass and midgrass prairie with areas of savanna and woodland. The vegetation is more diverse toward the south and eastern boundaries of the region in an area known as the Balcones Escarpment. This area is characterized by numerous canyons with deciduous woodland and forest.

The vegetation throughout the region has been heavily impacted by overgrazing of livestock and other developments. The prairies in the rolling plains, once dominated by side-oats grama, little false bluestem, and blue grama have been converted to grain fields or have been cleared for oil-well pads. Throughout the region, shrub species have been spreading as the native grasses decline due to overgrazing.

The primary sources used to develop descriptions and species lists are Barbour and Billings 1988; Bill Carr (pers. com.); Correll and Johnston 1970; McMahan et al. 1984; Riskin and Diamond 1988; Tharp 1939; Tharp 1991.

Dominant Ecological Communities

Upland Systems
> Deciduous Woodland
> Juniper–Oak Savanna
> Mesquite–Oak Savanna
> Mesquite Savanna
> Midgrass Prairie

Wetland Systems
> Floodplain Forest

UPLAND SYSTEMS

Deciduous Woodland

Along the southern boundary of the Central Texas region, the Balcones Escarpment provides conditions for woodlands and forests. The north- and east-facing slopes in the canyons of the Balcones Escarpment generally have the mesic conditions required for woodlands to develop. This community corresponds to Woodlands ERT.

Canopy

Characteristic Species

Fraxinus texensis	Texas ash
Juniperus ashei	Ashe's juniper
Prunus serotina	black cherry
Quercus buckleyi	Buckley's oak
Quercus fusiformis	plateau oak

Associates

Celtis laevigata	sugar-berry
Juglans major	Arizona walnut
Morus rubra	red mulberry

Quercus laceyi	Lacey's oak
Quercus muehlenbergii	chinkapin oak
Quercus sinuata	bastard oak
Tilia americana	American basswood
Ulmus americana	American elm
Ulmus crassifolia	cedar elm

Woody Understory

Aesculus pavia	red buckeye
Buddleja racemosa	wand butterfly-bush
Callicarpa americana	American beauty-berry
Cercis canadensis	redbud
Cornus drummondii	rough-leaf dogwood
Forestiera pubescens	stretchberry
Frangula caroliniana	Carolina buckthorn
Ilex decidua	deciduous holly
Ilex vomitoria	yaupon
Lindera benzoin	northern spicebush
Parthenocissus quinquefolia	Virginia-creeper
Prunus mexicana	bigtree plum
Ptelea trifoliata	common hoptree
Rhus aromatica	fragrant sumac
Rhus virens	evergreen sumac
Rubus trivialis	southern dewberry
Sideroxylon lanuginosum	gum bully
Smilax bona-nox	fringed greenbriar
Sophora affinis	Eve's necklacepod
Sophora secundiflora	mescal-bean
Ungnadia speciosa	Mexican-buckeye
Viburnum rufidulum	rusty blackhaw
Vitis monticola	sweet mountain grape
Vitis mustangensis	mustang grape
Zanthoxylum hirsutum	Texas hercules'-club

Herbaceous Understory

Anemone edwardsiana	plateau thimbleweed
Aquilegia canadensis	red columbine
Arisaema dracontium	greendragon
Aristolochia serpentaria	Virginia-snakeroot
Brickellia cylindracea	gravel-bar brickellbush
Carex edwardsiana	Edwards plateau sedge
Carex planostachys	cedar sedge
Chaetopappa effusa	spreading leastdaisy
Chasmanthium latifolium	Indian wood-oats
Desmodium psilophyllum	simple-leaf tick-trefoil
Dichanthelium acuminatum	tapered rosette grass
Dichanthelium oligosanthes	Heller's rosette grass
Dichanthelium pedicellatum	cedar rosette grass
Elymus virginicus	Virginia wild rye

Euphorbia roemeriana	Roemer's spurge
Galium texense	Texas bedstraw
Geum canadense	white avens
Hedeoma acinoides	slender false pennyroyal
Lespedeza texana	Texas bush-clover
Matelea edwardsensis	plateau milkvine
Matelea reticulata	netted milkvine
Muhlenbergia lindheimeri	Lindheimer's muhly
Muhlenbergia reverchonii	seep muhly
Muhlenbergia schreberi	nimblewill
Parietaria pensylvanica	Pennsylvania pellitory
Passiflora affinis	bracted passion-flower
Passiflora tenuiloba	bird-wing passion-flower
Polygala lindheimeri	shrubby milkwort
Salvia roemeriana	cedar sage
Schizachyrium scoparium	little false bluestem
Scutellaria drummondii	Drummond's skullcap
Scutellaria ovata	heart-leaf skullcap
Sporobolus asper	tall dropseed
Stipa leucotricha	Texas needle grass
Tinantia anomala	widow's-tears
Tradescantia edwardsiana	plateau spiderwort
Tragia ramosa	branched noseburn
Tridens spp.	fluff grass
Urtica chamidryoides	heart-leaf nettle
Verbesina virginica	white crownbeard
Viola missouriensis	Missouri violet

Juniper–Oak Savanna

Juniper–Oak Savanna occurs in the Edwards Plateau area where it occupies sloping sites with dry, shallow soil. It is composed of dense to very open stands of low trees and shrubs. The trees and shrubs are deciduous or broadleaf evergreen and needleleaf evergreen. This community corresponds to Küchler #86 and Woodlands ERT.

Canopy
Characteristic Species
Juniperus ashei	Ashe's juniper
Quercus fusiformis	plateau oak

Associates
Fraxinus texensis	Texas ash
Prosopis glandulosa	honey mesquite
Quercus buckleyi	Buckley's oak
Quercus sinuata	bastard oak

Woody Understory
Cercis canadensis	redbud
Diospyros texana	Texas persimmon
Echinocereus reichenbachii	lace hedgehog cactus
Echinocereus triglochidiatus	king-cup cactus

Mahonia trifoliata	Laredo Oregon-grape
Mimosa borealis	fragrant mimosa
Opuntia engelmannii	cactus-apple
Rhus aromatica	fragrant sumac
Rhus virens	evergreen sumac
Yucca rupicola	Texas yucca

Herbaceous Understory

Andropogon gerardii	big bluestem
Anemone berlandieri	ten-petal thimbleweed
Artistida longespica	red three-awn
Artistida purpurea	purple three-awn
Bouteloua curtipendula	side-oats grama
Bouteloua hirsuta	hairy grama
Bouteloua pectinata	tall grama
Buchloë dactyloides	buffalo grass
Carex planostachys	cedar sedge
Chaetopappa bellidifolia	white-ray leastdaisy
Chamaesyce angusta	black-foot sandmat
Chamaesyce fendleri	Fendler's sandmat
Dichanthelium oligosanthes	Heller's rosette grass
Dichanthelium pedicellatum	cedar rosette grass
Elymus canadensis	nodding wild rye
Eragrostis intermedia	plains love grass
Erioneuron pilosum	hairy woolly grass
Hedeoma acinoides	slender false pennyroyal
Hedeoma drummondii	Drummond's false pennyroyal
Hilaria belangeri	curly-mesquite
Leptochloa dubia	green spangletop
Lespedeza texana	Texas bush-clover
Marshallia cespitosa	puffballs
Melampodium leucanthum	plains blackfoot
Muhlenbergia reverchonii	seep muhly
Nassella leucotricha	Texas wintergrass
Panicum obtusum	blunt panic grass
Phyllanthus polygonoides	smartweed leaf-flower
Polygala alba	white milkwort
Schizachyrium scoparium	little false bluestem
Simsia calva	awnless bush-sunflower
Sorghastrum nutans	yellow Indian grass
Sporobolus asper	tall dropseed
Stillingia texana	Texas toothleaf
Tetraneuris linearifolia	fine-leaf four-nerve-daisy
Tetraneuris scaposa	stemmed four-nerve-daisy
Thelesperma filifolium	stiff greenthread
Thelesperma simplicifolium	slender greenthread
Tridens muticus	awnless fluff grass
Wedelia hispida	hairy creeping-oxeye

Mesquite–Oak Savanna

This is an open canopy community with a groundcover of tall to medium grasses. Tree density varies with local conditions. This community is common in the Llano uplift area. This community corresponds to Küchler #87 and Woodlands ERT.

Canopy

Characteristic Species

Prosopis glandulosa	honey mesquite
Quercus marilandica	blackjack oak
Quercus stellata	post oak

Associates

Aloysia ligustrina	beebrush
Celtis laevigata	sugar-berry
Diospyros texana	Texas persimmon
Juniperus ashei	Ashe's juniper
Quercus fusiformis	plateau oak
Ulmus crassifolia	cedar elm

Woody Understory

Aloysia gratissima	whitebrush
Mahonia trifoliata	Laredo Oregon-grape
Opuntia engelmannii	cactus-apple
Opuntia leptocaulis	Christmas cholla
Opuntia macrorhiza	twist-spine prickly-pear
Rhus aromatica	fragrant sumac
Rhus virens	evergreen sumac
Yucca constricta	Buckley's yucca
Yucca torreyi	Torrey's yucca
Zanthoxylum hirsutum	Texas hercules'-club

Herbaceous Understory

Agrostis elliottiana	Elliott's bent
Aphanostephus skirrhobasis	Arkansas dozedaisy
Aristida longespica	red three-awn
Aristida purpurea	purple three-awn
Astragalus nuttallianus	turkey-pea
Bothriochloa saccharoides	plumed beard grass
Bouteloua curtipendula	side-oats grama
Bouteloua hirsuta	hairy grama
Bouteloua rigidiseta	Texas grama
Buchloë dactyloides	buffalo grass
Callirhoe involucrata	purple poppy-mallow
Carex muhlenbergii	Muhlenberg's sedge
Carex planostachys	cedar sedge
Castilleja indivisa	entire-leaf Indian-paintbrush
Chaetopappa asteroides	Arkansas leastdaisy
Cnidoscolus texanus	Texas bull-nettle
Dichanthelium oligosanthes	Heller's rosette grass

Eragrostis intermedia	plains love grass
Eriogonum tenellum	tall wild buckwheat
Gaillardia pulchella	firewheel
Helenium amarum	yellowdicks
Helianthemum georgianum	Georgia frostweed
Hypericum drummondii	nits-and-lice
Lechea san-sabeana	San Saba pinweed
Linum hudsonioides	Texas flax
Lotus unifoliolatus	American bird's-foot-trefoil
Lupinus texensis	Texas lupine
Nassella leucotricha	Texas wintergrass
Nuttalanthus texana	Texas-toadflax
Panicum virgatum	wand panic grass
Phacelia patuliflora	sand scorpion-weed
Phlox drummondii	annual phlox
Schizachyrium scoparium	little false bluestem
Sedum nuttallianum	yellow stonecrop
Senecio ampullaceus	Texas ragwort
Sorghastrum nutans	yellow Indian grass
Spermolepis echinata	bristly scaleseed
Sporobolus asper	tall dropseed
Tripogon spicatus	American five-minute grass
Valerianella texana	Edwards plateau cornsalad
Xanthisma texanum	Texas sleepy-daisy

Mesquite Savanna

This is an open canopy community dominated by scattered broadleaf evergreen or deciduous shrubs and low trees with a dense to open grass groundcover. This community corresponds to Küchler #60 and #85 and Woodlands ERT.

Characteristic Species
Prosopis glandulosa	honey mesquite

Associates
Acacia greggii	long-flower catclaw
Juniperus ashei	Ashe's juniper
Juniperus pinchotii	Pinchot's juniper
Opuntia spp.	prickly-pear
Quercus fusiformis	plateau oak

Herbaceous Understory
Aphanostephus skirrhobasis	Arkansas dozedaisy
Aristida purpurea	purple three-awn
Astragalus nuttallianus	turkey-pea
Bothriochloa barbinodis	cane beard grass
Bouteloua curtipendula	side-oats grama
Bouteloua rigidiseta	Texas grama
Buchloë dactyloides	buffalo grass
Chaetopappa asteroides	Arkansas leastdaisy
Gaillardia pulchella	firewheel

Helenium amarum	yellowdicks
Hilaria belangeri	curly-mesquite
Hilaria mutica	Tobosa grass
Nuttalanthes texanus	Texas-toadflax

Midgrass Prairie

Midgrass Prairie occurs on upland areas in the Rolling Plains and Edwards Plateau. This community corresponds to Prairies ERT.

Characteristic Species
Bouteloua curtipendula	side-oats grama
Schizachyrium scoparium	little false bluestem

Associates
Artemisia ludoviciana	white sagebrush
Asclepias latifolia	broad-leaf milkweed
Astragalus missouriensis	Missouri milkvetch
Bothriochloa barbinodis	cane beard grass
Bouteloua gracilis	blue grama
Bouteloua hirsuta	hairy grama
Buchloë dactyloides	buffalo grass
Calylophus berlandieri	Berlandier's sundrops
Calylophus hartwegii	Hartweg's sundrops
Chamaesyce lata	hoary sandmat
Dalea aurea	golden prairie-clover
Dalea candida	white prairie-clover
Dalea enneandra	nine-anther prairie-clover
Dalea formosa	featherplume
Dalea frutescens	black prairie-clover
Digitaria californica	California crab grass
Elymus canadensis	nodding wild rye
Eriogonum alatum	winged wild buckwheat
Gaura villosa	woolly beeblossom
Hedyotis nigricans	diamond-flower
Helianthus maximiliani	Michaelmas-daisy
Hilaria belangeri	curly-mesquite
Juniperus ashei	Ashe's juniper
Juniperus pinchotii	Pinchot's juniper
Liatris punctata	dotted gayfeather
Machaeranthera pinnatifida	lacy tansy-aster
Melampodium leucanthum	plains blackfoot
Mirabilis albida	white four-o'clock
Monarda citriodora	lemon beebalm
Nassella leucotricha	Texas wintergrass
Panicum obtusum	blunt panic grass
Polygala alba	white milkwort
Prosopis glandulosa	honey msequite
Psoralidium tenuiflorum	slender-flower lemonweed
Quercus mohriana	Mohr's oak
Ratibida columnaris	red-spike Mexican-hat

Rhus microphylla	little-leaf sumac
Sorghastrum nutans	yellow Indian grass
Sphaeralcea angustifolia	copper globe-mallow
Sphaeralcea coccinea	scarlet globe-mallow
Sporobolus cryptandrus	sand dropseed
Thelesperma megapotamicum	Hopi-tea
Tidens muticus	awnless fluff grass
Tragia ramosa	branched noseburn
Ziziphus obtusifolia	lotebush

WETLAND SYSTEMS

Floodplain Forest

Floodplain Forest in this region occurs along perennial streams and in terraces of the Balcones Escarpment. Species composition varies from east to west with bald-cypress, sycamore, pecan, chinkapin oak, and Arizona walnut more important in the east, and live oak and sugar-berry dominant in the west. This community corresponds to Floodplain Forests ERT.

Canopy
Characteristic Species

Carya illinoinensis	pecan
Celtis laevigata	sugar-berry
Fraxinus pennsylvanica	green ash
Platanus occidentalis	American sycamore
Quercus fusiformis	plateau oak
Taxodium distichum	southern baldcypress
Ulmus americana	American elm
Ulmus crassifolia	cedar elm

Associates

Acer negundo	ash-leaf maple
Fraxinus berlandieriana	Mexican ash
Juglans major	Arizona walnut
Quercus buckeyi	Buckley's oak
Quercus macrocarpa	burr oak
Quercus muhlenbergii	chinkapin oak
Quercus virginiana	live oak
Salix nigra	black willow
Sapindus saponaria	wing-leaf soapberry
Tillia americana	American basswood

Woody Understory

Amorpha fruticosa	false indigo-bush
Ampelopsis arborea	peppervine
Cephalanthus occidentalis	common buttonbush
Cornus drummondii	rough-leaf dogwood
Ilex decidua	deciduous holly
Juglans microcarpa	little walnut

Parthenocissus quinquefolia	Virginia-creeper
Rhus trivialis	southern dewberry
Viburnum rufidulum	rusty blackhaw
Vitis mustangensis	mustang grape

Herbaceous Understory

Andropogon glomeratus	bushy bluestem
Aster drummondii	Drummond's aster
Boehmeria cylindrica	small-spike false nettle
Carex amphibola	eastern narrow-leaf sedge
Carex blanda	white sedge
Carex cephalophora	oval-leaf sedge
Carex emoryi	Emory's sedge
Carex microdonta	little-tooth sedge
Carex planostachys	cedar sedge
Chasmanthium latifolium	Indian wood-oats
Clematis pitcheri	bluebill
Dicliptera brachiata	branched foldwing
Elymus virginicus	Virginia wild rye
Eupatorium serotinum	late-flowering thoroughwort
Galium aparine	sticky-willy
Geum canadense	white avens
Malvaviscus drummondii	Texas wax-mallow
Melothria pendula	Guadeloupe-cucumber
Muhlenbergia lindheimeri	Lendheimer's muhly
Muhlenbergia schreberi	nimblewill
Muhlenbergia utilis	aparejo grass
Panicum virgatum	wand panic grass
Paspalum pubiflorum	hairy-seed crown grass
Poa arachnifera	Texas blue grass
Ranunculus macranthus	large buttercup
Ruellia drummondiana	Drummond's wild petunia
Sanicula canadensis	Canadian black-snakeroot
Setaria scheelei	southwestern bristle grass
Tinantia anomala	widow's-tears
Tradescantia gigantea	giant spiderwort
Valerianella stenocarpa	narrow-cell cornsalad
Verbesina virginica	white crownbeard
Viola missouriensis	Missouri violet

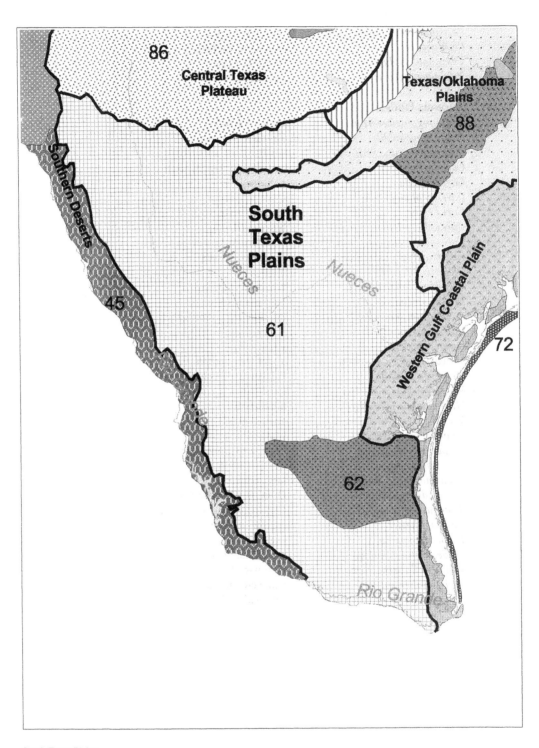

South Texas Plains

South Texas Plains

Introduction

The South Texas Plains region encompasses most of the southern tip of Texas. The northern boundary is the Balcones Escarpment. The region extends south to the tip of Texas at the mouth of the Rio Grande River near Brownsville. This region does not include the vegetation of the arid plains bordering the Rio Grande River to the west. A description of the vegetation found along the Rio Grande can be found in the Southern Deserts region. The vegetation of the Gulf Coast and barrier islands to the east is found in the Western Gulf Coastal Plains region.

The natural vegetation is characterized by grasslands and savanna–woodland communities. Before changes in the region, such as excessive grazing, conversion to cropland, and suppression of natural fires, the area was predominantly open grasslands with scattered trees and shrubs. Today, the area is increasingly dominated with the thorny scrub vegetation which was once limited to rocky upland areas.

Unique in this region are small groves of native palm which occur in southernmost Texas near Brownsville.

The primary sources used to develop descriptions and species lists are Barbour and Billings 1988; Bill Carr (pers. com.); Correll and Johnston 1970; McMahan et al. 1984; Texas Natural Heritage Program 1991; Thasp 1926.

Dominant Ecological Communities

Upland Systems
Mesquite–Granjeno Woodland
Mesquite–Blackbrush Shrub
Rio Grande Shrubland
Live Oak Savanna
South Texas Prairie
Wetland Systems
Floodplain Forest

UPLAND SYSTEMS

Mesquite–Granjeno Woodland

Mesquite–Granjeno Woodland occurs throughout the South Texas Plains. This woodland may occur within shrub-dominated areas where there is little or no distinction between canopy and understory layers. This community corresponds, in part, to Küchler #61 and Woodlands ERT.

Canopy
Characteristic Species
Celtis pallida	shiny hackberry (granjeno)
Prosopis glandulosa	honey mesquite

Associates
Acacia berlandieri	guajillo
Acacia greggii	long-flower catclaw
Acacia minuta	coastal scrub wattle

Acacia rigidula	chaparro-prieto
Aloysia gratissima	whitebrush
Colubrina texensis	hog-plum
Condalia hookeri	Brazilian bluewood
Eysenhardtia texana	Texas kidneywood
Forestiera angustifolia	Texas swamp-privet
Guajacum angustifolium	Texas lignumvitae
Karwinskia humboldtiana	coyotillo
Lycium berlandieri	silver desert-thorn
Opuntia engelmannii	cactus-apple
Opuntia leptocaulis	Christmas cholla
Opuntia macrorhiza	twist-spine prickly-pear
Parkinsonia aculeata	Jerusalem-thorn
Schaefferia cuneifolia	desert-yaupon
Sideroxylon lanuginosum	gum bully
Zanthoxylum fagara	lime prickly-ash
Ziziphus obtusifolia	lotebush

Herbaceous Understory

Acleisanthes anisophylla	oblique-leaf trumpets
Ambrosia confertiflora	weak-leaf burr-ragweed
Aphanostephus ramosissimus	plains dozedaisy
Aristida purpurea	purple three-awn
Bothriochoa saccharoides	plumed beard grass
Bouteloua rigidiseta	Texas grama
Bouteloua trifida	red grama
Cardiospermum corindum	faux-persil
Chloris ciliata	fringed windmill grass
Chloris cucullata	hooded windmill grass
Cnidoscolus texanus	Texas bull-nettle
Cocculus diverisifolius	snailseed
Croton dioicus	grassland croton
Florestina tripteris	sticky florestina
Gaillardia suavis	perfumeballs
Helianthus annuus	common sunflower
Helianthus ciliaris	Texas-blueweed
Heteropogon contortus	twisted tanglehead
Hilaria berlangeri	curly-mesquite
Isocoma drummondii	Drummond's jimmyweed
Liatris mucronata	cusp gayfeather
Machaeranthera pinnatifida	lacy tansy-aster
Nassella leucotricha	Texas wintergrass
Nyctaginia capitata	devil's-bouquet
Palafoxia texana	Texas palafox
Panicum hallii	Hall's panic grass
Pappophorum bicolor	pink pappus grass
Parthenium hysterophorus	Santa Maria feverfew
Passiflora foetida	scarlet-fruit

Salvia coccinea	blood sage
Schizachyrium scoparium	little false bluestem
Setaria macrostachya	plains bristle grass
Simsia calva	awnless bush-sunflower
Siphonoglossa pilosella	hairy tubetongue
Solanum elaeagnifolium	silver-leaf nightshade
Sporobolus pyramidatus	whorled dropseed
Tiquilia canescens	woody crinklemat
Tridens eragrostoides	love fluff grass
Tridens muticus	awnless fluff grass

Mesquite–Blackbrush Scrub

Mesquite–Blackbrush Scrub is widely represented in the region. It occurs in rocky, broken uplands and may grade into other shrub communities depending upon soil, slope, and moisture. This community corresponds, in part, to Küchler #61 amd Desert Lands ERT.

Characteristic Woody Species

Acacia rigidula	chaparro-prieto (blackbrush)
Prosopis glandulosa	honey mesquite

Associates

Acacia berlandieri	guajillo
Acacia greggii	long-flower catclaw
Acacia schaffneri	Schaffner's wattle
Aloysia gratissima	whitebrush
Castela erecta	goatbush
Celtis pallida	shiny hackberry
Condalia hookeri	Brazilian bluewood
Diospyros texana	Texas persimmon
Ephedra antisyphilitica	clapweed
Eysenhardtia texana	Texas kidneywood
Forestiera angustifolia	Texas swamp-privet
Guajacum angustifolium	Texas lignumvitae
Jatropha dioica	leatherstem
Koeberlinia spinosa	crown-of-thorns
Leucophyllum frutescens	Texas barometer-bush
Lycium berlandieri	silver desert-thorn
Mahonia trifoliata	Laredo Oregon-grape
Mammillaria heyderi	little nipple cactus
Opuntia engelmannii	cactus-apple
Opuntia leptocaulis	Christmas cholla
Rhus microphylla	little-leaf sumac
Schaefferia cuneifolia	desert-yaupon
Thelocactus setispinus	miniature-barrel cactus
Yucca treculeana	Don Quixote's-lace
Ziziphus obtusifolia	lotebush

Herbaceous Understory

Acleisanthes anisophylla	oblique-leaf trumpets
Acleisanthes longiflora	angel's trumpets
Acourtia runcinata	feather-leaf desert-peony
Amblyolepis setigera	huisache-daisy
Aphanostephus ramosissimus	plains dozedaisy
Aristida purpurea	purple three-awn
Bouteloua curtipendula	side-oats grama
Bouteloua gracilis	blue grama
Bouteloua hirsuta	hairy grama
Carex planostachys	cedar sedge
Centaurea americana	American star-thistle
Chamaesyce serpens	matted sandmat
Chloris ciliata	fringed windmill grass
Chloris cucullata	hooded windmill grass
Chloris verticillata	tumble windmill grass
Clematis drummondii	Texas virgin's-bower
Cordia podocephala	Texas manjack
Croton dioicus	grassland croton
Florestina tripteris	sticky florestina
Hedyotis nigricans	diamond-flower
Hymenopappus scabiosaeus	Carolina woollywhite
Isocoma drummondii	Drummond's jimmyweed
Liatris mucronata	cusp gayfeather
Oxalis dichondraefolia	peony-leaf wood-sorrel
Panicum hallii	Hall's panic grass
Pappophorum bicolor	pink pappus grass
Parthenium hysterophorus	Santa Maria feverfew
Phyllanthus polygonoides	smartweed leaf-flower
Pinaropappus roseus	white rock-lettuce
Polygala ovatifolia	egg-leaf milkwort
Ruellia nudiflora	violet wild petunia
Schizachyrium scoparium	little false bluestem
Senna roemeriana	two-leaved wild sensitive-plant
Setaria macrostachya	plains bristle grass
Setaria ramiseta	Rio Grande bristle grass
Siphonoglossa pilosella	hairy tubetongue
Thamnosma texanum	rue-of-the-mountains
Thymophylla pentachaeta	five-needle pricklyleaf
Tiquilia canescens	woody crinklemat
Tridens muticus	awnless fluff grass
Wedelia hispida	hairy creeping-oxeye

Rio Grande Shrubland

Most of the vegetation in the area around the mouth of the Rio Grande River has been altered through development or conversion to cropland. Some areas remain in the native subtropical evergreen shrubland or forest. This area, generally characterized by ebony blackbead (Texas ebony), also supports groves of Rio Grande palmetto. This community corresponds to Desert Lands ERT.

Characteristic Woody Species
Ehretia anacua	knockaway
Pithecellobium ebano	ebony blackbead

Associates
Acacia greggii	long-flower catclaw
Celtis pallida	shiny hackberry
Condalia hookeri	Brazilian bluewood
Harvardia pallens	haujillo
Leucaena pulverulenta	great leadtree
Opuntia leptocaulis	Christmas cholla
Phaulothamnus spinescens	devilqueen
Prosopis glandulosa	honey mesquite
Sabal mexicana	Rio Grande palmetto
Sideroxylon celastrinum	saffron-plum
Ulmus crassifolia	cedar elm
Yucca torreyi	Torrey's yucca
Zanthoxylum fagara	lime prickly-ash
Zizphus obtusifolia	lotebush

Herbaceous Understory
Bothriochloa barbinodis	cane beard grass
Buchloë dactyloides	buffalo grass
Cardiospermum corindum	faux-persil
Chloris pluriflora	multi-flower windmill grass
Cocculus diversifolius	snailseed
Eupatorium azureum	blue boneset
Malpighia glabra	wild crape-myrtle
Ruellia drummondiana	Drummond's wild petunia
Salvia coccinea	blood sage
Serjania brachycarpa	little-fruit slipplejack
Setaria leucopila	streambed bristle grass
Stachys drummondii	Drummond's hedge-nettle
Stellaria prostrata	prostrate starwort
Urtica chamidryoides	heart-leaf nettle
Verbesina microptera	Texas crownbeard

Live Oak Savanna

The trees of Live Oak Savanna are either evergreen or deciduous broadleaf and occur scattered or in groves. This community corresponds to Küchler #62 and Woodlands ERT.

Canopy
Characteristic Species
Prosopis glandulosa	honey mesquite
Quercus virginiana	live oak

Associates
Acacia rigidula	chaparro-prieto
Aloysia gratissima	whitebrush
Diospyros texana	Texas persimmon
Lycium berlandieri	silver desert-thorn

Maclura pomifera	osage-orange
Opuntia engelmannii	cactus-apple
Opuntia leptocaulis	Christmas cholla
Oxalis frutescens	shrubby wood-sorrel
Sideroxylon lanuginosum	gum bully
Zanthoxylum fagara	lime prickly-ash
Ziziphus obtusifolia	lotebush

Herbaceous Understory

Andropogon gerardii	big bluestem
Aristida purpurea	purple three-awn
Bouteloua gracilis	blue grama
Brachiaria ciliatissima	fringed signal grass
Callirhoe involucrata	purple poppy-mallow
Chamaecrista fasciculata	sleepingplant
Chamaesyce serpens	matted sandmat
Cnidoscolus texanus	Texas bull-nettle
Croton capitatus	hogwort
Elyonurus tripsacoides	pan-American balsamscale
Gaillardia aestivalis	lance-leaf blanket-flower
Helianthus argophyllus	silver-leaf sunflower
Heteropogon contortus	twisted tanglehead
Heterotheca subaxillaris	camphorweed
Muhlenbergia capillaris	hairy awn muhly
Panicum hallii	Hall's panic grass
Panicum virgatum	wand panic grass
Pappophorum bicolor	pink pappus grass
Paspalum monostachyum	gulf dune crown grass
Paspalum plicatulum	brown-seed crown grass
Schizachyrium scoparium	little false bluestem
Schrankia microphylla	little-leaf sensitive-briar
Senna roemeriana	two-leaved wild sensitive-plant
Solanum elaeagnifolium	silver-leaf nightshade
Sorghastrum nutans	yellow Indian grass
Sporobolus cryptandrus	sand dropseed
Tephrosia lindheimeri	Lindheimer's hoary-pea
Thelesperma nuecense	Rio Grande greenthread
Trachypogon secundus	one-sided crinkle-awn grass
Tridens elegans	Silveus' grass
Tridens flavus	tall redtop

South Texas Prairie

The South Texas Plains formerly supported a large area dominated by mid-height grasslands. The prairies, however, have been overgrazed resulting in an over-abundance of shrubs in this once grass-dominated area. Other factors in the elimination of grassland habitat include conversion to cropland and suppression of natural fires. This community corresponds to Prairies ERT.

Characteristic Species
Bothriochloa barbinodis	cane beard grass
Chloris pluriflora	multi-flower windmill grass

Associates
Ambrosia confertiflora	weak-leaf burr-ragweed
Aristida purpurea	purple three-awn
Bouteloua gracilis	blue grama
Buchloë dactyloides	buffalo grass
Chamaesyce serpens	matted sandmat
Chloris cucullata	hooded windmill grass
Cnidoscolus texanus	Texas bull-nettle
Hilaria belangeri	curly-mesquite
Pappophorum bicolor	pink pappus grass
Phyllanthus polygonoides	smartweed leaf-flower
Schizachyrium scoparium	little false bluestem
Senna roemeriana	two-leaved wild sensitive-plant
Setaria macrostachya	plains bristle grass
Thymophylla pentachaeta	five-needle pricklyleaf
Tridens muticus	awnless fluff grass

WETLAND SYSTEMS

Floodplain Forest

Floodplain Forest in the South Texas Plains is a deciduous forest which occurs along the major rivers. This community corresponds to Floodplain Forests ERT.

Canopy
Characteristic Species
Carya illinoinensis	pecan
Celtis pallida	shiny hackberry
Morus rubra	red mulberry
Prosopis glandulosa	honey mesquite
Ulmus americana	American elm
Ulmus crassifolia	cedar elm

Associates
Acer negundo	ash-leaf maple
Celtis laevigata	sugar-berry
Fraxinus berlandieriana	Mexican ash
Juglans nigra	black walnut
Platanus occidentalis	American sycamore
Quercus fusiformis	plateau oak
Quercus macrocarpa	burr oak
Salix nigra	black willow
Sapindus saponaria	wing-leaf soapberry

Woody Understory
Acacia minuta	coastal-scrub wattle
Ampelopsis arborea	peppervine

Baccharis neglecta	Roosevelt-weed
Diospyros texana	Texas persimmon
Mimosa pigra	black mimosa
Parkinsonia aculeata	Jerusalem-thorn
Rubus trivialis	southern dewberry
Sideroxylon lanuginosum	gum bully
Smilax bona-nox	fringed greenbriar
Toxicodendron pubescens	atlantic poison-oak
Vitis mustangensis	mustang grape

Herbaceous Understory

Ambrosia psilostachya	western ragweed
Carex brittoniana	Britton's sedge
Chasmanthium latifolium	Indian wood-oats
Chloracantha spinosus	Mexican devilweed
Clematis drummondii	Texas virgin's-bower
Cocculus diversifolius	snailseed
Dicliptera brachiata	branched foldwing
Elymus virginicus	Virginia wild rye
Malvaviscus drummondii	Texas wax-mallow
Nassella leucotricha	Texas wintergrass
Panicum virgatum	wand panic grass
Rivina humilis	rougeplant
Ruellia nudiflora	violet wild petunia
Salvia coccinea	blood sage
Urtica chamaedryoides	heart-leaf nettle
Verbesina virginica	white crownbeard

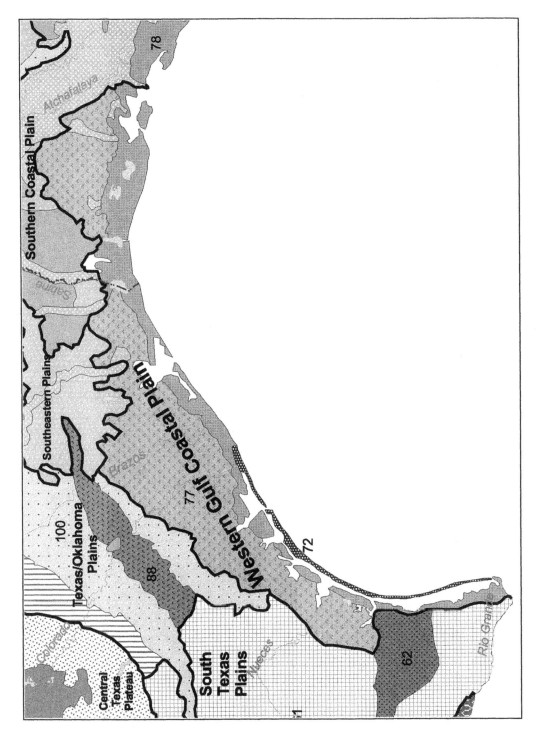

Western Gulf Coastal Plain

Western Gulf Coastal Plain

Introduction

The Western Gulf Coastal Plain is a long narrow grassy plain bordering the Gulf of Mexico along the Texas coast. The barrier islands off the coast are also included. The region supports both upland and lowland grasslands, evergreen woodlands, freshwater marshes, and salt marshes. Grasslands were once the characteristic vegetation of this region. The area is now mostly cultivated or heavily grazed, with natural vegetation occurring as fragmented remnants. Excessive grazing and other human disturbances have caused much of the area to be invaded by trees and brush such as mesquite, oaks, prickly pear, and acacias.

The primary sources used to develop descriptions and species lists are Barbour and Billings 1988; Bill Carr 1992 (pers. com.); Correll and Johnston 1970; Jones 1977; McMahan et al. 1984; Texas Natural Heritage program 1991; Tharp 1926.

Dominant Ecological Communities

Upland Systems
  Coastal Woodland
  Coastal Prairie
  Island Prairie
Wetland Systems
  Floodplain Forest
  Fresh Water Marsh
Estuarine System
  Salt Marsh

UPLAND SYSTEMS

Coastal Woodland

Coastal Woodland occurs throughout the Coastal Prairie. The composition of Coastal Woodland varies according to geographical distribution. The proximity of the woodland to Floodplain Forest, prairies, and other woodland communities such as Cross Timbers Woodland at the northeastern boundary of the region are important factors. This community corresponds to Woodlands ERT.

Canopy
Characteristic Species
  Carya illinoinensis   pecan
  Quercus stellata   post oak
  Quercus virginiana   live oak
Associates
  Celtis laevigata   sugar-berry
  Persea borbonia   red bay
  Quercus hemisphaerica   Darlington's oak
  Quercus laurifolia   laurel oak
  Quercus marilandica   blackjack oak
  Ulmus alata   winged elm

Woody Understory

Crateagus spp.	hawthorn
Ilex vomitoria	yaupon

Herbaceous Understory

Paspalum plicatulam	brown-seed crown grass
Schizachyrium scoparium	little false bluestem

Coastal Prairie

The Western Gulf Coastal Plain Region was once characterized by the tallgrass upland prairies. Woodlands of live oak may be mixed with the prairie. Most of the prairie today has been converted into farmland or heavily grazed. This community corresponds to Küchler #77 and Prairies ERT.

Characteristic Species

Andropogon gerardii	big bluestem
Schizachyrium scoparium	little false bluestem
Spartina spartinae	gulf cord grass
Sorghastrum nutans	yellow Indian grass

Associates

Andropogon glomeratus	bushy bluestem
Andropogon virginicus	broom-sedge
Aristida purpurea	purple three-awn
Bouteloua curtipendula	side-oats grama
Buchloë dactyloides	buffalo grass
Fimbristylis puberula	hairy fimbry
Muhlenbergia capillaris	hairy-awn muhly
Paspalum monostachyum	gulf dune crown grass
Paspalum plicatulum	brown-seed crown grass
Schizachyrium tenerum	slender false bluestem
Sporobolus indicus	smut grass

Island Prairie

Island Prairie is a midgrass to tallgrass prairie occurring on the partially stabilized dunes of the coastal barrier islands. This community corresponds to Küchler #72 and Prairies ERT.

Characteristic Species

Panicum amarum	bitter panic grass
Schizachyrium scoparium	little false bluestem
Uniola paniculata	sea-oats

Associates

Agalinis maritima	saltmarsh false foxglove
Ambrosia psilostachya	western ragweed
Aphanostephus skirrhobasis	Arkansas dozedaisy
Buchnera americana	American bluehearts
Chamaecrista fasciculata	sleepingplant
Chamaesyce bombensis	dixie sandmat
Croton capitatus	hogwort
Croton punctatus	gulf croton

Digitaria texana	Texas crab grass
Erythrina herbacea	red cardinal
Eupatorium betonicifolium	betony-leaf thoroughwort
Fimbristylis caroliniana	Carolina fimbry
Fimbristylis castanea	marsh fimbry
Helianthus argophyllus	silver-leaf sunflower
Hypericum hypericoides	St. Andrew's-cross
Ipomaea imperati	beach morning-glory
Ipomaea pes-caprae	bay-hops
Juncus spp.	rush
Limonium carolinianum	Carolina sea-lavendar
Machaeranthera phyllocephala	camphor-daisy
Muhlenbergia capillaris	hairy-awn muhly
Myrica cerifera	southern bayberry
Oenothera drummondii	beach evening-primrose
Opuntia engelmannii	cactus-apple
Panicum amarum	bitter panic grass
Panicum virgatum	wand panic grass
Paspalum monostachyum	gulf dune crown grass
Paspalum vaginatum	talquezal
Persea borbonia	red bay
Quercus hemisphaerica	Darlington's oak
Quercus virginiana	live oak
Sabal minor	dwarf palmetto
Sabatia arenicola	sand rose-gentian
Samolus ebracteatus	limewater brookweed
Senecio riddellii	Riddell's ragwort
Sesuvium portulacastrum	shoreline sea-purslane
Solidago sempervirens	seaside goldenrod
Spartina patens	salt-meadow cord grass
Trichoneura elegans	Siveus' grass
Vasevochloa multinervosa	Texas grass
Vigna luteola	piedmont cow-pea

WETLAND SYSTEMS

Floodplain Forest

Floodplain Forest occurs on floodplains and along bayous in the Western Gulf Coastal Plain. Evergreen Woodland and grassland species from adjacent communities contribute to the diversity of the floodplain forest. This community corresponds to Floodplain Forests ERT.

Canopy
Characteristic Species

Carya illinoinensis	pecan
Carya texana	black hickory
Fraxinus pennsylvanica	green ash
Platanus occidentalis	American sycamore
Populus deltoides	eastern cottonwood

Quercus nigra	water oak
Quercus phellos	willow oak
Quercus shumardii	Shumard's oak
Quercus virginiana	live oak
Salix nigra	black willow
Sapindus saponaria	wing-leaf soapberry
Ulmus americana	American elm

Associates

Celtis laevigata	sugar-berry
Liquidambar styraciflua	sweet-gum
Quercus lyrata	overcup oak
Quercus pagoda	cherry-bark oak
Ulmus crassifolia	cedar elm

Woody Understory

Aesculus pavia	red buckeye
Ampelopsis arborea	peppervine
Bignonia capreolata	crossvine
Campsis radicans	trumpet-creeper
Carpinus caroliniana	American hornbeam
Cornus drummondii	rough-leaf dogwood
Diospyros virginiana	common persimmon
Forestiera acuminata	eastern swamp-privet
Ilex decidua	deciduous holly
Ilex vomitoria	yaupon
Lonicera sempervirens	trumpet honeysuckle
Maclura pomifera	osage-orange
Ostrya virginiana	eastern hop-hornbeam
Parthenocissus quinquefolia	Virginia-creeper
Sambucus canadensis	American elder
Toxicodendron pubescens	Atlantic poison-ivy
Viburnum dentatum	southern arrow-wood

Herbaceous Understory

Arisaema dracontium	greendragon
Carex amphibola	eastern narrow-leaf sedge
Carex cephalophora	oval-leaf sedge
Carex cherokeensis	Cherokee sedge
Chasmanthium latifolium	Indian wood-oats
Chasmanthium laxum	Slender wood oats
Corydalis micrantha	small-flower fumewort
Dicliptera brachiata	branched foldwing
Elephantopus carolinianus	Carolina elephant's-foot
Elymus virginicus	Virginia wild rye
Nassella leucotricha	Texas wintergrass
Nemophila phacelioides	large-flower baby-blue-eyes
Ruellia caroliniensis	Carolina wild petunia
Senecio glabellus	cress-leaf ragwort
Stellaria media	common chickweed
Verbesina virginica	white crownbeard

Fresh Water Marsh

Fresh Water Marsh occurs in low, poorly drained areas where water is near or at the surface for prolonged periods of time. In this region, Fresh Water Marsh is usually found landward of Salt Marsh and Coastal Prairie. Emergent herbaceous plants that dominate the marsh are alligator weed and maiden-cane. Woody vegetation is almost completely lacking. This community corresponds to Nonestuarine Marshes ERT.

Characteristic Species

Alternanthera philoxeroides	alligator-weed
Panicum hemitomon	maiden-cane

Associates

Cabomba caroliniana	Carolina fanwort
Ceratophyllum demersum	coontail
Hydrocotyle verticillata	whorled marsh-pennywort
Lemna aequinoctialis	lesser duckweed
Nymphaea odorata	American white water-lily
Pontederia cordata	pickerelweed
Sagittaria papillosa	nipple-bract arrowhead
Typha latifolia	broad-leaf cat-tail

ESTUARINE SYSTEMS

Salt Marsh

Salt Marsh occurs along the Gulf coast and is influenced by tidal activity. The marshes that are subject to the daily ebb and flow of the tides will have *Juncus* species. This community corresponds to Küchler #78 and Estuarine Marshes ERT.

Characteristic Species

Distichlis spicata	coastal salt grass
Spartina alterniflora	saltwater cord grass
Spartina patens	salt-meadow cord grass
Spartina spartinae	gulf cord grass

Associates

Avicennia germinans	black mangrove
Batis maritima	turtleweed
Halodule beaudettei	shoalweed
Juncus effusus	lamp rush
Juncus roemerianus	Roemer's rush
Panicum hemitomon	maiden-cane
Panicum repens	torpedo grass
Panicum virgatum	wand panic grass
Paspalum vaginatum	talquezal
Phragmites australis	common reed
Ruppia maritima	beaked ditch-grass
Sagittaria lancifolia	bull-tongue arrowhead
Salicornia virginica	woody saltwort
Scirpus americanus	chairmaker's bulrush

Scirpus californicus	California bulrush
Spartina cynosuroides	big cord grass
Sporobolus virginicus	seashore dropseed
Typha domingensis	southern cat-tail
Zizaniopsis miliacea	marsh-millet

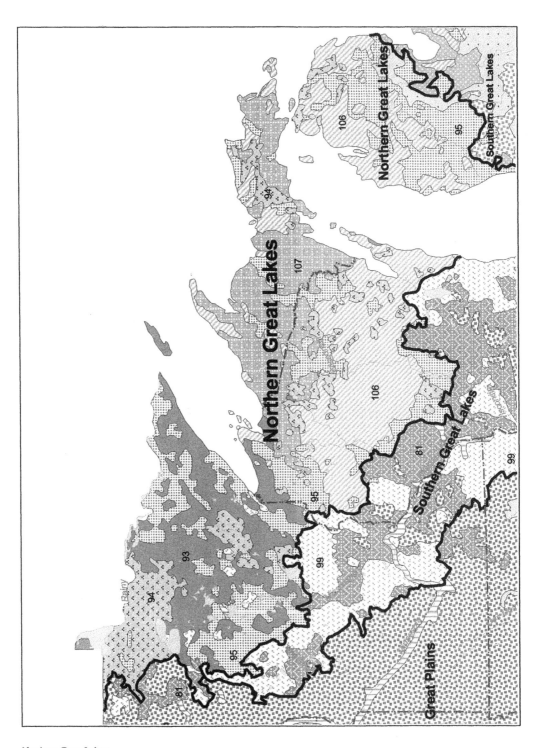

Northern Great Lakes

Northern Great Lakes

Introduction

The Northern Great Lakes region encompasses northern and upper peninsula Michigan, northern Wisconsin, and northeastern Minnesota. This region contains a narrow strip that crosses Michigan, Wisconsin, and Minnesota, known as the tension zone, that splits these states into northern and southern provinces. The northern and the southern zones in these states support distinctive flora, however, species overlap in the tension zone.

Spruce–fir forests dominate in northern Minnesota and the upper peninsula of Michigan. Northern Hardwoods–Conifer and Great Lakes Pine Forest dominate in Wisconsin and Michigan. Open canopy forests and barrens are also notable, especially the Pine Barrens. This vegetation type is becoming less common due to fire suppression. Wetland areas are important as habitat for migrating waterfowl. Forested bogs, shrub swamps, and wet meadows are common.

The primary sources used to develop descriptions and species lists are Aaseng et al. 1992; Barbour and Billings 1988; Chapman 1986; Curtis 1959.

Dominant Ecological Communities

Upland Systems
Northern Hardwoods–Conifer Forest
Great Lakes Spruce–Fir Forest
Great Lakes Pine Forest
Pine Barrens

Wetland Systems
Conifer Bog
Hardwood–Conifer Swamp
Northern Shrub Swamp
Open Bog
Northern Wet Meadow

UPLAND SYSTEMS

Northern Hardwoods–Conifer Forest

Northern Hardwoods–Conifer Forest occurs on dry to mesic sites and is frequently found on sandy-deep loam soils, but may also be found on coarser soils and on slopes. Species diversity is greater in the southern portions of this system. This community corresponds to Küchler #106 and #107 and Eastern Deciduous and Mixed Forests ERT.

Canopy
Characteristic Species

Abies balsamea	balsam fir
Acer rubrum	red maple
Acer saccharum	sugar maple
Betula alleghaniensis	yellow birch
Betula papyrifera	paper birch
Fagus grandifolia	American beech

Pinus strobus	eastern white pine
Quercus rubra	northern red oak
Tilia americana	American basswood
Tsuga canadensis	eastern hemlock

Associates

Acer pensylvanicum	striped maple
Carya ovata	shag-bark hickory
Fraxinus americana	white ash
Fraxinus nigra	black ash
Juglans cinerea	white walnut
Ostrya virginiana	eastern hop-hornbeam
Picea glauca	white spruce
Pinus resinosa	red pine
Populus grandidentata	big-tooth aspen
Prunus pensylvanica	fire cherry
Prunus serotina	black cherry
Quercus alba	northern white oak
Quercus ellipsoidalis	northern pin oak
Thuja occidentalis	eastern arborvitae
Ulmus americana	American elm
Ulmus rubra	slippery elm

Woody Understory

Acer spicatum	mountain maple
Corylus cornuta	beaked hazelnut
Dirca palustris	eastern leatherwood
Kalmia latifolia	mountain-laurel
Linnaea borealis	American twinflower
Lonicera canadensis	American fly-honeysuckle
Lonicera oblongifolia	swamp fly-honeysuckle
Prunus virginiana	choke cherry
Ribes lacustre	bristly black gooseberry
Taxus canadensis	American yew
Viburnum acerifolium	maple-leaf arrow-wood

Herbaceous Understory

Actaea pachypoda	white baneberry
Adiantum pedatum	northern maidenhair
Anemone quinquefolia	nightcaps
Aralia nudicaulis	wild sarsaparilla
Aralia racemosa	American spikenard
Aster macrophyllus	large-leaf aster
Athyrium filix-femina	subarctic lady fern
Botrychium virginianum	rattlesnake fern
Cardamine diphylla	crinkleroot
Circaea alpina	small enchanter's-nightshade
Claytonia caroliniana	Carolina springbeauty
Clintonia borealis	yellow bluebead-lily
Conopholis americana	American squawroot

Cornus canadensis	Canadian bunchberry
Dicentra canadensis	squirrel-corn
Dryopteris campyloptera	mountain wood fern
Epifagus virginiana	beechdrops
Erythronium americanum	American trout-lily
Galium triflorum	fragrant bedstraw
Gymnocarpium dryopteris	western oak fern
Hepatica nobilis	liverwort
Huperzia lucidula	shining club-moss
Lycopodium obscurum	prince-pine
Maianthemum canadense	false lily-of-the-valley
Maianthemum racemosum	feathery false Solomon's-seal
Mitchella repens	partridge-berry
Mitella nuda	bare-stem bishop's-cap
Oryzopsis asperifolia	white-grain mountain-rice grass
Osmorhiza claytonii	hairy sweet-cicely
Panax trifolius	dwarf ginseng
Polygonatum pubescens	hairy Solomon's-seal
Streptopus roseus	rosy twistedstalk
Trientalis borealis	American starflower
Trillium cernuum	whip-poor-will-flower
Trillium grandiflorum	large-flower wakerobin
Uvularia grandiflora	large-flower bellwort
Uvularia sessilifolia	sessile-leaf bellwort
Viola cucullata	marsh blue violet
Viola pubescens	downy yellow violet

Great Lakes Spruce–Fir Forest

Great Lakes Spruce–Fir Forest occurs in northern Minnesota, the upper peninsula of Michigan, on islands in the northern Great Lakes, and scattered in extreme northern areas of Wisconsin. The dominant species are balsam fir and white spruce. The spruce–fir forest is also known as the boreal forest due to its northern affinities. Cool temperatures, short growing season, abundant available moisture during the growing season, and deep snows in winter are the dominant climatic forces in the spruce–fir zone. Soils are generally sand and sandy loam. This community corresponds to Küchler #93 and Coniferous Forests ERT.

Canopy
Characteristic Species
Abies balsamea	balsam fir
Betula papyrifera	paper birch
Picea glauca	white spruce
Picea mariana	black spruce
Pinus strobus	eastern white pine
Thuja occidentalis	eastern arborvitae

Associates
Acer rubrum	red maple
Acer saccharum	sugar maple
Betula alleghaniensis	yellow birch
Carya cordiformis	bitter-nut hickory

Fraxinus americana	white ash
Fraxinus nigra	black ash
Fraxinus pennsylvanica	green ash
Larix laricina	American larch
Pinus banksiana	jack pine
Pinus resinosa	red pine
Populus balsamifera	balsam poplar
Populus grandidentata	big-tooth aspen
Populus tremuloides	quaking aspen
Prunus pensylvanica	fire cherry
Prunus serotina	black cherry
Quercus alba	northern white oak
Quercus ellipsoidalis	northern pin oak
Quercus rubra	northern red oak
Tilia americana	American basswood
Tsuga canadensis	eastern hemlock
Ulmus americana	American elm
Ulmus rubra	slippery elm

Woody Understory

Acer spicatum	mountain maple
Alnus incana	speckled alder
Carpinus caroliniana	American hornbeam
Cornus rugosa	round-leaf dogwood
Corylus cornuta	beaked hazelnut
Diervilla lonicera	northern bush-honeysuckle
Gaultheria hispidula	creeping-snowberry
Gaultheria procumbens	eastern teaberry
Ledum groenlandicum	rusty Labrador-tea
Linnaea borealis	American twinflower
Lonicera canadensis	American fly-honeysuckle
Lonicera hirsuta	hairy honeysuckle
Ostrya virginiana	eastern hop-hornbeam
Ribes cynosbati	eastern prickly gooseberry
Rubus idaeus	common red raspberry
Rubus parviflorus	western thimble-berry
Rubus pubescens	dwarf red raspberry
Sorbus americana	American mountain-ash
Sorbus decora	northern mountain-ash
Vaccinium angustifolium	late lowbush blueberry
Vaccinium myrtilloides	velvet-leaf blueberry

Herbaceous Understory

Actaea rubra	red baneberry
Anemone quinquefolia	nightcaps
Apocynum androsaemifolium	spreading dogbane
Aralia nudicaulis	wild sarsaparilla
Aster macrophyllus	large-leaf aster
Athyrium filix-femina	subarctic lady fern

Brachyelytrum erectum	bearded shorthusk
Carex arctata	drooping woodland sedge
Carex eburnea	bristle-leaf sedge
Carex pedunculata	long-stalk sedge
Carex pensylvanica	Pennsylvania sedge
Circaea alpina	small enchanter's-nightshade
Clintonia borealis	yellow bluebead-lily
Coptis trifolia	three-leaf goldthread
Cornus canadensis	Canadian bunchberry
Cypripedium arietinum	ram-head lady's-slipper
Dryopteris campyloptera	mountain wood fern
Equisetum arvense	field horsetail
Fragaria virginiana	Virginia strawberry
Galium triflorum	fragrant bedstraw
Goodyera oblongifolia	green-leaf rattlesnake-plantain
Gymnocarpium dryopteris	western oak fern
Hepatica nobilis	liverwort
Impatiens capensis	spotted touch-me-not
Iris lacustris	dwarf lake iris
Lycopodium clavatum	running ground-pine
Lycopodium obscurum	princess-pine
Maianthemum canadense	false lily-of-the-valley
Maianthemum racemosum	feathery false Solomon's-seal
Mitchella repens	partridge-berry
Mitella nuda	bare-stem bishop's-cap
Orthilia secunda	sidebells
Oryzopsis asperifolia	white-grain mountain-rice grass
Osmorhiza claytoni	hairy sweet-cicely
Phegopteris connectilis	narrow beech fern
Polygala paucifolia	gaywings
Polygonatum pubescens	hairy Solomon's-seal
Prenanthes alba	white rattlesnake-root
Pteridium aquilinum	northern bracken fern
Pyrola elliptica	shinleaf
Sanicula marilandica	Maryland black-snakeroot
Streptopus roseus	rosy twistedstalk
Trientalis borealis	American starflower
Trillium cernuum	whip-poor-will-flower
Viola blanda	sweet white violet
Viola conspersa	American dog violet
Viola pubescens	downy yellow violet

Great Lakes Pine Forest

Great Lakes Pine Forest occurs throughout the Northern Great Lakes region, mainly on dry sites. The dominant species are white pine, jack pine, and red pine. Pine forests, however, may have a mixture of hardwoods in the canopy especially red maple, quaking aspen, and oak. These forests are dependent upon periodic fires which help maintain the pines as dominants. This community corresponds to Küchler #95 and Eastern Deciduous and Mixed Forests ERT.

Canopy
Characteristic Species
Pinus banksiana	jack pine
Pinus resinosa	red pine
Pinus strobus	eastern white pine

Associates
Abies balsamea	balsam fir
Acer rubrum	red maple
Betula papyrifera	paper birch
Picea glauca	white spruce
Picea mariana	black spruce
Populus grandidentata	big-tooth aspen
Populus tremuloides	quaking aspen
Prunus serotina	black cherry
Quercus alba	northern white oak
Quercus coccinea	scarlet oak
Quercus ellipsoidalis	northern pin oak
Quercus rubra	northern red oak
Quercus velutina	black oak
Thuja occidentalis	eastern arborvitae

Woody Understory
Acer spicatum	mountain maple
Amelanchier stolonifera	running service-berry
Arctostaphylos uva-ursi	red bearberry
Comptonia peregrina	sweet-fern
Cornus rugosa	round-leaf dogwood
Corylus americana	American hazelnut
Corylus cornuta	beaked hazelnut
Diervilla lonicera	northern bush-honeysuckle
Gaultheria procumbens	eastern teaberry
Gaylussacia baccata	black huckleberry
Kalmia angustifolia	sheep-laurel
Lonicera canadensis	American fly-honeysuckle
Prunus pumila	sand cherry
Rubus flagellaris	whiplash dewberry
Vaccinium angustifolium	late lowbush blueberry
Vaccinium myrtilloides	velvet-leaf blueberry
Viburnum rafinesquianum	downy arrow-wood

Herbaceous Understory
Apocynum androsaemifolium	spreading dogbane
Aralia nudicaulis	wild sarsaparilla
Aster ciliolatus	Lindley's aster
Aster macrophyllus	large-leaf aster
Carex pensylvanica	Pennsylvania sedge
Cornus canadensis	Canadian bunchberry
Danthonia spicata	poverty wild oat grass
Deschampsia flexuosa	wavy hair grass
Fragaria virginiana	Virginia strawberry

Hieracium venosum	rattlesnake-weed
Lycopodium obscurum	princess-pine
Maianthemum canadense	false lily-of-the-valley
Melampyrum lineare	American cow-wheat
Oryzopsis pungens	short-awn mountain-rice grass
Pteridium aquilinum	northern bracken fern

Pine Barrens

Pine Barrens are open pine forests with a grassland type understory. It occurs in areas of very sandy or rocky soil. Historically, there were large areas in the northern Great Lakes covered with woodland/barren communities. Communities such as these are now rare due to the suppression of natural fires. Fires maintain the open canopy structure of such a community through the occasional killing and scarring of young trees. This community corresponds to Woodlands ERT.

Canopy
Characteristic Species
Pinus banksiana	jack pine
Pinus resinosa	red pine
Populus grandidentata	big-tooth aspen
Quercus ellipsoidalis	northern pin oak

Associates
Picea mariana	black spruce
Populus tremuloides	quaking aspen
Quercus macrocarpa	burr oak
Quercus palustris	pin oak
Quercus rubra	northern red oak

Woody Understory
Amelanchier stolonifera	running service-berry
Arctostaphylos uva-ursi	red bearberry
Ceanothus ovatus	redroot
Comptonia peregrina	sweet-fern
Corylus americana	American hazelnut
Corylus cornuta	beaked hazelnut
Diervilla lonicera	northern bush-honeysuckle
Gaylussacia baccata	black huckleberry
Salix bebbiana	long-beak willow
Salix humilis	small pussy willow
Vaccinium angustifolium	late lowbush blueberry
Viburnum rafinesquianum	downy arrow-wood

Herbaceous Understory
Andropogon gerardii	big bluestem
Carex pensylvanica	Pennsylvania sedge
Danthonia spicata	poverty wild oat grass
Lithospermum canescens	hoary puccoon
Lupinus perennis	sundial lupine
Pteridium aquilinum	northern bracken fern
Pulsatilla patens	American pasqueflower
Schizachyrium scoparium	little false bluestem

WETLAND SYSTEMS

Conifer Bog

Conifer Bog occurs throughout the Northern Great Lakes region but is most common in northern Minnesota. Black spruce is the most common tree species due to its ability to reproduce in the layers of sphagnum mosses which cover the forest floor. Conifer Bog develops on peatlands where the surface substrate becomes isolated from groundwater flow because of peat accumulation. The wetter areas in bog complexes are dominated by mosses and shrubs (see Open Bog) while the trees occur on drier sites such as crests and upper slopes of raised bogs. Most of the water and nutrients in the bogs come from rainfall. This results in low concentrations of dissolved nutrients which affect the growth rate of the vegetation. This community corresponds to Küchler #94 and Bogs and Fens ERT.

Canopy

Characteristic Species

Abies balsamea	balsam fir
Larix laricina	American larch
Picea mariana	black spruce
Thuja occidentalis	eastern arborvitae

Associates

Acer rubrum	red maple
Betula alleghaniensis	yellow birch
Betula papyrifera	paper birch
Fraxinus nigra	black ash
Picea glauca	white spruce
Pinus banksiana	jack pine
Pinus strobus	eastern white pine
Populus tremuloides	quaking aspen
Ulmus rubra	slippery elm

Woody Understory

Alnus incana	speckled alder
Andromeda polifera	bog-rosemary
Chamaedaphne calyculata	leatherleaf
Gaultheria hispidula	creeping-snowberry
Gaultheria procumbens	eastern teaberry
Ilex verticillata	common winterberry
Kalmia polifolia	bog laurel
Ledum groenlandicum	rusty Labrador-tea
Linnaea borealis	American twinflower
Nemopanthus mucronatus	catberry
Parthenocissus vitacea	thicket-creeper
Rubus pubescens	dwarf red raspberry
Vaccinium angustifolium	late lowbush blueberry
Vaccinium myrtilloides	velvet-leaf blueberry
Vaccinium oxycoccus	small cranberry
Vaccinium vitis-idaea	lingonberry

Herbaceous Understory

Aralia nudicaulis	wild sarsaparilla
Carex disperma	soft-leaf sedge
Carex trisperma	three-seed sedge
Clintonia borealis	yellow bluebead-lily
Coptis trifolia	three-leaf goldthread
Cypripedium acaule	pink lady's-slipper
Dicranum polysetum	feather moss
Dicranum undulatum	feather moss
Dryopteris campyloptera	mountain wood fern
Dryopteris cristata	crested wood fern
Equisetum fluviatile	water horsetail
Eriophorum vaginatum	tussock cotton-grass
Eriophorum virginicum	tawny cotton-grass
Hylocomium splendens	feather moss
Maianthemum canadense	false lily-of-the-valley
Maianthemum trifolium	three-leaf false Solomon's-seal
Monotropa uniflora	one-flower Indian-pipe
Osmunda cinnamomea	cinnamon fern
Pleurozium schreberi	feather moss
Polytrichum strictum	feather moss
Ptilium crista-castrensis	feather moss
Sarracenia purpurea	purple pitcherplant
Sphagnum fuscum	moss
Sphagnum magellanicum	moss
Sphagnum nemoreum	moss
Sphagnum recurvum	moss
Sphagnum rubellum	moss
Trientalis borealis	American starflower

Hardwood–Conifer Swamp

Hardwood–Conifer Swamp occurs on moist mineral soil or muck along floodplains, or as transitional communities between bogs and upland forests. This wetland type is similar to Hardwood Floodplain Forest, but beaver dams and wind-thrown logs rather than the annual or seasonal rise of the water determine the particular structure and composition of the forest. The most common characteristic species are black spruce, American larch (tamarack), and eastern arborvitae (white cedar); however, ash, birch, maple, and elm species are also common. This community corresponds to Swamp Forests ERT.

Canopy
Characteristic Species

Abies balsamea	balsam fir
Acer rubrum	red maple
Betula alleghaniensis	yellow birch
Fraxinus nigra	black ash
Larix laricina	American larch
Picea mariana	black spruce
Populus tremuloides	quaking aspen
Sorbus americana	American mountain-ash
Thuja occidentalis	eastern arborvitae

Associates
Betula papyrifera	paper birch
Fraxinus pennsylvanica	green ash
Pinus strobus	eastern white pine
Quercus macrocarpa	burr oak
Quercus rubra	northern red oak
Tsuga canadensis	eastern hemlock
Ulmus americana	American elm

Woody Understory

Alnus incana	speckled alder
Andromeda polifera	bog-rosemary
Chamaedaphne calyculata	leatherleaf
Gaultheria hispidula	creeping-snowberry
Gaultheria procumbens	eastern teaberry
Ledum groenlandicum	rusty Labrador-tea
Linnaea borealis	American twinflower
Nemopanthus mucronatus	catberry
Parthenocissus vitacea	thicket-creeper
Rubus pubescens	dwarf red raspberry
Toxicodendron vernix	poison sumac
Vaccinium angustifolium	late lowbush blueberry
Vaccinium myrtilloides	velvet-leaf blueberry
Vaccinium oxycoccus	small cranberry

Herbaceous Understory

Aralia nudicaulis	wild sarsaparilla
Athyrium filix-femina	subarctic lady fern
Caltha palustris	yellow marsh-marigold
Carex bromoides	brome-like sedge
Carex disperma	soft-leaf sedge
Carex leptalea	bristly-stalk sedge
Carex trisperma	three-seed sedge
Clintonia borealis	yellow bluebead-lily
Coptis trifolia	three-leaf goldthread
Cypripedium acaule	pink lady's-slipper
Dryopteris campyloptera	mountain wood fern
Dryopteris cristata	crested wood fern
Equisetum fluviatile	water horsetail
Eriophorum vaginatum	tussock cotton-grass
Eriophorum virginicum	tawny cotton-grass
Lycopus uniflorus	northern water-horehound
Maianthemum canadense	false lily-of-the-valley
Maianthemum trifolium	three-leaf false Solomon's-seal
Mentha arvensis	American wild mint
Mitella nuda	bare-stem bishop's cap
Onoclea sensibilis	sensitive fern
Osmunda cinnamomea	cinnamon fern
Osmunda claytoniana	interrupted fern

Pilea pumila	Canadian clearweed
Sarracenia purpurea	purple pitcher-plant
Scutellaria galericulata	hooded scullcap
Scutellaria lateriflora	mad dog scullcap
Symplocarpus foetidus	skunk-cabbage
Trientalis borealis	American starflower
Viola macloskeyi	smooth white violet

Northern Shrub Swamp

Northern Shrub Swamp is a common community along streams and around lakes where the soil is muck or peat. This community type is dominated by alder and is often referred to as alder thicket. Often shrub-dominated communities are considered a successional stage toward a forest community. Shrub swamps in this area, however, are relatively stable due to limiting conditions associated with stream or lakeside habitat that are unsuitable for tree growth. The Northern Shrub Swamp may also be caused by prior fires. This community corresponds to Shrub Swamps ERT.

Characteristic Woody Species

Alnus incana	speckled alder
Betula pumila	bog birch
Cornus stolonifera	redosier
Myrica gale	sweet gale
Ribes americanum	wild black currant
Spiraea alba	white meadowsweet
Viburnum nudum	possumhaw

Associates

Betula papyrifera	paper birch
Frangula alnus	alder-buckthorn
Fraxinus nigra	black ash
Larix laricina	American larch
Thuja occidentalis	eastern arborvitae
Toxicodendron vernix	poison sumac

Herbaceous Understory

Asclepias incarnata	swamp milkweed
Aster lanceolatus	white panicle aster
Aster puniceus	purple-stem aster
Bromus ciliatus	fringed brome
Calamagrostis canadensis	bluejoint
Campanula aparinoides	marsh bellflower
Carex lacustris	lakebank sedge
Carex prairea	prairie sedge
Carex stricta	uptight sedge
Chelone glabra	white turtlehead
Comarum palustre	purple marshlocks
Dryopteris cristata	crested woodfern
Eupatorium maculatum	spotted joe-pye-weed
Eupatorium perfoliatum	common boneset
Galium asprellum	rough bedstraw
Glyceria grandis	American manna grass
Impatiens capensis	spotted touch-me-not

Iris virginica	Virginia blueflag
Lycopus uniflorius	northern water-horehound
Mentha arvensis	American wild mint
Onoclea sensibilis	sensitive fern
Poa palustris	fowl blue grass
Polygonum sagittatum	arrow-leaf tearthumb
Rumex orbiculatus	greater water dock
Scirpus atrovirens	dark-green bulrush
Solidago canadensis	Canadian goldenrod
Solidago gigantea	late goldenrod
Thalictrum dasycarpum	purple meadow-rue
Thelypteris palustris	eastern marsh fern
Typha latifolia	broad-leaf cat-tail

Open Bog

Open Bog is one stage in the succession from open-water lakes to conifer bogs. A typical open bog has a surface layer of sphagnum mosses over a layer of dead and loosely compacted peat. The vegetation is dominated by shrubs, sedges, or cotton grass. Open Bog remains open due to periodic fluctuations in the water table correlated with weather cycles. Stunted black spruce, American larch (tamarack), and shrubs may be scattered over the area, but cover no more than 30 percent of the area. Surface layers in the bog may become dry enough to support fires which will destroy the advancing shrubs and forest. Pitcher-plants and sundews are some of the insectivorous species that are unique to Open Bog. Many areas that were once open bogs have been converted to cranberry marshes. Other disturbances, such as peat mining, have contributed to the decline of bogs and the habitat which supports bog species. This community corresponds to Bogs and Fens ERT.

Characteristic Species

Andromeda polifolia	bog-rosemary
Betula nana	swamp birch
Carex oligosperma	few-seed sedge
Carex trisperma	three-seed sedge
Chamaedaphne calyculata	leatherleaf
Drosera rotundifolia	round-leaf sundew
Eriophorum vaginatum	tussock cotton-grass
Kalmia polifolia	bog-laurel
Ledum groenlandicum	rusty Labrador-tea
Sarracenia purpurea	purple pitcherplant
Sphagnum cuspidatum	moss
Sphagnum magellanicum	moss
Vaccinium angustifolium	late lowbush blueberry

Associates

Arethusa bulbosa	dragon's-mouth
Betula pumila	bog birch
Calla palustris	water-dragon
Calopogon tuberosus	tuberous grass-pink
Carex lasiocarpa	woolly-fruit sedge
Carex limosa	mud sedge
Carex magellanica	boreal-bog sedge
Carex pauciflora	few-flower sedge
Comarum palustre	purple marshlocks

Cypripedium acaule	pink lady's-slipper
Dryopteris cristata	crested wood fern
Dulicheum arundinacea	three-way sedge
Eriophorum vaginatum	tussock cotton-grass
Eriophorum virginicum	tawny cotton-grass
Gaultheria hispidula	creeping-snowberry
Gaultheria procumbens	eastern teaberry
Iris versicolor	harlequin blueflag
Maianthemum trifolium	three-leaf false-Solomon's-seal
Menyanthes trifoliata	buck-bean
Nemopanthus mucronatus	catberry
Platanthera clavellata	green woodland orchid
Pogonia ophioglossoides	snake-mouth orchid
Rhynchospora alba	white beak sedge
Scheuchzeria palustris	rannoch-rush
Sphagnum capillifolium	moss
Sphagnum fuscum	moss
Sphagnum majus	moss
Sphagnum papillosum	moss
Sphagnum recurvum	moss
Sphagnum russowii	moss
Sphagnum teres	moss
Thelypteris palustris	eastern marsh fern
Trientalis borealis	American starflower
Utricularia cornuta	horned bladderwort
Vaccinium macrocarpon	large cranberry
Vaccinium myrtilloides	velvet-leaf blueberry
Vaccinium oxycoccos	small cranberry

Northern Wet Meadow

Northern Wet Meadow typically borders streams but can also be found on pond and lake margins and above beaver dams. The dominant species, depending upon the level of moisture, are sedges, bulrushes, or cat-tails. The soil is characteristically raw peat or muck and water is always plentiful. The stability of the meadow communities depends upon the degree of saturation. Periodically, fire burns the dryer areas inhibiting the invasion of shrub and tree species. This community corresponds to Meadowlands ERT.

Characteristics Species

Asclepias incarnata	swamp milkweed
Calamagrostis canadensis	bluejoint
Carex aquatilis	leafy tussock sedge
Carex haydenii	cloud sedge
Carex lacustris	lakebank sedge
Carex rostrata	swollen beaked sedge
Carex stricta	uptight sedge
Cicuta bulbifera	bulbet-bearing water-hemlock
Glyceria canadensis	rattlesnake manna grass
Phragmites australis	common reed
Poa palustris	fowl blue grass
Sagittaria latifolia	duck-potato

Scirpus acutus	hard-stem bulrush
Scirpus americanus	chairmaker's bulrush
Scirpus atrovirens	dark-green bulrush
Scirpus cyperinus	cottongrass bulrush
Scirpus fluviatilis	river bulrush
Scirpus heterochaetus	pale great bulrush
Scirpus tabernaemontani	soft-stem bulrush
Typha angustifolia	narrow-leaf cat-tail
Typha latifolia	broad-leaf cat-tail

Associates

Aster lanceolatus	white panicle aster
Aster puniceus	purple-stem aster
Bromus ciliatus	fringed brome
Calamagrostis stricta	slim-stem reed grass
Campanula aparinoides	marsh bellflower
Carex lanuginosa	woolly sedge
Chelone glabra	white turtlehead
Cicuta maculata	spotted water-hemlock
Equisetum arvense	field horsetail
Eupatorium maculatum	spotted joe-pye-weed
Eupatorium perfoliatum	common boneset
Euthamia graminifolia	flat-top goldentop
Impatiens capensis	spotted touch-me-not
Lycopus americanus	cut-leaf water-horehound
Lycopus uniflorus	northern water-horehound
Mentha arvensis	American wild mint
Onoclea sensibilis	sensitive fern
Polygonum amphibium	water smartweed
Polygonum sagittatum	arrow-leaf tearthumb
Scutellaria galericulata	hooded skullcap
Scutellaria lateriflora	mad dog skullcap
Solidago gigantea	late goldenrod
Solidago uliginosa	bog goldenrod
Spartina pectinata	freshwater cord grass
Spiraea alba	white meadowsweet
Thalictrum dasycarpum	purple meadow-rue
Thelypteris palustris	eastern marsh fern
Verbena hastata	simpler's-joy

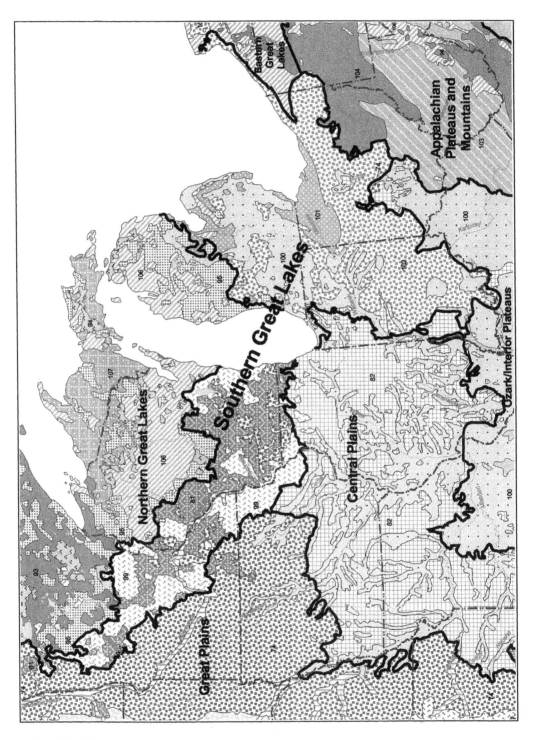

Southern Great Lakes

Southern Great Lakes

Introduction

The Southern Great Lakes region includes central Minnesota, the southern half of Wisconsin, Michigan, Indiana (excluding the very western portion), the northern half of Ohio, and a narrow band along Lake Erie which extends the region into a small area of Pennsylvania and eastern New York. The vegetation of the region is characterized by temperate deciduous forests. Maple–Basswood Forest is most common in Minnesota and Wisconsin. Beech–Maple and Oak–Hickory Forest are most common in Michigan, Indiana, and Ohio. Oak Barrens is a major vegetation type in Minnesota and Wisconsin which marks the transition between the forests in the east and the prairies to the west. A small but important number of prairie communities occur in the region. The prairies are generally mesic prairies, however, some occur on both very dry and wet areas.

Wetland systems include floodplain forests, shrub swamps, and wet meadows. Bogs occur but are not as prevalent as they are in the Northern Great Lakes Region.

Unique communities in the region include beech and dune communities associated with the Great Lakes. The beaches of the Great Lakes support a number of species which are rare on other lakes. The most important dune species along the Great Lakes is American beach grass (*Ammophila breviligulata*).

The primary sources used to develop descriptions and species lists are Barbour and Billings 1988; Braun 1950; Curtis 1959.

Dominant Ecological Communities
Upland Systems
Maple–Basswood Forest
Oak–Hickory Forest
Beech–Maple Forest
Oak Barrens
Mesic Prairie
Wetland Systems
Great Lakes Floodplain Forest
Shrub Swamp
Wet Meadow

UPLAND SYSTEMS

Maple–Basswood Forest

Maple–Basswood Forest extends from central Minnesota, which contains "Big Woods," the largest continuous area of Maple–Basswood Forest known, south into Wisconsin. The characteristic species are sugar maple, basswood, and beech. This forest occurs on fine-textured and well-drained loamy soils, also referred to as gray-brown forest soils. Maple–Basswood Forest is self-perpetuating in the absence of catastrophic disturbance and climate change. Natural stands of Maple–Basswood Forest are becoming rare because the soils on which the forest grows are suitable for cultivation and much of the area has been converted for crops. Grazing and lumbering has contributed to the loss of many of the ground layer species. This community corresponds to Küchler #99 and Eastern Deciduous and Mixed Forests ERT.

Canopy
Characteristic Species
Acer saccharum	sugar maple
Quercus rubra	northern red oak
Tilia americana	American basswood
Ulmus rubra	slippery elm

Associates
Betula papyrifera	paper birch
Carya cordiformis	bitter-nut hickory
Carya ovata	shag-bark hickory
Celtis occidentalis	common hackberry
Fagus grandifolia	American beech
Fraxinus americana	white ash
Fraxinus pennsylvanica	green ash
Fraxinus quadrangulata	blue ash
Gymnocladus dioicus	Kentucky coffeetree
Juglans cinerea	white walnut
Juglans nigra	black walnut
Populus grandidentata	big-tooth aspen
Prunus serotina	black cherry
Quercus alba	northern white oak
Quercus bicolor	swamp white oak
Quercus macrocarpa	burr oak
Quercus velutina	black oak
Ulmus americana	American elm
Ulmus thomasii	rock elm

Woody Understory
Acer negundo	ash-leaf maple
Carpinus caroliniana	American hornbeam
Celastrus scandens	American bittersweet
Cornus alternifolia	alternate-leaf dogwood
Corylus americana	American hazelnut
Dirca palustris	eastern leatherwood
Ostrya virginiana	eastern hop-hornbeam
Parthenocissus vitacea	thicket-creeper
Ribes cynosbati	eastern prickly gooseberry
Vitis vulpina	frost grape

Herbaceous Understory
Actaea pachypoda	white baneberry
Adiantum pedatum	northern maidenhair
Allium tricoccum	ramp
Amphicarpaea bracteata	American hog-peanut
Anemone quinquefolia	nightcaps
Arisaema triphyllum	jack-in-the-pulpit
Athyrium filix-femina	subarctic lady fern
Brachyelytrum erectum	bearded shorthusk
Cardamine concatenata	cut-leaf toothwort

Carex pedunculata	long-stalk sedge
Carex pensylvanica	Pennsylvania sedge
Caulophyllum thalictroides	blue cohosh
Circaea lutetiana	broad-leaf enchanter's-nightshade
Claytonia virginica	Virginia springbeauty
Cryptotaenia canadensis	Canadian honewort
Cystopteris fragilis	brittle bladder fern
Dicentra canadensis	squirrel-corn
Dicentra cucullaria	Dutchman's-breeches
Elymus hystrix	eastern bottle-brush grass
Enemion biternatum	eastern false rue-anemone
Erythronium americanum	American trout-lily
Galium aparine	sticky-willy
Galium concinnum	shining bedstraw
Galium triflorum	fragrant bedstraw
Geranium maculatum	spotted crane's-bill
Geum canadense	white avens
Hepatica nobilis	liverwort
Hydrophyllum virginianum	Shawnee-salad
Laportea canadensis	Canadian wood-nettle
Maianthemum racemosum	feathery false Solomon's-seal
Osmorhiza claytonii	hairy sweet-cicely
Phryma leptostachya	American lopseed
Podophyllum peltatum	may-apple
Polygonatum pubescens	hairy Solomon's-seal
Prenanthes alba	white rattlesnake-root
Sanguinaria canadensis	bloodroot
Sanicula odorata	clustered black-snakeroot
Smilax ecirrata	upright carrion-flower
Thalictrum dioicum	early meadow-rue
Trillium grandiflorum	large-flower wakerobin
Uvularia grandiflora	large-flower bellwort
Vernonia baldwinii	western ironweed
Viola cucullata	marsh blue violet
Viola pubescens	downy yellow violet

Oak–Hickory Forest

Oak–Hickory Forest covers extensive areas from Texas to Canada. Oak–Hickory Forest in the Southern Great Lakes Region is part of the northern division of the broader Oak–Hickory Forest association and occupies large areas across southern Michigan, Indiana, and central Ohio. Similar communities of predominantly oak species also occur in Minnesota and Wisconsin. White, red, and black oak and shagbark hickory are the most consistently occurring characteristic species. The Oak–Hickory Forest occurs on dry to dry-mesic sites. This community corresponds to Küchler #100 and Eastern Deciduous and Mixed Forests ERT.

Canopy

Characteristic Species

Carya alba	mockernut hickory
Carya glabra	pignut hickory
Carya ovata	shag-bark hickory
Quercus alba	northern white oak
Quercus macrocarpa	burr oak
Quercus rubra	northern red oak
Quercus shumardii	Shumard's oak
Quercus velutina	black oak

Associates

Acer rubrum	red maple
Acer saccharum	sugar maple
Carya cordiformis	bitter-nut hickory
Carya ovalis	red hickory
Fagus grandifolia	American beech
Fraxinus americana	white ash
Juglans nigra	black walnut
Pinus strobus	eastern white pine
Prunus serotina	black cherry
Quercus coccinea	scarlet oak
Quercus muehlenbergii	chinkapin oak
Tilia americana	American basswood
Tsuga canadensis	eastern hemlock
Ulmus americana	American elm

Woody Understory

Ceanothus americanus	New Jersey-tea
Corylus americana	American hazelnut
Hamamelis virginiana	American witch-hazel
Ostrya virginiana	eastern hop-hornbeam
Sassafras albidum	sassafras
Smilax tamnoides	chinaroot
Vitis aestivalis	summer grape

Herbaceous Understory

Actaea pachypoda	white baneberry
Adiantum pedatum	northern maidenhair
Brachyeletrum erectum	bearded shorthusk
Bromus pubescens	hairy woodland brome
Carex albursina	white bear sedge
Carex rosea	rosy sedge
Corallorrhiza maculata	summer coralroot
Galium triflorum	fragrant bedstraw
Hackelia virginiana	beggar's-lice
Lysimachia quadrifolia	whorled yellow loosestrife
Monotropa uniflora	one-flower Indian-pipe
Sanicula marilandica	Maryland black-snakeroot
Viola pubescens	downy yellow violet

Beech–Maple Forest

Beech–Maple Forest occurs throughout Indiana, Ohio, western Pennsylvania, and extends across the southern half of Michigan's lower peninsula. Beech–Maple occurs on rolling or slightly sloping topography in areas of medium moisture. On wetter sites, Beech–Maple grades into wet beech flats and on drier sites into oak-maple or oak-hickory. There may be conifers associated with Beech–Maple Forest, but they are confined to soils and topography unfavorable to many of the hardwoods. This community corresponds to Küchler #102 and Eastern Deciduous and Mixed Forests ERT.

Canopy
Characteristic Species
Acer saccharum	sugar maple
Fagus grandifolia	American beech
Liriodendron tulipifera	tuliptree
Quercus rubra	northern red oak

Associates
Acer rubrum	red maple
Aesculus glabra	Ohio buckeye
Carya cordiformis	bitternut hickory
Carya ovata	shagbark hickory
Fraxinus americana	white ash
Juglans nigra	black walnut
Nyssa sylvatica	black tupelo
Prunus serotina	black cherry
Quercus alba	northern white oak
Quercus macocarpa	burr oak
Tilia americana	American basswood
Ulmus americana	American elm
Ulmus rubra	slippery elm

Woody Understory
Asimina triloba	common pawpaw
Cornus alternifolia	alternate-leaved dogwood
Cornus florida	flowering dogwood
Diervilla lonicera	northern bush honeysuckle
Dirca palustris	eastern leatherwood
Evonymus atropurpureus	eastern wahoo
Hamamelis virginiana	American witch-hazel
Lindera benzoin	northern spicebush
Lonicera canadensis	American fly-honeysuckle
Ostrya virginiana	eastern hop-hornbeam
Ribes cynosbati	eastern prickly gooseberry
Sambucus canadensis	American elder
Smilax rotundifolia	horsebrier
Staphlea trifolia	American bladdernut
Viburnum acerifolium	maple-leaf arrow-wood
Vitis spp.	grape

Herbaceous Understory

Allium tricoccum	ramp
Aralia nudicaulis	wild sarsaparilla
Arisaema triphyllum	jack-in-the-pulpit
Asarum canadense	Canadian wild ginger
Cardamine concatenata	cut-leaf toothwort
Carex pensylvanica	Pennsylvania sedge
Carex plantaginea	broad scale sedge
Circaea lutetiana	broad-leaf enchanter's-nightshade
Claytonia virginica	Virginia springbeauty
Dicentra canadensis	squirrel-corn
Dicentra cucullaria	Dutchman's-breeches
Dryopteris carthusiana	spinulose wood fern
Enemion biternatum	eastern false rue-anemone
Epifagus virginiana	beechdrops
Erythronium americanum	American trout-lily
Galium aparine	sticky-willy
Geranium maculatum	spotted crane's-bill
Hepatica nobilis	liverwort
Hydrophyllum virginianum	Shawnee-salad
Maianthemum canadense	false lily-of-the-valley
Maianthemum racemosum	feathery false Solomon's-seal
Osmorhiza claytonii	hairy sweet-cicely
Phlox divaricata	wild blue phlox
Podophyllum peltatum	may-apple
Polygonatum biflorum	King Solomon's-seal
Sanguinaria canadensis	bloodroot
Streptopus amplexifolius	clasping twistedstalk
Trillium grandiflorum	large-flower wakerobin
Viola sororia	hooded blue violet

Oak Barrens

Oak Barrens occur as a transition between prairie and closed canopy oak forests and are common in central Minnesota and southern Wisconsin. The most common characteristic species is burr oak. Where the canopy is open, prairie species may dominate. Oak Barrens occur on dry to mesic sites, on sandy soils, and on bluffs or outwash terraces. Fires and grazing help maintain the canopy structure of woodlands, but on draughty sites with thin soil and steep slopes, they may persist without fire. This community corresponds to Küchler #81 and Woodlands ERT.

Canopy

Characteristic Species

Quercus alba	northern white oak
Quercus ellipsoidalis	northern pin oak
Quercus macrocarpa	burr oak
Quercus rubra	northern red oak
Quercus velutina	black oak

Associates

Acer negundo	ash-leaf maple
Betula papyrifera	paper birch

Carya ovata	shag-bark hickory
Fraxinus americana	white ash
Fraxinus pennsylvanica	green ash
Juglans nigra	black walnut
Juniperus viginiana	eastern red cedar
Populus balsamifera	balsam poplar
Populus grandidentata	big-tooth aspen
Populus tremuloides	quaking aspen
Prunus serotina	black cherry
Tilia americana	American basswood
Ulmus americana	American elm
Ulmus rubra	slippery elm

Woody Understory

Amelanchier spp.	service-berry
Amorpha canescens	leadplant
Ceanothus americanus	New Jersey tea
Cornus racemosa	gray dogwood
Corylus americana	American hazelnut
Parthenocissus vitacea	thicket-creeper
Prunus virginiana	choke cherry
Rhus glabra	smooth sumac
Ribes spp.	gooseberry
Rosa spp.	rose
Rubus spp.	raspberry
Salix humilis	small pussy willow
Toxicodendron radicans	eastern poison-ivy
Vitis riparia	river-bank grape
Zanthoxylum americanum	toothachetree

Herbaceous Understory

Amphicarpaea bracteata	American hog-peanut
Andropogon gerardii	big bluestem
Anemone cylindrica	long-head thimbleweed
Antennaria neglecta	field pussytoes
Apocynum androsaemifolium	spreading dogbane
Aquilegia canadensis	red columbine
Arabis lyrata	lyre-leaf rockcress
Aralia nudicaulis	wild sarsaparilla
Artemisia campestris	Pacific wormwood
Asclepias syriaca	common milkweed
Aster cordifolius	common blue wood aster
Aster linariifolius	flax-leaf white-top aster
Carex pensylvanica	Pennsylvania sedge
Comandra umbellata	bastard-toadflax
Coreopsis palmata	stiff tickseed
Dalea purpurea	violet prairie-clover
Dalea villosus	silky prairie-clover
Desmodium glutinosum	pointed-leaf tick-trefoil
Dichanthelium leibergii	Leiberg's rosette grass

Euphorbia corollata	flowering spurge
Fragaria virginiana	Virginia strawberry
Galium boreale	northern bedstraw
Galium concinnum	shining bedstraw
Geranium maculatum	spotted crane's-bill
Helianthemum canadense	long-branch frostweed
Helianthus occidentalis	few-leaf sunflower
Helianthus pauciflorus	stiff sunflower
Helianthus strumosus	pale-leaf woodland sunflower
Heliopsis helianthoides	smooth oxeye
Heuchera richardsonii	Richardson's alumroot
Hudsonia tomentosa	sand golden-heather
Koeleria macrantha	prairie Koeler's grass
Krigia biflora	two-flowered dwarf-dandelion
Lathyrus venosus	veiny vetchling
Lespedeza capitata	round-head bush-clover
Liatris aspera	tall gayfeather
Lithospermum canescens	hoary puccoon
Lupinus perennis	sundial lupine
Maianthemum racemosum	feathery false Solomon's-seal
Monarda fistulosa	Oswego-tea
Oenothera biennis	king's-cureall
Phlox pilosa	downy phlox
Physalis virginiana	Virginia ground-cherry
Polygonatum biflorum	King Solomon's-seal
Polygonella articulata	coastal jointweed
Potentilla arguta	tall cinquefoil
Prenanthes alba	white rattlesnake-root
Pteridium aquilinum	northern bracken fern
Rudbeckia hirta	black-eyed-Susan
Schizachyrium scoparium	little false bluestem
Smilax herbacea	smooth carrion-flower
Solidago nemoralis	gray goldenrod
Sorghastrum nutans	yellow Indian grass
Sporobolus heterolepis	prairie dropseed
Stipa spartea	porcupine grass
Tephrosia virginiana	goat's-rue
Tradescantia ohiensis	bluejacket
Viola cucullata	marsh blue violet
Viola pedata	bird foot violet

Mesic Prairie

Mesic Prairie is a grassland community characterized by a high species diversity on deep, fertile, and well-drained soils. Dominant grasses in the Mesic Prairie are big bluestem, little false bluestem, wand panic grass (switchgrass), yellow Indian grass, and prairie drop-seed. Forbs are also abundant but usually subdominant to grasses. Mesic Prairie is a fire-dependent community that succeeds to forest or woodland without periodic fires. This community corresponds to Küchler #74 and Prairies ERT.

Characteristic Species

Andropogon gerardii	big bluestem
Dodecatheon meadia	eastern shootingstar
Galium tinctorium	stiff marsh bedstraw
Helianthus grosseserratus	saw-tooth sunflower
Liatris punctata	dotted gayfeather
Lithospermum canescens	hoary puccoon
Oxalis violacea	violet wood-sorrel
Panicum virgatum	wand panic grass
Phlox pilosa	downy phlox
Ratibida pinnata	gray-head Mexican-hat
Rudbeckia hirta	black-eyed-Susan
Schizachyrium scoparium	little false bluestem
Solidago rigida	hard-leaf goldenrod
Sorghastrum nutans	yellow Indian grass
Sporobolus heterolepis	prairie drop-seed
Stipa spartea	porcupine grass
Tradescantia ohiensis	bluejacket

Associates

Ambrosia artemisiifolia	annual ragweed
Amorpha canescens	leadplant
Anemone cylindrica	long-head thimbleweed
Antennaria neglecta	field pussytoes
Apocynum androsaemifolium	spreading dogbane
Artemisia ludoviciana	white sagebrush
Asclepias ovalifolia	oval-leaf milkweed
Asclepias speciosa	showy milkweed
Asclepias syriaca	common milkweed
Aster ericoides	white heath aster
Aster laevis	smooth blue aster
Aster oolentangiensis	sky blue aster
Astragalus crassicarpus	ground-plum
Baptisia alba	white wild indigo
Baptisia bracteata	long-bract wild indigo
Bouteloua curtipendula	side-oats grama
Calystegia sepium	hedge false bindweed
Ceanothus americanus	New Jersey-tea
Cirsium discolor	field thistle
Cirsium flodmanii	Flodman's thistle
Comandra umbellata	bastard-toadflax
Coreopsis palmata	stiff tickseed
Dalea candida	white prairie-clover
Dalea purpurea	violet prairie-clover
Delphinium carolinianum	Carolina larkspur
Desmodium canadense	showy tick-trefoil
Desmodium illinoense	Illinois tick-trefoil
Dichanthelium leibergii	Leiberg's rosette grass
Dichanthelium oligosanthes	Hellen's rosette grass
Echinacea angustifolia	blacksamson

Elymus canadensis	nodding wild rye
Erigeron strigosus	prairie fleabane
Eryngium yuccifolium	button eryngo
Euphorbia corollata	flowering spurge
Fragaria virginiana	Virginia strawberry
Galium boreale	northern bedstraw
Helianthus occidentalis	few-leaf sunflower
Helianthus pauciflorus	stiff sunflower
Heuchera richardsonii	Richardson's alumroot
Koeleria macrantha	prairie Koeler's grass
Lactuca canadensis	Florida blue lettuce
Lathyrus venosus	veiny vetchling
Lespedeza capitata	round-head bush-clover
Liatris aspera	tall gayfeather
Liatris scariosa	devil's-bite
Lilium philadelphicum	wood lily
Monarda fistulosa	Oswego-tea
Pedicularis canadensis	Canadian lousewort
Pediomelum argophyllum	silver-leaf Indian-breadroot
Pediomelum esculentum	large Indian-breadroot
Physalis virginiana	Virginia ground-cherry
Potentilla arguta	tall cinquefoil
Prenanthes racemosa	purple rattlesnake-root
Ratibida columnifera	red-spike Mexican-hat
Rhus glabra	smooth sumac
Rosa arkansana	prairie rose
Silphium integrifolium	entire-leaf rosinweed
Silphium laciniatum	compassplant
Solidago canadensis	Canadian goldenrod
Solidago missouriensis	Missouri goldenrod
Solidago speciosa	showy goldenrod
Stipa spartea	porcupine grass
Viola pedatifida	crow-foot violet
Zigadenus elegans	mountain deathcamas
Zizia aptera	heart-leaf alexanders

WETLAND SYSTEMS

Great Lakes Floodplain Forest

Great Lakes Floodplain Forest occurs on seasonally inundated soils along the Mississippi River and other major rivers and streams. Silver maple, cottonwood, black willow, and American elm are among the most common characteristic species. Species in this forest type are tolerant of periodic inundation and are adapted to soil disturbances caused by scouring during flooding and by deposition of alluvium when river currents are slow. This community corresponds, in part, to Küchler #98 and Floodplain Forests ERT. This forest is also similar to Küchler's elm–ash forest (type #101) that covers low areas of ancient glacial lakebeds near Lakes Erie, Huron, and Michigan.

Canopy
Characteristic Species
Acer rubrum	red maple
Acer saccharinum	silver maple
Aesculus glabra	Ohio buckeye
Fraxinus pennsylvanica	green ash
Platanus occidentalis	American sycamore
Populus deltoides	eastern cottonwood
Salix nigra	black willow
Ulmus americana	American elm

Associates
Acer nigrum	black maple
Acer saccharum	sugar maple
Betula nigra	river birch
Carya cordiformis	bitter-nut hickory
Celtis occidentalis	common hackberry
Fagus grandifolia	American beech
Fraxinus americana	white ash
Fraxinus nigra	black ash
Fraxinus quadrangulata	blue ash
Gleditsia triacanthos	honey-locust
Gymnocladus dioica	Kentucky coffeetree
Juglans cinerea	white walnut
Juglans nigra	black walnut
Populus deltoides	eastern cottonwood
Populus tremuloides	quaking aspen
Prunus serotina	black cherry
Quercus alba	northern white oak
Quercus bicolor	swamp white oak
Quercus macrocarpa	burr oak
Quercus rubra	northern red oak
Tilia americana	American basswood
Ulmus rubra	slippery elm

Woody Understory
Asimina triloba	common pawpaw
Carpinus caroliniana	American hornbeam
Celastrus scandens	American bittersweet
Cephalanthus occidentalis	common buttonbush
Cercis canadensis	redbud
Clematis virginiana	devil's-darning-needles
Cornus alternifolia	alternate-leaf dogwood
Cornus amomum	silky dogwood
Crataegus spp.	hawthorn
Evonymus atropurpureus	eastern wahoo
Lindera benzoin	northern spicebush
Menispermum canadensis	Canadian moonseed
Morus rubra	red mulberry
Parthenocissus quinquefolia	Virginia-creeper

Parthenocissus vitacea	thicket-creeper
Salix amygdaloides	peach-leaf willow
Salix interior	sandbar willow
Sambucus canadensis	American elder
Smilax tamnoides	chinaroot
Symphoricarpos orbiculatus	coral-berry
Toxicodendron radicans	eastern poison-ivy
Vitis riparia	river-bank grape

Herbaceous Understory

Amphicarpaea bracteata	American hog-peanut
Apios americana	groundnut
Arisaema dracontium	greendragon
Aster lateriflorus	farewell-summer
Boehmeria cylindrica	small-spike false nettle
Camassia scilloides	Atlantic camas
Cardamine concatenata	cut-leaf toothwort
Carex typhina	cat-tail sedge
Chaerophyllum procumbens	spreading chevril
Cinna arundinacea	sweet wood-reed
Cryptotaenia canadensis	Canadian honewort
Diarrhena americana	American beakgrain
Echinocystis lobata	wild cucumber
Elymus virginicus	Virginia wild rye
Galium aparine	sticky-willy
Geum canadense	white avens
Impatiens capensis	spotted touch-me-not
Laportea canadensis	Canadian wood-nettle
Leersia virginica	white grass
Lycopus uniflorus	northern water-horehound
Lysimachia ciliata	fringed yellow-loosestrife
Mertensia virginica	Virginia bluebells
Muhlenbergia frondosa	wire-stem muhly
Onoclea sensibilis	sensitive fern
Osmorhiza claytoni	hairy sweet-cicely
Phlox divaricata	wild blue phlox
Pilea pumila	Canadian clearweed
Rudbeckia laciniata	green-head coneflower
Sanicula odorata	clustered black-snakeroot
Scutellaria lateriflora	mad dog skullcap
Sicyos angulatus	one-seed burr-cucumber
Solidago gigantea	late goldenrod
Stachys tenuifolia	smooth hedge-nettle
Teucrium canadense	American germander
Trillium nivale	dwarf white wakerobin
Trillium recurvatum	bloody-butcher
Trillium sessile	toadshade
Urtica dioica	stinging nettle

Shrub Swamp

Shrub Swamp occurs around lakes and ponds in the transition zone between sedge meadow and wet hardwood forests. Shrub Swamp is considered a successional community but may be quite stable when regularly fluctuating water levels retard forest development. The most important single species is the redosier, but willows collectively are of greater significance. Shrubs other than willows such as red raspberry, elderberry, and nanny-berry are common as well as numerous herbaceous species and woody vines. Shrub Swamp depends upon saturated soil and fires to maintain community structure and composition. This community corresponds to Shrub Swamps ERT.

Characteristic Woody Species

Cornus sericea	redosier
Salix bebbiana	long-beak willow
Salix discolor	tall pussy willow
Salix interior	sandbar willow
Salix lucida	shining willow
Salix petiolaris	meadow willow
Sambucus canadensis	American elder

Associates

Acer rubrum	red maple
Alnus incana	speckled alder
Alnus serrulata	brookside alder
Aronia melanocarpa	black chokeberry
Cephalanthus occidentalis	common buttonbush
Clematis virginiana	devil's-darning-needles
Cornus amomum	silky dogwood
Cornus racemosa	gray dogwood
Ilex verticillata	common winterberry
Parthenocissus vitacea	thicket-creeper
Ribes americanum	wild black currant
Rosa palustris	swamp rose
Rubus idaeus	common red raspberry
Salix amygaloides	peach-leaf willow
Salix eriocephala	Missouri willow
Salix nigra	black willow
Salix sericea	silky willow
Sambucus canadensis	American elder
Spiraea alba	white meadowsweet
Toxicodendron radicans	eastern poison-ivy
Toxicodendron vernix	poison-sumac
Ulmus americana	American elm
Vaccinium corymbosum	highbush blueberry
Viburnum dentatum	southern arrow-wood
Viburnum lentago	nanny-berry

Herbaceous Understory

Asclepias incarnata	swamp milkweed
Aster lanceolatus	white panicle aster
Aster puniceus	purple-stem aster
Calamagrostis canadensis	bluejoint

Carex lacustris	lakebank sedge
Carex prairea	prairie sedge
Cicuta maculata	spotted water-hemlock
Cirsium muticum	swamp thistle
Echinocystis lobata	wild cucumber
Equisetum sylvaticum	woodland horsetail
Eupatorium maculatum	spotted joe-pye-weed
Eupatorium perfoliatum	common boneset
Glyceria striata	fowl manna grass
Impatiens capensis	spotted touch-me-not
Lycopus americanus	cut-leaf water-horehound
Lycopus uniflorus	northern water-horehound
Onoclea sensibilis	sensitive fern
Poa palustris	fowl blue grass
Rumex orbiculatus	greater water dock
Solidago canadensis	Canadian goldenrod
Solidago gigantea	late goldenrod
Stachys palustris	woundwort
Thalictrum dasycarpum	purple meadow-rue
Thelypteris palustris	eastern marsh fern
Typha latifolia	broad-leaf cat-tail
Viola cucullata	marsh blue violet

Wet Meadow

Wet Meadow occurs along streams, adjacent to lakes, and in depressions and channels in glacial outwash. Sedges and rushes are dominant, but many other species are important in these meadows. Soils are wet mineral, muck, or shallow peat. Water in the meadows is abundant in the spring after heavy rains, but the water table is below the soil surface for much of the growing season. Wet meadows grade into wet prairies as moisture decreases. Wet meadow grades into cat-tail and reed marshes in wetter conditions. This community corresponds to Meadowlands ERT.

Characteristic Species

Asclepias incarnata	swamp milkweed
Calamagrostis canadensis	bluejoint
Carex aquatilis	leafy tussock sedge
Carex lacustris	lakebank sedge
Carex stricta	uptight sedge
Eleocharis acicularis	needle spike-rush
Eleocharis erythropoda	bald spike-rush
Eleocharis obtusa	blunt spike-rush
Eleocharis smallii	Small's spike-rush
Eupatorium maculatum	spotted joe-pye-weed
Juncus acuminatus	knotty-leaf rush
Juncus dudleyi	Dudley's rush
Juncus effusus	lamp rush
Juncus tenuis	poverty rush
Juncus torreyi	Torrey's rush
Peltandra virginica	green arrow-arum
Spartina pectinata	freshwater cord grass
Thalictrum dasycarpum	purple meadow-rue

Associates

Anemone canadensis	round-leaf thimbleweed
Angelica atropurpurea	purple-stem angelica
Aster lanceolatus	white panicle aster
Aster puniceus	purple-stem aster
Carex bebbii	Bebb's sedge
Carex cosmosa	bearded sedge
Carex crinita	fringed sedge
Carex frankii	Frank's sedge
Carex granularis	limestone-meadow sedge
Carex grayi	Gray's sedge
Carex haydenii	cloud sedge
Carex hystericina	porcupine sedge
Carex lanuginosa	woolly sedge
Carex lupulina	hop sedge
Carex rostrata	swollen beaked sedge
Carex scoparia	pointed broom sedge
Carex squarrosa	squarrose sedge
Carex tribuloides	blunt broom sedge
Carex vulpinoidea	common fox sedge
Chelone glabra	white turtlehead
Cicuta maculata	spotted water-hemlock
Cladium mariscoides	smooth saw-grass
Cyperus spp.	flat sedge
Dulichium arundinaceum	three-way sedge
Equisetum arvense	field horsetail
Eupatorium perfoliatum	common boneset
Galium obtusum	blunt-leaf bedstraw
Glyceria striata	fowl manna grass
Helianthus grosseserratus	saw-tooth sunflower
Impatiens capensis	spotted touch-me-not
Iris virginica	Virginia blueflag
Lathyrus palustris	marsh vetchling
Leersia oryzoides	rice cut grass
Lycopus americanus	cut-leaf water-horehound
Lycopus uniflorus	northern water-horehound
Mentha arvensis	American wild mint
Poa palustris	fowl blue grass
Sagittaria spp.	arrowhead
Salix bebbiana	long-beak willow
Salix discolor	tall pussy willow
Scirpus americanus	chairmaker's bulrush
Scirpus atrovirens	dark-green bulrush
Scirpus cyperinus	cottongrass bulrush
Scirpus fluviatilis	river bulrush
Scirpus lineatus	drooping bulrush
Solidago canadensis	Canadian goldenrod
Solidago gigantea	late goldenrod
Sparganium spp.	burr-reed

Stachys palustris	woundwort
Thelypteris palustris	eastern marsh fern
Typha angustifolia	narrow-leaf cat-tail
Typha latifolia	broad-leaf cat-tail
Viola cucullata	marsh blue violet

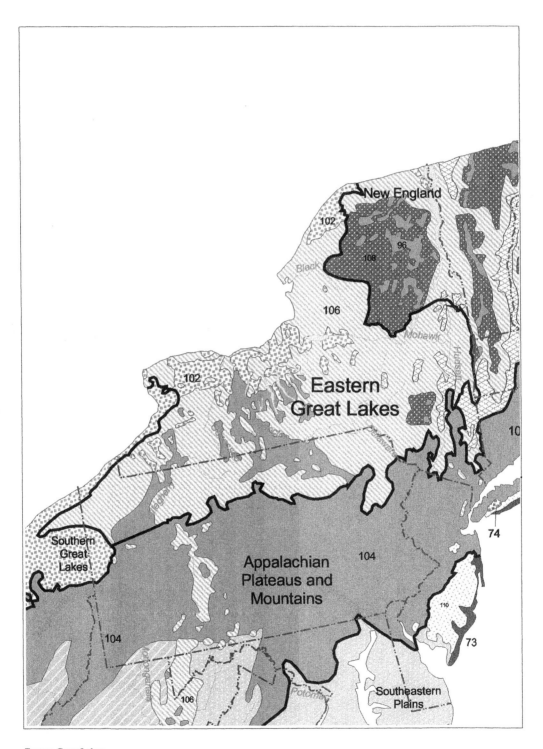

Eastern Great Lakes

Eastern Great Lakes

Introduction

The Eastern Great Lakes region encompasses northern Pennsylvania and most of New York state, except the Adirondack Mountain region and the coastal lowlands. The dominant forest communities in the Eastern Great Lakes are variants of northern hardwoods.

The primary sources used to develop descriptions and species lists are Barbour and Billings 1988; Eyre 1980; Reschke 1990; Smith (no date).

Dominant Ecological Communities

Upland Systems
Northern Hardwoods–Conifer Forest
Northern Oak Forest
Mesophytic Forest
Coniferous Woodland
Deciduous Woodland

Wetland Systems
Floodplain Forest
Deciduous Swamp Forest
Coniferous Swamp Forest
Shrub Bog/Fen
Shrub Swamp
Herbaceous Peatland Fen/Wet Meadow
Mineral Soil Marsh

UPLAND SYSTEMS

Northern Hardwoods–Conifer Forest

Northern Hardwoods–Conifer Forest occurs throughout the Eastern Great Lakes region as the major forest type. This broadly defined community has many regional and edaphic variants and plant associations within it. This community corresponds to Küchler #106 and Eastern Deciduous and Mixed Forests ERT.

Canopy
Characteristic Species
Acer rubrum	red maple
Acer saccharum	sugar maple
Betula papyrifera	paper birch
Fagus grandifolia	American beech
Picea rubens	red spruce
Pinus resinosa	red pine
Pinus strobus	eastern white pine
Populus tremuloides	quaking aspen
Tsuga canadensis	eastern hemlock

Associates

Abies balsamea	balsam fir
Betula alleghaniensis	yellow birch
Betula lenta	sweet birch
Prunus serotina	black cherry
Quercus rubra	northern red oak
Tilia americana	American basswood

Woody Understory

Acer pensylvanicum	striped maple
Acer spicatum	mountain maple
Amelanchier canadensis	Canadian service-berry
Epigaea repens	trailing-arbutus
Gaultheria procumbens	eastern teaberry
Kalmia angust∙folia	sheep-laurel
Lonicera canadensis	American fly-honeysuckle
Rhododendron maximum	great-laurel
Taxus canadensis	American yew
Vaccinium angustifolium	late lowbush blueberry
Vaccinium myrtilloides	velvet-leaf blueberry
Viburnum acerifolium	maple-leaf arrow-wood
Viburnum lantanoides	hobblebush
Viburnum nudum	possumhaw

Herbaceous Understory

Aralia nudicaulis	wild sarsaparilla
Carex pensylvanica	Pennsylvania sedge
Clintonia borealis	yellow bluebead-lily
Coptis trifolia	three-leaf goldthread
Cornus canadensis	Canadian bunchberry
Dicranum polysetum	moss
Dryopteris campyloptera	mountain wood fern
Dryopteris intermedia	evergreen wood fern
Epifagus virginiana	beechdrops
Huperzia lucidula	shining club-moss
Leucobryum glaucum	white moss
Maianthemum canadense	false lily-of-the-valley
Medeola virginiana	Indian cucumber-root
Melampyrum lineare	American cow-wheat
Mitchella repens	partridge-berry
Oryzopsis asperifolia	white-grain mountain-rice grass
Oxalis montana	sleeping-beauty
Pleurozium schreberi	feather moss
Polystichum acrostichoides	Christmas fern
Pteridium aquilinum	northern bracken fern
Streptopus roseus	rosy twistedstalk
Tiarella cordifolia	heart-leaf foamflower
Trientalis borealis	American starflower
Trillium erectum	stinking-Benjamin

Trillium undulatum	painted wakerobin
Uvularia sessilifolia	sissile-leaf bellwort
Viola rotundifolia	round-leaf yellow violet

Northern Oak Forest

Northern Oak Forest occurs throughout the Eastern Great Lakes region on a diversity of mostly well-drained sites. White oak, red oak, and black oak are the most common canopy species. Several species of hickory consistently occur in this forest type but seldom make up more than ten percent of the canopy cover. This community corresponds to Küchler #104 and Eastern Deciduous and Mixed Forests ERT.

Canopy
Characteristic Species
Quercus alba	northern white oak
Quercus prinus	chestnut oak
Quercus rubra	northern red oak
Quercus velutina	black oak

Associates
Acer rubrum	red maple
Betula lenta	sweet birch
Carya glabra	pignut hickory
Carya ovalis	red hickory
Carya ovata	shag-bark hickory
Fagus grandifolia	American beech
Fraxinus americana	white ash
Nyssa sylvatica	black tupelo
Pinus rigida	pitch pine
Pinus strobus	eastern white pine
Pinus virginiana	Virginia pine
Populus grandidentata	big-tooth aspen
Prunus serotina	black cherry
Quercus coccinea	scarlet oak
Quercus stellata	post oak
Tsuga canadensis	eastern hemlock

Woody Understory
Amelanchier arborea	downy service-berry
Cornus florida	flowering dogwood
Cornus foemina	gray dogwood
Corylus cornuta	beaked hazelnut
Galussacia baccata	black huckleberry
Gaultheria procumbens	eastern teaberry
Hamamelis virginiana	American witch-hazel
Kalmia latifolia	mountain-laurel
Ostrya virginiana	eastern hop-hornbeam
Prunus pensylvanica	fire cherry
Prunus virginiana	choke cherry
Quercus ilicifolia	bear oak
Rhododendron periclymenoides	pink azalea
Rubus idaeus	common red raspberry
Sassafras albidum	sassafras

Vaccinium angustifolium	late lowbush blueberry
Vaccinium pallidum	early lowbush blueberry
Vaccinium stamineum	deerberry
Viburnum acerifolium	maple-leaf arrow-wood

Herbaceous Understory

Aralia nudicaulis	wild sarsaparilla
Carex pensylvanica	Pennsylvania sedge
Chimaphila maculata	striped prince's-pine
Cimicifuga racemosa	black bugbane
Clintonia umbellulata	white bluebead-lily
Desmodium glutinosum	pointed-leaf tick-trefoil
Desmodium paniculatum	panicled-leaf tick-trefoil
Hepatica nobilis	liverwort
Hieracium venosum	rattlesnake-weed
Leucobryum glaucum	white moss
Maianthemum racemosum	feathery false Solomon's-seal
Prenanthes alba	white rattlesnake-root
Pteridium aquilinum	northern bracken fern
Solidago palmata	white goldenrod
Viola triloba	early blue violet
Waldsteinia fragarioides	Appalachain barren-strawberry

Mesophytic Forest

Mesophytic Forest occurs on moist, well-drained soils. In the Eastern Great Lakes region, this forest type is most commonly located at low elevations and on fertile, loamy soils. This is a diverse community with an abundance of spring wildflowers. This community corresponds to Küchler #102 and Eastern Deciduous and Mixed Forests ERT.

Canopy
Characteristic species

Acer rubrum	red maple
Acer saccharum	sugar maple
Betula lenta	sweet birch
Fagus grandifolia	American beech
Fraxinus americana	white ash

Associates

Carya cordiformis	bitter-nut hickory
Carya ovata	shag-bark hickory
Liriodendron tulipifera	tuliptree
Magnolia acuminanta	cucumber magnolia
Pinus strobus	eastern white pine
Prunus serotina	black cherry
Quercus alba	northern white oak
Quercus imbricaria	shingle oak
Quercus rubra	northern red oak
Tilia americana	American basswood
Tsuga canadensis	eastern hemlock
Ulmus americana	American elm
Ulmus rubra	slippery elm

Woody Understory

Acer pensylvanicum	striped maple
Amelanchier laevis	Allegheny service-berry
Asimina triloba	pawpaw
Carpinus caroliniana	American hornbeam
Cornus alternifolia	alternate-leaf dogwood
Cornus rugosa	round-leaf dogwood
Dievilla lonicera	northern bush-honeysuckle
Hamamelis virginiana	witch-hazel
Lonicera canadensis	American fly-honeysuckle
Ostrya virginiana	eastern hop-hornbeam
Rhododendron periclymenoides	pink azalea
Staphylea trifolia	American bladdernut
Vaccinium pallidum	early lowbush blueberry
Viburnum acerifolium	maple-leaf arrow-wood
Viburnum lantanoides	hobblebush

Herbaceous Understory

Actaea pachypoda	white baneberry
Allium tricoccum	ramp
Arisaema triphyllum	jack-in-the-pulpit
Asarum canadense	long-tail wildginger
Aster macrophyllus	large-leaf aster
Cardamine concantenata	cut-leaf toothwort
Cardamine diphylla	crinkleroot
Caulophyllum thalictroides	blue cohosh
Cimicifuga racemosa	black bugbane
Claytonia virginica	Virginia springbeauty
Clintonia umbellulata	white bluebead-lily
Collinsonia canadensis	richweed
Dicentra candensis	squirrel-corn
Dicentra cucullaria	Dutchman's-breeches
Disporum lanuginosum	yellow fairybells
Erythronium americanum	American trout-lily
Hepatica nobilis	liverwort
Maianthemum racemosum	feathery false Solomon's-seal
Mertensia virginica	Virginia bluebells
Mitchella repens	partridge-berry
Osmunda claytoniana	interrupted fern
Podophyllum peltatum	may-apple
Polystichum acrostichoides	Christmas fern
Prenanthes trifoliolata	gall-of-the-earth
Sanguinaria canadensis	bloodroot
Sanicula marilandica	Maryland black-snakeroot
Solidago caesia	wreath goldenrod
Thalictrum dioicum	early meadow-rue
Tiarella cordifolia	heart-leaf foamflower
Trillium erectum	stinking-Benjamin
Viola rotundifolia	round-leaf yellow violet

Coniferous Woodland

Coniferous Woodland occurs on flat summits and rocky ridges where the soil is well drained and often rocky or sandy. The species composition of the woodland varies according to local conditions. For example, where the area is warm and dry, deciduous species may be associates, but where it is cool, balsam fir will be a likely addition to the community. This community corresponds to Woodlands ERT.

Canopy

Characteristic Species

Picea rubens	red spruce
Pinus banksiana	jack pine
Pinus rigida	pitch pine
Pinus strobus	eastern white pine
Quercus prinoides	dwarf chinkapin oak

Associates

Abies balsamea	balsam fir
Carya ovata	shag-bark hickory
Juniperus communis	common juniper
Juniperus virginiana	eastern red cedar
Populus grandidentata	big-tooth aspen
Quercus prinus	chestnut oak
Sorbus americana	American mountain-ash

Woody Understory

Aronia melanocarpa	black chokeberry
Betula populifolia	gray birch
Comptonia peregrina	sweet-fern
Gaylussacia baccata	black huckleberry
Kalmia angustifolia	sheep-laurel
Ostrya virginiana	eastern hop-hornbeam
Quercus ilicifolia	bear oak
Vaccinium angustifolium	late lowbush blueberry
Vaccinium myrtilloides	velvet-leaf blueberry
Vaccinium pallidum	early lowbush blueberry
Viburnum nudum	possumhaw

Herbaceous Understory

Andropogon gerardii	big bluestem
Antennaria plantaginifolia	woman's-tobacco
Aralia nudicaulis	wild sarsaparilla
Campanula rotundifolia	bluebell-of-Scotland
Carex eburnea	bristle-leaf sedge
Carex pensylvanica	Pennsylvania sedge
Cornus canadensis	Canadian bunchberry
Cypripedium acaule	pink lady's-slipper
Danthonia spicata	poverty wild oat grass
Deschampsia flexuosa	wavy hair grass
Fragaria virginiana	Virginia strawberry
Gaultheria procumbens	eastern teaberry
Lechea mucronata	hairy pinweed

Lespedeza capitata	round-head bush-clover
Lespedeza hirta	hairy bush-clover
Lespedeza procumbens	trailing bush-clover
Lespedeza stuevei	tall bush-clover
Lupinus perennis	sundial lupine
Maianthemum canadensis	false lily-of-the-valley
Melampyrum lineare	American cow-wheat
Oryzopsis pungens	short-awn mountain-rice grass
Polygala nuttallii	Nuttall's milkwort
Pteridium aquilinum	northern bracken fern
Schizachyrium scoparium	little false bluestem
Sibbaldiopsis tridentata	shrubby-fivefingers
Solidago spathulata	coastal-dune goldenrod
Sorghastrum nutans	yellow Indian grass
Tephrosia virginiana	goat's-rue

Deciduous Woodland

Deciduous Woodland occurs on well-drained, shallow soils where the bedrock is close to the surface and there are numerous rock outcrops. This community type may also cover slopes and sandy soils. There are many variants of community composition depending upon soil fertility and moisture capacity. Tree species may occur in continuous stands with an open canopy or as clusters of trees within savannas. This community corresponds to Woodlands ERT.

Canopy
Characteristic Species

Acer saccharum	sugar maple
Carya ovata	shag-bark hickory
Fraxinus americana	white ash
Juniperus virginiana	eastern red cedar
Picea glauca	white spruce
Pinus strobus	eastern white pine
Quercus alba	northern white oak
Quercus macrocarpa	burr oak
Quercus prinus	chestnut oak
Quercus rubra	northern red oak
Tilia americana	American basswood

Associates

Abies balsamea	balsam fir
Betula papyrifera	paper birch
Carya glabra	pignut hickory
Prunus pensylvanica	fire cherry
Quercus muhlenbergii	chinkapin oak
Quercus velutina	black oak
Thuja occidentalis	eastern arborvitae
Ulmus americana	American elm
Ulmus thomasii	rock elm

Woody Understory

Acer pensylvanicum	striped maple
Acer spicatum	mountain maple
Cornus foemina	gray dogwood
Cornus rugosa	round-leaf dogwood
Juniperus communis	common juniper
Lonicera dioica	limber honeysuckle
Ostrya virginiana	eastern hop-hornbeam
Prunus pumila	Great Lakes sand cherry
Quercus prinoides	dwarf chinkapin oak
Rhamnus alnifolia	alder-leaf buckthorn
Rhus aromatica	fragrant sumac
Rhus glabra	smooth sumac
Ribes cynosbati	eastern prickly gooseberry
Rosa blanda	smooth rose
Rubus flagellaris	whiplash dewberry
Rubus idaeus	common red raspberry
Rubus occidentalis	black raspberry
Shepherdia canadensis	russet buffalo-berry
Staphylea trifolia	American bladdernut
Symphoricarpos albus	common snowberry
Toxicodendron radicans	eastern poison-ivy
Vaccinium angustifolium	late lowbush blueberry
Viburnum rafinesquianum	downy arrow-wood
Zanthoxylum americanum	toothachetree

Herbaceous Understory

Actaea pachypoda	white baneberry
Adiantum pedatum	northern maidenhair
Andropogon gerardii	big bluestem
Anemone cylindracea	long-head thimbleweed
Aquilegia canadensis	red columbine
Asarum canadense	Canadian wild ginger
Asclepias tuberosa	butterfly milkweed
Asplenium rhizophyllum	walking fern
Asplenium trichomanes	maidenhair spleenwort
Aster ciliolatus	Lindley's aster
Aster divaricatus	white wood aster
Aster ericoides	white heath aster
Aster macrophyllus	large-leaf aster
Athyrium filix-femina	subarctic lady fern
Botrychium virginianum	rattlesnake fern
Campanula rotundifolia	bluebell-of-Scotland
Carex aurea	golden-fruit sedge
Carex eburnea	bristle-leaf sedge
Carex pensylvanica	Pennsylvania sedge
Carex platyphylla	plantain-leaf sedge
Caulophyllum thalictroides	blue cohosh
Comandra umbellata	bastard-toadflax

Cystoperis bulbifera	bulblet bladder fern
Cystoperis fragilis	brittle bladder fern
Danthonia compressa	flattened wild oat grass
Danthonia spicata	poverty wild oat grass
Deparia acrostichoides	silvery false spleenwort
Desmodium glabellum	Dillenius' tick-trefoil
Desmodium paniculatum	panicled-leaf tick-trefoil
Dryopteris marginalis	marginal wood fern
Eleocharis elliptica	elliptic spike-rush
Elmus hystrix	eastern bottle-brush grass
Fragaria virginiana	Virginia strawberry
Geranium robertianum	herbrobert
Houstonia longifolia	long-leaf summer bluet
Hypericum gentianoides	orange-grass
Maianthemum canadense	false lily-of-the-valley
Maianthemum racemosum	feathery false Solomon's-seal
Monarda fistulosa	Oswego-tea
Panax quinquefloius	American ginseng
Panicum flexile	wiry panic grass
Panicum philadelphicum	Philadelphia panic grass
Penstemon hirsutus	hairy beardtongue
Phlox divaricata	wild blue phlox
Polygonatum pubescens	hairy Solomon's seal
Polypodium virginianum	rock polypody
Polystichum acrostichoides	Christmas fern
Pteridium aquilinum	northern bracken fern
Rudbeckia hirta	black-eyed-Susan
Sanguinaria canadensis	bloodroot
Sanicula marilandica	Maryland black-snakeroot
Schizachyrium scoparium	little false bluestem
Senecio pauperculus	balsam ragwort
Solidago caesia	wreath goldenrod
Solidago flexicaulis	zigzag goldenrod
Solidago hispida	hairy goldenrod
Solidago juncea	early goldenrod
Solidago ptarmicoides	prairie goldenrod
Sorghastrum nutans	yellow Indian grass
Thalictrum dioicum	early meadow-rue
Trichostema dichotomum	forked bluecurls
Trillium grandiflorum	large-flower wakerobin
Waldsteinia fragaroides	Appalachian barren-strawberry

WETLAND SYSTEMS

Floodplain Forest

Floodplain Forest occurs on floodplain soils which are deep, fertile, and mesic. This community occurs along rivers throughout the region. Soils are seasonally saturated as a result of overflow from a nearby water body, groundwater, or drainage from adjacent uplands. This community corresponds to Floodplain Forests ERT.

Canopy

Characteristic Species

Acer rubrum	red maple
Acer saccharum	sugar maple
Carya ovata	shagbark hickory
Fraxinus americana	white ash
Fraxinus nigra	black ash
Platanus occidentalis	American sycamore
Populus deltoides	eastern cottonwood
Quercus bicolor	swamp white oak
Salix nigra	black willow
Tilia americana	American basswood

Associates

Betula nigra	river birch
Carya cordiformis	bitter-nut hickory
Fraxinus pensylvanica	green ash
Juglans cinerea	butternut
Juglans nigra	black walnut
Quercus macrocarpa	burr oak
Ulmus americana	American elm

Woody Understory

Acer negundo	ash-leaf maple
Alnus serrulata	brookside alder
Cephalanthus occidentalis	common buttonbush
Clematis virginiana	devil's-darning-needles
Cornus amomum	silky dogwood
Lindera benzoin	northern spicebush
Parthenocissus quinquefolia	Virginia-creeper
Physocarpus opulifolius	Atlantic ninebark
Sambucus canadensis	American elder
Toxicodendron radicans	eastern poison-ivy

Herbaceous Understory

Ageratina altissima	white snakeroot
Arisaema dracontium	greendragon
Aster puniceus	purple-stem aster
Boehmeria cylindrica	small-spike false nettle
Impatiens capensis	spotted touch-me-not
Impatiens pallida	pale touch-me-not
Laportea canadensis	Canadian wood-nettle

Lobelia cardinalis	cardinal-flower
Lobelia siphilitica	great blue lobelia
Onoclea sensibilis	sensitive fern
Peltandra virginica	green arrow-arum
Polygonum virginianum	jumpseed
Saururus cernuus	lizard's-tail

Deciduous Swamp Forest

Deciduous Swamp Forest occurs in poorly drained depressions and along lakeshores and rivers in inorganic (mineral) soils. These areas are usually uniformly wet with minimal seasonal fluctuations in water levels. This community corresponds to Swamp Forests ERT.

Canopy
Characteristic Species

Acer rubrum	red maple
Acer saccharinum	silver maple
Betula lenta	sweet birch
Fraxinus nigra	black ash
Nyssa sylvatica	black tupelo
Pinus strobus	eastern white pine
Quercus bicolor	swamp white oak
Quercus palustris	pin oak
Salix nigra	black willow
Tsuga canadensis	eastern hemlock
Ulmus americana	American elm

Associates

Betula alleghaniensis	yellow birch
Carya cordiformis	bitter-nut hickory
Fraxinus americana	white ash
Fraxinus pennsylvanica	green ash
Juglans cinerea	white walnut
Picea mariana	black spruce
Pinus rigida	pitch pine
Quercus alba	northern white oak

Woody Understory

Alnus serrulata	brookside alder
Aronia melanocarpa	black chokeberry
Cephalanthus occidentalis	common buttonbush
Clethra alnifolia	coastal sweet-pepperbush
Cornus amomum	silky dogwood
Cornus sericea	redosier
Gaylussacia baccata	black huckleberry
Ilex verticillata	common winterberry
Lindera benzoin	northern spicebush
Parthenocissus quinquefolia	Virginia-creeper
Rhamnus alnifolia	alder-leaf buckthorn
Rhododendron periclymenoides	pink azalea

Rhododendron viscosum	clammy azalea
Toxicodendron radicans	eastern poison-ivy
Toxicodendron vernix	poison-sumac
Vaccinium angustifolium	late lowbush blueberry
Vaccinium corymbosum	highbush blueberry
Viburnum dentatum	southern arrow-wood

Herbaceous Understory

Carex intumenscens	great bladder sedge
Carex lacustris	lakebank sedge
Cinna arundinacea	sweet wood-reed
Cypripedium parviflorum	lesser yellow lady's-slipper
Dryopteris carthusiana	spinulose wood fern
Dryopteris cristata	crested wood fern
Glyceria striata	fowl manna grass
Impatiens capensis	spotted touch-me-not
Muhlenbergia racemosa	green muhly
Onoclea sensibilis	sensitive fern
Osmunda cinnamomea	cinnamon fern
Osmunda regalis	royal fern
Platanthera grandiflora	greater purple fringed orchid
Ranunculus recurvatus	blisterwort
Saururus cernuus	lizard's tail
Saxifraga pensylvanica	eastern swamp saxifrage
Scirpus cyperinus	cottongrass bulrush
Scutellaria galericulata	hooded skullcap
Symplocarpus foetidus	skunk-cabbage
Thelypteris palustris	eastern marsh fern
Trollius laxus	American globeflower

Coniferous Swamp Forest

Coniferous Swamp Forest occurs on sites with organic (peat) soils in poorly drained depressions. These areas are often spring fed and permanently or semipermanently saturated. It is often dominated by boreal coniferous trees such as black spruce, balsam fir, and American larch (tamarack) and influenced by acidic water. It may have a nearly closed to fairly open canopy with scattered shrubs. The herbaceous layer has a diversity of mosses, sedges, and forbs. This community corresponds to Swamp Forests ERT.

Canopy
Characteristic Species

Abies balsamea	balsam fir
Chamaecyparis thyoides	Atlantic white-cedar
Larix laricina	American larch
Picea mariana	black spruce
Picea rubens	red spruce
Pinus rigida	pitch pine
Pinus strobus	eastern white pine
Tsuga canadensis	eastern hemlock
Thuja occidentalis	eastern arborvitae

Associates

Acer rubrum	red maple
Betula alleghaniensis	yellow birch
Betula populifolia	gray birch
Fraxinus nigra	black ash
Nyssa sylvatica	black tupelo

Woody Understory

Alnus incana	speckled alder
Alnus serrulata	brookside alder
Amelanchier arborea	smooth serviceberry
Aronia melanocarpa	black chokeberry
Betula pumila	bog birch
Chamaedaphne calyculata	leatherleaf
Clematis virginiana	devil's-darning-needles
Clethra alnifolia	coastal sweet-pepperbush
Cornus sericea	redosier
Gaultheria hispidula	creeping-snowberry
Gaultheria procumbens	eastern teaberry
Ilex laevigata	smooth winterberry
Ilex verticillata	common winterberry
Kalmia angustifolia	sheep-laurel
Kalmia polifolia	bog-laurel
Ledum groenlandicum	rusty Labrador-tea
Lonicera oblongifolia	swamp fly-honeysuckle
Nemopanthus mucronatus	catberry
Pentaphylloides floribunda	golden-hardhack
Rhamnus alnifolia	alder-leaf buckthorn
Rhododendron maximum	great-laurel
Rhododendron viscosum	clammy azalea
Ribes hirtellum	hairy-stem gooseberry
Rubus pubescens	dwarf red raspberry
Spiraea alba	white meadowsweet
Toxicodendron vernix	poison-sumac
Vaccinium corymbosum	highbush blueberry
Vaccinium myrtilloides	velvet-leaf blueberry
Vaccinium oxycoccos	small cranberry

Herbaceous Understory

Aulacomnium palustre	moss
Calla palustris	water-dragon
Caltha palustris	yellow marsh-marigold
Carex bromoides	brome-like sedge
Carex disperma	soft-leaf sedge
Carex eburnea	bristle-leaf sedge
Carex interior	inland sedge
Carex intumescens	great bladder sedge
Carex lacustris	lakebank sedge

Carex leptalea	bristly-stalk sedge
Carex scabrata	eastern rough sedge
Carex stricta	uptight sedge
Carex trisperma	three-seed sedge
Cirsium muticum	swamp thistle
Clintonia borealis	yellow bluebead-lily
Coptis trifolis	three-leaf goldthread
Cornus canadensis	Canadian bunchberry
Cypripedium parviflorum	lesser yellow lady's-slipper
Dryopteris cristata	crested wood fern
Eriophorum spp.	cotton-grass
Geum rivale	purple avens
Glyceria striata	fowl manna grass
Gymnocarpium dryopteris	western oak fern
Hylocomium splendens	feathermoss
Maianthemum canadense	false lily-of-the-valley
Maianthemum stellatum	starry false Solomon's-seal
Mitchella repens	partridge-berry
Mitella nuda	bare-stem bishop's-cap
Onoclea sensibilis	sensitive fern
Osmunda cinnamomea	cinnamon fern
Osmunda claytonia	interrupted fern
Osmunda regalis	royal fern
Platanthera blephariglottis	white fringed orchid
Pogonia ophioglossoides	snake-mouth orchid
Pteridium aquilinum	northern bracken fern
Ptilium crista-castrensis	feathermoss
Sarracenia purpurea	purple pitcherplant
Saxifraga pensylvanica	eastern swamp saxifrage
Senecio aureus	golden ragwort
Solidago patula	round-leaf goldenrod
Sphagnum centrale	moss
Sphagnum russowii	moss
Sphagnum warnstorifii	moss
Symplocarpus foetidus	skunk-cabbage
Thalictrum pubescens	king-of-the-meadow
Thelypteris palustris	eastern marsh fern
Trichocolea tomentella	leafy liverwort
Trientalis borealis	American starflower
Trollius laxus	globeflower
Typha latifolia	broad-leaf cat-tail
Veratrum viride	American false hellebore

Shrub Bog/Fen

The Shrub Bog/Fen community type occurs on peat substrate in continually wet areas fed by springs. The nutrient level of the water that feeds the bogs and fens strongly influences the diversity and type of vegetation that occurs. Some areas may have a high pH (6.0-7.9) and support a greater variety of species than areas that are fed by nutrient-poor waters. Fen peatlands are influenced by calcareous water. In general, they are dominated by evergreen and deciduous shrubs although trees are often present. This community corresponds to Bogs and Fens ERT.

Characteristic Woody Species

Betula pumila	bog birch
Chamaedaphne calyculata	leatherleaf
Gaylussacia baccata	black huckleberry
Kalmia angustifolia	sheep-laurel
Kalmia polifolia	bog-laurel
Vaccinium corymbosum	highbush blueberry
Vaccinium macrocarpon	large cranberry
Vaccinium oxycoccus	small cranberry

Associates

Abies balsamea	balsam fir
Acer rubrum	red maple
Alnus incana	speckled alder
Andromeda polifolia	bog-rosemary
Cornus sericea	redosier
Decodon verticillatus	swamp-loosestrife
Ilex verticillata	common winterberry
Juniperus virginiana	eastern red cedar
Larix laricina	American larch
Nemopanthus mucronatus	catberry
Pentaphylloides floribunda	golden-hardhack
Picea mariana	black spruce
Pinus rigida	pitch pine
Pinus strobus	eastern white pine
Rhamnus alnifolia	alder-leaf buckthorn
Rhododendron maximum	great-laurel
Rhododendron viscosum	clammy azalea
Salix candida	hoary willow
Salix discolor	tall pussy willow
Toxicodendron vernix	poison-sumac
Vaccinium augustifolium	late lowbush blueberry
Vaccinium myrtilloides	velvet-leaf blueberry
Xyris montana	northern yellow-eyed-grass

Herbaceous Understory

Calla palustris	water-dragon
Calopogon tuberosus	tuberous grass-pink
Carex aquatilis	leafy tussock sedge
Carex atlantica	prickly bog sedge
Carex canescens	hoary sedge
Carex folliculata	northern long sedge
Carex lacustris	lakebank sedge
Carex limosa	mud sedge
Carex oligosperma	few-seed sedge
Carex rostrata	swollen beaked sedge
Carex stricta	uptight sedge
Carex trisperma	three-seed sedge
Drosera rotundifolia	round-leaf sundew
Eriophorum angustifolium	tall cotton-grass

Eriophorum vaginatum	tussock cotton-grass
Eriophorum virginicum	tawny cotton-grass
Glyceria canadensis	rattlesnake manna grass
Iris versicolor	harlequin blueflag
Lobelia kalmii	brook lobelia
Maianthemum trifolia	three-flower false Solomon's-seal
Osmunda cinnamomea	cinnamon fern
Osmunda regalis	royal fern
Peltandra virginica	green arrow-arum
Platanthera blephariglottis	white fringed orchid
Pogonia ophioglossoides	snake-mouth orchid
Rhynchospora alba	white beak sedge
Sarracenia purpurea	purple pitcherplant
Scirpus acutus	hard-stem bulrush
Sphagnum angustifolium	moss
Sphagnum capillifolium	moss
Sphagnum centrale	moss
Sphagnum fallax	moss
Sphagnum fimbriatum	moss
Sphagnum fuscum	moss
Sphagnum magellanicum	moss
Sphagnum nemoreum	moss
Sphagnum papillosum	moss
Sphagnum recurvuum	moss
Sphagnum rubellum	moss
Thelypteris palustris	eastern marsh fern
Triandenum virginicum	Virginia marsh-St. John's-wort
Utricularia cornuta	horned bladderwort

Shrub Swamp

Shrub Swamp occurs scattered throughout the region in poorly drained areas. This community is dominated by shrubs (at least 50 percent) with less coverage in trees, on permanently flooded or saturated mineral soils. Species composition is influenced primarily by water depth and pH. This community corresponds to Shrub Swamps ERT.

Characteristic Woody Species

Alnus incana	speckled alder
Alnus serrulata	brookside alder
Cephalanthus occidentalis	common buttonbush
Cornus amomum	silky dogwood
Cornus sericea	redosier
Decodon verticillatus	swamp-loosestrife
Rosa palustris	swamp rose
Sambucus canadensis	American elder
Vaccinium corymbosum	highbush blueberry

Associates

Aronia melanocarpa	black chokeberry
Betula populifolia	gray birch
Chamaedaphne calyculata	leatherleaf

Cornus foemina	gray dogwood
Gaylussacia baccata	black huckleberry
Ilex verticillata	common winterberry
Kalmia angustifolia	sheep-laurel
Lindera benzoin	northern spicebush
Lyonia ligustrina	maleberry
Physocarpus opulifolius	Atlantic ninebark
Rhododendron maximum	great-laurel
Rhododendron viscosum	clammy azalea
Salix bebbiana	long-beak willow
Salix discolor	tall pussy willow
Salix lucida	shining willow
Salix petiolaris	meadow willow
Spiraea alba	white meadowsweet
Spiraea tomentosa	steeplebush
Vaccinium macrocarpon	large cranberry
Vaccinium oxycoccos	small cranberry
Viburnum dentatum	southern arrow-wood

Herbaceous Understory

Caltha palustris	yellow marsh-marigold
Carex rostrata	swollen beaked sedge
Solanum dulcamara	climbing nightshade
Sphagnum spp.	moss
Symplocarpus foetidus	skunk-cabbage

Herbaceous Peatland Fen/Wet Meadow

Wet meadows and fens occur in peat soils that are fed by small springs or groundwater seepage. They have saturated soils and are seasonally or permanently flooded. Fens occur on higher or sloping ground and may contain trees and low shrubs. This community corresponds to Bogs and Fens ERT.

Characteristic Species

Argentina anserina	common silverweed
Aster umbellatus	parasol flat-top white aster
Caltha palustris	yellow marsh-marigold
Carex aquatilis	leafy tussock sedge
Carex flava	yellow-green sedge
Carex hystericina	porcupine sedge
Carex interior	inland sedge
Carex lacustris	lakebank sedge
Carex lanuginosa	woolly sedge
Carex leptalea	bristly-stalk sedge
Carex prariea	prairie sedge
Carex sterilis	dioecious sedge
Carex stricta	uptight sedge
Carex viridula	little green sedge
Chelone glabra	white turtlehead
Cladium mariscoides	smooth sawgrass

Cyperus bipartitus	shining flat sedge
Deschampsia cespitosa	tufted hair grass
Drepanpcladus revolvens	moss
Drosera rotundifolia	round-leaf sundew
Dryopteris cristata	crested wood fern
Eleocharis palustris	pale spike-rush
Eleocharis rostellata	beaked spike-rush
Equisetum arvense	field horsetail
Equisetum variegatum	varigated scouring rush
Eriophorum viridicarinatum	tassel cotton-grass
Eupatorium perfoliatum	common boneset
Geum rivale	purple avens
Liparis loeselii	yellow wide-lip orchid
Lobelia cardinalis	cardinal-flower
Lobelia kalmii	brook lobelia
Lycopus uniflorus	northern water-horehound
Lycopus virginicus	Virginia water-horehound
Mentha arvensis	American wild mint
Muhlenbergia glomerata	spike muhly
Muhlenbergia racemosa	green muhly
Osmunda cinnamomea	cinnamon fern
Osmunda regalis	royal fern
Panicum flexile	wiry panic grass
Parnassia glauca	fen grass-of-Parnassus
Platanthera dilatata	scentbottle
Pogonia ophioglossoides	snake-mouth orchid
Polygonum amphibium	water smartweed
Rhynchospora alba	white beak sedge
Rhynchospora capillacaea	needle beak sedge
Rudbeckia laciniata	green-head coneflower
Sarracenia purpurea	purple pitcherplant
Scirpus acutus	hard-stem bulrush
Scleria verticillata	low nut-rush
Senecio aureus	golden ragwort
Solidago ohioensis	Ohio goldenrod
Solidago patula	round-leaf goldenrod
Solidago uliginosa	bog goldenrod
Spiranthes lucida	shining ladies'-tresses
Symplocarpus foetidus	skunk-cabbage
Thalictrum pubenscens	king-of-the-meadow
Thelypteris palustris	eastern marsh fern
Triglochin palustre	marsh arrow-grass
Trollius laxus	American globeflower
Typha angustifolia	narrow-leaf cat-tail
Typha latifolia	broad-leaf cat-tail
Utricularia minor	lesser blatterwort

Associates

Acer rubrum	red maple
Acorus americanus	sweetflag
Amelanchier arborea	smooth shadbush
Angelica atropurpurea	purple-stem angelica
Aronia melanocarpa	black chokeberry
Calamagrostis canadensis	bluejoint
Carex eburnea	bristle-leaf sedge
Clematis virginiana	devil's-darning-needles
Cornus foemina	gray dogwood
Cornus sericea	redosier
Eupatorium maculatum	spotted joe-pye-weed
Myrica gale	sweet-gale
Myrica pensylvanica	northern bayberry
Pentaphylloides floribunda	golden-hardhack
Rhamnus alnifolia	alder-leaf buckthorn
Ribes hirtellum	hairy-stem gooseberry
Rubus pubescens	dwarf red raspberry
Salix candida	hoary willow
Spiraea alba	white meadowsweet
Tsuga canadensis	eastern hemlock
Vaccinium corymbosum	highbush blueberry
Vaccinium macrocarpon	large cranberry

Mineral Soil Marsh

Mineral Soil Marsh occurs along lake shores where the soil is gravelly, sandy, or muddy. These areas are permanently saturated and seasonally or periodically flooded. Where water is usually present and covering the soil, the community is known as an emergent marsh. Water depths in these areas may range from six inches to six feet. Woody species such as alder and willow may border the forb-dominated marshes. This community corresponds to Nonestuarine Marshes ERT.

Characteristic Species

Alisma plantago-aquatica	water plantain
Bidens frondosa	devil's-pitchfork
Calamagrostis canadensis	bluejoint
Carex lurida	shallow sedge
Carex prairea	prairie sedge
Carex tetanica	rigid sedge
Eleocharis acicularis	needle spike-rush
Eleocharis obtusa	blunt spike-rush
Equisetum arvense	field horsetail
Glyceria melicaria	melic manna grass
Juncus effusus	lamp rush
Lathyrus palustris	marsh vetchling
Leersia oryzoides	rice cut grass
Onoclea sensibilis	sensitive fern
Polygonum pensylvanicum	pinkweed
Sagittaria latifolia	duck-potato
Scirpus americanus	chairmaker's bulrush

Scirpus atrovirens	dark-green bulrush
Symplocarpus foetidus	skunk-cabbage
Typha latifolia	broad-leaf cat-tail
Zizania aquatica	Indian wild rice

Associates

Acorus americanus	sweetflag
Argentina anserina	common silverweed
Armoracia lacustris	lakecress
Asclepias incarnata	swamp milkweed
Campanula aparinoides	marsh bellflower
Carex aurea	golden-fruit sedge
Carex flava	yellow-green sedge
Carex granularis	limestone-meadow sedge
Carex lacustris	lakebank sedge
Carex schweinitzii	Schweinitz's sedge
Carex stricta	uptight sedge
Carex viridula	little green sedge
Carex vulpinoidea	commmon fox sedge
Cyperus bipartitus	shining flat sedge
Cyperus squarrosus	awned flat sedge
Decodon verticillatus	swamp-loosestrife
Dulichium arundinaceum	three-way sedge
Eleocharis intermedia	matted spike-rush
Eleocharis palustris	pale spike-rush
Epilobium strictum	downy willowherb
Equisetum fluviatile	water horsetail
Equisetum variegatum	varigated scouring-rush
Eupatorium perfoliatum	common boneset
Fimbristylis autumnalis	slender fimbry
Fraxinus pennsylvanica	green ash
Glyceria canadensis	rattlesnake manna grass
Gratiola neglecta	clammy hedge-hyssop
Heteranthera dubia	grass-leaf mud-plantain
Iris versicolor	Virginia blueflag
Juncus alpinoarticulatus	northern green rush
Juncus balticus	Baltic rush
Juncus canadensis	Canadian rush
Juncus nodosus	knotted rush
Juncus pelocarpus	brown-fruit rush
Juncus torreyi	Torrey's rush
Lobelia dortmanna	water lobelia
Lobelia kalmii	brook lobelia
Ludwigia palustris	marsh primrose-willow
Lythrum thrysiflora	thyme-leaf loosestrife
Nuphar lutea	yellow pond-lily
Nymphaea odorata	American white water-lily
Peltandra virginica	green arrow-arum
Polygonum amphibium	water smartweed
Ranunculus longirostris	long-beak water-crowfoot

Ranunculus reptans	lesser creeping spearwort
Rhynchospora capillacea	needle beak sedge
Salix petiolaris	meadow willow
Salix sericea	silky willow
Scirpus actus	hard-stem bulrush
Scirpus cyperinus	cottongrass bulrush
Scirpus fluviatilis	river bulrush
Scirpus tabernaemontani	soft-stem bulrush
Sium suave	water-parsnip
Sparganium americanum	American burr-reed
Sparganium eurycarpum	broad-fruit burr-reed
Thelypteris palustris	eastern marsh fern
Typha angustifolia	narrow-leaf cat-tail

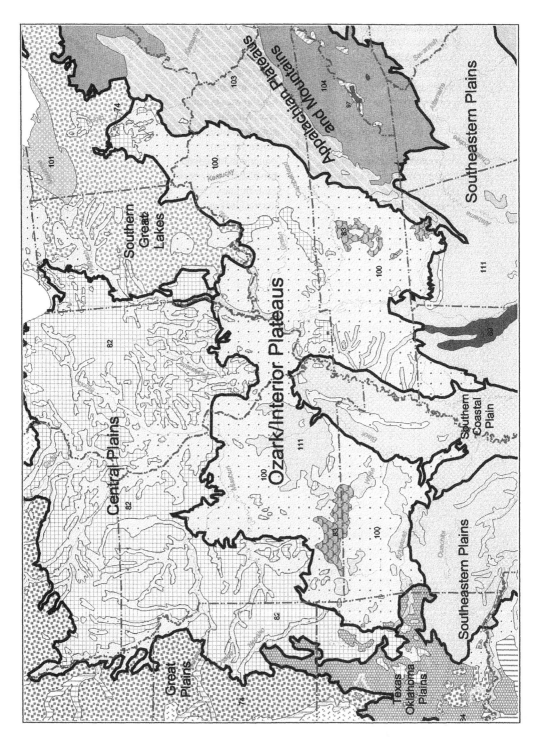

Ozark/Interior Plateaus

Ozark/Interior Plateaus

Introduction

This region encompasses most of Kentucky and Tennessee west of the Cumberland Plateau and portions of southwest Ohio, southern Indiana, extreme southern Illinois, southern Missouri, northern Arkansas, and extreme northeastern Oklahoma. It does not include small portions of extreme western Kentucky and Tennessee and the southeastern part of Missouri, known as the Bootheel region, which are part of the Southern Coastal Plain region.

At the time of settlement, most of this region was covered with deciduous forest. Oak–Hickory Forest dominates the upland landscape and covers much of the region. Oak–Pine Forest becomes important west of the Mississippi River in the Ozark Plateaus. Rich, mesophytic forests are mainly restricted to protected north slopes and ravines and areas with fertile, well-drained soils. Oak barrens, prairies, and glades are also important communities.

The primary sources used to develop descriptions and species lists are Braun 1950; Nelson 1980; Solecki 1980; Styermark 1963.

Dominant Ecological Communities

Upland Systems
Oak–Hickory Forest
Oak–Pine Forest
Western Mesophytic Forest
Oak Barrens
Bluestem Prairie
Glade

Wetland Systems
Floodplain Forest
Swamp Forest
Marsh
Wet Prairie

UPLAND SYSTEMS

Oak–Hickory Forest

Oak–Hickory Forest occurs throughout the region on dry to dry-mesic uplands. Dominance and species composition varies in response to topography, soils, geology, and disturbance history. This community corresponds to Küchler #100 and Eastern Deciduous and Mixed Forests ERT.

Canopy
Characteristic Species

Carya glabra	pignut hickory
Carya ovata	shag-bark hickory
Quercus alba	northern white oak
Quercus rubra	northern red oak
Quercus velutina	black oak

Associates

Acer rubrum	red maple
Acer sacharrum	sugar maple
Carya ovalis	red hickory
Carya texana	black hickory
Diospyros virginiana	common persimmon
Fraxinus americana	white ash
Fraxinus quadrangulata	blue ash
Nyssa sylvatica	black tupelo
Pinus echinata	short-leaf pine
Quercus coccinea	scarlet oak
Quercus falcata	southern red oak
Quercus marilandica	blackjack oak
Quercus muhlenbergii	chinkapin oak
Quercus shumardii	Shumard's oak
Quercus stellata	post oak
Ulmus rubra	slippery elm

Woody Understory

Amelanchier arborea	downy service-berry
Ceanothus americanus	New Jersey-tea
Cercis canadensis	redbud
Cornus drummondi	rough-leaf dogwood
Cornus florida	flowering dogwood
Cornus racemosa	gray dogwood
Parthenocissus quinquefolia	Virginia-creeper
Rhus aromatica	fragrant sumac
Symphoricarpos orbiculatus	coral-berry
Ulmus alata	winged elm
Vaccinium arboreum	tree sparkle-berry
Vaccinium pallidum	early lowbush blueberry
Vaccinium stamineum	deerberry
Vitis aestivalis	summer grape

Herbaceous Understory

Antennaria plantaginifolia	woman's-tobacco
Asplenium platyneuron	ebony spleenwort
Aster patens	late purple aster
Carex pensylvanica	Pennsylvania sedge
Clitoria mariana	Atlantic pigeonwings
Cunila marina	common dittany
Danthonia spicata	poverty white oat grass
Desmodium nudiflorum	naked-flower tick-trefoil
Galium concinnum	shining bedstraw
Galium pilosum	hairy bedstraw
Helianthus divaricatus	woodland sunflower
Helianthus hirsutus	whiskered sunflower
Hieracium gronovii	queendevil
Podophyllum peltatum	may-apple

Polystichum acrostichoides	Christmas fern
Prenanthes alba	white rattlesnake-root
Prenanthes altissima	tall rattlesnake-root
Ranunculus hispidus	bristly buttercup
Silene caroliniana	sticky catchfly
Solidago hispida	hairy goldenrod
Tradescantia virginiana	Virgina spiderwort

Oak–Pine Forest

In this region, Oak–Pine Forest occurs primarily in the Ozark Highlands or Ozark Plateaus. It occupies dry to xeric sites such as south and west facing slopes and ridgetops. In some ways it is similar to the Oak–Hickory forest type and shares many of the same species. However, pine is a dominant and conspicuous part of the overstory, sometimes occurring in pure stands, especially on acid soils. This community corresponds to Küchler #111 and Eastern Deciduous and Mixed Forests ERT.

Canopy
Characteristic Species
Carya texana	black hickory
Pinus echinata	short-leaf pine
Quercus alba	northern white oak
Quercus marilandica	blackjack oak
Quercus stellata	post oak
Quercus velutina	black oak

Associates
Carya ovalis	red hickory
Diospyros virginiana	common persimmon
Quercus coccinea	scarlet oak
Quercus falcata	southern red oak
Sassafras albidum	sassafras
Ulmus alata	winged elm

Woody Understory
Amelanchier arborea	downy service-berry
Ceanothus americanus	New Jersey-tea
Cornus florida	flowering dogwood
Frangula caroliniana	Carolina buckthorn
Hypericum hypericoides	St. Andrew's-cross
Ostrya virginiana	eastern hop-hornbeam
Rhus aromatica	fragrant sumac
Rhus copallinum	winged sumac
Rosa carolina	Carolina rose
Vaccinium arboreum	tree sparkle-berry
Vaccinium pallidum	early lowbush blueberry
Vaccinium stamineum	deerberry

Herbaceous Understory
Antennaria plantaginifolia	woman's-tobacco
Aster anomalus	many-ray aster

Aster linariifolius	flax-leaf white-top aster
Aster patens	late purple aster
Carex flaccosperma	thin-fruit sedge
Cunila marina	common dittany
Danthonia spicata	poverty wild oat grass
Dichanthelium dichotomum	cypress rosette grass
Helianthus divaricatus	woodland sunflower
Lespedeza procumbens	trailing bush-clover
Lespedeza repens	creeping bush-clover
Potentilla simplex	oldfield cinquefoil
Pteridium aquilinum	northern bracken fern
Ranunculus hispida	bristly buttercup
Silene virginica	fire-pink
Solidago buckleyi	Buckley's goldenrod
Solidago hispida	hairy goldenrod
Stylosanthes biflora	side-beak pencil-flower
Tradescantia virginiana	Virginia spiderwort
Viola pedata	bird-foot violet

Western Mesophytic Forest

Western Mesophytic Forest occurs throughout the region on rich, moist sites such as protected lower north slopes, deep ravines, and other areas of deep, rich, well-drained soils. It has the highest species diversity of the forested communities in this region. This community corresponds, in part, to Küchler #100 and Eastern Deciduous and Mixed Forests ERT.

Canopy
Characteristic Species

Acer saccharum	sugar maple
Carya cordiformis	bitter-nut hickory
Carya ovata	shag-bark hickory
Fagus grandifolia	American beech
Liriodendron tulipifera	tuliptree
Quercus alba	northern white oak
Quercus rubra	northern red oak
Tilia americana	American basswood

Associates

Aesculus glabra	Ohio buckeye
Celtis occidentalis	common hackberry
Fraxinus americana	white ash
Gymnocladus dioicus	Kentucky coffeetree
Juglans nigra	black walnut
Liquidambar styraciflua	sweet-gum
Magnolia acuminata	cucumber magnolia
Nyssa sylvatica	black tupelo
Platanus occidentalis	American sycamore
Prunus serotina	black cherry
Quercus shumardii	Shumard's oak
Ulmus rubra	slippery elm

Woody Understory

Asimina triloba	common pawpaw
Carpinus caroliniana	American hornbeam
Cercis canadensis	redbud
Cornus alternifolia	alternate-leaf dogwood
Cornus florida	flowering dogwood
Dirca palustris	eastern leatherwood
Hamamelis virginiana	American witch-hazel
Hydrangea arborescens	wild hydrangea
Lindera benzoin	northern spicebush
Ostrya virginiana	eastern hop-hornbeam
Sassafras albidium	sassafras
Staphylea trifolia	American bladdernut

Herbaceous Understory

Actaea pachypoda	white baneberry
Adiantum pedatum	northern maidenhair
Amphicarpaea bracteata	American hog-peanut
Amsonia tabernaemontana	eastern bluestar
Aquilegia canadensis	red columbine
Aralia racemosa	American spikenard
Arisaema triphyllum	jack-in-the-pulpit
Aruncus dioicus	bride's-feathers
Asarum canadense	Canadian wild ginger
Asplenium platyneuron	ebony spleenwort
Athyrium filix-femina	subarctic lady fern
Cardamine concatenata	cut-leaf toothwort
Chasmanthium latifolium	Indian wood-oats
Cimicifuga racemosa	black bugbane
Claytonia virginica	Virginia springbeauty
Cypripedium pubescens	greater yellow lady's-slipper
Diarrhena americana	American beakgrain
Dicentra canadensis	squirrel-corn
Dicentra cucullaria	Dutchman's-breeches
Dodecatheon meadia	eastern shootingstar
Dryopteris intermedia	evergreen wood fern
Dryopteris marginalis	marginal wood fern
Enemion biternatum	eastern false rue-anemone
Erythronium albidum	small white fawn-lily
Erythronium americanum	American trout-lily
Galearis spectabilis	showy orchid
Geranium maculatum	spotted crane's-bill
Hepatica nobilis	liverwort
Hydrastis canadensis	goldenseal
Hydrophyllum appendiculatum	great waterleaf
Hydrophyllum canadense	blunt-leaf waterleaf
Hydrophyllum virginianum	Shawnee-salad
Jeffersonia diphylla	twinleaf
Maianthemum racemosum	feathery false Solomon's-seal

Mitella diphylla	two-leaf bishop's-cap
Onoclea sensibilis	sensitive fern
Osmunda cinnamomea	cinnamon fern
Panax quinquefolius	American ginseng
Phacelia bipinnatifida	fern-leaf scorpion-weed
Phlox divaricata	wild blue phlox
Polygonatum biflorum	King Solomon's-seal
Polystichum acrostichoides	Christmas fern
Sanguinaria canadensis	bloodroot
Saxifraga virginiensis	early saxifrage
Silene virginica	fire-pink
Solidago caesia	wreath goldenrod
Solidago flexicaulis	zigzag goldenrod
Stylophorum diphyllum	celandine-poppy
Thalictrum thalictroides	rue-anemone
Tiarella cordifolia	heart-leaf foamflower
Trillium erectum	stinking-Benjamin
Trillium flexipes	nodding wakerobin
Trillium grandiflorum	large-flower wakerobin
Trillium recurvatum	bloody-butcher
Trillium sessile	toadshade
Viola canadensis	Canadian white violet
Viola pubescens	downy yellow violet
Viola sororia	hooded blue violet

Oak Barrens

Oak Barrens, a savanna-like community of scattered, open grown oaks with a variably open understory composed of prairie and woodland species originally occurred over a broad area. Except for scattered remnants, this community type is all but gone from the landscape. Opportunities exist for restorations that include reintroduction of fire as a management tool. This community corresponds to Küchler #82 and Woodlands ERT.

Canopy
Characteristic Species

Carya ovata	shag-bark hickory
Quercus alba	northern white oak
Quercus macrocarpa	burr oak
Quercus marilandica	blackjack oak
Quercus stellata	post oak

Associates

Carya texana	black hickory
Juniperus virginiana	eastern red cedar
Pinus echinata	short-leaf pine
Quercus velutina	black oak
Ulmus alata	winged elm

Woody Understory

Crataegus spp.	hawthorn
Forestiera ligustrina	upland swamp-privet
Rhus aromatica	fragrant sumac

Herbaceous Understory

Andropogon gerardii	big bluestem
Aster linearifolius	stiff-leaf aster
Baptisia bracteata	long-bract wild indigo
Blephilia ciliata	downy pagoda-plant
Bouteloua curtipendula	side-oats grama
Echinacea purpurea	eastern purple-coneflower
Gentiana puberulenta	downy gentian
Houstonia canadensis	Canadian summer bluet
Pedicularis canadensis	Canadian lousewort
Schizachyrium scoparium	little false bluestem
Silene regia	royal catchfly
Sorghastrum nutans	yellow Indian grass
Tephrosia virginiana	goat's-rue
Viola pedata	bird-foot violet

Bluestem Prairie

Bluestem Prairie once occupied millions of acres in this region. It occurred primarily as disjunct, variable-sized prairie openings in an otherwise forested area. The grass-dominated prairies were maintained by fire and grazing. The prairies ranged from wet to dry depending upon soil depth, moisture availability, aspect, and other factors. The prairies were primarily grass dominated with forbs scattered throughout. Trees and shrubs occurred only as isolated individuals or in groves. This community corresponds to Küchler #74 and #82 and Prairies ERT.

Characteristic Species

Andropogon gerardii	big bluestem
Aster novae-angliae	New England aster
Aster oolentangiensis	sky-blue aster
Aster pilosus	white oldfield aster
Baptisia alba	white wild indigo
Bouteloua curtipendula	side-oats grama
Chamaecrista fasciculata	sleepingplant
Chrysopsis mariana	Maryland golden-aster
Coreopsis lanceolata	lance-leaf tickseed
Dalea candida	white prairie-clover
Dalea purpurea	violet prairie-clover
Echinacea pallida	pale purple-coneflower
Eryngium yuccifolium	button eryngo
Euphorbia corollata	flowering spurge
Helianthus mollis	neglected sunflower
Heliopsis helianthoides	smooth oxeye
Lespedeza capitata	round-head bush-clover
Liatris aspera	tall gayfeather

Liatris pychnostachya	cat-tail gayfeather
Liatris spicata	dense gayfeather
Manfreda virginica	false aloe
Monarda fistulosa	Oswego-tea
Panicum virgatum	wand panic grass
Penstemon digitalis	foxglove beardtongue
Phlox pilosa	downy phlox
Physostegia virginiana	obedient-plant
Pycnanthemum tenuifolium	narrow-leaf mountain-mint
Ratibida pinnata	gray-head Mexican-hat
Rudbeckia hirta	black-eyed-Susan
Salvia azurea	azure-blue sage
Schizachyrium scoparium	little false bluestem
Silene regia	royal catchfly
Silphium compositum	kidney-leaf rosinweed
Silphium laciniatum	compassplant
Silphium terebinthinaceum	prairie rosinweed
Solidago nemoralis	gray goldenrod
Solidago rigida	hard-leaf goldenrod
Sorghastrum nutans	yellow Indian grass
Spartina pectinata	freshwater cord grass
Sporobolus heterolepis	prairie dropseed
Vernonia gigantea	giant ironweed
Veronicastrum virginicum	culver's-root

Glade

Glade is a naturally open, grass and/or forb dominated area where bedrock is at or near the surface of the ground. Glade communities may occur on slopes of hills or in topographically flat areas. They are usually dry and droughty during the growing season but may be locally saturated in the spring. Glades are usually small, but can be locally dominant in the central basin of Tennessee and in parts of Arkansas and Missouri in the Ozarks. Glade communities are often distinguished by bedrock type (i.e., sandstone, limestone, dolomite, shales, etc.). Glades are floristically diverse and often harbor many unusual and rare species. They share many species of the prairie and often grade into dry prairies where soil deepens. This community corresponds to Küchler #83 and Glades ERT.

Characteristic Species

Aristida longespica	red three-awn
Aster sericeus	western silver aster
Bouteloua curtipendula	side-oats grama
Castilleja coccinea	scarlet Indian-paintbrush
Cheilanthes lanosa	hairy lip fern
Crotonopsis elliptica	egg-leaf rushfoil
Dalea candida	white prairie-clover
Dalea purpurea	violet prairie-clover
Danthonia spicata	poverty wild oat grass
Diodia teres	poorjoe
Echinacea pallida	pale purple-coneflower
Eragrostis spectabilis	petticoat climber
Glandularia canadensis	rose mock vervain
Heliotropium tenellum	pasture heliotrope

Hypericum gentianoides	orange-grass
Lechea mucronata	hairy pinweed
Lechea tenuifolia	narrow-leaf pinweed
Liatris aspera	tall gayfeather
Liatris cylindracea	Ontario gayfeather
Liatris squarrosa	scaly gayfeather
Lithospermum canescens	hoary pucoon
Minuartia patula	Pitcher's stitchwort
Ophioglossum engelmannii	limestone adder's-tongue
Opuntia humifusa	eastern prickly-pear
Phlox pilosa	downy phlox
Sabatia angularis	rose-pink
Schizachyrium scoparium	little false bluestem
Sedum pulchellum	widow's cross
Senecio anonymus	Small's ragwort
Sorghastrum nutans	yellow Indian grass
Sporobolus asper	tall dropseed
Sporobolus heterolepis	prairie dropseed
Sporobolus vaginiflorus	poverty dropseed
Stylosanthes biflora	side-beak pencil-flower
Talinum parviflorum	sunbright
Tephrosia virginiana	goat's-rue

WETLAND SYSTEMS

Floodplain Forest

Floodplain Forest occurs throughout the region along large streams and rivers where floodplains have developed. The canopy composition is a variety of large deciduous trees, and the understory and herbaceous layers range from well developed to poorly developed, depending upon the duration of annual flooding. This community corresponds to Küchler #113 and Floodplains ERT.

Canopy
Characteristic species

Acer saccharinum	silver maple
Betula nigra	river birch
Carya laciniosa	shell-bark hickory
Celtis occidentalis	common hackberry
Liquidambar styraciflua	sweet-gum
Populus deltoides	eastern cottonwood
Quercus bicolor	swamp white oak
Quercus lyrata	overcup oak
Quercus macrocarpa	burr oak
Quercus pagoda	cherry-bark oak
Quercus palustris	pin oak
Ulmus americana	American elm

Associates

Acer negundo	ash-leaf maple
Acer rubrum	red maple
Carya cordiformis	bitter-nut hickory
Fraxinus pennsylvanica	green ash
Gleditsia triacanthos	honey-locust
Juglans nigra	black walnut
Nyssa sylvatica	black tupelo
Platanus occidentalis	American sycamore
Quercus michauxii	swamp chestnut oak
Salix nigra	black willow
Ulmus rubra	slippery elm

Woody Understory

Asimina triloba	common pawpaw
Bignonia capreolata	crossvine
Campsis radicans	trumpet-creeper
Carpinus caroliniana	American hornbeam
Cornus foemina	stiff dogwood
Ilex decidua	deciduous holly
Lindera benzoin	northern spicebush
Toxicodendron radicans	eastern poison-ivy
Vitis spp.	grape

Herbaceous Understory

Arundinaria gigantea	giant cane
Boehmeria cylindrica	small-spike false nettle
Carex grayi	Gray's sedge
Carex tribuloides	blunt broom sedge
Chelone glabra	white turtlehead
Cinna arundinacea	sweet wood-reed
Elymus virginicus	Virginia wild rye
Hymenocallis caroliniana	Carolina spider-lily
Impatiens capensis	spotted touch-me-not
Laportea canadensis	Canadian wood-nettle
Lobelia cardinalis	cardinal-flower
Lobelia siphilitica	great blue lobelia
Mertensia virginica	Virginia bluebells
Mimulus ringens	Allegheny monkey-flower
Monarda didyma	scarlet beebalm
Onoclea sensibilis	sensitive fern
Phacelia purshii	Miami-mist
Sanicula canadensis	Canadian black-snakeroot
Viola cucullata	marsh blue violet

Swamp Forest

Swamp Forest is uncommon and scattered throughout this region, usually associated with large floodplains of major rivers. Less common are upland swamps, usually associated with poorly drained depressions or sinkholes. Most swamps in this region are hardwood dominated although some can be dominated by southern bald-cypress. This community corresponds to Swamp Forests ERT.

Canopy
Characteristic species

Betula nigra	river birch
Nyssa aquatica	water tupelo
Populus heterophylla	swamp cottonwood
Quercus lyrata	overcup oak
Quercus palustris	pin oak
Salix nigra	black willow
Taxodium distichum	southern bald-cypress

Woody Understory

Cephalanthus occidentalis	common buttonbush
Decodon verticillatus	swamp-loosestrife
Ilex decidua	deciduous holly
Itea virginica	Virginia sweetspire
Rosa palustris	swamp rose

Herbaceous Understory

Ceratophyllum demersum	coontail
Iris virginica	Virginia blueflag
Lemna spp.	duckweed
Sagittaria latifolia	duck-potato
Saururus cernuus	lizard's-tail

Marsh

Marsh is primarily associated with floodplains of larger streams and rivers although it also occurs in upland depressions. These communities are shallowly flooded for much of the year. They are associated with Wet Prairie communities in dryer areas and Swamp Forest in deeper water areas. This community corresponds to Nonestuarine Marshes ERT.

Characteristic Species

Alisma subcordatum	Amercian water-plantain
Boltonia asteroides	white doll's-daisy
Carex frankii	Frank's sedge
Carex lupulina	hop sedge
Carex tribuloides	blunt broom sedge
Carex vulpinoidea	common fox sedge
Eleocharis acicularis	needle spike-rush
Eleocharis obtusa	blunt spike-rush
Hibiscus laevis	halherb-leaf rose-mallow
Hibiscus moscheutos	crimson-eyed rose-mallow

Iris brevicaulis	zigzag iris
Iris virginica	Virginia blueflag
Juncus effusus	lamp rush
Leersia virginica	white grass
Nelumbo lutea	Amercian lotus
Nuphar lutea	yellow pond-lily
Polygonum amphibium	water smartweed
Polygonum hydropiperoides	swamp smartweed
Polygonum pensylvanicum	pinkweed
Rumex verticillatus	swamp dock
Sagittaria latifolia	duck-potato
Scirpus americanus	chairmaker's bulrush
Scirpus atrovirens	dark-green bulrush
Scirpus tabernaemontani	soft-stem bulrush
Sparganium eurycarpum	broad-fruit burr-reed
Typha latifolia	broad-leaf cat-tail

Wet Prairie

Wet Prairie is often associated with marsh communities in floodplains of larger streams and in poorly drained depressions in the uplands. These grass-dominated wetlands are essentially eradicated now with most areas converted to cropland. This community corresponds to Wet Prairies and Grasslands ERT.

Characteristic Species

Agalinis purpurea	purple false foxglove
Asclepias incarnata	swamp milkweed
Bidens aristosa	bearded beggarticks
Bidens cernua	nodding burr-marigold
Carex arkansana	Arkansas sedge
Carex vulpinoidea	common fox sedge
Cephalanthus occidentalis	common buttonbush
Cicuta maculata	spotted water-hemlock
Eupatorium coelestinum	blue mistflower
Eupatorium maculatum	spotted joe-pye-weed
Eupatorium perfoliatum	common boneset
Helenium autumnale	fall sneezeweed
Helianthus angustifolius	swamp sunflower
Panicum virgatum	wand panic grass
Rudbeckia laciniata	green-head coneflower
Sium suave	hemlock water-parsnip
Solidago rugosa	wrinkle-leaf goldenrod
Spartina pectinata	freshwater cord grass
Spiraea alba	white meadowsweet
Tripsacum dactyloides	eastern mock grama
Verbena hastata	simpler's-joy

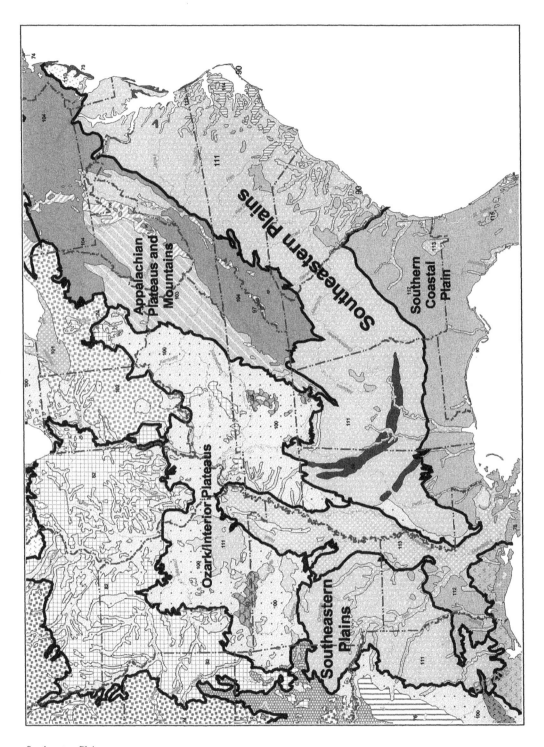

Southeastern Plains

Southeastern Plains

Introduction

The Southeastern Plains region extends from the Pine Barrens in New Jersey, south and east through the Coastal Plain and Piedmont physiographic regions of Virginia and North and South Carolina, and into the northern half of Georgia, Alabama, Mississippi, and Louisiana. The western boundary includes southern Arkansas and east Texas. Excluded from the region are the coastal zones of Georgia, Alabama, and Mississippi which are part of the Southern Coastal Plains region. The majority of the region is the relatively flat coastal plain and the remaining encompasses the Piedmont, an area of gently rolling hills bordering the Appalachian highlands.

Vegetation is very diverse, ranging from closed canopy forest to shrublands, savannas, grasslands, and many freshwater wetlands. The impact of human activity has caused dramatic changes to the natural communities. The suppression of natural fires has led to changes in the vegetation. The forests throughout the region are presently dominated by shortleaf or longleaf pine. The region was originally dominated by a mixture of oak and hickory, as well as pine. Forested lands, converted to agriculture and then abandoned, have been reseeded, either intentionally or naturally, with the now dominant pine species. Much of the forest land is intensively managed by wood products industries which make large investments in the establishment and growth of pine timber.

Other natural communities, especially wetlands, have been dramatically changed or reduced in number by drainage and development. Marshes, lakes, and swamps are often associated with the streams that wind gently through the sloping topography of the region. Wetland communities, not associated with rivers or nonalluvial wetlands, are Pocosin and Atlantic White-cedar Swamp. Wetlands associated with coastal conditions include Salt Shrub Thicket and Salt Marsh.

The primary sources used to develop descriptions and species lists are Barbour and Billings 1988; Forman 1979; Martin et al. In press; Nelson 1986; Schafale and Weakley 1990; U.S. Fish and Wildlife Service 1984; Wharton 1989.

Dominant Ecological Communities

Upland Systems
Oak–Hickory–Pine Forest
Southern Mixed Hardwoods Forest
Mesic Pine Forest
Xeric Pine Forest
Pine Barrens
Live Oak Woodland
Blackbelt Prairie

Wetland Systems
Southern Floodplain Forest
Cypress–Tupelo Swamp Forest
Atlantic White-cedar Swamp
Pocosin
Freshwater Marsh

Estuarine Systems
Salt Shrub Thicket
Salt Marsh

UPLAND SYSTEMS
Oak–Hickory–Pine Forest

Oak–Hickory–Pine Forest was originally the most extensive forest type in this region. This forest type occurs throughout the Piedmont and the Southeastern Coastal Plain on slopes, ridges, upland flats, and other dry to dry-mesic sites. The dominant trees are hickories, short-leaf pine, loblolly pine, northern white oak, and post oak. Examples of original oak–hickory–pine are rare because most of the area is now used for agriculture, urban development, or is occupied by pine stands after past agriculture. Examples of the original forest, however, may be found in fragmented areas along rivers or creeks. This community corresponds to Küchler #111 and Eastern Deciduous and Mixed Forests ERT.

Canopy
Characteristic Species

Carya alba	mockernut hickory
Carya carolinae-septentrionalis	southern shag-bark hickory
Carya glabra	pignut hickory
Pinus echinata	short-leaf pine
Pinus taeda	loblolly pine
Quercus alba	northern white oak
Quercus coccinea	scarlet oak
Quercus marilandica	blackjack oak
Quercus rubra	northern red oak
Quercus stellata	post oak
Quercus velutina	black oak

Associates

Acer rubrum	red maple
Carya cordiformis	bitter-nut hickory
Carya ovalis	red hickory
Carya ovata	shag-bark hickory
Carya pallida	sand hickory
Diospyros virginiana	common persimmon
Fraxinus americana	white ash
Juglans nigra	black walnut
Liquidambar styraciflua	sweet-gum
Liriodendron tulipifera	tuliptree
Nyssa sylvatica	black tupelo
Persea borbonia	red bay
Pinus strobus	eastern white pine
Pinus virginiana	Virginia pine
Prunus serotina	black cherry
Quercus falcata	southern red oak
Quercus muhlenbergii	chinkapin oak
Quercus shumardii	Shumard's oak

Woody Understory

Acer leucoderme	chalk maple
Aesculus sylvatica	painted buckeye
Amelanchier spp.	service-berry
Asimina parviflora	small-flower pawpaw

Calycanthus floridus	eastern sweetshrub
Carpinus caroliniana	American hornbeam
Cercis canadensis	redbud
Chionanthus virginicus	white fringetree
Cornus florida	flowering dogwood
Evonymus americanus	American strawberry-bush
Hypericum hypericoides	St. Andrew's-cross
Ilex decidua	deciduous holly
Ostrya virginiana	eastern hop-hornbeam
Oxydendron arboreum	sourwood
Rhododendron canescens	mountain azalea
Rhododendron periclymenoides	early azalea
Rhus aromatica	fragrant sumac
Symphoricarpos orbiculatus	coral-berry
Symplocos tinctoria	horsesugar
Toxicodendron pubescens	Atlantic poison-oak
Vaccinium arboreum	tree sparkle-berry
Viburnum acerifolium	maple-leaf arrow-wood
Viburnum prunifolium	smooth blackhaw
Viburnum rafinesquianum	downy arrow-wood
Viburnum rufidulum	rusty blackhaw

Herbaceous Understory

Agrimonia gryposepala	tall hairy grooveburr
Amianthium muscaetoxicum	flypoison
Andropogon virginicus	broom-sedge
Aristolochia serpentaria	Virginia-snakeroot
Aster dumosus	rice button aster
Aster paternus	toothed white-top aster
Aster solidagineus	narrow-leaf white-top aster
Aureolaria flava	smooth yellow false-foxglove
Carex albicans	white-tinge sedge
Carex nigromarginata	black-edge sedge
Chimaphila maculata	striped prince's-pine
Chrysopsis mariana	Maryland golden-aster
Coreopsis major	greater tickseed
Cunila marima	common dittany
Cypripedium acaule	pink lady's-slipper
Desmodium laevigatum	smooth tick-trefoil
Desmodium nudiflorum	naked-flower tick-trefoil
Desmodium obtusum	stiff tick-trefoil
Desmodium perplexum	perplexed tick-trefoil
Desmodium rotundifolium	prostrate tick-trefoil
Dichanthelium sphaerocarpon	round-seed rosette grass
Elephantopus spp.	elephant's foot
Euphorbia corollata	flowering spurge
Galium circaezans	licorice bedstraw
Galium pilosum	hairy bedstraw
Goodyera pubescens	downy rattlesnake-plantain

Hexastylis arifolia	little-brown-jug
Hexastylis virginica	Virginia heartleaf
Hieracium gronovii	queendevil
Hieracium venosum	rattlesnake-weed
Lespedeza repens	creeping bush-clover
Panicum spp.	panic grass
Phryma leptostachya	American lopseed
Piptochaetium avenaceum	black-seed spear grass
Polygonatum biflorum	King Solomon's-seal
Pycnanthemum flexuosum	Appalachian mountain-mint
Rhynchosia tomentosa	twining snout-bean
Scleria oligantha	little-head nut-rush
Solidago nemoralis	gray goldenrod
Strophostyles umbellata	pink fuzzy-bean
Stylosanthes bilflora	side-beak pencil-flower
Tephrosia virginiana	goat's-rue
Tipularia discolor	crippled-cranefly
Trillium catsbaei	bashful wakerobin
Uvularia perifoliata	perfoliate bellwort
Uvularia sessilifolia	sessile-leaf bellwort

Southern Mixed Hardwoods Forest

Southern Mixed Hardwoods Forest is restricted to coves and gorges in this region. American beech, northern white oak, red maple, and sweet-gum are the dominant species. Southern Mixed Hardwoods Forest is more mesic than Oak–Hickory–Pine Forest, yet it is distinct from forests along river bottomlands. Small remnant old-growth stands can be found on slopes, islands in swamps, and on a few upland flats. This community corresponds to Küchler #112 and Eastern Deciduous and Mixed Forests ERT.

Canopy
Characteristic Species

Acer rubrum	red maple
Acer saccharum	sugar maple
Carya cordiformis	bitter-nut hickory
Carya pallida	sand hickory
Fagus grandifolia	American beech
Juglans nigra	black walnut
Liquidambar styraciflua	sweet-gum
Liriodendron tulipifera	tuliptree
Quercus alba	northern white oak
Quercus pagoda	cherry-bark oak
Quercus michauxii	swamp chestnut oak
Quercus rubra	northern red oak
Quercus shumardii	Shumard's oak
Ulmus americana	American elm
Ulmus rubra	slippery elm

Associates

Acer barbatum	Florida maple
Carya alba	mockernut hickory
Carya glabra	pignut hickory

Celtis laevigata	sugar-berry
Juglans nigra	black walnut
Liquidambar styraciflua	sweet-gum
Magnolia grandiflora	southern magnolia
Magnolia tripetala	umbrella magnolia
Persea borbonia	red bay
Pinus echinata	short-leaf pine
Pinus elliottii	slash pine
Pinus palustris	long-leaf pine
Pinus taeda	loblolly pine
Populus heterophylla	swamp cottonwood
Quercus falcata	southern red oak
Quercus incana	bluejack oak
Quercus laevis	turkey oak
Quercus laurifolia	laurel oak
Quercus marilandica	blackjack oak
Quercus stellata	post oak
Quercus virginiana	live oak
Tilia americana	American basswood

Woody Understory

Acer negundo	ash-leaf maple
Aesculus pavia	red buckeye
Aesculus sylvatica	painted buckeye
Asimina parviflora	small-flower pawpaw
Asimina triloba	common pawpaw
Calycanthus floridus	eastern sweetshrub
Carpinus caroliniana	American hornbeam
Cercis canadensis	redbud
Cornus florida	flowering dogwood
Evonymus americanus	American strawberry-bush
Evonymus atropurpurea	eastern wahoo
Hamamelis virginiana	Amercian witch-hazel
Hydrangea arborescens	wild hydrangea
Ilex glabra	inkberry
Ilex opaca	American holly
Lindera benzoin	northern spicebush
Menispermum canadense	Canadian moonseed
Myrica cerifera	southern bayberry
Ostrya virginiana	eastern hop-hornbeam
Philadelphus inodorus	scentless mock orange
Sabal palmetto	cabbage palmetto
Staphylea trifolia	American bladdernut
Stewartia malacodendron	silky-camellia
Styrax grandifolia	big-leaf snowbell
Symplocos tinctoria	horsesugar
Viburnum rufidulum	rusty blackhaw

Herbaceous Understory

Actaea pachypoda	white baneberry
Adiantum pedatum	northern maidenhair
Anemone virginiana	tall thimbleweed
Asarum canadense	Canadian wild ginger
Aureolaria flava	smooth yellow false-foxglove
Campanulastrum americana	American-bellflower
Cimicifuga racemosa	black bugbane
Corydalis flavula	yellow fumewort
Cypripedium pubescens	greater yellow lady's-slipper
Delphinium tricorne	dwarf larkspur
Dicentra cucullaria	Dutchman's-breeches
Enemion biternatum	eastern false rue-anemone
Epifagus virginiana	beechdrops
Hepatica nobilis	liverwort
Heuchera americana	American alumroot
Hybanthus concolor	eastern green-violet
Lathyrus venosus	veiny vetchling
Mitchella repens	partridge-berry
Panax quinquefolius	American ginseng
Podophyllum peltatum	may-apple
Polystichum acrostichoides	Christmas fern
Sanguinaria canadensis	bloodroot
Scutellaria ovata	heart-leaf skullcap
Silene stellata	widow's frill
Thaspium trifoliatum	purple meadow-parsnip
Tiarella cordifolia	heart-leaf foamflower
Tipularia discolor	crippled-cranefly
Trillium cuneatum	little-sweet-Betsy
Trillium erectum	stinking-Benjamin

Mesic Pine Forest

In the coastal plain, Mesic Pine Forest may occur on flat or rolling land. These forests are often referred to as "Flatwoods." In recent years the suppression of fire has led to changes in community structure as hardwood species invade and share dominance. The pine forests of the region are divided into mesic and xeric categories. These two types share many of the same species. The Mesic Pine Forest has a denser and more diverse herbaceous layer than the Xeric Pine Forest. This community corresponds to Eastern Deciduous and Mixed Forests ERT.

Canopy
Characteristic Species

Pinus echinata	short-leaf pine
Pinus palustris	long-leaf pine
Pinus taeda	loblolly pine

Associates

Carya alba	mockernut hickory
Carya pallida	sand hickory
Gordonia lasianthus	loblolly-bay
Liquidambar styraciflua	sweet-gum

Magnolia virginiana	sweet-bay
Quercus falcata	southern red oak
Quercus incana	bluejack oak
Quercus laurifolia	laurel oak
Quercus marilandica	blackjack oak
Quercus nigra	water oak
Quercus stellata	post oak

Woody Understory

Cyrilla racemiflora	swamp titi
Ilex coriacea	large gallberry
Ilex glabra	inkberry
Kalmia hirsuta	hairy-laurel
Myrica cerifera	southern bayberry
Myrica inodora	odorless bayberry
Vaccinium tenellum	small black blueberry
Viburnum nudum	possumhaw

Herbaceous Understory

Andropogon gerardii	big bluestem
Anthaenantia villosa	green silkyscale
Aristida stricta	pineland three-awn
Aster paludosus	southern swamp aster
Aster tortifolius	dixie white-top aster
Aster walteri	Walter's aster
Chaptalia tomentosa	woolly sunbonnets
Dalea pinnata	summer-farewell
Elephantopus carolinianus	Carolina elephant's-foot
Gymnopogon brevifolium	short-leaf skeleton grass
Helianthus angustifolius	swamp sunflower
Helianthus radula	rayless sunflower
Lespedeza capitata	round-head bush-clover
Panicum virgatum	wand panic grass
Paspalum bifidum	pitchfork crown grass
Pteridium aquilinum	northern bracken fern
Pterocaulon virgatum	wand blackroot
Schizachyrium scoparium	little false bluestem
Solidago odora	anise-scented goldenrod
Sorghastrum nutans	yellow Indian grass
Tephrosia spicata	spiked hoary-pea
Tephrosia virginica	goat's-rue

Xeric Pine Forest

Xeric Pine Forest is the driest community in the Southeastern Plains region. This community is often dominated by long-leaf pine, turkey oak, and pineland three-awn (wiregrass). These xeric areas have been classified as "Sandhill Pine Forest," "Sandhill Pine Scrub," and "Xeric Sandhill Scrub." Xeric Pine Forests naturally experience frequent fires. This community corresponds to Eastern Deciduous and Mixed Forests ERT.

Canopy
Characteristic Species
Pinus palustris	long-leaf pine
Quercus laevis	turkey oak

Associate Species
Diospyros virginiana	common persimmon
Liquidamber styraciflua	sweet-gum
Nyssa sylvatica	black tupelo
Pinus echinata	short-leaf pine
Pinus taeda	loblolly pine
Quercus incana	bluejack oak
Quercus marilandica	blackjack oak

Woody Understory
Ceratiola ericoides	sand-heath
Cornus florida	flowering dogwood
Gaylussacia dumosa	dwarf huckleberry
Gaylussacia frondosa	blue huckleberry
Ilex vomitoria	yaupon
Licania michauxii	gopher-apple
Lyonia mariana	Piedmont staggerbush
Opuntia humifusa	eastern prickly-pear
Quercus margarettiae	sand post oak
Quercus pumila	runner oak
Sassafras albidum	sassafras
Toxicodendron pubescens	Atlantic poison-oak

Herbaceous Understory
Aristida stricta	pineland three-awn
Baptisia lanceolata	gopherweed
Chrysopsis gossypina	cottony golden-aster
Cnidoscolus stimulosus	ginger-rot
Dicranum condensatum	sandhill broom-moss
Dicranum spurium	broom moss
Eupatorium rotundifolium	round-leaf thoroughwort
Euphorbia ipecacuanhae	American-ipecac
Minuartia caroliniana	pine-barren stitchwort
Pluchea rosea	rosy camphorweed
Polygonella polygama	October-flower
Rudebeckia hirta	black-eyed-Susan
Schizachyrium scoparium	little false bluestem
Selaginella arenicola	sand spike-moss
Stillingia sylvatica	queen's-delight
Stipulicida setacea	pineland scaly-pink
Stylosanthes biflora	side-beak pencil-flower
Tephrosia virginiana	goat's rue

Pine Barrens

Pine Barrens is a dwarf forest dominated by pitch pine (*Pinus rigida*). This forest type occurs north of the Delaware Bay 100–200 feet above sea level in gently rolling terrain. The largest and most uniform pine barrens community occurs in New Jersey and covers nearly one and a half million acres. Fires are an important factor in maintaining the structure and composition of Pine Barrens. Toward the south, Pine Barrens is replaced by Live Oak Woodland. This community corresponds to Küchler #110 and Woodlands ERT.

Canopy
Characteristic Species

Pinus rigida	pitch pine

Associates

Pinus echinata	short-leaf pine
Quercus coccinea	scarlet oak
Quercus marilandica	blackjack oak
Quercus prinus	chestnut oak
Quercus stellata	post oak
Quercus velutina	black oak

Woody Understory

Arctostaphylos uva-ursi	red bearberry
Clethra alnifolia	coastal sweet-pepperbush
Comptonia peregrina	sweet-fern
Corema conradii	broom-crowberry
Epigaea repens	trailing-arbutus
Gaultheria procumbens	eastern teaberry
Gaylussacia baccata	black huckleberry
Gaylussacia dumosa	dwarf huckleberry
Gaylussacia frondosa	blue huckleberry
Ilex glabra	inkberry
Kalmia angustifolia	sheep-laurel
Kalmia latifolia	mountain-laurel
Leiophyllum buxifolium	sand-myrtle
Quercus ilicifolia	bear oak

Herbaceous Understory

Andropogon virginicus	broom-sedge
Cladonia atlantica	lichen
Cladonia caroliniana	lichen
Cladonia cristatella	lichen
Cladonia squamosa	lichen
Aureolaria pedicularia	fern-leaf yellow false-foxglove
Scleria triglomerata	whip nut-rush

Live Oak Woodland

Live Oak Woodland is associated with maritime conditions and is often referred to as "maritime forests." Live oak communities occur on the barrier islands and along the coast. Frequently, live oaks grow on old stabilized dunes and flats, protected from salt water and flooding and the most extreme salt spray. The trees are characterized by their asymmetric growth resulting from a combination of wind and salt spray. Sea-oats are important in the Live Oak Woodland due to their role in stabilizing the dunes. This community corresponds to Küchler #90 and Woodlands ERT.

Canopy

Characteristic Species

Pinus elliottii	slash pine
Quercus virginiana	live oak
Sabal palmetto	cabbage palmetto

Associates

Carya glabra	pignut hickory
Pinus palustris	long-leaf pine
Pinus taeda	loblolly pine
Quercus falcata	southern red oak
Quercus hemisphaerica	Darlington's oak
Quercus laevis	turkey oak
Quercus laurifolia	laurel oak
Quercus nigra	water oak

Woody Understory

Ampelopsis arborea	peppervine
Baccharis halimifolia	groundseltree
Berchemia scandens	Alabama supplejack
Bignonia capreolata	crossvine
Callicarpa americana	American beauty-bush
Campsis radicans	trumpet-creeper
Carpinus caroliniana	Amerian hornbeam
Cornus florida	flowering dogwood
Gelsemium sempervirens	evening trumpet-flower
Hamamelis virginiana	American witch-hazel
Ilex opaca	American holly
Ilex vomitoria	yaupon
Iva frutescens	Jesuit's-bark
Iva imbricata	seacoast marsh-elder
Juniperus virginiana	eastern red-cedar
Lyonia ferruginea	rusty staggerbush
Magnolia virginiana	sweet-bay
Myrica cerifera	southern bayberry
Myrica pensylvanica	northern bayberry
Opuntia humifusa	eastern prickly-pear
Osmanthus americana	devilwood
Parthenocissus quinquefolia	Virginia-creeper
Persea borbonia	red bay
Prunus caroliniana	Carolina laurel cherry

Rhus copallinum	winged sumac
Sabal minor	dwarf palmetto
Sassafras albidum	sassafras
Serenoa repens	saw-palmetto
Smilax bona-nox	fringed greenbrier
Toxicodendron radicans	eastern poison-ivy
Vitis rotundifolia	muscadine
Yucca aloifolia	aloe yucca
Yucca filamentosa	Adam's-needle
Zanthoxylum clava-herculis	Hercules'-club

Herbaceous Understory

Ammophila breviligulata	American beach grass
Andropogon virginicus	broom-sedge
Asplenium platyneuron	ebony spleenwort
Cakile edentula	American searocket
Cenchrus tribuloides	sand-dune sandburr
Chamaesyce bombensis	dixie sandmat
Chamaesyce polygonifolia	seaside spurge
Chasmanthium laxum	slender wood-oats
Cnidoscolus stimulosus	finger-rot
Conyza canadensis	Canadian horseweed
Croton punctatus	gulf croton
Dichanthelium commutatum	variable rosette grass
Diodia teres	poor-joe
Elephantopus nudatus	smooth elephant's-foot
Eustrachys petraea	pinewoods finger grass
Galium pilosum	hairy bedstraw
Houstonia procumbens	round-leaf bluet
Hydrocotyle bonariensis	coastal marsh-pennywort
Juncus roemerianus	Roemer's rush
Kostleltskya virginica	Virginia fen-rose
Mitchella repens	partridge-berry
Oenothera humifusa	seaside evening-primrose
Panicum amarum	bitter panic grass
Passiflora incarnata	purple passion-flower
Passiflora lutea	yellow passion-flower
Physalis walteri	Walter's ground-cherry
Piptochaetium avenaceum	black-seed spear grass
Schizachyrium scoparium	little false bluestem
Scleria triglomerata	whip nut-rush
Spartina alternifolia	salt water cord grass
Spartina patens	salt-meadow cord grass
Stenotaphrum secundatum	St. Augustine grass
Strophostyles helvula	trailing fuzzy-bean
Triplasis purpurea	purple sand grass
Uniola paniculata	sea-oats

Blackbelt Prairie

The Blackbelt region of Alabama and Mississippi has been converted to agriculture. The natural vegetation was probably a prairie and oak savanna mosaic. Küchler (#89) calls the Blackbelt a *Liquidamber-Quercus-Juniperus* community. It is likely that the control of fire and eradication of grazing animals resulted in succession of the prairie to forest vegetation. This community corresponds to Prairies ERT.

Characteristic Species

Andropogon gerardii	big bluestem
Andropogon virginicus	broom-sedge
Aristida purpurescens	arrow-feather three-awn
Aster dumosus	rice button aster
Aster patens	late purple aster
Aster pilosus	white oldfield aster
Chamaecrista fasciculata	sleepingplant
Dalea purpurea	violet prairie-clover
Echinacea purpurea	eastern purple-coneflower
Eupatorium altissimum	tall thoroughwort
Galactia volubilis	downy milk-pea
Gymnopogon brevifolius	short-leaf skeleton grass
Monarda fistulosa	Oswego-tea
Panicum anceps	beaked panic grass
Ratibita pinnata	gray-head Mexican-hat
Rudbeckia fulgida	orange coneflower
Sabatia angularis	rose-pink
Salvia lyrata	lyre-leaf sage
Schizachyrium scopraium	little false bluestem
Silphium asteriscus	starry rosinweed
Silphium laciniatum	compassplant
Sporobolus asper	tall dropseed
Sporobolus vaginiflorus	poverty dropseed
Tragia urticifolia	nettle-leaf noseburn
Tridens flavus	tall redtop

Associates

Asclepias verticillata	whorled milkweed
Aster undulatus	waxy-leaf aster
Bouteloua curtipendula	side-oats grama
Desmodium ciliare	hairy small-leaf tick-trefoil
Erigeron strigosus	prairie fleabane
Euphorbia corollata	flowering spurge
Gaillardia aestivalis	lance-leaf blanket-flower
Lobelia spicata	pale-spike lobelia
Panicum virgatum	wand panic grass
Pycnanthemum tenuifolium	narrow-leaf mountain-mint

WETLAND SYSTEMS
Southern Floodplain Forest

Southern Floodplain Forest occurs throughout the Coastal Plain along large and medium size rivers. There are many "zones" associated with floodplains based upon the length of time the soils are saturated. The dominants and associates in floodplain areas are determined, in large part, by the degree of moisture. The zone that encompasses the greatest area is saturated during the winter and spring which accounts for 20 to 30 percent of the year. In these areas, laurel oak is the dominant tree species. Willow oak, sweet-gum, green ash, and tuliptree may accompany laurel oak as a characteristic species. In higher areas, swamp chestnut oak and cherry-bark oak will be dominant and in lower areas, southern bald-cypress, water tupelo, and swamp tupelo are characteristic. This community corresponds to Küchler #113 and Floodplain Forests ERT.

Canopy
Characteristic Species

Carya aquatica	water hickory
Carya cordiformis	bitter-nut hickory
Carya ovata	shag-bark hickory
Fraxinus pennsylvanica	green ash
Juglans nigra	black walnut
Liquidambar styraciflua	sweet-gum
Liriodendron tulipifera	tuliptree
Pinus taeda	loblolly pine
Quercus laurifolia	laurel oak
Quercus lyrata	overcup oak
Quercus michauxii	swamp chestnut oak
Quercus phellos	willow oak

Associates

Acer rubrum	red maple
Betula nigra	river birch
Celtis laevigata	sugar-berry
Chamaecyparis thyoides	Atlantic white-cedar
Fraxinus americana	white ash
Fraxinus caroliniana	Carolina ash
Fraxinus profunda	pumpkin ash
Gleditsia aquatica	water-locust
Nyssa aquatica	water tupelo
Nyssa biflora	swamp tupelo
Nyssa sylvatica	black tupelo
Plantanus occidentalis	American sycamore
Populus deltoides	eastern cottonwood
Populus heterophylla	swamp cottonwood
Quercus falcata	southern red oak
Quercus nigra	water oak
Quercus pagoda	cherry-bark oak
Quercus shumardii	Shumard's oak
Taxodium distichum	southern bald-cypress
Ulmus alata	winged elm
Ulmus americana	American elm

Woody Understory

Acer negundo	ash-leaf maple
Alnus serrulata	brookside alder
Ampelopsis arborea	peppervine
Arundinaria gigantea	giant cane
Asimina triloba	common pawpaw
Berchemia scandens	Alabama supplejack
Bignonia capreolata	crossvine
Campsis radicans	trumpet-creeper
Carpinus carolinianus	American hornbeam
Cephalanthus occidentalis	common buttonbush
Clematis virginiana	devil's-darning-needles
Clethra alnifolia	coastal sweet-pepperbush
Cornus foemina	stiff dogwood
Cyrilla racemiflora	swamp titi
Evonymous americanus	American strawberry-bush
Ilex decidua	deciduos holly
Ilex opaca	American holly
Itea virginica	Virginia sweetspire
Lindera benzoin	northern spicebush
Lyonia ligustrinia	maleberry
Magnolia virginiana	sweet-bay
Parthenocissus quinquefolia	Virginia-creeper
Persea borbonia	red bay
Persea palustis	swamp bay
Planera aquatica	planertree
Salix nigra	black willow
Smilax rotundifolia	horsebrier
Staphylea trifolia	American bladdernut
Toxicodendron pubescens	Atlantic poison-oak
Toxicodendron radicans	eastern poison-ivy
Vaccinium elliottii	Elliott's blueberry
Vaccinium stamineum	deerberry
Viburnum acerifolium	maple-leaf arrow-wood
Viburnum dentatum	southern arrow-wood
Viburnum prunifolium	smooth blackhaw
Vitis aestivalis	summer grape
Vitis rotundifolia	muscadine

Herbaceous Understory

Arisaema draconteum	greendragon
Asclepias perennis	Aquatic milkweed
Asplenium platyneuron	ebony spleenwort
Aster lateriflorus	farewell-summer
Boehmeria cylindrica	small-spike false nettle
Cayaponia quinqueloba	five-lobe-cucumber
Chasmanthium latifolium	Indian wood-oats
Chasmanthium laxum	slender wood-oats
Commelina virginica	Virginia dayflower

Dioscorea villosa	wild yam
Galium aparine	sticky-willy
Hymenocallis caroliniana	Carolina spider-lily
Leersia lenticularis	catchfly grass
Lysimachia quadrifolia	whorled yellow-loosestrife
Matelea gonocarpos	angular-fruit milkvine
Onoclea sensibilis	sensitive fern
Osmunda cinnamomea	cinnamon fern
Osmunda regalis	royal fern
Oxalis violacea	violet wood-sorrel
Pilea pumila	Canadian clearweed
Polygonum virginianum	jumpseed
Senecio glabellus	cress-leaf ragwort
Spiranthes cernua	white nodding ladies'-tresses
Thelypteris palustris	eastern marsh fern
Tradescantia virginiana	Virginia spiderwort
Viola affinis	sand violet
Viola lanceolata	bog white violet
Woodwardia areolata	netted chain fern
Woodwardia virginica	Virginia chain fern
Zepharanthes atamasco	atamasco-lily

Cypress–Tupelo Swamp Forest

Cypress–Tupelo Swamp Forest occurs in the lowest and wettest parts of floodplains. Southern bald-cypress is the most typical tree species of this habitat. A unique feature of the bald-cypress and tupelo are their buttressed bases and "cypress knees." Water tupelo is most common where water is relatively deep and inundation periods are long. In shallower, less frequently inundated areas, swamp tupelo is abundant. This community corresponds, in part, to Küchler #113 and Swamp Forests ERT.

Canopy
Characteristic Species

Nyssa aquatica	water tupelo
Nyssa biflora	swamp tupelo
Taxodium ascendens	pond-cypress
Taxodium distichum	southern bald-cypress

Associates

Acer rubrum	red maple
Chamaecyparis thyoides	Atlantic white-cedar
Carya aquatica	water hickory
Fraxinus caroliniana	Carolina ash
Fraxinus profunda	pumpkin ash
Pinus serotina	pond pine
Pinus taeda	loblolly pine
Populus heterophylla	swamp cottonwood
Quercus nigra	water oak
Salix nigra	black willow

Woody Understory

Berchemia scandens	Alabama supplejack
Brunnichia ovata	American buckwheatvine
Campsis radicans	trumpet-creeper
Cephalanthus occidentalis	common buttonbush
Clethra alnifolia	coastal sweet-pepperbush
Crataegus marshallii	parsley hawthorn
Cyrilla racemiflora	swamp titi
Forestiera acuminata	eastern swamp-privit
Gaylussacia frondosa	blue huckleberry
Gordonia lasianthus	loblolly-bay
Ilex coriacea	large gallberry
Ilex decidua	deciduos holly
Ilex glabra	inkberry
Lyonia ligustrina	maleberry
Lyonia lucida	shinyleaf
Magnolia virginiana	sweet-bay
Myrica cerifera	southern bayberry
Myrica heterophylla	evergreen bayberry
Persea borbonia	red bay
Persea palustris	swamp bay
Planera aquatica	planertree

Herbaceous Understory

Azolla caroliniana	Carolina mosquito fern
Boehmeria cylindrica	small-spike false nettle
Carex gigantea	giant sedge
Centella asiatica	spadeleaf
Drosera capillaris	pink sundew
Dulichium arundinaceum	three-way sedge
Hydrocotyle verticillata	whorled marsh-pennywort
Lemna minor	common duckweed
Mayaca fluviatilis	stream bog-moss
Mikania scandens	climbing hempvine
Mitchella repens	partridge-berry
Peltandra virginica	green arrow-arum
Polygonum punctatum	dotted smartweed
Sarracenia flava	yellow pitcherplant
Sarracenia rubra	sweet pitcherplant
Saururus cernuus	lizard's-tail
Sphagnum spp.	moss
Utricularia subulata	zigzag bladderwort
Woodwardia areolata	netted chain fern

Atlantic White-cedar Swamp

White-cedar dominated swamp forests occur on deep peats, often over sandy substrates. This community type ranges from New England to northern Florida. Atlantic White-cedar Swamp usually occurs in large, even stands due to uniform regeneration following a fire. Extensive stands are found in the New Jersey Pine Barrens and the lower terraces of the North Carolina and Virginia coastal plains. This community corresponds to Swamp Forests ERT.

Canopy
Characteristic Species
Chamaecyparis thyoides	Atlantic white-cedar

Associates
Acer rubrum	red maple
Nyssa biflora	swamp tupelo
Pinus serotina	pond pine
Pinus taeda	loblolly pine
Taxodium ascendens	pond-cypress

Woody Understory
Cyrilla racemiflora	swamp titi
Gaylussacia frondosa	blue huckleberry
Gordonia lasianthus	loblolly-bay
Ilex coriacea	large gallberry
Ilex glabra	inkberry
Lyonia ligustrina	maleberry
Lyonia lucida	shinyleaf
Magnolia virginiana	sweet-bay
Myrica cerifera	southern bayberry
Myrica heterophylla	evergreen bayberry
Persea borbonia	red bay
Persea palustris	swamp bay

Herbaceous Understory
Drosera capillaris	pink sundew
Mayaca fluviatilis	stream bog-moss
Mitchella repens	partridge-berry
Peltandra virginica	green arrow-arum
Sarracenia flava	yellow pitcherplant
Sarracenia rubra	sweet pitcherplant
Sphagnum spp.	moss
Woodwardia areolata	netted chain fern

Pocosin

Shrub-dominated wetlands or pocosins are common features of the Coastal Plain. These peatlands are dominated by a dense, nearly impenetrable cover of evergreen and deciduous shrubs usually between three and nine feet tall. Scattered emergent trees, especially loblolly-bay, sweet-bay, and red bay, may invade a pocosin and become a "Bay forest." Pocosins are often divided into three types: low, medium, and high pocosins depending upon the percentage of shrub and tree cover. Pocosins are seasonally flooded or saturated, and extremely nutrient poor with nutrient input only from rainfall. Severe fires may periodically occur in pocosins. This community corresponds to Küchler #114 and Bogs and Fens ERT.

Characteristic Woody Species
Aronia arbutifolia	red chokeberry
Chamaedaphne calyculata	leatherleaf
Clethra alnifolium	summer sweet clethra
Cyrilla racemiflora	swamp titi

Ilex coriacea	large gallberry
Ilex glabra	inkberry
Leucothoe racemosa	swamp doghobble
Lyonia lucida	shinyleaf
Myrica cerifera	southern bayberry
Persea borbonia	red bay
Pinus serotina	pond pine
Rhododendron viscosum	clammy azalea
Smilax laurifolia	laurel-leaf greenbrier
Vaccinium corymbosum	highbush blueberry
Vaccinium fuscatum	black blueberry
Zenobia pulverulenta	honeycup

Associates

Acer rubrum	red maple
Gordonia lasianthus	loblolly-bay
Liquidambar styraciflua	sweet-gum
Magnolia virginiana	sweet-bay
Persea palustris	swamp bay
Pinus palustris	long-leaf pine
Vaccinium macrocarpon	large cranberry

Herbaceous Understory

Andropogon glomeratus	bushy bluestem
Carex stricta	uptight sedge
Sarracenia flava	yellow pitcherplant
Sphagnum spp.	moss
Woodwardia virginica	Virginian chain fern

Freshwater Marsh

Freshwater Marsh is located upstream from Salt Marsh and downstream from nontidal freshwater wetlands. It is characterized by near freshwater conditions, with the average salinity of 0.5 parts per trillion or below, except during periods of drought. Chesapeake Bay and the South Carolina coast have many freshwater marshes. Near the coast of North Carolina, swamps are more common than freshwater marshes. The characteristic vegetation of freshwater marshes consists of rushes, sedges, grasses, and cat-tails. Many of the marshes and associated swamps were diked, impounded, and converted to rice fields during the 18th and 19th centuries. Many of these impoundments still exist and provide habitat for waterfowl. This community corresponds to Nonestuarine Marshes ERT.

Characteristic Species

Carex stricta	uptight sedge
Centella asiatica	spadeleaf
Cladium mariscus	swamp saw-grass
Eleocharis fallax	creeping spike-rush
Eleocharis rosetellata	beaked spike-rush
Fuirena squarrosa	hairy umbrella sedge
Impatiens capensis	spotted touch-me-not
Iris virginica	Virginia blueflag
Kosteletzkya virginica	Virginia fen-rose
Lachnanthes carolina	Carolina redroot

Lemna perpusilla	minute duckweed
Limnobium spongia	American spongeplant
Ludwigia palustris	marsh primrose-willow
Lythrum alatum	wing-angle loosestrife
Myriophyllum aquaticum	parrot's feather
Myriophyllum heterophyllum	two-leaf water-milfoil
Nuphar lutea	yellow pond-lily
Nymphoides cordata	little floatingheart
Orontium aquaticum	goldenclub
Osmunda regalis	royal fern
Panicum hemitomon	maiden-cane
Peltandra virginica	green arrow-arum
Phragmites australis	common reed
Pontederia cordata	pickerelweed
Sagittaria graminea	grass-leaf arrowhead
Sagittaria latifolia	duck-potato
Saururus cernuus	lizard's-tail
Scirpus americanus	chairmaker's bulrush
Scirpus cyperinus	cottongrass bulrush
Scirpus robustus	seaside bulrush
Sium suave	hemlock water-parsnip
Sparganium americanum	American burr-reed
Spartina cynosuroides	big cord grass
Typha angustifolia	narrow-leaf cat-tail
Typha domingensis	southern cat-tail
Typha latifolia	broad-leaf cat-tail
Zizania aquatica	Indian wild rice
Zizaniopsis miliacea	marsh-millet

Associates

Asclepias incarnata	swamp milkweed
Boltonia asteroides	white doll's-daisy
Carex alata	broad-wing sedge
Cicuta maculata	spotted water-hemlock
Echinochloa walteri	long-awn cock's-spur grass
Eleocharis albida	white spike-rush
Eleocharis flavescens	yellow spike-rush
Hibiscus moscheutos	crimson-eyed rose-mallow
Hydrocotyle umbellata	many-flower marsh-pennywort
Lythrum lineare	saltmarsh loosestrife
Myrica cerifera	southern bayberry
Phragmites australis	common reed
Physostegia virginiana	obedient-plant
Pluchea foetida	stinking camphorweed
Rhynchospora corniculata	short-bristle horned beak rush
Saccharum giganteum	giant plume grass
Scirpus tabernaemontani	soft-stem bulrush
Setaria magna	giant bristle grass
Taxodium distichum	southern bald-cypress
Thelypteris palustris	eastern marsh fern

ESTUARINE SYSTEMS

Salt Shrub Thicket

Salt Shrub Thicket occurs in the high areas of salt marshes. These communities are especially common on the barrier islands. Nutrients in these communities are primarily provided by salt or brackish water as well as salt spray. Periodic salt water flooding prevents invasion by trees and other shrubs. Fire, overwash, and sand deposition may play a role in community composition. This community corresponds to Shrublands ERT.

Characteristic Species

Baccharis angustifolia	saltwater false willow
Baccharis glomeruliflora	silverling
Baccharis halimifoila	groundseltree
Borrichia frutescens	bushy seaside-tansy
Myrica cerifera	southern bayberry

Associates

Cynanchum angustifolium	Gulf coast swallow-wort
Iva frutescens	Jesuit's-bark
Juncus roemerianus	Roemer's rush
Juniperus virginiana	eastern red cedar
Solidago sempervirens	seaside goldenrod
Spartina patens	salt-meadow cord grass

Salt Marsh

Salt Marsh occurs along the margins of sounds and estuaries, along the backs of barrier islands, and near closed inlets with regular salt water tides. These areas are regularly flooded and dominated by salt-tolerant grasses. Saltwater cord grass (smooth cord grass) is the most common characteristic species. Marshes dominated by cord grass are perhaps the most widely known estuarine community. This community corresponds to Küchler #73 and Salt Marsh ERT.

Characteristic Species

Distichlis spicata	coastal salt grass
Juncus roemerianus	Roemer's rush
Spartina alterniflora	saltwater cord grass
Spartina patens	salt-meadow cord grass

Associates

Agalinis maritima	gerardia
Borrichia frutescens	bushy seaside-tansy
Juncus gerardtl	saltmarsh rush
Limonium carolinianum	Carolina sea-lavender
Plantago decipiens	seaside plantain
Salicornia virginica	woody saltwort
Triglochin maritimum	seaside arrow-grass

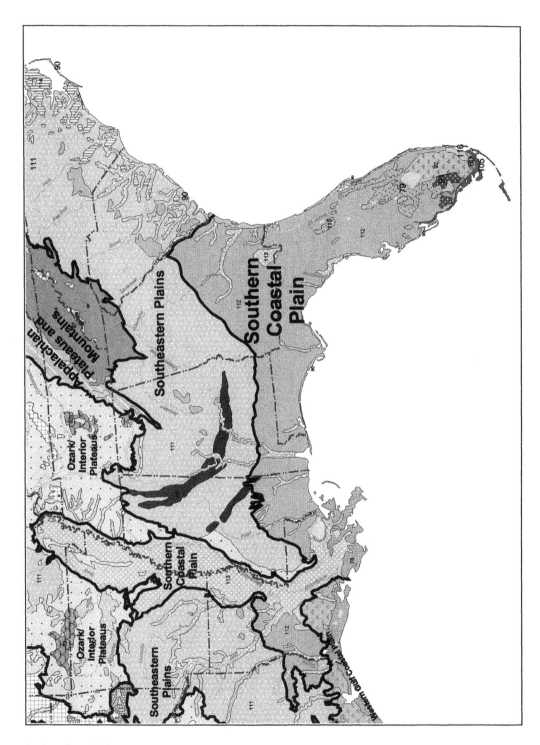

Southern Coastal Plain

Southern Coastal Plain

Introduction

This region occupies Florida, the southern portions of Georgia, Alabama, and Mississippi, and includes the Mississippi River Alluvial Plain from Lousiana north to extreme southern Illinois. It is underlain by limestone or alluvium, has sandy soils, and a low elevation. Fire and water have played a major role in plant community development in this region. Soils range from extensive organic development to soils with no organic content. The water regimes and natural flows, especially in Florida, have been highly altered by development. Extensive wetland areas have been drained and developed for agriculture and other purposes.

The vegetation ranges from dry, pine-dominated communities to swamp forests and wet prairies. This range of community types can sometimes occur over a small area. There are also extensive areas of other communities, such as the Sawgrass Marsh of the Everglades. The vegetation of south Florida has a strong tropical influence. Several communities are found only south of Lake Okeechobee. In some cases, community types occur throughout the region, but have some distinctive plant species in the southern tip of Florida and the Keys.

The primary sources used to develop descriptions and species lists are Barbour and Billings 1988; Florida Natural Areas Inventory and Department of Natural Resources (no date); Martin et al. In press; Myers and Ewel 1990; Deborah White (pers. com.).

Dominant Ecological Communities

Upland Systems
Southern Mixed Hardwoods Forest
Rockland Hammock
Maritime Hammock
Sandhill
Pine Flatwoods
Pine Rockland
Scrub
Dry Prairie
Beach Dune

Wetland Systems
Southern Swamp Forest
Baygall Forest
Dome Swamp Forest
Floodplain Marsh
Coastal Grassland
Sawgrass Marsh
Marl Prairie
Wet Prairie
Upland Marsh (Basin/Depression Marshes)

Estuarine Systems
Mangrove
Salt Marsh

UPLAND SYSTEMS
Southern Mixed Hardwoods Forest

These forests are sometimes called hammocks. They are closed canopy forests in ravines, on slopes, and upland rolling hills. A wide variety of species occur in these mesic forests with topography, moisture, and other abiotic factors determining species composition on a particular site. Species with northern affinities drop out on a southern gradient and are generally absent in peninsular Florida. This community corresponds to Küchler #111 and #112 and Eastern Deciduous and Mixed Forests ERT.

Canopy
Characteristic Species
Carya glabra	pignut hickory
Fagus grandifolia	American beech
Liquidambar styraciflua	sweet-gum
Magnolia grandiflora	southern magnolia
Morus rubra	red mulberry
Oxydendrum arboreum	sourwood
Persea borbonia	red bay
Pinus glabra	spruce pine
Pinus taeda	loblolly pine
Prunus caroliniana	Carolina laurel cherry
Quercus hemisphaerica	Darlington's oak
Quercus michauxii	swamp chestnut oak
Quercus nigra	water oak
Quercus virginiana	live oak

Associates
Carya pallida	sand hickory
Celtis laevigata	sugar-berry
Diospyros virginiana	common persimmon
Fraxinus americana	white ash
Liriodendron tulipifera	tuliptree
Magnolia pyramidata	pyramid magnolia
Quercus alba	northern white oak
Quercus austrina	bluff oak
Tilia americana	American basswood
Ulmus alata	winged elm
Ulmus americana	American elm

Woody Understory
Acer barbatum	Florida maple
Aralia spinosa	devil's-walkingstick
Callicarpa americana	American beauty-berry
Calycanthus floridus	eastern sweetshrub
Carpinus caroliniana	American hornbeam
Cercis canadensis	redbud
Chionanthus virginicus	white fringetree
Cornus florida	flowering dogwood
Dirca palustris	eastern leatherwood
Evonymus americana	American strawberry-bush

Hamamelis virginiana	American witch-hazel
Ilex ambigua	Carolina holly
Ilex opaca	American holly
Juniperus virginiana	eastern red cedar
Magnolia ashei	Ashe's magnolia
Osmanthus americanus	devilwood
Ostrya virginiana	eastern hop-hornbeam
Sabal palmetto	cabbage palmetto
Sebastiania fruticosa	Gulf sebastain-bush
Smilax bona-nox	fringed greenbrier
Smilax pumila	sarsaparilla-vine
Stewartia malacodendron	silky-camellia
Styrax grandifolius	big-leaf snowbell
Symplocos tinctoria	horsesugar
Vaccinium arboreum	tree sparkle-berry
Zanthoxylum clava-herculis	Hercules'-club

Herbaceous Understory

Actaea pachypoda	white baneberry
Adiantum pedatum	northern maidenhair
Campanula spp.	bellflower
Goodyera pubescens	downy rattlesnake-plantain
Hepatica nobilis	liverwort
Hexastylis arifolia	little-brown-jug
Mitchella repens	partridge-berry
Passiflora lutea	yellow passion-flower
Polygonatum biflorum	King Solomon's-seal
Polystichum acrostichoides	Christmas fern
Sanicula spp.	black-snakeroot
Trillium spp.	wakerobin
Uvularia spp.	bellwort

Rockland Hammock

Rockland Hammock is found in extreme southern Florida. This species-rich closed canopy forest of broad-leaved tropical and semitropical trees does not normally flood or burn and occurs on limestone that is exposed or near the surface. This community corresponds to Eastern Deciduous and Mixed Forests ERT.

Canopy
Characteristic Species

Coccothrinax argentata	Florida silver palm
Eugenia foetida	box-leaf stopper
Eugenia rhombea	red stopper
Guapira discolor	beeftree
Myrcianthes fragrans	twinberry

Associates

Acer rubrum	red maple
Bursera simaruba	gumbo-limbo
Capparis cynophallophora	Jamaican caper
Capparis flexuosa	falseteeth

Celtis laevigata	sugar-berry
Chrysophyllum oliviforme	satinleaf
Coccoloba diversifolia	tietongue
Conocarpus erectus	button mangrove
Exostema caribaeum	Carribbean princewood
Exothea paniculata	butterbough
Ficus aurea	Florida strangler fig
Gouania lupuloides	whiteroot
Guajacum sanctum	holywood lignumvitae
Krugiodendron ferreum	leadwood
Lysiloma latisiliquum	false tamarind
Metopium toxiferum	Florida poisontree
Morus rubra	red mulberry
Persea borbonia	red bay
Piscidia piscipula	Florida fishpoison-tree
Quercus laurifolia	laurel oak
Quercus pumila	running oak
Quercus virginiana	live oak
Roystonea elata	Florida royal palm
Sabal palmetto	cabbage palmetto

Woody Understory

Acacia pinetorum	pineland acacia
Alvaradoa amorphoides	Mexican alvaradoa
Ampelopsis arborea	peppervine
Amyris elemifera	sea torchwood
Annona glabra	pond-apple
Ardisia escallonoides	island marl-berry
Baccharis angustifolia	softwater false willow
Baccharis glomeruliflora	silverling
Baccharis halimifolia	groundseltree
Berchemia scandens	Alabama supplejack
Bourreria ovata	Bahama strongbark
Bourreria succulenta	bodywood
Byrsonima lucida	long key locust-berry
Caesalpinia bonduc	yellow nicker
Caesalpinia crista	gray nicker
Callicarpa americana	American beauty-berry
Calyptranthes pallens	pale lidflower
Calyptranthes zuzygium	myrtle-of-the-river
Canella winteriana	pepper-cinnamon
Catesbaea parviflora	small-flower lilythorn
Cephalanthus occidentalis	common buttonbush
Chiococca alba	West Indian milkberry
Chrysobalanus icaco	icaco coco-plum
Cissus verticillata	seasonvine
Citharexylum fruticosum	Florida fiddlewood
Coccoloba uvifera	seaside-grape

Colubrina arborescens	greenheart
Colubrina cubensis	Cuban nakedwood
Colubrina elliptica	soldierwood
Cordia globosa	curaciao-bush
Cordia sebestena	large-leaf geigertree
Cornus foemina	stiff dogwood
Crossopetalum rhacoma	maiden-berry
Cupania glabra	Florida toadwood
Dalbergia brownii	Brown's Indian-rosewood
Dalbergia ecastaphyllum	coinvine
Dodonaea viscosa	Florida hopbush
Drypetes diversifolia	milkbark
Drypetes lateriflora	Guianna-plum
Erithalis fruticosa	blacktorch
Erythrina herbacea	red-cardinal
Eugenia axillaris	white stopper
Eugenia confusa	red-berry stopper
Ficus citrifolia	wild banyantree
Forestiera segregata	Florida swamp-privet
Guettarda eliptica	hammock velvetseed
Gyminda latifolia	West Indian false box
Gymnanthes lucida	oysterwood
Hamelia patens	scarletbush
Hippocratea volubilis	medicine-vine
Hippomane mancinella	manchineel
Hypelate trifoliata	inkwood
Ilex cassine	dahoon
Ilex krugiana	tawny-berry holly
Jacquinia keyensis	joewood
Lantana involucrata	button-sage
Magnolia virginiana	sweet-bay
Manilkara jaimiqui	wild dilly
Maytenus phyllanthoides	Florida mayten
Myrica cerifera	southern bayberry
Myrsine floridana	guianese colicwood
Parthenocissus quinquefolia	Virginia-creeper
Pisonia aculeata	devil's-claw pisonia
Pisonia rotundata	smooth devil's-claws
Pithecellobium unguis-cati	cat's-claw blackbead
Prunus myrtifolia	West Indian cherry
Pseudophoenix sargentii	Florida cherry palm
Psidium longipes	mangroveberry
Psychotria ligustrifolia	Bahama wild coffee
Psychotria nervosa	Seminole balsamo
Randia aculeata	white indigo-berry
Reynosia septentrionalis	darling-plum
Rhus copallinum	winged sumac
Salix caroliniana	coastal-plain willow
Sapindus saponaria	wing-leaf soapberry

Savia bahamensis	Bahama-maidenbush
Schaefferia frutescens	Florida boxwood
Schoepfia chrysophylloides	island beefwood
Serenoa repens	saw-palmetto
Sideroxylon celastrinum	saffron-plum
Sideroxylon foetidissimum	false mastic
Sideroxylon reclinatum	Florida bully
Sideroxylon salicifolium	white bully
Simarouba glauca	paradise-tree
Smilax spp.	greenbrier
Solanum donianum	mullein nightshade
Solanum erianthum	potato-tree
Swietenia mahagoni	West Indian mahogany
Tetrazygia bicolor	Florida clover-ash
Thrinax morrisii	Key thatch palm
Thrinax radiata	Florida thatch palm
Tournefortia hirsutissima	chiggery-grapes
Tournefortia volubilis	twining soilderbush
Trema lamarckianum	pain-in-back
Trema micranthum	Jamaican nettletree
Vallesia glabra	pearlberry
Viburnum obovatum	small-leaf arrow-wood
Vitis labrusca	fox grape
Vitis rotundifolia	muscadine
Ximenia americana	tallow-wood
Zamia pumila	coontie
Zanthoxylum clava-herculis	Hercules'-club
Zanthoxylum fagara	lime prickly-ash

Herbaceous Understory

Eupatorium villosum	Florida Keys thoroughwort
Solanum bahamense	Bahama nightshade

Maritime Hammock

Maritime Hammock is a low salt-pruned forest that occurs interior to the dune system. This is the stable dune community of the coast. The trees form a dense canopy which allows for humus buildup and moisture retention. This community grades into coastal scrub toward the ocean. Soils are sandy and well drained. This community corresponds to Küchler #90 and Eastern Deciduous and Mixed Forests ERT.

Canopy
Characteristic Species

Ilex vomitoria	yaupon
Magnolia grandiflora	southern magnolia
Persea borbonia	red bay
Quercus virginiana	live oak
Sabal palmetto	cabbage palmetto

Associates

Bursera simaruba	gumbo-limbo
Capparis spp.	caper tree
Celtis laevigata	sugar-berry
Ficus aurea	Florida strangler fig
Ilex opaca	American holly
Juniperus virginiana	eastern red cedar
Metopium toxiferum	Florida poisontree
Myrsine floridana	guianese colicwood
Prunus serotina	black cherry
Quercus hemisphaerica	Darlington's oak
Simarouba glauca	paradise-tree

Woody Understory

Ardisia escallonioides	island marl-berry
Callicarpa americana	American beauty-berry
Coccoloba uvifera	seaside-grape
Erythrina herbacea	red-cardinal
Lyonia ferruginea	rusty staggerbush
Myrica cerifera	southern bayberry
Osmanthus americanus	devilwood
Psychotria nervosa	Seminole balsamo
Rivina humilis	rougeplant
Serenoa repens	saw-palmetto
Sideroxylon tenax	tough bully
Sideroxylon foetidissimum	false mastic
Smilax auriculata	ear-leaf greenbrier
Symphoricarpos albus	common snowberry
Vaccinium darrowi	Darrow's blueberry
Vitis munsoniana	little muscadine
Zamia pumila	coontie
Zanthoxylum clava-herculis	Hercule's-club

Herbaceous Understory

Salvia lyrata	lyre-leaf sage
Verbesina virginica	white crownbeard

Sandhill

Sandhill, also called high pine, is a vegetation type that occurs on rolling hills of sand throughout Florida north of Lake Okeechobee and into southern Georgia and southern Alabama. It is generally an open, long-leaf pine forest with a grass and oak shrub understory. Pineland three-awn (wiregrass), *Aristida stricta*, is the characteristic ground cover species and is important in facilitating low intensity ground fires. This vegetation type is highly tolerant of and requires fire on a regular basis. Slash pine has been brought in to replace long-leaf pine on many sites. This community corresponds, in part, to Küchler #115 and Eastern Deciduous and Mixed Forests ERT.

Canopy
Characteristic Species
Pinus elliottii	slash pine
Pinus palustris	long-leaf pine
Quercus laevis	turkey oak

Associates
Carya alba	mockernut hickory
Diospyros virginiana	common persimmon
Quercus falcata	southern red oak
Quercus geminata	sand live oak
Quercus incana	bluejack oak
Quercus margarettiae	sand post oak
Quercus marilandica	blackjack oak
Quercus stellata	post oak
Sassafras albidum	sassafras
Vaccinium arboreum	tree sparkle-berry

Woody Understory
Asimina incana	woolly pawpaw
Gaylussacia dumosa	dwarf huckleberry
Gaylussacia frondosa	blue huckleberry
Gelsemium spp.	trumpet-flower
Ilex glabra	inkberry
Licania michauxii	gopher-apple
Opuntia spp.	prickly-pear
Quercus minima	dwarf live oak
Quercus pumila	runner oak
Rhus copallinum	winged sumac
Rubus cuneifolius	sand blackberry
Smilax laurifolia	laurel-leaf greenbrier
Vitis labrusca	fox grape
Vitis rotundifolia	muscadine

Herbaceous Understory
Andropogon gerardii	big bluestem
Andropogon ternarius	split-beard bluestem
Andropogon virginicus	broom-sedge
Aristida stricta	pineland three-awn
Aster spp.	aster
Aureolaria flava	smooth yellow false-foxglove
Balduina angustifolia	coastal-plain honeycomb-head
Berlandiera subacaulis	Florida greeneyes
Chamaecrista fasciculata	sleepingplant
Croton argyranthemus	healing croton
Dalea pinnata	summer farewell
Dyschoriste oblongifolia	oblong-leaf snakeherb
Eriogonum tomentosum	dog-tongue wild buckwheat
Galactia volubilis	milk peas
Indigofera spp.	indigo

Lechea spp.	pinweed
Liatris pauciflora	few-flower gayfeather
Liatris tenuifolia	short-leaf gayfeather
Muhlenbergia capillaris	hair-awn muhly
Pityopsis graminifolia	narrow-leaf silk-grass
Polanisia tenuifolia	slender-leaf clammeyweed
Pteridium aquilinum	northern bracken fern
Rhynchosia spp.	snout-bean
Sanicula spp.	black snakeroot
Schizachyrium scoparium	little false bluestem
Solidago spp.	goldenrod
Sorghastrum nutans	yellow Indian grass
Sporobolus junceus	wire grass
Stillingia sylvatica	queen's-delight
Stylosanthes biflora	side-beak pencil-flower
Tephrosia virginiana	goat's-rue

Pine Flatwoods

Flatwoods range from open forests of scattered pines with little understory to dense pine stands with a rather dense undergrowth of grasses (particularly *Aristida*), saw palmettos, and other low shrubs. Flatwoods occur on level topography with acidic sands. The dominant canopy species is usually *Pinus elliottii*, but can be other pine species depending upon latitude, soils, hydroperiod, and fire frequency. This community corresponds to Eastern Deciduous and Mixed Forests ERT.

Canopy
Characteristic Species

Pinus elliottii	slash pine

Associates

Coccothrinax argentata	Florida silver palm
Pinus palustris	long-leaf pine
Quercus chapmanii	Chapman's oak
Quercus geminata	sand live oak
Quercus inopina	sandhill oak
Quercus myrtifolia	myrtle oak
Quercus virginiana	live oak

Woody Understory

Befaria racemosa	tarflower
Garberia heterophylla	garberia
Gaylussacia dumosa	dwarf huckleberry
Hypericum spp.	St. John's-wort
Ilex glabra	inkberry
Kalmia hirsuta	hairy-laurel
Licania michauxii	gopher-apple
Lyonia ferruginea	rusty staggerbush
Lyonia fruticosa	coastal-plain staggerbush
Lyonia lucida	shinyleaf
Myrica cerifera	southern bayberry
Persea humilis	silk bay

Quercus pumila	runner oak
Sabal etonia	scrub palmetto
Serenoa repens	saw-palmetto
Vaccinium arboreum	tree sparkle-berry
Vaccinium elliottii	Elliott's blueberry
Zamia pumila	coontie

Herbaceous Understory

Agalinis spp.	false foxglove
Aristida spp.	three-awn
Aster paternus	toothed white-topped aster
Chrysopsis spp.	golden aster
Cladonia spp.	lichen
Lachnocaulon spp.	bogbutton
Lechea spp.	pinweed
Panicum abscissum	cut-throat grass
Pterocaulon virgatum	wand blackroot
Solidago spp.	goldenrod
Verbesina virginica	white crownbeard
Xyris spp.	yellow-eyed-grass

Pine Rockland

This is an open canopy forest of slash pine with a shrub, palm, and grass understory. It occurs on limestone outcrops and is maintained by fire. This community corresponds to Küchler #116 and Woodlands ERT.

Canopy
Characteristic Species

Alvaradoa amorphoides	Mexican alvaradoa
Annona glabra	pond-apple
Bursera simaruba	gumbo-limbo
Chrysophyllum oliviforme	satinleaf
Citharexylum fruticosum	Florida fiddlewood
Conocarpus erectus	button mangrove
Diospyros virginiana	common persimmon
Erithalis fruticosa	blacktorch
Exothea paniculata	butterbough
Ficus aurea	Florida strangler fig
Guapira discolor	beeftree
Hypelate trifoliata	ironwood
Lysiloma latisiliquum	false tamarind
Metopium toxiferum	Florida poisonwood
Persea borbonia	red bay
Pinus elliottii	slash pine
Piscidia piscipula	Florida fishpoison-tree
Psidium longipes	mangroveberry
Quercus geminata	sand live oak
Quercus virginiana	live oak

Sabal palmetto	cabbage palmetto
Sideroxylon foetidissimum	false mastic
Sideroxylon salicifolium	white bully
Thrinax morrisii	Key thatch palm
Thrinax radiata	Florida thatch palm

Woody Understory

Acacia farnesiana	sweet acacia
Acacia pinetorum	pineland acacia
Amorpha herbacea	cluster-spike indigo-bush
Ardisia escallonoides	island marl-berry
Asimina reticulata	netted pawpaw
Baccharis angustifolia	saltwater false willow
Baccharis glomeruliflora	silverling
Baccharis halimifolia	groundseltree
Befaria racemosa	tarflower
Bourreria cassinifolia	smooth strongbark
Byrsonima lucida	long key locust-berry
Caesalpinia pauciflora	few-flower holdback
Callicarpa americana	American beauty-berry
Catesbaea parviflora	small-flower lilythorn
Cephalanthus occidentalis	common buttonbush
Chamaecrista fasciculata	sleepingplant
Chrysobalanus icaco	icaco coco-plum
Coccoloba diversifolia	tietongue
Coccoloba uvifera	seaside-grape
Coccothrinax argentata	Florida silver palm
Colubrina arborescens	greenheart
Colubrina cubensis	Cuban nakedwood
Crossopetalum rhacome	maiden-berry
Crotalaria spp.	rattlebox
Croton linearis	grannybush
Dodonaea viscosa	Florida hopbush
Drypetes diversifolia	milkbark
Echites umbellata	devil's-potato
Eugenia axillaris	white stopper
Eugenia foetida	box-leaf stopper
Ficus citrifolia	wild banyantree
Forestiera segregata	Florida swamp-privet
Guettarda elliptica	hammock velvetseed
Guettarda scabra	wild guave
Gymnanthes lucida	oysterwood
Hippomane mancinella	manchineel
Ilex cassine	dahoon
Ilex glabra	inkberry
Ilex krugiana	tawny-berry holly
Jacquinia keyensis	joewood
Lantana involucrata	button sage
Lyonia fruticosa	coastal-plain staggerbush

Magnolia virginiana	sweet-bay
Manilkara jaimiqui	wild dilly
Myrica cerifera	southern bayberry
Myrcianthes fragrans	twinberry
Myrsine floridana	guianese colicwood
Pisonia rotundata	smooth devil's-claws
Pithecellobium keyense	Florida Keys blackbead
Psychotria nervosa	Seminole balsamo
Quercus minima	dwarf live oak
Quercus pumila	running oak
Randia aculeata	white indigo-berry
Reynosia septentrionalis	darling-plum
Rhus copallinum	winged sumac
Salix caroliniana	coastal-plain willow
Sambucus canadensis	American elder
Serenoa repens	saw-palmetto
Sideroxylon celastrinum	saffron-plum
Sideroxylon salicifolium	white bully
Sideroxylon reclinatum	Florida bully
Smilax spp.	greenbrier
Solanum donianum	mullein nightshade
Solanum erianthum	potato-tree
Symphoricarpos albus	common snowberry
Tetrazygia bicolor	Florida clover-ash
Trema lamarckianum	pain-in-back
Trema micranthum	Jamaican nettletree
Toxicodendron radicans	eastern poison-ivy
Vaccinium myrsinites	shiny blueberry
Ximenia americana	tallow-wood
Zamia pumila	coontie

Herbaceous Understory

Andropogon virginicus	broom-sedge
Anemia adiantifolia	pineland fern
Aristida spp.	three-awn
Eupatorium villosum	Florida Keys thoroughwort
Muhlenbergia spp.	muhly

Scrub

This is generally a shrub community dominated by a variety of evergreen oak species. It occurs on well-drained, sandy soils throughout Florida and into southern Georgia. Scrub may or may not have a pine overstory. Although the herbaceous layer is usually very sparse, the mid-story can form a dense thicket. Natural fire cycles are longer than in other pine-dominated communities in this region and fire events are intense. Scrub supports numerous endemics, several of which are federally listed. This community corresponds to Küchler #115 and Shrublands ERT.

Canopy
Characteristic Species
Carya floridana	scrub hickory
Ceratiola ericoides	sand-heath
Pinus clausa	sand pine
Quercus chapmanii	Chapman's oak
Quercus geminata	sand live oak
Quercus inopina	sandhill oak
Quercus myrtifolia	myrtle oak

Associates
Juniperus virginiana	eastern red-cedar
Lyonia ferruginea	rusty staggerbush
Lyonia fruticosa	coastal-plain staggerbush
Magnolia grandiflora	southern magnolia
Persea borbonia	red bay
Persea humilis	silk bay
Pinus elliottii	slash pine
Piscidia piscipula	Florida fishpoison-tree
Quercus virginiana	live oak
Sabal etonia	scrub palmetto
Serenoa repens	saw-palmetto

Woody Understory
Ampelopsis arborea	peppervine
Baccharis halimifolia	groundseltree
Caesalpinia bonduc	yellow nicker
Chrysoma pauciflosculosa	woody goldenrod
Dalbergia ecastaphyllum	coinvine
Echites umbellata	devil's-potato
Erythrina herbacea	red-cardinal
Forestiera segregata	Florida swamp-privet
Garberia heterophylla	garberia
Ilex opaca	scrub holly
Ilex vomitoria	yaupon
Lantana involucrata	button-sage
Limonium spp.	sea-lavender
Myrcianthes fragrans	twinberry
Myrica cerifera	southern bayberry
Myrsine floridana	guianese colicwood
Opuntia stricta	erect prickly-pear
Osmanthus americanus	devilwood
Pithecellobium keyense	Florida Keys blackbead
Pithecellobium unguis-cati	cat-claw blackbeard
Randia aculeata	white indigo-berry
Sideroxylon celastrinum	saffron-plum
Sideroxylon tenax	tough bully
Smilax auriculata	ear-leaf greenbrier
Sophora tomentosa	yellow necklacepod

Toxicodendron radicans	eastern poison-ivy
Vaccinium arboreum	tree sparkle-berry
Vitis munsoniana	little muscadine
Ximenia americana	tallow-wood
Yucca aloifolia	aloe yucca
Zanthoxylum clava-herculis	Hercules'-club

Herbaceous Understory

Alternanthera flavescens	yellow joyweed
Ambrosia hispida	coastal ragweed
Andropogon floridanus	Florida bluestem
Andropogon glomeratus	bushy bluestem
Asclepias curtissii	Curtis's milkweed
Cladonia evansii	lichen
Cladonia leporina	lichen
Cladonia prostrata	lichen
Cladonia subtenuis	lichen
Croton glandulosus	vente-conmigo
Dichanthelium sabulorum	hemlock rosette grass
Euphorbia cyathophora	fire-on-the-mountain
Eustachys petraea	pinewoods finger grass
Flaveria floridana	Florida yellowtops
Galactia spp.	milk-pea
Heterotheca subaxillaris	camphorweed
Hydrocotyle bonariensis	coastal marsh-pennywort
Ipomoea pes-caprae	bay-hops
Lechea cernua	nodding pinweed
Oenothera humifusa	seaside evening-primrose
Palafoxia feayi	Feay's palafoxia
Physalis angustifolia	coastal ground-cherry
Rhynchospora megalocarpa	sandy-field beak sedge
Schizachyrium sanguineum	crimson false bluestem
Trichostema dichotomum	forked bluecurls

Dry Prairie

This is a treeless grass, saw palmetto, and low shrub vegetation type. It occurs on moderately to poorly drained soils. A fire frequency of one to four years appears necessary to maintain this vegetation type. This community corresponds to Küchler #79 and Prairies ERT.

Characteristic Herbaceous Species

Andropogon virginicus	broom-sedge
Aristida purpurascens	arrow-feather three-awn
Aristida spiciformis	bottlebrush three-awn
Axonopus spp.	carpet grass
Eragrostis spp.	love grass
Hyptis alata	clustered bush-mint
Liatris spp.	gayfeather
Lilium catesbaei	southern red lily

Panicum virgatum	wand panic grass
Polygala spp.	milkwort
Pterocaulon virgatum	wand blackroot
Sabatia spp.	rose-gentian
Schizachyrium scoparium	little false bluestem
Solidago spp.	goldenrod
Sorghastrum nutans	yellow Indian grass

Characteristic Woody Species

Asimina spp.	pawpaw
Ilex glabra	inkberry
Lyonia ferruginea	rusty staggerbush
Lyonia lucida	shinyleaf
Myrica cerifera	southern bayberry
Quercus pumila	runner oak
Serenoa repens	saw-palmetto
Vaccinium myrsinites	shiny blueberry

Beach Dune

This community is the foredune or upper beach zone along the coast. It is a dynamic community due to the wind and wave action that continually moves the coastal sands. Dunes are sandy, xeric soils. The vegetation must tolerate exposure to salt spray, blowing sand, and direct sunlight. The shrubs are scattered and do not become dominant in the community. This community corresponds to Sand Dunes ERT.

Characteristic Herbaceous Species

Alternanthera flavescens	yellow joyweed
Atriplex pentandra	crested saltbush
Cakile edentula	American searocket
Canavalia rosea	bay-bean
Cenchrus spp.	sandburr
Chamaesyce bombensis	dixie sandmat
Chamaesyce polygonifolia	seaside spurge
Cnidoscolus stimulosus	finger-rot
Commelina erecta	white-mouth dayflower
Croton punctatus	gulf croton
Distichlis spicata	coastal salt grass
Eustachys petraea	pinewoods finger grass
Helianthus debilis	cucumber-leaf sunflower
Honkenya peploides	seaside sandplant
Hudsonia tomentosa	sand golden-heather
Hydrocotyle bonariensis	coastal marsh-pennywort
Ipomoea imperati	beach morning-glory
Ipomoea pes-caprae	bay-hops
Lathyrus japonicus	sea vetchling
Oenothera humifusa	seaside evening-primrose
Okenia hypogaea	burrowing-four-o'clock
Panicum amarum	bitter panic grass
Paspalum distichum	jointed crown grass

Physalis walteri	Walter's ground-cherry
Polygonum glaucum	seaside knotweed
Sesuvium portulacastrum	shoreline sea-purslane
Solidago sempervirens	seaside goldenrod
Spartina patens	salt-meadow cord grass
Sporobolus virginicus	seashore dropseed
Suaeda linearis	annual seepweed
Uniola paniculata	sea-oats

Characteristic Woody Species

Argusia gnaphalodes	sea-rosemary
Dalbergia ecastophyllum	coinvine
Iva imbricata	seacoast marsh-elder
Myrica cerifera	southern bayberry
Myrica pensylvanica	northern bayberry
Scaevola plumieri	gullfeed
Sophora tomentosa	yellow necklacepod
Suriana maritima	bay-cedar

WETLAND SYSTEMS

Southern Swamp Forest

For the purposes of this book, the Southern Swamp Forest includes many types of forested wetlands including bottomland hardwood, cypress-tupelo, sloughs, and other forests of the floodplain. These forests are closed canopy and have a variety of understories that range from dense shrub to a mix of herbs and grasses and, on some sites, very little cover. This community corresponds to Küchler #113 and Swamp Forests ERT.

Canopy
Characteristic Species

Acer negundo	ash-leaf maple
Acer rubrum	red maple
Acer saccharinum	silver maple
Annona glabra	pond-apple
Betula nigra	river birch
Carya aquatica	water hickory
Carya glabra	pignut hickory
Catalpa bignonioides	southern catalpa
Catalpa speciosa	northern catalpa
Celtis laevigata	sugar-berry
Chamaecyparis thyoides	Atlantic white-cedar
Chrysobalanus icaco	icaco coco-plum
Fagus grandifolia	American beech
Ficus aurea	Florida strangler fig
Fraxinus caroliniana	Carolina ash
Fraxinus pennsylvanica	green ash
Fraxinus profunda	pumpkin ash
Gleditsia aquatica	water-locust
Gordonia lasianthus	loblolly-bay

Halesia spp.	silverbell
Juniperus virginiana	eastern red cedar
Liquidambar styraciflua	sweet-gum
Magnolia grandiflora	southern magnolia
Nyssa aquatica	water tupelo
Nyssa ogeche	Ogeechee tupelo
Nyssa sylvatica	black tupelo
Persea borbonia	red bay
Pinus elliotti	slash pine
Pinus glabra	spruce pine
Pinus palustris	long-leaf pine
Pinus serotina	pond pine
Pinus taeda	loblolly pine
Platanus occidentalis	American sycamore
Populus deltoides	eastern cottonwood
Populus heterophylla	swamp cottonwood
Quercus laurifolia	laurel oak
Quercus lyrata	overcup oak
Quercus michauxii	swamp chestnut oak
Quercus nigra	water oak
Quercus phellos	willow oak
Quercus virginiana	live oak
Roystonea elata	Floridian royal palm
Sabal palmetto	cabbage palmetto
Taxodium distichum	southern bald-cypress
Ulmus americana	American elm

Woody Understory

Alnus serrulata	brookside alder
Ampelopsis arborea	peppervine
Aronia arbutifolia	red chokeberry
Aster carolinianus	climbing aster
Berchemia scandens	Alabama supplejack
Bignonia capreolata	crossvine
Carpinus caroliniana	American hornbeam
Cephalanthus occidentalis	common buttonbush
Clethra alnifolia	coastal sweet-pepperbush
Cliftonia monophylla	buchwheat-tree
Cornus foemina	stiff dogwood
Crataegus marshallii	parsley hawthorn
Cyrilla racemiflora	swamp titi
Diospyros virginiana	common persimmon
Ficus citrifolia	wild banyantree
Forestiera acuminata	eastern swamp-privet
Gelsemium sempervirens	evening trumpet-flower
Ilex cassine	dahoon
Ilex coriacea	large gallberry
Ilex decidua	deciduous holly
Ilex glabra	inkberry

Ilex myrtifolia	myrtle dahoon
Ilex vomitoria	yaupon
Itea virginica	Virginia sweetspire
Leucothoe axillaris	coastal doghobble
Leucothoe racemosa	swamp doghobble
Lyonia ferruginea	rusty staggerbush
Lyonia lucida	shinyleaf
Magnolia virginiana	sweet-bay
Myrica cerifera	southern bayberry
Myrica heterophylla	evergreen bayberry
Myrsine floridana	guianese colicwood
Persea palustris	swamp bay
Planera aquatica	planertree
Rhapidophyllum hystrix	needle palm
Rhododendron vaseyi	pink shell azalea
Rhododendron viscosum	clammy azalea
Rubus argutus	saw-tooth blackberry
Sabal minor	dwarf palmetto
Salix caroliniana	coastal-plain willow
Salix nigra	black willow
Sambucus canadensis	American elder
Smilax bona-nox	fringed greenbrier
Smilax glauca	sawbrier
Smilax laurifolia	laurel leaf greenbrier
Smilax walteri	coral greenbrier
Toxicodendron radicans	eastern poison-ivy
Vaccinium arboreum	tree sparkle-berry
Vaccinium corymbosum	highbush blueberry
Viburnum nudum	possumhaw
Viburnum obovatum	small-leaf arrow-wood
Vitis aestivalis	summer grape
Vitis rotundifolia	muscadine
Vitis shuttleworthii	calloose grape
Wisteria frutescens	American wisteria

Herbaceous Understory

Acrostichum danaeifolium	leather fern
Bacopa spp.	water hyssop
Canna flaccida	bandanna-of-the-everglades
Carex spp.	sedge
Cladium mariscus	swamp sawgrass
Crinum americanum	seven-sisters
Juncus effusus	lamp rush
Leersia virginica	white grass
Lemna spp.	duckweed
Limnobium spongia	American spongeplant
Ludwigia palustris	marsh primrose-willow
Nymphoides aquatica	big floatingheart
Oplismenus setarius	short-leaf basket grass

Osmunda regalis	royal fern
Panicum rigidulum	red-top panic grass
Peltandra spp.	arrow-arum
Pistia stratiotes	water-lettuce
Polygonum punctatum	dotted smartweed
Pontederia cordata	pickerelweed
Sagittaria spp.	arrowhead
Saururus cernuus	lizard's-tail
Thalia geniculata	bent alligator-flag
Thelypteris palustris	eastern marsh fern
Zizaniopsis miliacea	marsh-millet

Baygall Forest

This is a closed canopy evergreen forest found in poorly drained, peat-filled, shallow depressions. This community type typically forms at the base of a slope where soils are acidic (pH 3.5–4.5). This community corresponds to Swamp Forests ERT.

Canopy
Characteristic Species

Chamaecyparis thyoides	Atlantic white-cedar
Gordonia lasianthus	loblolly-bay
Ilex decidua	deciduos holly
Ilex myrtifolia	myrtle dahoon
Liquidambar styraciflua	sweet-gum
Nyssa biflora	swamp tupelo
Persea palustris	swamp bay
Taxodium ascendens	pond cypress

Woody Understory

Aronia arbutifolia	red chokeberry
Clethra spp.	sweet-pepperbush
Ilex cassine	dahoon
Ilex coriacea	large gallberry
Itea virginica	Virginia sweetspire
Leucothoe axillaris	coastal doghobble
Leucothoe racemosa	swamp doghobble
Lyonia ligustrina	maleberry
Lyonia lucida	shinyleaf
Magnolia virginiana	sweet-bay
Myrica cerifera	southern bayberry
Myrica inodora	odorless bayberry
Rhapidophyllum hystrix	needle palm
Smilax laurifolia	laurel leaf greenbrier
Toxicodendron radicans	eastern poison-ivy
Vitis spp.	grape

Herbaceous Understory

Osmunda cinnamomea	cinnamon fern
Woodwardia areolata	netted chain fern
Woodwardia virginica	chain fern

Dome Swamp Forest

Dome Swamp Forest (cypress dome or cypress hammock) is a forested wetland type that occurs in depressions. Tree height increases toward the center where the water is deeper. This gives these areas a dome appearance. These communities typically form in sandy flatwoods and sinkholes where sand has partially filled them. This community corresponds to Swamp Forests ERT.

Canopy
Characteristic Species

Acer barbatum	Florida maple
Annona glabra	pond-apple
Gordonia lasianthus	loblolly-bay
Persea palustris	swamp bay
Taxodium ascendens	pond-cypress

Associates

Acer rubrum	red maple
Nyssa biflora	swamp tupelo

Woody Understory

Cephalanthus occidentalis	common buttonbush
Clethra alnifolia	coastal sweet-pepperbush
Cyrilla racemiflora	swamp titi
Decodon verticillatus	swamp-loosestrife
Hypericum spp.	St. John's-wort
Ilex cassine	dahoon
Ilex coriacea	large gallberry
Ilex glabra	inkberry
Itea virginica	Virginia sweetspire
Leucothoe racemosa	swamp doghobble
Lyonia lucida	shinyleaf
Magnolia virginiana	sweet-bay
Myrica cerifera	southern bayberry
Salix caroliniana	coastal-plain willow
Smilax laurifolia	laurel-leaf greenbrier
Smilax walteri	coral greenbrier
Toxicodendron radicans	eastern poison-ivy
Vaccinium spp.	blueberry

Herbaceous Understory

Bacopa spp.	water hyssop
Cladium mariscus	swamp sawgrass
Lachnanthes caroliniana	Carolina redroot
Ludwigia palustris	marsh primrose-willow
Nymphoides aquatica	big floatingheart
Osmunda cinnamomea	cinnamon fern
Panicum hemitomon	maiden-cane
Peltandra spp.	arrow-arum
Saururus cernuus	lizard's-tail
Sphagnum spp.	sphagnum moss

Thalia geniculata	bent alligator-flag
Woodwardia areolata	netted chain fern
Woodwardia virginica	chain fern

Floodplain Marsh

Floodplain Marsh occurs in river floodplains and is dominated by shrubs and herbaceous plants. This community corresponds to Nonestuarine Marshes ERT.

Characteristic Herbaceous Species

Amaranthus australis	southern amaranth
Amphicarpum muhlenbergianum	perennial goober grass
Cabomba caroliniana	Carolina fanwort
Cladium mariscus	swamp sawgrass
Coreopsis spp.	tickseed
Crinum americanum	seven-sisters
Eleocharis equisetoides	horsetail-spike-rush
Eleocharis vivipara	viviparous spike-rush
Juncus acuminatus	knotty-leaf rush
Juncus effusus	lamp rush
Lachnanthes caroliniana	Carolina redroot
Leersia hexandra	southern cut grass
Ludwigia repens	creeping primrose-willow
Luziola fluitans	southern water grass
Nymphaea mexicana	banana water-lily
Nymphaea odorata	American white water-lily
Osmunda regalis	royal fern
Orontium aquaticum	goldenclub
Panicum hemitomon	maiden-cane
Panicum repens	torpedo grass
Peltandra virginica	green arrow-arum
Polygonum punctatum	dotted smartweed
Polygonum sagittatum	arrow-leaf tearthumb
Pontederia cordata	pickerelweed
Reimarochloa oligostachya	Florida reimar grass
Rhynchospora inundata	narrow-fruit horned beak sedge
Rhynchospora tracyi	Tracy's beak sedge
Sacciolepis striata	American glenwood grass
Sagittaria latifolia	duck-potato
Salicornia spp.	saltwort
Scirpus spp.	bulrush
Sesuvium spp.	sea-purslane
Spartina bakeri	bunch cord grass
Sporobolus virginicus	seashore dropseed
Thalia geniculata	bent alligator-flag
Thelypteris palustris	eastern marsh fern
Typha spp.	cat-tail
Utricularia foliosa	leafy bladderwort
Vigna luteola	Piedmont cow-pea

Characteristic Woody Species

Cephalanthus occidentalis	common buttonbush
Decodon verticillatus	swamp-loosestrife
Glottidium vesicarium	bagpod
Hibiscus grandiflorus	swamp rose-mallow

Coastal Grassland

Coastal Grassland occurs in low flat areas behind the foredunes. It is often covered by salt water during storms. This community corresponds to Wet Prairies and Grasslands ERT.

Characteristic Herbaceous Species

Andropogon glomeratus	bushy bluestem
Andropogon spp.	bluestem
Cenchrus spp.	sandburr
Cirsium horridulum	yellow thistle
Croton glandulosus	vente-conmigo
Cyperus spp.	sedge
Eragrostis spp.	love grass
Eustachys petraea	pinewood finger grass
Hydrocotyle spp.	pennywort
Ipomoea imperati	beach morning-glory
Ipomoea pes-caprae	bay-hops
Muhlenbergia spp.	muhly
Oenothera spp.	evening primrose
Panicum amarum	bitter panic grass
Pentalinon luteum	hammock vipertail
Physalis spp.	ground-cherry
Salicornia spp.	saltwort
Schizachyrium sanguineum	crimson false bluestem
Sesuvium spp.	sea-purslane
Spartina patens	saltmeadow cord grass
Sporobolus spp.	dropseed
Uniola paniculata	sea-oats
Chamaesyce garberi	Garber's sandmat

Characteristic Woody Species

Baccharis halimifolia	groundseltree
Borrichia frutescens	bushy seaside-tansy
Iva imbricata	seacoast marsh-elder
Myrica cerifera	southern bayberry
Opuntia stricta	erect prickly-pear
Waltheria indica	basera-prieta

Sawgrass Marsh

Sawgrass Marsh is a predominant community type in the Florida Everglades. Sawgrass forms a nearly impenetrable mat on sites with deep organic soils. This community corresponds to Küchler #92 and Non-estuarine Marshes ERT.

Characteristic Herbaceous Species

Bacopa spp.	water hyssop
Cladium mariscus	swamp sawgrass
Crinum americanum	seven-sisters
Eleocharis elongata	slim spike-rush
Eleocharis spp.	spike-rush
Hymenocallis spp.	spider-lily
Muhlenbergia spp.	muhly
Nymphaea odorata	American shite water-lily
Panicum hemitomon	maiden-cane
Pontederia cordata	pickerelweed
Rhynchospora spp.	beak sedge
Sagittaria spp.	arrowhead
Utricularia spp.	bladderwort
Ludwigia repens	creeping primrose-willow

Characteristic Woody Species

Cephalanthus occidentalis	common buttonbush
Salix caroliniana	coastal-plain willow

Marl Prairie

Marl Prairie occurs on seasonally flooded sites on alkaline soils with limestone or marl near the surface. The vegetation is sparse with widely scattered dwarf cypress. This community corresponds to Küchler #80 and #91 and Wet Prairies and Grasslands.

Characteristic Herbaceous Species

Aletris spp.	colicroot
Aristida purpurascens	arrow-feather three-awn
Aster spp.	aster
Cladium mariscus	swamp sawgrass
Crinum americanum	seven-sisters
Cyrtopodium punctatum	cowhorn orchid
Eleocharis spp.	spike-rush
Eragrostis spp.	love grass
Eupatorium capillifolium	dogfennel
Hyptis alata	clustered bush-mint
Leersia spp.	cutgrass
Muhlenbergia capillaris	hair-awn muhly
Nymphaea odorata	American white water-lily
Polygala spp.	milkwort
Pontederia cordata	pickerelweed
Rhynchospora colorata	narrow-leaf white-top
Rhynchospora microcarpa	southern beak sedge
Schizachyrium rhizomatum	Florida false bluestem
Schoenus nigricans	black bog-rush
Spartina bakeri	bunch cord grass
Thalia geniculata	bent alligator-flag

Characteristic Woody Species

Taxodium ascendens	pond-cypress

Wet Prairie

This is a flat, poorly drained treeless community on soils saturated or underwater 50 to 150 days each year. The species in this community type vary with soils, fire, and hydroperiod. Species have a tolerance for both flooding and dry periods. This community corresponds to Wet Prairies and Grasslands ERT.

Characteristic Herbaceous Species

Aletris spp.	colicroot
Aristida spp.	three-awn
Cladium mariscus	swamp sawgrass
Coreopsis spp.	tickseed
Ctenium aromaticum	toothache grass
Drosera spp.	sundew
Eleocharis spp.	spike-rush
Eragrostis spp.	love grass
Eriocaulon spp.	pipewort
Helenium spp.	sneezeweed
Helianthus spp.	sunflower
Muhlenbergia capillaris	hair-awn muhly
Panicum hemitomon	maiden-cane
Panicum spp.	panic grass
Rhexia spp.	meadow beauty
Rhynchospora colorata	narrow-leaf white-top
Rhynchospora tracyi	Tracy's beak sedge
Rudbeckia hirta	black-eyed-Susan
Sabatia spp.	rose-gentian
Saccharum giganteum	giant plume grass
Sarracenia spp.	pitcherplant
Setaria corrugata	coastal bristle grass
Spartina bakeri	bunch cord grass
Verbesina chapmanii	Chapman's crownbeard
Xyris spp.	yellow-eyed-grass

Characteristic Woody Species

Hypericum fasciculatum	peel-bark St. John's-wort
Myrica cerifera	southern bayberry

Upland Marsh

Upland Marsh occurs in basins or depressions located outside the floodplain, such as old lake beds, ponds, and sinkholes. The frequency of fire determines the degree of shrub invasion. Hydroperiod can vary from 50 days to all year. This community corresponds to Nonestuarine Marshes ERT.

Characteristic Herbaceous Species

Bidens bipinnata	Spanish-needles
Eleocharis spp.	spike-rush
Eupatorium capillifolium	dogfennel
Hydrocotyle spp.	marsh-pennywort
Juncus effusus	lamp rush

Lachnanthes caroliniana	Carolina redroot
Leersia spp.	cut grass
Ludwigia palustris	marsh primrose-willow
Ludwigia repens	creeping primrose-willow
Luziola fluitans	southern water grass
Nelumbo lutea	American lotus
Panicum hemitomon	maiden-cane
Panicum spp.	panic grass
Phragmites australis	common reed
Pontederia cordata	pickerelweed
Sagittaria spp.	arrowhead
Thalia geniculata	bent alligator-flag
Utricularia spp.	bladderwort
Woodwardia spp.	chain fern
Xyris spp.	yellow-eyed-grass

Characteristic Woody Species

Baccharis spp.	false willow
Cephalanthus occidentalis	common buttonbush
Hypericum spp.	St. John's-wort
Myrica cerifera	southern bayberry
Salix caroliniana	coastal-plain willow
Salix spp.	willow
Sambucus canadensis	American elder

ESTUARINE SYSTEMS

Mangrove

Mangrove is a shrub community composed of any of the following species: white mangrove, black mangrove, red mangrove, and button mangrove. Mangrove occurs in marine and estuarine tidal areas along the southern peninsula in a variety of soils that are saturated with brackish water and under water during high tide. This community corresponds to Küchler #105 and Mangroves ERT.

Canopy
Characteristic Species

Avicennia germinans	black mangrove
Borrichia frutescens	bushy seaside-tansy
Conocarpus erectus	button mangrove
Laguncularia racemosa	white mangrove
Rhizophora mangle	red mangrove

Herbaceous Understory

Batis maritima	turtleweed
Distichlis spicata	coastal salt grass
Eleocharis spp.	spike-rush

Juncus roemerianus	Roemer's rush
Salicornia spp.	saltwort
Sesuvium spp.	sea-purslane
Spartina spartinae	gulf cord grass

Salt Marsh

Salt Marsh is an intertidal coastal community type consisting of salt-tolerant grasses, rushes, sedges, and other halophytic herbs. This community corresponds to Küchler #78 and Salt Marsh ERT.

Characteristic Herbaceous Species

Acrostichum aureum	golden leather fern
Aster tenuifolius	perennial saltmarsh aster
Batis maritima	turtleweed
Boltonia diffusa	small-head doll's-daisy
Cladium mariscus	swamp saw-grass
Distichlis spicata	coastal salt grass
Heliotropium curassavicum	seaside heliotrope
Juncus effusus	lamp rush
Juncus roemerianus	Roemer's rush
Limonium carolinianum	Carolina sea-lavender
Monanthochloe littoralis	shore grass
Paspalum distichum	jointed crown grass
Paspalum vaginatum	talquezal
Pluchea spp.	camphorweed
Salicornia bigelovii	dwarf saltwort
Salicornia virginica	woody saltwort
Scripus spp.	bulrush
Sesuvium portulacastrum	sea-purslane
Solidago sempervirens	seaside goldenrod
Spartina alterniflora	saltwater cord grass
Spartina bakeri	bunch cord grass
Spartina cynosuroides	big cord grass
Spartina patens	salt-meadow cord grass
Spartina spartinae	gulf cord grass
Sporobolus virginicus	seashore dropseed
Suaeda linearis	annual seepweed
Suaeda maritima	herbaceous seepweed
Typha spp.	cat-tail

Characteristic Woody Species

Avicennia germinans	black mangrove
Baccharis halimifolia	groundseltree
Borrichia arborescens	tree seaside-tansy
Borrichia frutescens	bushy seaside-tansy
Conocarpus erectus	button mangrove
Iva frutescens	Jesuit's-bark
Laguncularia racemosa	white mangrove
Lycium carolinianum	Carolina desert-thorn

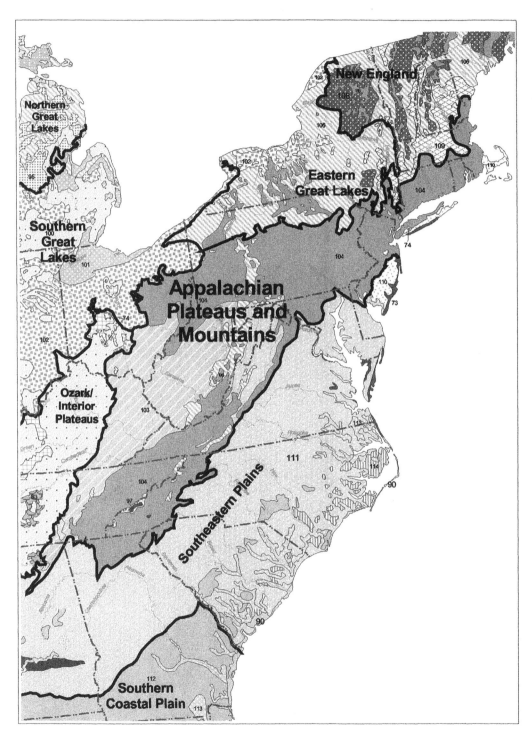

Appalachian Plateaus and Mountains

Appalachian Plateaus and Mountains

Introduction

This region occupies a long and relatively narrow area encompassing the central and southern Appalachian Mountains and the Cumberland and Appalachian plateaus. For the most part, this region is rugged and mountainous. Elevations range from about 1300–1500 feet and go up to 6711 feet at the summit of Mt. Mitchell. Oak-chestnut once covered much of the region and is being replaced by predominantly oak forests with the loss of the American chestnut to the blight. The diverse topography has given rise to many kinds of ecological communities in this region.

The primary sources used to develop descriptions and species lists are Barbour and Billings 1988; Braun 1950.

Dominant Ecological Communities

Upland Systems
Northern Conifer Forest
Northern Hardwoods Forest
Appalachian Oak Forest
Mesophytic Forest
Pine–Oak Forest
Wetland Systems
Floodplain Forest

UPLAND SYSTEMS
Northern Conifer Forest

In this region, Northern Conifer Forest is restricted to the highest elevations and is more widespread in the northern part of the region. In the Southern Appalachians, Fraser's fir is endemic and restricted to the highest mountain slopes and peaks where it replaces balsam fir, which is characteristic of more northern mountain peaks. This community corresponds to Küchler #96 and #97 and Coniferous Forests ERT.

Canopy
Characteristic Species

Abies balsamea	balsam fir
Abies fraseri	Fraser's fir
Betula alleghaniensis	yellow birch
Picea rubens	red spruce
Pinus strobus	eastern white pine

Associates

Acer spicatum	mountain maple
Betula papyrifera	paper birch
Prunus pensylvanica	fire cherry
Sorbus americanus	American mountain-ash
Tsuga canadensis	eastern hemlock

Woody Understory

Ilex montana	mountain holly
Menziesia pilosa	minniebush
Rhododendron catawbiense	catawba rosebay
Rubus alleghaniensis	allegheny blackberry
Rubus canadensis	smooth blackberry
Rubus idaeus	common red raspberry
Vaccinium corymbosum	highbush blueberry
Vaccinium erythrocarpum	southern mountain-cranberrry
Viburnum lantanoides	hobblebush

Herbaceous Understory

Ageratima altissima	white snakeroot
Aster acuminatus	whorled wood aster
Aster divaricatus	white wood aster
Athyrium filix-femina	subarctic lady fern
Carex pensylvanica	Pennsylvania sedge
Chelone lyonii	pink turtlehead
Circaea alpina	small enchanter's-nightshade
Clintonia borealis	yellow bluebead-lily
Dryopteris camplyoptera	mountain wood fern
Huperzia lucidula	shining club-moss
Maianthemum canadense	false lily-of-the-valley
Oxalis montana	sleeping-beauty
Solidago glomerata	clustered goldenrod
Streptopus roseus	rosy twistedstalk
Viola macloskeyi	smooth white violet

Northern Hardwoods Forest

Northern Hardwoods Forest occurs throughout this region at higher elevations but usually below Northern Conifer Forest. It is generally found on mesic sites with high rainfall, abundant fog, and low temperatures. This community corresponds to Küchler #106 and Eastern Deciduous and Mixed Forests ERT.

Canopy

Characteristic Species

Acer saccharum	sugar maple
Aesculus flava	yellow buckeye
Betula alleghaniensis	yellow birch
Fagus grandifolia	American beech

Associates

Acer rubrum	red maple
Fraxinus americana	white ash
Magnolia acuminata	cucumber magnolia
Prunus serotina	black cherry
Tilia americana	American basswood

Woody Understory

Acer pensylvanicum	striped maple
Acer spicatum	mountain maple
Amelanchier laevis	allegheny service-berry
Cornus alternifolia	alternate-leaf dogwood
Hydrangea arborescens	wild hydrangea
Ostrya virginiana	eastern hop-hornbeam
Rhododendron catawbiense	catawba rosebay
Sorbus americana	American mountain-ash
Viburnum lantanoides	hobblebush

Herbaceous Understory

Actaea pachypoda	white baneberry
Ageratina altissima	white snakeroot
Anemone quinquefolia	nightcaps
Arisaema triphyllum	jack-in-the-pulpit
Arnoglossum muehlenbergii	great Indian-plantain
Aster cordifolius	common blue wood aster
Aster divaricatus	white wood aster
Athyrium filix-femina	subarctic lady fern
Cardamine concatenata	cut-leaf toothwort
Carex debilis	white-edge sedge
Carex pensylvanica	Pennsylvania sedge
Caulophyllum thalictroides	blue cohosh
Cimicifuga americana	mountain bugbane
Claytonia caroliniana	Carolina springbeauty
Collinsonia canadensis	richweed
Dryopteris goldiana	Goldie's wood fern
Dryopteris intermedia	evergreen wood fern
Dryopteris marginalis	marginal wood fern
Erythronium umbilicatum	dimpled trout-lily
Hydrophyllum canadense	blunt-leaf waterleaf
Hydrophyllum virginianum	Shawnee-salad
Impatiens pallida	pale touch-me-not
Laportea canadensis	Canadian wood-nettle
Lilium superbum	Turk's-cap lily
Maianthemum racemosum	feathery false Solomon's-seal
Monarda didyma	scarlet beebalm
Osmorhiza claytonii	hairy sweet-cicely
Phacelia bipinnatifida	fern-leaf scorpion-weed
Prenanthes altissima	tall rattlesnake-root
Rudbeckia laciniata	green-head coneflower
Stellaria pubera	great chickweed
Streptopus roseus	rosy twistedstalk
Trillium erectum	stinking-Benjamin
Trillium grandiflorum	large-flower wakerobin
Trillium luteum	yellow wakerobin
Viola blanda	sweet white violet
Viola canadensis	Canadian white violet
Viola hastata	halberd-leaf yellow violet
Viola rostrata	long-spur violet

Appalachian Oak Forest

Appalachian Oak Forest is widespread throughout this region and has many variants. It occupies lower elevation slopes and ridges that are well drained and range from dry to dry-mesic. Before chestnut blight eliminated the American chestnut, it was one of the most important canopy trees in this community and the region. This community corresponds to Küchler #104 and Eastern Deciduous and Mixed Forests ERT.

Canopy
Characteristic Species
Castanea dentata	American chestnut
Quercus alba	northern white oak
Quercus coccinea	scarlet oak
Quercus prinus	chestnut oak
Quercus velutina	black oak

Associates
Acer rubrum	red maple
Carya glabra	pignut hickory
Carya alba	mockernut hickory
Liriodendron tulipifera	tuliptree
Nyssa sylvatica	black tupelo
Pinus echinata	short-leaf pine
Pinus rigida	pitch pine
Quercus rubra	northern red oak
Quercus stellata	post oak

Woody Understory
Amelanchier arborea	downy service-berry
Clethra acuminata	mountain sweet-pepperbush
Cornus florida	flowering dogwood
Corylus cornuta	beaked hazalnut
Epigaea repens	trailing-arbutus
Gaultheria procumbens	eastern teaberry
Gaylussacia baccata	black huckleberry
Hamamelis virginiana	American witch-hazel
Kalmia latifolia	mountain-laurel
Oxydendrum arboreum	sourwood
Prunus pensylvanica	fire cherry
Pyrularia pubera	buffalo-nut
Quercus ilicifolia	bear oak
Rhododendrum calendulaceum	flame azalea
Rhododendrum maximum	great-laurel
Sassafras albidum	sassafras
Vaccinium corymbosum	highbush blueberry
Vaccinium stamineum	deerberry
Viburnum acerifolium	maple-leaf arrow-wood

Herbaceous Understory
Aureolaria laevigata	entire-leaf yellow false-foxglove
Chimaphila maculata	striped prince's-pine
Coreopsis major	greater tickseed

Galax urceolata	beetleweed
Goodyeara pubescens	downy rattlesnake-plantain
Heuchera longifolia	long-flower alumroot
Hieracium venosum	rattlesnake-weed
Lysimachia quadrifolia	whorled yellow-loosestrife
Maianthemum racemosum	feathery false Solomon's-seal
Medeola virginiana	Indian cucumber-root
Melanthium parviflorum	Appalachian bunchflower
Pedicularis canadensis	Canadian lousewort
Polygonatum biflorum	King Solomon's-seal
Prenanthes trifoliolata	gall-of-the-earth

Mesophytic Forest

In this region, Mesophytic Forest is well developed. It is the richest and most diverse forest type occurring in the southern Appalachians and Cumberland Mountains. Dominance in these forests is often shared by many species and numerous variants occur. This community corresponds to Küchler #103 and Eastern Deciduous and Mixed Forests ERT.

Canopy
Characteristic Species

Acer saccharum	sugar maple
Aesculus flava	yellow buckeye
Betula lenta	sweet birch
Fagus grandifolia	American beech
Fraxinus americana	white ash
Liriodendron tulipifera	tuliptree
Magnolia acuminata	cucumber magnolia
Quercus alba	northern white oak
Tilia americana	American basswood
Tsuga canadensis	eastern hemlock

Associates

Acer rubrum	red maple
Betula alleghaniensis	yellow birch
Carya cordiformis	bitter-nut hickory
Carya ovata	shag-bark hickory
Halesia carolina	Carolina silverbell
Juglans cinerea	white walnut
Juglans nigra	black walnut
Prunus serotina	black cherry
Quercus rubra	northern red oak
Ulmus rubra	slippery elm

Woody Understory

Acer pensylvanicum	striped maple
Acer spicatum	mountain maple
Asimina triloba	common pawpaw
Carpinus caroliniana	American hornbeam
Cornus florida	flowering dogwood
Hydrangea arborescens	wild hydrangea

Leucothe axillaris	coastal doghobble
Magnolia fraseri	Fraser's magnolia
Magnolia tripetala	umbrella magnolia
Ostrya virginiana	eastern hop-hornbeam
Rhododendrum maximum	great-laurel

Herbaceous Understory

Actaea pachypoda	white baneberry
Adiantum pedatum	northern maidenhair
Ageratina altissima	white snakeroot
Arisaema triphyllum	jack-in-the-pulpit
Asarum canadense	Canadian wild ginger
Aster cordifolius	common blue wood aster
Cardamine concatenata	cut-leaf toothwort
Caulophyllum thalictroides	blue cohosh
Cimicifuga racemosa	black bugbane
Clintonia umbellata	white bluebead-lily
Dicentra canadensis	squirrel-corn
Dicentra cucullaria	Dutchman's-breeches
Dryopteris intermedia	evergreen wood fern
Galium triflorum	fragrant bedstraw
Hepatica nobilis	liverwort
Hydrophyllum canadense	blunt-leaf waterleaf
Hydrophyllum virginianum	Shawnee-salad
Impatiens capensis	spotted touch-me-not
Impatiens pallida	pale touch-me-not
Laportea canadensis	Canadian wood-nettle
Meehania cordata	Meehan's-mint
Mitchella repens	partridge-berry
Mitella diphylla	two-leaf bishop's-cap
Osmorhiza claytonii	hairy sweet-cicely
Panax quinquefolius	American ginseng
Podophyllum peltatum	may-apple
Polystichum acrostichoides	Christmas fern
Sedum ternatum	woodland stonecrop
Stellaria pubera	great chickweed
Thalictrum clavatum	mountain meadow-rue
Thelypteris noveboracensis	New York fern
Tiarella cordifolia	heart-leaf foamflower
Trillium erectum	stinking-Benjamin
Trillium grandiflorum	large-flower wakerobin
Viola canadensis	Canadian white violet
Viola rostrata	long-spur violet
Viola rotundifolia	round-leaf yellow violet

Pine–Oak Forest

Pine and oak dominated forests occur throughout this region on infertile and excessively well-drained, sandy or gravelly sites on upper slopes and other exposed or south-facing ridges. Some sites can be dominated exclusively by pine. This community corresponds, in part, to Küchler #110 and Eastern Deciduous and Mixed Forests ERT.

Canopy
Characteristic Species
Pinus echinata	short-leaf pine
Pinus pungens	table mountain pine
Pinus rigida	pitch pine
Pinus virginiana	Virginia pine
Quercus coccinea	scarlet oak
Quercus prinus	chestnut oak

Associates
Acer rubrum	red maple
Castanea dentata	American chestnut
Nyssa sylvatica	black tupelo
Oxydendrum arboreum	sourwood
Robinia hispida	bristly locust
Robinia pseudo-acacia	black locust
Sassafras albidum	sassafras
Tsuga caroliniana	Carolina hemlock

Woody Understory
Castanea pumila	allegheny-chinkapin
Comptonia peregrina	sweet-fern
Epigea repens	trailing-arbutus
Gaultheria procumbens	eastern teaberry
Gaylussacia baccata	black huckleberry
Gaylussacia ursina	bear huckleberry
Kalmia latifolia	mountain-laurel
Leucothoe recurva	red-twig doghobble
Quercus ilicifolia	bear oak
Rhododendron catawbiense	catawba rosebay
Rhododendron maximum	great-laurel
Smilax glauca	sawbrier
Symplocos tinctoria	horsesugar
Vaccinium pallidum	early lowbush blueberry
Vaccinium stamineum	deerberry

Herbaceous Understory
Chimaphila maculata	striped prince's-pine
Coreopsis major	greater tickseed
Cypripedium acaule	pink lady's-slipper
Galax urceolata	beetleweed
Melampyrum lineare	American cow-wheat
Pteridium aquilinum	northern bracken fern
Schizachyrium scoparium	little false bluestem
Tephrosia virginiana	goat's-rue
Uvularia puberula	mountain bellwort
Xerophyllum asphodeloides	eastern turkeybeard

WETLAND SYSTEMS

Floodplain Forest

Floodplain Forest occurs primarily along the larger streams and rivers where a relatively large valley bottom has developed. Floodplain forests can vary considerably in composition depending upon substrate, drainage, and duration of flooding. This community corresponds to Floodplain Forests ERT.

Canopy
Characteristic Species
Acer saccharinum	silver maple
Betula nigra	river birch
Celtis laevigata	sugar-berry
Fraxinus americana	white ash
Fraxinus pennsylvanica	green ash
Liquidambar styraciflua	sweet-gum
Platanus occidentalis	American sycamore
Ulmus americana	American elm

Associates
Acer rubrum	red maple
Carya cordiformis	bitter-nut hickory
Carya ovata	shag-bark hickory
Juglans nigra	black walnut
Liriodendron tulipifera	tuliptree
Quercus bicolor	swamp white oak
Quercus michauxii	swamp chestnut oak
Quercus pagoda	cherry-bark oak

Woody Understory
Acer negundo	ash-leaf maple
Alnus serrulata	brookside alder
Arundinaria gigantea	giant cane
Asimina triloba	common pawpaw
Bignonia capreolata	crossvine
Campsis radicans	trumpet-creeper
Carpinus caroliniana	American hornbeam
Cornus amomum	silky dogwood
Cornus florida	flowering dogwood
Evonymous americanus	American strawberry-bush
Ilex opaca	American holly
Leucothoe recurva	red-twig doghobble
Lindera benzoin	northern spicebush
Menispermun canadense	Canadian moonseed
Parthenocissus quinquefolia	Virginia creeper
Physocarpus opulifolius	Atlantic ninebark
Rhododendron maximum	great-laurel
Toxicodendron radicans	eastern poison-ivy
Xanthorhiza simplicissima	shrub yellowroot

Herbaceous Understory

Arisaema triphyllum	jack-in-the-pulpit
Aster divaricatus	white wood aster
Boehmeria cylindrica	small-spike false nettle
Carex laxiflora	broad loose-flower sedge
Chasmanthium latifolium	Indian wood-oats
Corydalis flavula	yellow fumewort
Cryptotaenia canadensis	Canadian honewort
Dichanthelium dichotomum	cypress rosette grass
Elymus hystrix	eastern bottle-brush grass
Elymus virginicus	Virginia wild rye
Impatiens pallida	pale touch-me-not
Laportea canadensis	Canadian wood-nettle
Lobelia cardinalis	cardinal-flower
Mertensia virginica	Virginia bluebells
Osmunda cinnamomea	cinnamon fern
Polemonium reptans	greek-valerian
Polygonum virginianum	jumpseed
Rudbeckia laciniata	green-head coneflower
Senecio aureus	golden ragwort
Solidago caesia	wreath goldenrod
Stellaria pubera	great chickweed
Verbesina alternifolia	wingstem

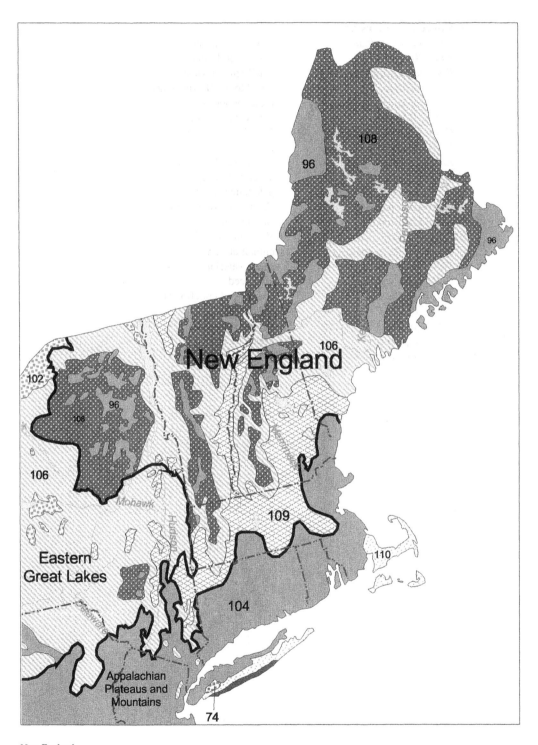

New England

New England

Introduction

The New England region encompasses Maine, New Hampshire, Vermont, Massachusetts, and eastern New York. This region is primarily vegetated with closed canopy forest. The most common upland community type is the Northern Hardwoods–Conifer Forest. On higher elevations, exposed sites, and cool flats, spruce-fir forests occur, often mixed with yellow and white birch. Oak, oak–hickory, and oak–pine forests are not extensive but do occur on dry slopes, especially in the southern part of the region. Pine and spruce forests and woodlands occupy narrow zones along the coast as well as specialized habitats inland.

The dominant wetland community in New England is swamp dominated by red maple or conifers such as spruce, fir, or white-cedar. Other wetland types which occur, but are generally localized or small in extent, include shrub swamps, marshes, bogs, fens, and estuarine salt and tidal marshes.

Specialized or localized communities, usually very limited in extent, include alpine ridges and krummholz at the highest elevations; heath, deciduous, and coniferous barrens or woodlands; pine forests of red and/or white pine; coastal dune/shoreline communities; and many others.

The primary sources used to develop descriptions and species lists are Barbour and Billings 1988; Maine Natural Heritage Program 1991; Reschke 1990.

Dominant Ecological Communities

Upland Systems
Northeastern Spruce–Fir Forest
Northern Hardwoods–Conifer Forest
Oak–Pine–Hickory Forest
Beech–Maple Forest
Coniferous Woodland
Deciduous Woodland

Wetland Systems
Floodplain Forest
Deciduous Swamp Forest
Coniferous Swamp Forest
Forested Bog
Shrub Bog/Fen
Shrub Swamp
Herbaceous Bog/Fen
Mineral Soil Marsh

Estuarine Systems
Brackish Tidal Marsh
Freshwater Tidal Marsh

UPLAND SYSTEMS
Northeastern Spruce–Fir Forest

Northeastern Spruce–Fir Forest occurs on high-elevation slopes in mountainous areas such as the Adirondack high peaks and northern Appalachians (the Green and White Mountains). It also occurs along the coast of Maine where white and black spruce are common, and on moist, poorly drained to well-drained soils on low flats. This is a broadly defined forest type with many variants and segregates dependant on elevation, exposure, and drainage. Spruce–fir forests include spruce flats, balsam flats, maritime spruce-fir, spruce slopes, and mountain-fir forests. This community corresponds to Küchler #96 and Coniferous Forests ERT.

Canopy
Characteristic Species

Abies balsamea	balsam fir
Picea glauca	white spruce
Picea mariana	black spruce
Picea rubens	red spruce

Associates

Acer rubrum	red maple
Betula alleghaniensis	yellow birch
Betula papyrifera	paper birch
Pinus strobus	eastern white pine
Prunus serotina	black cherry
Sorbus americana	American mountain-ash
Tsuga canadensis	eastern hemlock

Woody Understory

Acer pensylvanicum	striped maple
Acer spicatum	mountain maple
Alnus viridis	sitka alder
Gaultheria hispidula	creeping-snowberry
Kalmia angustifolia	sheep-laurel
Ledum groenlandicum	rusty Laborador-tea
Lonicera canadensis	American fly-honeysuckle
Nemopanthus mucronatus	catberry
Prunus pensylvanica	fire cherry
Ribes glandulosum	skunk currant
Rubus idaeus	common red raspberry
Rubus pubescens	dwarf red raspberry
Sorbus americana	American mountain-ash
Vaccinium angustifolium	late lowbush blueberry
Vaccinium myrtilloides	shiny blueberry
Viburnum lantanoides	hobblebush
Viburnum nudum	possumhaw

Herbaceous Understory

Aralia nudicaulis	wild sarsaparilla
Aster acuminatus	whorled wood aster
Athyrium filix-femina	subarctic lady fern

Clintonia borealis	yellow bluebead-lily
Coptis trifolia	three-leaf goldthread
Cornus canadensis	Canadian bunchberry
Dalibarda repens	robin-run-away
Dryopteris campyloptera	mountain wood fern
Dryopteris carthusiana	spinulose wood fern
Huperzia lucidula	shining club moss
Maianthemum canadense	false lily-of-the-valley
Oxalis montana	sleeping-beauty
Solidago macrophylla	large-leaf goldenrod
Streptopus amplexifolius	clasping twistedstalk

Northern Hardwoods–Conifer Forest

Northern Hardwoods–Conifer Forest is the most common forest community in the New England region. It occupies lower and mid-elevation slopes and well-drained upland and lowland flats. This broadly defined community has many regional and edaphic variants and plant associations within it. The dominant species include hardwoods (sugar maple, beech, and yellow birch) and conifers (hemlock, spruce, and white pine). The characteristic species composition varies according to available moisture, elevation, and soil conditions. This community corresponds to Küchler #108 and Eastern Deciduous and Mixed Forests ERT.

Canopy
Characteristic Species

Acer rubrum	red maple
Acer saccharum	sugar maple
Betula alleghaniensis	yellow birch
Carya cordiformis	bitter-nut hickory
Fagus grandifolia	American beech
Fraxinus americana	white ash
Picea rubens	red spruce
Pinus resinosa	red pine
Pinus strobus	eastern white pine
Tsuga canadensis	eastern hemlock

Associates

Abies balsamea	balsam fir
Betula lenta	sweet birch
Betula papyrifera	paper birch
Carya cordiformis	bitter-nut hickory
Juglans cinerea	white walnut
Picea glauca	white spruce
Populus tremuloides	quaking aspen
Prunus serotina	black cherry
Quercus rubra	northern red oak
Tilia americana	American basswood
Ulmus americana	American elm

Woody Understory

Acer pensylvanicum	striped maple
Acer spicatum	mountain maple
Amelanchier canadensis	Canadian service-berry

Carpinus caroliniana	American hornbeam
Cornus alternifolia	alternate-leaf dogwood
Corylus cornuta	beaked hazelnut
Epigaea repens	trailing-arbutus
Gaultheria procumbens	eastern teaberry
Hamamelis virginiana	American witch-hazel
Kalmia angustifolia	sheep-laurel
Kalmia latifolia	mountain-laurel
Lonicera canadensis	American fly-honeysuckle
Ostrya virginiana	eastern hop-hornbeam
Taxus canadensis	American yew
Vaccinium angustifolium	late lowbush blueberry
Vaccinium myrtilloides	velvet-leaf blueberry
Viburnum acerifolium	maple-leaf arrow-wood
Viburnum dentatum	southern arrow-wood
Viburnum lantanoides	hobblebush
Viburnum nudum	possumhaw

Herbaceous Understory

Aralia nudicaulis	wild sarsaparilla
Aster macrophyllus	large-leaf aster
Carex pensylvanica	Pennsylvania sedge
Clintonia borealis	yellow bluebead-lily
Cornus canadensis	Canadian bunchberry
Coptis trifolia	three-leaf goldthread
Cypripedium acaule	pink lady's-slipper
Deparia acrostichoides	silver false spleenwort
Dryopteris campyloptera	mountain wood fern
Dryopteris goldiana	Goldie's wood fern
Dryopteris intermedia	evergreen wood fern
Huperzia lucidula	shining club-moss
Lycopodium obscurum	princess-pine
Maianthemum canadense	false lily-of-the-valley
Medeola virginiana	Indian cucumber-root
Melampyrum lineare	American cow-wheat
Mitchella repens	partridge-berry
Oryzopsis asperifolia	white-grain mountain-rice grass
Oxalis montana	sleeping-beauty
Polystichum acrostichoides	Christmas fern
Pteridium aquilinum	northern bracken fern
Streptopus roseus	rosy twistedstalk
Tiarella cordifolia	heart-leaf foamflower
Trientalis borealis	American starflower
Trillium erectum	stinking-Benjamin
Trillium undulatum	painted wakerobin
Uvularia sessilifolia	sessile-leaf bellwort
Viola rotundifolia	round-leaf yellow violet

Oak–Pine–Hickory Forest

Oak–Pine–Hickory Forest occurs on south- and west-facing slopes and ridgetops where the soil is sandy or rocky. Limited moisture and poor soil formation are important factors in maintaining the characteristic mix of oak, pine, and hickory in the canopy. This community corresponds to Eastern Dedicuous and Mixed Forests ERT.

Canopy
Characteristic Species
Carya glabra	pignut hickory
Carya ovalis	red hickory
Carya ovata	shag-bark hickory
Pinus rigida	pitch pine
Pinus strobus	eastern white pine
Quercus alba	northern white oak
Quercus coccinea	scarlet oak
Quercus prinus	chestnut oak
Quercus rubra	northern red oak
Quercus velutina	black oak

Associates
Acer rubrum	red maple
Betula lenta	sweet birch
Fagus grandifolia	American beech
Fraxinus americana	white ash
Prunus serotina	black cherry
Tilia americana	American basswood
Tsuga canadensis	eastern hemlock

Woody Understory
Amelanchier arborea	downy service-berry
Cornus florida	flowering dogwood
Cornus foemina	stiff dogwood
Corylus cornuta	beaked hazelnut
Gaultheria procumbens	eastern teaberry
Gaylussacia baccata	black huckleberry
Hamamelis virginiana	American witch-hazel
Kalmia latifolia	mountian-laurel
Ostrya virginiana	eastern hop-hornbeam
Prunus virginiana	choke cherry
Quercus ilicifolia	bear oak
Rubus idaeus	common red raspberry
Vaccinium angustifolium	late lowbush blueberry
Vaccinium pallidum	early lowbush blueberry
Viburnum acerifolium	maple-leaf arrow-wood

Herbaceous Understory
Aralia nudicaulis	wild sarsaparilla
Carex pensylvanica	Pennsylvania sedge
Chimaphila maculata	striped prince's-pine
Cimicifuga racemosa	black bugbane
Desmodium glutinosum	pointed-leaf tick-trefoil

Desmodium paniculatum	panicled-leaf tick-trefoil
Hepatica nobilis	liverwort
Maianthemum racemosum	feathery false Solomon's-seal
Prenanthes alba	white rattlesnake-root
Pteridium aquilinum	northern bracken fern
Smilax rotundifolia	horsebrier
Solidago bicolor	white goldenrod

Beech–Maple Forest

Beech–Maple Forest occurs on sheltered, low to mid-elevation sites and on east or north exposures. The soil is generally loamy, high in fertility, moist, and well drained. The most characteristic species include sugar maple, beech, and basswood. This community corresponds to Küchler #102 and Eastern Deciduous and Mixed Forests ERT.

Canopy
Characteristic Species

Acer saccharum	sugar maple
Fagus grandifolia	American beech
Fraxinus americana	white ash
Tilia americana	American basswood

Associates

Betula alleghaniensis	yellow birch
Betula lenta	sweet birch
Carya cordiformis	bitter-nut hickory
Juglans nigra	black walnut
Prunus serotina	black cherry
Quercus rubra	northern red oak
Ulmus americana	American elm

Woody Understory

Acer pensylvanicum	striped maple
Carpinus caroliniana	American hornbeam
Cornus alternifolia	alternate-leaf dogwood
Hamamelis virginiana	American witch-hazel
Ostrya virginiana	eastern hop-hornbeam

Herbaceous Understory

Actaea pachypoda	white baneberry
Allium tricoccum	ramp
Arisaema triphyllum	jack-in-the-pulpit
Asarum canadense	Canadian wild ginger
Carex plantaginea	broad scale sedge
Carex platyphylla	plantain-leaf sedge
Caulophyllum thalictroides	blue cohosh
Dicentra cucullaria	Dutchman's-breeches
Dryopteris goldiana	Goldie's wood fern
Impatiens pallida	pale touch-me-not
Maianthemum racemosum	feathery false Solomon's-seal
Panax quinquefolius	American ginseng
Polystichum acrostichoides	Christmas fern
Sanguinaria canadensis	bloodroot

Coniferous Woodland

Coniferous Woodland occurs on gravelly slopes, sandy and rocky soils, dunes, and rocky areas. These sites generally have nutrient-poor soils and are excessively well drained. In the coastal areas and along lake shores, pitch pine is dominant in the south, while jack pine is dominant in the north part of the region. Fire may have played a role in maintaining the coniferous woodland communities. Woodlands and barrens will be less distinct as the forest canopy closes in areas where fire is suppressed. This community corresponds to Woodlands ERT.

Canopy
Characteristic Species
Picea mariana	black spruce
Picea rubens	red spruce
Pinus banksiana	jack pine
Pinus resinosa	red pine
Pinus rigida	pitch pine
Pinus strobus	eastern white pine

Associates
Abies balsamea	balsam fir
Acer rubrum	red maple
Betula papyrifera	paper birch
Betula populifolia	gray birch
Corema conradii	broom-crowberry
Picea glauca	white spruce
Populus grandidentata	big-tooth aspen
Populus tremuloides	quaking aspen
Quercus coccinea	scarlet oak
Quercus rubra	northern red oak
Thuja occidentalis	eastern arborvitae

Woody Understory
Aronia melanocarpa	black chokeberry
Chamaedaphne calyculata	leatherleaf
Comptonia peregrina	sweet-fern
Corema conradii	broom-crowberry
Cornus foemina	stiff dogwood
Gaultheria procumbens	eastern teaberry
Gaylussacia baccata	black huckleberry
Kalmia angustifolia	sheep-laurel
Lonicera dioica	limber honeysuckle
Quercus ilicifolia	bear oak
Rhamnus alnifolia	alder-leaf buckthorn
Ribes cynosbati	eastern prickly gooseberry
Rubus idaeus	common red raspberry
Rubus occidentalis	black raspberry
Staphylea trifolia	American bladdernut
Toxicodendron radicans	eastern poison-ivy
Vaccinium angustifolium	late lowbush blueberry
Vaccinium myrtilloides	shiny blueberry
Vaccinium pallidum	early lowbush blueberry

Herbaceous Understory

Aralia nudicaulis	wild sarsaparilla
Asplenium trichomanes	maidenhair spleenwort
Aster macrophyllus	large-leaf aster
Botrychium virginianum	rattlesnake fern
Carex pensylvanica	Pennsylvania sedge
Carex platyphylla	plantain-leaf sedge
Cypripedium acaule	pink lady's-slipper
Danthonia spicata	poverty wild oat grass
Deschampsia flexuosa	wavy hair grass
Dryopteris marginalis	marginal wood fern
Fragaria virginiana	Virginia strawberry
Geranium robertianum	herbrobert
Hudsonia tomentosa	sand golden-heather
Maianthemum canadense	false lily-of-the-valley
Maianthemum racemosum	feathery false Solomon's-seal
Melampyrum lineare	American cow-wheat
Polypodium virginanum	rock polypody
Pteridium aquilinum	northern bracken fern
Sanicula marilandica	Maryland black-snakeroot
Solidago caesia	wreath goldenrod
Thalictrum dioicum	early meadow-rue
Trillium grandiflorum	large-flower wakerobin

Deciduous Woodland

Deciduous Woodland occurs on shallow soils over bedrock, usually with numerous outcrops. Trees occur scattered or in groves, but never with a closed canopy. The characteristic species vary according to soil type and moisture availability. Deciduous woodlands may need fire in certain areas to maintain the distinct woodland composition. This community corresponds to Woodlands ERT.

Canopy

Characteristic Species

Carya glabra	pignut hickory
Fraxinus americana	white ash
Juniperus virginiana	eastern red cedar
Ostrya virginiana	eastern hop-hornbeam
Pinus strobus	eastern white pine
Quercus alba	northern white oak
Quercus prinus	chestnut oak
Quercus rubra	northern red oak

Associates

Acer saccharum	sugar maple
Carya ovata	shag-bark hickory
Quercus macrocarpa	burr oak
Quercus muhlenbergii	chinkapin oak
Thuja occidentalis	eastern arborvitae
Tilia americana	American basswood

Woody Understory

Acer pensylvanicum	striped maple
Acer spicatum	mountain maple
Cornus foemina	stiff dogwood
Cornus rugosa	round-leaf dogwood
Gaultheria procumbens	eastern teaberry
Gaylussacia baccata	black huckleberry
Juniperus communis	common juniper
Kalmia angustifolia	sheep-laurel
Lonicera dioica	limber honeysuckle
Quercus ilicifolia	bear oak
Quercus prinoides	dwarf chinkapin oak
Rhamnus alnifolia	alder-leaf buckthorn
Rhus glabra	smooth sumac
Ribes cynosbati	eastern prickly gooseberry
Rubus idaeus	common red raspberry
Rubus occidentalis	black raspberry
Staphylea trifolia	American bladdernut
Toxicodendron radicans	eastern poison-ivy
Vaccinium angustifolium	late lowbush blueberry
Viburnum rafinesquianum	downy arrow-wood
Zanthoxylum americanum	toothachetree

Herbaceous Understory

Actaea pachypoda	white baneberry
Adiantum pedatum	northern maidenhair
Antennaria plantaginifolia	woman's-tobacco
Asarum canadense	Canadian wild ginger
Asplenium rhizophyllum	walking fern
Asplenium trichomanes	maidenhair spleenwort
Aster divaricatus	white wood aster
Aster macrophyllus	large-leaf aster
Athyrium filix-femina	subarctic lady fern
Botrychium virginianum	rattlesnake fern
Carex eburnea	bristle-leaf sedge
Carex pensylvanica	Pennsylvania sedge
Carex platyphylla	plantain-leaf sedge
Caulophyllum thalictroides	blue cohosh
Cystopteris bulbifera	bulblet bladder fern
Cystopteris fragilis	brittle bladder fern
Deparia acrostichoides	silver false spleenwort
Deschampsia flexuosa	wavy hair grass
Dryopteris marginalis	marginal wood fern
Elymus hystrix	eastern bottle-brush grass
Fragaria virginiana	Virginia strawberry
Geranium robertianum	herbrobert
Hieracium venosum	rattlesnake-weed
Lycopodium sabinifolium	savin-leaf ground-pine
Maianthemum canadense	false lily-of-the-valley

Maianthemum racemosum	feathery false Solomon's-seal
Oryzopsis racemosa	black-seed mountain-rice grass
Panax quinquefolius	American ginseng
Penstemon hirsutus	hairy beardtongue
Polygonatum pubescens	hairy Solomon's-seal
Polystichum acrostichoides	Christmas fern
Pteridium aquilinum	northern bracken fern
Sanguinaria canadensis	bloodroot
Sanicula marilandica	Maryland black-snakeroot
Schizachyrium scoparium	little false bluestem
Sibbaldiopsis tridentata	shrubby-fivefingers
Solidago caesia	wreath goldenrod
Solidago flexicaulis	zigzag goldenrod
Thalictrum dioicum	early meadow-rue
Trillium grandiflorum	large-flower wakerobin
Waldsteinia fragarioides	Appalachian barren-strawberry

WETLAND SYSTEMS

Floodplain Forest

Floodplain Forest occurs on low terraces of river floodplains and river deltas throughout the region. These areas are seasonally flooded, usually in the spring. The soils are generally mineral soils. The composition of Floodplain Forest changes in relation to flood frequency and elevation of floodplain terraces along large rivers. The characteristic species are dominant throughout the region, but some of the associate species become characteristic in specific areas. Cottonwood, for example, occurs in the floodplain forest in northern New York as a characteristic species, but not in Maine. This community corresponds to Floodplain Forests ERT.

Canopy
Characteristic Species

Acer rubrum	red maple
Acer saccharinum	silver maple
Juglans cinerea	white walnut
Salix nigra	black willow

Associates

Carya cordiformis	bitter-nut hickory
Fraxinus americana	white ash
Fraxinus nigra	black ash
Fraxinus pennsylvanica	green ash
Nyssa sylvatica	black tupelo
Platanus occidentalis	American sycamore
Populus deltoides	eastern cottonwood
Quercus bicolor	swamp white oak
Quercus macrocarpa	burr oak
Tilia americana	American basswood
Ulmus americana	American elm
Ulma rubra	slippery elm

Woody Understory

Clematis virginiana	devil's-darning-needles
Lindera benzoin	northern spicebush
Parthenocissus quinquefolia	Virginia-creeper
Toxicodendron radicans	eastern poison-ivy

Herbaceous Understory

Ageratina altissima	white snakeroot
Allium tricoccum	ramp
Anemone quinquefolia	nightcaps
Caulophyllum thalictroides	blue cohosh
Erythronium americanum	American trout-lily
Impatiens capensis	spotted touch-me-not
Matteuccia struthiopteris	ostrich fern
Onoclea sensibilis	sensitive fern
Polygonum virginianum	jumpseed
Sanguinaria canadensis	bloodroot
Solidago canadensis	Canadian goldenrod

Deciduous Swamp Forest

Deciduous Swamp Forest occurs in poorly drained basins and along lake shores, streams, and rivers throughout the region. The characteristic species include red maple, black and green ash, swamp white oak, and American elm. Species composition varies according to moisture and soil conditions. Soils are hydric, usually saturated, and inorganic. This community corresponds to Swamp Forests ERT.

Canopy
Characteristic Species

Acer rubrum	red maple
Fraxinus nigra	black ash
Fraxinus pennsylvanica	green ash
Quercus bicolor	swamp white oak
Ulmus americana	American elm

Associates

Acer saccharinum	silver maple
Betula alleghaniensis	yellow birch
Carya cordiformis	bitter-nut hickory
Fraxinus americana	white ash
Juglans cinerea	white walnut
Pinus rigida	pitch pine
Pinus strobus	eastern white pine
Quercus alba	northern white oak
Quercus coccinea	scarlet oak
Salix nigra	black willow
Tsuga canadensis	eastern hemlock

Woody Understory

Aronia melanocarpa	black chokeberry
Cornus sericea	redosier
Gaylussacia baccata	black huckleberry

Ilex verticillata	common winterberry
Lindera benzoin	northern spicebush
Nemopanthus mucronatus	catberry
Parthenocissus quinquefolia	Virginia-creeper
Rhododendron periclymenoides	pink azalea
Toxicodendron radicans	eastern poison-ivy
Vaccinium angustifolium	late lowbush blueberry
Vaccinium corymbosum	highbush blueberry
Viburnum dentatum	southern arrow-wood
Viburnum nudum	possumhaw

Herbaceous Understory

Carex intumescens	great bladder sedge
Carex lacustris	lakebank sedge
Dryopteris carthusiana	spinulose wood fern
Dryopteris cristata	crested wood fern
Glyceria striata	fowl manna grass
Impatiens capensis	spotted touch-me-not
Laportea canadensis	Canadian wood-nettle
Onoclea sensibilis	sensitive fern
Osmunda cinnamomea	cinnamon fern
Osmunda claytoniana	interrupted fern
Osmunda regalis	royal fern
Scirpus cyperinus	cottongrass bulrush
Scutellaria galericulata	hooded skullcap
Symplocarpus foetidus	skunk-cabbage
Thalictrum pubescens	king-of-the-meadow
Thelypteris palustris	eastern marsh fern

Coniferous Swamp Forest

Coniferous Swamp Forest occurs in stagnant basins or bordering small streams. Coniferous swamps are distinguished from upland coniferous forests by their hydric soils, lower elevation, and greater importance of mosses in the ground layer. Coniferous swamps include both basin swamps, with somewhat stagnant water, and seepage swamps on gentle slopes, where soils remain saturated through groundwater seepage. In the Adirondacks, these swamps are often found in drainage basins occasionally flooded by beavers. Some coniferous swamps are supplied with water containing a high concentration of minerals from the substrate. The high-mineral content sets them apart from Forested Bog which occurs in peat soils and which is fed with relatively mineral-poor water. This community corresponds to Swamp Forests ERT.

Canopy
Characteristic Species

Abies balsamea	balsam fir
Picea glauca	white spruce
Picea mariana	black spruce
Picea rubens	red spruce
Thuja occidentalis	eastern arborvitae

Associates

Acer rubrum	red maple
Larix laricina	American larch

Woody Understory

Alnus incana	speckled alder
Clethra alnifolium	summer sweet clethra
Gaultheria hispidula	creeping-snowberry
Ilex laevigata	smooth winterberry
Ilex verticillata	common winterberry
Sorbus americana	American mountain-ash
Vaccinium corymbosum	highbush blueberry
Viburnum nudum	possumhaw

Herbaceous Understory

Coptis trifoila	three-leaf goldthread
Dalibarda repens	robin-run-away
Dryopteris campyloptera	mountain wood fern
Osmunda cinnamomea	cinnamon fern
Oxalis montana	sleeping-beauty
Sphagnum spp.	moss

Forested Bog

Forested Bog develops mainly on accumulated peat, fine grained marl, and organic muck. American larch (tamarack) and black spruce are very common characteristic species. Bogs get most of their moisture through rainfall rather than more nutrient-rich streams or rivers. The lack of nutrients in the peat contributes to the slow growth of trees in bogs. This community corresponds to Bogs and Fens ERT.

Canopy

Characteristic Species

Larix laricina	American larch
Picea mariana	black spruce
Pinus strobus	eastern white pine
Thuja occidentalis	eastern arborvitae

Associates

Abies balsamea	balsam fir
Acer rubrum	red maple
Betula alleghaniensis	yellow birch
Chamaecyparis thyoides	Atlantic white-cedar
Fraxinus nigra	black ash
Pinus banksiana	jack pine
Pinus rigida	pitch pine

Woody Understory

Alnus incana	speckled alder
Alnus serrulata	brookside alder
Aronia melanocarpa	black chokeberry
Betula pumila	bog birch
Chamaedaphne calyculata	leatherleaf
Cornus sericea	redosier
Gaultheria procumbens	eastern teaberry
Ilex verticillata	common winterberry
Kalmia polifolia	bog-laurel

Ledum groenlandicum	Laborador tea
Lonicera oblongifolia	swamp fly-honeysuckle
Nemopanthus mucronatus	catberry
Pentaphylloides floribunda	golden-hardhack
Rhamnus alnifolia	alder-leaf buckthorn
Rubus pubescens	dwarf red raspberry
Spiraea alba	white meadowsweet
Toxicodendron vernix	poison-sumac
Vaccinium corymbosum	highbush blueberry
Vaccinium oxycoccus	small cranberry

Herbaceous Understory

Caltha palustris	yellow marsh-marigold
Carex lacustris	lakebank sedge
Carex leptalea	bristly-stalk sedge
Carex stricta	uptight sedge
Carex trisperma	three-seed sedge
Coptis trifolia	three-leaf goldthread
Dryopteris cristata	crested wood fern
Osmunda regalis	royal fern
Sarracenia purpurea	purple pitcherplant
Solidago patula	round-leaf goldenrod
Thelypteris palustris	eastern marsh fern

Shrub Bog/Fen

The Shrub Bog/Fen community type occurs on peat substrate. These areas are generally fed by fog, precipitation, and water drainage, but may also be fed by springs. There are many characteristic shrubs which depend on specific conditions. Bog-rosemary, leatherleaf, bog-laurel, and highbush blueberry are among the most common characteristic species occurring in both nutrient-poor and nutrient-rich areas. Shrub bogs and fens often occupy a band that is a transition or succession between forested and open bogs. This community corresponds to Bogs and Fens ERT.

Characteristic Woody Species

Acer rubrum	red maple
Alnus incana	speckled alder
Andromeda polifolia	bog-rosemary
Chamaedaphne calyculata	leatherleaf
Gaylussacia baccata	black huckleberry
Kalmia angustifolia	sheep-laurel
Kalmia polifolia	bog-laurel
Ledum groenlandicum	rusty Labrador-tea
Salix candida	sage willow
Vaccinium corymbosom	highbush blueberry
Vaccinium oxycoccos	small cranberry

Associates

Aronia melanocarpa	black chokeberry
Betula pumila	bog birch
Clethra alnifolia	coastal sweet-pepperbush
Cornus sericea	redosier

Decodon verticillatus	swamp-loosestrife
Ilex laevigata	smooth winterberry
Ilex verticillata	common winterberry
Larix laricina	American larch
Lonicera oblongifolia	swamp fly-honeysuckle
Myrica gale	sweet gale
Nemopanthus mucronatus	catberry
Pentaphylloides floribunda	golden hardhack
Picea mariana	black spruce
Pinus rigida	pitch pine
Pinus strobus	eastern white pine
Rhamnus alnifolia	alder-leaf buckthorn
Rubus chamaemorus	cloudberry
Toxicodendron vernix	poison-sumac
Vaccinium angustifolium	late lowbush blueberry
Vaccinium macrocarpon	large cranberry

Herbaceous Understory

Calla palustris	water-dragon
Carex aquatilis	leafy tussock sedge
Carex canescens	hoary sedge
Carex exilis	coastal sedge
Carex lacustris	lakebank sedge
Carex oligosperma	few-seed sedge
Carex trisperma	three-seed sedge
Drosera intermedia	spoon-leaf sundew
Drosera rotundifolia	round-leaf sundew
Eriophorum vaginatum	tussock cotton-grass
Eriophorum virginicum	tawny cotton-grass
Geocaulon lividum	false toadflax
Iris versicolor	harlequin blueflag
Maianthemum trifolium	three-leaf false Solomon's-seal
Menyanthes trifoliata	buck-bean
Osmunda cinnamomea	cinnamon fern
Osmunda regalis	royal fern
Peltandra virginica	green arrow-arum
Rhynchospora alba	white beak sedge
Sarracenia purpurea	purple pitcherplant
Scheuchzeria palustris	rannoch-rush
Scirpus acutus	hard-stem bulrush
Scirpus cespitosus	tufted bulrush
Sphagnum angustifolim	moss
Sphagnum centrale	moss
Sphagnum cuspidatum	moss
Sphagnum fallax	moss
Sphagnum fimbriatum	moss
Sphagnum magellanicum	moss
Sphagnum majus	moss
Sphagnum nemoreum	moss

Sphagnum papillosum	moss
Sphagnum rubellum	moss
Thelypteris palustris	eastern marsh fern
Triadenum virginicum	Virginia marsh-St. John's-wort
Woodwardia virginica	Virginia chain fern

Shrub Swamp

Shrub Swamp occurs on mineral soils in areas that are flooded too long and too deeply for trees to tolerate or as successional transition zones between forested and open wetlands. Often shrub swamps occur around the margins of ponds and bogs where the water may reach a depth of six inches or more. This community corresponds to Shrub Swamps ERT.

Characteristic Woody Species

Alnus incana	speckled alder
Cornus amomum	silky dogwood
Cornus sericea	redosier
Myrica gale	sweet gale

Associates

Alnus serrulata	brookside alder
Cephalanthus occidentalis	common buttonbush
Cornus racemosa	gray dogwood
Decodon verticillatus	swamp loosestrife
Lindera benzoin	northern spicebush
Lyonia ligustrina	maleberry
Rhododendron viscosum	clammy azalea
Salix bebbiana	long-beak willow
Salix discolor	tall pussy willow
Salix lucida	shining willow
Salix petiolaris	meadow willow
Spiraea alba	white meadowsweet
Spiraea tomentosa	steeplebush
Vaccinium corymbosum	highbush blueberry
Viburnum dentatum	southern arrow-wood
Viburnum nudum	possumhaw

Herbaceous Understory

Saxifraga pensylvanica	eastern swamp saxifrage

Herbaceous Bog/Fen

The Herbaceous Bog/Fen community type is an open wetland on organic soils of peat or muck. The dominant species include many sedges, grasses, and rushes. Shrubs and trees may also occur, but woody species cover is less than 50 percent of the area. This community corresponds to Bogs and Fens ERT.

Characteristic Species

Acorus americanus	sweetflag
Angelica atropurpurea	purple-stem angelica
Aster umbellatus	parasol flat-top white aster
Aulacomnium palustre	moss

Calamagrostis canadensis	bluejoint
Carex aquatilis	leafy tussock sedge
Carex diandra	lesser tussock sedge
Carex flava	yellow-green sedge
Carex hystericina	porcupine sedge
Carex livida	livid sedge
Carex magellanica	boreal-bog sedge
Carex oligosperma	few-seed sedge
Carex prairea	prairie sedge
Carex stricta	uptight sedge
Cladium mariscoides	swamp saw-grass
Drosera rotundifolia	round-leaf sundew
Dulichium arundinaceum	three-way sedge
Eleocharis rostellata	beaked spike-rush
Equisetum arvense	field horsetail
Eupatorium maculatum	spotted joe-pye-weed
Iris versicolor	harlequin blueflag
Juncus stygius	moor rush
Menyanthes trifoliata	buck-bean
Muhlenbergia glomerata	spiked muhly
Osmunda regalis	royal fern
Parnassia glauca	fen grass-of-Parnassus
Pogonia ophioglossoides	snake-mouth orchid
Rhynchospora alba	white beak rush
Sarracenia purpurea	purple pitcher-plant
Scheuchzeria palustris	rannoch-rush
Scirpus atrovirens	dark-green bulrush
Spiraea alba	white meadowsweet
Thalictrum pubescens	king-of-the-meadow
Thelypteris palustris	eastern marsh fern
Tomenthypnum nitens	moss
Typha angustifolia	narrow-leaf cat-tail
Typha latifolia	broad-leaf cat-tail
Utricularia minor	lesser bladderwort

Associates

Andromeda polifolia	bog rosemary
Cornus racemosa	gray dogwood
Cornus sericea	redosier
Machaerium revolvens	moss
Myrica pensylvanica	northern bayberry
Pentaphylloides floribunda	golden hardhack
Rhamnus alnifolia	alder-leaf buckthorn
Salix candida	sage willow
Scorpidium scorpioides	moss
Sphagnum contortum	moss
Sphagnum subsecundum	moss
Vaccinium macrocarpon	large cranberry

Mineral Soil Marsh

Mineral Soil Marsh occurs in areas periodically inundated with standing or slow-moving water on mineral soils with little or no peat accumulation. Species composition varies according to available moisture. Some emergent marshes are permanently saturated, while others are seasonally flooded. Emergent marshes may have water up to six feet deep and support species such as yellow pond-lily and American white water-lily. Other marsh communities may occur along the shores of lakes and ponds, and along river or stream channels where disturbance prevents peat accumulation. This community corresponds to Nonestuarine Marshes ERT.

Characteristic Species

Acorus americanus	sweetflag
Alisma plantago-aquatica	water plantain
Armoracia lacustris	lakecress
Calamagrostis canadensis	bluejoint
Caltha palustris	yellow marsh-marigold
Campanula aparinoides	marsh bellflower
Carex lacustris	lakebank sedge
Carex stricta	uptight sedge
Cyperus squarrosus	awned flat sedge
Dulichium arundinaceum	three-way sedge
Eleocharis acicularis	needle spike-rush
Eleocharis obtusa	blunt spike-rush
Eleocharis palustris	pale spike-rush
Fimbristylis autumnalis	slender fimbry
Glyceria canadensis	rattlesnake manna grass
Gratiola neglecta	clammy hedge-hyssop
Heteranthera dubia	grass-leaf mud-plantain
Iris versicolor	harlequin blueflag
Juncus canadensis	Canadian rush
Leersia oryzoides	rice cut grass
Lobelia dortmanna	water lobelia
Ludwigia palustris	marsh primrose-willow
Lythrum thrysiflora	thyme-leaf loosestrife
Nuphar lutea	yellow pond-lily
Nymphaea odorata	American white water-lily
Peltandra virginica	green arrow-arum
Polygonum amphibium	water smartweed
Polygonum pensylvanicum	pinkweed
Ranunculus longirostris	long-beak water-crowfoot
Ranunculus reptans	lesser creeping spearwort
Scripus acutus	hard-stem bulrush
Scripus atrovirens	dark-green bulrush
Scripus cyperinus	cottongrass bulrush
Scripus microcarpus	red-tinge bulrush
Scripus tabernaemontanii	soft-stem bulrush
Sparganium eurycarpum	broad-fruit burr-reed
Typha angustifolia	narrow-leaf cat-tail
Typha latifolia	broad-leaf cat-tail
Veratrum viride	American false hellebore
Zizania aquatica	Indian wild rice

ESTUARINE SYSTEMS

Brackish Tidal Marsh

Brackish Tidal Marsh occurs in coastal tidal areas where the salinity ranges from 0.5 to 18 parts per trillion and where water is less that six feet deep at high tide. Within this broadly defined marsh community, there may be up to three distinct zones: broad mudflats, marshes dominated by tall graminoids and muddy gravel, or rock shores. Each of these zones supports a different type of herbaceous vegetation. This community corresponds to Estuarine Marshes ERT.

Characteristic Species

Cardamine longii	Long's bittercress
Lilaeopsis chinensis	eastern grasswort
Limosella australis	Welsh mudwort
Sagittaria calycina	hooded arrowhead
Sagittaria latifolia	duck-potato
Samolus valerandi	seaside brookweed
Sium suave	hemlock water-parsnip
Spartina alterniflora	saltwater cord grass
Zizania aquatica	Indian wild rice

Freshwater Tidal Marsh

Freshwater Tidal Marsh occurs in shallow bays, shoals, and the mouths of tributaries of large tidal river systems where the water is usually fresh. Salinity in these marshes is less than 0.5 parts per trillion. Freshwater Tidal Marsh has up to three distinctive zones: mudflats, marshes, and muddy or rocky shores, each with a characteristic species composition. This community corresponds to Estuarine Marshes ERT.

Characteristic Species

Crassula aquatica	water pygmyweed
Eriocaulon parkeri	estuary pipewort
Limosella australis	Welsh mudwort
Lobelia cardinalis	cardinal-flower
Samolus valerandi	seaside brookweed
Scirpus pungens	three-square
Typha angustifolia	narrow-leaf cat-tail
Zizania aquatica	Indian wild rice

Ecological Restoration Types*

* See Chapter 7 for a general description of Ecological Restoration Types and sections included
  for each type.

Table of Contents

Coniferous Forests

Introduction

The distinctive feature of coniferous forests is that they are dominated by conifers or gymnosperms (cone-bearing plants). Conifers include the evergreen tree species commonly called pines, spruces, firs, larches (tamaracks), cedars, cypresses, hemlocks, and yews. While conifers are considered less advanced, evolutionarily speaking, they collectively dominate large areas and fill important ecological niches. Reasons for dominance of conifers may be tolerance of water stress or drought stress and higher nutrient-use efficiency.

An important use of conifers is for timber. Conifers are more widely used as lumber than are deciduous hardwood trees. Thus, the management of our coniferous resources has been the subject of much research and experimentation. While clearcutting may be the primary threat to coniferous forests, acid rain can substantially lower the pH of the soil, interfering with the nutrients available to the trees. Trees weakened by acid rain then become more vulnerable to insect damage and disease.

Extensive conifer forests, known collectively as taiga or boreal forests, stretch across most of Canada and Alaska and dip down into the Great Lakes states and New England to about 45°N latitude. The taiga is a monotonous forest, with crowded trees of modest stature and low species diversity in the overstory and understory. This low-elevation forest extends over impressive distances, interrupted often with lakes, ponds, and moss-covered bogs.

Coniferous forests extend south of the taiga, at moderate to high elevations in mountain chains. The subalpine forests, just below the alpine tundra, is similar to the taiga, but different species dominate. The montane zones below this are usually covered by a richer diversity of conifers, and the forests are often more open and the trees larger, more productive, and longer lived than those of the taiga and subalpine forests. Below the montane zones lie vegetation types dominated by broadleaved trees, scrub, or grassland. Exceptions to this pattern are the mountain ranges that penetrate north into the tiaga; in these areas, the surrounding low-elevation vegetation is still conifer dominated.

Site Conditions

Soils

Taiga soils mostly belong to the spodsol soil order. These are acidic soils (pH 3–5) whose upper layers have been leached of clay, organic matter, and many nutrients. Conifers are characterized by shallow root systems which lie above zones of nutrient and clay accumulation. However, nutrient uptake is improved by mycorrhizae. All taiga conifers possess mycorrhizae and there is experimental evidence that tree growth and survival would be minimal without them. Soils are acidic, in part, because of leaching and because conifer foliage is acidic. Soils beneath montane and coast forests are generally either podzolic or young and weakly developed (inceptisol order). If podzolic, they are not as acidic as taiga soils.

While creating a variety of forest types indigenous to New York, Hellman (1984) found that pines (*Pinus* spp.) may be able to tolerate a higher pH than commonly thought. He noted that pines, spruces (*Picea* spp., other than black), and fir (*Abies* spp.) generally grew well on a dolomitic escarpment despite early concern that they would be inhibited by the high pH. Physical properties such as texture and permeability are often more important than pH.

Topography

The topography of coniferous forests is quite variable as their development seems to be more dependent upon soil and nutrient attributes. Slope and aspect, however, contribute to soil development. Much of the area where these communities occur was scoured and shaped by glaciers in ages past, and many of its landforms bear witness to this. The topography is mainly flat plains with occasional low hills or gentle mountains, countless lakes and bogs in shallow depressions, and the great piles of stones called moraines. It is important to restore these communities in places that have similar topography to nearby, natural examples. Slope, elevation, and aspect determine the species of coniferous forests. These factors must be considered when selecting which community and species to use for restoration.

Moisture and hydrology

Near the base of the Rocky Mountains precipitation is only about ten to twenty inches, but increases to about forty inches on the higher slopes. It occurs mostly in winter and as snow at higher elevations. The eastern side of the mountains, bordering the Great Plains, is drier than the western slopes. With increasing altitude, the length of the growing season shortens, and temperatures in both winter and summer are lower.

In the Sierra-Cascade forests, precipitation amounts are greater and temperatures are cooler at higher elevations. Both winter and summer are warmer, and rainfall is more seasonal (mostly in winter) at lower elevations.

The precipitation of the Pacific forests ranges from a moderate 30 inches in the south to as much as 300 inches on Washington's Olympic Peninsula. Often there is more rainfall in the winter. Humidity is high in most areas.

The moisture and hydrology of coniferous forests differ depending upon the individual community. Some of these community types are significantly drier than other forest types. However, some develop into very rich, mesic areas. It is worthwhile to compare the moisture levels and hydrology of natural, or reference, communities to potential restoration sites to determine suitability.

Establishment

Site preparation

If the restoration site has been severely altered, then some earth moving may be useful to restore more natural contours. Many conifers are extremely adaptable to poor soil conditions and can prosper in degraded areas. This is one reason they are often used for mine reclamation.

Planting considerations

Root dips have often been used to protect conifer seedlings prior to and following planting in revegetated hot, dry areas of the Intermountain West. Sloan (1994) found that several varieties of dips did not significantly improve seedling survival, shoot growth, or root growth of native conifer species under dry growing conditions in both field and greenhouse experiments. Sloan concludes that what is important is the careful handling of seedlings prior to planting—sheltering them from the sun and the wind–and the proper planting of each seedling.

Dense lichen and moss growth may mechanically inhibit tree regeneration. Seeds or seedlings will need to be planted in holes dug through this layer of bryophyte growth. Seeds of ponderosa pine (*Pinus ponderosa*) are known to do best when the litter layer is thin or absent, and in relatively open sunlight. Seedlings of white fir (*Abies concolor*) can tolerate denser shade and a deeper litter layer. Tree planters need to be familiar with light and soil requirements of all species being planted. Many of the dominant and characteristic species of coniferous forest communities (see Appendix A) are commercially available (see Appendix D).

Bare-root material is typically used (especially on larger sites) as it is more economical, quicker to plant, and easier to handle. While the mortality may be higher than that of container stock or balled and burlapped plants, the lower costs and reduced labor associated with bare-root material make it a better choice.

Spacing considerations

As a general rule, coniferous forest trees should be planted in an open "parkland" style spacing arrangement. This will allow for filling in when planted trees reach reproductive maturity. The uniform row plantings typical of pine plantations that are used for timber production should be avoided. The goal should be to create a more heterogenous landscape with structural diversity that will benefit other species of plants and animals as well.

Temporal spacing for different plants and layers

Restoration (especially landscape-scale restoration) typically depends heavily upon natural succession. Planting in phases allows the restorationist to assess previous plantings and make necessary adjustments along the way. Reseeding and/or replanting may need to be done for three to five years to ensure proper development of desired vegetation types.

Singer (cited in Jellinek 1995) found that by removing smaller competing trees, the growth rate of second growth trees (hemlock [*Tsuga canadensis*] and hardwoods in northern Wisconsin and Michigan) can be increased by 50 to 100 percent. This technique for accelerating the development of sec-

ond growth forests would probably be useful in other forest types as well. By removing competition and encouraging individual trees, the establishment of a canopy may be achieved more quickly.

Maintenance and Management

Nutrient requirements

Conifers are generally efficient users of nutrients. Many of the nutrients in coniferous forests may be locked up in litter. Nutrient cycling is often slowed by mosses if they are present in the ground layer. Mineral cycling is typically low.

Fertilizer pellets will likely improve survival of plantings. Pellets should be placed in individual planting holes. A broad-scale fertilizer application is generally not needed. However, some soil amendments may be useful on severely degraded sites. One such amendment that appears to improve growth and survival is the inoculation of mycorrhizal fungi on the site or on the roots of individual plantings. Mycorrhizae form a symbiotic relationship and help plants to accumulate nutrients.

Water requirements and hydrology

Temperature and precipitation are major environmental gradients that influence the range and abundance of conifer species, especially along an altitudinal gradient. Many conifers have a high tolerance for water stress and/or drought stress which gives them a competitive edge against less tolerant species. Supplemental watering of newly planted seeds or seedlings for the first few seasons can greatly improve survival and subsequent forest establishment.

Disturbance

Ponderosa pine (*Pinus ponderosa*) forests often have a rather open, parklike aspect. The open physiognomy and pure dominance have apparently been maintained by natural surface fires with a frequency of one fire every three to ten years. Fire suppression policies over the past sixty years have resulted in denser stands, more brush, loss of grass, and the invasion of other conifers. Many foresters and ecologists recognize that disruption of the historic pattern of frequent fires in ponderosa pine forests has resulted in major ecological changes, including increasingly severe wildfires and insect and disease epidemics (Everett 1994, Gray 1992, Weaver 1943). Historically, fires, at intervals averaging five to thirty years in most areas, thinned small trees and helped produce open, park-like, fire-resistant stands (Arno 1988). What was once ponderosa pine forests now consists of fir thickets impacted by insect and disease epidemics due to fire exclusion and logging of overstory pines. Also, large stand-destroying wildfires, formerly rare in the open ponderosa pine forests, have become common in the dense stands that have developed as a result of fire exclusion (Arno et al. 1995). Agee (1993) provides a detailed account of fire ecology in the Pacific Northwest.

It is usually necessary to begin restoration treatment in today's dense stands with a "low thinning" that removes excess understory and weaker overstory trees that cannot be safely killed in an underburn (Arno et al. 1995). Ponderosa pine and western larch are fire resistant and long lived, commonly surviving 400 to over 600 years in conjunction with frequent underburns. Cutting without subsequent burning is ineffective for forest restoration, because only fire can efficiently remove large proportions of the small understory trees (often thousands per acre) that occupy many stands and which, if not removed, will develop into thickets of stressed trees (ibid.). In addition, fire produces unique changes in soil chemistry such as increasing pH and rapid oxidation of nutrients, both of which may significantly influence nutrient availability.

Once thinning and burning are completed, open stands can be maintained with trees of many ages and sizes, which can be perpetuated in a healthy condition with selective cutting and underburning at intervals of about fifteen to thirty years. Arno et al. (1995) found that one year after a burn, a luxuriant undergrowth of grasses and herbs, including abundant flowering of arnica (*Arnica cordifolia*) and other species that seldom flower in untreated stands, attests to the stimulating effects of fire.

The boreal forest is a disturbance (disclimax) forest. Small areas are often disturbed by windthrow from winter storms or by fire from summer dry-lightning strikes. Many conifers have serotinous or semiserotinous cones and regenerate well after fire. Succession often proceeds through meadow, then broadleaf tree seral stages before reaching an even-aged spruce-fir climax in a span of about 300 years.

In the Great Lakes region, which is an ecotone between the boreal forest and deciduous forest, it is believed that pine is seral to hardwood. Pine may form an edaphic climax on sites too dry to support the hemlock-hardwoods.

Eastern Deciduous and Mixed Forests

Northern Hardwoods–Conifer Forest A-249, A-283, A-381
Great Lakes Pine Forest A-253
Maple–Basswood Forest A-265
Oak–Hickory Forest A-267, A-305
Beech–Maple Forest A-269, A-384
Northern Oak Forest A-285
Mesophytic Forest A-286, A-373
Oak–Pine Forest A-307
Western Mesophytic Forest A-308
Oak–Hickory–Pine Forest A-320
Southern Mixed Hardwoods Forest A-322, A-342
Mesic Pine Forest A-324
Xeric Pine Forest A-325
Rockland Hammock A-343
Maritime Hammock A-346
Sandhill A-347
Pine Flatwoods A-349
Northern Hardwoods Forest A-370
Appalachian Oak Forest A-372
Pine–Oak Forest A-374
Oak–Pine–Hickory Forest A-383

Introduction

Eastern deciduous and mixed forest communities have a well-developed canopy as well as a subcanopy, shrub layer, and herbaceous layer on the forest floor. While these communities are canopied, natural openings from tree falls, severe weather, and other natural disturbances may occur.

Mixed forests are those with coniferous (evergreen) species co-dominating with deciduous trees. Forest communities dominated by coniferous species are discussed in the coniferous forest restoration type. Most of these forests have been cut for timber. Because this is a region of deep, fertile soils, the forests have been cut and the land used for agriculture in many areas. The older forests today are little more than a century old. Braun (1950), Spurr and Barnes (1980), Perry (1994), and Yahner (1995) provide excellent overviews of eastern deciduous forests.

Site Conditions

Soils

Two soil orders characterize the eastern deciduous forest. The alfisol order extends from north of Virginia and Kentucky to well into an ecotone area with the taiga (Barbour et al. 1987). Alfisols are also known as gray-brown podzolics; they are acidic (pH 5-6), but not as acidic as taiga podzols, partly because of the pH 4 of spruce and pine. Diversity and activity of soil biota are high. The ultisol order dominates the southern part of the forest (ibid.). Ultisols have not been glaciated; hence, they are older and more weathered than alfisols. Ultisols are slightly more acidic alfisols, and considerably lower in fertility.

Soils of eastern deciduous and mixed forests range from sandy and well drained to clayey and poorly drained. Many are loams (with approximately equal amounts of sand, silt, and clay). Forests with well-drained soils typically have faster rates of litter decomposition and, hence, a thinner layer of leaf litter than forests with poorly drained soils (Yahner 1995).

Topography

Eastern deciduous and mixed forests develop on a variety of topographic settings, from flat to steeply sloping. When restoring areas where the terrain is steeply sloping, erosion control needs to be considered. Slope, elevation, and aspect determine the species composition of these communities. Through its influence on water availability, aspect can also directly influence important functions such as primary productivity.

Moisture and hydrology

Eastern deciduous and mixed forest communities range from xeric (dry) to mesic (moist) depending upon the topography and soils present. The major source of water in the eastern deciduous forest is precipitation in the form of rain or snow that percolates into the soil (Spurr and Barnes 1980). Not all precipitation, however, reaches the forest floor; some hits the foliage and evaporates back into the atmosphere. In fact, about seventy percent of the total precipitation is recycled into the atmosphere from the forest and the amount of water entering the forest soil is lessened further by surface runoff. The forest soil is usually very wet and at full water capacity near the beginning of the growing season. Soil moisture becomes depleted but is replenished by rainfall as the growing season progresses. In general, an average of twenty-four to forty inches of rainfall are needed for forest tree growth during the growing season in northern and southern states, respectively (Yahner 1995).

Moisture in the forest soil can vary depending upon the degree of slope and the position on the slope, which in turn, can influence the distribution of plants and animals. Sites with steep slopes usually have relatively dry soils compared to sites lower down the slope; hence, chestnut oak (*Quercus prinus*) predominates on upper, dryer sites, whereas tulip poplar (*Liriodendron tulipifera*) is abundant on lower moister sites (Gleason and Cronquist 1963).

The southern tip of Florida lies near enough to the tropics to share the tropical wet-dry climate of the Caribbean. In this region, rain falls mainly from late spring to mid-fall, and the winter months are usually dry. Year-round mild temperatures make possible the growth of many tropical plants found nowhere else in the continental United States.

Establishment

Site preparation

It is generally best to begin with a degraded shrubland or woodlot (if one is present) rather than clearing the site all at once. To release such a woodland can be slow and difficult. Begin removing exotics and making improvements where the trees are largest and the shade is greatest, and work outward toward the edge. The idea is to move from the best toward the worst to allow existing rootstocks and seeds of native plants the best chance to regenerate (Scott 1992).

Litter accumulation as branches, leaves, and even entire trees is a natural part of any forest ecosystem. Litter builds and insulates the soil, holds moisture, and prevents erosion.

Planting considerations

Many of the dominant and characteristic trees, shrubs, and herbaceous plants of eastern deciduous and mixed forests (see Appendix A) are available commercially (see Appendix D). Brenner and Simon (1984) found that direct seeding of native mast-producing trees on surface mines can accelerate establishment of deciduous communities by at least ten years. They planted black walnut (*Juglans nigra*), northern red oak (*Quercus rubra*), and American beech (*Fagus grandifolia*) at a depth of four-tenths of an inch on a surface mine reclaimed with fescue (*Festuca elatior*) and crown vetch (*Coronilla varia*). The percent germination was twelve and sixteen percent for northern red oak and American beech, respectively; and twenty-five percent for black walnut. Seedlings of all species had survived on the site for six years.

Schaal (1996) described a practical and effective method for planting white oak (*Quercus alba*) seedlings. First, loosen the soil and remove any existing plants, and then lay down a three–four inch thick layer of coarse woodchip mulch with a diameter of twelve inches. Next, plant a single acorn just below the surface of the mulch and lightly water the site. Because white oak acorns begin to germinate roots in the fall, that is the best time to plant. Then in the following spring, just before the buds swell, work a small amount of 12-12-12 granular fertilizer into the soil around the seedling. After replenishing the mulch with another three to four inch thick layer of woodchips, water the planting area thoroughly. This technique may be useful for seed planting other oaks as well.

Reber and Pope (1995) found that northern red oak seedlings inoculated with ectomycorrhizal *Pisolithus tinctorius* were taller and had greater root, stem, and foliage biomass than with the other treatments. Treatments with high levels of nitrogen and phosphorus produced the largest seedlings, but mycorrhizal colonization decreased significantly in these fertile soils.

Most oaks have deep penetrating roots and the genus has adapted to a wide range of soil conditions. Hickories also have deep taproots that enable them to adapt well to dry soils. Hickories usually appear in a developing woods after the oaks have begun to bear acorns, because the acorns attract squirrels which then bring in the hickory nuts (Scott 1992).

Browsing deer and other forest animals can have a negative impact on restoration efforts and many types of chemical repellents are available. Some are systemic and are placed directly into the nursery soil. Hardwood trees may have to be protected from deer by enclosing them in tree shelters which are available in a variety of sizes and styles (some are photodegradable). Tree shelters have proven to be very effective for protecting the trees from browsing animals, and they have the added advantages of significantly lowering mortality and enhancing the growth rate of the trees by acting as miniature greenhouses (Bainbridge and Fidelibus 1994). While tree shelters are beneficial to young, newly planted trees, Carey and Robertson (1995) reported that songbirds sometimes become trapped in tree shelters. Blakeman (1996) found that inserting stems of common reed (*Phragmites australis*) in tree shelters allows any bird that enters to escape by walking up the stalk.

Pineland three-awn, also known as wiregrass (*Aristida stricta*), is a dominant grass in the understory of the mesic and xeric pine forest communities of the southeastern plains. Cole (1995) recommends transplanting wiregrass during seasons of higher rainfall (September to January in Florida); maintaining a distance of no more than fifty miles between collection and transplanting sites; and using a scale appropriate to the labor- and time-intensive methods required for this work. Walters et al. (1994) found that allozyme variation within and among sample populations of pineland three-awn from three regions of the species' range reveal noticeable genetic variation. They recommend specifying locally-adapted pineland three-awn genotypes for restoration projects.

Bisset (1995) broadcast seeds of pineland three-awn across a one-foot thick layer of clay-sand over-burden placed over quartz mine-tailings. The result was a good stand of the grass, mixed with other native species. This suggests that this grass may be used to establish cover on surface-mined sites.

Barber (1994) began growing pineland three-awn seedlings from seeds collected throughout Florida. Production trials with bare root plants showed that seed sources with a ten-year history of summer burns on a three-year rotation yielded the best germination. Barber (1994) found that the best time to burn pineland three-awn is April through June with seed collection in November. Outcalt (1995) directly seeded pineland three-awn grass seed across different site conditions under existing South Carolina pine stands. Neither prior cultivation nor the removal of seed from stalks before sowing affected seedling densities after one year. Seedling densities were greatest on moist sites having low density of pine overstory and lowest on dry sites with denser tree cover.

Spacing considerations

The spacing of tree plantings in the restoration of eastern deciduous and mixed forests should be open and "parkland"-like to allow for filling when trees reach reproductive maturity. The existing seedbank (if one is present) and the proximity to intact forest should also be considered. If no seedbank is present or the distance to intact forest is great, then higher densities of trees should be planted. If the site has not been severely altered, volunteer (or pioneer) recruitment can be expected.

Temporal spacing for different plants and layers

In converting an old field to a forest, early pioneer trees such as dogwood (*Cornus* spp.), hawthorne (*Crataegus* spp.), red cedar (*Juniperus* spp.), sassafras (*Sassafras albidum*), and wild cherry (*Prunus serotina*) will serve to shade and weaken the weedy herbaceous plants present. As more shade is provided, more typical forest trees and shrubs such as tulip poplar (*Liriodendron tulipifera*), red maple (*Acer rubrum*), ash (*Fraxinus* spp.), viburnum (*Viburnum* spp.), spicebush (*Lindera benzoin*), and witch hazel (*Hamamelis virginiana*) begin to appear and eventually shade out the earlier hawthorne, red cedar, and wild cherry. Next, dominant species such as oaks (*Quercus* spp.), hickory

(*Carya* spp.), or beech (*Fagus grandifolia*) can be established. Then ferns and wildflowers that characterize the forests will have a suitable habitat. In many eastern forests, succession may be accelerated by initially planting tulip poplar. Tulip poplars rapidly grow very tall; they must to survive, because their seeds do not sprout in the dense shade of a mature forest.

Restoration (especially landscape-scale restoration) typically depends heavily on natural succession. Planting in phases allows the restorationist to assess previous plantings and make necessary adjustments along the way. Reseeding and/or replanting may need to be done for three to five years to ensure proper development of desired vegetation types.

Eastern deciduous and mixed forests typically have four levels. Below the canopy are the understory trees, smaller species that tolerate the shade of the canopy species. Below the understory trees are the shrubs. Next is the groundcover of herbaceous plants and ferns. Lowest of all are the mosses. Each layer has adapted to progressively diminishing levels of light, so herbaceous plants, which need the least, are both the last to arrive in a developing woods and the first to green up in the spring (Scott 1992).

Matt Singer (cited in Jellinek 1995) found that by removing smaller competing trees, the growth rate of second growth mesophytic forest may be increased from 50 to 100 percent. Sally Dahir (cited in Jellinek 1995) discovered that existing hemlock (*Tsuga canadensis*)-hardwood old-growth forests in northern Wisconsin and Michigan's Upper Peninsula form gaps in the canopy at a rate of seven percent of stand area per decade, indicating a complete forest turnover every 140 years.

Robertson and Robertson (1995) planted white pines (*Pinus strobus*) along a border where the invasive exotics, oriental bittersweet (*Celastrus orbiculatus*) and porcelainberry vines (*Ampelopis brevipedunculata*) had formerly "sealed off" the edge. The pines were placed strategically to cast as much shade as possible in the interior of the woods and establish a light barrier of native trees. However, the use of noninvasive, nonindigenous ornamental conifers like Norway spruce (*Picea abies*) was also suggested because their branching habit maintains shade along the entire length of the trunk, right down to ground level (ibid.). Some good native

edge species to encourage are are tulip poplar, dogwood, redbud (*Cercis canadensis*), red maple, sugar maple (*Acer saccharum*), white ash (*Fraxius americana*), sassafras, or birch (*Betula* spp.). In addition to pines, shadbush (*Amelanchier* spp.), sumac (*Rhus* spp.), viburnum, and blueberry (*Vaccinium* spp.), or huckleberry (*Gaylussacia* spp.) are naturally edge inhabitants. Native vines that also occur in this area are greenbrier, blackberry or black raspberries (*Rubus* spp.), trumpet creeper (*Campsis radicans*), and poison ivy (*Toxicodendron radicans*).

Maintenance and Management
Weed control

Several exotic trees, shrubs, and vines invade eastern deciduous and mixed forests. Exotic trees compete with native trees. Tree-of-heaven (*Ailanthus altissima*), princess-tree (*Paulownia tomentosa*), and Norway maple (*Acer platanoides*), for example, all grow readily in disturbed areas, and often invade disturbed forests in the eastern United States. Norway maples have very large, thick leaves which emerge early in the spring, creating a deep shade that severely inhibits native woodland wildflowers. It is also strongly allelopathic; a toxic substance is released into the soil by the roots that suppresses the growth of competing species. This explains why Norway maple can dominate entire woodlands unless controlled. Norway maple can be easily recognized in the fall as its leaves do not turn yellow until mid-November when most other woodland trees are bare. This is a good time to identify trees for removal.

Bird cherry (*Prunus avium*) and corktree (*Phellodendron japonicum*) have become pests in some eastern U.S. forests and woodlands. Black locust (*Robinia pseudoacacia*), while native, is considered a weed by some restorationists due to its ability to establish and reproduce in disturbed areas. However, in the long term, black locust poses little threat to biological integrity of a community as it is shade intolerant and will not spread into areas with well-established canopies.

Seedlings and small saplings up to two inches in diameter can be pulled up using a weed wrench. Young trees can be cut down and herbicide applied to the stump. Large trees can be girdled (spring is a good time to do this). However, girdling alone will not control Norway maple; unless herbicide is applied to the wound, it will soon be calloused over.

Exotic buckthorns (*Rhamnus cathartica* and *R. frangula*) are effectively controlled by applying herbicide to freshly cut stumps. Privet (*Ligustrum* spp.) and bush honeysuckle (*Lonicera* spp.) are also problematic in eastern forests and may be controlled using the same technique. When treating the stump, cut it close to the ground and be sure to cover the outside bark and cambium layer as well as the root collar at the soil line or resprouting may occur.

Oriental bittersweet (*Celastrus orbiculatus*) can occupy degraded forests and woodlands to the exclusion of other mid-story and groundlayer plants. It is likely that oriental bittersweet is hybridizing with the native bittersweet (*C. scandens*) in parts of its range, making the problem more complicated. Busciano (1995) found that herbicides may be effectively used to control heavy infestations of this invasive vine. Robertson and Robertson (1995) found that during the growing season, extensive but shallowly rooted bittersweet plants could be manually uprooted.

Japanese honeysuckle (*Lonicera japonica*) can be removed by uprooting and removing the vines by hand. Fire may be an effective means of control in some instances (Barden 1992, Barden and Matthews 1980). Robertson and Robertson (1995) found that in a forest in southeastern Pennsylvania, an area that was burned remained remarkably free of exotic vegetation and spicebush (*Lindera benzoin*) resprouted from its roots, but Japanese honeysuckle also resprouted. The effect of fire will obviously be temporary, but the reduced competition has given the native plants that were introduced into the charred area a chance to become established (ibid.). Herbicides appear to be most effective during flowering (April to June) (Romney et al. 1976). A combination of burning and/or cutting followed by herbicide application may be most effective.

Japanese barberry (*Berberis thunbergii*) and multiflora rose (*Rosa multiflora*) generally must be dug out with a shovel or spade. A potential biological control organism (*Megastigmus aculeatus*) may eventually play an important role in controlling multiflora rose. Multiflora rose proved so attractive to birds that it is now credited with the northern spread of cardinals and mockingbirds.

Macartney rose (*Rosa bracteata*) is an invasive shrub in the southeastern United States. Gordon et al. (1982) and Gordon and Scifries (1977) found that prescribed burning removes more than ninety percent of topgrowth, but that this rose resprouts vigorously. Herbicides also effectively killed topgrowth, but left a tangle of dead canes that provide trellis-like supports for resprouting canes. Mechanical methods such as shredding may spread the infestation, since fragments of canes readily reroot. Cutting and/or burning in early spring and applying herbicide between June and September may reduce Macartney rose by ninety percent up to two years after treatment (Gordon et al. 1982). These methods may also prove effective for multiflora rose.

Robertson and Robertson (1995) found that cutting porcelainberry stems (in the winter) just above the soil surface and also five feet above the ground created a gap to prevent porcelainberry root sprouts from reaching the hanging, dead vines during the next growing season.

Anyone who has restored an area where kudzu (*Pueraria lobata*) had previously invaded knows that it can be a challenge. As kudzu "suffocates" nearly all vegetation, often herbicides are applied and sometimes the entire area is cleared before revegetation efforts begin. Kudzu, an aggressive herbaceous vine, is very difficult to control. Often burning and herbicide are required to control kudzu.

Fire is an effective control method for garlic mustard (*Alliaria petiolata*), but it is not always feasible at a restoration site. Manual removal is most effective; however, it is labor intensive. Herbicides have been used to control stilt grass (*Microstegium vimineum*) (Robertson and Robertson 1995), however, this vigorous grass is nearly impossible to completely control in eastern United States forests and woodlands where it competes with native herbaceous plants. Robertson and Robertson (1995) stated that controlled burning is an effective means of managing stilt grass.

The fact that the land was once farmed may be a factor in the spread and persistence of noxious weed. Farm fertilizers that raise the level of nutrients, like phosphorus and nitrogen in the soil, act to stimulate the growth of fast-growing exotic species and strengthen their grip on native vegetation.

Cogongrass (*Imperata cylindrica*), a perennial grass native to southeast Asia, has become a serious problem in pine forests, sandhill, and pine flatwoods communities of the southeastern United States. It spreads by both seed and rhizomes and, once established, has the ability to displace other vegetation on forests and rangelands, as well as on reclaimed phosphate mines and other disturbed areas (Shilling and Gaffney 1995). Only a combination of techniques, applied in an integrated way, can effectively control cogongrass (ibid.). Mowing cogongrass must be done consistently over two or more years to deplete the starch reserves that support the growth of new shoots. Tilling knocks down new shoots and cuts and helps dry out the rhizomes. Deep tillage is important since cogongrass rhizomes rarely resprout from depths greater than six inches (ibid.). Complete control of cogongrass requires repeated tilling until there is no regrowth—typically three tillings every year for three years.

Only a few herbicides have proven effective in controlling cogongrass. Shilling and Gaffney (1995) found that an application of imazapyr or glyphosate in September provided seventy to eighty percent control up to one year after a single treatment. Both of these herbicides have some drawbacks, however. Both kill all plants in the treated area and may remain active in the soil inhibiting the establishment of desirable species. Shilling and Gaffney (1995) stated that the key to long-term control of cogongrass is replacing it with a competitive plant community capable of closing ranks and resisting reinvasion. Establishing new species in cogongrass-infested areas is difficult, because cogongrass secretes allelopathic chemicals, has an extensive system of rhizomes, and creates a dense canopy. Burning followed by tilling and herbicide applications should contain cogongrass long enough to give restorationists a chance to establish the species chosen to compete successfully with cogongrass over the long-term.

Nutrient requirements

Soil nutrients are vital to the growth of plants. A ready source of nutrients is organic matter –freshly fallen leaves, twigs, stems, bark, and flowers–which fall to the forest floor, decompose, and are recycled in the eastern deciduous forest (Aber 1990, Bockheim 1990). Conversely, nutrients are lost from the forest as trees are removed for timber and other forest products (Lorimer 1990). Deciduous trees in the eastern forest typically require more nutrients for growth than conifers, which are often capable of thriving on nutrient-poor soils. The amount of nutrients needed by deciduous or coniferous trees is

much less, however, than that required by most lawns in suburbia.

Water requirements and hydrology

These forests are generally moist due to the large amount of shade provided by the canopy and subcanopy trees. The constant buildup of litter on the forest floor by falling leaves, branches, logs, and dying herbaceous plants builds organic matter which holds moisture and also helps prevent erosion. Soil water is critical to nutrient cycling in the forest ecosystem (Spurr and Barnes 1980, Bockheim 1990). As plants absorb water through root systems, nutrients suspended in the water are transported into the plant. For trees to grow properly, there must be an optimal balance between the amount of water and airspace in the soil.

Disturbances

Tree falls by lightning, fire, disease, or wind are constant in these eastern deciduous and mixed forest types. While these disturbances typically affect only small and isolated areas within a forest, they may be responsible for resetting the process of plant succession over large areas. This is not as common under natural circumstances as clearcutting by humans is in resetting succession to its earliest stages.

Hammocks consist of evergreen, tropical hardwood trees. These are "tree islands," surrounded by other vegetation types, usually pineland or swamp. Hammocks are often found where fire protection is given (pinelands burn frequently), on deeper soils. "High" hammocks are drier and "low" hammocks are wetter, with different species growing in each type. Pinelands are maintained by frequent fire, hammocks are not.

Johnson (1994b) rejects the idea that the revegetation of the high pine sandhills of Florida requires the removal of dominant hardwood species. Historical records show that hardwood species such as turkey oak (*Quercus laevis*) and bluejack oak (*Q. incana*) did occur in these communities in the past. Those hardwoods that occur naturally with longleaf pine (*Pinus palustris*) should be identified and considered when using prescribed burning as a management tool (ibid.).

For additional information concerning fire in this restoration type, see United States Forest Service (1978).

Case studies

The Pennypack Ecological Trust developed three major restoration objectives for the restoration of Overlook Woods, a forested four-acre tract of mixed mesophytic woods in southeastern Pennsylvania (Robertson and Robertson 1995). These objectives were control exotic vegetation, reintroduce native shrubs and herbs, and plant canopy-gaps with seedlings and saplings of native trees that will begin regeneration; introduce several extirpated native species such as black cohosh (*Cimicifuga racemosa*) and bellwort (*Uvularia sessilifolia*); and exclude exotics by creating dense shade. A published flora (Rhoads and Klein 1993) was used to develop a list of plants native to the area. Plants to be used were chosen based on the habitats present, sunny woods-edges and rocky woodlands. The plan included the eradication of exotic trees, shrubs, vines, and herbs prior to planting native species. Survival of native herbs was high, possibly due to the use of systemic animal repellent and weeding to reduce competition.

Western Deciduous and Mixed Forests

Mixed Evergreen Forest A-11
Mojave Montane Forest A-67
Western Ponderosa Forest A-79
Aspen–Lodgepole Pine Forest A-103
Blackhills Pine Forest A-167
Eastern Ponderosa Forest A-169
Oak–Hickory Forest A-200, A-209

Introduction

Western deciduous and mixed forests have closed to relatively open canopies. Conifers are sometimes the major dominant, however, broadleaved deciduous trees (especially oaks and aspen) also play an important role in these communities. Fire is a natural disturbance that serves to maintain their structure and composition. These communities are often transitional between coniferous forests or open scrub or grassland communities. See also **coniferous forests**, **desert lands**, and **eastern deciduous and mixed forests** restoration types.

Site conditions

Soils

Western deciduous and mixed forests soils are classified in the Ultisols and Mollisols orders (U.S. Geological Survey 1970, U.S. Department of Agriculture 1975). County soil surveys contain soil maps, descriptions of soil map units, and other information necessary to identify western deciduous and mixed forests restoration sites based upon soil characteristics. Bear in mind that past land uses influence soil compaction and water-holding capacity.

Topography

Western deciduous and mixed forests develop on a variety of topographic settings, from flat to steeply sloping. When restoring areas where the terrain is steeply sloping, erosion control needs to be considered. Slope, elevation, and aspect determine the species composition of these communities.

Through its influence on water availability, aspect can also directly influence important functions such as primary productivity.

Moisture and hydrology

Most of these communities receive have a mean annual precipitation of eight to twelve inches (Barbour and Billings 1988). Precipitation is equally distributed throughout the year, although some places receive peak precipitation in summer. The relative humidity is sixty to eighty percent throughout the year in the deciduous forest.

Establishment

Site preparation

It is best to begin with with an existing forest rather than clearing the site all at once. To release such a forest can be slow and difficult. Begin removing exotics and making improvements where the trees are the largest and the shade is greatest and work outward toward the edge. The idea is to move from the best toward the worst to allow existing rootstocks and seeds of native plants the best chance to regenerate (Scott 1992).

Litter accumulation of branches, leaves, and even entire trees is a natural part of any forest ecosystem. Litter builds and insulates the soil, holds moisture, and prevents erosion.

Planting considerations

Before planting begins, several important considerations must be addressed: (1) which species to use, (2) type of plant materials to use (seeds, seedlings, saplings, cuttings, container stock), (3) source of plant materials, (4) density of plantings (spacing pattern), and (5) plant protection (screening, fencing, stakes, repellent). Once these factors are addressed revegetation efforts can begin. Planting should not be a one-time event. Supplemental planting for the first three to five seasons can help ensure survival of plants to maturity and establishment of desired vegetation type.

Many of the dominant and characteristic trees, shrubs, and herbaceous plants of western deciduous and mixed forests (see Appendix A) are available commercially (see Appendix D). Methods of establishing vegetation at a restoration site include direct seeding, planting bare-root stock, planting container stock, and/or planting balled and burlapped stock. A combination of materials may be most appropriate for a site. For example, it may be best to use container stock on steeper, more erosion-prone areas and seeds and bare-root stock on more level ground. Bare-root material is generally the best choice (especially for larger sites) as it is more economical, quicker to plant, and easier to handle. While the mortality may be higher than that of container stock or balled and burlapped plants, the lower costs and reduced labor associated with bare-root material make it a better choice.

Spacing requirements

Spacing should be at densities similar to those found in natural examples of these communities. However, it is often best to use a higher density as some mortality can be expected. It is likely that some volunteer species recruitment will take place, especially if there is a seedbank and/or nearby intact forests.

Temporal spacing for different plants and layers

Restoration (especially landscape-level restoration) typically depends heavily on natural succession. Planting in phases allows the restorationist to assess previous plantings and make necessary adjustments along the way. Reseeding and/or replanting for a few seasons following initial planting ensures development of desired vegetation types.

Maintenance and Management

Nutrient requirements

If the soil is found to be deficient in important nutrients (as compared to a more natural reference community), soil amendments or fertilizer may need to be applied to the site. Several factors influence the decision to use fertilizer: (1) the nutrient needs of the plant species to be planted; (2) known nutrient deficiencies in the soil; (3) effect of fertilizers on soils; (4) necessity of refertilization; (5) cost; and (6) available soil water (U.S.D.A. Forest Service 1979).

Water requirements and hydrology

While these communities should become self-sustaining, supplemental watering for a few months after planting will improve survival and development of trees and shrubs, especially in the summer.

Disturbances

Tree falls by lightning, fire, disease, or wind are constant in these western deciduous and mixed forests. While these disturbances typically affect only relatively small and isolated areas within a forest, they may be responsible for resetting the process of plant succession over large areas, although this is not as common, under natural circumstances, as clearcutting by humans in resetting succession to its earlier stages.

Over a four-year period, Crow et al. (1994) investigated the dynamics between oak (*Quercus* spp.) recruitment and fire management on a fifteen-acre mixed forest/grassland site in Lucas County, Iowa. Fire stimulated oak growth and sprouting and while the oaks did not rapidly colonize within the study area, they may persist and eventually dominate those communities subjected to periodic burning (ibid.).

Woodlands

Introduction

Woodlands used broadly here, include communities commonly known as savannas and barrens. Early explorers called any area that was useless for farming because of thin, nutrient-poor soil "barrens." They generally have a parkland appearance with sparse individual trees or some groves and a herbaceous layer in which grasses and/or heaths dominate. There are many geographical variations of these community types in almost every part of North America. Leach and Ross (1995) define a savanna as "...any area where scattered trees and/or shrubs and other large persistent plants occur over a continuous and permanent groundlayer usually dominated by herbs, usually graminoids." Risser (cited in Anderson and Anderson 1995) emphasized that even though we have accumulated a considerable amount of information on the ecology of savannas and barrens we do not always use this knowledge in an effective manner.

The Midwest Oak Ecosystem Recovery Plan (Leach and Ross 1995) fashioned after the United States Fish and Wildlife Service endangered species recovery plans offers a management plan which identifies the steps necessary to conserve savanna/ woodland biodiversity.

Haney and Apfelbaum (1994) presented a classification system of midwestern oak savannas based upon moisture conditions and upon a series of hydrological and geographical gradients. This serves as a valuable ordination for savanna communities given their extensive range from Texas to Manitoba. Each description of a savanna type precedes a description of the consequences of controlling fire in these communities.

A woodland is an open forest: an environment still dominated by trees, but where the trees provide only between thirty and seventy percent cover. Because a woodland is an open community with ample light, it usually has many shrubs, forbs, and grasses in the understory. The pinyon–juniper woodland of the American Southwest and the Great Basin and the evergreen-oak woodland of the California chaparral are the most extensive North American woodlands. For additional information on the ecology of woodlands and savannahs, see Barbour (1988) and Barbour and Major (1977).

Site conditions

Soils

Woodlands may have occurred on many different soil types, including "wetland," "prairie," and "forest" soils. Curtis (1959) noted that savannas often occupied the border between prairie and forest. Dyksterhuis (1957) noted that savannas occurred on prairie or forest sites and were sometimes of anthropogenic origin. The soils of woodlands are often sandy. They are typically porous and permit free movement of water and air but have low capacity for holding either water or nutrients. Due to the good drainage, soil temperature fluctuates greatly. Soil development is minimal in these xeric, fire-maintained communities.

The typical soil of pine barrens is best described as dirty sand. It is a well-drained, nutrient-poor, podzol soil. The surface is covered with litter of pine needles, cones, and undecomposed oak leaves. The soil is too acidic (pH 4) for earthworms or bacteria.

Topography

Woodland communities develop under a variety of circumstances. Physical characteristics such as well-drained soils and/or presence of rock at or near the soil surface appear to be more important than topography in determining the community structure and composition. As a result these communities are dry and fire plays a significant role in reducing fuel loads and resetting plant succession. Topography is quite variable and ranges from flat or level to steep mountainside barrens.

Moisture and hydrology

In spite of adequate rainfall and a generally humid climate, these communities are subject to punishing drought because of the inability of sand to retain moisture. Excessive dryness also makes fire a constant presence. Most eastern woodlands have a mild winter climate, hot humid summers, and precipitation of ten to fifteen inches spread evenly throughout the year. Because of the high precipitation, many soils are nutrient poor (the frequent rains carry the nutrients below the reach of plant roots) and the water table is often high near the surface.

These communities do not hold moisture. Any water that moves into the community is from rain or surface runoff from upslope. Water is quickly used by plants adapted to living in these xeric communities. Water that is not used drains rapidly through the porous soils or over the rocky surface to adjacent areas that may be more mesic.

Oak savannas and woodlands grew across a broad range of hydrologic and moisture conditions, from dry to moist (Leach and Ross 1995). Most high quality remnants of oak savanna and woodlands are on dry, often sandy, nutrient-poor soils. Many former savannas on richer and more mesic soils were converted to agriculture or, in the absence of fire, were invaded and altered by fire-intolerant species (Curtis 1959, McCune and Cottam 1985).

Establishment

Site preparation

Restoration of woodlands typically occurs in existing woodlands, however, some woodlands have been created on previously cleared sites. Site preparation for the restoration of woodland communities will involve earthwork only if the site has been drastically altered and natural contours need to be restored. Usually site preparation involves specifications for the removal of unsuitable materials/debris. This may include the removal of some woody vegetation (living or dead) that has increased the fuel load. This will allow a fire regime to be restored. The control of invasive exotic species is also part of the site preparation. If seeding or planting is to be conducted, then the surface of planting areas will need to be prepared.

Planting considerations

The restoration of woodlands often does not require planting or requires only supplemental plantings such as the reintroduction of rare species. Most woodland restoration projects involve the improvement of ecological conditions so that natural potential vegetation will develop at the restoration site. Of course, management using prescribed burning is a key ingredient in woodland restoration.

Spacing considerations

Trees in woodland communities are often spaced so that there is no canopy overlap, in fact, there are often relatively large openings with just shrubby and herbaceous plants. Most savanna communities are classified by percent of canopy coverage, which is typically low (five to twenty percent). So, knowing the optimum percent coverage of the canopy can guide spacing schemes. Herbaceous plants must be established at suitable microhabitats within these woodlands. As successful establishment is somewhat experimental in communities with extreme conditions (i.e., xeric and full-insolation), some mortality should be expected. However, successfully established plants will colonize areas or disperse propagules to nearby "safe sites."

Temporal spacing for different plants and layers

Restoration, especially landscape-scale restoration, depends heavily upon natural succession. Planting in phases allows the restorationist to assess previous plantings and make necessary adjustments along the way. Reseeding and/or replanting may need to be done for three to five years until plants become well established.

Most woodland communities are relatively open so most plantings can be made at one time. Supplemental plantings may be needed to achieve the desired species diversity, but grasses are typically very successful after initial establishment.

Maintenance and Management

Weed control

See weed control section in **prairies** and **grasslands** restoration types.

Nutrient requirements

Most plant species of woodlands have low nutrient requirements and the sites where they occur are often considered poor quality because of their sandy, rocky nature. Restoration of severely degraded sites may involve some soil amendments or fertilizer applications if the soils have been depleted of certain elements. A comparison of soil tests between degraded and natural woodlands can help identify deficiencies.

Water requirements and hydrology

Woodland communities are typically xeric and have low water requirements. Watering may be helpful during initial establishment of woodland species, but is generally not needed as the plant species inhabiting these communities are well-adapted to dry conditions.

Water quickly cycles through woodland communities. This occurs by one of two mechanisms; water either percolates downward through the porous soils or is deferred to adjacent land as surface runoff. The flora of these woodlands use water efficiently as it is scarce here.

Disturbances

The juniper and juniper–pinyon woodlands of the intermountain region between the Sierra Nevada, Cascade, and Rocky Mountain ranges have undergone many changes. Open, sparse, savanna-like western juniper (*Juniperus occidentalis*) woodlands occurred on low-sagebrush flats with shallow soils, and on rocky ridges, where there was insufficient fuel to carry a fire (Miller 1995). During the last 100 years, however, western juniper has increased both in distribution and density throughout its range. Western juniper has invaded sagebrush-grassland, aspen groves, and riparian communities. Grazing promotes the spread of juniper by reducing fine fuels, thus decreasing fire frequencies and increasing the density of sagebrush, a nurse plant for young junipers (ibid.). Lack of fire is partly responsible for the increase in western juniper densities. Restoration of juniper woodlands to shrub steppe or aspen communities requires burning and/or tree cutting.

Frequent fires play a major role in maintaining woodland communities. The oaks and pines are generally more fire tolerant than the other trees and shrubs, and in many areas have just a ground layer of fire-tolerant grasses as an understory. Oaks that have adapted to frequently burned areas include post oak (*Quercus stellata*), blackjack oak (*Q. marilandica*), and bear oak (*Q. ilicifolia*). Pitch pine (*Pinus rigida*) and shortleaf pine (*P. echinata*) are able to sprout from the trunk after a fire.

Because savannas include trees and shrubs, restorationists must use methods of prescribed burning that are different from those used on prairies. Nagel (1996) suggested the following strategies and precautions:

- Firebreaks should be mowed, as they are in prairies, and then cleared of leaf litter and grass by using leaf blowers or rakes. Managers should avoid ninety degree corners to go around obstacles when designing a firebreak.
- Living or dead trees located near a fire line can also create problems if they catch fire. Nagel (1996) suggests cutting down any potentially troublesome dead snags or live, hollow trees prior to burning. Trees or branches that are already lying in or across the fire line should be removed or cut so that full width of the firebreak is maintained. In addition, logs on steep slopes can roll downslope across firebreaks as they continue to burn.
- Woody debris falling from standing or burnt-out trees poses a danger to crews and could fall into nearby roads or powerlines. Also, there is potential for smoke-related problems to occur even hours after the burn. More staff time is needed to mop up and monitor a savanna burn.
- Red cedars (*Juniperus virginiana*) are highly flammable and thickets of them are capable of carrying a crown fire. Nagel recommends keeping fire lines away from cedar thickets, if possible.

Once the desired canopy coverage is achieved, fire intervals can be lengthened. Pruka (1995) believes that the presence of certain light-dependent understory species indicates a recent closure of the canopy. He provides a list of indicator species for oak savannas and open oak woodlands in southern Wisconsin. He stated that wooded sites containing sufficient populations of these species have the highest potential for recovery if properly managed through the use of prescribed burns, mechanical canopy thinning, and other techniques.

Case studies

Missouri is currently carrying some of the largest oak ecosystem restoration projects anywhere, and may serve as a model for restorationists planning landscape-scale projects in other areas (Bader 1996). The Proceedings of the North American Conference on Savannas and Barrens (Fralish et al. 1994) contains several significant contributions to the knowledge, restoration, and management of woodland communities. For additional information on the restoration and management of woodlands and savannahs, see Anderson and Anderson (1995).

Desert Lands

Introduction

Desert lands are dry communities often referred to as arid or semiarid lands. The plants and animals that inhabit these extreme communities have many adaptations for survival. High temperatures, low rainfall, and low nutrient levels characterize these areas. Fire is not uncommon in communities where fuel accumulates and many species are fire tolerant or even require it for their survival. Without intervention, desert areas disturbed by human activities such as off-road vehicle recreation and mining may take decades or centuries to recover (Bainbridge and Virginia 1990).

Restoration of desert communities is usually quite a challenge. As Ross (1985) reported, "In the west we have been interested in plants that can resist harsh environments–radical changes in temperature, arid conditions, salinity. But with the new disturbances, there are other harsh conditions resulting from toxic and hazardous wastes, introduction into the surface environment of heavy metals, changes in salt concentrations and alkalinity, and changes in water-holding capacity of the substrate." These changes in the surface environment must also be evaluated before restoration begins.

Gelt (1993) stated that in Arizona the Sonoran saltbrush scrub is an ecosystem that is now considered rare. Researchers with the University of Arizona and the Desert Botanical Garden are studying methods to hasten the return of native plant species such as mesquite (*Prosopis* spp.), creosote-bush (*Larrea tridentata*), and saltbush (*Atriplex* spp.) to former farmlands. Findings indicate that building water catchments, applying coarse, woody mulch, and repeatedly planting a portion of a field every third year serves as an effective revegetation strategy that replicates natural succession in an abandoned field.

The introduction of cattle and disruption of burn cycles on the High Desert shrub-steppe of the Pacific Northwest brought about changes that no one clearly understands how to reverse. Grazing, for example, not only removes bunch-grasses, increasing the interval between fires, but also results in the breaking up of a cryptobiotic crust that forms on the soil surface. This crust is composed of lichens and cyanobacteria. Destruction of the cryptobiotic crust opens up the ecosystem to invasion by weeds–especially cheatgrass (*Bromus tectorum*). A new trajectory is established–one that inevitably leads to the loss of many native species (Jordan 1995b).

Over the past 120 years, the grasslands of the northern Chihuahuan Desert have been converted to shrublands dominated by species such as mesquite (*Prosopis* spp.). Unfortunately, attempts to restore these grasslands have often been unsuccessful despite substantial expenditures of money, labor, and fossil fuels (Fredrickson et al. 1996). For additional information on the ecology of desert lands, see Baars (1995), Bahre (1991), Benson (1969), Benson and Darrow (1981), and Hasting and Turner (1965).

Site conditions

Soils

Desert land soils are highly alkaline, may have high salt concentrations, and often have a layer of cementlike caliche at fairly shallow depth. The caliche makes growth difficult for most plants. Caliche, which is a form of calcium, can be broken down to liberate calcium to the soil by treating with iron sulfate or sulfur. If caliche is not present, gypsum applied to the soil surface and lightly worked in will provide calcium.

An often overlooked characteristic of arid and semiarid ecosystems is the presence of microbiotic (or cryptogamic) soil crusts, delicate sybioses of cyanobacteria, lichens, and mosses from a variety of taxa (Fleischner 1994). The essential role of these microbiotic crusts in nutrient cycling of arid ecosystems has been increasingly documented. Crusts perform the major share of nitrogen fixation in desert ecosystems (Rychert et al. 1978). The availability of nitrogen in the soil is a primary limiting factor on biomass production in deserts. In the Great Basin Desert, at least, it is second in importance only to the lack of moisture (James and Jurinak 1978). Microbiotic crusts in arid ecosystems have been correlated with increased organic matter and available phosphorus (Kleiner and Harper 1977) and increased soil stability (Kleiner and Harper 1972, Rychert et al. 1978). Crusts also play an important role in ecological succession because they provide favorable sites for the germination of vascular plants (St. Clair et al. 1984).

Desert plants are generally drought tolerant, have high oxygen demands, and are susceptible to many nursery pathogens. They commonly benefit from a rapid-draining, porous soil mix (Bainbridge

et al. 1995). Additionally, most desert plants form or require symbiotic associations with vesicular arbuscular mycorrhizal fungi (VAM). Inoculating plants with appropriate fungi may increase survival and growth of many species. The fungal association improves phosphorous, water, and nutrient uptake (Allen 1988, Allen 1992). Studies have shown that transplants that received VAM inoculations grew faster and outcompeted weedy annuals more effectively than uninoculated plants (Allen and Allen 1984). Inoculation adds another complication to the task of providing nutrients to young plants. The fungi require nitrogen, but may be inhibited by phosphorous, so it is important to provide these nutrients in ratios that optimize growth of seedlings and mycorrhizae (Hayman 1982). Mycorrhizal inoculants for restoration projects are becoming more widely available from commercial sources.

Some desert plant communities include mounds or dunes that may require stabilization. Tiszler et al. (1995) found that in building mesquite mounds, straw crimped into the sand with shovels or coir (coconut-husk fiber) netting covering the windward two-thirds of the mound were the most effective sand stabilizers. The cost of crimped straw was one-tenth the cost of coir netting.

Bainbridge (1996) found that placing straw, sticks, or brush upright in the soil–vertical mulching–can reduce erosion and aid in the revegetation of bare areas that result from construction, excessive vehicle traffic, or recreational activities. Placing these materials in an upright position can slow the movement of water and improve its penetration into the soil. They also provide safe sites for seeds to catch and sprout, serve as windbreaks, offer shade and cover for seedlings, and, as they decay, provide a source of below-ground organic matter to restore the soil (ibid.). In addition, vertical mulches are biodegradable and can be more effective than mechanical treatments. Bainbridge (1996) suggests experimenting with bundles of seed-laden native grass straw.

Saltbush shrub communities occur where the soil and available groundwater have high degrees of either salt or alkali. If the land has been previously used for agriculture or mining, the upper soil horizons may remain as impermeable sandy clays. Coarser textured sandy loams allow proper rainwater infiltration that favors natural and artificial revegetation (Jackson et al. 1991).

Topography

While most desert lands are relatively flat, some include areas of mounds or dunes. Microtopography is of utmost importance as it creates areas of suitable seed germination and establishment (safe sites). The microtopography of desert land communities must be considered in the restoration of desert land communities. Comparison with the nearest intact (or relatively intact) similar community will help the restorationist determine proper elevations and microtopographic relief to be used in the project.

Moisture and hydrology

Desert land communities are very dry communities and water is often a limiting factor that directly influences species composition and structure. As the soils are typically well drained, moisture must be quickly utilized and/or stored by plants that inhabit these xeric communities. Desert plants are well adapted to these conditions in natural areas, however, in degraded areas that are being restored, irrigation and supplemental watering can mean the difference between success and failure.

In many desert lands, precipitation is very low and irregular, temperatures are high, and strong drying winds are common. Such a climate severely limits plant growth, and here the desert vegetation of widely spaced shrubs and succulent cacti is found.

Where chaparral communities occur, the climate is generally mild in both winter and summer, and precipitation of twelve to forty inches occurs mainly as winter rain. Summers are hot and dry, and the plants are very tolerant of water stress.

In the cold desert, rainfall amounts range from five to twenty inches, mostly in winter. Summers are hot and winters moderately cold. Frost may occur in any month of the year. Rain is mainly as erratic winter rains in the Mojave Desert, winter and summer rains in the Sonoran Desert, and primarily as summer storms in the Chiuahuan Desert. In some areas, no appreciable rain may fall for two years. Winters are moderate with occasional frost, except for the Mojave, which has cold winters. Summers are very hot, with extremely high evaporation (and therefore, high water stress for plants).

Desert lands are well drained and water is generally scarce as soils are porous, solar radiation is intense, winds are strong, and annual rainfall is slight. Water cycles through these ecosystems very quickly and must be efficiently used by plants and animals inhabiting desert communities. Washes and arroyos develop where precipitation runs across the soil surface cutting into the desert floor.

Establishment

Site preparation

See **case studies** of this restoration type.

Tiszler et al. (1995) noted that local records suggest that large accretion dunes, which formed around mesquite (*Prosopis glandulosa* var. *torreyana*) trees, covered sizeable areas of the Imperial Valley (Yuha Desert, California) before it was developed for agricultural purposes. These "mesquite mounds" are not as abundant today.

Uresk and Yamamoto (1994) tested a variety of native and nonnative herbaceous and shrubby plants in amended (gypsum, saw dust, perlite, straw, and vermiculite) or untreated bentonite mine spoil. Fourwing saltbush (*Atriplex canescens*) had the greatest survival in all treatments and did the best in perlite-amended plots. Other species did very poorly in all treatments.

Zamora (1994) studied the ability of six native species of wild buckwheat (*Eriogonum* spp.) to revegetate degraded, semiarid sites in the northern intermountain region. When planted in locations environmentally similar to their native habitats, these plants have the capability to work with other early successional species in stabilizing disturbed soil areas. In particular, snow buckwheat (*Eriogonum niveum*) develops a deep taproot, which makes it useful in revegetating coarsely-grained mine deposits. Virginia (1990) provides a discussion of the role of woody legumes in desert restoration.

Planting considerations

Harsh desert climates, intense solar radiation, high temperatures, high winds, low rainfall, low soil fertility, and intense herbivore pressure can limit transplant success unless plants are carefully prepared and protected after planting. Newly planted seedlings are often the most succulent plants available, and rodents, rabbits, reptiles, domestic livestock, and insects find them especially attractive (Bainbridge et al. 1995). This herbivory can severely damage and even kill seedlings unless they are protected.

Many strategies have been developed to protect plants from environmental pressures. These include treeshelters, rock mulches, plant collars, and animal repellents. All of these may be effective when properly used and matched with site conditions and herbivore species. Their advantages and disadvantages should be carefully considered, and alternatives should be tested before they are applied on a larger scale. Bainbridge et al. (1995) provides a review of treeshelters, rock mulches, plant collars, and animal repellents.

Fredrikson et al. (1996) presented methods using both water and animals to deposit viable seeds into favorable microsites. They have developed gully seeders that disperse seeds during peak rainfall events. The seeders have a movable element—either a vane or a piece of wood–that is placed in the water channel. When sufficient water flows through the channel, the vane or piece of wood moves, releasing the stopper from the upturned bottle of seed.

Domestic and native animals have also been used to disperse seed. Barrow and Havstad (1992) fed steers gelatin capsules containing fourwing saltbush (*Atriplex canescens*), side-oats grama (*Bouteloua curtipendula*), stiff panicum (*Panicum hirsutum*), and alkali sacaton (*Sporobolus airoides*). Other researchers have noted that seedlings do emerge from the feces of cattle, sheep, deer, and rabbits in temperate pastures (Welch 1985). Fredrickson et al. (1996) believe that fourwing saltbush seed hoarded by native kangaroo rats will germinate and become established. In fact, germination rates could be quite high, because the seed is exposed to a fungal inoculum while being carried in the cheek pouches of the kangaroo rats (ibid.).

Karpiscak (1980) found that reestablishment of the formerly dominate saltbush (*Atriplex* spp.) and creosote-bush was highly variable in abandoned agricultural fields of southern Arizona. Only a few sites adjacent to undisturbed vegetation are recovering rapidly.

Conditions favorable for seed germination and seedling establishment are infrequent and unpredictable in the desert, making direct seeding an ineffective restoration strategy (Cox et al. 1982, Barbour 1968). Fortunately, many desert shrubs are easy to grow in a nursery and respond well to transplanting (Bainbridge et al. 1995). Many desert plants would benefit greatly from deeper (taller) containers than are commonly used (Bainbridge 1994, Felker et al. 1988, Holden 1992). Containers should be designed to encourage the seedling to form an extensive and vigorous root system that can be protected until the seedling is planted. Growth is commonly limited by water or nutrients. A large, active root system will increase uptake of water and nutrients. Bainbridge et al. (1995) provides an excellent review of the common container types used by restorationists.

Downs et al. (1995) found that herbicide treatments of perennial and annual forbs prior to early spring seeding will increase sagebrush (*Artemisia tridentata*) germination and establishment.

The size of the seed or cutting, growth rate, susceptibility to disease, temperature preferences, and desired outplanting size must all be considered. In general, larger seedlings will survive better than smaller ones, however, larger seedlings will cost more. Ortega-Rubio et al. (1995) found that members of the Agavaceae and Cactaceae families had the highest survivorship rate (between seventy and ninety-five percent), while the members of the Burseraceae, Euphorbiaceae, and Fouqueriaceae families showed the lowest (between twelve and one-half and twenty percent).

Aldon and Doria (1995) found that four-wing saltbush (*Atriplex canescens*) can be grown under saline conditions to reclaim farm land abandoned due to brackish groundwater intrusion. It is possible to germinate manually-scarified fourwing saltbush seeds at seventy degrees Fahrenheit and obtain forty percent germination with twenty-five percent saline water. Transplanted seedlings are vulnerable to saline conditions until they are seventy days old. Once established under saline conditions, no more than one-half of the shrub's ground height should be harvested annually or death may result. Natural precipitation should be relied upon for sustaining the plant in the long run.

Seed propagation of bitterbush (*Purshia tridentata*) is difficult. Kituku et al. (1995) found that stem propagation may be the most effective method. Cuttings from seven individual bitterbush populations across south-central Wyoming were collected in early summer, transplanted in the greenhouse, and subjected to winter conditions. After forty-five to sixty days, sixty percent of the cuttings had root growth of three-fourths of an inch or more, and roughly one-half of the plants resumed spring growth following winter simulation.

Spacing considerations

Woodward and Cornman (1996) randomly assigned shrubs to planting locations on a forty-seven by forty-seven inch grid on the restoration site. Spacing will need to be assessed on a site-specific basis. For most desert land plantings, spacing will depend upon suitable microhabitats for particular species. If plants are placed in unsuitable microhabitats, they may be too wet or too dry. Experimental plantings may be useful to determine appropriate planting zones. Comparisons with nearby desert land communities may prove useful in delineating zones of particular vegetation types.

Desert plants should be widely spaced to conserve water by allowing each plant room to spread its roots through a large volume of soil. Mulching with organic or stone mulch will also help conserve water by slowing evaporation from the soil.

Temporal spacing for different plants and layers

Survival rates can be improved if outplanting is properly timed. Honey mesquite (*Prosopis glandulosa*) survived best when planted in mid-spring to midsummer (Bainbridge et al. 1995). Planting in phases allows the restorationist to assess previous plantings and make necessary adjustments along the way. Reseeding and or replanting may need to be done for three to five years to ensure development of desired vegetation types. Restoration, especially landscape-scale restoration, typically depends heavily on natural succession.

Maintenance and Management

Weed control

Cheatgrass (*Bromus tectorum*) is an invasive weed in desert land communities. It is extremely difficult to control. Unfortunately, it thrives on fire. Prescribed burning is not an effective control mechanism for cheatgrass.

The spread of German ivy (*Senecio mikanioides*) into new areas, including San Diego, now makes it a primary concern in coastal areas (Sigg 1994). German ivy threatens the existence of coastal canyon vegetation and is nearly impossible to eradicate once established.

California coastal vegetation is also threatened by the succulent perennials, sea fig (*Carpobrotus chilensis*) and the hottentot fig (*C. edulis*). Bowler (1996) found that dedicated hand pulling the plants along with their root and vegetative suckers can control these invasive sea figs. Native vascular plant species gradually recolonize areas cleared of sea fig. Sea figs or other iceplants can be effectively controlled by mechanical means.

Jackson et al. (1991) found that abandoned farmlands in the Sonoran Desert lowlands of Arizona have become dominated by the short-lived shrub burroweed (*Isocoma tenuisecta*). In all their experimental plots, uneven weed infestations of narrow-leaf goosefoot (*Chenopodium desiccatum*), filaree (*Erodium cicutarium*), and London rocket (*Sisymbrium imbrio*) occurred. Russian thistle or tumbleweed (*Salsola kali*) was also a problematic weed.

Nutrient requirements

As the availability of nitrogen in the soil is a primary limiting factor on biomass production in deserts, any deficits must be taken into consideration. Fertilizer may need to be used for initial establishment of desert plants, however, it must be used moderately or applied systematically as fertilizer may encourage unwanted weeds. Another problem with adding nutrients is that high nutrient levels may inhibit root growth and contribute to a low root-to-shoot ratio. Low root-to-shoot ratio is detrimental to plant survival. Nutrients can be quickly leached from desert soils by excessive watering or heavy rains and may necessitate extra fertilization. Microbiotic crusts play an important role in nutrient cycling of arid ecosystems (see **Soils**).

Water requirements and hydrology

Desert plants are generally drought tolerant (up to 50–60 bars). Survival of desert plants on arid sites depends upon root growth potential and the ability of the root system to gain access to soil moisture and generate new roots. In disturbed arid sites, growth is commonly limited by water or nutrients. Watering will likely be required for initial establishment of desert vegetation. Seedlings may need a prolonged period of moist conditions in order to withstand dry periods.

Woodward and Cornman (1996) installed a drip irrigation system to provide supplemental water during establishment of shrubs in the restoration of a blackbrush community in Texas. Only 12 of the

112 shrubs they planted died even though south Texas had experienced a major drought during the summer. The authors concluded that restoration of the blackbrush community is possible especially when a drip irrigation system can be employed.Yamashita and Manning (1995) found that saltbush survival was affected most significantly by irrigation, although weed control and fertilizing are also having positive effects.

Particles of superabsorbent polyacrylamide polymers can be useful when planting. Polyacrylimide is a type of plastic that repeatedly absorbs and releases moisture and does not biodegrade. The polymers retain water that would normally drain from the soil, eliminating fluctuations in moisture level. Polyacrylamide has much potential in the restoration of degraded lands where the soil has lost some of its natural water-retaining capabilities. The particles or dry crystals should be mixed with the soil in the area to be planted. Be sure you purchase polyacrylamides and not polyacrylates. Polyacrylates only absorb and do not release liquid.

Water is rapidly cycled through desert ecosystems and hydrology is not typically a concern of desert restoration efforts. However, if the site has been severely altered, grading to the original topography will probably restore the flow of surface water.

Disturbances

Wilson et al. (1995) studied the postfire regeneration of shrubs associated with the giant saguaro cactus (*Carnegiea gigantea*). They theorize that the long-term health of the cactus depends upon the restoration or regeneration of these woody species. Their analysis of basal and branch resprouting supports the idea that some shrubs do not regenerate following spring fires. For the most part, desert land communities are significantly influenced by periodic fire and have evolved to accomodate its occurrence.

Case studies

While building mesquite mounds in the Yuha Desert (California), Tiszler et al. (1995) planted eight, seven-month-old, ten-inch tall mesquite (*Prosopis glandulosa* var. *torreyana*) seedlings on each mound (five of which were protected with tree shelters of varying heights). Water-retaining gypsum soil moisture blocks were placed beneath the sand dunes. The mesquite responded well to outplanting, with some individuals more than doubling in height (from ten to twenty-two inches) in six weeks. The plants appeared to grow equally well within and outside tree shelters, although initially the tallest plants grew in tall (twenty-four inches) shelters and the more highly branched plants were found either in short (twelve inches) shelters or in the open. While no seedlings died, several had to be replanted when their roots became exposed.

Jackson et al. (1991) presented the results of the restoration of abandoned farmland in Pinal County, Arizona. They recognized that recovery to a predisturbance condition, whether by natural recovery or active restoration, may be impossible. Tillage had mixed formerly distinct soil horizons, changing the dynamics of water infiltration and availability, thus reducing the already minuscule chances for seedlings to become established. The leveling of hundreds of contiguous square miles has transformed the area from a network of dendritic ephemeral streams into a cartesian grid system of roads and ditches. The nearby Santa Cruz River has been diverted and channelized to prevent it from flooding the valley. As a result, it no longer recharges the shallow water tables that some plants depended upon. Critical to any restoration plan for this area was an accurate assessment not only of its historical plant and animal communities, but also of its current potential to support them. Historical aerial photos taken between 1936 and 1979 in combination with an intimate knowledge of the soils and vegetation of the site allowed them to piece together its history. They found fragments of ironwood (*Olneya teosota*) and a large (three-foot diameter) trunk of blue paloverde (*Cercidium floridum*) on the study site, suggesting greater water availability in the past. Imperfect leveling of older fields meant that some microtopography was still in place to provide safe sites for new seedlings. One portion of the site that still received some channel runoff from the Santa Cruz was found to be recovering rapidly with mesquite (*Prosopis velutina*), desert saltbush, wolfberry (*Lycium* spp.), and globemallow (*Sphaeralcia* spp.)—all native to the area—now abundant. The goal of Jackson et al. (1991) was to find out whether it is ecologically feasible to establish perennial shrubs on a typical abandoned field that showed no signs of recovery. They chose two barren areas (less than one-tenth percent cover) within the 640-acre study site and planted seeds of the native species found growing in nearby desert scrub. Jackson et al. (1991) used simple water catchments to concentrate rainfall, enhance water infiltration through tillage, and modify the environment around the seed and seedling. They hoped to ameliorate the adverse conditions preventing natural seed establishment. They experimented with various combinations of straw mulch and living mulches of weeds or annual grasses on a series of plots seeded with desert saltbush, creosote-bush, mesquite, and wolfberry. Seeds of the perennial creosote-bush, desert salt-bush, linear-leaved saltbush (*Atriplex canescens* var. *linearis*), mesquite, wolfberry, and globemallow were used in the revegetation seed mix. All of the seed for the study was donated by local native seed companies or collected by Desert Botanical Garden volunteers. Summer rains were better than average in 1990, producing good crops of seeds from warm-season ephemerals such as six-weeks grama (*Bouteloua barbata*) and three-awn grama (*B. aristoides*). The grass seed was collected, in some cases, by sweeping the seeds off the desert floor with brooms and dustpans before it could be taken by harvester ants.

Jackson et al. (1991) found that natural reveg-etation was more rapid and yielded greater cover-ages where the upper one to twelve inches of the soil profiles contained coarser-textured horizons (sandy loams). Rainwater can infiltrate sandy loams more quickly than the relatively impermeable sandy clays found in the barren sites. They suspect that their experimental barren sites have remained nearly free of vegetation due to their finer-textured, less perme-able soils. Because summer rains are less predict-able than winter rains and are accompanied by high-evaporation conditions, they planted during the win-ter rainy season in early January. To increase the amount of water available to the plantings, they used rainwater running off the barren, impermeable soil surface to supplement their experimental plots where its flow was arrested by low berms arranged on the contour. Then, using a border disk commonly used for making cotton rows, they created two rows of nine parallel berms separated by water catchment areas fifty feet wide. Each berm was 100 feet long and sixteen inches high. They used a stiff-tined "ripper" to create an eight foot wide seed bed (the actual experimental plot) on the uphill side of each berm. The ripper disturbed the soil to a depth of four to six inches and left large (two to four inch diam-eter) clods on the surface. At each site, the seed bed next to each berm was broadcast seeded with a mixture of perennial shrubs. Half of each plot was designated at random and seeded with Mediterra-nean grass (*Schismus barbata*), six-weeks grama, three-awn grama, and Indian wheat (*Plantago insularis*), all common ephemerals of the lowland deserts. Mediterranean grass, the only exotic spe-cies, was planted because it is completely natural-ized to the area and was the only available cool-season grass. Each combination of mulch and weed-ing treatments (four in all) was randomly assigned to four replicate berms at each site.

Jackson et al. (1991) concluded that the water catchments worked well. Overall, seeds of desert saltbush germinated and survived best in the mulched, weeded plots. The number of saltbush seedlings in all treatments was substantially higher than it would be in mature stands. Annual grasses did not come up in sufficient quantity to have any effect on either weed suppression or desert salt brush establishment. The primary concept in the project was by getting any sort of plants established, even weeds, that would eventually improve the chances for natural recovery by increasing soil organic matter and water holding capacity. They found that while weeding may not be necessary, mulch greatly improves the survival of plantings. Major changes in soil proper-ties and regional hydrology made natural plant es-tablishment all but impossible. Even in undisturbed deserts, seedling establishment occurs only infre-quently, when there is unusually plentiful rainfall. Not only is moisture necessary for immediate ger-mination and growth, but seedlings must grow large enough to tolerate the drier, normal conditions (Went 1948; Jordan and Nobel 1979, 1981; West et al. 1979). It is likely that any restoration in deserts relying on direct seeding and rainfall will fail in most years (Jackson et al. 1991). Water catchment systems and mulching will increase the chances for seedling establishment. Repeatedly planting a frac-tion of the restoration site until a good rainfall occurs is a good strategy in desert land restoration. Depend-ing upon the life of the seeds in the soil, an area might need to be reseeded every third year or so. Seeding, mulching, and catchment techniques would all need to be tailored to such a long-term approach (Jackson et al. 1991). This approach mimics the process of natural succession.

For additional information on the restoration and management of desert land, see Allen (1988), Aronson et al. (1993a,b), and Bainbridge (1990).

Shrublands

Salt Shrub Thicket A-338
Scrub A-352

Introduction

Some of the factors that promote shrubland development, as opposed to woodlands or grasslands, at a given site include sandy soils, hydric conditions, droughty conditions, fire, windthrow, and low nutrient levels. It is usually a combination of some of these factors that leads to the development of shrubland community types. Native shrubs have been used to hasten natural plant succession on disturbed sites (Dehgan et al. 1987, Martin and Sick 1995, and Slagle 1992) and have been highly recommended for this purpose by Foote and Jones (1989).

Salt shrub thickets are low, evergreen shrub associations that develop just upland of high salt marsh (see **estuarine marsh** restoration type). They also occur on the sound side of barrier islands (see **sand dunes** restoration type). The environment where salt shrub thickets occur is usually rather windy, possesses high salinity, and has infertile soils.

Scrub is generally a shrub community dominated by a variety of evergreen oak species and sand pine. Although the herbaceous layer is usually very sparse, the midstory layer can form a dense thicket. Today, unfortunately, scrub communities are disappearing at an alarming rate. Some have been converted to citrus groves, but most are falling prey to housing developments, golf courses, and urban encroachment in general (Myers 1990).

Site Conditions

Soils

The soils of salt shrub thickets have a high sand content. They are sometimes moist but may be well drained. Some peat is usually present. The soils of salt shrub thickets are relatively infertile and have a high salt content.

Scrub soils are sandy, infertile, and well drained. Practically all scrub soils are entisols (soils with little or no horizon development) that fall under the classification of Quartzipsamments, which are entisols derived from quartz sand (Myers 1990). Soils supporting scrub are excessively well drained, siliceous sands practically devoid of silt, clay, and organic matter and thus low in nutrients.

Topography

The terrain of scrub and salt shrub thickets ranges from relatively flat to rolling and sometimes includes dunes or swales.

Moisture and hydrology

Although scrub soils are excessively well drained, drought stress may not be a common occurrence (Myers 1990). Even though the majority of fine roots of scrub species are shallow (probably to facilitate nutrient capture), these species also have deep "sinker" roots that tap soil moisture at considerable depths.

Salt shrub thickets receive water input from precipitation and as salt spray from the ocean. They are typically well drained but often not far above the water table.

Establishment

Planting considerations

Many landscape restoration projects "fail" due to improper seedbed preparation, poor seeding/plant techniques, inappropriate vegetation selection, or poor timing. Before planting begins, several important considerations must be addressed: (1) which species to use, (2) type of plant materials to use (seeds, seedlings, saplings, cuttings, container stock), (3) source of plant materials, (4) density of plantings (spacing pattern), and (5) plant protection (screening, fencing, stakes, repellent). Once these factors are addressed, revegetation efforts can begin. Planting should not be a one-time event. Supplemental planting for the first three to five seasons can help ensure survival of plants to maturity and establishment of desired vegetation type.

New plantings may survive better if they are mulched with wood chips or other organic materials, but keep the mulch away from the stems of woody plants as they are sensitive to soil around their trunks. Some native shrubland species, especially annual wildflowers, can be easily started from seed. The best sowing time is in fall or summer, just before the rains are expected.

Clewell and Poppleton (in Robertson 1983) reported successful establishment of sand pine scrub communities using mulch transferred from natural woodlands. Attempts to "seed" whole communities by transferring mulch or topsoil or both from intact communities to disturbed sites appears to be quite effective.

Older false-willow (*Baccharis neglecta*) plants are difficult to transplant because they have long, woody, and brittle taproots that are hard to remove intact from the rocky site where they frequently grow (Nokes 1986). Smaller plants, less than three feet, may be moved if cut back and kept watered.

Nokes (1986) recommended the following methods for propagating the false-willow. The light, feathery seeds are dispersed by wind and should be collected by hand in the fall when they are no longer green, but before they completely dry out and blow away. Spread the achenes on a screen or table to air-dry a few days before storage. Seeds may be stored for at least two years in sealed containers held at thirty-six to forty degrees Fahrenheit. If held at an average temperature of seventy degrees Fahrenheit, length of viability is reduced to one year (Vories 1981). Seeds may be sown immediately after collection in a greenhouse or cold frame or stored over winter and planted outdoors in the spring after all danger of frost is past. The seed flat must contain light, well-drained soil, and the seeds should be lightly covered. Fresh seeds will germinate in seven to fifteen days. Some species of *Baccharis* can be propagated by cuttings that are taken from the current season's growth and placed under a mist.

Southern wax-myrtle (*Myrica cerifera*) is important for reforestation purposes because of its nitrogen-fixing root nodules, which contain actinomycetes that work to improve the soil. Southern wax-myrtle plants up to four feet have been successfully bare-root transplanted in the winter (Nokes 1986). Cut back the tops by two-thirds to a maximum height of two feet. Trim the roots and then place the plant into a container. Transplanting is more successful on cool damp days and if the roots are treated with a root stimulator like IBA. Balled and burlapped plants may be transplanted in the summer if placed in a shady location, pruned slightly, and kept moist. Pruning to remove dead limbs or promote bushy growth should be done in spring.

Nokes (1986) recommended the following methods for propagating southern wax-myrtle. Collect the fruit in the fall when it turns bluish green. The fruit can be beaten from the bush onto a drop cloth or collected by hand with clippers. The waxy covering should be left on the seeds to be stored over winter, but removed before sowing or stratification. Soaking the seeds in a dilute mixture of lye and warm water (one tenth of a fluid ounce to three and one-half quarts water) is an effective way to dissolve the coating. Store seeds in sealed containers in the refrigerator. Germination of southern wax-myrtle is delayed by a dormant embryo. Pretreat seeds by sowing outdoors in the fall or stratifying in moist peat for sixty to ninety days at thirty-four to forty-one degrees Fahrenheit. Natural germination occurs in the spring following seedfall. Outdoor sowing should be delayed until late fall or early winter to avoid frost damage to early germinating seedlings. Transplant seedlings from the flat when they are two inches tall. Mulch outdoor seedbeds over winter. Southern wax-myrtle will root from softwood and semihardwood cuttings two to three inches long. Some growers find that a healed cutting roots best. Treat the cutting with IBA and place under mist. After rooting occurs, reduce the frequency of misting and harden-off. Fall-rooted cuttings should remain in a coldframe over winter.

Spacing considerations

For many shrubland plantings, spacing will depend upon suitable microhabitats for particular species. If plants are placed in unsuitable microhabitats, the habitat may be too wet or too dry. Experimental plantings may be useful to determine appropriate planting zones. Comparisons with nearby natural scrub communities may prove useful in delineating zones of particular vegetation types. Plantings need to be spaced appropriately in order to effectively control erosion and provide the desired amount of plant cover. Exact spacing must be determined on a site-specific basis.

Temporal spacing for different plants and layers

Restoration (especially landscape-level restoration) typically depends heavily upon natural succession. Planting in phases allows the restorationist to assess previous plantings and make necessary adjustments along the way. Reseeding and/or replanting may need to be done for a few seasons following initial planting to ensure development of desired vegetation types.

Many factors affect revegetation, whether it is natural succession or restoration. Most likely it will be some of both. Shrubland species can be planted all at once. However, some of the desired species may not be initially successful because of source of plant material, timing of planting, herbivory, parasites, nutrient, or water requirements.

Maintenance and Management

Nutrient requirements

Nutrients in salt shrub thickets are primarily provided by salt or brackish water as well as salt spray. The evergreen nature of many scrub species is assumed to be an adaptation that helps retain nutrients (Monk 1966).

The needs of most shrubland plants will be supplied by an organic mulch, which slowly feeds nutrients to the soil. Adding a time-released fertilizer tablet to the planting hole will also help supply nutrients required for establishment.

Water requirements and hydrology

The greatest effect of droughty conditions on scrub communities is probably the limitation of regeneration by seed. Extended dry periods following fire have been shown to limit sand pine establishment. Supplemental watering will likely increase survival rates of plantings or germination of seeds. After plants are established, the necessary water will come from precipitation, sea spray, and low lying fog.

Particles of superabsorbent polyacrylamide polymers can be useful when planting. Polyacrylimide is a type of plastic that repeatedly absorbs and releases moisture and does not biodegrade. The polymers retain water that would normally drain from the soil, eliminating fluctuations in moisture level. Polyacrylamide has much potential in the restoration of degraded lands where the soil has lost some of its natural water retaining capabilities. The particles or dry crystals should be mixed with the soil in the area to be planted. Be sure you purchase polyacrylamides and not polyacrylates. Polyacrylates only absorb and do not release liquid.

Disturbances

Because scrub burns, it is considered a *pyrogenic* ecosystem—one maintained by high intensity, infrequent fires, loosely termed *catastrophic* or *stand-replacing* (Myers 1990). Fire in scrub follows a reasonably predictable sequence of events. After a lengthy fire-free period of fuel accumulation, an intense fire occurs, facilitated by severe burning conditions. The slow accumulation of fuel reduces the likelihood of a reburn for several decades (ibid.).

Because of fragmentation and isolation from other ecosystems, most scrubs are no longer self-maintaining (Myers 1990). Fires no longer sweep across the landscape as they once did, and these pyric ecosystems must now be burned using prescribed fire. The fires need to be applied in such a manner that various stages of development are maintained within isolated fragments; otherwise, species with special habitat requirements would be adversely affected and perhaps eliminated (ibid.)

Windthrow may be as important as fire in maintaining scrub, especially in coastal areas. Other disturbances in scrub that may create "open" sites include blowouts in coastal scrubs, pits and mounds created by tree falls, clusters of dead trees caused by lightning strikes, and burrows and mounds created by animals, particularly the gopher tortoise (*Gopherus polyphemus*).

Mechanical methods such as broad-scale cutting, scraping, and chopping also can be used to maintain scrub habitat. These methods may mimic the natural regeneration process in scrub.

Case studies

Researchers at Kennedy Space Center/Merritt Island National Wildlife Refuge are using combinations of mechanical treatment and burning to restore unburned scrub (Schmalzer 1994). In stands dominated by myrtle oak (*Quercus myrtifolia*), sand live oak (*Q. geminata*), Chapman's oak (*Q. chapmanii*), saw palmetto (*Serenoa repens*), and ericaceous shrubs such as staggerbush (*Lyonia ferruginea*), the shrubs resprout vigorously when they are cut and then burned. See also Schmalzer and Hinkle (1987).

Meadowlands

Mountain Meadow A-27, A-44, A-86
Meadow Steppe A-53
Dry Mountain Meadow A-108
Wet Mountain Meadow A-109
Wet Meadow A-174, A-194, A-278
Northern Wet Meadow A-261

Introduction

Meadowlands are communities that, for various reasons, have practically no woody trees or shrubs. Meadows may be created by conditions too wet to support trees or may be the result of avalanche or fire activity. A shallow water table is the most important factor in explaining the occurrence and distribution of wet meadows. Meadows also occur under drier conditions, the latter even in gravelly soils. Environmental conditions such as high summer temperatures, intensity of solar radiation, and rocky soils prohibit the growth of trees in dry meadows. Wet meadows fall on the line between wetland and terrestrial systems. Many are true wetlands, however, others are not. Trees and shrubs do not readily invade these communities due to wet soils, rocky soils, elevation, and climate. Meadows are dominated by perennial herbaceous plants. The sedge and grass families are important in many meadows, but a great diversity of forbs and wildflowers also abound here. Disturbance such as heavy livestock grazing or draining severely alters the species composition and structure of these communities. Natural disturbances such as fire, wind, and herbivores have a role in maintenance of wet meadows.

So-called "meadow" plantings or landscape designs have become common in recent years. Many of these resemble natural meadows but may contain many nonnative plants. The concept of a meadow community (whether typical meadow, prairie, grassland, or a mix of exotic and native wildflowers) can seriously reduce maintenance costs as they only need mowing once a year. There is great potential for their incorporation into roadside areas, powerline right-of-ways, parks, golf courses, and corporate landscapes. They also provide habitat for many species of birds, butterflies, other insects, and small mammals.

Site Conditions

Soils

The soils of dry meadows are mostly entisols. The entisols have little soil profile development, and their traits reflect the character of the parent material more than that of the macroenvironment. Some are composed of soils that have developed on glacial outwash depsoits.

Soils of wet meadowlands vary considerably from poorly drained hydric to dry, gravelly soils. Soils of the Southern Great Lakes wet meadows (A-278) are wet mineral, muck, or shallow peat. Typically, wet meadow soils consist of two soil orders: mollisols and inceptisols. Mollisols are base-rich, fine-textured soils characteristic of grasslands. Inceptisols are soils of recent origin.

Where soils have been eroded or washed away, imported soil may be required to establish the desired community. In these degraded sites, a nurse crop (of noninvasive legumes and/or grasses that hold soil together) will improve the site for future plantings.

Topography

Meadows generally occur in broad and relatively flat bottoms of valleys. They also occur on slopes and other gradients where suitable conditions prevail.

Microrelief affects meadow vegetation markedly, because (a) it will determine how late into the summer snow will lie on the ground, first covering vegetation, then supplying it with water; (b) it will determine the degree of protection that the plants will receive from ice-blasting winter winds; and (c) it will have a bearing on the depth of the talik (topsoil), and hence on the amount of root space and the adequacy of drainage (Barbour et al. 1987). Ridge tops support very little plant cover because winter winds remove protecting snow cover. Without the snow, abrasive ice crystals carried by the wind can prune plants back to the ground or, at least, winter winds can dessicate twig tissue. In summer, ridge soils dry rapidly. In contrast, north-facing slopes may accumulate snow drifts that remain well into summer, providing melt water to wet meadows.

Moisture and hydrology

Meadowlands are quite variable as far as moisture and hydrology. As Cooper (1990) found in Rocky Mountain National Park, Colorado, the water table is highest in May and lowest in fall and early winter. The growing season usually lasts only three months (late May to late August), and water tables drastically fluctuate among years depending upon the depth of the snowpack. Thus, plants and soils have developed under a hydrological regime involving large annual variability.

Meadows receive water as precipitation (surface runoff) and wet meadows often have shallow water tables (subsurface runoff). Overbank flooding, in meadows adjacent to streams, is also an important source of water.

Establishment

Site preparation

Meadows occurring on slopes are more prone to erosion problems, especially if they have been grazed. Whitall (1995) found that check dams constructed of concrete-filled burlap bags and aprons of loosely-piled rock reduced erosion in the largest gullies. For smaller gullies and rills, a series of wooden energy-dissipators were made from abandoned fencing on-site and installed in the gully soil.

Planting considerations

Blankets and cylinders of coir (processed fiber of coconut tree) have been used with a great deal of success to hold plants in place in wetland restoration and streambank stabilization projects. Other benefits of this material are the high tensile strength when wet, durability (up to ten years), and it is 100 percent organic matter. Whitall (1995) tested three types of mulch (straw, bark chips, and black silvicultural matting) for revegetating a degraded meadow slope. She found that bark mulch was the most effective.

Seeds require suitable sites in the soil/seedbed and weather conditions conducive to germination. Seeding is generally the best method of restoring meadows, although some species that are more difficult to establish from seed may need to be planted as plugs or container stock. In many meadows the planting holes can be made with the appropriate tool (such as a dibble bar or hoedad). In some meadows, however, the soil is rocky and suitable microhabitats (i.e., pockets of accumulated soil) must be found when installing plants. Natural revegetation may yield some desired vegetation if a seedbank is present.

Spacing considerations

For most meadow plantings, spacing will depend upon suitable microhabitats for particular species. If plants are placed in unsuitable microhabitats, they may be too wet or too dry. Experimental plantings may be useful to determine appropriate planting zones. Comparisons with nearby natural meadows may prove useful in delineating zones of particular vegetation types.

Temporal spacing for different plants and layers

In degraded dry meadows, initially establishing the more vigorous species such as grasses (e.g., *Deschampsia cespitosa*) or shrubs can be advantageous. If the reasons for degradation have been removed and these more vigorous species become established, then other herbaceous plants can be gradually introduced. By revegetating the area in phases, the chances of success are greatly improved. Reseeding and/or replanting may need to be done for three to five years to ensure development of desired vegetation types.

Some of the desired species may not initially be successful because of source material, planting time, herbivory, parasites, amount of light, nutrient requirements, and/or water requirements. Unsuccessful plantings must be assessed and attempted at a later date, preferably once the reason for initial failure has been determined.

Maintenance and Management

Weed control

In some meadow steppe (A-53) communities,

grazing and fire have impacted community composition and structure. In these areas exotics such as Kentucky bluegrass (*Poa pratensis*) may dominate. Past and present land uses influence the plant species composition and density.

Nutrient requirements

Past land uses greatly influence present nutrient levels. Nutrients present in the soil should be assessed and fertilizer or other soil amendments may need to be applied before revegetation begins. Berendse et al. (1992) found that nutrients may need to be exported through hay or sod removal to restore meadows that were previously used as pasture.

Water requirements and hydrology

Once plantings have become well established, watering will not be required, however, seeds and seedlings may need irrigation or supplemental watering to grow and develop to maturity. Particles of superabsorbent polyacrylamide polymers can be useful when planting. Polyacrylimide is a type of plastic that repeatedly absorbs and releases moisture and does not biodegrade. The polymers retain water that would normally drain from the soil, eliminating fluctuations in moisture level. Polyacrylamide has much potential in the restoration of degraded lands where the soil has lost some of its natural water retaining capabilities. The particles or dry crystals should be mixed with the soil in the area to be planted. Be sure you purchase polyacrylamides and not polyacrylates. Polyacrylates only absorb liquid and do not release it.

Most dry meadows are well drained and water flows rapidly through the system. Restoring the natural water flow patterns (input) and drainage (output) may be difficult in degraded meadows. Success of restoration, however, depends upon reestablishing the hydrologic regime to the extent possible given land-use changes within the area.

Disturbances

Fire and grazing have played a role in the development of dry meadows. Both may be important, however, overgrazing and improperly timed burning have caused degradation of some meadows. Properly managed grazing regimes may even be beneficial to seed germination as seeds are compressed into the soil improving the chances of germination and subsequent establishment.

Mooberry (1984) found that in Pennsylvania mowing around mid-June (or as close as possible depending on how wet the soil is) can effectively control woody plant invasion. Poison ivy (*Toxicodendron radicans*), multiflora rose (*Rosa multiflora*), honeysuckle (*Lonicera japonica*), and oriental bittersweet (*Celastrus orbiculatus*) can also be controlled, though not eliminated, by annual mowing.

Case studies

For additional information on the restoration and management of meadowlands, see Murn (1993) and Schuster and Zuck (1986).

Glades

Glade A-312

Introduction

Glades are naturally open, grass- and/or forb-dominated areas where bedrock is at or near the surface of the ground. See also the Woodlands Ecological Restoration Type.

Site Conditions

Soils

The soils of glades are thin and often have a large amount of rock and gravel. Areas within glades may be only bedrock at the surface and plants may become established in crevices where soil has been deposited. The glade community (A-312) discussed here has soils with a neutral or slightly alkaline pH. Glade communities are often distinguished by bedrock type (i.e., sandstone, limestone, dolomite, shale, serpentine, etc.). At the edges of glades where the soil gets deeper, prairie or woodland communities may be present.

Topography

The terrain of glades may be relatively steep or very level; it depends upon the landscape in which it developed. Conditions conducive to glade development are not limited by topography.

Moisture and hydrology

Glade communities are very dry. They are usually dry and droughty during the growing season, but may be locally saturated in the spring.

Different substrates have various storage capacities for ground water. Some soils, like gravel and sand, have a much greater storage capacity than loam, clay, and other fine-grained soils. Likewise, limestone, dolomite, and sandstone hold more water than other kinds of less permeable bedrock such as metamorphic or igneous rocks. Within the bedrock, water is stored in cracks and dissoved spaces.

Establishment

Site preparation

Glades are easiest to restore in areas where there are existing degraded glades. However, the glade community may also be created by scraping the upper layers of soil off of bedrock. In fact, some disturbed/degraded areas have plant species typical of glades due to soil removal.

Planting considerations

Before planting begins, several important considerations must be addressed: (1) which species to use, (2) type of plant materials to use (seeds, seedlings, saplings, cuttings, container stock), (3) source of plant materials, (4) density of plantings (spacing pattern), and (5) plant protection (screening, fencing, stakes, repellent). Once these factors are addressed, revegetation efforts can begin. Planting should not be a one-time event. Supplemental planting for the first three to five seasons can help ensure survival of plants to maturity and establishment of desired vegetation type.

Seeds require suitable sites in soil or seedbeds. Also weather conditions must be appropriate for germination and subsequent establishment. Natural revegetation may yield some desired species provided the seedbank is intact.

Spacing considerations

For most glade plantings, spacing will depend upon suitable microhabitats for particular species. If plants are placed in unsuitable microhabitats, they may be too wet or too dry. Experimental plantings may be useful to determine appropriate planting zones. Comparisons with nearby natural glades may prove useful in delineating zones of particular vegetation types.

Temporal spacing for different plants and layers

Restoration (especially landscape-level restoration) typically depends heavily upon natural succession. Planting in phases allows the restorationist to assess previous plantings and make necessary adjustments along the way. Reseeding and/or replanting may need to be done for a few seasons following initial planting to ensure development of desired vegetation types.

Maintenance and Management

Weed control

Queen Anne's lace (*Daucus carota*), oxeye daisy (*Chysanthemum leucanthemum*), spotted knapweed (*Centaurea maculosa*), and thistles (*Cirsium* spp. and *Carduus* spp.) are common problematic weeds of glades. Manual removal is most appropriate in glade communities. However, a combination of methods may be more effective. See the weed control section in the **prairie** restoration type.

Nutrient requirements

Obligate glade species compete best in nutrient-poor, xeric soils, and resist colonization by plants with higher nutrient needs by further reducing available nitrogen levels (Tilman 1987) and inhibiting nitrogen-fixing bacteria and algae (Rice 1972, 1974). By starting with low-nutrient soils and planting species that further reduce levels of available nutri-ents, the restorationist would be, in effect, accomplishing what Luken (1990) refers to as the resources-depletion method for maintaining a plant community in an early successional stage.

Water requirements and hydrology

Supplemental watering (during dry periods) of plantings for the first three to five seasons, or until they are well established and reproducing naturally, will greatly improve survival and establishment of desired vegetation. The community should, however, be monitored for weeds during the irrigation stages as greater moisture availability may trigger weed outbreaks. Once plantings have become well established and are reproducing, supplemental watering is no longer needed.

Disturbances

Fire plays an important role in keeping glade communities open by controlling the invasion of woody species such as eastern red-cedar (*Juniperis virginiana*), redbud (*Cercis canadensis*), oaks (*Quercus* spp.), and elms (*Ulmus* spp.).

Grasslands

Coastal Mountain Grassland A-16
California Prairie A-41
Palouse Prairie A-53
Wheat grass–Blue grass Prairie A-108
Galleta–Three-Awn Shrubsteppe A-117
Grama–Tobosa Shrubsteppe A-118
Grama–Galleta Steppe A-119
Desert Grassland A-136, A-146, A-159

Introduction

The terms prairies and grasslands are used interchangeably by many ecologists. The ecological communities known as California prairie, Palouse prairie, and wheat grass-blue grass prairie are listed here as grasslands (even though there are many similarities with prairies). The term prairies is reserved for the short, mixed, and tallgrass prairies and their derivatives to the east (see **Prairies** section). Grasslands are generally found in areas where there is not enough rain to support trees. Grasslands are characterized by dominant grasses, although many forbs are also present and may seem dominant when in bloom. Shrubs and trees may also be found in grasslands, but usually only as individuals or groves growing near creeks or where the water table is within reach of their roots.

Urbanization, farming practices, livestock grazing, and the introduction of exotic plant species–particularly Mediterranean species–have all contributed to the decline of native grassland communities (Bornstein 1985).

Wilson et al. (1995) discussed the once large prairies of the Willamette Valley in western Oregon. Wetland prairies occurred on poorly drained areas of the valley floor, while upland prairies and oak savannas occurred on better-drained soils along valley margins. Frequent burning by indigenous people reduced the abundance of shrubs and trees, favored grasses such as tufted hairgrass (*Deschampsia cespitosa*) on wetter sites and red fescue (*Festuca rubra*) on upland sites, and promoted a rich variety of native forbs. Frequent flooding also helped keep some sites in an open, early successional condition. However, today these areas have been greatly altered due to cessation of burn-

ing, less frequent flooding, and development for agriculture and urban uses. Many grasslands and prairies have suffered similar change.

Wilson et al. (1995) listed three challenges to resource managers trying to conserve and restore these natural areas and the species they support. First, succession of prairies to shrub- and tree-dominated communities precludes a purely protectionist approach and compels active intervention. Second, widespread invasion of exotic species has significantly altered these prairies. Third, prescribed burning is not always feasible because of strict smoke-management regulations in the area.

Site Conditions

Soils

The soils in the galleta–three-awn shrubsteppe are sandy and covered by a microphytic crust. Grazing breaks up the crust causing dunes to form and allowing shrubs to invade and increase.

St. John (1996) described a recently invented imprinter that places inoculum of beneficial fungi (mycorrhizae) in the root zone, while on the same pass creating soil surface heterogeneity, improving infiltration and soil aeration, and distributing and pressing seeds into the soil. Restorationists have long understood that ecosystem function depends heavily upon mycorrhiza (Miller 1985). However, until now, introduction of inoculum into the root zone has not been economically feasible.

The restorationist must take into account that past land uses influence soil compaction and water-holding capacity. If these sources of degradation are not removed, revegetation efforts may fail.

Topography

Grassland communities occur on valley bottoms and mountain tops or ridges, where the topography is typically flat. However, these grasslands are very extensive in some regions and occur in a variety of topographic positions and elevations.

Moisture and hydrology

The communities are typically dry and have well-drained soils. Grasslands may also occupy wetter areas along washes or streams.

Water usually cycles relatively rapidly through arid grassland communities where the soils are well-drained and evaporation is high. However, moisture gradients do occur within large grasslands and seeps or washes may cause changes in species composition and structure.

Establishment

Site preparation

Wilson (1995) created a small grassland on a disturbed site that was previously a parking lot. The area was deeply plowed with a ripping-shank, then seeded with a nonpersistent grass covercrop. The following season, ripping was repeated and an effort was made to control weeds using a combination of spot applied herbicide and hand pulling. The following spring, the site was ripped again, raked, and covered with clear plastic sheeting to use solar heating to further reduce a seedbank dominated by noxious nonnative grasses and forbs. The plastic remained in place until mid-September, then it was removed and native grass seed was sown directly on the site. Two more grasses, California brome (*Bromus carinatus*) and blue wild rye (*Elymus glaucus*) were introduced as seedlings, sown in plug-flats early in August, and set out on one and one-half foot centers in October. Several other grasses and forbs will be introduced later.

Danielsen (1996) found that plots that were tilled twice and weeded prior to planting were most successful. This suggests that eliminating competition from weeds prior to planting is a key to successfully establishing native grasses.

Planting considerations

Studies conducted at several reserves in California show promise of considerable reestablishment of natives following cessation of grazing (Pitschel 1984). So, if the restoration site was previously utilized for livestock grazing, it would be worthwhile to exclude grazing from an area and see if native vegetation reappears.

Danielsen (1996) collected seeds of native grasses in California from April through August, then sowed them in flats during September and October. By December and January the seedlings had grown large enough to be outplanted. Newly planted grasses were watered through March (no summer watering) and the plots were weeded as necessary.

Spacing considerations

Native bunch grasses should be relatively widely spaced to allow for as much natural regeneration as possible once plantings begin to produce seed. Planting grass plugs on one and one-half to three-foot centers is good for initial establishment.

Temporal spacing for different plants and layers

Supplemental plantings or replantings will likely be needed in grassland restorations. Planting is often completed in phases with different plots and species being planted at different times. This allows the restorationist to experiment and determine the best times to plant seeds or sprigs.

Maintenance and Management

Weed control

A diversity of pernicious weeds plague the grassland communities of the western United States. The majority of these were introduced from the Mediterranean region or Europe as forage for livestock or erosion control.

Yellow-star thistle (*Centaurea solstitialis*), an annual, is a relative of the knapweeds (*C. maculosa, C. repens, C. jacea,* and *C. nigra*), bachelor's button (*C. cyanus*), and dusty miller (*C. cineria*). We have only one American species of knapweed (*C. americana*) and it is a native of the Southwest. Dremann (1996) observed that yellow-star thistle can be controlled by shading it out. Native bunchgrasses such as tussock grass (*Nassella pulchra*) can withstand invasion by star thistle even when completely surrounded by the weed. As a result, maintaining shade-producing standing grasses or its thatch, especially during May and June, may be the most cost-effective method of controlling star thistle (ibid.). If grasses must be cut or grazed, Dremann recommends waiting until after May 15 to do so and then to leave at least five to six inches of stubble to shade the ground. Dremann also found that applying wheatstraw mulch killed the yellow-star thistle in direct proportion to its depth (five inches provided 100 percent control).

The exotic leguminous shrub, Scotch broom (*Cytisus scoparius*) competes with and shades out native prairie vegetation along the rocky coastline of the West Coast. Scotch broom can tolerate warm, dry summers and grow on thin soils in exposed rock crevices where native shrubs cannot grow (Leopold and Wright 1995). It readily spreads into meadow environments and resprouts when cut. Leopold and Wright (1995) found that the most effective way to eliminate Scotch broom was to cut it and paint the stumps with glyphosate herbicide. The cut brush was then piled and burned. Seeds persist in the seedbank and seedlings will likely have to be pulled each year. Fires seem to promote sprouting of Scotch broom from stump or root and accelerates seed germination (ibid.). Another invasive shrub— Nootka rose (*Rosa nutkana*), may be controlled the same way (ibid.).

Russian thistle (*Salsola kali*) may be a problematic weed on severely degraded or contaminated sites (Smith 1995). It can be controlled by successful establishment of perennial bunchgrasses (ibid.).

Reed canarygrass (*Phalaris arundinacea*) is a cool-season perennial grass that often forms dense, monotypic stands in wetlands and disturbed uplands. Techniques such as solarization and hydrologic control may prove effective for its control (Wilson et al. 1995).

Himalayan blackberry (*Rubus discolor*) can form impenetrable thickets in grassland communities and disturbed forests of the western United States. Mechanical cutting and direct application of herbicide to the cut stems may be the most effective method to control Himalayan blackberry. Fire may cause an initial decline, however, most *Rubus* spp. respond positively to burning so it should not be used where blackberry infestations are abundant.

Tarweed (*Holocarpa virgata*) is a native plant that has increased to pest levels on California range-lands. The reasons for this include overgrazing, introduction of annual forbs and grasses, and grassland fire control—all tending to replace or reduce the native perennial grasses that originally kept tarweed in check. Perrier et al. (1981) found that herbicides were most effective when applied in April, prior to stem elongation. Mowing was most effective just before flowering (in August), presumably when carbohydrates are concentrated in stems and leaves, and root reserves are minimal. Application of fertilizer in the spring tended to discourage tarweed by encouraging growth of annual grasses that compete for soil water. Deferred fall and winter grazing also encouraged annual grasses at the expense of tarweed.

Gorse (*Ulex europeus*) is a thorny, nonnative shrub that is present along both the east and west coasts of the United States. Amme (1984) noted that gorse was a major problem during the restoration of a coastal grassland in northern California.

A combination of herbicide treatment, cutting, and burning may be most effective in controlling South American pampas grass (*Cortaderia selloana*).

German ivy (*Senecio mikanioides*) can be eradicated with properly timed applications of herbicides.

Mowing during April and May reduced wild oats (*Avena barbata*) and ripgut brome (*Bromus diandrus*), while mowing during summer dramatically reduced the exotic composite, yellow star-thistle (*Centaurea solstitialis*) (Danielsen 1996). Although some star-thistle will seed out after an August mowing, the seedlings that result tend to cluster around the parent plants, making then easy to spot and remove.

Nutrient requirements

Nutrient levels are affected by past and present land uses and must be evaluated to determine if fertilizers or soil amendments are needed prior to revegetation.

Water requirements and hydrology

Smith (1995) found that an application of poly-acrylamide can improve the moisture-retaining ability of degraded soils within the root zone. This can significantly enhance the growth and establishment of bunchgrasses (ibid.).

Unless a grassland community receives excessive artificial water input, as from urban stormwater or landfill leakage, hydrologic restoration is not needed.

Disturbances

Fire suppression throughout all of the Southwest grasslands has resulted in the mass invasion of junipers, creosotebush, acacia, mesquites, and other shrubs into vast areas where purely grassland occurred at the time of the Spanish conquest (Wilson 1984). Overgrazing has also played a role in this degradation of the southwestern rangeland. In California, much coastal perennial grass prairie has been replaced by coastal sage chaparral, as a result of fire suppression and overgrazing. Consequently, the communities we see today are very different from the original vegetation.

Case Studies

The restoration objectives for the St. Johns Landfill Grassland was to plant the area with a mixture of species representative of plant communities native to northwest Oregon (Wilson 1995). The site consists of a closed landfill that was once a shallow pond or marsh connected to other floodplain wetlands and slough channels at the confluence of the Willamette and Columbia Rivers in Portland. The communities chosen were mesic prairie (for the nearly flat ridgetops), xeric prairie (for the steep side slopes), and shrubs, trees, and mesic prairie (for the valley bottoms) (ibid.). A variety of native grasses were hand seeded, but failed due to an existing competitive seedbank. A sterile cover crop was used on each area followed by various seedbank control strategies the following year. The second fall, native grasses were established on site where seedbank management techniques had proved successful. Only two native grasses, the pioneer native grass, California brome (*Bromus carinatus*) and the slower-growing Idaho fescue (*Festuca idahoensis*) were seeded as a mix using a drill and no-till techniques.

Wilson (1995) found that just two months after planting, California brome produced a stand of twenty to thirty percent cover on well-drained soils. Native grass germination was highest on the plot that was solarized (covered with plastic) to control weeds. Solarized and herbicide application plots were the most weed free. Wilson also found that swathing (side cutting) reduced undesirable grass seedlings.

Bornstein (1985) described the restoration of a one-acre meadow in California. The site was dominated by the spring-flowering California poppies (*Eschscholtzia californica*) followed by weeds in the summer and fall. The clay soil had poor drainage and high salinity. The goals of the project were to establish a more authentic meadow by increasing species diversity and improving soil texture and structure, decrease maintenance (particularly weeding), decrease irrigation, and extend the blooming season. First, the area was rototilled and organic matter was added. Eight species of perennial bunchgrasses were directly seeded in subplots. Germination was poor and seedling survival low due to heavy weed competition. Subsequent plant materials were propagated in a greenhouse. Twenty-five species of bunchgrasses were successfully propagated from seed, transplanted to three-inch jiffy pots, and planted throughout the winter rainy season in irregular stands.

Weeds, especially weedy grasses, were a continual problem. A mulch of woodchips, applied in stages, was effective in reducing the weed population and in retaining moisture. In two years, fifteen species of California native bunchgrasses had been successfully established. The original goals were partially achieved, but maintenance was still intensive.

Danielsen (1996) described the restoration of a few acres of native California grassland at Mount Diablo State Park, California. A major obstacle was the conversion of a Mediterranean annual grassland dominated by exotic oats (*Avena* spp.), brome (*Bromus* spp.), and ryegrass (*Lolium* spp.) into a diverse assemblage of native grasses and forbs. Seeds of native grasses were collected from April through August, then sowed in flats during September and October. By December and January the seedlings had grown large enough to be outplanted. Newly planted grasses were watered through March (no summer watering) and the plots were weeded as necessary.

Danielsen found that plots tilled twice and weeded prior to planting were the most successful. This suggests that eliminating competition from weeds prior to planting is a key to successfully establishing native grasses. Results of this project include healthy plots of pine bluegrass (*Poa secunda* var. *secunda*), melic grass (*Melica californica* and *M. imperfecta*), purple needlegrass (*Nassella pulchra*), foothill needlegrass (*N. lepida*), California brome (*Bromus carinatus*), woodland brome (*B. laevipes*), and blue wild rye (*Elymus glaucus*). Forbs successfully established in areas of smaller grasses such as pine bluegrass and melic grass were dove weed (*Eremocarpus setigerus*), spikeweed (*Hemizonia fitchii*), blue-eyed grass (*Sisyrinchium bellum*), *Hesperevax sparsiflora* var. *sparsiflora*, and fireweed (*Epilobium* spp.). California poppy (*Eschscholzia californica*), narrow-leaf milkweed (*Asclepias fascicularis*), and tarweed (*Holocarpha heermannii*) thrived on all the plots.

Of the several species of native forbs, only clarkia (*Clarkia ungiculata, C. purpurea*, and several subspecies), yarrow (*Achillea millefolium*), and California buttercup (*Ranunculus californicus*) grew well when sown directly or as transplanted seedlings. Many of the transplants drowned during a wet winter as the soil around them settled, forming depressions that held water.

Seed has been harvested from most of the plots and three-year-old plants are as tall as natural stands (Danielsen 1996). After the plots are successfully established, plantings will be expanded beyond the grids by dividing mature plants, direct sowing seed, and continuing to raise seedlings for fall planting. The ultimate goal is to have the approximately ten acres of former agricultural field converted to a grassland by 2002.

For additional information on the ecology of these communities, consult Collins (1976, 1979), Kearney and Peebles (1960), Leake et al. (1993), Shreve and Wiggins (1951), Turner and Brown (1982), Weaver and Albertson (1956), and Whitfield and Anderson (1983).

For additional information on the restoration and management of grasslands, see Stromberg and Kephart (1996).

Prairies

Shortgrass Prairie A-127, A-182
Foothills Prairie A-85, A-172
Mixed Grass Prairie A-172, A-191
Northern Shortgrass Prairie A-173
Sagebrush–Bluestem Prairie A-181
Sandhill Prairie A-182, A-190
Bluestem Prairie A-188, A-199, A-214, A-311
Western Bluestem Prairie A-189
Mixed Grass–Shrub Prairie A-192
Fayette Prairie A-215
Blackland Prairie A-216
Midgrass Prairie A-227
South Texas Prairie A-236
Coastal Prairie A-242
Island Prairie A-242
Mesic Prairie A-272
Blackbelt Prairie A-330
Dry Prairie A-354

Introduction

Prairies are open, grass-dominated expanses with deep, fertile soils. Prairie communities occur predominantly in midwestern North America, however, outliers and smaller "prairie patches" or "prairie remnants" are found at both the eastern and western extremes of the prairie region. Grazing pressure and drought have caused many changes in the original vegetation. Prairie communities have subtle structural differences based upon their topographic-soil moisture gradients.

Thompson (1992) stated that an ideal restored prairie community in the botanical sense alone would require an area close to 200 acres (it would require closer to 1000 acres to restore the full fire-bison interactions of the original tallgrass prairie) and contain at least 200 plant species from propagules obtained within a fifty-mile radius of the restoration site.

The restoration of prairies has been the focus of much research. The *Tallgrass Restoration Handbook for Prairies, Savannas, and Woodlands* (Packard and Mutel 1997) is an excellent source of information concerning the restoration of the tallgrass prairie ecosystem.

Site Conditions

Soils

The native prairie was a naturally self-perpetuating plant community that helped to build and maintain the fertility and ideal physical properties of some of the world's most productive soils (Hayden 1945). Prairie communities not only protect against erosion, but also enrich the chemical and physical properties of the soil (e.g., promotion of aggregate formation and stability of structure) (Jastrow 1987). Data indicate that for large-scale and long-term production, humans have been unable to mimic the sustained productivity of the native prairie with maintenance of soil properties, much less soil improvement.

Schramm and Kalvin (1978) suggested the use of prairie plant communities to ameliorate severely degraded soils, such as in strip mine reclamation and road construction cut/fill areas. Dancer (1985) stated that plants on disturbed land reclaimed with topsoil are subject to unusually severe water stress during drought because of the destruction of soil structure during handling. In addition to nutrient deficiencies, occasional toxicities from newly exposed minerals confound the influence of drought stress on disturbed coal-mined land, making management more difficult (ibid.).

Prairie-influenced soils typically had deep (greater than eighteen inches), dark, nutrient-rich surface horizons and are often classified in the order Mollisols in the United States Department of Agriculture system of soil classification (United States Department of Agriculture 1975). Prairie soil is typically alkaline. County soil surveys contain soil maps, descriptions of soil map units, and other information necessary to identify potential prairie restoration sites based upon soil characteristics. Due to extensive cultivation, accelerated loss of topsoil may have changed the surface soil characteristics of former prairie soils (Fenton 1983), particularly on steep slopes. As you go from the tallgrass prairie westward to the shortgrass plains, soil fertility decreases, as does rainfall.

St. John (1996) described a recently invented imprinter that places inoculum of beneficial fungi (mycorrhizae) in the root zone, while on the same pass creating soil surface heterogeneity, improving infiltration and soil aeration, and distributing and pressing seeds into the soil. Restorationists have long understood that ecosystem function depends heavily upon mycorrhiza (Miller 1985). However, until now, introduction of inoculum into the root zone has not been economically feasible.

Topography

The topography of most prairie communities is relatively flat or gently rolling, however, some prairie communities occur on hillslopes or steep grades. The topography of a prairie community (along with moisture) determine the subtle structural and species composition changes that contribute to the diversity of these communities.

Moisture and hydrology

In the tallgrass prairie region, rainfall is lower than in the adjacent forests, and steadily decreases toward the west. Amounts vary from forty inches in the east to fifteen inches in the shortgrass plains of the west, and it occurs mainly as summer rain. Winters can be very cold and summers very hot, with hot, drying winds.

Although prairies are typically very dry, moisture gradients and areas where the soil is not well drained often occur. These areas are represented by changes in species composition. The water table is usually a few to several yards below the soil surface. Some deeply rooted prairie grasses may reach the water table, but the primary source of water is in the form of precipitation.

Establishment

Site preparation

Site preparation is usually the most labor-intensive phase of prairie restoration. Existing vegetation should be carefully evaluated. It may be worthwhile to conduct a burn and wait one season to see what species appear or install an exclosure (to protect the area from grazing) and see if native plants appear. Hubner and Leach (1995) found that prairie parsley (*Polytaenia nuttallii*) reappeared following brush cutting and burning at a site on which the species had been formerly documented. Many other examples exist of rare species reappearing after proper management is implemented on degraded habitats. McClain (1986) cautions against attempting prairie restoration on recently cropped fields because of the possibility of herbicide carryover, which can damage prairie plant seeds and seedlings. Such fields should be allowed to lie fallow for at least one year.

Fast establishment of prairie plants from seed has been obtained with a firm, weed-free seedbed (Betz 1986, Lekwa 1984, Rock 1981). It is often necessary to begin seedbed preparation in the fall prior to sowing seed in spring. If the site was used the previous year for row crops, shallow spring disking two or three times (the last time just before planting prairie seeds) is recommended. On former pasture or areas with a mixture of annual and perennial weeds, it may be necessary to kill the existing sod completely before attempting to seed in prairie species. One possible approach involves late-summer mowing to a height of about twelve inches, followed by fall plowing to a depth of at least eight inches and shallow spring disking (at two- to three-week intervals) up to the time of planting (Betz 1986, Lekwa 1984). Betz (1986) and Rock (1981) obtained good results when spring disking was followed by harrowing (to level the area) and using a roller or cultipacker to firm the seedbed just prior to and just after planting seed. These approaches should be used on relatively level sites as seed may be lost in heavy rains.

No-till chemical (herbicide) seedbed preparation is possible for both small- and large-scale prairie restoration. Lekwa (1984) reported initial success in prairie plant establishment (especially grasses) using glyphosate to kill existing vegetation a week to eighteen days before planting prairie seed. Several combinations of tillage and herbicide application have been used in restoration projects.

Pauly (1984) treated an area with herbicide in September and burned it the following April. In June, herbicide was again applied and three weeks later seeds of thirty-three species of prairie plants were hand broadcast at a rate of four pounds/acre forbs and seven pounds/acre grasses. Seeds were set with a toothed drag that stirred up the top one-eighth to one-fourth inch of soil, but did not penetrate the sod of dead grasses. In the first year, smooth oxeye (*Heliopsis helianthoides*) bloomed and twelve species bloomed during the second growing season.

In general, most prairie restoration sites are prepared by disking or tilling. However, sites that contain some native grasses and/or forbs may not need to be prepared. Planting, weed control, and other management practices may proceed. The determination of whether or not to start with a prepared site is subjective and should be made by someone familiar with native plant communities in the region. The *Tallgrass Restoration Handbook for Prairies, Savannas, and Woodlands* (Packard and Mutel 1997) contains excellent detailed information about site preparation for restoring prairie communities.

Planting considerations

Prairie plants are most commonly planted as seeds, however, some herbaceous plants (especially rare or uncommon plants) should be planted from container-grown stock. Many prairie restoration projects have utilized a combination of seeds and transplants to establish a prairie community. The more plants that can be established using seeds, the more economical prairie restoration becomes. The best results are often obtained by drilling seed directly into the soil. Lathrop (1995) found that hand broadcasting successfully established six species of Nebraska grasses. Furthermore, native prairie seedlings did quite well, especially Canada wild rye (*Elymus canadensis*), foxtail dalea (*Dalea leporina*), prairie clovers (*Dalea candida, D. purpureum*), sunflowers (*Helianthus* spp.), asters (*Aster* spp.), cordgrass (*Spartina pectinata*), and slender wheat-

grass (*Agropyron trachycaulum*) (Lathrop 1995).

For restoration of prairie communities using seed, consideration must be given to seed treatment prior to planting, the timing of the planting, the amount or rate of seeding, the method of sowing seed, and whether to sow an annual cover crop at the same time (Thompson 1992). Local seeds may be hand collected for use in prairie restoration or collected on a larger scale with specialized equipment. Merchant and Olsen (1995) found that seed samples collected using a flail-vac stripper contained an average of fourteen and one-half native species, compared with twenty-eight species in samples collected by a combine. This difference is attributed to the greater number of forb species found in combine-harvested samples. Depending upon the collection site, legal permission may be necessary to collect seed, and a general rule of thumb is not to take more than one-fifth to one-third of the seed crop for common species in a given year and to take only a few individual seeds, if any, from rare plants.

Hanson and Lipke (1995) found that broadcasting seed onto snow is an effective way to establish tallgrass prairie species on small restoration sites. The authors note two advantages in winter broadcasting–the distribution and amount of seed can be easily observed resulting in a more even application, and activity during the winter lessens the management workload of spring and early summer.

As changes in topography (which are generally slight) dictate species composition changes, two ways to determine what to plant where are: (1) comparison with the topographic/species composition changes in an adjacent natural prairie and (2) experimental transects in which a species is planted along a transect to determine the optimum topography (microtopography) for a particular species. Using both techniques prior to full-scale plantings will help improve survival rates.

Prairie plants are often difficult to trasplant due to their extensive root systems. Seeding for prairie restoration is usually most successful in early to late spring, when the weather has begun to warm (June in the North; earlier in the South). You can also sow in the fall, especially to prepare dormant seeds for germination with the alternate freezing and thawing of winter. Or you can stratify the seeds indoors over the winter and sow them in spring. This will give the best germination and avoid loss of seed to spring runoff. Mix the seed with some sand to permit more even sowing (the sand will also show where you've sown). Rake the seeds in lightly and roll the area. Machinery designed especially for grassland planting is useful for large areas.

Seed obtained from a commercial source has usually been pretreated by one or more of the following processes: stratification (cold storage at about thirty-four degrees Fahrenheit for six to sixteen weeks), scarification (abrasion of the seed coat by chemical or mechanical means), and inoculation (introduction of nitrogen-fixing bacteria to prairie legume seeds). If seed is collected from local prairie remnants, proper conditioning is essential to achieve adequate germination in the first year (Shramm 1978). The majority of prairie grasses and forbs will germinate the first season after planting when they have been dry or moist stratified (seed mixed with sand or vermiculite, moistened for damp stratification) and stored in plastic containers at thirty-two to forty-one degrees Fahrenheit for at least six weeks (Rock 1981, Schramm 1978).

Many prairie restorationists have experimented with various grass to forb planting ratios. In Wisconsin, Liegel and Lyon (1985) planted separate plots with 60:40 and 40:60 grass to forb seeding ratios. Five years later, the 60:40 grass to forb planting had an 82:18 composition of grasses to forbs (cover). The 40:60 grass to forb planting had a 64:36 grass to forb composition. Planting ratios appear to be directly related to the resulting mixture of grasses and forbs, so by adjusting this ratio, the desired species composition may be achieved.

Kearns (1985) experimented with transplanting prairie plants in Wisconsin and found that growth and survival of grasses and forbs were similar in spring, summer, and fall. Kearns (1985) concluded that most prairie species can survive transplanting at any time during the growing season if done under proper weather conditions of cool temperatures, high humidity, and low solar radiation. Both mechanical and manual sod transplants resulted in survival of most species of native and exotic forbs, grasses, and sedges. Root pruning was found to increase survival in lupine (*Lupinus perennis*) and puccoon (*Lithospermum carolinense*). Root pruning in fall showed slightly higher survival rates than did spring root pruning.

Zajicek-Traeger et al. (1985) found that in highly disturbed areas, seedling establishment of prairie forbs and their adaptation to a low-maintenance landscape may depend upon mycorrhizal fungi. They found that mycorrhizal colonization of native prairie forbs improves seedling growth at low phosphorus levels in the greenhouse. Introduction of mycorrhizal fungi may play an important role in the establishment of prairie forbs on disturbed sites. Mycorrhizal inoculants for restoration projects are available from commercial sources.

For more detailed information on planting of prairies, see Bragg and Sutherland (1988), Henderson (1995), and Shirley (1994).

Spacing considerations

Prairie grasses often have a tendency to form clumps and many spread by rhizomes, so spacing should be adequate to allow some filling in. Given the fact that prairie grasses are the dominant plants of prairie communities, forbs should be planted "here and there" within the grasses and in the appropriate microhabitat (i.e., xeric vs. mesic).

Temporal spacing for different plants and layers

In grassland communities such as prairies, the vertical structure is relatively simple and most species can be planted at the same time. Liegel and Luthin (1984) experimented with spring and fall prairie plantings in Wisconsin. Under similar conditions, they planted one area in late May and another in late November. They found that although almost twice as many species had been planted in the spring, more were found in the fall planting (thirty-five vs. twenty-six). More weeds and less prairie grasses were found in the fall planting. Results indicated that asters are best planted in the fall immediately after collecting. Hefty (1985) found that in Wisconsin prairie, some grasses germinating after mid-July will survive the winter.

Because of the effort required for site preparation and initial maintenance, year-to-year variations in weed seed germination, year-to-year differences in availability, variety, and viability of prairie seed, and the benefit gained from experience along the way, large restoration sites should be divided into several years of plantings (Ahrenhoerster and Wilson 1981, Shramm 1978).

Maintenance and Management

Weed Control

Once established, prairies can dominate a site (in the absence of major disturbances other than fire), preventing invasion by many weeds (exceptions include aggressive exotics such as smooth brome [*Bromus inermis*] and leafy spurge [*Euphorbia esula*]). A combination of factors contribute to the value of native prairie communities for weed control. These include: (1) species diversity of the prairie community, (2) the low fertility requirement of prairie species, (3) most prairie plants are perennials, and (4) prairie plants are well adapted to local conditions (Thompson 1992). In most cases, after establishment of prairie vegetation, periodic prescribed burns and/or occasional mowing are all that is needed to keep weeds under control. However, if an area has been severely degraded by grazing or agriculture, weed control will be more challenging until the biological integrity of the community is reestablished.

To control smooth brome (*Bromus inermis*), Grilz and Romo (1995) recommend evaluating the existing vegetation and seedbank for its natural revegetation potential. A spring burn followed by an application of glyphosate herbicide, although it may damage some native species, will effectively control brome (ibid.). Additional applications of herbicide or manual removal of individual plants will likely be necessary. Smooth brome increases when mowed, but spring or summer burning decreases its density.

Queen Anne's lace (*Daucus carota*) frequently invades prairie plantings in the upper Midwest. A deep-rooted biennial, which is also capable of producing thousands of seeds, it often displaces prairie seedlings (Kilgour and Kilgour 1995). They found that many plants cut to ground layer did not resprout. This was especially true when plants were cut while the seeds were fat, but still green–only twenty percent resprouted. No matter when plants were cut or at what height, the plants that did resprout produced at most a few hundred seeds (ibid.). This is far fewer seeds than uncut plants, which commonly produce 3,000–5,000 seeds per plant. Parsnip (*Pastinaca sativa*) can also be controlled by cutting.

Prickly pear cactus (*Opuntia* spp.) is a native cactus that can invade and become a pest as a result of overgrazing. Burning may be an effective control, however, prickly pear cacti generally grow in areas with little or no fuel loads (Smith and Ueckert 1983). Herbicide treatments have not proved effective for control. Smith and Ueckert (1983) achieved the best results by applying herbicide immediately after a prescribed burn.

Musk thistle (*Carduus nutans*) is a pest in many prairie and savanna communities. A combination of herbicide treatments and two introduced insects (*Rhinocyllus conicus* and *Ceuthorhynchidius horridus*) effectively controlled musk thistle (Trubble and Kok 1980a, 1980b). They also found that herbicide application while primary blooms were in the late bud or early bloom stage did not significantly reduce survival or reproduction of these two weevils. This suggests that herbicide application may be used to prevent production of viable seed, leaving the weevils to damage or destroy the existing plant. Canadian thistle (*Cirsium arvense*), which is actually European, is the most aggressive thistle. It is a perennial that spreads by stolons as well as wind-blown achenes. Bull thistle (*Cirsium vulgare*) is also a common pest of open lands. Cutting off the flower before thistle has a chance to go to seed can offer some control.

Crested wheatgrass (*Agropyron cristatum* and *A. desertorum*) is an introduced bunchgrass that invades native grasslands and forms monotypic stands in the western United States and Canada. Romo et al. (1994) found that burning or mowing followed by an application of glyphosate was more effective than either treatment without herbicide. Applications of herbicide may need to be repeated until seed reserves in the soil are depleted.

Choi and Pavlovic (1994) tested fire, herbicide, and sod/topsoil removal treatments to rid razed residential areas of exotic vegetation within the Indiana Dunes National Seashore. Herbicide applications and sod removal effectively reduced bluegrass (*Poa pratensis*) and quackgrass (*Agropyron repens*). Burning, however, actually increased the amount of bluegrass.

Smooth sumac (*Rhus glabra*), although a North American native, can invade prairies and rapidly transform them into brushland. Aldous (1929, 1934) found that cutting in mid-June (in Kansas) led to the most effective control. However, this presents a problem in communities such as northern prairies, where summer fire or mowing would set back warm-season perennial grasses. Kline (1982) effectively combined spring burning with late summer or fall cutting of sumac in Wisconsin. Waller (1982) cut sumac in June and treated the stumps with a glyphosate solution and none of these stems resprouted. Martin (1981) burned a mesic prairie in Wisconsin in April, then controlled resprouting by mowing twice annually between mid-June and late August for two years. Grasses and forbs released by the cutting helped suppress resprouting and provided a good fuel base for maintenance burns.

Late spring burning is usually effective in controlling yellow sweet clover (*Melilotus officinalis*), if done prior to or during flowering. Schwarzmeier (1984) found that sweet clovers (*M. alba* and *M. officinalis*) can be controlled using a refined mow/burn prescription in Wisconsin. First, note the earliest date that the average sweet clover stem bases (zero to five inches) receive nearly full shade (May 9 in Wisconsin). Next, burn seven to ten days after this date. Alternatively, mow at two to three inches one to five weeks after the full shade date.

Prevention is an important consideration in protecting sites from weeds. Horses spread sweet clover and exotic grasses in their manure and may need to be excluded from prairies. Seed can also be carried on equipment such as mowers.

One of the worst weeds in the prairies of the northern Plains is leafy spurge (*Euphorbia esula*). Ranchers, farmers, and federal agencies spend more than $100 million dollars every year to keep it from spreading beyond the five million acres it already infests. There is even a newsletter (Leafy Spurge News) concerning the control of leafy spurge as well as a CD-ROM containing over 400 papers about leafy spurge and its control (Spencer 1995). Hull-Sieg (1994) found that fall herbicide followed by spring burning worked best in controlling both seed germination and plant density.

Biocontrol applications using black-dot (*Aphthona nigriscutis*) or brown-dot (*A. cyparissiae*) (flea) beetles can significantly reduce leafy spurge given the proper setting. Harris (1994) found that black-dot beetles are most effective in prairie sites with needle-and-thread (*Stipa comata*) and porcupine grass (*S. spartea*) present, spurge stems less than twenty-eight inches tall with densities less than sixty stems/square meter, and soils having more than sixty percent sand content. The best locations for brown-dot beetles have green needle-grass (*Stipa viridula*), spurge stems taller than twenty inches, with densities between 50 and 125 stems/square meter, and soils forty to sixty percent sand. Van Fleet (1994) found that in Wyoming county-wide elimination of leafy spurge increased tenfold over two years following the introduction of *A. nigriscutis* flea beetles to over 100 sites in the western part of the state. Fellows (1995) found that the establishment of *Aphthona* beetles used to control leafy spurge (*Euphorbia esula*) can be increased when the release sites are burned the preceding fall or spring.

It was once thought that the soil of old fields was simply too thin to support vegetation other than broomsedge (*Andropogon virginicus*), little bluestem (*Schizachyrium scoparium*), or beardgrass (*Andropogon gyrans*); but recent research has shown that *Andropogon* is allelopathic. That is, its roots exude a chemical that inhibits other plant growth, a fact that could be put to good use by restorationists who are battling invasive aliens (Scott 1992).

Nutrient requirements

Since most prairie plants have relatively low nutrient requirements, addition of fertilizers often encourages weeds. However, on severely depleted sites amendments may be needed, especially sites that have been cultivated or grazed for many years. On newly prepared sites (disked and seeded) a mulch will help germination and stabilization of the site. The type of mulch that is used should be prairie litter and not contain viable hay or straw grasses used in agricultural practices; although the use of sterile forms of straw is a possibility.

Water requirements and hydrology

Most prairie plants have low water requirements. Watering should be done with caution as it may encourage weeds. However, during initial establishment of transplants, watering may increase survival. O'Keefe (1996) concluded that June and July rainfall may well be one of the biggest factors in the success of a prairie restoration effort—even in the high-rainfall Midwest. In fact, even in a normal-rainfall year (eight inches) you could improve the restoration success by forty percent if you could irrigate three inches a month for June and July to bring the total up to fourteen inches (ibid.).

Disturbances

Prairie communities evolved under the influence of frequent fires and prairie plants can compete with exotic perennial species and woody plants under management regimes that include regularly scheduled controlled fires. Burning has several beneficial effects, primarily the removal of accumulated dead plant material. Removal of the dead mulch associated with prairie plants leads to earlier warming of the soil surface in spring and exposes soil for germination of new seedlings (Hulbert 1986, Lekwa 1984). Periodic fires can actually stimulate the flowering and productivity of many prairie species, especially grasses (Richards and Landers 1973, Hill and Platt 1975, Hulbert 1986). Howe (1995) concluded that the composition of presettlement prairies, which were subject to summer fires set by lightning, may have significantly differed from that of modern prairies maintained by prescribed fires in other seasons.

Romo et al. (1993) found that the density of western snowberry (*Symphoricarpos occidentalis*) increased two- to threefold in the first two growing seasons following a single burn. Although the optimum interval between burns to enhance species diversity, productivity, and flowering is unknown (Hulbert 1986), generally recommended action is to plan for a burn once every three to five years after the prairie is well established (Lekwa 1984). Heckert (1994) recommends prescribed burning to manage small prairie fragments, while large prairie fragments must incorporate burned and unburned areas to ensure the availability of suitable habitat for grassland breeding birds, which are experiencing significant population declines in the Midwest.

The optimum frequency and timing of burning will be site specific, particularly with respect to which weed species are a problem on the site (e.g., late spring burning is more effective in controlling bromegrass [*Bromus* spp.]) (Thompson 1992). The consensus is that spring (late March or April) usually provides optimum conditions for burning; the ground is cool and moist, many prairie plants are still dormant, weather conditions are more reliably favorable, few birds have begun nesting, reptiles and amphibians are still hibernating, and good responses by prairie species are usually observed (Bragg 1978, Schramm 1978). Benning and Bragg (1993) measured the growth and flowering of big

bluestem that they had burned earlier that year at four-day intervals between April 6 and May 20. They noted increases in growth and reproduction in bluestem burned after May 12, leading them to conclude that responses to burning can vary over periods as short as four days.

In areas where burning is not possible because of proximity to buildings or other structures, haying can serve as a substitute. In order to most closely simulate some of the important effects of fire (e.g., warming of the soil surface in spring), removal of cuttings by raking and baling is recommended (Diboll 1986, Lekwa 1984). Thompson (1992) advises mowing to a height of five to twelve inches in late June or early July the year after planting, followed by a similar treatment in May or June of the second year. O'Keefe (1995a) speculates that additional mowing can enhance the success of a newly-planted prairie. Bragg and Sutherland (1988) and Ross and Vanderpoel (1991) both suggest that mowing for more than the first year encourages the establishment of prairie species. Boettcher and Bragg (1988), on the other hand, found that regular mowing of mature, native prairies reduces species diversity and increases the number of aggressive and introduced species.

Proper planning is of utmost importance for controlled burning. Hulbert (1978) and McClain (1986) recommend the following considerations:

- Construction of firebreaks where natural firebreaks are not present. Roads, ditches, ponds, and creeks can serve as "natural" firebreaks. Constructed firebreaks can be grassy areas mowed periodically during the growing season. Disked, plowed, or raked areas forming two- to three-yard strips surrounding the burn area can also serve as a firebreak. In areas particularly susceptible to erosion, it is possible to create firebreaks without tillage (see Hulbert 1978).

- The number of persons required to contain the fire will depend upon the size of the proposed burn and the amount of accumulated fuel in the area. Use a crew of no fewer than four persons, one to set the main fire, two to control the fire, and one to monitor the backfire.

- Wind velocity should be less than fifteen miles per hour (but strong enough that the wind direction is stable), fine fuel moisture (dry weight) between seven and twenty percent, the relative humidity (surface) between twenty-five and sixty-five

percent, and the temperature less than eighty degrees Fahrenheit (O'Keefe 1995b, Pauly 1988).
■ It may be necessary to obtain an open burning permit from local agencies.

Perfect burn days rarely occur, yet "perfect" conditions are required for a burn to be effective, yet manageable and safe (O'Keefe 1995b). The United States Forest Service uses a program known as BEHAVE, that models fire behavior, as a tool for planned burning (Rothermel et al. 1986). O'Keefe (1995b) discovered an excellent method to schedule prescribed burns. He found that by analyzing previous weather conditions, a number of possible burn days could be scheduled that would guarantee a certain percent chance of getting the required burn. O'Keefe (1995b) further stated that chances of getting good burn conditions can be increased by scheduling burns for the early evening hours, when the chance of suitable conditions is at its greatest. See also Donoghue and Johnson (1975) and Pauly (1988).

Proper grazing management is necessary on sites that will be used for cattle or bison. Historically, bison played a major role in maintaining the prairie environment and its diversity of species. Hickman et al. (1995) found that the greatest diversity on sites subjected to (1) grazing by bison; (2) seasonal grazing by cattle; or (3) periodic burning. They caution that the different effects of bison and cattle grazing may have had more to do with grazing management than with grazing habitats and preferences of these herbivores.

Case Studies

Curtis Prairie was established at the University of Wisconsin Arboretum in 1936 with approximately 200 single-species plantings, using any one of 47 different prairie species (Sperry 1994). Surveys done in 1982 and 1990 showed that the planting method was a successful restoration strategy, producing a gradual development of mixed prairie. A total of twenty-three species showed good establishment following their planting in 1936, including big bluestem (*Andropogon gerardii*), Indian grass (*Sorghastrum nutans*), rattlesnake master (*Eryngium yuccifolium*), white indigo (*Baptisia leucantha*), rosinweed (*Silphium integrifolium*), and bushclover (*Lespedeza capitata*).

Collins (1985) described the restoration of a five-acre hillside pasture to prairie in Illinois. Presettlement survey notes indicated that the area was a mixture of prairie and savanna. First, the site was partially burned and monitored to determine the amount of native plant species present. A few hardy native plants were observed, including black-eyed susan (*Rudbeckia hirta*) and flowering spurge (*Euphorbia corollata*). Two spring-flowering prairie forbs present in significant numbers were prairie violet (*Viola pedatifida*) and blue-eyed grass (*Sisyrinchium albidum*). Each year, one acre was cleared of brush and planted to prairie using a mixture of one-year old transplants grown on the site and cold-stratified seed collected locally. The planting site was lightly disked prior to planting to provide a seedbed. Seeding ratios were set to approximate the composition of hill prairie remnants within a twenty-mile radius of the site. Special attention was given to the establishment of spring-flowering plants such as shooting star (*Dodecatheon meadia*), hoary puccoon (*Lithospermum canescens*), and prairie smoke (*Geum triflorum*). Also many rare plants recorded by early botanists as part of the county flora (but now extirpated) were reintroduced to the site. These include pasque flower (*Anemone patens wolfgangiana*), Hill's thistle (*Cirsium hillii*), prairie buttercup (*Ranunculus rhomboideus*), and prairie dandelion (*Agoseris cuspidata*).

For additional information on the ecology of prairies, see Risser et al. (1981), Weaver (1954, 1965, 1968), and Weaver and Albertson (1956).

Sand Dunes

Beach Dune A-355

Introduction

Sand dune habitats occur wherever loose, shifting sands build up into mounds called dunes. Dunes are also found on dredge material (spoil) islands where sand and silt have been deposited and formed mounds. In time, the barren mounds become vegetated with hardy dune plants that help stabilize the sand. The row of dunes nearest the ocean (primary dunes) is directly exposed to the full effects of ocean beach environmental forces such as strong winds, moving sand, and intense sun. Harsh physical forces characterize the dune habitat. Strong winds are a nearly constant force year around. Sand is constantly being moved by the wind causing dunes to shift and migrate, often covering vegetation. Winds carry salt spray from the ocean to the dunes. The salt spray, heaviest during storms, periodically kills back shrub and tree branches facing the ocean. Behind the primary dunes, salt spray causes such plants to grow stunted and to appear sculptured. The few species adapted to living in this habitat, particularly in the primary dunes, are highly specialized and quite successful.

Although some sand dunes occur on mainland beach areas, they mostly develop on barrier islands or peninsulas. Barrier beaches are perhaps the most dynamic coastal land masses along the open-ocean coast. Fournier (1996) stated that undisturbed coastal sand dune ecosystems are very dynamic and heal quite well after storms, but, where human disturbance and development occur, much of that resilience is lost. Hurricanes are a very real part of this coastal system. Even a beach that appears to be stable is constantly changing, with periods of erosion balanced over time by periods of deposition. Sand dunes constantly change as the forces of wind and wave action mold them. Many sand dunes have been severely altered or lost due to development and/or shoreline stabilization. The demand for development and recreational uses of these areas is intense; the result is an alarmingly high rate of habitat loss and the decline of many sand dune plant and animal species. With the loss of the sand dune habitat, the beach cannot effectively absorb large

waves, nor can it supply the sand needed to adjust the beach profile during storms. Without the natural process of sand removal and replenishment, erosion occurs, and developed property ultimately requires protection. Costly engineering projects provide only temporary solutions. Seawalls, groins, bulkheads, jetties, and other human-made structures all seek to "harden" beach sands to keep them from shifting or moving. These projects invariably fail because the natural forces exerted upon the beach cannot be changed.

Sand dunes are interrupted only by tidal inlets. Barrier islands are often linked by the longshore sediment transport system, so that an engineering action taken in any one beach area can have major impacts on adjacent downdrift beaches. Barrier islands are typically low lying, flood prone, and underlain by easily erodible, unconsolidated sediments. Natural sand dunes are formed by winds blowing on shore over the beach, transporting sand landward. Vegetation creates a natural barrier against sea attack and dunes provide a reservoir of beach sand during severe storms and thus help prevent flood and wave damage to adjacent property. In areas where substantial dunes exist, the poststorm beach width can be greater than the prestorm width (National Research Council 1990).

Attempts have been made to mimic nature by promoting the function of artificial dunes. Artificial dunes have been created in many countries around the world, as well as in the United States (ibid.). States where large-scale dune construction has occurred include North Carolina, Texas, Florida, and New Jersey.

The *Dune Restoration and Revegetation Manual* (Salmon et al. 1982) is an excellent source of information concerning the restoration and management of sand dunes.

Site Conditions

Soils

The soils of sand dune communities, as the

name implies, are composed almost entirely of course and fine sand. Sand plays an important role in this community and may need to be managed with fences or artificial barriers during restoration. Koske and Gemma (1992) found that American beach grass (*Ammophila breviligulata*) thrived in areas where sand was actively moving or accumulating, before giving ground to secondary species. Plant succession on sand dunes at Cape Cod, Massachusetts was observed to be highly dependent on mycorrhizal populations (ibid.). Inoculating plantings with mycorrhizae may improve success rates and survival of plants.

Newton (1989) found that sand erosion (substrate instability) was possibly the main factor in determining success or failure of propagules. Dredge spoils with shell debris and larger sized particles serve to armor a site and prevent wind erosion. Although blowing sand is an important natural component of sand dune communities, some type of erosion control may be necessary. Of the three types of erosion control fencing used by Newton (1989) (wood-slat, plastic sand, and hay bale), wood-slat material was the best overall choice.

Although the climate is generally mild, soils are porous and nutrient poor and plants must tolerate salt spray and occasional inundation by salt water. Most plants adapted to the seashore have very large and intricate root systems.

Topography

Sand dunes range from relatively flat (just after a hurricane or tropical storm) to small ridges and mounds of sand that may be a few yards high and several yards wide. The dunes are shaped by the wind and wave action that continually moves the coastal sands.

Moisture requirements and hydrology

The water supply on sand dune habitats is scarce. The plants have many characteristics similar to those of desert plants. Rain may be plentiful along the dunes and barrier beaches, but the sand is so porous that the water is only briefly available to plants, while the hot sun and ocean wind continually dry their tissues. Water quickly seeps down from the surface, seeking the natural water table which may be six yards or more below (Spitsbergen 1980).

The water table is generally a few to several yards below the sand dunes (depending upon their maturity) due to close proximity to mean sea level. Saltwater (in the form of salt spray) is almost constantly being blown into the dunes. Precipitation and occasional flooding also provide inflows of water to sand dune habitats. However, due to the intense sunlight, constant winds, and sandy soils, these communities remain very xeric.

Establishment

Site preparation

The process of dune building is a natural one that takes many years and then may be washed away by a hurricane or storm only to start over again. The process of dune building in restoration, however, may be accelerated by forming a mosaic of small ridges and mounds as found in natural sand dune communities. Bowman (1994) reconstructed the beach area of Long Island's Fire Island (New York) using dredged offshore beach sand. The fill was then graded into protective dunes sixteen feet wide and four feet high, and planted with plugs of American beach grass (*Ammophila breviligulata*). Vegetation plays an important role in stabilizing sand dunes and must be quickly established if dunes are constructed artificially.

Planting considerations

Newton (1989) found that sprigs were more likely to survive than seeds, and if erosion cannot be adequately controlled at a site, then sprigs should be used. The freshness of the plant material used on a project can greatly effect the results. Newton (1989) found that when material was collected nearby, heeled-in on the same day, and quickly planted and irrigated, a high survival rate resulted.

Starting with plants rather than seeds will help establish dune vegetation more quickly. Seeding is difficult because of the dry soil and harsh conditions. Most dune plants are impossible to transplant because of their extensive root systems. A combination of seeding and planting may improve chances of success. Seedlings should be planted with a rootball of soil for their nutrition and stability. Plants without a good rootball can easily be blown over in a strong wind.

Newton (1989) states that direct seeding, under natural conditions (i.e., moving sand), requires a soil amendment to stabilize the substrate. If the substrate is stabilized by artificial means, seeding (with species that grow well from seeds) would be more economical (Newton 1989). The seed mix application rate and ratio of each species within the mix is critical for success. If the application rate is too low, there will be too few plants to control erosion (ibid.). If the rate is too high, competition among plants will reduce the size and vigor of surviving plants (ibid.). The cover of a natural dune community is fifty to sixty-five percent (Duebendorfer 1985).

Spacing considerations

For most beach dune plantings, spacing will depend upon suitable microhabitats for particular species. If plants are placed in unsuitable microhabitats, they may be too wet or too dry. Experimental plantings may be useful to determine appropriate planting zones. Comparisons with nearby natural dune vegetation may prove useful in delineating zones of particular vegetation types.

Temporal spacing for different plants and layers

The highly-adapted plant species in sand dune habitats often occur in large numbers. The native plants that dominate the sand dunes are the first organisms to establish, unless the area has been severely altered and weeds invade. American beach grass (*Ammophila breviligulata*), a primary sand dune pioneer species, spreads three to six feet annually (Fournier 1996). As dunes mature and advance seaward, sand accumulation slows. The established American beach grass no longer receives its needed nutrients. This results in a decline in the health of plants in older stands and stimulates natural plant succession (ibid.).

Newton (1989) noted that some sand dune species (*Solidago, Artemisia*) must be planted at a certain time of the year to achieve optimum survival rates. Planting in phases allows the restorationist to assess previous plantings and make necessary adjustments along the way. Reseeding and/or replanting may need to be done for three to five years to ensure establishment of desired vegetation types.

It is difficult to restore sand dune areas. It may require several years of gradually reclaiming open sand or poor soil, getting the pioneer forbs and grasses established, then gradually adding shrubs (in some areas). This process establishes more favorable microclimates for plant growth and makes successive plantings possible.

Maintenance and Management
Weed control

While severely altered or degraded sand dunes often contain weeds, sand dunes generally do not have the problems with exotics found in other communities. However, the impact of exotics can be overwhelming as they compete with the relatively low diversity of native dune vegetation and, over time, may alter the structure of the community. Some exotic plants, such as the invasive European beach grass (*Ammophila arenaria*), have been introduced to sand dunes to control erosion and stabilize dunes. The amount of exotic plant species removed from sand dune habitats may need to be limited at any one time to avoid serious erosion. Salt spray rose (*Rosa rugosa*), shore juniper (*Juniperus conferta*), and Japanese sedge (*Carex kobomugi*) are species from east Asia that have been used to stabilize dunes. Several other exotic species (from coastal areas outside of the United States) that are planted in dune areas can escape into adjacent dune communities. However, these "escapes" are often local and do not present serious threats to the dune ecosystem. Some exotic weeds that are invasive in dune communities include sea blite (*Suaedea maritima*), orache (*Atriplex patula*), beach wormwood (*Artemisia stelleriana*), dusty miller (*Centaurea cineria*), and horn poppy (*Glaucium flavum*).

Mattoni (1989) found that the disturbance at El Segundo sand dunes (California) caused a shift in succession to an exotic iceplant (*Mesembryanthemum* spp.) savannah with two Australian species of acacia (*Acacia cyclops, A. retinoides*). The reason for success of these *Acacia* species in both disturbed and undisturbed dune habitats may be its capacity of fixing nitrogen in a soil type that is severely nitrogen limited (ibid.). Cowan (1995) found that in revegetating the dunes and swales on California's Monterey Bay after a 1990 freeze killed much of the exotic African ice plant (*Carpobrotus edulis*), intensive weed control is still required to kill ice plant, European beach grass (*Ammophila arenaria*), and ripgut brome (*Bromus diandrus*).

Fertilizers and hay bales (used for mulch or sand fences) while beneficial for some aspects of sand dune restoration have been documented to promote weed density and diversity (e.g., Newton 1989, Mattoni 1989).

Nutrient requirements

Salt spray provides the main source of nutrients on the primary dunes. Salt spray falls on the blades and leaves of dune plants or settles on dune surfaces. Inorganic materials and water contained in the spray quickly leach downward and percolate into the sandy soil. Since nutrient minerals quickly leach out of the sand, these dunes have low fertility. Nitrogen is particularly scarce in dune soil since little or no decaying plant and animal matter accumulates to enrich the habitat. Inorganic nutrients and water are at a premium, especially in the primary dunes, and plant productivity is relatively low. Cycling of organic to inorganic material, so prevalent in decomposition processes in other nearby communities, is greatly reduced on primary dunes.

Native dune plants are adapted to these extreme conditions; they send roots deep into the sand to catch moisture and absorb nutrients. Many dune plants also absorb moisture and nutrient materials directly from salt spray collected on the surface of their blades (Spitsbergen 1980).

Newton (1989) and Mattoni (1989) found that applying fertilizer in dune restoration may produce more robust plants. However, there are some drawbacks of fertilization. Fertilizer, applied at the time of seeding, has a toxic effect on germination (Roberts and Bradshaw 1985, Sheldon and Bradshaw 1977). Fertilization after germination will alleviate this effect, but may not be logistically and economically possible. Newton (1989) noted that at some sites plants that received fertilizer were much more robust, thereby compensating for losses during germination. The major drawback of fertilization is the associated increase in weed growth. The advantages of increased growth of the preferred species, which yield greater soil stabilization, must be weighed against the disadvantage of increased weed growth (Newton 1989).

The hardiest grass and shrub dune species will not need fertilization after a few years, but it will help get them started. Another way to provide plant nutrients and to improve soil is by mulching. Any organic material makes a useful mulch, including seaweed (preferably with the salt washed off). Coarser, fibrous materials will last longer and remain in place better in the wind.

Water requirements and hydrology

While dune plants are well adapted to extreme conditions where water is at a premium, watering plantings in the early phases of establishment may be necessary. There is, however, a trade-off since too much watering may encourage weeds. Irrigation of a site is used for two reasons: to stabilize the substrate and to foster propagule growth (Newton 1989).

Particles of superabsorbent polyacrylamide polymers can be useful when planting. Polyacrylimide is a type of plastic that repeatedly absorbs and releases moisture and does not biodegrade. The polymers retain water that would normally drain from the soil, eliminating fluctuations in moisture level. Polyacrylamide has much potential in the restoration of degraded lands where the soil has lost some of its natural water-retaining capabilities. The particles or dry crystals should be mixed with the soil in the area to be planted. Be sure you purchase polyacrylamides and not polyacrylates. Polyacrylates only absorb and do not release liquid.

The hydrology of the sand dune community is typical of the ocean/beach interface. The water table generally occurs several meters below the dunes. For dune plants, water is obtained by salt spray, occasional flooding, and precipitation. However, some dune plants send down deep root systems in order to reach the water table.

Case Studies

While several cases of sand dune restoration projects are well documented, the majority of the projects had major goals of stabilization and coastal erosion control rather than restoration. The roots of dune plants form a matted growth that penetrates the interior of a dune, helping to hold and stabilize the sand.

In 1984, a twenty-three acre artificial dune system was created from dredge spoils at Buhne Point, California to protect the King Salmon area from tidal erosion (Newton 1989). The goals of the project were to revegetate with native species while stabilizing the sand substrate. Prior to the project, most of the site was destabilized and devoid of vegetation except for the parallel ridges that delineate the northern and southern boundaries of the project. The revegetation design used both seeds and vegetative propagules, with the propagules placed in areas most prone to wind erosion. Wood-slat fencing and hay bales were used to minimize sand blowing onto and within the project site. A "foredune" mix, dominated by sprawling perennials, was used on more stable areas and on ridges defining the north and south boundaries. Many species (*Artemisia, Fragaria, Poa, Tanacetum*) installed as vegetative propagules had high survival rates, while a couple (*Calystegia, Lathyrus*) had low survival rates. Survival rates for *Elymus* were highly variable (fifty-three to eighty-seven percent) and were likely due to proximity to irrigation and length of time held. The desired amount of coverage (approximately fifty percent) of plants with normal stature was achieved during Phase I, however, excessive coverage (eighty percent or more) was produced in Phase II and had to be reconciled.

Thirty acres of dunes and swales on California's Monterey Bay are being restored to native vegetation after a 1990 freeze killed much of the exotic African ice plant (*Carpobrotus edulis*) growing on the dunes, threatening their stability (Cowan 1995). Revegetation teams planted native species such as beach sagewort (*Artemisia pycnocephala*), lizard tail (*Eriophyllum staechadifolium*), and coastal buckwheat (*Eriogonum latifolium*). However, intensive weed control is still required to kill ice plant, European beach grass (*Ammophila arenaria*), and ripgut brome (*Bromus diandrus*).

For additional information about the ecology and restoration of sand dunes, see Clark 1976, Duebendorfer 1985, Fournier 1996, Godfrey and Godfrey 1976, Graetz 1973, Koske and Gemma 1992, Mattoni 1989, Moreno-Casasola and Espejel 1986, Newton 1989, Perkins 1974, Ranwell 1975, Salmon et al. 1982, Spitsbergen 1980, Statler 1993a, Statler 1993b, and Wiedemann (1984).

Floodplain Forests

Introduction

The major difference in swamp forests and floodplain forests is that swamp forests hold water longer than floodplain (or bottomland hardwood) forests. Floodplain forests are temporarily flooded to great depths, but the water easily runs off. Although water depth in floodplain forests may be very deep for a shorter period as a result of river flooding, such floodplain forests typically are flooded during the trees' dormant period. If flooded for very long during the growing season, many species would die and eventually the area would become either marshlike (as sunlight entered a formerly shaded area) or dominated by still more water-tolerant trees (Weller 1994).

Floodplain forests occur as an ecotone between aquatic and upland ecosystems but have distinct vegetation and soil characteristics. They are uniquely characterized by the combination of high species diversity, high species densities, and high productivity (Johnson and McCormick 1979). They process large fluxes of energy and materials from upstream systems. Continuous interactions occur between riparian, aquatic, and upland terrestrial ecosystems through exchanges of energy, nutrients, and species.

Floodplain forests are one of the dominant types of riparian ecosystems in the United States. Historically, the term bottomland hardwood forest (or bottomland hardwoods) has been used to describe the vast forests that occur on river floodplains of the southeastern United States (Mitsch and Gosselink 1986). Huffman and Forsythe (1981) suggest that the term now applies to floodplain forests of the eastern and central United States as well. Estimates of the total area of floodplain forests are quite variable. Although they represent one of the most abundant types of wetland in the United States, they are being lost at an alarming rate to other uses such as agriculture and housing.

Floodplain forests also play significant roles in nutrient cycling. Peterjohn and Correll (1984) found that a fifty-yard-wide riparian forest in an agricultural watershed near the Chesapeake Bay in Maryland removed an estimated eighty-nine percent of the nitrogen and eighty percent of the phosphorus that entered it from upland runoff, groundwater, and precipitation. Elder and Mattraw (1982) and Elder (1985) discovered that, in Florida, the floodplain forests were nutrient transformers rather than nutrient sinks. These riparian forests were net importers of inorganic forms of nutrients and net exporters of organic forms (Elder 1985).

For additional information on ecology and management of floodplain forests, see Johnson and McCormick (1979), Brinson et al. (1981), Clark and Benforado (1981), and Wharton et al. (1982).

Site Conditions

Soils

The soils of floodplains are rich and deep due to deposition of upstream soil that occurs during flooding. In heavily scoured areas, the soils are higher in sand and rocks. In areas where stream velocity is slower, silt and clay accumulate. The soils within the root zone become periodically saturated during the growing season. The soils of the New England floodplain forests are generally mineral soils.

The type of alluvial soil in the floodplain, its capacity to hold water (storage capacity), and the size of the water-holding deposits all help to determine the depth of surface water and the time the floodplain will be under water (Mitsch and Gosselink 1986). Some of the most important characteristics of floodplain soils are soil aeration, organic and clay content, and nutrient content. All of these characteristics are influenced by the flooding and subsequent dewatering of these ecosystems; the characteristics of the soil, in turn, greatly affect the structure and function of the plant communities that are found in the riparian ecosystem (ibid.).

The organic content of floodplain soils is usually intermediate (two to five percent) compared with the highly organic peats (twenty to sixty percent) on one extreme and upland soils (four-tenths to one and five-tenths percent) on the other (Wharton et al. 1982). The alternating aerobic and anaerobic conditions slow down, but do not eliminate decomposition. In addition, the high clay content provides a degree of protection against decomposition of litter and other organic matter on the floodplain. Wharton et al. (1982) suggest that five percent organic content of soils is a good indicator of periodically flooded alluvial floodplains as opposed to the more permanently flooded blackwater swamps of the southeastern United States.

After only a few days of flooding, anaerobic conditions develop in floodplain soils. Once dewatering occurs in the floodplain, aerobic conditions quickly return. Aeration is important for most floodplain species and extended inundation can cause mortality. Soil aeration is dramatically curtailed by flood water, but it is also affected by several other characteristics of the soil, including texture, amount of organic matter, permeability to water flow level of groundwater, and degree of compaction. When floodplain soils are compacted, the air-filled pores dramatically decrease, thereby decreasing soil aeration (Patrick 1981).

Topography

Floodplain forests occur along river terraces. They are adjacent to rivers and streams and directly affected by their actions (i.e., flooding, scouring, washing in propagules, soil deposition). The terrain is typically level, although floodplain forests may also be established at the lower edge of steeper slopes. As the elevation increases from the floodplain, more upland tree species are encountered. The complex microrelief does not always show a distinct change from one zone to the next. The levee next to the stream is often one of the most diverse parts of the floodplain due to fluctuations in its elevation.

Moisture and hydrology

Soils of floodplain forests are often seasonally saturated as a result of overflow from a nearby water body, groundwater, or drainage from adjacent up-

lands. Floodplain forests are hydrologically open to periodic flooding.

Floodplain forests are seasonally flooded, usually in the spring. The habitat may be inundated or saturated by surface or groundwater periodically during the growing season. The flooding water and subsequent groundwater levels are the main determinants of the type and productivity of vegetation found in the riparian zone (the land adjacent to the stream or other water body, that is, at least periodically, influenced by flooding). Flooding waters also bring nutrient-rich sediments to the floodplain, and serve as a primary agent for long-term aggradation and degradation of the floodplain (Mitsch and Gosselink 1986).

Typically, midwestern and eastern United States riparian ecosystems are flooded for several days to several weeks during the spring thaw, while flooding of southern United States bottomland forests is for longer periods and is less apt to be only in the spring. Flooding duration is particularly related to the drainage area of the stream. Bedinger (1981) found that sites that had larger drainage areas had correspondingly longer times in standing water in two Arkansas basins. This is because small watersheds have rapid runoff and sharp flood peaks, while large watersheds have flood peaks that are less sudden and longer lasting.

Periodic flooding usually contributes to higher productivities in at least three ways (Brinson et al. 1981).

■ Flooding provides an adequate water supply for the vegetation. This is particularly important in arid riparian wetlands.
■ Nutrients are supplied and favorable alteration of soil chemistry results from the periodic overbank flooding. These alterations include nitrification, sulfate reduction, and nutrient mineralization.
■ The flowing water offers a more oxygenated root zone than if the water were stagnant. The periodic "flushing" also carries away many waste products of soil and root metabolism, such as carbon dioxide and methane.

Establishment

Site preparation

If the restoration site is severely eroded, the streambanks will need to be stabilized by planting deeply rooted woody vegetation and creating a buffer zone to dissipate surface runoff. Reintroducing riffles and meanders to the stream is often required to achieve a naturalistic ecological community. Large woody debris plays an important role in stream and river ecosystems. Maser and Sedell (1994) provide an excellent detailed discussion of the ecology of wood in streams, rivers, estuaries, and oceans.

Planting considerations

Nokes (1986) found that ashes (*Fraxinus* spp.) can be propagated without a great deal of difficulty. The samaras can be collected when they are no longer green, but when they have returned a light yellow color with crisp-textured wings. Samaras should not be collected from the ground because they are likely to be of poor quality and insect infested. Dewinging before sowing is not necessary. If seeds are being stored, they should be sealed in containers in a cool dry place. Green ash (*Fraxinus pennsylvanica*) remained viable for seven years when dried to seven to ten percent of its fresh weight and stored at fifty-eight degrees Fahrenheit (United State Department of Agriculture 1974).

The seeds may be sown outdoors in the fall after collection in well-tilled and mulched beds (Nokes 1986). Or ash seeds may be stored over winter, then stratified in moist sand or perlite for thirty to sixty days at fifty-eight degrees Fahrenheit, and then sown in a greenhouse or outdoors after all danger of frost has past. Plant the seeds one-fourth to three-fourths of an inch deep in well-drained soil. Seedlings benefit from light shade the first season. Germination rates are variable depending upon species and seed source.

Several sources have indicated that ash is difficult or impossible to root from cuttings. There may be some potential, however, in cuttings from young ashes that have not yet flowered.

Birch (*Betula* spp.) seeds must be collected as soon as the "cones" are full grown and beginning to turn brown, but before they dry completely and open to disperse the seeds from April through June (Nokes 1986). The strobiles can be collected when still slightly green by picking or stripping the tree and then spread out to air dry until they have opened. Seed quality varies from year to year and location to location. Sow seeds immediately or store in sealed containers at temperatures slightly above freezing (United States Department of Agriculture 1974).

Seeds may be sown in the fall after collection or stored and stratified for spring sowing (Nokes 1986). Seeds gathered when mature and sown immediately in well-prepared beds of sandy loam with organic matter (peat moss or compost) often germinate in the highest percentages. The seeds should be pressed firmly into the soil and covered lightly with fine sand. Do not bury the seeds. The bed should be kept moist and shaded during the first growing season. For spring sowing, stratify the seeds thirty to sixty days at fifty-eight degrees Fahrenheit, remove in early spring and plant as recommended above. Successful rooting of cuttings are reported to occur less than fifty percent of the time.

Blankets and cylinders of coir (processed fiber of coconut tree) have been used with a great deal of success to hold plants in place in wetland restoration and streambank stabilization projects. Other benefits of this material are the high tensile strength when wet, durability (up to ten years), and it is 100 percent organic matter (Santha 1994).

Spacing considerations

George Palmiter, a pioneer in naturalistic river management, found that just a few trees strategically placed may be even more effective than a solid stand (Jordan 1984). This is a good point especially since seeds will be washed in by flooding. If a large floodplain is being restored, it is better to plant trees in phases with initial plantings widely spaced to allow for natural filling in. Mowing may be necessary to reduce competition. If erosion is a problem, plantings may need to be denser in erosion-prone areas. Biotechnical erosion control and/or the use of coir blankets may be needed in extreme areas.

Temporal spacing for different plants and layers

The dominant tree species must be established first, then characteristic shrubs and herbaceous plants can be introduced. If intact floodplain forests occur upstream, flooding will likely deposit many seeds or propagules of desired species. Deliberate planting, however, will not only accelerate this natural process but will help achieve desired species composition. Restored floodplain forests are usually quickly colonized by trees, shrubs, and herbs due to the high nutrient content. Some early colonizers will be nonnative species that may need to be controlled depending upon their invasiveness.

Maintenance and Management

Weed control

Several exotic shrubs such as the bush honeysuckles (*Lonicera mackkii, L. morrowii,* and *L. tartarica*), privets (*Ligustrum* spp.), autumn olives (*Eleagnus angustifolia* and *E. umbellata*), and salt cedar (*Tamarix* spp.) have changed the character and successional characteristics of many floodplain and riparian ecosystems. These woody shrubs can be manually removed with a weed wrench or can be controlled by cutting and applying herbicide to the stump.

Nutrient requirements

Nutrients are typically abundant in floodplain forest communities due to several processes. The high clay content of the soils results in higher concentrations of nutrients such as phosphorus, which has a higher affinity for clay particles than for sand or silt particles. The high organic content results in higher concentrations of nitrogen than would be found in upland soils with low organic content (Patrick 1981). The sediment deposited by rivers is fine-grained clays and silt. Nutrients such as phosphorus are likely to be deposited in greater amounts than if the material were coarse grained.

Flooding causes soils to be in a highly reduced oxidation state and often causes a shift in pH. This can lead to both greater availability of certain nutrients and also to an accumulation of potentially toxic compounds in the soil (Mitsch and Gosselink 1986). The periodic wetting and drying of floodplain soils is important in the release of nutrients from leaf litter. The generally high concentrations of nutrients and the relatively good soil texture during dry periods suggest that the major limiting factor in riparian ecosystems is the physical stress of inadequate root oxygen during flooding rather than the inadequate supply of any mineral (ibid.).

Water requirements and hydrology

Floodplain species tolerate periods of both flooding and drought. Supplemental watering of initial plantings will likely increase their survival. Some riparian species that depend upon groundwater will benefit from irrigation during dry periods until their roots have grown deep enough to utilize groundwater.

Kondolf and Micheli (1995) argue for more intensive evaluation procedures for the increasing number of stream restoration efforts. Post-construction evaluation strategies adapted to specific goals must be incorporated into each project design. Designers must also extend themselves further into both past and future by viewing historic channel conditions as potential design models and by monitoring over at least ten years, including surveys after each flood event that exceeds a specified threshold (ibid.).

The direct influence of the river is crucial to floodplain forests. If either is altered, the other will surely change in time because floodplains and their rivers are in a continual dynamic balance between the building of structure and the removal of structure (Mitsch and Gosselink 1986). In the long term, floodplains result from the combination of the deposition of alluvial materials (aggradation) and downcutting of surface geology (degradation) over many years (ibid.). Alluvial materials are deposited on the inside curves of rivers (point bars) and from overbank flooding. The resulting floodplain is composed of alluvial sediments (or alluvium) that can range from 30–270 feet thick. The alluvium, derived from the river over many thousands of years, generally progresses from gravel or coarse sand at the bottom to fine-grained material at the surface (Bedinger 1981). Degradation of floodplains occurs when the supply of sediments is less than the outflow of sediments, a condition that could be naturally caused with a shift in climate or synthetically with the construction of an upstream dam (Mitsch and Gosselink 1986).

Case Studies

For additional information on the restoration and management of flood plain forests, see Koebel (1995), Natural Resources Conservation Service (1996), Messina and Conner (1997), and Trettin (1997).

Bogs and Fens

Introduction

Bogs are rather widely distributed over the humid regions of the earth. They are best developed and most abundant in the cold northern (boreal) forested regions of North America, Europe, and Asia. They differ sharply from swamps and marshes in physical and chemical characteristics and community composition. However, bog communities throughout the northern hemisphere are similar to each other in terms of function, structure, and even species composition. Bogs and, to some extent, fens belong to a major class of freshwater wetlands called peatlands or mires. Called moors in Europe and muskegs in Canada, bogs are peat deposits. The water table in bogs is generally high, yet bogs have no significant inflow or outflow streams. Bogs support acidophilic (acid-loving) vegetation, particularly mosses (Mitsch and Gosselink 1986). Fens, on the other hand, are open wetland systems that generally receive some drainage from surrounding mineral soils. Fens are often covered by grasses, sedges, or reeds (ibid.) and may be considered intermediate between bogs and marshes. However, they are typically early successional stages of bog communities. Rich fens, high in calcium, give rise to white cedar (*Thuja occidentalis*) swamps where peat accumulation is limited (Crum 1992). But fens of lesser richness build up enough peat to alter the movement and chemistry of water, eventually to the extent that they are transformed into acid bogs receiving water and nutrients only from the atmosphere. In the change from mineral-rich to mineral-poor peatlands, sphagnum moss (*Sphagnum* spp.) plays an important role. Peatlands range from sedgy intermediate fens through wet *Sphagnum* lawns, or poor fens, to open bogs covered by a low shrub, hummock-hollow complex, and eventually, black spruce (*Picea mariana*) bog (ibid.).

Most bogs have several features in common. They usually develop where drainage is very poor; all have cushion-like vegetation, and all have an accumulation of peat that forms an impenetrable layer that seals off any groundwater input. Bogs are nutrient poor (although many nutrients may be present, they are mostly unavailable to plants). Most at some time have (or have had) a marginal semifloating mat of vegetation, usually sphagnum moss and heaths (usually leatherleaf [*Chamaedaphne calyculata*]). Fens are structurally similar to bogs. The main difference is that fens are calcareous (i.e., basic as opposed to acidic), receive input from groundwater, and are nutrient rich.

Raised bogs or high moors develop on drier ridges surrounding dystrophic lakes and bogs on flat, poorly drained areas (Dansereau and Segadas-Vianna 1952). Sphagnum moss becomes established in depressions on drier ridges. The spongy sphagnum with its tremendous water-holding capacity retains more and more precipitation. As a result mesophytic trees are replaced by trees and shrubs that live in a wet, acidic environment. The sphagnum accumulates, often to a height of over one yard, and rises above the adjoining ground in large, cushionlike mounds. As the mounds increase in size, they join. The whole area becomes a raised bog surrounded by a swampy moat. Like the flat bog, the raised bog is eventually occupied by concentric zones of vegetation interrupted by bog pools and puddles.

Many classification schemes exist for peatland communities. According to Mitsch and Gosselink (1986), classification using hydrologic and chemical conditions usually defines three general types of peatland: (1) minerotrophic (true fens), (2) ombrotrophic (raised bogs), and (3) transition (poor fens). Ombrotrophic bogs are isolated from mineral-bearing groundwater and thus display lower pH, lower nutrients, lower minerals, and more dominance by mosses than do minerotrophic fens.

Life in the bog lake is restricted in number of species, but the organisms present may be abundant. A variety of aquatic insects are represented including dragonflies, damselflies, stoneflies, beetles, and flies. Fish are few in number or absent, but amphibians, especially frogs, are universally present. Warblers are the most abundant birds of the bog forests and the bog lemming is a common small mammal.

The bog turtle (*Clemmys muhlenbergii*) is considered to be America's rarest turtle (Nickens 1994). Efforts are underway to maintain or restore the bog habitats needed by the bog turtle which is protected in each of the twelve eastern states where it is known to occur. The bog turtle's breeding habitat of wet, open-canopy glades, once maintained by beavers, bison, or elk, is now being managed by humans by damming and tree girdling.

Many studies have focused on bogs because of their vast area in temperate climates, the economic importance of peat as a fuel and soil conditioner, and their unique biota and successional patterns. Bogs are also important for the anthropological and ecological archives they contain. The pollen deposited year after year, century after century, provides a continuous datable record of climate and vegetation. Many bogs show alternating layers of sphagnum and sedge peat which indicate changed conditions in the past (Curtis 1959).

Many areas that were once open bogs have been converted to cranberry marshes, which completely destroys the original community in the process. Other disturbances, such as peat mining, have contributed to the decline of bogs and the habitat that supports bog species. Peat is formed at a very slow rate. Various estimates in the Great Lakes region range from 100 to 800 years per twelve inches. Peat mining is not sustainable. Crum (1992) postulated that regeneration is unlikely in the Great Lakes region. Because sphagnum will not grow where the water table is lowered, owing to difficulties in getting capillary water, the early stages of succession cannot be repeated (ibid.). Crum (1992) further states that as a natural resource, peat should indeed be used, but peatlands have some values greater than monetary. Environmental impact studies during and after exploitation should include monitoring the

water level and quality in adjacent peatlands, as well as measuring the rate of outflow and its effect on silting and streambank erosion below the area of discharge (ibid.).

Anderson and Kurmis (1981) and Green (1983) found that the revegetation of mined peatlands is variable. Recovery is very slow if drainage ditches have lowered the water table and there is a low concentration of nutrients. If the ditches are blocked, however, the higher water table will promote a good recovery within ten years. Trees will typically colonize spoil banks because of their deeper aerated zone. Green (1983) suggests that aluminum toxicity may be important in restricting plant colonization, but the primary limiting factor seems to be the position of the water table.

While some "reclamation" has focused on natural communities, many mined peatlands are used for growing a variety of agricultural crops such as carrots, radishes, beets, potatoes, onions, tulips, lettuce, celery, mint, wild rice, blueberries, and strawberries. In addition, many peatlands and other wetland communities in the northeastern United States are exhibiting increases in acidity due to acid rain. Rain is normally somewhat acid (pH 5.6), but in the last three decades, it has become much more acid, at least in areas subject to atmospheric pollution (Crum 1992). Acid rain in the northeastern United States can be traced, in large part, to sulphur dioxide pollution produced by the coal- and oil-burning industries of the Ohio Valley (ibid.).

Site Conditions

Soils

In bogs, organic debris is only partially decomposed and forms peat. Peat is formed from sedges, shrubs, and trees but also sphagnum moss. Water adjacent to bog vegetation may be acid or alkaline, but water with sphagnum moss, the dominant plant of bogs, is always acid. The soils of bogs are organic soils of peat or muck. Peat, according to one of several definitions, is a soil formed under water-soaked, anaerobic conditions, accumulated to a depth greater than twelve inches drained and eighteen inches undrained and having an organic content of at least twenty percent and an ash content of less than fifty percent (Crum 1992). Its physical and chemical properties are based upon its content of organic matter, minerals, gas, and water, and that content varies with vegetational origin and degree of decomposition (ibid.). Peat can be classified according to vegetational origin (aquatic, sedge, wood, herbaceous, or *Sphagnum*) or degree of decomposition (fibric, hemic, or sapric). The principal components of peat, in addition to water, are cellulose, lignin, and colloidal humic substances resulting from their incomplete decay (ibid.).

The accumulation of peat in bogs is determined by the production of litter (from primary production) and the destruction of organic matter (decomposition) (Mitsch and Gosselink 1986). As with the primary production, the rate of decomposition in peat bogs is generally low due to (1) waterlogged conditions, (2) low temperatures, and (3) acid conditions (Moore and Bellamy 1974). Besides leading to peat accumulation, slow decomposition also leads to slower nutrient recycling in an already nutrient-limited system. As pH decreases, the fungal component of the decomposer food web becomes more important relative to bacterial populations (Mitsch and Gosselink 1986).

Peatlands can also develop by the water soaking of well-drained slopes. Blanket bogs can occupy a sloping terrain because of watertight soils such as fine-grained and compacted clay (Crum 1992). Coarse soils are permeable at the surface, but downward percolation may bring about podzolization, a process by which oxides of iron and aluminum leach downward and accumulate as a hardpan (ibid.). Pan formation is favored by an acid litter produced by conifers or heath plants. The accumulation of litter in itself causes water retention and oxygen-poor conditions that limit decomposition and favor bog initiation (ibid.).

Fen peats are alkaline due to the high level of calcium and magnesium carbonates present in their spring waters. Fens develop under the influence of mineral-rich ground or surface water (Crum 1992). But, as peat accumulates, a fen receives reduced groundwater input, sphagnum growth increases, nutrient availability decreases, and acidity increases (lower pH), favoring bog development. However, not all fens develop into bogs. Some are successional stages of swamps or marshes. Also, disturbances (fires, flooding) may set back succession prolonging the existence of some fen communities.

Many terrestrial plants utilize fungi in order to absorb minerals from the soil. Mycorrhizae have a limited significance in wet, acid peatlands, but heaths

and orchids are mycorrhizal, and mycorrhizae have been demonstrated in tamarack, black spruce, and other woody plants of bogs and fens (Crum 1992). Some mycorrhizal fungi also protect the host from soil toxins (ibid.). The result of this symbiosis is that both host and fungus can exploit habitats unsuited to either alone. It is interesting that one family characteristic of peatlands, sedges, have no such associations.

Forested bog develops mainly on accumulated peat, fine grained marl, and organic muck. The soil of shrub bogs is generally hydric organic. Hellmann (1984) found that a high pH caused by the leaching of calcium and magnesium from the dolomitic escarpment on which a created forested bog was situated caused black spruce to be stunted and chlorotic. So, to simulate bog conditions, pine and spruce chips (donated by a local utility company) were brought in to provide an organic growing base. Hellmann concluded that the "slow, acidic decay is creating a favorable habitat, the spruces are thriving, and other bog species are becoming increasingly successful."

Topography

Bogs and fens occur in low wet areas that are generally flat, but sometimes there are minor ridges or swales present. However, there are some notable exceptions such as "raised bogs," "blanket bogs," or "domed or sloped bogs." Topographic variations may be correlated with differences in climate, water source, and water movement (Crum 1992). Some flat fens develop into raised bogs of slight convexity. Some raised bogs of oceanic Britain exceed one mile in width and fifteen or even thirty feet in height (Crum 1992). The centers are nearly flat and the marginal slopes gentle. When raised bogs develop on slopes they are called "eccentric bogs."

Favorable humid climates allow peat to literally "blanket" very large areas far from the site of original peat accumulation. Peat in these areas can generally advance on slopes of up to eighteen degrees; extremes of twenty-five degree slopes covered by blanket bogs have been noted in western Ireland (Moore and Bellamy 1974). A bog or fen may develop on a slope where water accumulates or flows rather continuously. In a few cases, wetland plant associations form on hillsides where groundwater discharges, so that mosses, sedges, and cattails can be seen in patches on otherwise dry road cuts or hillsides.

Slope bogs are similar to blanket bogs but limited in extent by topography. Usually small, they probably originate as a result of infilling at the base of a slope followed by waterlogging and upslope invasion by peatland vegetation (Crum 1992). Blanket bogs are more extensive, covering depressions as well as irregular terrain. In Wisconsin, fens sometimes occur on hillsides overlooking existing or extinct glacial lakes (Curtis 1959). Nonetheless, most bogs and fens are found in relatively level areas which contain depressions, whether in valleys or on mountain tops. Crum (1992) provides descriptions of types of peatland as categorized by surface topography.

Moisture and hydrology

The majority of bog wetlands are completely saturated for a large part, if not all, of the year. Many have open water areas that may be very deep. The wetness or dryness of bogs determines which vegetation types develop. In general, the wetter the bog, the more likely it will be dominated by sphagnum or sphagnum and sedges. Dryer bogs are more likely to be dominated by shrubs and trees. Fens can be succeeded by coniferous swamps or by bogs, depending upon water movement and quality. Conditions favorable to the continued growth of *Sphagnum* lead toward a bog sequence (Crum 1992).

Drainage ditches have had a serious impact upon peatland vegetation by promoting the growth of trees and shrubs at the expense of the more hydrophilous sedges and herbs (Glaser 1987). In eastern Minnesota, ditching has enhanced the growth of various shrubs (particularly bog birch [*Betula pumila*]) and trees.

Bogs and fens occur where precipitation exceeds potential evapotranspiration, creating a net water surplus (Mitsch and Gosselink 1986). There must also be some impediment, if only lack of relief, that prevents the rapid drainage of the surplus water (ibid.). Large expanses of the boreal and northern deciduous zones of Europe, Siberia, and North America have humid conditions with flat terrain and a high groundwater level, an ideal environment for bog or fen formation (Walter 1973). Where the groundwater level is eighteen inches or more below the surface, forest vegetation is possible; for higher water tables, bogs and fens may dominate (Mitsch and Gosselink 1986).

The only water source for many bogs is rainwater. Bogs raise above the groundwater by an accumulation of peat. Fens, however, differ in that they receive groundwater. In a fen, the constant flow of water is usually provided by an artesian spring. Some bog wetlands actually may draw water from lower levels by capillary action in the organic substrate.

Bogs can develop either from aquatic ecosystems through the filling of a basin or from terrestrial ecosystems through the spreading of blanket bogs through wooded areas (called paludification) (Mitsch and Gosselink 1986). It should be emphasized that not all lakes and forests in the boreal and northern deciduous zones are destined to become bogs and fens. Mitsch and Gosselink (1986) discussed the three major processes of bog formation.

■ Flow-through succession—development of a bog from a lake basin that originally had continuous inflow and outflow of surface and groundwater.

■ Quaking bog succession—involves the filling of the basin from the surface, creating a quaking bog. These bogs develop only in small lakes that have little wave action; they receive their name from the quaking of the entire surface that can be caused by walking on the floating mat.

■ Paludification—bog evolution occurs when blanket bogs exceed the basin boundaries and encroach on formerly dry land. This process of paludification can be brought about by climatic change, by beaver dams, by logging of forests, or by natural advancement of a peatland (Moore and Bellamy 1974).

Shrub bogs are wet most of the year, but standing water may be present only in the winter or spring. Adequate moisture is important in maintaining shrub bog vegetation. The site must be wet enough to discourage larger woody growth (i.e., trees) but not too wet for shrubs.

Establishment

Site preparation

Amon and Briuer (1993) created a fen near Beaver Creek, Ohio. Groundwater was introduced by excavating to groundwater levels. With groundwater less than three feet below the site surface, enough soil was removed to cause groundwater to be at, or very near, the surface. Since the layer of clay-like mineral soil probably would not permit good flow of groundwater, the soil was removed down to a layer that appeared to be a more heterogenous till. The organic soil, mixed with some of the less organic overburden, was stockpiled. The clay layers were removed to a nearby upland for disposal. With the excavation of the clay layer completed, the best-quality organic soil was returned to the site to establish the necessary surface elevation compatible with desired hydrologic conditions.

Water entering the fen is direct rain or groundwater. A ditch was cut into the downslope end of the swale as an outlet and then extended over a hundred meters toward the natural wetland. During the winter months, most of the swale became saturated with water. It became necessary to even out the surface after the soil settled to maintain a downward slope toward the exit ditch.

Phillips (1985) created a montane bog of the southern Appalachian Mountains, a northern boreal community featuring large cranberry (*Vaccinium macrocarpon*) and associated bog species in North Carolina. The bog was constructed by removing the native clay soil to a depth of twenty-four inches and then backfilling with a mixture of predominantly rotted oak leaves, two bales of peat moss, and a dusting of cottonseed meal. Then the area was planted with sphagnum moss (*Sphagnum* sp.) which, when established, provided a suitable growing mat for herbaceous species. Supplemental watering was required as a part of the maintenance plan, as well as weeding, and curbing the spread of marsh fern (*Thelypteris* sp.) several times during the year.

Sollenberger (1984) used extensive grading and the addition of twelve to twenty-four inches of gravel covered by a meter of calcareous peat to construct a fen in Illinois. This constructed fen was then artificially irrigated to provide a constant flow of water.

Planting considerations

Bog plants are primarily planted as plugs, bareroot plantings, or divisions of rhizomes, tubers, corms, bulbs, or of vegetative portions for some species. Seeding is always experimental as germination sites may not be present. The hydrology and soil water chemistry must be appropriate for any bog species. Species must be carefully chosen keeping in mind the successional stage of the bog being restored (see Jones and Foote 1990).

Glattstein (1994) found that rose pogonia (*Pogonia ophioglossoides*) is quite amenable to

cultivation if it is given appropriate growing conditions. It has fibrous roots that grow just beneath the surface, be it sphagnum or sandy, peaty soil. These stoloniferous roots are a significant means of increase, readily producing new plantlets. Their shallow placement also makes this plant very susceptible to drought; it dies rather rapidly when conditions become even briefly dry. In cultivation, it thrives in sandy, peaty, acid conditions with a constant supply of water. Summer heat does not seem to affect rose pogonia. Many peatland shrubs can be propagated by taking cuttings.

In the absence of disturbances, the shift in dominance to woody plants is usually rapid, but the exact rate is considerably influenced by the size of the local rabbit populations (Curtis 1959). The willows (*Salix* spp.), dogwoods (*Cornus* spp.), and bog birches (*Betula* spp.) are all highly palatable to these animals and their annual growth may be completely pruned back during the winter.

Spacing considerations

For most bog plantings, spacing will depend upon suitable microhabitats for particular species. If plants are placed in unsuitable microhabitats, they may be too wet or too dry. Experimental plantings may be useful to determine appropriate planting zones. Comparisons with nearby bogs or fens may prove useful in delineating zones of particular vegetation types.

Temporal spacing for different plants and layers

The hydrology must be in place before restoration can begin. In a bog, saturated soils or water levels very near the surface are important. Fens are groundwater supplied and must have an outlet for the water. Once the hydrology has been established and stabilized, planting may begin. There should also be a healthy growth of sphagnum moss prior to the establishment of sedges, shrubs, or trees. Bog plants will likely colonize a suitable site without planting. However, if it is a long distance to the nearest natural bog, then planting may accelerate recovery.

Larger peatlands may contain bog complexes or several bog communities in various stages of development. These areas should be divided into separate planning units as they will require different management techniques. Older parts of the bog, at the outer margin of the mat, support an open forest of black spruce. *Sphagnum* continues to be abundant in this forested area, but leatherleaf is replaced by the more shade-tolerant rusty Labrador-tea. The bog surface usually has a slight convexity and the slope drainage has some significance in the establishment of a black spruce stand.

Bogs are sphagnum moss-dominated communities. So, the first step in restoring a bog is to establish sphagnum at the site. Unless a new bog is being created, sphagnum mosses are likely already present at the site.

Maintenance and Management

Weed control

Bog plants are highly specialized and few weeds persist in this harsh environment with waterlogged conditions, acid waters, low nutrients, and extreme temperatures. Jorgensen and Nauman (1994) found that the disturbance caused by the presence of commercial cranberry beds can extend into the adjacent wetlands. They found that the relative abundance of certain native species increased closer to the disturbance. The only exotic reported was devil's-paintbrush (*Hieracium aurantiacum*), a native of Europe, which can be controlled by mechanical means.

Cat-tails (*Typha* spp.) and common reed (*Phragmites communis*), more indicative of fens and marshes, may become established in disturbed bogs. Fens are more nutrient rich and weeds of wet prairies and grasslands, marshes, and/or swamps may pose a threat. See the weed control section of the following restoration types: wet prairies and grasslands, nonestuarine marshes, and estuarine marshes.

Nutrient requirements

Bog waters are extremely low in nutrients and form acidic peats. Available nitrogen, potassium, and phosphorus are tied up in the peat. These waters have a high carbon dioxide content and possess traces of hydrogen sulfide, the result of activity by sulfur bacteria. Rainwater is the principle source of incoming nutrients in bogs.

Fen peats, on the other hand, are not acidic. Fens receive nutrients from sources other than precipitation, usually through groundwater movement. Sollenberger (1984) stated that the growth of fen

indicator species is probably not dependent upon the high pH of fens. These species also seem to grow well on neutral soils. It is probable that these species claim fens as their niche because they more efficiently utilize nutrients available in high pH environments (ibid.).

Soil water chemistry is one of the most important factors in the development and structure of the bog ecosystem (Heinselman 1970). Factors such as pH, mineral concentration, available nutrients, and cation-exchange capacity influence the vegetation types and productivity. Conversely, the plant communities influence the chemical properties of the soil water (Gorham 1967). In few wetland types is the interdependence between chemistry and ecosystem productivity as apparent as in northern peatlands (Mitsch and Gosselink 1986).

The pH of peatlands generally decreases as organic content increases with development from a minerotrophic fen to an ombrotrophic bog (Mitsch and Gosselink 1986). Bogs commonly have a pH of about three to four, poor fens four to six, and richer fens about six to seven and one-half (Crum 1992). This range of pH is equal to the difference between vinegar and sea water. Fens are dominated by minerals from surrounding soils, while bogs rely on a sparse supply of minerals from precipitation. Bogs are exceedingly deficient in available plant nutrients; fens, with groundwater and surface water sources, have considerably more nutrients (Moore and Bellamy 1974). While bogs have a shortage of all available nutrients, phosphorus and potassium are more important limiting factors than nitrogen. Goodman and Perkins (1968) found that potassium was the major limiting factor for growth of tussock cotton-grass (*Eriophorum vaginatum*) in a bog in Wales, while Heilman (1968) found that levels of phosphorous and to a lesser extent, potassium, were deficient in black spruce (*Picea mariana*) foliage in a sphagnum bog in Alaska. Bog formation in its latter stages is essentially limited by the input of nutrients by precipitation.

The ability of roots to take up nutrients is impaired at pH levels less than five. Temperatures too low for metabolic processes, as well as the lack of oxygen, further reduce nutrient uptake (Crum 1992). Macronutrients are generally available for uptake at pH four to nine, but some of the micronutrients show much narrower ranges of availability. The total content of most minerals decreases along a fen-to-bog gradient. Most of them have their greatest availability below pH seven and one-half, but restricted availability at four and one-half to five (ibid.).

Prior to litter fall in autumn, some bog plants, notably cotton grass (*Eriophorum* spp.), translocate nutrients back to organs surviving over the winter. This process provides nutrient reserves for the following year's growth and seedling establishment (Mitsch and Gosselink 1986). Bog litter has been demonstrated to release potassium and phosphorus, often the most limiting nutrients, more rapidly than other nutrients. This adaptation keeps these nutrients in the upper layers of peat (Moore and Bellamy 1974).

Leaf assays of bog heaths show phosphorus and potassium at levels inadequate for plants of other habitats, but the amounts of iron and manganese are much more than adequate (Crum 1992). A high content of manganese in bog plant tissues suggests that tolerance to that mineral, generally toxic in all but slight amounts, bestows a competitive advantage (ibid.).

Another adaptation to nutrient deficiency in bogs is the ability of carnivorous plants to trap and digest insects. This special feature is seen in several insectivorous bog plants, including pitcher-plant (*Sarracenia* spp.), sundew (*Drosera* spp.), butterwort (*Pinguicula* spp.), and bladderwort (*Utricularia* spp.). Some bog plants also carry out symbiotic nitrogen fixation. The bog myrtle (*Myrica gale*) and alder (*Alnus* spp.) develop root nodules characteristic of nitrogen fixers and have been shown to fix atmospheric nitrogen in bog environments (Mitsch and Gosselink 1986). Bog plants have also adapted to being overgrown by peat mosses. These plants must raise their shoot bases by elongating their rhizomes or by developing adventitious roots (ibid.). Trees such as pine, birch, and spruce are often severely stunted due to the moss growth and poor substrate; they grow better on bogs where vertical growth of moss has ceased (ibid.).

Peatland communities depend upon the chemical environment and the availability of some nutrients. Nutrients become locked up in the accumulating peat. Nutrient cycling depends upon decay, and as decay is limited, so are the available supplies of phosphorus, nitrogen, and potassium (Crum 1992). The effective use of atmospheric input, scant as it is, makes possible the continued development from fen to bog (ibid).

Amon and Bruier (1993) applied aqueous fertil- izer containing ammonium nitrate to leaves and stems of plants in selected parts of a created fen during late spring and early summer. They found that it did not affect growth or coloration of the plants, leading to the conclusion that nitrogen limitation was not likely. They observed that higher rates of nitrogen fixation occurred in the wetter areas during the daylight hours. Denitrification, the loss of nitrate to atmospheric nitrogen, was also greatest in the wettest areas.

Water requirements and hydrology

Bogs and fens are saturated much of the year. While the plants characteristic of these communities are stressed by these conditions, they are well adapted to life in a wet environment. The species composition and vegetation structure of bogs and fens is determined by the relative amount of water present in a bog. Sphagnum lawns or sphagnum-sedge-dominated bogs (open bogs, herbaceous bogs) occur on the wettest sites while forested bogs occur on dryer sites. A disturbance such as fire or flooding can set back succession from a forested bog to a sphagnum lawn.

Amon and Briuer (1993) found that groundwater flow into a created fen was sufficient for fen development. Water levels rose and fell normally in response to rainfall, and anaerobic soil zones and dense vegetation developed.

An artificial irrigation system may be used in fen construction to simulate a spring (Sollenberger 1984). Sollenberger (1984) designed a constructed fen by grading the site and adding twelve to twenty-four inches of gravel covered by a meter of calcareous peat. Irrigation lines were placed in the gravel layer. These lines were intentionally punctured so that water would be absorbed by the peat. The amount of water entering the system could be controlled. Obviously, this type of fen creation could be carried out only on a small area.

Disturbances

While bogs and fens are typically very wet environments, peat burns well when it has experienced drought conditions. Bog fires can be hard to control especially when they burn underground. Fire has been used to deepen bog wetlands in some north-central states, but it is very difficult to direct and extinguish (Linde 1969). Historical evidence from core samples of bog peats provides evidence that fire has played a role in some peatland communities.

Leatherleaf sometimes covers the hummock-hollow topography of the open bog. Because its spreading rootstocks are protected by wet sphagnum, leatherleaf survives superficial burns, but more severe burns favor an ingress of blueberries (Crum 1992). Indians and white settlers used to burn bogs to encourage blueberries. More severe fires, destroying the hummocks in which trees and shrubs are rooted, cause a reversion to a wet sedge fen or even standing water. If natural or artificial drainage has lowered the water table, a considerable volume of the surface peat layers may be removed by fire resulting in an open bog or sedge meadow community (Curtis 1959).

Curtis (1959) found that when fens are protected from external disturbance, they tend to develop into shrublands. In natural systems, fire probably maintained the fen. The fire burned the mulch and top growth of the community with little danger to the peat beneath because of the steady water supply. In herbaceous fens, an increase in shrub density often causes a decrease in orchid flower production (Curtis 1946), so shrubs may need to be controlled if a site is being managed for particular rare species. Amon and Briuer (1993) burned a created fen in early December to give slow-growing plants a chance to receive light in the spring. Preliminary data show that the burning released the nutrient phosphate, which is being flushed away from the site by the high water table.

Case Studies

Researchers at Wright State University (Dayton, Ohio) created a fen in the 1000-acre Beaver Creek wetlands along Big Beaver Creek (Amon and Briuer 1993). The goals were to demonstrate methods needed to create a fen and to test the possibility of using free-flowing groundwater to hydrate the created wetland. Initial excavation, which began in October 1991, resulted in a swale approximately ten feet wide by forty feet long. Groundwater flow into the fen was sufficient. Water levels rose and fell normally in response to rainfall. Anaerobic soil zones and dense vegetation developed.

Amon and Bruier (1993) observed that excavation should not proceed without adequate assessment of the groundwater topography. Microscale water table irregularities do occur. Seepage points can be seen during excavation and the permeability of the substrate can be assessed. The success or failure of a groundwater-driven wetland depends upon the ability of the groundwater to enter the excavation. Drying of the newly created soil surface occurred prior to plant establishment, but was not problematic. The soil rapidly developed anaerobic zones after planting, as indicated by repeated core samples. Plant establishment from plugs, transplants, seedbank, and planted seed was very successful for the wetland. Data analyses are planned for geological, chemical, and biological observations of the fen over several seasons.

As a part of a long-range study, simulation models were used in central Michigan to investigate the feasibility of using a rich fen community for disposal of treated sewage (Kadlec and Tilton 1979). Models were designed to (1) give long-term predictions on the ability of the wetland to maintain its structural and functional integrity and (2) design an optimal disposal scheme for the wastewater (Dixon 1974). A large-scale simulation model, analogous to chemical process models used by chemical engineers, was developed as part of that study (Parker and Kadlec 1974).

For additional information on the restoration and management of peat lands, see Messina and Conner (1997), Shear et al. (1992), Shear et al. (1996), and Trettin (1997).

Swamp Forests

Hardwood–Conifer Swamp A-257
Deciduous Swamp Forest A-293, A-389
Coniferous Swamp Forest A-294, A-390
Swamp Forest A-315
Cypress–Tupelo Swamp Forest A-333
Atlantic White-cedar Swamp A-334
Southern Swamp Forest A-356
Baygall Forest A-359
Dome Swamp Forest A-360

Introduction

Swamp forests are wetlands dominated by woody trees or shrubs with relatively deep and often extensive water that prevents invasion of herbaceous plants. They usually occur in basins formed by rivers. The major difference in swamp forests and floodplain forests is that swamp forests hold water longer than floodplain (or bottomland hardwood) forests. Floodplain forests are temporarily flooded to great depths but the water runs off easily. Although water depth in floodplain forests may be very deep for a shorter period as a result of river flooding, such bottomlands typically are flooded during the trees' dormant period. If flooded for very long during the growing season, many species would die and eventually the area would become either marshlike (as sunlight entered a formerly shaded area) or dominated by still more water-tolerant trees (Weller 1994).

For additional information on the ecology of swamp forests, see Allen et al. (1989) and Wharton et al. (1982).

Site Conditions

Soils

For creating forested wetlands, many practitioners have had good success using "wetland humus" (topsoil salvaged from wetlands that are being developed). This technique is often effective as propagules and seed banks are transported along with the soil. However, be sure to not use soil salvaged from wetland areas containing nuisance plants such as purple loosestrife (*Lythrum salicaria*) and common reed (*Phragmites australis*). Also, do not let soil sit long as cat-tails (*Tyha* spp.) will invade it.

Dome swamp forests (A-360) are usually underlain by an impermeable clay layer and sometimes by a hardpan, a layer of consolidated and relatively impermeable material (Mitsch and Gosselink 1986). Both layers impede downward drainage, although there is often some exchange of groundwater with the surrounding upland as infiltration radiates outward from the dome rather than vertically (Heimburg 1984).

Topography

Deciduous swamp forests in the northeastern United States have characteristic "mound-and-pool" microtopography, also referred to as "pit-and-mound" or "hummock-and-hollow" (Golet et al. 1993). This situation increases the diversity of the plant community, since mounds provide establishment sites with a relatively wide range of flooding frequencies and moisture conditions. The height of mounds in natural wetlands apparently depends upon both hydrology and vegetation, with the wettest forested swamps having the most pronounced microrelief (ibid.). Reproducing a diverse microtopography is the key to restoration of wetland forests.

In natural forested wetlands, fallen trees or tip-ups are an integral part of the dynamics of the ecosystem. They contribute to the mound-and-pool topography that plays an important role in the development of wetland vegetation. Restorationists need to keep this in mind when restoring or enhancing these community types. In fact, in newly constructed swamp forests, some practitioners have even used existing logs and large branches to create structural diversity.

The transition to adjacent lands should be gradual. A gradual transition will increase the likelihood of providing a range of water levels that will support a variety of wetland vegetation types.

Moisture and hydrology

The major loss of water from dome swamp forests (A-360) is evapotranspiration. Infiltration is rapid during the dry season, but relatively slow during the wet summers, when water levels surrounding the dome forests are also high (Heimburg 1984).

Dome swamp forests have standing water that is mostly the result of rainfall and surface inflow, with little or no groundwater inflow.

Establishment

Site preparation

Grading is crucial to the success of swamp forest restoration. Mound-and-pool microrelief should be created to provide adequate sites for plants that characterize the communities. The transition to adjacent plant communities should be gradual to discourage erosion and create a naturalistic continuum.

Barry et al. (1996) hydroseeded a mix of three herbaceous wetland species (fox sedge [*Carex vulpinoidea*]; blue vervain [*Verbena hastata*]; and swamp milkweed [*Asclepias incarnata*]) in the fall to create a ground cover and stabilize the soil the following spring before planting trees.

Penfound (1952) defined the freshwater deep swamp as having "fresh water, woody communities with water throughout most or all of the growing season." Swamp forests are diverse communities due in part to the microtopography and the associated moisture conditions and differences in flooding frequencies present on hummocks or mounds. These mounds increase the margin of tolerance for plants placed in fluctuating hydrological conditions.

Planting considerations

Barry et al. (1996) found that a major advantage of the mound-and-pool microrelief is that high quality, expensive woody transplants can be placed on the mounds with some assurance that they will not be killed by flooding, a major cause of plant mortality in created wetlands.

Many oaks are difficult to transplant because of deep taproots that develop at an early age. The root system of pin oak spreads more than that of most oaks, making it easier to transplant and therefore, the most widely offered and planted oak species (Glattstein 1994). Swamp white oak has a deep taproot and is difficult to transplant.

The seeds of swamp forest trees, including cypress and tupelo, require oxygen for germination. This means that occasional drawdowns, if only at relatively infrequent intervals, are necessary for persistence of these deepwater swamps (Mitsch and Gosselink 1986). Otherwise, continuous flooding will ultimately lead to an open-water pond, although individual cypress trees may survive for a century or more. Cypress seeds and seedlings require very moist, but not flooded, soil for germination and survival (Mattoon 1915). After germination, cypress seedlings survive by rapid vertical growth to keep above the rising waters

Nokes (1986) found that ashes (*Fraxinus* spp.) can be propagated without a great deal of difficulty. The samaras can be collected when they are no longer green, but when they have turned a light yellow color with crisp-textured wings. Samaras should not be collected from the ground because they are likely to be of poor quality and insect infested. Dewinging before sowing is not necessary. If seeds are being stored, they should be sealed in containers in a cool dry place. Green ash (*Fraxinus pennsylvanica*) remained viable for seven years when dried to seven to ten percent of its fresh weight and stored at fifty-eight degrees Fahrenheit (United States Department of Agriculture 1974).

The seeds may be sown outdoors in the fall after collection in well-tilled and mulched beds (Nokes 1986). Ash seeds may be stored over winter, then stratified in moist sand or perlite for thirty to sixty days at fifty-eight degrees Fahrenheit, and then sown in a greenhouse or outdoors after all danger of frost has past. Plant the seeds one-fourth to three-fourths of an inch deep in well-drained soil. Seedlings benefit from light shade the first season. Germination rates are variable depending upon species and seed source.

Several sources have indicated that ash is difficult or impossible to root from cuttings. There may be some potential, however, in cuttings from young ashes that have not yet flowered.

Birch (*Betula* spp.) seeds must be collected as soon as the "cones" are full grown and beginning to turn brown, but before they dry completely and open to disperse the seeds from April through June (Nokes 1986). The strobiles can be collected when still slightly green by picking or stripping the tree and then spread out to air dry until they have opened. Seed quality varies from year to year and location to location. Sow seeds immediately or store in sealed containers at temperatures slightly above freezing (United States Department of Agriculture 1974).

Seeds may be sown in the fall after collection or stored and stratified for spring sowing (Nokes 1986). Seeds gathered when mature and immediately sown in well-prepared beds of sandy loam with organic matter (peat moss or compost) often germinate in the highest percentages. The seeds should be pressed firmly into the soil and covered lightly with fine sand. Do not bury the seeds. The bed should be kept moist and shaded during the first growing season. For spring sowing, stratify the seeds thirty to sixty days at fifty-eight degrees Fahrenheit, remove in early spring, and plant as recommended above. Successful rooting of cuttings are reported to occur less than fifty percent of the time.

Spacing considerations

As a general rule, trees should be planted on about ten-foot centers. They should not be planted on regular intervals but instead should be staggered to appear more natural. The microrelief and location of mounds will also dictate the best places to plant trees. If a seed bank is present or an intact swamp forest is nearby, recruitment of desired species can be expected. However, planting will accelerate plant succession and ensure that desired species are present on the site.

Temporal spacing for different plants and layers

Bald cypress trees may live for centuries and achieve great sizes. Virgin stands of cypress were typically from 400 to 600 years old (Mattoon 1915). Most of the virgin stands of cypress in the United States were logged over the last century, however, and few individuals over 200 years old remain (Mitsch and Gosselink 1986). Mature bald cypress are typically 100–130 feet high and three to five feet in diameter.

Many mature swamp forests are devoid of any understory vegetation. Even when enough light is available for understory vegetation, it is difficult to generalize about its composition (Mitsch and Gosselink 1986). There can be a dominance of woody shrubs, of herbaceous vegetation, or of both. Fetterbush (*Lyonia lucida*), wax myrtle (*Myrica cerifera*), and Virginia willow (*Itea virginica*) are common shrubs and small trees in nutrient-poor cypress domes (ibid.). Understory species in higher nutrient river swamps include buttonbush (*Cephalanthus occidentalis*) and Virginia willow. Some continually flooded cypress swamps that have high concentrations of dissolved nutrients in the water develop dense mats of duckweed (*Lemna* spp., *Spirodela* spp.) or mosquito fern (*Azolla* spp.) on the water surface most of the year.

Maintenance and Management

Nutrient requirements

Mitsch and Gosselink (1986) found that wide ranges of acidity, dissolved substances, and nutrients are found in the waters of swamp forests. Some facts should be noted from this wide range of water quality.

■ Swamp forests do not necessarily have acid water.

■ Nutrient conditions vary from nutrient-poor conditions in rainwater swamps to nutrient-rich conditions in alluvial river swamps.

■ An alluvial river swamp often has water quality very different from that of the adjacent river.

Many deepwater wetlands, particularly alluvial river swamps, are "open" to river flooding and other inputs of neutral and generally well-mineralized waters (Mitsch and Gosselink 1986). The pH of many alluvial swamps in the southeast is six to seven, and there are high levels of dissolved ions. Dome swamp forests, on the other hand, are primarily fed by rainwater and have acidic waters, usually in the pH range of three and one-half to five, due to humic acids produced within the swamp.

Nutrient inflows, often coupled with hydrologic conditions, are a major influence on swamp productivity. Swamps can be nutrient sinks whether the nutrients are a natural source or are artificially applied.

Water requirements and hydrology

Swamp forests are truly aquatic systems. They require water to persist, although permanent water is not needed. Many swamp forests occur on soggy soils but are only seasonally or periodically inundated.

Various natural and unnatural factors affect the hydrology of swamp forest communities. These include seasonal fluctuations in water levels, different underlying soils, adjacent land uses, and beaver dams. An assessment of the current hydrology and any proposed changes in the hydrology should be an integral component of the restoration plan for these community types.

Restorationists must decide whether or not to install water-control structures or attempt to mimic a naturally occurring, closed-basin depressional wetland fed by groundwater. Both have advantages and disadvantages. A water-control structure allows the land manager to dictate water levels which can be useful in the control of certain wetland weeds. However, water-control structures require high initial cost as well as continuous maintenance. The choice will likely depend upon a number of factors such as project goals (i.e., future use of the site). For many wetland restoration projects, one of the primary goals is to create a naturalistic wetland that is self-sustaining. In this case, a water-control structure is not appropriate unless it is temporary. However, if the site is to be used for water quality improvement (as are many constructed wetlands), educational purposes, water conservation, irrigation, or wildlife habitat, then a control structure may be needed.

Swamp forest productivity is closely tied to its hydrologic regime. Highest productivities result in systems that are neither very wet nor too dry, but have seasonal hydrologic pulsing. Stillwater swamps receive their major inputs of water and nutrients from rainfall and are thus poorly nourished.

Disturbances

When swamp forests have been drained, or when their dry period is dramatically extended, they can be invaded by pine (*Pinus* spp.) or hardwoods.

Ewel and Mitsch (1978) investigated the effects of fire on a cypress dome in Florida. They found that fire had a "cleansing" effect on the dome, selectively killing almost all of the pines and hardwoods and yet killing relatively few pond cypress, suggesting a possible advantage of fire to some shallow cypress ecosystems in eliminating competition that is less water tolerant.

Case Studies

For additional information on the restoration and management of swamp forests, see Brandel (1985), Barry et al. (1996), Deitz et al. (1996), Messina and Conner (1997), Trettin (1997), and Wheeler et al. (1995).

Riparian Forests and Woodlands

Riparian Forest A-17, A-42, A-86, A-127
Riparian Woodland A-76, A-136, A-146, A-163
Desert Oasis Woodland A-75

Introduction

This Ecological Restoration Type includes the western riparian forests and woodlands and desert oasis woodlands. These riparian forests of the semi-arid grasslands and arid western United States differ from those found in the humid eastern and southern United States. The predominant ecosystems of western uplands are grasslands, deserts, or other nonforested ecosystems, so the riparian zone is a conspicuous feature of the landscape (Brinson et al. 1981). The riparian zone in arid regions is also narrow and sharply defined in contrast with the wide alluvial valleys of the southeastern United States.

The status of woody-deciduous riparian habitats has caused special concern because of their disproportionate importance in supporting wildlife diversity (Maser et al. 1984), especially migratory songbirds (Bock et al. 1993, Dobkin 1994). Riparian trees and shrubs not only provide protection, roosting areas, and favorable microclimates for many species, but they also provide standing dead trees and "snags" in streams that provide habitat value for both terrestrial and aquatic animals (Brinson et al. 1981). The riparian vegetation also shades the stream, stabilizes the streambank with tree roots, and produces leaf litter, all of which support a greater variety of aquatic life in the stream.

Many of these communities have been degraded by agriculture, groundwater pumping, livestock grazing, urbanization, and alteration of rivers and streams. Few high quality examples exist. Li et al. (1994) found that abundance of trout was diminished due to elevated stream temperature with increased solar input when riparian canopies are disturbed by grazing. Other impacts include sreambank erosion, excessive stream siltation, and stream intermittency in severely grazed riparian systems that previously flowed year round (ibid.). Cattle overgrazing can cause streams to become intermittent through lowering of the water table due to reduced interaction of the stream channel with riparian vegetation, low-

ered water permeability of riparian soils due to compaction, and dewatering (ibid.). Channel structure becomes more simple when riparian vegetation is removed, because inputs of large woody debris and their influence on channel structure are diminished (Gregory et al. 1991). Because livestock congregate in riparian ecosystems, which are among the biologically richest habitats in arid and semiarid regions, the ecological costs of grazing are magnified in these sites (Fleischner 1994). Fleischner (1994) also found that in western North America, grazing has reduced the density and biomass of many plant and animal species. Reduced biodiversity has aided the spread of exotic species, interrupted ecological succession, impeded the cycling of the most important limiting nutrient (nitrogen), changed habitat structure, and disturbed community organization. Zasoski and Edmonds (1986) found that riparian plant communities function as effective filters for many industrial and agricultural effluents from uplands.

California fan palms (*Washingtonia filifera;* indicator of desert oasis woodlands) were historically more numerous and widespread, but as the climate became increasingly more arid, they became more restricted in range (Larson 1970). Today they grow only where sufficient permanent water supplies are available, clustered about an occasional hillside seep or in a canyon or rare wash area with springs or subterranean water supplies (Larson 1970).

Briggs et al. (1994), in an assessment of the effectiveness of riparian revegetation in Arizona, found that the greatest limitation of riparian revegetation is that it often does not address the causes of degradation. When it does not, it is likely that the riparian system will continue to be unstable. Such instability will often hamper the establishment of artificially planted vegetation to the same degree as it has natural regenerative processes (Briggs et al. 1994).

Site Conditions

Soils

The soils of western riparian forests and woodlands are shallow, poorly drained, and subject to flooding. These soils developed as a direct result of annual periods of inundation. Increased soil salinity may be a problem in degraded riparian areas (Briggs et al. 1994) and should be determined before revegetation/restoration begins.

California fan palms (*Washingtonia filifera*) are able to tolerate fairly high concentrations of alkali (Larson 1970). Alkali may produce a crust on moist soils, preventing the growth of many plant species. However, some plants (halophytes) require alkali conditions.

Topography

The topography is fairly level in western riparian forests and woodlands that occur along stream terraces of canyons. In the arid areas where western riparian forests and woodlands occur, the vegetation is very different from adjacent upland vegetation and topographic gradients can dramatically change throughout the valley. Streamside vegetation at higher elevations (nearer the headwaters) does not form as distinct an ecological community but instead grades quickly into the upland vegetation type. For these areas, refer to the appropriate upland restoration type. Narrower canyons tend to be more moist (mesic), while broader canyons are drier (xeric).

Moisture and hydrology

Since western riparian forests and woodlands are driven by periodic flooding, moisture varies quite a bit depending upon season. These communities occur in close proximity to a stream or river and are distinctly influenced by it. They naturally experience flooding in the spring and become very dry in the summer. The native plants in these ecological communities are well adapted to these extremes. However, native transplanted trees and shrubs, cuttings, or other propagules will likely require irrigation for successful establishment. Sacchi and Price (1992) found that the major cause of mortality in arroyo willow (*Salix lasiolepis*) was a shortage of soil moisture. Biotic factors such as herbivory and competition played only minor roles in the dynamics of the species' ability to populate a riverine area.

Desert oasis woodlands occur at heads or in the bottoms of canyons and arroyos where permanent springs or seeps occur. Subsurface runoff is an important component of western riparian forests and woodlands. Summertime streamflow represents groundwater discharge, while flows during springtime primarily result from snowmelt at higher elevations (Ehleringer et al. 1992). Western riparian forests and woodlands are seasonally flooded and inundated part of the year. Many western United States rivers and streams have been altered and/or impounded, which has disrupted the natural cycle of flooding events both spatially and temporally. Nelson and Anderson (1994) found that natural colonization by understory and other herbaceous plants may not occur when surface soils in the riparian zone are hydrologically disconnected from the river. The river introduces and irrigates plants via the flood cycle. Before the hydrology was altered, these rivers and streams flowed year-around (perennial). Many historically perennial river systems in the western United States now flow intermittently or only following heavy runoff. Bren (1993) illustrates how changes in flooding frequency directly changes the surrounding vegetation types. Generally, a major flooding event can be expected about every twenty years in perennial and semiperennial streams.

Periodic flooding usually contributes to higher productivities in at least three ways (Brinson et al. 1981).

■ Flooding provides an adequate water supply for the vegetation. This is particularly important in arid riparian wetlands.

■ Nutrients are supplied and favorable alteration of soil chemistry results from the periodic overbank flooding. These alterations include nitrification, sulfate reduction, and nutrient mineralization.

■ The flowing water offers a more oxygenated root zone than if the water were stagnant. The periodic "flushing" also carries away many waste products of soil and root metabolism, such as carbon dioxide and methane.

For additional information on the ecology of riparian forests and woodlands, see Brinson et al. (1981).

Establishment

Site Preparation

One of the most important considerations during site preparation is the removal of exotic or pest plants. Not only do these foreign species have to be removed prior to planting trees, but they have to be controlled following planting or they will out-compete many of the native tree species (Sweeney 1993). Because the widespread use of herbicides in riparian areas should be avoided, mechanical removal is advocated (see **Weed Control**). Increased soil salinity plays an important role in site degradation and must be addressed before developing a restoration plan (Briggs et al. 1994).

Grazing livestock should be excluded from the restoration site prior to revegetation (preferably for several seasons). Attempts at artificial revegetation of riparian shrubs and trees in the western United States have often failed, in part because of a lack of understanding of establishment requirements and also because of continued grazing by livestock (Bryant and Skovlin 1982). Glinski (1977) demonstrated that cattle grazing of small seedlings prevented cottonwood (*Populus fremontii*) regeneration in a southern Arizona riparian zone. He concluded that long-term grazing could eliminate or reduce the upper canopy by preventing the establishment of saplings. Carothers et al. (1974) noted the lack of cottonwood in grazed areas along the Verde River, Arizona. Prevention of seedling establishment due to grazing and trampling by livestock has transformed a variety of riparian areas in the southwestern United States into even-aged, nonreproducing vegetative communities (Carothers 1977, Szaro 1989). In Oregon, grazing retarded succession in the willow-balsam poplar (*Salix-Populus balsamifera*) community and little, if any, regeneration of alders (*Alnus*) or poplars (Kauffman et al. 1983) occurred. Davis (1977) concluded that livestock grazing was "probably the major factor contributing to the failure of riparian communities to propagate themselves."

In severely eroded sites, the first objective may be to prevent or slow further erosion. A method known as biotechnical erosion control (BEC) is being used by a growing number of landscape architects (Northcutt 1994). Northcutt (1994) stated that BEC offers environmentally sound solutions to erosion and sediment problems, using dormant nonrooted cuttings (typically willow, dogwood, spiraea, and poplar) to build living structures that protect the surface from erosion by wind or water and stabilize the soil several feet below the surface. BEC can be used to control erosion on slopes, to repair gullies, to stabilize headcuts, and to control shallow landslides or slumps. It is probably most popular for stabilizing stream banks and protecting lakeshores (Northcutt 1994). For more information on BEC, refer to Gray and Leiser 1982, Northcutt 1994, and Schiechtl 1980.

Planting considerations

One difference in the western riparian forests and the eastern and midwestern floodplain forests is the general absence of oak (*Quercus* spp.) (Brinson et al. 1981). Also, few species are common to both regions, due to the differences in climate. However, species of cottonwood and willow occur in both regions.

Desert oasis woodlands (A-75) are dominated by California fan palm (*Washingtonia filifera*) and California sycamore (*Platanus racemosa*). The palms are fire tolerant, but the understory species are not. Periodic fire opens the understory allowing seedlings to establish. The reduced understory also removes competition for water.

In desert oasis woodlands, California fan palms are the dominant plant species, but they are joined by other plants as well. These plant species depend upon numerous factors including the terrain and type of soil. Some water plants grow where permanent standing water is available. The palms grow around the water source and beyond them in a transition zone from the oasis to desert. Larson (1970) states that species richness in desert oasis woodlands is low with an average of only eleven species of plants per oasis, and that included the plants of the three zones–water, oasis, and oasis-desert transition. An oasis may support one or two palm trees or several hundred. When present in great numbers, they provide a distinctively humid microhabitat for a number of animals (ibid.).

Species recommendations for riparian plantings must, by necessity, be very site specific. Other sources should be consulted to accurately determine which species are native to the specific area and their relative abundance. Some primary sources that provide species lists and/or vegetation characterizations for riparian forests and woodlands and desert oasis woodlands are Bahre 1991, Barbour and Major 1977, Beatley 1976, Bender 1982, Bendix (1994), Cottam and Evans 1945, Gardener 1951, Leopold 1951, Lowe 1964, MacMahon 1988, Ohmart and Anderson 1982, Szaro 1989, Szaro and Pase 1983, Warner and Hendrix 1984, and Winegar 1977.

Because transplanting large trees is not feasible

(either economically or physically) for most land-owners, streamside plantings usually involve small (<eighteen inches) to intermediate (three to five feet) size seedlings (Sweeney 1993). Small seedlings are more desirable because they can usually be purchased from state or federal government at about one-tenth the cost of the intermediate-size plants. However, survivorship of small seedlings is often low when they are stressed by factors such as soil moisture, competition with other plants (native and exotic), and predation by herbivores (Sweeney 1993). Since the importance of each of these mortality factors greatly diminishes as the tree gets larger, it behooves the landowner to either start with an intermediate-size seedling (if economically possible) or to start with a small seedling and try to promote the fastest possible vertical growth. This can be accomplished by implementing irrigation, reducing competition, mulching, and/or using fertilizer. Another method of optimizing vertical seedling growth for a given level of nutrients and moisture is to install "tree shelters" when the trees are planted (Sweeney 1993). Tree shelters, which were originally developed in England, are tall plastic tubes installed around seedlings to protect them from vertebrate herbivores (e.g., deer, rabbits, mice, etc.) and improve growth by creating a "greenhouse-like" effect around each tree (Tuley 1985).

Hujik and Griggs (1995a) found that cuttings of Fremont cottonwood (*Populus fremontii*) and Goodding willow (*Salix gooddingii*) had survival rates in excess of seventy percent and arroyo willow (*S. lasiolepis*) cuttings had survival rates over ninety percent. Drake and Langel (1995) used a hydraulic jet to effectively plant riparian tree cuttings (live stakes) to reduce streambank erosion. The hydraulic jet allows cuttings to be planted in a six-foot-deep hole, instead of the usual two- to three-foot-deep holes typically used and sometimes washed away. In 1994, Drake and Langel planted approximately 4,000 dormant sandbar willows (*Salix interior*) on the sandy, eroding banks of Clear Creek in east-central Iowa. They found nearly all the cuttings alive by the end of the first growing season, at which point they had grown three feet on average. More importantly, when they excavated around these cuttings, it was found that the cuttings had rooted along their entire underground portions–both above and below the water table. Since the roots extend more than two feet below the deepest pool level, they

expect to have erosion-resistant thickets of willows by the end of 1995. See Drake and Langel (1995), for more information concerning the use of hydraulic jetting for planting cuttings.

Ischinger and Shafroth (1995) found that human-induced damage of cottonwood roots will promote root-suckering. However, more experiments are needed to develop techniques that maximize root-suckering and insure long-term survival of the young saplings.

Lippitt et al. (1994) provides an excellent discussion of techniques for collecting, processing, and storing seeds for revegetation projects in the western United States.

Blankets and cylinders of coir (processed fiber of coconut tree) have been used with a great deal of success to hold plants in place in wetland restoration and streambank stabilization projects. Other benefits of this material are the high tensile strength when wet, durability (up to ten years), and organic matter content (100 percent).

Many trees and shrubs of riparian forests are adapted to flooding and depend upon inundation to fulfill requirements for seed germination. Snyder and Miller (1992) reported that extensive flooding along the Colorado River in 1983–84 resulted in considerable natural reproduction of seedlings. Some trees and shrubs of desert wash areas, such as smokethorn (*Psorothamnus spinosus*), have seeds that require abrasion before germination can occur, and this action takes places in the wash during floods (Larson 1970).

Most ashes (*Fraxinus* spp.) up to four inches in diameter can be easily transplanted in winter if the site permits the taking of a good ball of earth with the roots. Transplanted trees should be pruned back and watered.

Nokes (1986) found that ashes (*Fraxinus* spp.) can be propagated without a great deal of difficulty. The samaras can be collected when they are no longer green, but when they have turned a light yellow color with crisp-textured wings. Samaras should not be collected from the ground because they are likely to be of poor quality and insect infested. Dewinging before sowing is not necessary. If seeds are being stored, they should be sealed in containers in a cool dry place. Green ash (*Fraxinus pennsylvanica*) remained viable for seven years when dried to seven to ten percent of its fresh weight and stored at fifty-eight degrees Fahrenheit (United

States Department of Agriculture 1974).

The seeds may be sown outdoors in well-tilled and mulched beds in the fall after collection (Nokes 1986). Or ash seeds may be stored over winter, stratified in moist sand or perlite for thirty to sixty days at fifty-eight degrees Fahrenheit, and then sown in a greenhouse or outdoors after all danger of frost has past. Plant the seeds one-fourth to three-fourths of an inch deep in well-drained soil. Seedlings benefit from light shade the first season. Germination rates are variable depending upon species and seed source.

Several sources have indicated that ash is difficult or impossible to root from cuttings. There may be some potential, however, in cutting from young ashes that have not yet flowered.

Birch (*Betula* spp.) seeds must be collected as soon as the "cones" are full grown and beginning to turn brown, but before they dry completely and open to disperse the seeds from April through June (Nokes 1986). The strobiles can be collected when still slightly green by picking or stripping the tree and then spread out to air dry until they have opened. Seed quality varies from year to year and location to location. Sow seeds immediately or store in sealed containers at temperatures slightly above freezing (United States Department of Agriculture 1974).

Seeds may be sown in the fall after collection or stored and stratified for spring sowing (Nokes 1986). Seeds gathered when mature and immediately sown in well-prepared beds of sandy loam with organic matter (peat moss or compost) often germinate in the highest percentages. The seeds should be pressed firmly into the soil and lightly covered with fine sand. Do not bury the seeds. The bed should be kept moist and shaded during the first growing season. For spring sowing, stratify the seeds thirty to sixty days at fifty-eight degrees Fahrenheit, remove in early spring and plant as recommended above. Successful rooting of cuttings are reported to occur less than fifty percent of the time.

The California fan palm (*Washingtonia filifera*) has annual crops of seeds that remain dormant on the moist alkaline soils and may require prolonged rains to leach inhibiting materials from the seeds or the soil before germination can occur (Larson 1970). Coyotes eat fallen California palm fruit and since palm seeds are found in large numbers in their droppings, they are considered to be important disseminators of these seeds (Larson 1970).

Spacing considerations

Competition (space, light, nutrients, moisture, etc.) with other plants and density of herbivores during the first two to five years following planting are very important factors that effect seedling survivorship and growth (Sweeney 1993). Mechanical weed abatement (mowing and hand weeding) and mulching can greatly reduce plant competition and improve seedling growth. Sweeney (1993) found that seedlings planted in rows with about a three-yard spacing worked well. This accommodated quick and effective mowing of large areas with tractors and provides proper spacing between seedlings to minimize strong competition among neighboring plants during their first six to eight years of life. Herbiciding the area immediately around seedlings is also effective. Herbicide application must be done with caution in riparian areas because of their proximity to both the stream and to shallow groundwater sources providing a portion of the streamflow (Sweeney 1993).

Sweeney (1993) found that mowing areas planted with seedlings at least twice per year (May and August) for the first three to five years is the minimum necessary treatment in order to avoid almost complete mortality. Mowing probably increases survivorship, because it keeps light levels nonlimiting and prevents the seedlings from being physically overgrown by competing herbaceous plants and vines (Sweeney 1993). However, mowing probably does little to reduce competition for nutrients and moisture because seedlings are still surrounded with low-lying plants and their root systems.

Temporal spacing for different plants and layers

The establishment of canopy species is primary to restoration of western riparian forests and woodlands. After the canopy has been established (ten to fifteen years), shrubs and woody understory and herbaceous species should be planted, although some natural revegetation will likely occur. Any native species that appears and is a natural component of the ecological community being restored should be preserved. Exotic plants and nondesirable species (such as some pioneer species) should be removed to reduce competition and encourage growth of native plants.

The high density and species diversity of wild-life in riparian areas have been attributed to the many structural layers of riparian vegetation and to its ecotonal nature (Rucks 1978, Thomas et al. 1979). The physical structure of ecosystems, in-cluding vegetation stratification, is often changed by livestock grazing. Rucks (1978) stated that livestock grazing causes the replacement of shrub-nesting birds with species showing no preference for vertical vegetation structure. Riparian zones can be managed for nongame species richness by main-taining high structural diversity of vegetation (Shulz and Leininger 1991).

In desert oasis woodlands, California fan palms (*Washingtonia filifera*) may live as long as 200 years, although most probably die before 150 years of age. Optimum conditions for germination and growth of seedlings must occur only once every hundred years for stands to reproduce themselves (Larson 1970).

Natural establishment of willows and cotton-woods is most prevalent following spring runoff on the course-textured sediments of gravel bars (Busse 1989). So, planting of seeds or propagules may be most successful during that season.

Maintenance and Management

Weed control

Streamside areas that have been historically clearcut for agriculture generally become immedi-ately dominated by foreign plant species. Saltcedar or tamarisk (*Tamarix* spp.) is a serious weed in many areas in the western United States (Larson 1970, MacMahon 1988, Neill 1993, Nelson and Andersen 1994, Snyder and Miller 1992). *Tamarix* spp. are native to Europe and Asia and were introduced into North America where they have prospered. They grow where water is readily available and they have become dominant along many southwestern streams and rivers. *Tamarix* ssp. have crowded out native vegetation. So prolific and so wasteful of water are *Tamarix* ssp. that attempts to eradicate them are often unsuccessful. Neill (1993) provides a history of tamarisk introduction into California, its physical and morphological attributes, and some reviews of efforts to control it. Russian olive (*Elaeagnus angustifolia*) is pioneering along the Colorado River (Snyder and Miller 1992). MacMahon (1988) and

Hickman (1993) stated that giant reed (*Arundo donax*) is a problem along some streams and rivers, and in springs and seeps.

The most efficient and cost effective method is to mow or cut off by hand the stem of each undesired plant at ground level during the spring just after the plants have leafed out (i.e., after a substantial amount of stored energy has been expended by the plant) (Sweeney 1993). Although most plants will sprout stems from the cut-off root stump, mowing or cut-ting once or twice again during the subsequent summer can weaken and kill many of the plants.

Bendix (1994) speculates that fire may play as much an ecological role in the riparian environment as on the surrounding hillslopes. If so, fire may serve as a valuable tool to control weeds as well as maintain natural riparian forests along southwestern rivers and streams. Fire may play an important role in desert oasis woodlands. The thatch and fallen leaves of the California fan palms are highly flam-mable. If a fire starts in a desert oasis (due to humans or lightning), the thatch, ground debris, and some of the smaller plants growing in the area are burned. The palms are relatively fire tolerant and usually survive (Larson 1970). A fire can benefit the desert oasis by removing dead debris and leaving clear, moist soil in which new plants may grow. Fire also removes small shrubby species that shade and pre-vent optimum development of young, small palms (Larson 1970).

Grazing destabilizes plant communities by aid-ing the spread and establishment of exotic species, such as tamarisk (*Tamarix*) (Ohmart and Anderson 1982, Hobbs and Huenneke 1992). Livestock help spread exotic plant species by (1) dispersing seeds in fur and dung; (2) opening up habitat for weedy species, such as cheatgrass (*Bromus tectorum*) (Gould 1973, Mack 1981), that thrive in disturbed areas; and (3) reducing competition from native species by eating them. As D'Antonio and Vitousek (1992) pointed out, alien grass invasions in North America have been most severe in the arid and semiarid West, where invasion by many species (including *Bromus tectorum, B. rubens, B. mollis, B. diandrus, Taeniatherum asperum,* and *Avena* spp.) was associated with grazing.

Bertin Anderson (cited in Howald 1996) con-cluded that anthropogenic activities such as agricul-ture, grazing, and dam construction have modified some desert riparian habitats to such a degree that they are no longer able to support native cottonwoods and willows. He concludes that restoration with natives may be infeasible in such circumstances, and that restoration attempts may actually reduce the value of these ecosystems as habitat for native birds.

Nutrient requirements

While applying fertilizer is not always necessary, Briggs et al. (1994) found that applying fertilizer can increase survival of trees and shrubs. A soil analysis should be conducted to determine whether or not soil nutrients are depleted at the revegetation/restoration site.

Water requirements and hydrology

Flooding is important to maintain the integrity of western riparian forests and woodlands. The hydrologic regime should mimic the natural flooding cycle, if possible. Irrigation will likely be required where the stream systems have been altered. Irrigation overflow from agricultural fields sustained Fremont cottonwoods (*Populus fremontii*) along a revegetated portion of the lower Colorado River (Nelson and Andersen 1994). During the initial phases of revegetation, the use of a drip irrigation system for the first two summers following planting increased survival (Briggs et al. 1994). Ehleringer et al. (1992) found that many mature riparian tree species obtain their water from groundwater and not stream water. Griggs (1994) found that Fremont cottonwood and willows (*Salix* spp.) will grow without irrigation if their roots can reach the water table.

Allowing accumulation of woody debris in streams is an important part of the interaction between the terrestrial and aquatic communities present in western riparian forests and woodlands. Debris not only provides food and shelter for aquatic organisms, but also naturally effects the flow of the stream or river. Groundwater decline and channel instability were cited by Briggs et al. (1994) as important factors in site degradation that must be addressed before developing a mitigation plan. Allowing a period in which natural changes in channel stability and groundwater level can be monitored should be advantageous to restoration efforts in western riparian forests and woodlands.

Kondolf and Micheli (1995) argue for more intensive evaluation procedures for the increasing number of stream restoration efforts. Post-construction evaluation strategies adapted to specific goals must be incorporated into each project design. Designers must also extend themselves further into both past and future by viewing historic channel conditions as potential design models and by monitoring over at least ten years, including surveys after each flood event that exceeds a specified threshold (ibid.).

Case Studies

Fleischner (1994) found that attempts at restoration of livestock-damaged ecosystems have offered both good and bad news. Riparian areas often show rapid recovery upon removal of livestock, but more xeric uplands demonstrate little inherent capacity for healing. At a Sonoran Desert spring, Warren and Anderson (1987) documented dramatic recovery of marsh and riparian vegetation within five years of livestock removal. All nine aspects of trout habitat studied along Summit Creek, Idaho, improved within two years of livestock removal (Keller and Burnham 1979). Mahogany Creek, Nevada, also showed major improvement in fisheries habitat after only two years of exclosure (Dahlem 1979). Beaver and waterfowl returned to Camp Creek, Oregon, within nine years of cattle exclosure (Winegar 1977). However, the aquatic component of riparian systems often is the quickest to show improvement. Szaro and Pase (1983) observed extremely limited recovery of a cottonwood-ash-willow association in Arizona after four years. Knopf and Cannon (1982) noted that a willow community was slower to heal than the adjacent stream. Ten to twelve years was insufficient for the former.

The United States General Accounting Office (1988) recently reviewed riparian restoration efforts on Bureau of Land Management and United States Forest Service lands in the West and concluded (1) that even severely degraded habitats can be successfully restored and (2) that successful restoration to date represents only a small fraction of the work that needs to be done. They noted that successful techniques varied considerably from site to site, and that many sites could repair themselves, given respite from overuse by livestock. Successful riparian restoration efforts are summarized by the United States General Accounting Office (1988) and Chaney et al. (1990).

Goose Flats, along the Colorado River near Blythe, Arizona, was successfully revegetated with greater than twenty percent of planted material (which is high for degraded riparian communities) (Briggs et al. 1994). Loss of riparian vegetation at this site had been caused by several factors, including manipulation of natural river flow by impoundment structure construction, accumulation of sodium and other salts in the soils of the floodplain, and competition from salt cedar (*Tamarix chinensis*) (Briggs et al. 1994). Prior to planting, a bulldozer was used to clear salt cedar and other nonnative species. Several months later, plants that resprouted were sprayed with glycophosphate herbicide. A fence was installed around the site to control beaver and off-road vehicles. Holes fifteen inches in diameter by eight feet deep to the water table were augered at a density of 100 holes per acre. The holes were loosely refilled to provide a path to the water table for irrigation water. A drip irrigation system was installed and the plantings were irrigated during the first two summers following planting, and fertilizer was applied to all trees. These rather extensive efforts paid off as a total of 4,172 trees and shrubs (seven species) had been established. The ten-year survival rate was predicted to be around seventy percent.

Along the ungrazed reaches of Meadow Creek (a tributary of the Grand Ronde River, Oregon), over 120 plant species were present on gravel bars created by a spring flood three years prior to sampling (Kauffman et al. 1995). At the onset of the study, the tree/shrub component of this riparian ecosystem was in a severely degraded condition because of excessive utilization by cattle, elk, and deer. Densities of riparian-obligate shrubs and trees (willows, cottonwood, and alder) were far below potential. In addition, fewer than eight percent of the willows were in a reproductive state (that is, were producing catkins), and the mean height of black cottonwood (*Populus trichocarpa*) was only thirteen inches. Two years after the cessation of grazing by cattle, the height and crown volume of black cottonwood had increased 106 and 800 percent, respectively. Crown volume of alder and willows had increased 200 and 288 percent, respectively (Case 1995). The overall density of shrubs increased fifty percent, or approximately one shrub for every thirty feet of streambank. If current rates of recovery continue, artificial reintroduction of willows and cottonwoods will not be necessary on this site. The rapid recovery and establishment of riparian shrubs following cessation of livestock grazing is a good example of successful passive riparian restoration.

Pyle (1995) planned a restoration project that included a rest from grazing to increase survival of young shrubs and trees, planting of willow to facilitate their establishment on sites where seed sources are depleted, and using prescribed burning to foster establishment and regeneration of aspen and other woody-riparian shrubs. A total of 4,000 cuttings of willow from local stock–mainly Lemmon's willow (*Salix lemmonii*), Booth's willow (*S. boothii*), and Geyer's willow (*S. geyeriana*)–were planted at seven sites along semiperennial streams during March and April, 1991–94. Since then, a survey of 560 willows planted in 1991–92 indicated that twenty-one percent survived the third growing season at two sites. Sixty-one willows planted in spring following a prescribed burn the previous fall had a survival rate of eighty-one percent two growing seasons after the fire. An examination of forty-six mature willows on the burned site indicated that seventy-eight percent resprouted and survived the second growing season.

For additional information on the restoration and management of riparian forests, see Averitt et al. (1994) and Warner and Hendrix (1984).

Shrub Swamps

Cedar–Alder Swamp A-17
Northern Shrub Swamp A-259
Shrub Swamp A-277, A-298, A-394

Introduction

Shrubs swamp communities possess interesting and diverse biota, however, many of these areas have been drained and the remaining examples are often degraded. Muck farming and clearcutting have also played major roles in the degradation of these communities. The draining of these communities has often been accredited to the fact that they are breeding grounds for mosquitoes and biting flies. See also the swamp forests and nonestuarine marshes Ecological Restoration Types.

Site Conditions

Soils

Northern shrub swamp (Northern Great Lakes region) is a common community along streams and around lakes where the soil is muck or peat.

Cedar–alder (*Thuja-Alnus*) swamps develop on peat muck soils underlain by glacial lake beds.

Topography

These communities are usually very flat. They may occur in a valley or on mountain tops, but fully develop only where drainage is poor and saturation occurs for a large portion of the year.

Moisture and hydrology

Even during drought, these communities often contain water, either standing or near the soil surface. The plant species inhabiting shrub swamps are tolerant and even adapted to these very wet conditions. Moisture is a key factor in the development and perpetuation of shrub swamp communities as they are often flooded too long and too deeply for trees to become established. Cedar–alder swamps (*Thuja plicata-Alnus rubra*) occur where there is a high water table, or even standing surface water, for all or a portion of the year.

The hydrology of shrub swamps may be a result of surface runoff accumulating in areas where the drainage is poor or input from an adjacent lake or stream. The water table is often not far below the soil surface. Some shrub swamps have outflows while others hold water.

Establishment

Site preparation

Site preparation must involve the following four components:
- A grading plan.
- Specifications for soil amendments and preparing the surface for revegetation.
- Specifications for the removal of unsuitable materials/debris.
- Specifications for weed control.

Shrub swamps are wet and mucky environments and the operation of earth-moving equipment can prove difficult. Site preparation should be done at a time when the site is driest. Creating an uneven pit-and-mound microtopography in the basin is a good way to provide variable hydrologic conditions for different plant species.

Planting considerations

White cedar or eastern arborvitae (*Thuja occidentalis*) has valuable uses aside from wetland plantings. It is very similar to the oriental arborvitae (*Thuja orientalis*) commonly used in landscape plantings. White cedar could be used at the edge of woodlands to "seal off" the edge and discourage weeds and light penetration. It also performs well in a variety of habitats from rocky ledges and talus slopes to mucky blackwater swamps. It would be a good candidate for hydroseeding or biotechnical erosion control on steep slopes. There is much potential for this native conifer in landscape planning and restoration.

Many shrub swamp species can be propagated by cuttings. Brookside alder (*Alnus serrulata*) may be increased by separating root sprouts, underground stems, and suckers (Nokes 1986). It is also easily transplanted from the wild in the winter. Saplings of common buttonbush (*Cephalanthus occidentalis*) may be dug in the spring. Large clumps of older plants may be separated in early spring (Taylor and Hamblin 1963).

Nokes (1986) recommended the following methods for propagating brookside alder (*Alnus serrulata*). Gather strobiles (conelike fruits) in early fall before they completely dry and disperse the nutlets or drop from the plants. Seeds should be thoroughly air dried before storage. Storage in sealed containers is adequate for over winter. For longer storage, refrigeration is recommended. Brookside alder seeds held in sealed containers at near freezing have been known to retain their viability for as long as ten years. Seeds may be sown outdoors in well-prepared seedbeds in late fall. Gently press the seeds into the soil and lightly cover with washed sand. The seedbed should be mulched over winter, and the mulch later removed as germination begins. Seeds sown indoors or in the spring must be stratified for sixty to ninety days at 34–40°F. Stratification should be planned so the seeds are ready for spring planting. Seed germination and seedling survival during hot weather is poor. Do not allow the seedbed or container to dry out.

Nokes (1986) recommended the following methods for propagating common buttonbush (*Cephalanthus occidentalis*). Gather seeds in late summer or early fall before the heads dry out and break apart. Separate the seeds by putting the fruits in a bag and lightly beating them. Fresh seeds will germinate without pretreatment in thirty to forty days. Germination rates vary greatly and generally are not high. Plant seeds in a greenhouse or outdoors after the soil has warmed and all danger of frost is past. Buttonbush is easily rooted from both softwood and hardwood cuttings. Semi-hardwood cuttings taken in late July and treated with IBA rooted in three to four weeks in high percentages. These were planted in sand and kept under intermittent mist. Rooting percentages for buttonbush are usually high. Stock plants held in a greenhouse and routinely cut back provide a source of easily rooted cutting material. Shrubs should be planted on mounds in the basin.

Spacing considerations

Shrub swamp plantings need to be appropriately spaced in order to effectively control erosion and provide the desired amount of plant cover. Exact spacing must be determined on a site-specific basis. Shrubs should be planted on mounds in the basin.

Temporal spacing for different plants and layers

Many factors affect revegetation, whether it is natural succession or restoration. Most likely it will be some of both. Shrub swamp species can be planted all at once. However, some of the desired species may not be successful initially because of source of plant material, timing of planting, herbivory, parasites, or nutrient or water requirements. Therefore, survival may need to be monitored and additional planting may need to be done.

Maintenance and Management
Weed control

Brandel (1985) found that whitetail deer pose a major hurdle to restoration of cedar stands. Exclosures may need to be constructed if damage to plantings is excessive.

Past and present land uses influence the plant species composition and density. Nonnative plants are generally not problematic in shrub swamps. However, nonnative species discussed in the nonestuarine marshes restoration types may also invade shrub swamps and may need to be controlled.

Nutrient requirements

Past land uses directly influence the nutrient levels. In shrub swamps that have relatively consistent inflows and outflows, nutrients are not a limiting factor. However, in stillwater shrub swamps nutrient-poor conditions may occur.

Water requirements and hydrology

Shrub swamp species tolerate periods of both flooding and drought. Supplemental watering of initial plantings will likely increase their survival. Some shrub swamp species that depend upon groundwater will benefit from irrigation during dry periods until their roots have grown deep enough to utilize groundwater.

In created shrub swamps, a device to control the water levels may be useful for management. As the hydrology of these communities may be dependent upon surface runoff, adjacent areas that are well-drained (artificially or naturally) may have adverse affects on the establishment of swamp-loving vegetation. If the quantity and quality of water discharged into a wetland has been altered, the restorationist must decide if a different community type might be more appropriate for the site than the post-disturbance type. Restoration goals should be to reestablish natural water flow patterns to the extent possible given land-use changes within the watershed.

Case Studies

For additional information on the restoration and management of shrub swamps, see Brandel (1985) and Wheeler et al. (1995).

Nonestuarine Marshes

Introduction

Nonestuarine marshes occur in areas where water naturally accumulates and supports a wide array of submersed and emergent aquatic plants. Marshes are typically dominated by emergent grasses (graminoids) and sedges. These communities range from small potholes less than an acre in size to the immense sawgrass monocultures of the Florida Everglades. Nonestuarine marshes have received much attention due to the fact that they are important staging areas for waterfowl. Waterfowl nest in northern freshwater marshes, winter in southern marshes, and rest in other marshes during their migrations.

Nutrients play an important role in the ecological health of marshes. Marshes are "sinks" for nutrients washed in through surface run-off as well as from groundwater sources. Many nonestuarine marshes receive run-off from livestock operations and farm fields. This run-off has resulted in excessive nutrient loading. Wood (1995) found that in the Pacific Northwest, low dissolved oxygen and lethal pH levels (seasonally exceeding 10) upset the ecological balance of lakes and wetlands resulting in shifts in species composition and abundance. Comparatively, existing bulrush (or tule) marshes maintain summer pH just above the neutral level, locking nutrients into the bottom of the marsh and purifying the waters (ibid.). Nonestuarine marshes are valuable as wildlife habitat and have been tested extensively as sites that can assimilate nutrients from human domestic wastes.

The major factors influencing marsh formation and development are the dense growth of a diversity of plants that capture sunlight and build plant biomass; slow water movement; and basins sealed by accumulation and settling of silts, clays, and organic matter. Water movement may not occur in marshes with perched basins, but water loss via evaporation and transpiration and inflow from rainfall may cycle water fairly rapidly through the system. These factors contribute to the fact that marshes are generally very productive (high energy), nutrient-rich wetlands.

A number of excellent references on nonestuarine marshes exist. The reader is referred to Good et al. (1978), Weller (1994), Gore (1983), Prince and D'Itri (1985), Mitsch and Gosselink (1993), van der Valk (1988), and Galawitsch and van der Valk (1995).

Site conditions

Soils

Marsh soils are generally poorly drained as a result of settling and accumulation of organic matter. Because of this fact, newly created marshes initially may drain at a faster rate than natural ones, however, over time, silts and clays will "seal" the basin. It may be decades before the amounts of soil organic matter in restored and mitigated wetlands is equal that in natural ones (Sumner and Kentula 1994). Nonestuarine marshes have mineral soils rather than peat (as in bogs) and are generally rich in mineral nutrients and organic materials and therefore, productive. Marsh soils often have high soil calcium, medium or high nutrient loading rates, high productivity, and high soil microbial activity that leads to rapid decomposition, rapid recycling, and nitrogen fixation (Mitsch and Gosselink 1993). Peat may or may not occur. Groff (1994) found that adding organic amendments to disturbed mineral soils is not always beneficial, especially in areas where soils are always saturated. However, he does recommend the use of organic fertilizers for seasonally-saturated or upland soils.

Prairie pothole marshes are found in greatest abundance in moraines of undulating glacial till. These potholes contain rich soils. Soil salts, even in low concentrations, determine the species present at a site. Because many inland marshes are potholes that collect water that leaves only by evaporation, salts may become concentrated during periods of low precipitation, adversely affecting the growth of salt-intolerant species (Mitsch and Gosselink 1986).

Topography

These community types occur in relatively low, flat areas where a basin or depression exists. Even slight changes in topography imply transition to uplands or deepwater aquatic systems. Small freshwater marshes are sometimes created on steeper slopes by roadcuts, manmade impoundments, or the damming activities of beavers (*Castor canadensis*). The most extensive marshlands occur in flat valley bottoms while upland depression marshes are relatively small.

Moisture and hydrology

Nonestuarine marshes are truly aquatic systems. They may not be totally inundated year-round, but the water table is at or above the soil surface for most of the year. The variable moisture levels in these communities contribute to the highly variable species structure and composition. Even when standing water is not present, organic soils may hold sufficient water to promote germination or to sustain the growth of emergent aquatic or "water-loving" plants (hydrophytes).

The flooding regime determines the character of nonestuarine marshes. Water levels are controlled by the balance between precipitation and evapotranspiration, especially for those marshes in small watersheds with restricted throughflow (Mitsch and Gosselink 1986). Many marshes dry down seasonally, but the plant species found there reflect the hydric conditions found during most of the year. A marsh may have sources of water other than direct precipitation (ibid.). Some marshes intercept groundwater supplies. Other marshes have "sealed" or "perched" basins that are not influenced by ground-

water. The water level of these marshes is highly variable. Although usually less than twenty percent of the water impounded in freshwater wetlands enters the groundwater supply, this water may be an important contribution to the watershed (Weller 1994). Overflowing lakes supply water and nutrients to adjacent marshes. Riverine marshes are supplied by the rise and fall of the adjacent river. However, as nonestuarine marshes depend upon rainfall, which is unpredictable, water regimes vary spatially and temporally. Wetlands are generally classified according to their water regime as ephemeral, temporary, seasonal semipermanent, and permanent.

Transpiration loss often exceeds that due to evaporation—and is greater with tall than with short emergent plants (Weller 1994). In some North Dakota prairie potholes, the combined loss of water during the summer through evaporation and transpiration was thirty inches, exceeding the annual rainfall of twelve inches (Eisenlohr et al. 1972). Thus, runoff was essential for maintenance of the water level, and snowfall, early spring rains, and carryover were essential for maintenance of levels under dry conditions. Areas with water high in dissolved salts that also have high summer evaporation become alkaline (sulfate) or saline (chloride) wetlands, with a corresponding change in vegetation and animal life (Stewart and Katrund 1972).

The average slope of the Florida Everglades is only one and eight-tenths inches per mile, so the freshwater from Lake Okeechobee flows slowly southward across the land as a broad sheet during the wet season (Mitsch and Gosselink 1986). The freshwater head prevents salt intrusion from the Gulf of Mexico in the south. During the dry season, the surface water dries up and is found only in the deepest holes. These deep holes are crucial habitat for many wading birds and alligators (*Alligator mississippiensis*).

Establishment

Site preparation

As nonestuarine marshes are usually wet, site preparation can prove difficult unless the project is a creation of a marsh where one had not previously existed. Many marsh restoration projects take place on pre-existing wetlands that may not require intensive site preparation. In both restored and created marshes the hydrology is the most crucial physical characteristic related to the success of the project.

The primary consideration when preparing the site is the hydrology. If the hydrology is not conducive to the development of desired marsh vegetation (to determine this may require experimentation), then the hydrology will have to be restored first. Grading to the proper elevations and restoring the hydrology will accelerate revegetation efforts. If the hydrology is already appropriate, vegetation enhancement may proceed.

Preconstruction planning involving detailed contour mapping of prospective sites is essential (Verry 1985). Site observations during natural flooding periods also are useful because contour maps may not provide the precision essential in water-level regulation. In general, the contours of marshes are slightly sloping and the basins are relatively shallow. The creation of sinuous margins or islands is beneficial to breeding birds.

Earth-moving equipment is commonly used for contouring and can be used to create water depths associated with the desired plant community. Islands, bays, and other structural features can be created during construction if soil character and shoreline protection are considered. Where such work is done on areas with a rich seed bank, soil should be moved off site and returned as topsoil because of its organic content and seed bank. Dams and levees often are built of borrowed soil taken from the water side to create some deepwater sites or to facilitate installation of water control structures. When basin shape is modified by scraping, a barren substrate may be created that must be reseeded or await the natural processes of seed transport, germination, and local seed production (Kelting and Penfound 1950). Returning topsoil also may reduce invasions by exotics. Modifying shallow wetlands to create open pools and deeper water can be done by dewatering and a bulldozer, by drag-line movement of basin substrate, or by blasting with explosives (Strohmeyer and Fredrickson 1967, Mathiak 1965).

Because so many plants and invertebrates of marshes survive as a result of tolerant seeds or eggs, and because these organisms are distributed by wind, birds, and other animals, diversity seems to come quickly to marshes (Weller 1994). A restoration site that is situated in a low area may have a seed bank representative of a marsh, indicating a former wetland. Once a water supply is returned to an area

it may be worthwhile to wait and see what germinates.

Planting considerations

In marsh communities, and most wetlands, different species often occur in zones on slight gradients. This is usually dependent on average water levels. For example, the narrow-leaf cat-tail (*Typha angustifolia*) is more flood tolerant than the broad-leaf cat-tail (*T. latifolia*), and may grow in water up to three feet deep. Marshes are dominated by angiosperms, or flowering plants, as opposed to bogs which are dominated by sphagnum mosses.

Muskrat (*Ondatra zibethicus*), nutria (*Myocastor coypus*), beaver, birds and waterfowl have major influences on the structure and composition of marsh communities. For example, muskrats and nutria are credited with "eat-outs" that decimate marsh vegetation. However, muskrats and nutria keep some aggressive marsh plants, such as cat-tails, in check by feeding on the tubers. Beavers create marshes and ponds where none existed before and land managers are forced to decide if beavers are compatible with other management goals. Birds and waterfowl serve as dispersal agents for a variety of marsh plants and also keep certain aquatic plants under control.

Cages or "in-marsh exclosures" may be required around planting plots to prevent carp and other wildlife from uprooting the newly planted vegetation. Muskrats can chew their way into cages made of plastic snowfence, so wire fence should be used. Muskrats have been known to eat nearly all initial plantings in some restoration projects. Cat-tail (*Typha* spp.) tuber feeding by muskrats and food searching by fish such as carp can cause floatation of the buoyant stocks. Once vegetation is well-established the exclosures may be used for another plot at the restoration site or used in another restoration project.

Blankets and cylinders of coir (processed fiber of coconut tree) have been used with a great deal of success to hold plants in place in wetland restoration and streambank stabilization projects. Other benefits of this material is the high tensile strength when wet, durability (up to ten years), and organic matter/ content (100 percent).

Emergent plants should be established prior to submersed vegetation as it may decrease water turbidity by reducing wave action and by trapping some of the suspended sediments.

Bailey (1985) listed the optimum water levels for some common wetland species:

0–6 inches
    skunk cabbage (*Symplocarpus foetidus*)
0–12 inches
    buttonbush (*Cephlanthus occidentalis*)
    river scouring rush (*Equisetum fluviatile*)
    narrow-leaf cat-tail (*Typha angustifolia*)
    broad-leaf cat-tail (*Typha latifolia*)
    large-fruited burreed (*Sparganium eurycarpum*)
    mud plantain (*Alisma plantago-aquatica*)
    pickerelweed (*Pontederia cordata*)
    water arum (*Peltandra virginica*)
    softstem bulrush (*Scirpus validus*)
    marsh marigold (*Caltha palustris*)
    lizard tail (*Saururus cernuus*)
    crowfoot (*Ranunculus septentrionalis*)
0–24 inches
    duck potato (*Sagittaria latifolia*)
6–15 inches
    water pepper (*Polygonum hydropiperoides*)
6–24 inches
    common bladderwort (*Utricularia vulgaris*)
    milfoil (*Myriophyllum verticillatum*)
12–24 inches
    water smartweed (*Polygonum amphibium*)
    yellow pond lily (*Nuphar luteum*)
    American white water lily (*Nymphaea odorata*)
    coontail (*Ceratophyllum demersum*)

These planting depths were used by Bailey (1985) in Ohio and serve as useful guidelines for a few common marsh species. However, species and appropriate planting depths need to be determined on a site-specific basis. We suggest reviewing literature that contains this type of information, personal communication with local or regional native plant suppliers, observing nearby natural wetlands, and experimental plantings. To determine optimum planting depth in a particular wetland, plant a transect or strip of one species along a wetland gradient and over time, the vigor of the plants will reveal the proper depth for optimum growth.

Most aquatic perennials can be proagated by division of rhizomes. Arrowhead is easy to transplant. Tubers or plants should be planted one yard apart in water from a few inches to one yard deep between the beginning of March and the end of July.

These species thrive along the shore or in shallow water and protect the soil against erosion.

Lizard's tail can be propagated by pressing small pieces of the rhizome into the mud in very moist areas or under shallow water. Rhizomes are the method the plant itself uses to form large colonies.

Seeding and planting of vegetation in nonestuarine marshes is not always successful. Natural seedbanks may prove more successful in vegetation establishment if they are present and water levels can be manipulated to induce germination.

Bailey (1985) used plant materials propagated from seed and root stalks. Each species was planted in large, natural-looking clumps. Those species that tend to be aggressive such as yellow pond lily (*Nuphar luteum*) and American white water lily (*Nymphaea odorata*) were contained in planter boxes (ibid.).

Bremholm and van der Valk (1994) examined the response of uptight sedge (*Carex stricta*) seedlings to treatments of compost, top soil, or fertilizer. In the greenhouse, all amendments improved seedling growth—a finding confirmed by the field studies. However, plots treated with both compost and topsoil also produced significant numbers of unwanted volunteer species.

Thullen and Eberts (1995) described methods for successfully germinating seeds of hard-stem bulrush (*Scirpus acutus*). Seeds were collected, scarified, and cold stratified, then incubated in light- and temperature-controlled regimes under a fourteen-hour photoperiod. Germination rates over ninety-seven percent were reported for seed exposed to a fifty-eight to seventy-seven degrees Fahrenheit regime after a twelve-week cold period. No significant germination differences were observed between seeds collected from a dry river bed or from a flood control pond.

Spacing considerations

As a general rule, groups of ten to forty aquatic emergent plants should be planted in widely spaced plots, often in containment structures (to stabilize sediment until vegetation takes over) or exclosures (to prevent wildlife from uprooting or eating plantings). As plants become established, they often spread vegetatively as well as sexually (by seed) and colonize areas between plots.

Temporal spacing for different plants and layers

Prairie potholes exhibit a characteristic five- to twenty-year cycle of dry marsh, regenerating marsh, degenerating marsh, and lake (Weller and Spatcher 1965, van der Valk and Davis 1978), which is related to periodic drought. During drought years, standing water disappears. Buried seed in the exposed mudflats germinate to grow a cover of annuals (*Bidens, Polygonum, Cyperus, Rumex*) and perennials (*Typha, Scirpus, Sparganium, Sagittaria*). When rainfall returns to normal, the mudflats are inundated. Annuals disappear, leaving only the perennial emergent species. Submerged species also reappear (*Potamogeton, Najas, Ceratophyllum, Myriophyllum, Chara*). While this phenomenon has been well documented in prairie potholes, similar long-term cycles occur in other freshwater marshes.

Maintenance and management

Weed control

With several introduced weeds such as water-hyacinth (*Eichhornia crassipes*), hydrilla (*Hydrilla verticillata*), and purple loosestrife (*Lythrum salicaria*), wetlands are impossible to keep open without some type of control measure. Water-hyacinth, Eurasian water milfoil (*Myriophyllum spicatum*), and hydrilla have cost millions of dollars in control programs. Grass carp, or white amur (*Ctenopharyngodon idella*), are widely used throughout the South for weed control. Some state agencies allow such introductions only in closed basins, but use of such biological control agents have been successful in reducing populations of the exotic water weed hydrilla. While natural reproduction of grass carp in lakes is unlikely (as it naturally spawns in streams), sterile hybrids have been developed for weed-control efforts. Grass carp may compete with waterfowl and other fish in marshes as it utilizes submerged water plants and invertebrates (Gasaway and Drda 1977).

Some native plants such as cat-tails (*Typha* spp.) and willows (*Salix* spp.) may become aggressive in created or restored wetlands and require management. Water-level manipulation is often effective for these, however, this is not possible in many wetlands and other mechanical or chemical control methods will be needed.

Common reed (*Phragmites australis*) may be a problematic weed in some nonestuarine marsh communities. For a discussion of its control, see the Weed Control section of the Estuarine Marshes restoration type. Common reed is distributed worldwide and is native to some North American marsh communities and an invasive weed in others.

One of the worst weeds of nonestuarine marshes may be purple loosestrife (*Lythrum salicaria*). Anderson (1995) stated that there is not enough quantitative evidence to support the common perception that purple loosestrife causes species decline within native plant communities. Five European insect species have been recently approved by the United States Department of Agriculture for control of purple loosestrife (Anonymous 1995). They are *Hylobius tranversovittatus*, a root-mining beetle; two leaf-eating beetles, *Galerucella pusilla* and *G. calmariensis*; and two flower-eating beetles, *Nanophyes brevis* and *N. marmoratus*. The initial phase of insect release and colonization is expected to be completed by 1996.

Nutrient requirements

Flow of water through a marsh is the primary source of its nutrients. Some marshes intercept groundwater supplies. Groundwater flows are usually rather small sources of nutrients. Surface inflow, where it occurs, is usually a major source; it varies widely, sometimes providing many times the needs of the vegetation (Mitsch and Gosselink 1986). Excessive nutrients from stormwater and/or sewage can have a negative impact upon freshwater marshes (Chow-Fraser and Lukasik 1995). Dissolved inorganic nitrogen and phosphorus, the elements that most often limit plant growth, often vary seasonally from very low concentrations in the summer, since plants take them up as rapidly as they become available, to high concentrations in late winter, when plants are dormant but mineralization continues in the soil (Mitsch and Gosselink 1986).

To create a freshwater marsh in Ohio, Bailey (1985) adjusted the soil and water to a near neutral pH to allow for optimum levels of nutrient uptake and microbial activity. Marshlands thrive best at a six and one-half to seven pH (ibid.). Marsh peat is saturated with bases and as a result, the pH is close to neutral. Because nutrients are plentiful, productivity is higher than in bogs, bacteria are active in nitrogen fixation and litter decomposition, and turnover rates are high (Mitsch and Gosselink 1986). Peat accumulation results from high production rates, not from inhibition of decomposition by low pH (as occurs in bogs) (ibid.).

Fire has been used by some managers to release soil nutrients in marshes. Some others have experimented with the use of fertilizers to enrich marshes. This practice seems wasteful, since marshes seem to be nutrient sinks anyway, but some positive short-term responses (increase in plant growth) have been noted (Weller 1994). The use of fire or fertilizers in marshlands should be carefully evaluated prior to implementation.

Water requirements and hydrology

Submerged aquatic plants require light for germination, growth, and development. In some areas, high turbidity reduces light penetration and limits the growth of submersed vegetation and the seedlings of emergent plants. Water level (and sometimes flow) fluctuations are important for the germination and survival of many marsh plants.

Chow-Fraser and Lukasik (1995) found that drastically altered water levels may be responsible for the demise of some freshwater marshes. A restoration project may have to be designed to utilize current and not historical water levels in a marsh in order to successfully establish a diverse self-perpetuating community of aquatic plants.

Marshes may be formed in any basin that will hold water long enough for the germination and survival of semiaquatic plants. Basins probably hold water poorly when they are new, unless they are created in fine silt or clay that seals easily, but buildup of organic matter helps to fill the many pores (Weller 1994). Eventually, basins collect groundwater, rainfall, snowmelt, or floodwater from watershed, river, or lake.

The use of a water control device in marsh restoration may be especially useful in managed or urban areas, marsh creations, or if wildlife habitat is a goal. Water-level management is generally feasible only when a water-control structure has been constructed, either to restore a marsh or to aid in water-level stability. Water-control structures can improve chances of successful marsh management in areas with modified stream flow, lowered water tables, increased sedimentation due to wind or water erosion, and increased disturbance or consumption of vegetation because of artificially high levels of wildlife or domestic livestock.

A common problem is a control structure that cannot regulate levels within the necessary precision of one to two inches. Control structures should be located in such a way that complete drainage of the basin, a method for creating suitable plant seedbeds through oxidation and decomposition or for eliminating carp or muskrats, is possible. Water-level regulation may be used to flood out dense or weedy vegetation established during dry periods, or to dry out the marsh for revegetation. Drawdowns for revegetation may be (A) complete, when major restoration is essential in an open marsh; or (B) partial, when vegetation needs to be encouraged or herbivores discouraged (Weller 1994). Timing of the drawdown is important; too early a drawdown may induce emergents and willows that become weedy and too late a drawdown may favor valuable but annual species that do tolerate much flooding. Reflooding should be a gradual process (several weeks), to avoid floatation of emergents, direct scouring of other plants, or plant mortality due to the turbidity of muddy waters.

Water levels should be reduced to meadowlike depths (water at or near the soil surface) to encourage vegetative propagation of emergents and germination and growth of submergents in early summer. Management of muskrats, nutrias, beaver, or other herbivores is a valuable but more difficult task, since it depends upon the efficiency of population regulation (Weller 1994).

Case Studies

During the initial plantings for the restoration of Cootes Paradise Marsh, Chow-Fraser and Lukasik (1995) found that ten percent of the broad-leaf cattails (*Typha latifolia*) and twenty-four percent of the duck-potatoes (*Sagittaria latifolia*) planted in eight inches of water had become established from initial plantings. None of the emergent plants at depths greater than twelve inches survived. Installing silt screen around exclosures ameliorated some of the effects of deep water. None of the arrowheads in the unscreened exclosure survived, whereas all of those in the exclosure surrounded by siltscreen survived despite a water depth of greater than fifteen inches. Many plants grew vegetative shoots from runners and in one year more than 100 plants were counted in exclosures where only thirty had been planted.

Many of these plants also set seed. Similar results were obtained in shallower water (six inches) where the survival rate of cat-tails planted in the screened exclosure was more than double that of cat-tail planted in the unscreened exclosure.

Schreiber and Dinsmoore (1995) surveyed seven natural and fourteen restored wetlands in northwestern Iowa to determine the structure of their breeding bird community. Their results showed that despite an increase in the number of bird species breeding on restored wetlands, the communities still differ from those of natural wetlands. Common natural wetland species, including the marsh wren (*Cisothorus palustris*) and Virginia rail (*Rallus limicola*), were absent from the restored wetlands. However, mallards (*Anas discors*) nested more often on restored wetlands than on natural wetlands.

For additional information on the restoration and management of nonestuarine marshes, see Wheeler et al. (1995).

Wet Prairies and Grasslands

Wet Prairie A-205, A-316, A-364
Coastal Grassland A-362
Marl Prairie A-363

Introduction

Wet prairies and grassland communities are mostly treeless, grass-dominated communities that have water near or above the soil surface for a large part of the year. They are different from the communities included in the prairies and grasslands restoration types in that they are more mesic and often hydric (wetland). However, some of the information presented in the prairies and grasslands sections may be useful here.

Wet prairies and grasslands often occur at the edge of marsh communities. Shallow marshes grade into seasonally flooded, meadow-like areas that dry early and are covered by herbaceous vegetation. The vegetation is composed primarily of annuals and low-profile emergents such as spike-rush, rushes, and water-tolerant grasses (Weller 1994).

Site Conditions

Soils

The soils of wet prairie and grassland communities are typically poorly drained mineral soils. In the Southern Coastal Plain, wet prairie communities develop over peat and marl and each soil type supports distinct communities. Marl is a calcitic mud generated by precipitation of calcium carbonate by the blue-green algal mat in seasonally inundated, short-hydroperiod marshes (Browder et al. 1994).

Topography

The soil surface elevation is a function of the underlying bedrock topography and accumulations of soil. Bedrock near the surface is constantly being dissolved at the same time soil is developing and accumulating. Other variables (such as fire, hydroperiod, vegetation) influence the rates at which these processes occur (Gunderson 1994).

The overall topographic gradient is relatively level, however, microtopography is very important. Slight changes in topography can have significantly different hydrologic regimes and consequently different species composition may occur.

Moisture and hydrology

Marl prairies and coastal grasslands are drier than wet prairies. Floods and droughts are not uncommon in wet prairie and grassland communities. Dramatic shifts in type of vegetation community and composition can occur during floods and droughts.

The soil surface of wet prairies and grasslands is usually just above the permanent water table, but at certain times during the year (or after heavy rains) standing water may be present. Wet prairie and grassland communities are poorly drained. Often a hardpan occurs near the soil surface and impedes drainage. A hardpan is a hardened layer in the soil, cemented by iron oxide, silica, organic matter, or some other substance.

In south Florida, the summer rains cause water levels to rise and reach annual maxima by September and October in most years (Gunderson 1994). As rainfall decreases through the fall and winter months, water levels decline and reach annual minima during the spring months. Gunderson (1990) studied water level records in five wetland communities of the Everglades National Park and found that historical hydropatterns did not differ statistically among community types, due to the high year-to-year variability. All wet prairie and grassland communities are inundated for at least some period during the year. Marl prairie is inundated on average three to seven months/year and mean depths average four inches (Olmsted et al. 1980).

Establishment

Site preparation

The hydrology must be in place before the soil surface is prepared for planting or seeding. It may be worthwhile to wait one full season after the hydrology has been restored before initiating revegetation plans. Once the hydrology is restored, species com-position will begin to change to more typically aquatic species due to natural processes. However, the speed at which this take place may be very slow, the relative abundances of the species present may not represent the desired community composition, and/or nonnatives may need to be controlled.

If the community is somewhat intact, it may be best to use supplemental plantings to enhance exist-ing desired vegetation. However, if the species composition is far from desired and nonnatives are plentiful, an appropriate planting surface will need to be prepared. In wet prairie and grassland situa-tions, it may be difficult to operate large equipment due to wetness; therefore, the site may need to be prepared when the soils are least inundated. The least disturbing technique to prepare the planting surface would be no-till methods such as a combina-tion of mowing and insulation (covering the surface with a layer of dark plastic) or a mowing and burning or herbicide treatment. While the use of herbicides in wetlands is often discouraged, some herbicides claim to be safe for such purposes. It is important to check local regulations before using any chemical pesticides. If tilling is used to prepare a planting surface, it should be very shallow, turning over only the top few inches of soil.

Planting considerations

Marl prairies occurring on peat soils are domi-nated by spike-rush, beak sedge (*Rhynchospora* spp.), and maidencane (*Panicum hemitomon*). Wet prairies over marl substrates occur in the southern Everglades at localities where bedrock elevations are slightly higher and hydroperiods slower (Gunderson 1994). These marl prairies are domi-nated by muhly grass (*Muhlenbergia* spp.) and swamp-sawgrass (*Cladium mariscus*). Beak rush is common in the lower, wetter areas of the marl prairies. Olmstead and Loope (1984) reported over 100 species, the majority of which are herbaceous plants; yet these comprise less than one percent of the ground cover (Olmstead et al. 1980). They found from nine to twelve species per square yard. Typi-cally the grassy vegetation is less than three feet tall.

The marl prairies have undergone changes in hydrologic and fire regime, with resulting changes in vegetation. One such shift has been the increasing abundance of muhly grass (Atwater 1954).

Unlike butterflyweed (*Asclepias tuberosa*),

which has a deep tap root, swamp milkweed (*A. incarnata*) has a more shallow, fibrous root system. This allows easy transplantation of this species. Queen-of-the-prairie (*Filipendula rubra*) can be propagated by dividing the rhizomes. Iris (*Iris* spp.) plants expand to form clumps. These can be divided in early spring or after flowering. They must be promptly replanted and not allowed to dry out.

The rhizome of pickerelweed (*Pontederia cordata*) can be separated and replanted to propa-gate more plants. In addition, the small seeds float on water to create more distant new colonies.

Powdery thalia (*Thalia dealbata*) survives the winter best under twenty-four inches of water. In mild regions, propagate by dividing the roots and planting them under one to six inches of water.

Spacing considerations

As a general rule, most wet prairie and grassland communities have one or a few dominant species that comprise around ninety percent of the total plant cover. These species should be planted to allow for natural spreading (eighteen to thirty-six inches). Less common associates should be planted here and there wherever suitable microhabitats exist or can be created.

Temporal spacing for different plants and layers

In these grass-dominated communities, the ver-tical structure is relatively simple and most species can be planted at the same time. Because of the effort required for site preparation and initial mainte-nance, year-to-year variations in weed seed germi-nation, year-to-year differences in availability, vari-ety and viability of prairie seed, and the benefit gained from experience along the way, large resto-ration sites should be divided into several years of plantings (Ahrenhoerster and Wilson 1981, Shramm 1978).

Maintenance and Management

Weed control

As in sawgrass marshes, wet prairies in the Everglades have been invaded by cajeput tree (*Melaleuca quinquenervia*) within the last fifty years. The southeastern coastal prairies have been invaded by Australian pine (*Casuarina equisetifolia*) since the 1950s (Egler 1952). Some parts of the Everglades have shown significant invasion by cajeput tree and Australian pine. Both of these invasive trees require intensive control programs (Randall and Marinelli 1996).

Nutrient requirements

The amount of nutrients and salinity in the water influences the distribution of the vegetation. Most plants of wet prairies and grasslands are adapted to low nutrient levels. The relationship between water chemistry and nutrient levels is dynamic. Nutrients tend to be more available for growth at certain pH and dissolved oxygen levels and temperatures. Nutrients, in turn, like calcium and magnesium, can moderate the pH of waters by acting as buffers that neutralize acids. In limestone waters that are thick with plants using up the available carbon dioxide, white calcium carbonate, called travertine or marl, can precipitate out and cover the plants and substrate (Caduto 1990).

Disturbances

The disturbances that influence wet prairies and grasslands include fire, frosts, and animal activity (Davis 1943, Loveless 1959, Craighead 1971). Fires play a major role in these communities. Severe fires have been a part of the ecology of the Everglades throughout the existence of current vegetation patterns, as evidenced by the presence of charcoal throughout the basal peats of the wetlands (Cohen 1984, Gleason and Stone 1994). Freezes influence the sensitive species. Alligators are the most important animal effecting these communities (Craighead 1971). Alligators influence soil elevations by the physical sculpting associated with their movement activity (Gunderson 1994). Alligator nests may also be incipient tree islands, because they serve as foci for colonization by swamp hardwoods, that then may increase in size to larger tree islands (ibid.).

Case Studies

For additional information on the restoration and management of wet prairies and grasslands, see Davis and Ogden (1994), Galatowitsch and van der Valk (1995), Murn (1993), Stewart and Katrund (1972), Thompson (1992), and van der Valk (1988).

Vernal Pools

Vernal Pools A-43

Introduction

Vernal pools are formed by the accumulation of winter rains on top of a hardpan layer, which results in a perched water table that is exposed in depressions of shallow soil and surrounded by mounds of deeper soils. As water evaporates in spring, showy wildflowers appear around the margins of the pools and swales, while hydrophytes grow in the water. The upland areas of highest relief usually are dominated by introduced grasses and forbs, while inundated areas contain successions of mostly native forbs and grasses that are endemic and adapted to life in this specialized habitat (Thorp and Leong 1995). For more detailed information on the ecology of vernal pools, see Jain (1976), Holland (1978a,b), Jain and Moyle (1984), Zedler (1987), and Ikeda and Schisling (1990).

Vernal pools apparently have been a part of the California landscape for at least tens of thousands of years, as judged from the wide and, in many cases, disjunct distribution and large number of species that are restricted to the habitat (Holland and Jain 1977). In the past, these pools were a common feature in most areas of the Central Valley, but agricultural expansion, mineral extraction, and residential development have eliminated most of these areas. The best remaining pools are found on higher, older terraces. Vernal pools, and associated endangered species, also face threats from vineyard and effluent irrigation (Waaland 1990).

Because of extreme environmental conditions, the vernal pool is a difficult place for living organisms to inhabit. For a time the area is submerged; then follows a period of progressively drier conditions, until finally the bottom is dry. During the time that water is present, the inhabitants of vernal pools must complete their reproductive activity. These same species must also survive during the dry period. Later the depression fills with water and the cycle begins again.

Three kinds of vernal pools exist in the Great Central Valley of California: valley pools, pools of volcanic areas, and terrace pools. Valley pools, most common in low places of the San Joaquin Valley, occur in basins or valleys in saline or alkaline soils (Schoeherr 1992). Pools of volcanic areas are found throughout the state. They are typified by those on the Vina Plains, in Tehama County, and Mesa De Colorado, in Riverside County. Floristically, they are similar to the terrace pools. Terrace pools occur on some of the oldest soils in the state, ancient flood terraces on higher ground.

While vernal pools also occur in the eastern deciduous forest, these are quite different from the California vernal pools discussed here. In the east, vernal pools occur in forested and open areas. These habitats are very important habitats for breeding amphibians and rare animals such as fairy shrimp and burrowing crayfish. There have been several efforts to preserve and/or restore these unique temporary wetlands.

Site Conditions

Soils

Because of the high rates of evaporation, soils are usually alkaline. Stone (1957) concluded that "of the many factors which may be collectively responsible for vernal pools, low precipitation, poor drainage, and high evaporation are factors which are equally responsible in the genesis of alkali soils." Thus soil moisture and salinity could be used to classify vernal pools. The basic soil types in which vernal pools develop are terrace soils, soils derived from volcanic mudflows, and on lava-capped mesas. Vernal pool soils may be clayey or sandy; the ability to hold water is derived from impervious hardpans or claypans.

Topography

The microrelief usually differs by only several inches. Associated with vernal pools on terrace soils are peculiar mounds known as "mima mounds," up to six feet in height (Schoenherr 1992). The terrain has a rolling, mounded appearance with vernal pools in the low spots. There are many theories about the origin of these mounds, including wind deflation of old stream channels and the piling of soil around old shrub fields. There are several places in the world typified by mima topography, but no satisfactory explanation exists for the mounds because of the diversity of climates and geologic processes in the localities where they occur (Schoenherr 1992).

One of the most conspicuous features of vernal pool vegetation is the concentric arrangement of species in relation to topography. At certain times during the year, topographical maps of individual pools may be prepared on the basis of zonation of various plants (Holland and Jain 1977). These maps are as accurate as maps prepared with transit and stadia. Specialization by pool taxa for portions of the topographical gradient is apparent. Lin (1970) used three zones to describe this concentric pattern. Species richness varies in relation to the topographical gradient. It is highest at the pool margin, slightly lower in the adjacent grassland, and considerably lower in the pool (Holland and Jain 1977).

Moisture and hydrology

Vernal pools are temporary wetlands. They hold water for a relatively short period and some years they are dry year-round. Impervious hardpans or claypans inhibit subsurface drainage of rain water and result in water tables perched near the surface. The water in these pools evaporates during spring months, resulting in a concentration of ions in the pool water and soil.

Vernal pools, sometimes called hogwallows, are formed by runoff during winter rains, as the water does not percolate into the clay soil. Water remains standing to form pools of various sizes, which gradually disappear by evaporation in the spring. Year-to-year fluctuations in precipitation do provide a highly variable water regime.

Establishment

Site preparation

Purer (1939) described the physiographical features of vernal pools in San Diego County, California in some detail. The pools are more or less oval in shape, with the long axis usually in a north-south direction, and they vary in size from 700 square yards to less than one square yard. Pool depth in the center varies in the range of six to twenty-four inches.

The proper microtopography must first be established on the restoration site. This includes mounds and pools. Vernal pools are characterized by a semirolling landscape, with hummocks of annual grassland vegetation surrounding the pools. The pools are ecological islands with a different annual flora. Soil texture, moisture, salinity, and microrelief are important determinants of vernal pool vegetation. These parameters must be considered during plant species selection and installation. The primary manipulation of the physical environment is the excavation of shallow basins.

Vernal pools are small depressions in valley grasslands that fill with water during the winter and become dry during the spring and summer. As they dry, various annual plants flower in concentric rings. The vegetation of these pools is commonly described in terms of "zones." The zones are usually organized from pool bottom to mound top. The pools support alkali or salt-associated plants. No trees, shrubs, or stem succulents are known to occur in vernal pools.

Planting considerations

The plants of vernal pools are unique. Holland and Jain (1977) compiled a list of 101 species considered typical of vernal pools. Of these, more than seventy percent were native ephemerals, and over half were endemic to California. Two genera of rare grasses, *Orcuttia* and *Neostapfia*, are found only in association with vernal pools, and five species of *Orcuttia* are endemic to California. Many of the vernal pool species are rare or endangered. Habitat destruction due to development and grazing has caused the disappearance of many areas of vernal pools.

Those involved in the restoration and management of California vernal pool ecosystems must consider the intimate relationships between the indigenous spring wildflowers and the small ground-nesting bees that inhabit these ecosystems. Thorp and Leong (1995) discovered that the annual cycles of host-specific pollinating bees and particular wildflower genera such as *Blennosperma, Lasthenia, Limnanthes,* and *Downingia,* are closely synchronized. Maintaining these cycles through the preservation of vernal pools is the preferred way of ensuring showy wildflower displays since it keeps all elements of the habitat intact. Vernal pool restorationists should transplant or monitor bee pollinators even though it is not required.

These ground-nesting bees (such as *Andrena [Diandrena] blennospermatis*) are endemic to vernal pool habitats and are even more limited in distribution than their pollen host plants (Thorp and Leong 1995). The andrenid bees carry out pollination, a vital function of the community in specialized association with their host plants. Unlike other bees, such as bumble bees and honey bees, andrenid bees probably are able to fly less than one-half mile. As a result of the continuing fragmentation of vernal pool habitats, many remaining pools are separated by distances greater than one-half mile.

Vernal pools contain many organisms listed on federal and state (California) endangered or threatened lists; these include the cryptic fairy and tadpole shrimp. Prevalent taxon-specific approaches to vernal pool habitat conservation tend to be biased toward the preservation of high-profile species such as these. Thorp and Leong (1995) suggest that a community or ecosystem-based approach to vernal pool habitat conservation will be more effective in preserving and protecting both well-known and relatively unknown species within vernal pool habitats.

In vernal pools, the soil is variable; however, planting holes can be made with the appropriate tool (such as a planting bar, dibble, or hoedad). In some vernal pools the soil is stony and suitable microhabitats (i.e., pockets of accumulated soil) must be found when installing plants. Rabbit grazing is a persistent but low-level natural disturbance in vernal pools (Zedler and Black 1989). Grazing appeared to be especially severe in those artificial pools that were surrounded by dense chaparral (ibid.). If the local rabbit population is dense, protective fencing may need to be used until plants become well-established.

The success and establishment of individual plants after their seed reaches a given area is highly dependent upon the weather of the particular season. Natural regeneration in degraded vernal pools depends upon the seedbank present and the proximity of natural pools. Many vernal pool plants are available as plugs or seeds from California native plant nurseries (see Appendix D).

Granholm (1989) reported that seeds should be raked, drilled, or otherwise worked into the soil surface. Broadcast seeding and hydro-seeding are discouraged, because they are expensive, require far more seeds, and are vulnerable to seed predation. In addition, hydro-seeding often reduces seed viability for wetland plants by prematurely wetting seeds.

Spacing considerations

A sharp ecotone exists between grassland and pool vegetation. In surrounding grassland, cover may be 100 percent, however, most pools have a characteristically low total cover, frequently less than fifteen to thirty percent (Holland and Jain 1977). Zedler and Black (1989) found that natural vernal pools in San Diego County, California averaged about forty-four percent. Vernal pool plants must be planted in the appropriate pool zone to become established.

Temporal spacing for different plants and layers

Most vernal pool species can be planted at the same time, however, supplemental planting or seeding may be required to achieve desired composition.

Maintenance and Management

Weed control

Unlike adjacent grassland and marshes, the vernal pool habitat has resisted invasions by nonnative species. Holland and Jain (1977) found that vernal pools contain only five to ten percent introduced species. These introduced species come primarily from South America and Africa.

Holland (1978a,b) found that introduced taxa were present in low cover in most pools. Plants such as smooth cat's ear (*Hypochoeris glabra*), brome (*Bromus mollis*), and long-beak stork's-bill (*Erodium botrys*) were observed late in the summer. They were probably plants from seed produced earlier that year in the surrounding grassland. Another commonly encountered introduced species was pimpernel (*Anagallis arvensis*). Mechanical removal or increasing water levels can be used to effectively control vernal pool weeds. As many weeds move in from adjacent grassland, their elimination from these areas would benefit pool habitats.

Nutrient requirements

Adjacent upland areas are important to vernal pool habitats. The upland aspects are interconnected to the pools through nutrient flows, surface and subsurface water flows, food webs, and other biotic and abiotic processes. Failure to restore both landscape features is likely to cause disruptions in these processes.

Water requirements and hydrology

While vernal pool plants require water to grow and reproduce, many are adapted to the harsh pool habitat that may be dry in particular years and inundated in others. Supplemental watering will likely increase initial plant establishment and discourage weeds.

Restored vernal pools must be constructed so that they are able to hold water as natural pools do. Comparisons with water levels and fluctuations in natural vernal pools are useful for assessing hydrologic regimes. Species composition within and between pools is strongly influenced by the duration of standing water. Zedler and Black (1989) found that artificial pools could support vernal pools floras provided they had the appropriate water durations and depths (30 to 100 days and 4–8 inches deep) and were constructed on the same soils as those of natural pools. Much of the water lost from vernal pools is subsurface which takes place at a very slow rate (ibid.). Evidently the subsoils beneath the artificial pools have a less well-developed network of cracks and macropores through which water movement can occur. Vernal pools may be thought of as microhabitats with two repeating natural disturbances—seasonal inundation and seasonal drought (ibid.).

Case Studies

Zedler and Black (1989) created forty vernal pools in San Diego County, California. To accelerate revegetation of artificial pools, an inoculum of vernal pool plants was introduced. This inoculum was collected from adjacent natural pools, by raking up the plant material with a small amount of the surface soil. This procedure resulted in some disruption to the natural pools. Plant cover in the artificial pools was seven percent in the second year of the project whereas nearby natural pools have an average of fourty-four percent. After two years, the project was relatively successful in terms of days of standing water, numbers of species, plant size, and other readily quantifiable parameters. Zedler and Black (1989) caution that successful creation of

vernal pool habitat should not be used to mitigate for the losses of natural pools. The close proximity of natural vernal pools was likely a major advantage for the project.

Pritchett and Ferren (1990) documented the enhancement, restoration, or creation of nine vernal pools at an urban twelve-acre site in Santa Barbara, California. The project included debris clean-up and public education. One pool was enhanced by constructing two dams to prevent drainage through ditches and low areas. The dams were used to increase the depth and duration of flooding, decrease the abundance of weedy introduced species, and increase native vernal pool species. Two pools were restored by removing piles of soil and other debris and by excavating the pools to a greater depth than they were before the project. Excavated material was deposited at a disturbed upland to form an observational area. The restored pools exhibited an increase in native wetland species. Six pools were created by excavating depressions into an upland dominated by introduced grasses and underlain by a clay subsoil. Three of the six pools were inoculated with a seedbank obtained from surface scrapings of natural vernal pools in the vicinity; as an experimental comparison, the remaining three pools were not inoculated. Three years of monitoring indicate that vegetative cover is establishing well in the restored and created pools. Naturalized grasses, however, persist in the restored and enhanced pools and dominate areas where the soil was disturbed and where flooding was insufficient to favor the growth of vernal pool plants. Plant cover and invertebrate animal populations in the inoculated created pools featured high cover and abundances of native vernal pool species, but created pools that were not inoculated supported few vernal pool species. Recovery of native species in the enhanced and restored pools was slow. Pritchett and Ferren (1990) concluded that excavation of depressions into a clay subsoil and subsequent inoculation with a seedbank from existing vernal pools can result in short-term creation of new vernal pool habitat, but the long-term persistence of such new habitats is uncertain.

Thorp and Leong (1995) stated that it is likely that artificial creation of vernal pools for compensatory mitigation actually results in a net loss of specialized ground-nesting bee populations which are crucial to the reproductive success of native vernal pool plants. A lack of specialized bee pollinators may result in decreased levels of seed production in vernal pool plants. Increasing the density of pools at one site to compensate for the loss of vernal pool habitat elsewhere is another example of actions that can produce a net loss of upland bee nesting habitats. Waaland (1990) studied many vernal pools in Santa Rosa, California and concluded that vernal pool creation is not a recommended mitigation, although it may have potential to compensate for loss of low-priority sites.

Mangroves

Mangrove A-365

Introduction

The term mangrove can mean both the group of woody species collectively known as "mangroves" or the ecological communities in which they are dominant. Mangrove wetlands are located in tropical and subtropical regions such as the south and Gulf coasts of Florida and Hawaii. These coastal forested wetlands are dominated by halophytic trees such as red and black mangroves. Mangrove wetlands are influenced by a wide range of salinity and tidal conditions.

The importance of mangroves in exporting organic matter to adjacent coastal food chains, in providing physical stability to certain shorelines to prevent erosion, in protecting inland areas from severe damage during hurricanes and tidal waves, and in functioning as sinks for nutrients and carbon has been well established. There is an extensive literature on the mangrove on a worldwide basis– possibly more than 3,000 titles (Por and Dor 1984). Mangroves are intriguing ecosystems that occur worldwide and possess many unique features. Compared to primary production estimates from other ecosystems, including agricultural ecosystems (Odum 1971), mangroves are among the world's most productive ecosystems. Healthy mangrove ecosystems appear to be more productive than sea grass, marsh grass, and most other coastal systems. Some of the major values of mangrove ecosystems are shoreline stabilization and storm protection, habitat value to wildlife, importance to threatened and endangered species, value to sport and commercial fisheries, aesthetics, and economic products.

It is no longer accepted dogma that mangroves are "land-builders" that are gradually encroaching on the sea (Mitsch and Gosselink 1986). In most cases, mangrove vegetation has a passive role in the accumulation of sediments and the vegetation usually follows, not leads, the land building that is caused by current and tidal energies. It is only after the substrate has been established that the vegetation contributes to land building by slowing erosion and by increasing the rate of sediment accretion (Lugo 1980).

Estimates of the total area occupied by mangrove wetlands in Florida vary widely between 430,000 acres to over 500,000 acres (Odum et al. 1982). Ninety percent of Florida's mangroves are located in the four southern counties of Lee, Collier, Monroe, and Dade. Much of the area covered by mangroves in Florida is presently owned by federal (Everglades National Park), state, or county governments, or by nonprofit organizations such as the National Audubon Society.

Historically, mangrove wetlands have been held in low regard as they are generally impenetrable, hot, and mosquito infested. It is possible that more acres of mangrove worldwide have been obliterated by humans in the name of "reclamation" than any other type of coastal environment (Odum et al. 1982). Destruction of mangrove forests in Florida has occurred by deliberate destruction such as land filling, diking, and flooding. Destruction also occurs due to introduction of fine particulate material and pollution damage, particularly oil spills.

The estuarine marshes of temperate mid- and high latitudes gives way to its analog, the mangrove (also called mangals, mangrove swamps, mangrove forests), in tropical and subtropical regions of the world. Mangroves are found along tropical and subtropical coastlines throughout the world, usually between 25°N and 25°S (Mitsch and Gosselink 1986). Their limit in the northern hemisphere generally ranges from 24°N to 32°N, depending upon local climate and the southern limits of freezing weather (ibid.). In the United States, mangrove wetlands are primarily found along the Atlantic and Gulf coasts of Florida south of 28-29°N latitude (ibid.).

Site Conditions

Soils

Mangroves are found on a wide variety of substrates including fine, inorganic muds, muds with a high organic content, peat, sand, and even rock and dead coral, if there are sufficient crevices for root attachment. The most productive mangrove ecosystems develop along coasts or in estuaries that have fine-grained muds composed of silt, clay, and a high percentage of organic matter (Odum et al. 1982). Anaerobic sediments pose no problems for mangroves and exclude competing vascular plant species. Mangrove trees function as stabilizers of the soil, which is continually shifting due to tides, currents, and winds. Mangroves in Florida often modify the underlying substrate through peat deposition. It is not unusual to find layers of mangrove peat several meters thick underlying well-established mangrove ecosystems such as those along the southwest coast of Florida (Odum et al. 1982).

Zieman (1972) presented an interesting argument suggesting that mangrove peat may be capable of dissolving underlying limestone rock, since carbonates may dissolve at pH 7.8. Through this process, shallow depressions might become deeper and the overlying peat layer thicker without raising the surface of the forest floor.

If mangrove peat is dried in the presence of oxygen, highly acidic soils (pH 3.5–5) can develop. This "cat clay" problem has greatly complicated the conversion of mangrove regions to agricultural land in Africa and southeast Asia.

Although information on mangrove peats and soils is limited, Odum et al. (1982) provided the following generalizations:

- Mangroves can grow on a wide variety of substrates including mud, sand, rock, and peat.
- Mangrove ecosystems appear to flourish on fine-grained sediments that are usually anaerobic and may have a high organic content.
- Mangrove ecosystems that persist for some time may modify the underlying substrate through peat formation. This appears to occur only in the absence of strong physical forces.
- Mangrove peat is primarily formed by red mangroves and predominantly consists of root material.
- Red mangrove peats may reach thicknesses of several meters, have a relatively low pH, and may be capable of dissolving underlying layers of limestone.
- When drained, dried, and aerated, mangrove soils usually experience dramatic increases in acidity due to the oxidation of reduced sulfur compounds. This greatly complicates their conversion to agriculture.

Topography

The terrain where mangroves develop is relatively level to slightly rolling. Topography is of utmost importance since only a few inches affects the amount and duration of tidal flushing. Mangroves also develop further inland along river banks.

Moisture and hydrology

Mangrove is a truly aquatic community. The soils are saturated and under water during high tide. Proximity to terrestrial ecosystems, the ocean, and human activities are all important in determining overall water quality. Surface waters associated with mangroves are characterized by (1) a wide range of salinities from virtually fresh water to above forty parts per trillion, (2) low macronutrient concentrations (particularly phosphorus), (3) relatively low dissolved oxygen concentrations, and (4) frequently increased water color and turbidity (Odum et al. 1982).

Tides import nutrients, aerate the soil water, and stabilize soil salinity in mangroves. The resulting salinity is important to the mangroves in eliminating competition from freshwater species. In areas with high evaporation rates, the action of the tides helps to prevent soil salinities from reaching concentrations that might be lethal to mangroves. In some locations, tides bring salt water up the estuary against the outward flow of freshwater and allow mangroves to become established well inland. Tides transport nutrients and relatively clean water into mangrove ecosystems and export accumulations of organic carbon and reduced sulfur compounds. Tides also aid in the dispersal of mangrove proagules and detritus.

Establishment

Site preparation

The surface of the restoration site must have elevations approximating those of nearby natural mangroves. Mangroves are not as particular as estuarine marsh plants about exact elevations. Operation of earth moving equipment may be difficult on some substrates.

Mangroves have a series of remarkable adaptations which enable them to flourish in an environment characterized by high temperatures, widely fluctuating salinities, and shifting, anaerobic substrates. Most mangrove species are facultative halophytes that compete very poorly with other plants in freshwater conditions, but have a decided advantage when brackish or saline conditions are present. Mangroves have the ability to both prevent salt from entering the plant at the roots (salt exclusion) and to excrete salt from the leaves (salt secretion).

Planting considerations

Four species of aquatic grasses occur on bay and creek bottoms adjacent to mangrove forests (Odum et al. 1982). Turtle grass (*Thalassia testudinum*) and manatee grass (*Syringodium filiforme*) are two tropical sea grasses which occur in waters with average salinities about twenty parts per trillion. Shoal grass (*Halodule wrightii*) is found at somewhat lower salinities and beaked ditch-grass (*Ruppia maritima*) is a freshwater grass which can tolerate lower salinities.

Davis (1940) is generally credited with the best early description of plant zonation in Florida mangroves. Odum et al. (1982) stated that zonation of mangrove species appear to be controlled by the interplay of physical and chemical factors with interspecific competition and possibly, through tidal sorting of propagules. Lugo and Snedaker (1974) provided a useful classification system based on mangrove forest physiognomy. They identified six major community types resulting from different geological and hydrological processes. Each type has its own characteristic set of environmental variables such as soil type and depth, soil salinity range, and flushing rates. The community types are as follows:

■ *Overwash mangrove forests*—these islands are frequently overwashed by tides and thus have high rates of organic export. All species of mangrove may be present, but red mangroves usually dominate. Maximum height of the mangroves is about twenty-three feet.

■ *Fringe mangrove forests*—mangroves form a relatively thin fringe along waterways. These forests are best defined along shorelines whose elevations are higher than mean high tide. Maximum height of mangroves is about thirty-three feet.

■ *Riverine mangrove forests*—this community type includes the tall flood plain forests along flowing waters such as tidal rivers and creeks. Although a shallow berm often exists along the creek bank, the entire forest is usually flushed by daily tides. All three species of mangroves are present, but red mangroves (with noticeably few, short prop roots) predominate. Mangroves may reach heights of sixty to sixty-five feet.

■ *Basin mangrove forests*—these forests occur inland in depressions channeling terrestrial runoff toward the coast. Close to the coast, they are influenced by daily tides and are usually dominated by red mangroves. Moving inland, the tidal influence lessens and dominance shifts to black and white mangroves. Trees may reach fifty feet in height.

■ *Hammock forests*—hammock mangrove communities are similar to the basin type except that they occur on ground that is slightly elevated (two to four inches) relative to surrounding areas. All species of mangroves may be present. Trees rarely exceed sixteen feet in height.

■ *Scrub or dwarf forests*—this community type is limited to the flat coastal fringe of south Florida and the Florida Keys. All three species are present. Individual plants rarely exceed five feet in height, except where they grow over depressions filled with mangrove peat. Many of these tiny trees are forty or more years old. Nutrients appear to be limiting although substrate (usually limestone marl) must play a role.

Salt tolerant herbaceous plants occur where slight increases in elevation exist and where sufficient light filters through the mangrove canopy. Carter et al. (1973) list the following as examples of the mangrove community: leather ferns (*Acrostichum aureum* and *A. danaeifolium*), Spanish bayonet (*Yucca aloifolia*), spider lily (*Hymenocallis latifolia*), sea blite (*Suaeda linearis*), chaff flower (*Alternanthera ramosissima*), samphire (*Philoxerus vermicularis*), bloodleaf (*Iresine celosia*), pricklypear cactus (*Opuntia stricta*), marsh elder (*Iva frutescens*), rubber vine (*Rhabdadenia biflora*), the lianas (*Ipomoea tuba* and *Hippocratea volubilis*), and a variety of bromeliads (Bromeliaceae).

Hammocks occur just upland of the mangrove zone. Typical trees in both communities include the fan palm (*Thrinax radiata*), button mangrove, manchineel (*Hippomane mancinella*), and previously, mahogany (*Swietenia mahagoni*).

Blankets and cylinders of coir (processed fiber of coconut tree) have been used with a great deal of success to hold plants in place in wetland restoration and streambank stabilization projects. Other benefits of this material is the high tensile strength when wet, durability (up to ten years), and its 100 percent organic matter content.

There must be adequate protection from wave action for the development of mangroves. Mitsch and Gosselink (1986) found several physiographic settings that protect mangroves, including (1) protected shallow bays, (2) protected estuaries, (3) lagoons, (4) the leeward sides of peninsulas and islands, (5) protected seaways, (6) behind spits, and (7) behind offshore shell or shingle islands. Unvegetated coastal and barrier dunes usually result where this protection does not exist, with mangroves often found behind the dunes (Chapman 1976b).

Red mangrove often is found growing in continually flooded coastal waters below normal low tide. Above the low tide level, but well within the intertidal zone, full-grown red mangrove predominate with well-developed prop roots. The area behind these red mangroves is dominated by black mangroves with numerous pneumatophores. Flooding occurs only during high tides. Button mangrove often forms a transition zone between the mangrove zones and upland ecosystems. Mangroves can be found several miles inland along riverbanks where there is less tidal action. Lugo (1981) found that these inland mangroves depend upon storm surges and "are not isolated from the sea but critically dependent on it as a source of fresh sea water."

Odum et al. (1982) found that direct herbivory of mangrove leaves, leaf buds, and propagules is moderately low, but highly variable from one site to the next. Identified grazers of living plant parts (other than wood) include the white-tailed deer (*Odocoileus virginianus*), the mangrove tree crab (*Aratus pisonii*), and insects including beetles, larvae of lepidopterans (moths and butterflies), and orthopterans (grasshoppers and crickets).

The tides provide a subsidy for the movement and distribution of the seeds of several mangrove species. Mangroves share two common reproductive strategies: dispersal by means of water and vivipary (Rabinowitz 1978a). Vivipary means that the embryo develops continuously while attached to the parent tree and during dispersal. This is apparently an adaptation for seedling success where shallow anaerobic water and sediments would other-

wise inhibit germination. The seedling eventually falls and often will root if it lands on sediments or will float and drift in currents and tides if it falls into the sea. After a time, if the floating seedling becomes stranded and the water is shallow enough, it will attach to the sediments and root. Often the seedling becomes heavier with time, rightens to a vertical position, and develops roots if the water is shallow (Mitsch and Gosselink 1986). Floating seeds is a valuable characteristic for the dispersal of mangrove species.

Savage (1972) mentioned that on the Florida gulf coast, red mangrove propagules mature and fall from the tree from July to September. Black mangroves bear fruit from August to November, while white mangroves bear fruit from July to October (Loope 1980). The propagules of the three species of Florida mangroves are easy to differentiate. Apparently, there is an obligate dispersal time for all Florida mangroves, i.e., a certain period of time must elapse during dispersal for germination to be complete and after which seedling establishment can take place. Rabinowitz (1978a) estimates the obligate dispersal period at approximately eight days for white mangroves, fourteen days for black, and forty days for red. Her estimate for viable longevity of the propagules is thirty-five days for white mangroves and 110 days for black. Davis (1940) reports viable propagules of red mangrove that kept floating for twelve months.

Rabinowitz (1978a) also concluded that black and white mangroves require a standing period of five days or more above the influence of tides to take hold into the soil. As a result, these two species are usually restricted to the higher portions of the mangrove ecosystem where tidal effects are infrequent. The white mangrove, which has the smallest propagule, has the highest rate of seedling mortality. The black mangrove has an intermediate mortality rate, while the red mangrove, with the largest propagule, has the lowest seedling mortality rate (Odum et al. 1982).

McMillan (1975) found that seedlings of black and white mangroves survived short-term exposures to 80 parts per trillion and 150 parts per trillion sea water if they were grown in a soil with a moderate clay content. They failed to survive these salinities, however, if they were grown in sand. A soil with seven to ten percent clay appeared to be adequate for increased protection from hypersaline conditions.

An extensive body of literature exists concerning mangrove planting techniques in Florida (Savage 1972, Carlton 1974, Pulver 1976, Teas 1977, Goforth and Thomas 1979, Lewis 1979).

Spacing considerations

Mangroves (especially red mangrove) typically have high reproductive success in suitable habitats. For this reason wider spacing can be used as filling in will occur when individuals reach reproductive maturity. Spacing of plants is a site-specific consideration and must take into account size of project, availability of plant materials, and human resources to install them. Teas (1977) suggested planting unrooted propagules and established seedlings three feet apart and three-year-old nursery trees four feet apart.

Temporal spacing for different plants and layers

After the dominant mangrove species have been established and a canopy is formed, herbaceous understory plants can be installed. Light is usually the most serious limiting factor underneath existing mangrove canopies. Rabinowitz (1978b) suggested that species with short-lived propagules must become established in an area that already has adequate light levels either due to tree fall or some other factor. In contrast, red mangrove seedlings can become established under an existing, dense canopy and then, due to their superior embryonic reserves, are able to wait for months for tree fall to open up the canopy and present an opportunity for growth.

Maintenance and Management

Weed control

Most weeds have a difficult time competing with the characteristic species of mangroves. Also, many plants of the communities have propagules that are distributed by ocean currents and therefore are found in many places where mangrove communities occur.

Nutrient requirements

Mangroves require a greater percentage of their energy for maintenance rather than for growth in both low-salinity and high-salinity conditions (Mitsch and Gosselink 1986). Nutrients are supplied to mangrove wetlands from tidal action as well as drainage from terrestrial systems. The greatest productivity appears to be where influences from both terrestrial nutrient sources and tidal factors are collectively optimal, not where either factor by itself is maximum. Nutrient limitation has not been a problem in mangrove restorations where proper site selection and preparation have been practiced. Mangrove ecosystems are best developed in environments flushed adequately and frequently by seawater and exposed to high nutrient concentrations (Lugo and Snedaker 1974). Mangrove ecosystems tend to act as a sink (net accumulator) for various elements including macronutrients such as nitrogen and phosphorus, trace elements, and heavy metals (Odum et al. 1982). Nitrogen fixation within mangrove wetlands may provide much of the nitrogen needed for growth (ibid.). Zuberer and Silver (1978) have emphasized the importance of nitrogen-fixing bacteria in the zone around mangrove roots.

Water requirements and hydrology

Mangrove wetlands are typically saturated and influenced by the tide. Mangrove species can tolerate fresh water, however, salt water gives them a competitive advantage over other plant species. There is a wide annual variation of salinity in mangroves.

When the prop roots or pneumatophores of mangroves are continuously flooded by stabilizing the water levels, those mangroves that have submerged pneumatophores or prop roots soon die (Macnae 1963, Day 1981).

While mangroves are intertidal, a large tidal range is not necessary (as it is in estuarine marshes). Most mangroves are found in tidal ranges from two to ten feet and more; the mangrove tree species can tolerate a wide range of inundation frequencies (Chapman 1976b). It is important to recognize that the intertidal zone in most parts of Florida changes seasonally; there is a tendency for sea level to be higher in the fall than in the spring (Provost 1974).

Disturbances

Egler (1952) felt that the impacts of fire and hurricanes made conventional succession impossible in the mangroves of Florida. The mangrove's successional dynamics appear to involve a combination of (1) peat accumulation balanced by tidal export, fire, and hurricanes over years and decades, and (2) advancement or retreat of zones according to the fall or rise of sea level over centuries. Disturbances of mangrove ecosystems range from small area disturbances (lightning strikes) to large area disturbances (sea level change, hurricane damage) and may cause succession to regress to an earlier stage. There is some evidence in south Florida that hurricanes occur on a fairly regular basis, creating a pattern of cyclic succession. Major hurricanes occur about every twenty to twenty-five years in south Florida (Lugo and Snedaker 1974). Coincidently, mangrove ecosystems appear to reach their maximum levels of productivity in about the same period of time (ibid.). Possibly other physical perturbations may influence mangrove succession including incursions of freezing temperatures into central Florida, periodic droughts causing unusually high soil salinities (Cintron et al. 1978), and fire spreading into the upper zones of mangrove forests from terrestrial sources.

Most fires in the Florida mangroves are initiated by lightning and consist of small circular openings in the mangrove canopy (Taylor 1980). Fire may also play a role in limiting the inland spread of mangroves. Taylor (1981) pointed out that Everglades fires appear to prevent the encroachment of red and white mangroves into adjacent herbaceous communities.

Any process, natural or artificial, which coats the aerial roots with fine sediments or covers them with water for extended periods has the potential for mangrove destruction (Odum et al. 1982).

Case Studies

Lewis et al. (1979) found that where mangroves have been destroyed, they can be replanted or suitable alternate areas can be planted.

Properly designed plantings are usually seventy-five to ninety percent successful, although the larger the transplanted tree, the lower its survival rate (Teas 1977). Important considerations in transplanting mangroves are: (1) to plant in the intertidal zone and avoid planting at too high or too low an elevation, (2) to avoid planting where the shoreline energy is too great, (3) to avoid human vandalism, and (4) to avoid accumulations of dead sea grass and other wrack.

Additional sources of information on the restoration and management of mangroves are found in Lewis et al. (1979) and Teas (1977).

Estuarine Marshes

Salt Marsh A-18, A-46, A-245, A-338, A-366
Brackish Tidal Marsh A-397
Freshwater Tidal Marsh A-397

Introduction

Estuarine marshes are transitional zones between the sea and uplands and are a dominant feature of many coastal areas. These marshes are low-lying meadows that are frequently inundated by tidewater from the sea or saturated by floodwater draining from the uplands. They contain grasses, sedges, rushes, and other plants especially adapted for growth in a semiaquatic environment. The physical and biological processes that create estuarine marshes operate very slowly. Centuries often are required for a shallow embayment to be filled with sediment and colonized with emergent plants (Chabreck 1988). Studies at Barnstable Marsh in Massachusetts traced the developmental process of the marsh and disclosed that over 3,000 years were required for it to reach its present size (Redfield 1967).

The loss of California's coastal wetland habitat is largely due to filling and dredging for urban developments (Zedler 1992). Two types of hydrologic alterations have also had devastating effects on the intertidal salt marsh. These are decreased tidal influence and augmented freshwater inflows (ibid.). Both alterations have occurred and in some locations, the salt marsh has been replaced by brackish cat-tail and bulrush marshes. Contamination by wastewater or sewage is also a problem for estuarine marshes. Nordby and Zedler (1991) believe that, in California, the unusual flows of nonsaline water are more to blame than the contaminants carried in the sewage. In addition, lowered soil salinities may be caused by inflows from street drains. Invasions of exotic species occur where soil salinities are reduced, with the extent related to the amount (volume and flow frequency) of freshwater inflow (Zedler et al. 1990). The analysis of tidal inflow and freshwater flow will assist in determining the success of a restoration project or the alteration necessary for a successful project.

Chabreck (1988) found that the loss of estuarine marshes mostly involves conversion of marsh to open water, but in many instances, marshes have been diked and drained or covered with spoil to form dry land. Ditching, canal dredging, and spoil disposal were prominent activities for many years and have affected most coastal regions (ibid.). The immediate loss of marshes to canal dredging was only part of the problem. After construction, canal banks have continued to erode and additional marsh is lost as the channels widen. In addition, many ditches and canals permit saltwater to move further inland, thus destroying vegetation in freshwater marsh zones. Restoration of such degraded areas will initially involve restoring as nearly as possible the original elevations and drainages within the marsh.

Estuarine marsh ecosystems are networks of complex relationships of interdependent organisms. A vital component of this network is cord grass (*Spartina* spp.). The ability of this grass to convert solar energy into biomass makes estuarine marshes one of the most productive ecosystems on earth. However, these estimates often greatly vary within and among sites and seasonally. Estuarine marshes provide a broad base for a trophic pyramid that supports myriad creatures from diatoms, copepods, nematodes, oysters, and mussels to bluefish, osprey, and herons.

For additional information on the ecology of estuarine marshes, see Adam (1990), Alexander et al. (1986), Ranwell (1975), Redfield (1967, 1972), Teal (1986), Stout (1984), and Selisker and Gallagher (1983).

Site Conditions

Soils

The mineral component of marsh soils consists of silt and clay particles of a wide range of sizes (Chabreck 1988). They reflect the type of material eroded from the upland watershed and deposited by rivers or washed from the floor of the sea and deposited by waves and tides. The organic component of marsh soil is composed of the remains of plants. The amount of plant material accumulating on the soil surface is influenced by the nature of the plant communities, frequency of flooding, and magnitude of tidal currents (Chabreck 1988). As plant remains fall to the marsh surface, they mix with mineral sediment being added from alluvial or marine sources. The ratio of organic matter to sediment determines the type of marsh soil and varies considerably from one location to another (ibid.). Several studies have shown that soils of constructed estuarine marshes had less than half the organic matter content of natural marshes (Craft et al. 1986, 1988; Lindau and Hossner 1981; Langis et al. 1991). A portion of the organic material produced by marsh plants is transported by tidal flushing to surrounding waters, providing a source of organic detritus that becomes a part of the estuarine food web (Teal 1962).

Organic soils are classified as peats or mucks based on the stage of decomposition. Peat contains partially decomposed plant parts, has a brown color, and usually consists of more than fifty percent organic matter (Chabreck 1988). Muck is usually dark gray or black and well decomposed with none of the plant parts identifiable (ibid.). The organic content of muck usually ranges from fifteen to fifty percent. Soil with less than fifteen percent organic matter is classified as mineral soil (Dachnowski-Stokes 1940).

In certain Atlantic and Gulf coastal regions, brackish marsh soil consisting of noncalcareous, heavy clay may form high acidity when subjected to prolonged drying. The Dutch recognize this condition in soils of acid meadows called *katterklei*, referred to as cat clays in the United States (Gallagher 1980). Restoration of marshland once cat clays form is a costly and time-consuming process. Heavy liming offers only a temporary solution to the problem, and corrective measures require alternate flooding and draining of the land over a period of several years (Chabreck 1988).

An extremely important consideration in the restoration of estuarine marshes is the physical properties of the soil. A soil analysis should be conducted prior to revegetation of a restored estuarine marsh to determine soil type, nutrient content, acidity, and salinity. Mechanical operations such as grading, shaping, and planting are generally easier on sandy soils than silt or clay because of the greater bearing capacity and trafficability of sand (Broome 1990). Salinities of interstitial water may become too high for plant growth, especially in depressions

that do not drain at low tide. Clay or other restrictive layers in dredged material may cause perched water tables, also resulting in concentrated soil solutions due to evaporation (Broome 1990). Grading old dredged material disposal sites may expose surfaces with high residual salt concentrations (ibid.). Residual salt may have previously accumulated in a ponded area and then been covered with spoil during a subsequent dredging operation (ibid.).

Broome (1990) found that favorable conditions for plant growth occur when seepage from adjacent uplands produces low salinity in the soil water. Such seepage may also provide a supply of plant nutrients to enhance growth. The chemical properties of the soil should be determined to insure the availability of plant nutrients and test for the possible presence of toxic contaminants. Soil problems may be encountered when mitigation projects involve grading upland sites to elevations suitable for intertidal habitat. Cuts into the B and C horizon of soils in the southeastern United States expose a new surface that is likely to be acid and deficient in plant nutrients. An alternative is to strip and stockpile the topsoil, grade the site, and replace the topsoil to bring the surface to the correct intertidal elevation. Topsoil is higher in organic matter and has better chemical and physical properties but also may be nutrient deficient.

For more detailed information on marsh soils, see Dachnowski-Stokes 1940, Edelman and van Staveren 1958, Gallagher 1980, Lytle and Driskell 1954, and Phleger 1977.

Topography

The variation in elevations of estuarine marshes is slight, generally less than ten feet. Estuarine marshes are situated near sea level and tidal action is a major influence on the physiognomy. The terrain is relatively flat and gently sloping toward the ocean. Mudflats and sandflats occur at extreme low water, where exposure to the air is brief; salt marsh vegetation develops where soils are more often exposed than inundated–usually above mean sea level. Slope and tidal level are important factors that affect the areal extent of the intertidal zone, the zonation of plant species, and drainage and erosion potential (Broome 1990). The elevation of the soil surface must be graded to the elevation that provides the hydrologic regime to which the desired plant species are adapted.

Slopes in the range of one to three percent are preferable. Elevation zones occupied by a particular plant species should be determined from observation of nearby natural marshes or from trial plantings on the site. A surveyor's level may be used to relate the elevation limits and water level of a natural marsh to a restoration site (Broome 1990). Marking the upper and lower limits of the water levels of an adjacent marsh may be a useful reference. In regularly flooded saline areas, the vertical range of saltwater cord grass (*Spartina alterniflora*) is from about mean sea level to mean high tide and saltmeadow cordgrass (*S. patens*) occupies the zone from mean high tide to the storm high tide line (ibid.).

Two important physical characteristics of salt marshes are the meandering creeks and the pond holes or pans. The creeks are the drainage pathways that carry water back out to the sea. These creeks are formed by complex processes. Some channels are formed by water deterred by irregularities on the surface. The river itself forms the primary channel. After they are formed, they may be deepened by scouring and heightened by a steady accumulation of organic matter and silt. In conjunction with these processes, the heads of the creeks erode and small branch creeks develop. Lateral erosion and undercutting causes the banks to cave in in places, cutting off smaller channels. The distribution and pattern of the creek system play an important role in the drainage of the surface water as well as the drainage and movement of the water in the subsoil.

The pond holes or pans occurring in tidal marshes are typically round or elliptical depressions. During high tides, they are submerged and at low tide, they remain filled with salt water. In shallow pans, the water may evaporate and leave an accumulating concentration of salt on the mud. These depressions occur naturally as the marsh develops. Early plant colonization is irregular and bare spots on the salt become surrounded by vegetation. The surrounding vegetation causes the adjacent areas to become slightly raised and a pond hole is formed. If a pond hole or pan becomes connected to a creek, it may drain and become vegetated. Other pans, especially in sandy marshes, are derived from creeks. Vigorous growth of marsh plants can partially dam a creek or lateral erosion may block the channel. Thus, water remains when the tide retreats inhibiting plant growth. Often a series of pans form on the upper

reaches of a single creek. A pan caused by the death of small patches of vegetation due to excess salinity or inadequate drainage is called a rotten-spot pan.

Pond holes or pans may support submerged plants if the depth and the salinity are appropriate. Pools with a firm bottom and sufficient depth to retain tidal water support dense growths of widgeon grass or beaked ditch-grass (*Ruppia maritima*) with long, threadlike leaves and small black triangular seeds relished by waterfowl. Shallow depressions in which water evaporates are covered with a heavy algal crust and crystallized salt. The edges of these salt flats may be invaded by glasswort (*Salicornia* spp.), coastal spike grass (*Distichlis spicata*), Carolina sea-lavender (*Limonium carolinianum*), or dwarf saltwater cord grass (*Spartina alterniflora*).

While many natural tidal wetlands are dissected by a network of tidal channels and creeks, most constructed marshes begin with smoothly graded terrain. Observations of such wetlands in southern California suggest that functioning is impaired by topographic homogeneity (Zedler 1992). Some shortcomings of constructed marshes may be correctable by excavating tidal creeks. Newling and Landin (1985) reported greater use by fish, benthos, and shorebirds where tidal channels were purposely created in restored marshes.

Moisture and hydrology

Tidal action in estuaries serves as an environmental moderator by stabilizing soil moisture and salinity daily or twice daily (Zedler 1992). Water levels are the major factor influencing plant and animal communities. The changes in range and character of tides are primarily influenced by changing positions of the moon in relation to the earth and sun.

The tidal range at any location varies from day to day, month to month, and year to year. The range of tidal fluctuation also varies regionally (National Ocean Service 1986a, 1986b). Estuarine marshes bordering the Gulf of Mexico are subjected to a daily tidal range of less than twenty-four inches (Marmer 1954). However, in some areas along the coasts of New England and the Pacific Northwest, tides may fluctuate twelve to sixteen feet daily (Hedgpeth and Obrebski 1981, Whitlatch 1982). Tides within a locality may respond in different ways; south of Cape Cod, the range is three to five feet and north of Cape Cod, the range is ten to thirteen feet (Whitlatch 1982).

Tidal wetlands have water regimes dictated by ocean tides. Typically this event occurs twice every twenty-four hours, but it varies from one to three times a day in various areas at different times of the year (Chabreck 1988, Weller 1994).

All wetlands are a function of their hydrology. In estuarine marshes, the two main hydrologic controls are tidal flushing and stream flow. In estuaries where these controls have been altered, the vegetation types often change to those more typical of brackish waters. Tidal flushing and stream flow must be restored before revegetation begins. Grading the restoration site to the correct elevation will insure proper flushing and flow as well as soil moisture and salinity regimes.

Severity of the wave climate is an important factor that affects initial establishment and long term stability of marshes. Four shoreline characteristics (average fetch, longest fetch, shore configuration, and grain size of sediments) are useful indicators of wave climate severity (Knutson et al. 1981, Knutson and Inkeep 1982). Planting success is inversely related to fetch, the distance over water that wind blows to generate waves. The shoreline configuration or shape is a subjective measure of the shorelines' vulnerability to waves. Grain size of beach sand is also related to wave energy. Fine-grained sands generally indicate low energy, while courser textured sand indicates high energy. This is affected by the texture of sand available in a particular environment. Knutson et al. (1981) developed a numerical site evaluation form for rating potential success using these four indicators.

Establishment

Site preparation

Before initiating a restoration or creation project in estuarine marshes, the environmental impact on existing habitat must be evaluated. Exposure to waves and currents that might cause erosion also need to be considered. Most sites will need to be graded to obtain the correct elevation relative to the tide level. This can be difficult in estuarine marshes as variation in elevation is usually slight throughout the marsh. The correct elevation will provide the hydrologic regime to which the desired plants are adapted. Construction should be completed well in advance of optimum planting dates.

Prior to restoration of a seventy-acre freshwater tidal swamp, Bowers (1995) conducted a historical study of the marsh spanning the past 100 hundred years, assessed current conditions, and developed a plan to restore mid-marsh vegetation over thirty to fifty percent of the existing mudflats (thirty acres). Since the extensive mudflats were too low to support mid-marsh vegetation, fill material was used to achieve proper elevations. Containment structures (twenty feet by twenty feet by two feet) were used to stabilize the sediment while the plants became established. The original structures were constructed using brush bundles; however, even with the addition of burlap liners, the brush bundle containment structure did not stabilize the sediment. Bowers (1995) found that bales of hay of Atlantic coastal panic grass (*Panicum amarum* var. *amarulum*) provided the required stabilization. While Atlantic coastal panic grass is not native to the site, it is possible that other varieties of baled hay may be available (or grown in advance) for marsh restoration projects.

A particular concern in designing the Sonoma Baylands Project (San Francisco Bay, California) was determining the proper elevation for dredged material above Mean Higher High Water at which natural tidal slough channels would develop, thus increasing the rate of marsh development (Williams and Florsheim 1994).

Garbisch and Garbisch (1994) recommend clearing shoreline trees and shrubs that shade locations better suited to the development of tidal marsh communities–a more effective stabilizing technique.

Planting considerations

All species of marsh plants are adapted for growth within a specific range of water salinity and exhibit best germination and growth where that range prevails. Studies have identified the optimum salinity range of many marsh plants and have established a procedure for identifying prevailing salinity regimes and marsh-types boundaries by the presence or absence of key plant associations (Penfound and Hathaway 1938, Chabreck 1970, 1972).

Chabreck (1970), in a survey of Louisiana coastal marshes, reported that diversity of plant species decreased as water salinity increased. He found that ninety-three species were encountered in freshwater marsh, intermediate marsh had fifty-four species, brackish marsh had forty species, and salt marsh contained only seventeen species. Most plants adapted for growth in highly saline environments (halophytes) also grew in less saline areas. However, most species in freshwater marsh are unable to survive highly saline conditions, hence the greater number of species at lower salinities.

Unfortunately, disastrous oil spills have been the impetus for some estuary restoration projects. As a result, much research has focused on the role of restoration in improving the quality of these areas by reestablishing salt marsh. Revegetating oiled coastal wetlands is a difficult task. Matsil and Feller (1996) stated that the challenge was to reaerate the peat in order to promote growth of heterotrophic bacteria, that would then degrade the petroleum hydrocarbons. They planted saltwater cord grass with thirty gram doses (one film-canister-full) of time-release fertilizer with iron under each plant (Bergen and Levandowski 1994). The purpose of the fertilizer was to promote growth of both the cord grass and the bacteria (under these stressful conditions, Matsil and Feller [1996] found that survival of cordgrass was only about 50 percent without fertilizer, but essentially 100 percent with it). Cord grass acts as a pump to provide oxygen to its root zone (Teal and Kanwisher 1966). Matsil and Feller (1996) hoped that its growth would reaerate the root zone and, in conjunction with the fertilizer, would ameliorate the conditions limiting growth of the bacteria. Results based on Total Petroleum Hydrocarbon (TPH) reductions so far suggest that this is working. Broome (1990) suggested selecting the plant species adapted to environmental conditions at the site, using vigorous transplants or seedlings of local origin, planting at a spacing that will provide cover in a reasonable length of time, and fertilizing with nitrogen and phosphorus to enhance initial growth.

Numerous plantings have been utilized in salt marshes for shoreline protection and dredge-spoil stabilization. The principle species used were saltwater cord grass (*Spartina alterniflora*) on the Atlantic and Gulf coasts and California cord grass (*S. foliosa*) on the west coast. Survival of plants used for shoreline protection was largely influenced by sediment grain size, wave severity, and shoreline configuration (Chabreck 1988). Greatest survival occurred where fine sediment was present, wave height was minimal, and plantings were made in a sheltered cove (Knutson et al. 1981). Saltwater cord

grass has been established on dredge spoil by direct seeding and transplanting. Success was obtained by direct seeding only in the upper portion of the intertidal zone and at higher elevations where seedlings could become firmly rooted. However, transplants were successful throughout the spoil area except where waves were severe or salinity excessive (greater than 45 parts per trillion). Hoeger (1995) was able to establish cord grass using coconut-fiber pallets planted with cord grass. These fiber pallets are useful in disturbed areas where the makeup of the soil or wave energy prevents nearby cord grass plants from naturally reproducing. Established stands of saltwater cord grass were more vigorous where salinities remained about 20 parts per trillion (Chabreck 1988). The cost of establishing stands of *Spartina* spp. was greatly reduced where direct seeding could be used (Seneca 1980, Mason 1980).

California cord grass is not a ready invader of newly created sites. It has limited establishment ability and must be transplanted with great care (Zedler 1984, 1986). Transplantation may need to be delayed until tidal flushing lowers salinity to at least fifty to sixty parts per trillion (Zedler 1992). Zedler (1992) recommends using test plantings because little is known of the requirements of individual plant species, and almost nothing of the interactions among factors that reduce plant growth. Therefore, an initial investment in experimental transplantation can identify suitable soils and elevations for individual sites.

Wildlife may cause problems during revegetation. Bowers (1995) found that Canada geese (*Branta canadensis*) removed tubers of arrow-arum (*Peltandra virginica*) and young plants within weeks after planting. Goose-exclusion fences can be made of wooden stakes with lines of string placed every twelve inches above the ground. While these fences are insubstantial, they provide effective barriers to hungry geese. Nutria (*Myocastor coypus*) and muskrats (*Ondatra zibethicus*) have big appetites for marsh plants as well and can greatly reduce biomass in certain areas. New plantings may need to be protected until well established. Taylor and Grace (1995) found that the species richness of three marsh communities along an elevation and salinity gradient was unaffected in plots containing nutria compared to those protected from the herbivore. However, lower amounts of biomass negatively reduces the accumulation of organic matter in the marsh, thus influencing its microbial and faunal components.

Although marsh plants display wide tolerance levels to salt concentrations, seeds of practically all species germinate best at low salinity (Chabreck 1988). Palmisano (1971) found that even most halophytes germinate best in distilled water.

Broome (1990) provides detailed information regarding the propagation of saltwater cord grass, salt-meadow cord grass, Roemer's rush, and coastal salt grass. He outlines methods of collecting and germinating seeds, field digging of plants, handling seedlings, transplanting techniques, and fertilizer recommendations.

For additional information on planting marsh plants, see Knutson et al. 1981 and Woodhouse 1979.

Spacing considerations

Estuarine marsh plantings need to be appropriately spaced in order to effectively control erosion and provide the desired amount of plant cover. Exact spacing must be determined on a site-specific basis. Jontos and Allan (1984) planted salt-meadow cord grass and saltwater cord grass on thirty-six inch centers as six-inch plugs taken from an adjacent marsh and reported new plants sprouting between clumps the next season.

Broome (1990) provides detailed information regarding the proper spacing of plantings of saltwater cord grass, salt-meadow cord grass, Roemer's rush, and coastal salt grass.

Temporal spacing for different plants and layers

Most estuarine marsh plants can be planted initially, however, it may be more effective if less abundant forbs are installed after dominants, such as cord grass and saltwort, have become established.

Broome (1990) found that the optimum planting dates for intertidal vegetation in the southeastern United States are between April 1 and June 15. Dates earlier than April 1 increase the likelihood of storm damage before the plants have taken root and field-dug plants are hard to obtain. Dates later than June 15 limit the length of the growing season available for plants to become established.

California cord grass marshes support vascular plant and algal growth all year in southern Califor-

nia, although growth rates are lower during winter (Zedler 1992). Algal productivity is high, relative to Atlantic and Gulf of Mexico cord grass marshes, a consequence of hypersaline soils, lower cordgrass production, and more open canopies (Zedler 1980).

Maintenance and Management

Weed control

The common reed (*Phragmites australis*), which has a nearly cosmopolitan range, is considered a serious weed in many wetland communities. Its spread into estuarine marshes where it was not historically known is of great concern to those attempting restoration in these environments. Common reed is tolerant of salt, but also invades freshwater marshes, swamps, and wet shores, and disturbed wet area (such as ditches, canals, and landfills) expanding its already wide range. Marks et al. (1994) provides a concise presentation of the species's ecology and of its response to a wide variety of mechanical, chemical, and hydrological management tools. Marks et al. (1994) found two effective methods of control, annual cutting in midsummer and the application of glyphosate herbicide after the tasseling stage in late summer. Other techniques showing some success include plastic mulching, altering water and salinity levels, and handpulling. Rozsa and Orson (1995) found that by simply reintroducing tidal flushing, the vegetation at two sites (Connecticut) changed from monocultures of common reed to stands of native cordgrass (*Spartina* spp.). Prescribed burning is not recommended unless fire can reach the underground or submerged rhizomes. No biological controls have been identified.

Matsil and Feller (1996) identified some factors that limit the growth and spread of common reed. With the exception of fire, which increases the shoot-density of common reed by removing litter, they found that landfill disturbances that create gaps in the stand commonly result in decreased density. These gaps are filled in by fast-growing, early successional trees, which can eventually shade out the common reed. They also discovered that sand provides neither sufficient available nutrients nor adequate moisture for optimal growth of common reed. On landfill sites in New York where the sand cover is in excess of four feet, common reed roots cannot reach water and nutrients. On such sites, common reed is either absent, or grows sparsely and is stunted.

Matsil and Feller (1996) arrived at a realistic goal in dealing with common reed. They do not plan to totally eradicate common reed, which dominates many acres of New York City parkland, but to seek a balance, where less productive fill sites are manipulated for the benefit of regional biodiversity and adjacent natural communities.

In southern California, reduced tidal inflows and increased freshwater inflows act together to shift habitat type from tidal creeks to shallow-water brackish lagoons (Zedler 1992). Sedimentation fills former creeks and salts are diluted in both the water and soils. The impact of reduced salinity is that marsh habitats are quickly invaded by brackish marsh plants such as southern cat-tail (*Typha domingensis*). Perennial pepperweed (*Lepidium latifolium*) is rapidly invading brackish marshes of California's coast as well as interior marshes. Atlantic cordgrasses, such as saltwater cord grass (*Spartina alterniflora*), are considered to be invasive perennial halophytes on the west coast (Simenstad and Thom 1995). Management predictions should not be based on the behavior of this species on the Atlantic and Gulf coasts where it is native and where the regional geologic and tidal regimes are different from Pacific conditions (ibid.). However, consistent species patterns have been noted across its range relative to its sediment-trapping role in the conversion from open mudflat to higher tidal marsh (ibid.). These changes may have adverse effects on navigation and fisheries by converting mudflats to seas of grass (Howald 1996).

The establishment of aggressive weeds has produced detrimental effects in coastal marshes. Their rapid spread has reduced stands of native plants in many areas. Also, wildlife species dependent on native plants and unable to utilize the exotic species are affected by the change (Chabreck 1988). Some exotic plants in the tidal freshwater and brackish marshlands of the Gulf Coast are water hyacinth (*Eichhornia crassipes*), alligatorweed (*Alternanthera philoxeroides*), and elephantsear (*Colocasia antiquorum*). Eurasian watermilfoil (*Myriophyllum spicatum*) and hydrilla (*Hydrilla verticillata*) are abundant along the Gulf and Atlantic coasts. Waterchestnut (*Trapa natans*), another exotic species, is common in the northeast.

Regulating water levels (flooding or drying) and/or controlled burning have been used to control weeds in marshes. Where these methods are not possible or effective, herbicides can be useful. However, any herbicide used must comply with state and local regulations and extreme caution should be taken to ensure wildlife, aquatic organisms, or native plants are not harmed.

Nutrient requirements

Natural estuarine marshes are productive ecosystems and generally contain an abundance of nutrients necessary for vigorous plant growth. In upland systems soil nutrients are primarily derived from the weathering of parent materials from which the soils were formed. However, estuarine marshes were formed mostly from alluvium, and subsurface deposits usually contain lower concentrations of nutrients than surface soils (Lytle and Driskell 1954). The major sources of most nutrients are rivers and seawater. Rivers transport rich sediment from their drainage basin during flood stages and fertilize downstream marshes. In marshes subject to tides, nutrients from seawater are added during each flood cycle (Phleger 1977).

Covin and Zedler (1988) demonstrated that the salt marsh in southern California is nitrogen limited, and in areas of mixed woody saltwort and California cord grass, woody saltwort is the superior competitor for nitrogen. Zalejko (1989) found low levels of nitrogen fixation in one cord grass marsh in San Diego Bay; if those results are typical for California wetlands, this helps to explain why nitrogen is limiting. Since seawater is low in nitrogen, supplies must come from the watershed. Levels of inorganic nitrogen in the channels and marsh soils are low except during sewage spills or other inflow events (Zedler and Nordby 1986). Broome (1990) stated that in the southeastern United States possible sources of nutrients for the marsh were seepage from adjacent sandy upland soils, which included fertilized lawns and septic tank drainage fields, tidal water, and sediments deposited by tides and waves. Over time, nutrients and organic matter accumulate in the soil and plant biomass.

The quality of the substrate and water often differ for natural and constructed wetlands. The substrate of excavated sites may become acid upon exposure to air, as accumulated sulfides are oxidized to sulfuric acid (Zedler 1992). Excavation can inadvertently uncover toxic materials if the former wetland had been used as a landfill. Restoration sites may also be downstream of contaminated waters. In urban areas, there is the additional problem that marshes can accumulate trash, which smothers vegetation. Such locations will not support high-quality wetlands.

Organic sediments are a basic feature of natural wetlands. It has been suggested that organic matter influences nearly every aspect of wetland ecosystem functioning, by changing sediment porosity and water-holding capacity, by influencing nutrient dynamics, by altering the growth rates and nutrient content of plants, and by influencing the species composition and abundance of invertebrates associated with the sediments (Zedler 1992). The microbes, plants, and animals, in turn, affect the rate of organic matter accumulation in wetland sediments. For most wetlands, the development of organic sediment occurs over time periods that are closer to centuries or millennia, than to years or decades (ibid.). Comparisons of constructed and restored wetlands in California (Langis et al. 1991), North Carolina (Craft et al. 1986, 1988), and Texas (Lindau and Hossner 1981) indicate that sediment organic matter and nitrogen levels are lower than in natural reference wetlands.

A number of studies have shown that nitrogen is a limiting factor in the growth of saltwater cordgrass in natural marshes (Sullivan and Daiber 1974, Valiela and Teal 1974, Broome et al. 1975). Experiments in natural cordgrass marshes have shown that nitrogen additions increase foliar nitrogen and plant biomass (e.g., Covin and Zedler 1988). Valiela et al. (1984) have shown that nutrient additions affect a wide range of ecosystem functions, e.g., a twenty-four percent increase in decomposition of vascular plant material following enrichment. Cord grass responded strongly to treatments with both nitrogen and organic matter added, and it grew slowly where neither was added (Zedler 1992). After six months of growth, the highest total stem length (with alfalfa plus N fertilizer added) was three and four-tenths times the lowest (with rototilling but no amendments). In North Carolina, Broome and Seneca (1985) found that applying nitrogen and phosphorus fertilizers to the soil was essential for growth and establishment of *Spartina* species.

Water requirements and hydrology

Most North American estuaries develop a salinity gradient from the coastal inlet toward the inland extremes. In Mediterranean-type regions, however, there are few rainfall events during the year and streamflows are more variable (Zedler 1992). As a result, channels are marine in character during most of the year. Occasional floods briefly lower the water salinity of channels from seawater (thirty-four parts per trillion) to brackish, making the system intermittently estuarine (ibid.). In other regions, freshwater plays a larger role in ecosystem structure and function. The salinity regime may vary greatly both temporally and spatially (e.g., Schroeder et al. 1992).

Cord grass is clearly dependent on tidal flushing; it has been eliminated from at least one site in southern California that shifted from being continuously tidal to infrequently tidal sometime after 1939 (Zedler 1986). An eight-month period without tidal flushing nearly decimated the light-footed clapper rail (*Rallus longirotris levipes*) population, its invertebrate foods, and its protective cord grass canopy at Tijuana Estuary (Zedler and Nordby 1986).

Zedler (1992) stated that the many problems of restored marshes can be grouped as hydrologic, chemical, and biological. Of these, hydrologic problems are the most often documented, in part because they are easier to see, and in part, because the impacts on the entire system are dramatic. Zedler (1992) lists the following common hydrologic and topographic problems:

■ overexcavation
■ underexcavation
■ unexpected sediment accumulation
■ shifting of tidal access during site construction with resultant erosion (undercutting transplants)
■ deposition (smothering transplants)

Perhaps the most obvious hydrologic/topographic error is that constructed marshes rarely "look like" natural marshes (Zedler 1992). The complex networks of small tidal creeks are always lacking and the broad, flat marsh plains are compressed or absent. In their place, bulldozers often create a collage of habitat types that would rarely abut one another—steep-sided islands and deep, straight channels. Isolated habitats can pose serious problems for the biota that need transitions between habitats (ibid.).

Disturbances

Plant succession, the replacement of one plant community by another, is an important process in the development of estuarine marshes. Natural, as well as unnatural, disturbances often take place that cause a "setback" in plant succession and result in changes in ecological communities. Brackish marsh along the Gulf Coast that remains undisturbed for a decade will contain ecological communities dominated by salt-meadow cord grass (Chabreck 1988). A major disturbance such as fire often hinders succession and permits a rapidly growing species, such as chairmaker's bulrush (*Scirpus americanus*), to dominate the burned area (ibid.). Chairmaker's bulrush is typically subdominant and, if the area is undisturbed for several years, salt-marsh cordgrass will gradually replace chairmaker's bulrush (ibid.). Properly timed, periodic burning, however, will maintain chairmaker's bulrush as the dominant species. Burning is often used in brackish marsh containing mixed stands of salt-meadow cord grass and chairmaker's bulrush to regulate species composition (Chabreck 1988). Fall burns favor growth of chairmaker's bulrush, an important wildlife food plant and without burning, the plant is eliminated by the more dominant cord grass.

Freshwater and intermediate marshes can be burned when encroachment of woody plants must be controlled; however, burning of salt marsh and spoil deposits should be avoided (Chabreck 1988). The mixture of woody shrubs on spoil deposits provides important food and cover for wildlife and is killed by fire (Soil Conservation Service 1984). Before initiating a burning program in coastal marshes, consider that the organic matter deposited by plants is a major source of material for marsh building. Fire will not only destroy the standing crop of vegetation, but when dry marsh is burned, organic material that has accumulated on the soil surface for many years may also be lost (Hackney and de la Cruz 1981, Soil Conservation Service 1984).

For additional information on the use of fire in marshes, see Hackney and de la Cruz (1981) and Soil Conservation Service (1984).

Case Studies

Belaire and Templet (1995) built a salt marsh habitat for whooping cranes (*Grus americana*) along Mesquite Bay in the Aransas National Wildlife Refuge. Over 50,000 cubic yards of dredged sand and silt were applied to a thirteen-acre site at elevations similar to those of known crane habitats. Three-fourths of the site was then stabilized with plantings of saltwater cord grass (*Spartina alterniflora*), while the remainder received a mixed planting including turtleweed (*Batis maritima*), coastal salt grass (*Distichlis spicata*), and salt-meadow cordgrass (*Spartina patens*). Since 1991, three whooping crane sightings have been reported, as has the presence of numerous other marsh waterfowl species.

Jontos and Allan (1984) restored 370 square yards of salt marsh along a tidal stream tributary to Long Island Sound in Greenwich. After removing rock fill and construction debris and controlling the spread of common reed, areas were selected for planting based on elevation. Salt-meadow cord grass and saltwater cord grass were planted on thirty-six inch centers as six-inch plugs taken from an adjacent marsh. The removal of large rock fragments resulted in a very uneven soil surface. The deeper pockets, which are subject to daily flooding, were replanted with saltwater cordgrass, while those areas at or above mean high water received plugs of salt-meadow cord grass. Nursery stock salt-spray rose (*Rosa rugosa*) and northern bayberry (*Myrica pennsylvanica*) were planted in areas of transition from marsh to upland. Granular fertilizer (10-10-10) was broadcast under all planting stock.

After two months, limited recolonization of disturbed areas by volunteer species had taken place. All the planted saltwater cord grass and salt-meadow cord grass had survived and were growing well. Fiddler crab colonies returned to the restored area from undisturbed marsh. By the end of the growing season, saltwater cord grass had added up to four inches of growth. New plants are sprouting between transplanted clumps. Although no indication of growth among planted salt-meadow cord grass was apparent, survival appeared close to one hundred percent.

For additional information on the restoration and management of tidal marshes, see Bourn and Cottam (1950), Bowers (1995), Seneca (1980), Seneca and Broome (1992), and Zedler (1984, 1990, 1992, 1996).

APPENDIX C

Woody and Herbaceous Plant Matrices

This appendix contains woody and herbaceous plant matrices that provide seven categories of information for each plant species in Appendix A. The matrices were compiled using state and regional floras, references on gardening and landscaping, and ecological studies of individual species as well as the authors' own observations. This information is intended to assist the restorationist and landscape designer in selecting appropriate species for a given planning unit or site. The Wildflower Research Center in Austin, Texas provides lists of native plants recommended for landscaping for each state.

The seven categories of information are plant type, environmental tolerance, aesthetic value, wildlife value, flower color, bloom period, and suggested landscape uses. Letter codes are used in each category, alone or in combination, to characterize each species.

The user will find that some of the categories are very closely related. For example, short perennials (plant type) often make excellent groundcovers (landscape use); annuals (plant type) that tolerate shade and flooding (environmental tolerances) may be suitable to rapidly stabilize wooded streambanks (landscape use).

Each category and the letter codes used in the matrices are described below. A "—" denotes we could find no documented, significant or otherwise, known value. An "X" indicates the category is not applicable to that species.

PLANT TYPE

Plant type refers to the life cycle (e. g., annual, biennial, perennial) as well as plant height and other physical attributes that help characterize plant species. Height codes for both woody and herbaceous plants refer to the maximum height achieved in favorable habitat. Woody plants are perennial species that generally increase annually in size up to their maximum height. Those which achieve the greatest height comprise the canopy or overstory in a forest community. Understory trees reach a maximum height at maturity that is beneath that of canopy species, but taller than shrub species. Shrubs comprise the lower layer in a forest community, or may be the sole woody component in a shrub or scrub community. For herbaceous plants, the 'small' code includes species with trailing, reclining, or procumbent habits. It should be noted that species will sometimes exceed the maximum height when used in a managed landscape where competition is reduced and other favorable factors exist.

Since woody plants are perennial, the life cycle code applies only to herbaceous plants. Herbaceous plants may complete all stages of their life cycle in a single season or live for many years. The life cycle code refers to the normal growth pattern exhibited within a species' natural range. The life cycle outside a species' natural range may change in response to climate. For example, a perennial species such as California poppy will often act as an annual when used in colder climates outside its native range.

The letter code 'G' is used to distinguish grasses from forbs, because many landscape situations call for the use of grasses or forbs, or a balanced combination of both. Included in this delineation are sedges, rushes, and other grass-like plants.

Additional letter codes are used for miscellaneous codes that help to further characterize each species.

Woody Plants—Height Codes

LC   large canopy tree (over 75 feet)
SC   small canopy tree (50 to 75 feet)
LU   large understory tree (35 to 50 feet)
SU   small understory tree (20 to 35 feet)
LS   large shrub (12 to 20 feet)
MS   medium shrub (6 to 12 feet)
SS   small shrub (less than 6 feet)
WV   woody vine (height variable)
S   shrub (maximum height unknown)

Herbaceous Plants—Height Codes

L   large (over 4 feet)
M   medium (2 to 4 feet)
S   small (less than 2 feet)
HV   herbaceous vine (height variable)

Herbaceous Plants—Life Cycle Codes

A   annual—plants that live only one season, producing flowers and seeds the first season.
B   biennial—plants that live two seasons, producing flowers and seeds in both the first and second seasons or only the second season.
P   perennial—plants that live more than two seasons and generally produce flowers and seeds beginning the second season.

Miscellaneous Codes

AQ   aquatic—includes emergent plants with floating leaves and free-floating plants
C   cactus
c   coniferous (evergreen)
e   evergreen (not coniferous)
G   grass or grass-like plant (grasses, sedges, and rushes)
m   moss

ENVIRONMENTAL TOLERANCE

The tolerance of a plant species to specific environmental conditions can greatly influence its use in a managed landscape. For example, species which tolerate drought, "D," are often suitable for use in xeriscaping. Those which tolerate flooding, "F," may be suitable for streambank restoration, and those which tolerate wet soil, "W," may be suitable for wetland restoration or water gardening. Those which tolerate shade, "s," can be grown beneath mature trees, while species which tolerate both shade and sun, "Ss," are often best suited for use in woodland and savannah restoration or in gardens that receive shade part of the day. Plants tolerant of salt, "Na," may be used in coastal, brackish, or saline conditions.

D	drought	Na	salt	s	shade
F	flood	S	sun	W	wet soil

AESTHETIC VALUE

This category refers to the notable ornamental value of a species. While this category tends to be subjective, it is based on generally recognized aesthetic attributes. As botanists, we can find almost any physical characteristic to be of aesthetic interest or value. As landscapers, we look for characteristics which distinguish a plant and make it outstanding in a setting, such as a showy flower or striking autumn coloration. Many wildflowers that produce stunning displays when grown in drifts or masses are identified with the "F" code. This category is intended as a guide only. Since one person's weed is another's rose, we encourage all users of this manual to become familiar with the species native to their region and to draw their own conclusions in this regard.

A	autumn coloration	F	flowers
B	bark, twigs, or stems	f	fruit
D	dried specimens	L	foliage

WILDLIFE VALUE

It is important to recognize that nearly all plants, native or not, provide cover for wildlife species, from minute insects to large mammals. This category is intended to identify specific groups of animals that use native plants for food. Many butterflies and moths, "B," are nectar feeders, while their caterpillars can be very host-specific and feed more extensively on sap or other plant parts. Hummingbirds are also nectar feeders, while ground birds and songbirds rely more extensively on seeds. Mammals use many plant parts for food, with browsing mammals relying most heavily on stems and leaves. Browsing mammals include rabbits and deer; large mammals include muskrat, raccoon, and beaver; small mammals include squirrels, mice, and moles; ground birds include grouse and quail; songbirds include finches and orioles; waterfowl and shorebirds include herons, ducks, and geese. Providing habitat for these groups will also attract larger predatory wildlife such as hawks, owls, and foxes.

The benefits of native plants to wildlife have long been recognized by biologists, but such information remains largely undocumented for many species. Most information on plant use by wildlife is derived from food habit studies of particular wildlife species. In the past, such studies primarily focused on game mammals, while other species received little attention. With the increasing popularity of gardening to attract butterflies, hummingbirds, songbirds, and other desirable wildlife, more information on uses of specific plants is becoming available. Where information was available for a genus or a particular species, we often inferred wildlife value for other closely related species having similar attributes.

B	butterflies and moths	h	honeybees	S	songbirds
C	caterpillars	L	large mammals	W	waterfowl, shorebirds
G	upland ground birds	M	browsing mammals	U	unknown
H	hummingbirds	m	small mammals		

FLOWER COLOR

Often, a species will exhibit a range of flower color, or the color may grade from one shade to another. Many species will therefore have more than one color code.

B	blue	P	pink
Br	brown	Pu	purple
Bu	burgundy	R	red
G	green	W	white
Gr	gray	Y	yellow
L	lavender	yg	yellow-green
O	orange		

BLOOM PERIOD

The period of time during which a species produces flowers may encompass one season or a range of seasons. The bloom period may also vary from one region to another in response to climatic and other factors. For most species, a range of bloom periods is provided.

Sp	spring	F	fall
Su	summer	W	winter

LANDSCAPE USE

Landscape use is closely related to the other categories and refers to the various ways a plant can be successfully used in a designed landscape. There is little information identifying specific landscape uses for most native species; therefore, this information was derived from analyzing the information in the other categories. The use of letter codes was somewhat arbitrary and we hope it will not be confusing to the user. Some of the codes are so closely related that with perfect hindsight, we might have combined them; for example, quick cover, "C," and erosion control, "E." The accent specimen, "A," code is reserved for plants which, because of their showy flowers or other outstanding ornamental appeal, can be used alone as a focal point in a landscape design.

A   accent specimen (outstanding aesthetic quality)
B   wooded buffer, border, woodland, or savannah, western riparian and other partially open sites
b   bird and butterfly garden
C   quick cover
E   erosion control
F   foundation planting (suitable for use around building foundations)
G   groundcover (low cover)
g   garden bed or border
H   hedge
I   interior forest (moderate to deep shade)
M   meadow, includes deserts, glades, prairies, and other dry to mesic, open sites such as along roadsides
N   nut or fruit crop (for human consumption)
P   physical barrier (dense, thorny, stinging)
R   rocks (able to grow on rocks in dry or wet sites)
S   streambank, shoreline including sandy beaches, dunes subject to flooding
s   screening (dense growth habit or evergreen)
T   shade tree
W   areas of wet soil including wet meadow, swamp, bog, marshy pond, or lake margins
wg water garden

Woody Plant Matrix

PLANT LIST	Plant Type	Env. Tol.	Aesthetic Value	Wildlife Value	Flower Color	Bloom Period	Landscape Uses
Abies amabilis Pacific silver fir	cLC	Ss	L	GLMmS	Pu	Sp	ABIs
Abies balsamea balsam fir	cSC	FSs	L	GLMmS	PuRY	Sp	ABIsW
Abies concolor white fir	cLC	DSs	BL	GLMmS	R	SpSu	ABITs
Abies fraseri Fraser's fir	cSC	DS	L	GLMmS	—	SpSu	ABIs
Abies grandis grand fir	cLC	FS	L	GLMmS	Y	Sp	ABIs
Abies lasiocarpa subalpine fir	cLC	Ss	L	MmS	BPu	Su	ABIs
Abies magnifica California red fir	cLC	Ss	BfL	GLMmS	Pu	Sp	ABIs
Abies procera noble fir	cLC	S	L	GLMmS	PPu	—	As
Acacia angustissima prairie wattle	SS	Ss	FL	GMm	WY	SpSu	BM
Acacia berlandieri guajillo	eLS	DS	F	HhM	W	WSp	ABHMs
Acacia constricta mescat wattle	MS	DS	FB	GhMm	Y	Su	ABM
Acacia farnesiana mealy wattle	LS	DFS	F	GhMm	OY	Su	ABM
Acacia greggii long-flower catclaw	LS	DS	F	GhMm	WY	SpSu	ABMP
Acacia minuta coastal-scrub wattle	LS	DS	F	GhMm	Y	SpSu	ABM
Acacia neovernicosa trans-pecos wattle	MS	DS	F	GhMm	Y	Su	ABM
Acacia pinetorum pineland wattle	SS	DS	F	GhMm	Y	Sp	BI
Acacia rigidula chaparro-prieto	LS	DS	F	GhMm	WY	Sp	ABEP
Acacia schaffneri Schaffner's wattle	SU	DS	B	GhMm	OY	Sp	AB
Acacia tortuosa poponax	MS	DS	BF	GhMm	OY	Sp	AB
Acamptopappus shockleyi Shockley's goldenhead	SS	DS	F	M	Y	SpF	BM
Acer barbatum Florida maple	SU	FS	ABf	GHmS	yg	Sp	ST
Acer circinatum vine maple	SU	FSs	A	LM	PuR	Sp	A
Acer glabrum Rocky Mountain maple	SU	S	AL	M	yg	SpSu	AE

PLANT LIST	Plant Type	Env. Tol.	Aesthetic Value	Wildlife Value	Flower Color	Bloom Period	Landscape Uses
Acer grandidentatum canyon maple	SC	S	A	M	Y	Sp	BIT
Acer leucoderme chalk maple	SU	DS	AL	GHmS	yg	Sp	HMS
Acer macrophyllum big-leaf maple	SC	FSs	ABFL	GhLMmS	Y	Sp	ABST
Acer negundo ash-leaf maple	LU	DFS	AL	CGMmSW	yg	Sp	SWT
Acer nigrum black maple	LC	s	AL	GLmS	yg	Sp	IST
Acer pensylvanicum striped maple	LU	s	AB	GMmS	Y	Sp	AI
Acer rubrum red maple	LC	FS	AL	HmSW	R	Sp	AIWT
Acer saccharinum silver maple	LC	FS	B	HmSW	RY	Sp	CISWT
Acer saccharum sugar maple	LC	Ss	ABf	GHmS	yg	Sp	AST
Acer spicatum mountain maple	SU	Ss	A	GMmS	yg	Sp	AI
Adenostoma fasciculatum common chamise	eMS	S	BF	MmS	W	SpSu	BMs
Adenostoma sparsifolium redshank	eLS	S	B	MmS	W	Su	BMs
Aesculus californica California buckeye	LS	DS	BFfL	m	W	Su	AB
Aesculus flava yellow buckeye	LC	FSs	ABFL	m	Y	Sp	AIT
Aesculus glabra Ohio buckeye	LU	DFSs	AFfL	hm	Y	Sp	AbIT
Aesculus pavia red buckeye	LS	Ss	FL	m	R	Sp	ABS
Aesculus sylvatica painted buckeye	LS	s	F	m	PRyg	Sp	AI
Agave lechuguilla lechuguilla	eLS	DS	FfL	G	Ryg	Su	AMP
Agave neomexicana New Mexico century-plant	eLS	DS	FfL	G	yg	Su	AMP
Agave palmeri Palmer's century-plant	eLS	DS	FfL	G	GPu	Su	AMP
Agave parryi Parry's century-plant	eLS	DS	FfL	G	Ryg	Su	AMP
Agave parviflora small-flower century-plant	eLS	DS	FfL	G	PW	Su	AMP
Agave scabra rough century-plant	eLS	DS	FfL	G	Y	Su	AMP
Agave schottii Schott's century-plant	eMS	DS	FfL	G	Y	Su	AMP
Agave utahensis Utah century-plant	eLP	DS	FfL	G	Y	SpSu	AMP

PLANT LIST	Plant Type	Env. Tol.	Aesthetic Value	Wildlife Value	Flower Color	Bloom Period	Landscape Uses
Ageratina wrightii Wright's snakeroot	SS	DS	—	BS	W	F	bM
Allenrolfea occidentalis iodinebush	MS	FNaS	—	U	G	Su	W
Alnus arrulata smooth alder	LU	FS	B	GLMmS W	PuR	Sp	W
Alnus incana speckled alder	LS	FS	BL	GLMmS W	Pu	Sp	SW
Alnus oblongifolia Arizona alder	LU	Fs	AL	GLMS	RY	Sp	ES
Alnus rhombifolia white alder	SC	FSs	ABL	GLMS	G	FW	S
Alnus rubra red alder	SC	S	ABL	GLMS	R	Sp	IS
Alnus serrulata brookside alder	LS	FSW	AL	GMmS W	Y	Sp	SW
Alnus viridis sitka alder	MS	FSsW	BFL	GLMS	G	Sp	SW
Aloysia gratissima whitebrush	MS	DSs	FL	h	W	SpSuF	ABFHMR
Aloysia ligustrina beebrush	MS	DSs	FL	h	PuW	SpSuF	ABFHMR
Aloysia wrightii Wright's beebrush	MS	DS	FL	hM	W	Sp	ABFHMR
Alvaradoa amorphoides Mexican alvaradoa	eSC	Ss	—	U	GY	WF	AI
Ambrosia deltoidea triangle burr-ragweed	SS	DS	—	GMmS W	G	Sp	BM
Ambrosia dumosa white burrobush	SS	DS	—	GMmS W	W	SpSu	BM
Amelanchier alnifolia Saskatoon service-berry	SS	DS	FfL	GLMmS	W	Sp	BIN
Amelanchier arborea downy service-berry	SU	DS	ABF	GLMmS	W	Sp	ABINS
Amelanchier canadensis Canadian service-berry	LU	Ss	AFL	GLMmS	W	Sp	ABINS
Amelanchier laevis Allegheny service-berry	SU	Ss	AFL	GLMmS	W	Sp	BINS
Amelanchier pallida pale service-berry	SS	DS	AFL	GLMmS	W	Sp	BINS
Amelanchier stolonifera running service-berry	SS	S	Ff	GLMmS	W	Sp	B
Amelanchier utahensis Utah service-berry	SU	DS	Ff	LMmS	PW	SpSu	Ab
Amorpha canescens leadplant	SS	DSs	F	B	BPu	SpSu	bEgM
Amorpha fruticosa false indigo-bush	LS	S	F	GS	BPu	Su	AES
Amorpha herbacea cluster-spike indigo-blue	eSS	D	—	GS	Pu	—	E

PLANT LIST	Plant Type	Env. Tol.	Aesthetic Value	Wildlife Value	Flower Color	Bloom Period	Landscape Uses
Ampelopsis arborea peppervine	WV	DFSs	AfL	GLmS	G	Su	IW
Ampelopsis cordata heart-leaf peppervine	WV	Fs	AfL	GLmS	G	Sp	IS
Amyris elemifera sea torchwood	eSU	Ss	F	U	W	WSpSu F	AI
Andromeda polifolia bog-rosemary	eSS	FSs	FL	m	PW	Sp	BCW
Anisacanthus thurberi Thurber's desert-honeysuckle	SS	DS	B	M	BOR	Sp	M
Annona glabra pond-apple	eSC	FSW	F	mS	W	Sp	NW
Aplopappus laricifolius turpentine bush	SS	NaS	F	U	Y	F	BM
Aralia spinosa devil's-walkingstick	SU	Ss	FfL	GMS	W	Su	ABI
Arbutus arizonica Arizona madrone	eLU	S	B	GMmS	WP	Sp	A
Arbutus menziesii Pacific madrone	eLC	S	ABFL	GMmS	W	Sp	AB
Arctostaphylos canescens hoary manzanita	eMS	DS	B	GLMmS	WP	WSp	BM
Arctostaphylos glandulosa Eastwood's manzanita	eMS	DS	BF	GLMmS	WP	WSp	ABM
Arctostaphylos glauca big-berry manzanita	eLS	DS	BF	GLMmS	W	WSp	ABEM
Arctostaphylos manzanita big manzanita	eSU	DS	BF	GLMmS	PW	WSp	ABEM
Arctostaphylos myrtifolia ione manzanita	eSS	DS	BF	GLMmS	PW	WSP	BM
Arctostaphylos nevadensis pinemat manzanita	eSS	S	BfL	GLMmS	PW	Sp	BCGI
Arctostaphylos nummularia Fort Bragg manzanita	eSS	S	BL	GLMms	W	WSp	BM
Arctostaphylos parryana pineland manzanita	eMS	DS	BL	GLMmS	PW	Sp	B
Arctostaphylos patula green-leaf manzanita	eMS	Ss	BL	GLMmS	P	Sp	B
Arctostaphylos pringlei pink-bract manzanita	eSU	DS	BF	GLMmS	W	W	E
Arctostaphylos pungens Mexican manzanita	eMS	DS	BL	GLMm	PW	WSp	BEMN
Arctostaphylos stanfordiana Stanford's manzanita	eMS	S	BL	GLMmS	PPu	WSp	BM
Arctostaphylos tomentosa hairy manzanita	eSS	S	BL	GLMmS	PW	WSp	BM
Arctostaphylos uva-ursi red bearberry	eSS	DS	BFfL	hLMm	PW	Sp	bGM
Arctostaphylos viscida white-leaf manzanita	eMS	DS	BFL	GLMmS	P	WSp	AB

PLANT LIST	Plant Type	Env. Tol.	Aesthetic Value	Wildlife Value	Flower Color	Bloom Period	Landscape Uses
Ardisia escallonoides island marl-berry	eSU	DSs	BFf	S	W	SpF	BAI
Argusia gnaphalodes sea-rosemary	eMS	DS	F	U	W	WSpSuF	EgS
Aronia arbutifolia red chokeberry	MS	DFSs	AFfL	GLMmS	W	Sp	AgNW
Aronia melanocarpa black chokeberry	MS	FSs	AFfL	GLMmS	W	SpSu	AgNW
Artemisia arbuscula dwarf sagebrush	eSS	DSs	L	GMm	Br	Su	BM
Artemisia bigelovii flat sagebrush	eSS	DS	BL	GMm	Y	SuF	BgM
Artemisia californica coastal sagebrush	eSS	DS	L	GLMm	PuY	SuF	M
Artemisia cana hoary sagebrush	eSS	DS	BL	M	WY	Su	ABgM
Artemisia filifolia silver sagebrush	SS	DS	L	GLMm	WY	SpF	ABEgM
Artemisia frigida prairie sagebrush	SS	DS	L	GM	Y	SuF	BM
Artemisia nova black sagebrush	SS	DNaS	L	GLMm	BrG	SuF	M
Artemisia pedatifida big-foot sagebrush	SS	DS	L	GLMm	BrG	SpSu	M
Artemisia rothrockii timberline sagebrush	eSS	S	BL	GLMm	PuY	SuF	BM
Artemisia spinescens bud sagebrush	eSS	DNaS	BL	GLMm	PY	SpSu	GgM
Artemisia tridentata big sagebrush	eMS	DNaS	BL	GLMm	BrY	SuF	BM
Artemisia tripartita three-tip sagebrush	SS	DS	L	GLMm	Br	Su	M
Arundinaria gigantea giant cane	eLPG	DFSs	BL	MS	G	SpSu	BEIMPSsW
Asimina incana woolly pawpaw	MS	Ss	F	Lm	W	Sp	IN
Asimina parviflora small-flower pawpaw	MS	Ss	F	Lm	Pu	Sp	IN
Asimina reticulata netted pawpaw	SS	S	F	Lm	W	Sp	I
Asimina triloba common pawpaw	SU	Ss	FfL	Lm	R	Sp	BINS
Aster carolinianus climbing aster	WV	FW	F	GMmS	PPu	WSpSuF	W
Atriplex canescens four-wing saltbush	eSS	DNaS	f	GMm	—	SuF	EM
Atriplex confertifolia shadscale	SS	DNaS	f	GLMmSW	—	SpSu	M
Atriplex gardneri Gardner's saltbush	MS	DNaS	f	GLMmSW	Br	SuF	BMW

PLANT LIST	Plant Type	Env. Tol.	Aesthetic Value	Wildlife Value	Flower Color	Bloom Period	Landscape Uses
Atriplex hymenelytra desert-holly	SS	DS	fL	GMsSW	—	SuF	M
Atriplex lentiformis quailbush	MS	DNaS	f	GLMmS BW	—	SuF	M
Atriplex nuttallii Nuttall's saltbush	SS	DNaS	f	GLMmS W	—	Su	M
Atriplex polycarpa cattle-spinach	SS	DFNa S	f	GLMmS W	—	SuF	M
Atriplex spinifera spinescale	SS	DNaS	—	GLMmS W	—	SuF	M
Avicennia germinans black mangrove	LS	NaSW	f	h	WY	Su	W
Baccharis angustifolia saltwater false willow	MS	FNaS W	F	M	WY	WSpSu F	W
Baccharis glomeruliflora silverling	LS	FNaS W	F	M	WY	WSpSu F	W
Baccharis halimifolia groundseltree	MS	DFNa S	AFfL	SW	GW	F	ESW
Baccharis neglecta Roosevelt-weed	MS	—	—	M	Br	—	—
Baccharis pilularis coyotebrush	SS	DS	F	M	WY	SuF	BE
Baccharis salicifolia mulefat	LS	FNaS W	F	MS	WR	—	BES
Baccharis sarothroides rosinbush	SS	DNaS	BFf	M	WY	SuF	AEM
Baccharis sergiloides squaw's false willow	SS	DS	F	M	WY	SPSUF	BM
Befaria racemosa tarflower	eMS	DSS	F	U	PW	WSp	ABgI
Berchemia scandens Alabama supplejack	WV	DFS	AfL	GMmS	W	Sp	BW
Betula alleghaniensis yellow birch	LC	Ss	AB	CGLMm S	yg	Sp	ABI
Betula glandulosa scrub birch	SS	FSW	L	CGLMm S	G	Sp	BW
Betula lenta sweet birch	SC	DSs	A	CGLMm S	yg	Sp	AIT
Betula nana swamp birch	SS	FSW	L	CGLMm S	yg	SP	W
Betula nigra river birch	SC	DFS	ABL	C	Br	Sp	ABSsT
Betula occidentalis water birch	SU	FSW	ABL	M	G	Sp	ABSW
Betula papyrifera paper birch	SC	DS	ABL	CGMmS	yg	Sp	ABI
Betula populifolia gray birch	LU	FS	ABL	CMmS W	yg	SpF	BCEW
Betula pumila bog birch	LS	FSW	A	CGmSW	yg	Sp	W

PLANT LIST	Plant Type	Env. Tol.	Aesthetic Value	Wildlife Value	Flower Color	Bloom Period	Landscape Uses
Bignonia capreolata crossvine	eWV	S	F	H	OR	Su	BI
Borrichia arborescens tree seaside-tansy	SS	FNaSW	F	U	Y	WSpSuF	SW
Borrichia frutescens bushy seaside-tansy	SS	FNaSW	F	U	Y	SpF	SMW
Bourreria cassinifolia smooth strongbark	eSU	DSs	F	U	W	WSpSuF	ABI
Bourreria ovata Bahama strongbark	eSU	Ss	F	U	W	WSpSuF	ABI
Bourreria succulenta bodywood	eSU	DSs	F	U	W	WSpSuF	ABI
Brickellia californica California brickellbush	SS	FDS	F	G	GY	Su	BRS
Brickellia laciniata split-leaf brickellbush	SS	D	—	G	GWY	—	—
Brunnichia ovata American buckwheatvine	WV	F	—	GMmS	G	Su	B
Buddleja racemosa wand butterfly-bush	SS	DS	F	M	Y	Sp	—
Buddleja scordioides escobilla butterfly-bush	SS	DS	F	M	GW	SpSuF	bRM
Bursera microphylla elephant tree	LS	DS	B	U	W	Su	ABM
Bursera simaruba gumbo-limbo	SC	DS	B	U	G	SpSu	ABI
Byrsonima lucida long key locust-berry	eLU	DFS	Ff	U	PPuW	WSpSuF	AIB
Caesalpinia bonduc yellow nicker	WV	DFNaS	F	U	Y	WSpSuF	B
Caesalpinia crista gray nicker	SS	DNaS	F	U	Y	SpSu	BSM
Caesalpinia pauciflora few-flower nicker	SS	DSs	F	U	Y	Su	BFI
Calliandra eriophylla fairy-duster	SS	DS	F	GLM	PuW	Sp	AM
Callicarpa americana American beauty-berry	MS	DS	FfL	GLMmS	PPu	SpSu	ABFgHM
Calocedrus decurrens incense-cedar	cLC	DSs	FfL	M	—	—	ABIST
Calycanthus floridus eastern sweetshrub	MS	FNaSs	F	U	BrR	SpSu	ABfg
Calyptranthes pallens pale lidflower	eSU	s	B	U	Br	SpSu	BFI
Calyptranthes zuzygium myrtle-of-the-river	eSU	s	BL	U	W	Su	BI
Campsis radicans trumpet-creeper	WV	DFS	FL	HT	OR	SuF	BbSs
Canella winteriana pepper-cinnamon	eSC	Ss	BFf	U	PuR	SuF	AI

PLANT LIST	Plant Type	Env. Tol.	Aesthetic Value	Wildlife Value	Flower Color	Bloom Period	Landscape Uses
Canotia holacantha crucifixion-thorn	LS	DS	B	U	W	Su	BM
Capparis cynophallophora Jamaican caper	SC	Ss	L	GMmS	W	SpSu	AI
Capparis flexuosa falseteeth	SC	Ss	L	GMmS	PW	SpSu	AI
Carnegia gigantea saguaro	C	DS	B	GMmS	W	Sp	AM
Carpinus caroliniana American hornbeam	LU	FS	AB	SW	R	Sp	ABISW
Carya alba mockernut hickory	LC	DFSW	ABL	GLMmS BW	yg	Sp	AINT
Carya aquatica water hickory	LC	FSW	B	GLMmS BW	BY	Sp	ITW
Carya carolinae-septentrionalis southern shag-bark hickory	LC	DFS	ABL	GLMmS BW	yg	Sp	INT
Carya cordiformis bitter-nut hickory	LC	FSs	AL	GLMmS BW	yg	Sp	IT
Carya floridana scrub hickory	SC	DS	AL	GLMmS BW	yg	Sp	BI
Carya glabra pignut hickory	LC	DS	AL	GLMmS BW	yg	Sp	IT
Carya illinoinensis pecan	LC	FS	AL	GLMmS B	yg	Sp	INST
Carya laciniosa shell-bark hickory	LC	S	AL	GLMmS B	yg	Sp	INT
Carya ovalis red hickory	LC	DFS	AL	GLMmS BW	yg	Sp	IT
Carya ovata shag-bark hickory	LC	DFS	ABL	GLMmS BW	yg	Sp	INT
Carya pallida sand hickory	LC	S	AB	GLMmS BW	B	Sp	IT
Carya texana black hickory	SC	DFSW	AL	GLMmS BW	—	Sp	ITW
Casasia ligustrina seven-year-apple	eSU	D	—	U	W	WSpSuF	BS
Cassiope mertensiana western moss-heather	SS	DS	FfL	GLMmS	W	Su	GMNR
Castanea dentata American chestnut	LC	Ss	Ff	GMmS	Y	Su	AINT
Castanea pumila Allegheny-chinkapin	SU	DS	Ff	GMmS	Y	Su	BIN
Castanopsis chrysophylla golden chinkapin	eLC	DS	BF	GMS	W	Su	BNS
Castanopsis sempervirens Sierran chinkapin	eMS	DS	F	GMS	W	Su	Bs
Castela emoryi thorn-of-Christ	MS	DS	Ff	U	PuR	Sp	gM
Castela erecta goatbush	SS	DS	Ff	U	OR	Sp	BM

PLANT LIST	Plant Type	Env. Tol.	Aesthetic Value	Wildlife Value	Flower Color	Bloom Period	Landscape Uses
Catalpa bignonioides southern catalpa	LU	FS	Ff	C	W	Su	ABT
Catalpa speciosa nothern catalpa	LC	FSW	Ff	C	W	Sp	ASTW
Catesbaea parviflora small-flower lilythorn	MS	DSs	f	U	W	SpSuF	BPS
Ceanothus americanus New Jersey-tea	SS	DSs	FL	GhMmS	W	SpSu	BFgM
Ceanothus cordulatus mountain whitethorn	SS	DS	BF	GhLMm	W	SpSu	BgP
Ceanothus crassifolius snowball	MS	DS	F	GhLMm	W	SpSu	Bg
Ceanothus cuneatus wedge-leaf buckbrush	MS	DS	F	GhLMm	W	SpSu	BMP
Ceanothus dentatus sandscrub	eSS	S	F	GhLMm	B	SpSu	BgM
Ceanothus diversifolius pinemat	eSS	Ss	FL	GhLMm	B	SpSu	BgI
Ceanothus fendleri Fendler's buckbrush	SS	Ss	F	GhLMm	W	SpSu	BFgR
Ceanothus fresnensis fresnomat	SS	S	F	GhLMm	B	—	BM
Ceanothus greggii Mojave buckbrush	SS	DS	F	GhLMm	W	SpSu	BHMR
Ceanothus herbaceous prairie redroot	SS	FSs	FfL	GHhMmS	W	Su	BbgM
Ceanothus impressus Santa Barbara buckbrush	eSS	Ss	F	GhLMm	B	—	BgM
Ceanothus integerrimus deerbrush	MS	Ss	BF	GhLMm	BW	SuF	ABg
Ceanothus jepsonii muskbush	SS	S	F	GhLMm	BW	—	B
Ceanothus leucodermis jackbrush	S	S	—	GMm	BW	SuF	—
Ceanothus oliganthus explorer's-bush	MS	DS	F	GhLMm	BPuW	—	Bg
Ceanothus palmeri cuyamaca-bush	MS	DS	BF	GhLMm	BW	Sp	Bg
Ceanothus parryi ladybloom	eMS	Ss	F	GhLMm	B	—	Bgs
Ceanothus parvifolius cattlebush	SS	Ss	F	GhLMm	B	—	Bg
Ceanothus pinetorum Kern River buckbrush	SS	Ss	F	GhLMm	BW	—	BI
Ceanothus prostratus squawcarpet	SS	Ss	BFL	GhLMm	B	SpSu	BGI
Ceanothus sanguineus Oregon teatree	LS	Ss	BF	GhLMm	W	SpSu	ABg
Ceanothus spinosus redheart	eSU	DS	BF	GhLMm	BW	Sp	Bgs

PLANT LIST	Plant Type	Env. Tol.	Aesthetic Value	Wildlife Value	Flower Color	Bloom Period	Landscape Uses
Ceanothus thyrsiflorus bluebrush	eSU	Ss	F	GhLMm	BW	Sp	BgIS
Ceanothus tomentosus ionebush	MS	DS	BFL	GhLMm	B	—	Bg
Ceanothus velutinus tobacco-brush	eSS	Ss	FL	GhLMm	W	Su	BgI
Ceanothus verrucosus barranca-bush	SS	DS	F	GhLMm	W	Su	BgM
Celastrus scandens American bittersweet	WV	DFSs	AfL	GmS	G	Sp	BI
Celtis laevigata sugar-berry	SC	FSs	AB	S	G	Sp	ABISTW
Celtis occidentalis common hackberry	LC	DFSs	B	BmS	yg	Sp	AbNST
Celtis pallida shiny hackberry	eLS	DS	B	BGhLMS	W	—	BMs
Cephalanthus occidentali common buttonbush	MS	FSW	FL	BFhMmSW	W	Su	SWwg
Ceratiola ericoides sand-heath	eMS	DS	FL	U	RY	SpSuF	BFHM
Cercis canadensis redbud	SU	DFSs	ABFL	Hh	P	Sp	ABbgI
Cercocarpus intricatus little-leaf mountain-mahogany	S	DS	—	GM	—	—	B
Cercocarpus ledifolius curl-leaf mountain-mahogany	eSU	DS	L	GM	—	Su	BS
Cercocarpus montanus alder-leaf mountain-mahogany	MS	DS	fL	GM	G	Sp	B
Chamaebatia foliolosa Sierran mountain-misery	eSS	s	L	U	W	—	GI
Chamaebatiaria millifolium fernbush	SS	DS	FL	U	W	Su	FgMR
Chamaecyparis lawsoniana Port Orford-cedar	cLC	Ss	L	LMS	R	Sp	ABFH
Chamaecyparis nootkatensis Alaska-cedar	cLC	S	L	LMS	BY	Sp	A
Chamaecyparis thyoides Atlantic white-cedar	cLU	FS	L	LMS	Ryg	Sp	wg
Chamaedaphne calyculata leatherleaf	eSS	FSW	L	GMS	W	SpSu	W
Chilopsis linearis desert-willow	SU	DS	F	hS	PPu	Su	A
Chimaphila umbellata pipsissewa	eSP	s	L	U	P	Su	GI
Chiococca alba West Indian milkberry	eSS	Ss	L	U	WY	WSpSuF	ABI
Chionanthus virginicus white fringetree	SU	Ss	AFL	S	W	Sp	ABFg
Chrysobalanus icaco icaco coco-plum	eSU	Na	—	U	G	Sp	Is

PLANT LIST	Plant Type	Env. Tol.	Aesthetic Value	Wildlife Value	Flower Color	Bloom Period	Landscape Uses
Chrysoma pauciflosculosa woody goldenrod	SS	DS	—	U	Y	SpSu	BG
Chrysophyllum oliviforme satinleaf	eSC	DSs	L	U	G	WF	As
Chrysothamnus depressus long-flower rabbitbrush	SS	DS	F	MmS	Y	Su	M
Chrysothamnus nauseosus rubber rabbitbrush	MS	DFNaS	FL	hLMmS	WY	SuF	Eg
Chrysothamnus paniculatus dotted rabbitbrush	SS	DS	F	MmS	Y	Su	BM
Chrysothamnus parryi Parry's rabbitbrush	SS	DS	F	MmS	Y	Su	BM
Chrysothamnus pumilis rabbitbrush	SS	DS	BF	MmS	Y	Su	BM
Chrysothamnus teretifolius needle-leaf rabbitbrush	SS	DS	F	LMmS	Y	Su	BM
Chrysothamnus vaseyi Vasey's rabbitbrush	SS	DS	F	MmS	Y	Su	BM
Chrysothamnus viscidiflorus green rabbitbrush	MS	DS	BF	MmS	Y	Su	M
Cissus verticillata seasonvine	WV	—	—	U	GW	SpSu	—
Citharexylum fruticosum Florida fiddlewood	eSU	Ss	F	U	W	WSpSuF	BIS
Clematis columbiana Columbian virgin's-bower	WV	S	F	U	BPu	Su	B
Clematis ligusticifolia deciduous traveler's-joy	WV	DS	F	M	W	SpSuF	Bg
Clematis virginiana devil's-darning-needles	WV	DFSs	FL	S	W	SuF	BW
Cleome isomeris bladder-pod spider-flower	S	—	—	U	Y	WSpSuF	—
Clethra acuminata mountain sweet-pepperbush	MS	s	AFL	C	W	Su	BFI
Clethra alnifolia coastal sweet-pepperbush	MS	FSsW	AFL	BGmSW	PW	SuF	bIWwg
Cliftonia monophylla buckwheat-tree	eSU	FSW	Ff	hM	R	WSp	AWs
Coccoloba diversifolia tietongue	eSC	FS	fL	U	W	Su	BEs
Coccoloba uvifera seaside-grape	eLS	S	L	U	W	Su	ABFSs
Coccothrinax argentata Florida silver palm	eSU	DS	L	S	W	Su	ABs
Coleogyne ramosissima blackbrush	SS	DS	F	M	WY	Sp	E
Colubrina arborescens greenheart	eSU	DS	—	L	G	SuF	Is
Colubrina cubensis Cuban nakedwood	eSU	Ss	—	L	G	Su	Is

PLANT LIST	Plant Type	Env. Tol.	Aesthetic Value	Wildlife Value	Flower Color	Bloom Period	Landscape Uses
Colubrina elliptica soldierwood	eSU	Ss	B	L	yg	SpF	AIs
Colubrina texensis hog-plum	MS	DS	—	L	yg	Sp	H
Comptonia peregrina sweet-fern	SS	DS	L	GMm	Br	Sp	EI
Condalia ericoides javelin-bush	SS	DS	—	U	—	—	M
Condalia hookeri Brazilian bluewood	eSU	Ss	FL	S	G	Sp	Hs
Condalia spathulata squawbush	eMS	DS	—	GM	G	—	Ms
Conocarpus erectus button mangrove	eSU	FNaSW	L	U	G	WSpSuF	Ws
Cordia globosa curaciao-bush	eSS	Ss	F	U	W	WSpSuF	BI
Cordia sebestena large-leaf geigertree	eSU	NaS	F	U	R	WSpSuF	ABFs
Corema conradii broom-crowberry	eSS	Ss	FL	U	BrPu	Sp	BEs
Cornus alternifolia alternate-leaf dogwood	SU	FSs	AFf	BGMmSW	W	Sp	ABFIS
Cornus amomum silky dogwood	MS	FS	ABFL	BMmSW	W	Su	BEgISW
Cornus drummondii rough-leaf dogwood	MS	Ss	L	GLMmS	W	Sp	B
Cornus florida flowering dogwood	LU	DSs	AFf	BGLMmS	WY	Sp	ABI
Cornus foemina stiff dogwood	LS	FS	ABF	BGLMmS	W	Sp	BW
Cornus glabrata smooth-leaf dogwood	SU	FSs	A	GLMmS	W	Sp	BS
Cornus nuttalli Pacific flowering dogwood	SU	s	FfL	GLMmSW	W	Sp	BI
Cornus racemosa gray dogwood	LS	DFSs	ABF	BGMmSW	W	Sp	BbFg
Cornus rugosa round-leaf dogwood	MS	Ss	AfL	BGLMmS	W	Su	BI
Cornus sericea redosier	MS	FSsW	ABL	BLMmSW	W	SpSu	BSW
Corylus americana American hazelnut	MS	DSs	Ff	GLMmS	BrR	Sp	ABIN
Corylus cornuta beaked hazelnut	SS	DSs	AL	GLMmS	BrR	Sp	BHNs
Coryphantha cornifera rhinoceros cactus	C	DS	BF	U	Y	SpSu	gM
Coryphantha macromeris nipple beehive cactus	C	DS	BF	U	PuR	SpSu	gM
Coryphantha ramillosa whiskerbush	C	DS	BF	U	PPu	SpSu	gM

PLANT LIST	Plant Type	Env. Tol.	Aesthetic Value	Wildlife Value	Flower Color	Bloom Period	Landscape Uses
Coryphantha recurvata Santa Cruz beehive cactus	C	DS	BF	U	RY	SpSu	gM
Coryphantha scheeri long-tubercle beehive cactus	C	DS	BF	U	RY	SpSu	gM
Coryphantha sulcata pineapple cactus	C	DS	BF	U	RY	SpSu	gM
Coursetia glandulosa rosary baby-bonnets	Ss	DSs	BF	U	PW	SpSuF	gM
Crataegus crus-galli cock-spur hawthorn	LS	DFS	BFfL	GhLMm SW	W	Sp	BbP
Crataegus douglasii black hawthorn	LS	DFS	BFfL	GhLMm SW	W	Sp	BbFS
Crataegus marshallii parsley hawthorn	LS	FS	BFfL	GhLMm SW	W	Sp	b
Crataegus spathulata little-hip hawthorn	LS	Ss	L	GhLMm SW	W	Sp	Bb
Crataegus viridis green hawthorn	LS	FSsW	FfL	GhLMm SW	W	Sp	ISW
Crossopetalum rhacoma maiden-berry	SU	s	—	U	GR	SpSu	I
Crossosoma bigelovii ragged rockflower	SS	DS	—	U	PuW	Sp	—
Croton linearis grannybush	SS	DSs	—	GmS	—	WSpSuF	BMS
Cupania glabra Florida toadwood	eSU	Ss	—	U	—	F	s
Cupressus arizonica Arizona cypress	cSC	DS	BfL	Mm	Y	Sp	AEs
Cupressus bakeri Modoc cypress	cSC	DS	L	Mm	—	—	AHs
Cupressus forbesii Tecate cypress	c	DS	BL	Mm	—	—	AHs
Cupressus glabra Arizona smooth cypress	cSC	DNaS	BL	Mm	—	—	AHs
Cupressus goveniana Gowen cypress	cSC	S	BL	Mm	—	—	AHs
Cupressus macnabiana MacNab's cypress	cLU	DS	BL	Mm	yg	Sp	AHs
Cupressus macrocarpa Monterey cypress	cSC	NaS	BL	Mm	—	—	AHs
Cupressus nevadensis Paiute cypress	cLU	S	L	Mm	Y	WSp	AHs
Cupressus pygmaea Mendocino cypress	cSC	DS	L	Mm	—	—	AHs
Cupressus sargentii Sargent's cypress	cMS	DS	L	Mm	—	—	AHs
Cupressus stephensonii Cuyamaca cypress	c	S	L	Mm	—	—	AHs
Cyrilla racemiflora swamp titi	eSU	DFS	AFL	hMSW	W	Su	bgW

PLANT LIST	Plant Type	Env. Tol.	Aesthetic Value	Wildlife Value	Flower Color	Bloom Period	Landscape Uses
Dalbergia brownii Brown's Indian-rosewood	SS	DSs	F	U	PW	—	BgP
Dalbergia ecastaphyllum coinvine	MS	DSW	Ff	U	PW	Sp	BSW
Dalea formosa featherplume	SS	DS	F	m	Pu	SpSuF	AM
Dasylirion leiophyllum green sotol	SS	DS	L	GM	WY	Su	AM
Dasylirion wheeleri common sotol	SS	DS	L	GM	WY	Su	AM
Decodon verticillatus swamp-loosetrife	LS	FSW	F	m	P	Su	Wwg
Diervilla lonicera northern bush-honeysuckle	SS	DSs	AFL	BHS	R	Su	bI
Diospyros texana Texas persimmon	eLU	DS	BL	BGLMmS	GW	Sp	ABNs
Diospyros virginiana common persimmon	SC	DFS	BLf	BGLMmS	Y	Su	BNT
Diplacus longiflorus southern bush-monkey-flower	eSS	S	—	U	PY	SpSu	—
Dirca palustris eastern leatherwood	SS	Ss	AL	GMS	Y	Sp	IS
Dodonaea viscosa Florida hopbush	eSU	FS	—	U	yg	WF	AWs
Drypetes diversifolia milkbark	eSU	S	—	U	yg	Su	Is
Drypetes lateriflora Guiana-plum	eSU	Ss	—	U	yg	WSp	Is
Echinocactus horizonthalonius devil's-head	C	DS	BF	mS	PY	Sp	AgM
Echinocactus polycephalus cotton-top cactus	C	DS	BF	mS	Y	Sp	AgM
Echinocactus texensis horse-crippler	C	DS	BF	mS	PuR	Sp	AgM
Echinocereus chloranthus brown-spine hedgehog cactus	C	DS	BF	mS	BrGR	Sp	AgM
Echinocereus coccineus scarlet hedgehog cactus	C	DS	BF	mS	R	Sp	AgM
Echinocereus engelmannii saints cactus	C	DS	BF	mS	PPu	Sp	AgMR
Echinocereus enneacanthus pitaya	C	DS	BF	mS	Pu	Sp	AgMR
Echinocereus fendleri pink-flower hedgehog cactus	C	DS	BF	mS	PuR	Sp	AgMR
Echinocereus pectinatus Texas rainbow cactus	C	DS	BF	mS	LPuY	Sp	Agm
Echinocereus reichenbachii lace hedgehog cactus	C	DS	BF	mS	PPu	SpSu	AgM
Echinocereus rigidissimus rainbow hedgehog cactus	C	DS	BF	mS	—	Sp	AgM

PLANT LIST	Plant Type	Env. Tol.	Aesthetic Value	Wildlife Value	Flower Color	Bloom Period	Landscape Uses
Echinocereus stramineus strawberry hedgehog cactus	C	DS	BF	mS	—	Sp	AgM
Echinocereus triglochidiatus king-cup cactus	C	DS	BF	mS	R	Sp	AgM
Echites umbellata devil's-potato	Wv	Ss	F	U	W	WSpSuF	B
Ehretia anacua knockaway	eSC	Ss	AB	GhLMS	W	Sp	ABITs
Elaeagnus commutata American silver-berry	MS	DNaS	BF	GMS	Y	Sp	AMW
Encelia californica California brittlebush	SS	DS	F	U	PuY	Sp	gM
Encelia farinosa goldenhills	SS	DS	FL	U	Y	Sp	gM
Encelia virginensis virgin river brittlebush	SS	DSs	L	U	Y	Sp	M
Ephedra antisyphilitica clapweed	eSS	DS	B	GLM	Y	Sp	BgM
Ephedra californica California joint-fir	SS	DS	B	GLMm	—	Sp	gM
Ephedra nevadensis Nevada joint-fir	SS	DS	B	GLM	—	Sp	gM
Ephedra torreyana Torrey's joint-fir	SS	DS	B	GLM	Y	Sp	gM
Ephedra trifurca long-leaf joint-fir	MS	DS	B	GLM	Y	Sp	gM
Ephedra viridis Mormon-tea	SS	DS	B	GLM	BG	SpSu	gM
Epigaea repens trailing-arbutus	eSS	DSs	FL	GMm	PW	Sp	BGgI
Epithelantha micromeris ping-pong-ball cactus	C	DS	B	S	P	—	gM
Eriastrum densifolium giant woolstar	P	—	—	U	BYW	Su	—
Ericameria cooperi Cooper's heath-goldenrod	SS	DS	F	M	Y	—	M
Ericameria cuneata cliff heath-goldenrod	eSS	DS	F	M	Y	—	MR
Ericameria gilmanii white-flower heath-goldenrod	S	DS	F	M	WY	SuF	M
Ericameria laricifolia turpentine-bush	eSS	DS	F	M	Y	F	BGM
Ericameria linearifolius narrow-leaf heath-goldenrod	S	DS	F	M	Y	Sp	M
Eriodictyon angustifolium narrow-leaf yerba-santa	eMS	DS	F	M	W	Sp	BgMs
Eriodictyon californicum California yerba-santa	eMS	DS	F	M	BW	Sp	gMs
Eriodictyon crassifolium thick-leaf yerba-santa	eMS	DS	F	M	Pu	Sp	gMs

PLANT LIST	Plant Type	Env. Tol.	Aesthetic Value	Wildlife Value	Flower Color	Bloom Period	Landscape Uses
Eriodictyon trichocalyx hairy yerba-santa	MS	DS	F	M	BW	Sp	gMs
Eriogonum fasciculatum eastern Mojave wild buckwheat	SS	DS	—	GhLMmS	W	SuF	BgM
Eriogonum heermannii Heermann's wild buckwheat	SS	DS	—	GLMmS	Y	Su	BM
Eriogonum heracleoides Parsnip-flower wild buckwheat	SS	DS	—	GLMmS	W	Su	BM
Eriogonum microthecum slender wild buckwheat	SS	DS	—	GLMmS	PWY	Su	BM
Eriogonum niveum snow wild buckwheat	SS	L	—	GLMmS	W	Su	BgM
Eriogonum ovalifolium cushion wild buckwheat	SS	S	BL	GLMmS	RW	Su	gMR
Eriogonum umbellatum sulphur-flower wild buckwheat	SS	DS	FL	GLMmS	Y	Su	CgMR
Eriogonum wrightii bastard-sage	SS	DS	F	GLMmS	PW	Su	BEGR
Erithalis fruticosa blacktorch	SC	Ss	—	U	W	WSpSuF	BIS
Erythrina flabelliformis coral-bean	SU	DS	F	U	R	Sp	ABH
Erythrina herbacea red-cardinal	LS	S	F	U	R	SpSuF	ABHT
Escobaria orcuttii Orcutt's fox-tail cactus	C	DS	B	mS	—	—	gM
Escobaria tuberculosa white-column fox-tail cactus	C	DS	B	mS	—	—	gM
Escobaria vivipara spinystar	C	DS	B	Ms	P	—	gM
Eugenia axillaris white stopper	eSU	Ss	Ff	U	W	SuF	ABIs
Eugenia confusa red-berry stopper	eSU	Ss	Ff	U	W	Su	ABIs
Eugenia foetida box-leaf stopper	eSU	Ss	Ff	U	W	SuF	ABIs
Eugenia rhombea red stopper	eSU	Ss	Ff	U	W	SuF	ABFIs
Eupatorium azureum blue boneset	MS	Ss	F	BGS	B	WSpSu	Bb
Evonymus americanus American strawberry-bush	MS	Ss	f	GmS	Pu	Su	I
Evonymus atropurpureus eastern wahoo	SU	FSs	AfL	GS	R	Su	BSs
Exostema caribaeum Caribbean princewood	eSC	Ss	—	U	W	Su	BFIs
Exothea paniculata butterbough	eSC	Ss	F	U	W	SpSu	ABFIs
Eysenhardtia orthocarpa Tahitian kidneywood	SU	DS	—	U	W	Sp	—

PLANT LIST	Plant Type	Env. Tol.	Aesthetic Value	Wildlife Value	Flower Color	Bloom Period	Landscape Uses
Eysenhardtia texana Texas kidneywood	LS	DS	—	h	W	SpSu	Ab
Fagus grandifolia American beech	LC	Ss	AB	GLMmS	yg	Sp	AIT
Fallugia paradoxa Apache-plume	MS	DS	Ff	M	W	Su	E
Fendlerella utahensis Utah-fendlerbush	SS	DS	B	M	W	Su	BM
Ferocactus cylindraceus California barrel cactus	C	DS	BF	mS	Y	SpSu	AgRM
Ferocactus hamatacanthus turk's head	C	DS	BF	mS	Y	SpSu	AgM
Ferocactus wislizeni candy barrel cactus	C	DS	BF	mS	OY	SpSu	AgM
Ficus aurea Florida strangler fig	eSC	DSs	BL	mS	—	SpSuF W	ABIs
Ficus citrifolia wild banyantree	eSU	FS	B	mS	R	WSpSu F	ABs
Flourensia cernua American tarwort	SS	DS	—	U	Y	WF	M
Forestiera acuminata eastern swamp-privit	LS	FSsW	L	GMSW	G	Sp	SW
Forestiera angustifolia Texas swamp-privet	eSU	NaS	—	GMS	G	Sp	BMs
Forestiera ligustrina upland swamp-privet	MS	DSs	L	GMSW	G	Su	BgS
Forestiera pubescens stretchberry	MS	FSs	—	GMmS	G	Sp	BES
Forestiera segregata Florida swamp-privet	eSU	FS	L	GMSW	Y	Sp	BIs
Fouquieria splendens ocotillo	LS	DS	BF	S	OR	WSpSu F	AH
Frangula alnus alder-buckthorn	MS	S	L	h	G	Sp	H
Frangula betulifolia birch-leaf buckthorn	MS	—	L	GLMmS	—	—	—
Frangula californica California coffee berry	eSS	DS	L	GLMmS	G	Su	B
Frangula caroliniana Carolina buckthorn	LS	FSsW	FL	GLMmS	W	SpSu	ABS
Frangula purshiana cascara sagrada	LS	S	—	GLMmS	G	—	—
Frankenia salina alkali sea-heath	SS	S	L	U	P	—	—
Fraxinus americana white ash	LC	DFS	A	GLMmS W	Pu	Sp	IT
Fraxinus anomala single-leaf ash	SC	DS	L	GMmNS	G	Sp	A
Fraxinus berlandieriana Mexican ash	SC	F	L	GLMmS	G	Sp	AT

PLANT LIST	Plant Type	Env. Tol.	Aesthetic Value	Wildlife Value	Flower Color	Bloom Period	Landscape Uses
Fraxinus caroliniana Carolina ash	LU	FS	L	GLMmS W	yg	Sp	W
Fraxinus dipetala two-petal ash	MS	DS	F	GLMmS W	W	Sp	B
Fraxinus latifolia Oregon ash	SC	S	L	GLMmS W	—	Sp	T
Fraxinus nigra black ash	SC	FS	AL	GLMmS W	Pu	Sp	bSW
Fraxinus pennsylvanica green ash	LC	DFS	A	GLMmS W	Pu	Sp	IT
Fraxinus profunda pumpkin ash	LC	FSW	L	GLMmS W	—	—	W
Fraxinus quadrangulata blue ash	SC	DSs	AL	GLMmS W	Pu	Sp	BIST
Fraxinus texensis Texas ash	LU	DS	AL	GLMmS	Y	Sp	AT
Fraxinus velutina velvet ash	SC	S	L	GLMmS	yg	Sp	T
Fremontodendron californicum California flannelbush	eSU	S	—	U	Y	Sp	—
Garberia heterophylla garberia	eMS	DS	F	U	P	Su	BIs
Garrya buxifolia dwarf silktassel	eSS	DS	L	GM	—	—	—
Garrya elliptica wavy-leaf silktassel	eMS	S	FL	GM	—	W	s
Garrya flavescens ashy silktassel	eMS	DS	L	GM	—	W	s
Garrya fremontii bearbush	eMS	DS	L	GM	—	—	s
Garrya ovata lindheimer silktassel	eLS	DSs	L	GM	—	Sp	BR
Garrya veatchii canyon silktassel	SS	DS	L	GM	—	W	—
Garrya wrightii Wright's silktassel	eMS	DSs	L	GM	—	SpSu	BRs
Gaultheria hispidula creeping-snowberry	eSS	s	FfL	GLMmS	W	Sp	FGIW
Gaultheria procumbens eastern teaberry	eSS	DFSs	L	GLMm	W	SpSu	BGIW
Gaultheria shallon salal	eSS	—	L	GLMmS	W	—	BG
Gaylussacia baccata black huckleberry	SS	DSs	L	GLMmS	W	Sp	GN
Gaylussacia dumosa dwarf huckleberry	SS	FS	L	GLMmS	PW	Su	N
Gaylussacia frondosa blue huckleberry	SS	DSs	L	GLMmS	PuW	SpSu	N
Gaylussacia ursina bear huckleberry	SS	Ss	fL	GLMmS	GPW	Sp	BFIN

PLANT LIST	Plant Type	Env. Tol.	Aesthetic Value	Wildlife Value	Flower Color	Bloom Period	Landscape Uses
Gelsemium sempervirens evening trumpet-flower	eWV	Ss	FL	GM	Y	Sp	Gs
Gleditsia aquatica water-locust	LU	FSW	—	GhLM	—	Su	AW
Gleditsia triacanthos honey-locust	SC	FDS	AL	GhLM	yg	Su	AT
Glossopetalon spinescens spiny greasebush	SS	DS	—	U	W	—	R
Glottidium vesicarium bagpod	MS	FS	—	U	Y	SpSuF	—
Gordonia lasianthus loblolly-bay	eLU	FSW	FL	U	W	SpF	As
Gouania lupuloides whiteroot	WV	Fs	—	U	YG	SpSu	BI
Grayia spinosa spiny hop-sage	SS	DNaS	L	U	—	—	M
Guajacum angustifolium Texas lignumvitae	eMS	S	FL	U	PuW	SpSuF	As
Guajacum sanctum holywood lignumvitae	SU	DSs	BFL	U	BPu	SpSu	AI
Guapira discolor beeftree	SU	DSs	—	U	yg	Su	AI
Guettarda elliptica hammock velvetseed	SC	S	F	U	W	Su	BI
Guettarda scabra wild guave	SU	DSs	—	U	W	Su	BI
Gutierrezia microcephala small-head snakeweed	SS	DS	—	GLMm	Y	SuF	—
Gyminda latifolia West Indian false box	eSU	Ss	—	U	W	Su	BIs
Gymnanthes lucida oysterwood	eSU	S	—	U	yg	SpF	ABFIs
Gymnocladus dioicus Kentucky coffeetree	LC	FS	AL	mS	GW	Su	AST
Halesia carolina Carolina silverbell	LU	Ss	AF	U	W	Sp	ABIg
Hamamelis virginiana American witch-hazel	SU	Ss	ABF	MmS	Y	F	ABbISW
Hamelia patens scarletbush	eSS	DS	F	U	R	SpSuFW	BF
Haploesthes greggii false broomweed	SS	DNaS	—	U	GY	—	B
Havardia pallens haujillo	MS	DSs	F	U	WY	Su	B
Hazardia squarrosa saw-tooth bristleweed	SS	DS	—	GLMmS	Y	F	g
Hechtia texensis Texas false agave	MS	DS	—	U	—	—	—
Heteromeles arbutifolia California-Christmas-berry	eMS	DS	FfL	GmS	W	Su	A

PLANT LIST	Plant Type	Env. Tol.	Aesthetic Value	Wildlife Value	Flower Color	Bloom Period	Landscape Uses
Hibiscus denudatus paleface	SS	DS	FfL	U	LW	Sp	A
Hibiscus grandiflorus swamp rose-mallow	LS	FSW	F	U	P	SpSu	—
Hippocratea volubilis medicine-vine	WV	—	—	U	—	SpSu	—
Hippomane mancinella manchineel	SU	FSs	—	U	W	Sp	BI
Holodiscus boursieri Boursier's oceanspray	MS	DSs	BL	U	—	SuF	B
Holodiscus discolor hillside oceanspray	SS	DS	F	U	W	Su	A
Holodiscus dumosus glandular oceanspray	SS	DS	F	U	PW	—	AH
Hydrangea arborescens wild hydrangea	SS	s	FL	GMS	W	Su	BgI
Hymenoclea monogyra single-whorl cheesebush	SS	FS	—	U	—	—	E
Hymenoclea salsola white cheesebush	SS	FDS	—	U	—	Sp	—
Hypelate trifoliata inkwood	SC	Ss	—	U	G	SpSu	BI
Hypericum fasciculatum peel-bark St. John's-wort	eLS	FW	F	U	Y	SpSu	AW
Hypericum hypericoides St. Andrew's-cross	SS	FS	FL	U	Y	SpF	SW
Ilex ambigua Carolina holly	LS	s	fL	GLMmS	W	Sp	BIS
Ilex cassine dahoon	eSU	sW	fL	GLMmS W	W	Su	ABISsW
Ilex coriacea large gallberry	eMS	FSsW	fL	GLMmS W	W	Sp	ABISsW
Ilex decidua deciduos holly	SU	DFSs	BfL	GLMmS W	GW	Sp	BISW
Ilex glabra inkberry	eMS	FSs	fL	GLmSW	W	Sp	FHWs
Ilex krugiana tawny-berry holly	eSU	FSs	fL	GLMS	W	SpSu	AIs
Ilex laevigata smooth winterberry	MS	FSs	AfL	GLMmS W	W	Sp	SW
Ilex montana mountain holly	LS	s	A	GLMmS	W	Sp	ABI
Ilex myrtifolia myrtle dahoon	eSU	SsW	fL	GLMmS	W	Sp	ABSsW
Ilex opaca American holly	eLU	SsW	fL	GLMmS W	W	Sp	AIsW
Ilex verticillata common winterberry	MS	FSs	fL	GLMmS W	W	Su	BSW
Ilex vomitoria yaupon	eLS	DFSs	BfL	GLMmS W	W	Sp	AFHPsW

PLANT LIST	Plant Type	Env. Tol.	Aesthetic Value	Wildlife Value	Flower Color	Bloom Period	Landscape Uses
Isocoma acradenia alkali jimmyweed	SS	DNaS	F	U	Y	SuF	gM
Isocoma menziesii jimmyweed	SS	—	F	U	Y	—	—
Isocoma pluriflora southern jimmyweed	SS	DS	F	U	Y	Su	gM
Isocoma tenuisecta shrine jimmyweed	SS	DS	F	U	Y	SuF	M
Itea virginica Virginia sweetspire	MS	DFSs W	AFL	GmSW	W	Su	gSW
Iva frutescens Jesuit's-bark	MS	FS	—	U	W	SuF	W
Iva imbricata seacoast marsh-elder	SS	NaS	—	U	W	SuF	W
Jacquinia keyensis joewood	eLS	FNaS	FL	U	WY	SuF	AEIs
Jamesia americana five-petal cliffbush	MS	DS	A	U	W	Su	ABH
Jatropha cardiophylla sangre-de-cristo	SS	DS	—	U	—	Su	BM
Jatropha dioica leatherstem	SS	DS	—	U	PW	Su	BM
Juglans californica Southern California walnut	SU	FSs	L	LS	Y	Sp	BS
Juglans cinerea white walnut	LC	FS	AL	mS	yg	Sp	IN
Juglans major Arizona walnut	LU	FS	f	m	yg	Sp	BNT
Juglans microcarpa little walnut	SU	FS	L	m	G	Sp	SBN
Juglans nigra black walnut	LC	DFS	AL	m	yg	Sp	ANPT
Juniperus ashei Ashe's juniper	cSU	DS	fL	GLMmS	—	—	ABRM
Juniperus californica California juniper	cLS	DS	L	GLMmS	—	—	BEMP
Juniperus communis common juniper	cSU	DS	L	GLmS	Y	Sp	CEGg
Juniperus deppeana alligator juniper	cSC	DS	B	GLMmS	—	WSp	BM
Juniperus erythrocarpa red-berry juniper	cLS	DS	L	GLMmS	—	—	BM
Juniperus flaccida drooping juniper	cSU	DSs	BL	GLMmS	—	—	ABMR
Juniperus monosperma one-seed juniper	cSC	DS	BL	GLMmS	Y	Sp	ABM
Juniperus occidentalis western juniper	cLU	DS	BL	GLMmS	—	—	BMPR
Juniperus osteosperma Utah juniper	cLS	DS	L	GLMmS	—	—	BP

PLANT LIST	Plant Type	Env. Tol.	Aesthetic Value	Wildlife Value	Flower Color	Bloom Period	Landscape Uses
Juniperus pinchotii Pinchot's juniper	cSU	DS	L	GLMmS	—	—	B
Juniperus scopulorum Rocky Mountain juniper	cSU	DS	L	GLMmS	Y	Sp	ABM
Juniperus virginiana eastern red-cedar	cSC	DSs	fL	GLMmS	PuY	Sp	ABCEFHIPs
Kalmia angustifolia sheep-laurel	eSS	FSs	FL	Gm	PR	Su	BW
Kalmia hirsuta hairy-laurel	SS	FS	—	Gm	—	—	W
Kalmia latifolia mountian-laurel	eLS	DFSs	LF	GLMm	PW	Sp	BgW
Kalmia polifolia bog-laurel	eSS	FSsW	LF	Gm	P	Su	W
Karwinskia humboldtiana coyotillo	eSU	DS	L	U	—	—	A
Koeberlinia spinosa crown-of-thorns	SU	DS	—	GM	GW	Sp	EHP
Krameria grayi white ratany	SS	DS	FL	GM	Pu	SpSuF	EM
Krascheninnikovia lantata winterfat	SS	DFNaS	f	Mm	G	SpSuF	ES
Krugiodendron ferreum leadwood	eSC	Ss	L	U	G	Su	BFI
Laguncularia racemosa white mangrove	SC	FNaSsW	BL	U	W	WSpSuF	W
Lantana involucrata button-sage	eMS	DS	F	U	L	WSpSuF	BEFs
Larix laricina American larch	cSC	FS	L	GMmS	RY	Sp	BSW
Larix lyallii subalpine larch	cSC	S	L	GMmS	—	Sp	—
Larix occidentalis western larch	cLC	S	BL	GMmS	—	Sp	—
Larrea tridentata creosote-bush	eSS	DS	L	m	Y	SpSu	A
Ledum glandulosum glandular Labrador-tea	eSS	—	—	U	W	Sp	—
Ledum groenlandicum rusty Labrador-tea	eSS	FS	FL	BGS	W	Sp	W
Leiophyllum buxifolium sand-myrtle	eSS	DFSs	FL	GmS	W	Sp	CGg
Lepidospartum latisquamum scalebroom	MS	DS	—	U	—	—	—
Leptodactylon pungens granite prickly-phlox	SS	DS	F	U	PW	SpSu	GgR
Leucaena pulverulenta great leadtree	SC	F	L	U	W	Sp	AST
Leucophyllum frutescens Texas barometer-bush	MS	S	FL	U	PPuW	Su	AH

PLANT LIST	Plant Type	Env. Tol.	Aesthetic Value	Wildlife Value	Flower Color	Bloom Period	Landscape Uses
Leucophyllum minus Big Bend barometer-bush	MS	—	FL	U	PPu		AH
Leucothoe axillaris coastal doghobble	eMS	SW	AFL	U	W	WSp	AsW
Leucothoe racemosa swamp doghobble	eSS	SsW	AFL	M	W	Sp	AFHW
Leucothoe recurva red-twig doghobble	eLS	Fs	ABFL	U	W	Sp	ASs
Licania michauxii gopher-apple	eSS	DS	L	RS	W	Sp	B
Lindera benzoin northern spicebush	MS	DS	AFf	CGMSB	Y	Sp	BIN
Linnaea borealis American twinflower	eSS	Ss	F	M	PuW	SpSu	C
Liquidambar styraciflua sweet-gum	LC	FS	AL	GLmSW	yg	Sp	AISTW
Liriodendron tulipifera tuliptree	LC	Ss	AFL	HhLmS	yg	Su	AIT
Lithocarpus densiflorus tan-oak	eLC	Ss	BL	—	W	—	BIs
Lonicera albiflora white honeysuckle	WV	DS	F	GS	W	—	—
Lonicera canadensis American fly-honeysuckle	SS	DSs	Ff	BGHmS	O	Sp	bNW
Lonicera ciliosa orange honeysuckle	WV	s	F	GHS	OY	—	AI
Lonicera conjugialis purple-flower honeysuckle	SS	s	Ff	Hh	Pu	Su	I
Lonicera dioica limber honeysuckle	SS	DFSs	Ff	BGHmS	Y	Sp	bgM
Lonicera hirsuta hairy honeysuckle	WV	S	Ff	BH	Y	Su	I
Lonicera hispidula pink honeysuckle	eMS	—	F	GS	RY	Su	As
Lonicera interrupta chaparral honeysuckle	SS	DS	Ff	—	Y	—	Bg
Lonicera involucrata four-line honeysuckle	MS	s	F	GS	GPuR	Su	H
Lonicera oblongifolia swamp fly-honeysuckle	SS	FS	Ff	BH	W	Sp	W
Lonicera sempervirens trumpet honeysuckle	WV	s	F	—	ROY	Sp	AI
Lonicera subspicata Santa Barbara honeysuckle	SS	DS	Ff	—	Y	—	Bg
Lycium andersonii red-berry desert-thorn	MS	DNaS	F	GHM	PuW	WSpSu	BbM
Lycium berlandieri silver desert-thorn	MS	DNaS	BF	GHM	BPu	SpSuF	BMR
Lycium carolinianum Carolina desert-thorn	SS	NaSW	F	MW	BPu	SpSu	BgSW

PLANT LIST	Plant Type	Env. Tol.	Aesthetic Value	Wildlife Value	Flower Color	Bloom Period	Landscape Uses
Lycium cooperi peachthorn	MS	DS	B	GMmW	W	SpSuF	BMP
Lycium fremontii Fremont's desert-thorn	MS	DS	Ff	GHM	PuW	WSp	BgMN
Lycium pallidum pale desert-thorn	SS	DS	BLf	M	W	WSP	MPR
Lycium richii Santa Catalina desert-thorn	MS	DS	Ff	GMmW	Pu	WSp	BMP
Lyonia ferruginea rusty staggerbush	SU	DFS	—	mS	W	—	SW
Lyonia fruticosa coastal-plain staggerbush	MS	DFS	—	mS	W	Sp	SW
Lyonia ligustrina maleberry	LS	DFSW	AL	mS	W	SpSu	SW
Lyonia lucida shinyleaf	eSS	FSsW	FL	mS	W	Sp	IW
Lyonia mariana Piedmont staggerbush	SS	SsW	F	U	PW	SpSu	BM
Lysiloma latisiliquum false tamarind	eSC	S	L	U	W	SpSu	ABFIs
Lysiloma watsonii little-leaf false tamarind	LS	—	—	U	W	—	—
Maclura pomifera osage-orange	LC	DFS	f	GLMm	G	Su	H
Magnolia acuminanta cucumber magnolia	LC	Ss	FfL	MmS	yg	Sp	ABIT
Magnolia ashei Ashe's magnolia	SU	Ss	FfL	MmS	W	Sp	A
Magnolia fraseri Fraser's magnolia	SU	s	FL	MmS	YW	Sp	ABI
Magnolia grandiflora southern magnolia	eSC	Ss	FfL	MmS	W	SpSu	AHsT
Magnolia pyramidata pyramid magnolia	SU	Fs	FL	MmS	W	Sp	ABT
Magnolia tripetala umbrella magnolia	SU	Fs	FfL	MmS	W	Sp	ABI
Magnolia virginiana sweet-bay	eLS	FSs	FL	MmS	W	SpSu	ASsW
Mahonia aquifolium holly-leaf Oregon-grape	eSS	DS	fL	GS	Y	Sp	GN
Mahonia fremontii desert Oregon-grape	eMS	DS	L	LM	Y	Su	As
Mahonia haematocarpa red Oregon-grape	eMS	DS	L	GS	Y	Su	As
Mahonia nervosa Cascade Oregon-grape	eSS	s	fL	GS	Y	Sp	GIN
Mahonia repens creeping Oregon-grape	eSS	s	FL	GMS	Y	Sp	EG
Mahonia trifoliata Laredo Oregon-grape	eMS	DS	L	GhS	GY	Sp	Hs

PLANT LIST	Plant Type	Env. Tol.	Aesthetic Value	Wildlife Value	Flower Color	Bloom Period	Landscape Uses
Malosma laurina laurel-sumac	MS	S	FL	GhLMmS	—	—	Bg
Malpighia glabra wild crape-myrtle	SS	Ss	Ff	U	PPu	SpSuF	Bg
Malus ioensis prairie crabapple	LU	S	FL	GLMmS	PW	Sp	B
Malvaviscus drummondii Texas wax-mallow	MS	Ss	F	U	R	WSpSuF	BFg
Mammillaria grahamii Graham's nipple cactus	C	DS	BFf	U	PPuW	—	GgMR
Mammillaria heyderi little nipple cactus	C	DS	BF	U	PPu	—	M
Mammillaria mainiae counter-clockwise nipple cactus	C	DS	BFf	U	RW	—	GgMR
Mammillaria pottsii rat-tail nipple cactus	C	DS	BFf	U	PuR	—	M
Mammillaria tetrancistra corkseed cactus	C	DS	BF	U	PPu	—	M
Mammillaria viridiflora green-flower nipple cactus	C	DS	BF	U	PPu	—	M
Mammillaria wrightii Wright's nipple cactus	C	DS	BF	U	P	—	M
Manilkara jaimiqui wild dilly	SU	DS	—	h	Y	SpF	BF
Maytenus phyllanthoides Florida mayten	eSU	DS	f	U	GW	WSp	BFs
Menispermum canadense Canadian moonseed	WV	FSs	AL	GLmS	Wyg	Su	s
Menodora scabra rough menodora	SS	DS	FL	M	Y	SpSuF	B
Menodora spinescens spiny menodora	SS	DS	—	U	W	Su	ABP
Menziesia ferruginea fool's-huckleberry	LS	FSs	FL	U	GPu	Su	BS
Menziesia pilosa minniebush	SS	DFSs	L	U	PW	SpSu	W
Metopium toxiferum Florida poisontree	SC	Ss	—	U	W	SpSu	B
Mimosa biuncifera cat-claw mimosa	MS	DS	F	GLMm	PW	Su	EH
Mimosa borealis fragrant mimosa	SS	S	F	GLMm	P	Sp	A
Mimosa dysocarpa velvet-pod mimosa	SS	DS	F	GLMm	P	Su	A
Mimosa pigra black mimosa	SS	F	F	GLMm	P	—	A
Mortonia sempervirens Rio Grande saddlebush	SS	DS	—	U	W	SpSuF	E
Morus microphylla Texas mulberry	SU	DS	f	GLMS	G	Sp	N

PLANT LIST	Plant Type	Env. Tol.	Aesthetic Value	Wildlife Value	Flower Color	Bloom Period	Landscape Uses
Morus rubra red mulberry	LU	DFSs	AfL	GLmS	yg	Sp	INS
Myrcianthes fragrans twinberry	eSU	DNaSs	FL	U	W	—	ABs
Myrica californica Pacific bayberry	eSU	FSs	BL	U	—	Sp	AsW
Myrica cerifera southern bayberry	eSU	FSsW	BL	GLMSW	yg	Sp	AFsW
Myrica gale sweetgale	eSS	FSW	L	GLMSW	—	—	AsW
Myrica heterophylla evergreen bayberry	eLS	FSsW	L	GLMSW	—	WSp	AsW
Myrica inodora odorless bayberry	eSU	DF	L	GLMSW	G	Sp	As
Myrica pensylvanica northern bayberry	MS	DFS	fL	GLMSW	yg	Sp	BEFNSW
Myrsine floridana guianese colicwood	eSU	DSs	L	U	RW	WSp	BFIs
Nemopanthus mucronatus catberry	MS	WFSs	f	U	Y	Sp	BIW
Nolina erumpens foothill bear-grass	eSS	DS	FL	U	PW	SpSu	AM
Nolina microcarpa sacahuista bear-grass	eSS	DS	L	U	PuW	SpSu	AM
Nolina texana Texas bear-grass	eSS	DS	L	U	W	SpSu	AM
Nyssa aquatica water tupelo	LC	FSW	L	GhMS	G	Sp	Wwg
Nyssa biflora swamp tupelo	LC	FSW	L	GhLMmSW	—	Sp	Wwg
Nyssa ogeche Ogeechee tupelo	MC	FSsW	f	GhLMmSW	—	Sp	SW
Nyssa sylvatica black tupelo	LC	DFSW	AL	GhLMmSW	W	Sp	ABISTW
Oemleria cerasiformis oso-berry	LS	—	—	S	G	Sp	—
Olneya tesota desert-ironwood	eLU	S	FL	HhMS	BPu	Su	AMs
Oplopanax horridus devil's-club	MS	Fs	—	U	—	Su	BPS
Opuntia acanthocarpa buck-horn cholla	C	DS	BF	GMmS	R	SpSu	AMP
Opuntia basilaris beaver-tail cactus	C	DS	BF	GLMmS	PR	Sp	gMR
Opuntia bigelovii teddy-bear cholla	C	DS	BF	GMmS	yg	SpSu	AMP
Opuntia chlorotica clock-face prickly-pear	C	DS	B	GMmS	Y	SpSu	M
Opuntia echinocarpa golden cholla	C	DS	BF	GMmS	Y	SpSu	AMP

PLANT LIST	Plant Type	Env. Tol.	Aesthetic Value	Wildlife Value	Flower Color	Bloom Period	Landscape Uses
Opuntia engelmannii cactus-apple	C	DS	B	GMmS	Y	Sp	AM
Opuntia erinacea oldman cactus	C	DS	BF	GMmS	RY	SpSu	AMP
Opuntia fragilis pygmy prickly-pear	C	DS	B	S	yg	SpSu	MR
Opuntia fulgida jumping cholla	C	DS	BFf	GMmS	PPuR W	Su	AM
Opuntia gosseliniana violet prickly-pear	C	DS	B	GLMmS	—	SpSu	M
Opuntia humifusa eastern prickly-pear	C	DS	BF	GLMmS	Y	SpSu	BgPRS
Opuntia imbricata tree cholla	C	DS	B	GLMmS	Pu	SpSu	M
Opuntia kleiniae candle cholla	C	DS	BF	GLMmS	Pu	SpSu	AM
Opuntia kunzei devil's cholla	C	DS	B	GLMmS	—	SpSu	M
Opuntia leptocaulis Christmas cholla	C	DS	BF	GLMmS	yg	Su	M
Opuntia littoralis coastal prickly-pear	C	DS	B	GLMmS	—	SpSu	M
Opuntia macrocentra purple prickly-pear	C	DS	B	GLMmS	Y	SpSu	AM
Opuntia macrorhiza twist-spine prickly-pear	C	DS	B	GLMmS	—	SpSu	M
Opuntia parishii matted cholla	C	DS	B	GLMmS	RY	SpSu	M
Opuntia phaeacantha tulip prickly-pear	C	DS	B	GLMmS	—	SpSu	M
Opuntia polyacantha hair-spine prickly-pear	C	DS	B	GLMmS	—	SpSu	M
Opuntia ramosissima darning-needle cholla	C	DS	BF	GLMmS	Pu	SpSu	AP
Opuntia santa-rita Santa Rita prickly-pear	C	DS	B	GLMmS	—	SpSu	M
Opuntia schottii dog cholla	C	DS	B	GLMmS	—	SpSu	M
Opuntia spinosior walkingstick cactus	C	DS	B	GLMmS	PPuW Y	Sp	M
Opuntia tetracantha Tucson prickly-pear	C	DS	B	GLMmS	—	SpSu	M
Opuntia tunicata thistle cholla	C	DS	BF	GLMmS	yg	SpSu	AM
Opuntia versicolor stag-horn cholla	C	DS	B	GLMmS	RY	SpSu	M
Opuntia whipplei rat-tail cholla	C	DS	B	GLMmS	—	SpSu	M
Osmanthus americanus devilwood	eSU	F	FfL	U	W	Sp	SWs

PLANT LIST	Plant Type	Env. Tol.	Aesthetic Value	Wildlife Value	Flower Color	Bloom Period	Landscape Uses
Ostrya virginiana eastern hop-hornbeam	LU	DSs	AL	GmS	R	Sp	AFI
Oxalis frutescens shrubby wood-sorrel	S	—	—	U	Y	SpSuF	—
Oxydendron arboreum sourwood	LU	Ss	AFL	GmS	W	Su	A
Pachycereus schottii senita	C	DS	B	U	—	—	M
Pachystima myrsinites myrtle boxleaf	eSS	S	F	Gm	G	SpSu	BG
Parkinsonia aculeata Jerusalem-thorn	SU	NaS	BL	hMm	Y	SpSuF	AH
Parkinsonia florida blue palo-verde	SU	DS	B	hMm	Y	SpSu	A
Parkinsonia microphylla yellow palo-verde	MS	DS	BF	hMm	Y	Sp	A
Parkinsonia texana Texas palo-verde	SU	DS	B	hMm	Y	SpSu	A
Parthenium argentatum guayule	SS	DS	—	U	W	—	M
Parthenium incanum mariola	SS	DS	—	U	W	—	M
Parthenocissus quinquefolia Virginia-creeper	WV	DFSs	ALF	GLmS	W	Su	BIW
Parthenocissus vitacea thicket-creeper	WV	DSs	AL	GLS	G	Su	R
Peniocereus greggii night-blooming-cereus	C	DS	B	GS	W	Su	A
Pentaphylloides floribunda golden-hardhack	SS	DFS	FL	GLMmS	Y	Su	BFgHW
Persea borbonia red bay	eLU	FS	L	S	Y	Sp	sW
Persea humilis silk bay	eSU	DS	L	S	Y	Sp	Is
Persea palustris swamp bay	eSU	FSW	L	S	—	Sp	sWwg
Petradoria pumila grassy rock-goldenrod	P	DS	F	U	Y	Su	M
Petrophyton caespitosum rockmat	SS	DS	—	U	W	Sp	gMR
Peucephyllum schottii Schott's pygmy-cedar	MS	DS	—	U	Y	—	M
Phaulothamnus spinescens devilqueen	MS	—	—	U	—	F	—
Philadelphus inodorus scentless mock orange	SS	Ss	F	m	W	Sp	BS
Philadelphus lewisii Lewis' mock orange	MS	Ss	F	m	W	Su	AB
Philadelphus microphyllus little-leaf mock orange	SS	DS	FL	m	W	Su	AM

PLANT LIST	Plant Type	Env. Tol.	Aesthetic Value	Wildlife Value	Flower Color	Bloom Period	Landscape Uses
Phyllodoce breweri red mountain-heath	SS	Ss	FL	u	P	Su	BGI
Phyllodoce empetriformis pink mountain-heath	eSS	SW	FL	u	R	—	GgsW
Physocarpus capitatus Pacific ninebark	MS	DFSW	BF	GmW	W	—	BMSs
Physocarpus malvaceus mallow-leaf ninebark	MS	DSs	L	M	W	Su	BM
Physocarpus monogynus mountain ninebark	SS	DS	FL	Gm	PW	Su	BMs
Physocarpus opulifolius Atlantic ninebark	MS	DFS	Bf	GmW	W	SpSu	BMSs
Picea engelmannii Engelmann's spruce	cLC	Ss	L	MmS	PuR	Su	A
Picea glauca white spruce	cSC	FSs	L	GLMmS	Pu	Sp	AHPSs
Picea mariana black spruce	cSC	FSs	L	GLMmS	R	Su	SsW
Picea pungens blue spruce	cLC	FS	L	GLMmS	yg	Sp	AHPs
Picea rubens red spruce	cSC	S	L	GLMmS	R	—	AMs
Picea sitchensis sitka spruce	cLC	FSs	L	GLMmS	R	—	As
Pickeringia montana stingaree-bush	eMS	DS	F	U	Pu	SpSu	Ps
Pinus albicaulis white-bark pine	cSC	—	L	GMmS	X	X	s
Pinus aristata bristle-cone pine	cLU	DS	BL	GMmS	X	X	AFs
Pinus attenuata knob-cone pine	cSU	DS	BL	GLMmS	X	X	Is
Pinus balfouriana fox-tail pine	cSC	—	L	GLMmS	X	X	I
Pinus banksiana jack pine	cSC	DS	L	GMmS	X	X	Ps
Pinus cembroides Mexican pinyon	cLU	S	L	GLMmS	X	X	As
Pinus clausa sand pine	cMC	DS	L	GMmS	X	X	s
Pinus contorta lodgepole pine	cSC	DNaS	B	GLMmS	X	X	Es
Pinus coulteri Coulter's pine	cLC	DSs	BL	GLMmS	X	X	ABIsT
Pinus echinata short-leaf pine	cLC	DS	BL	GLMmS	X	X	gTsW
Pinus edulis two-needle pinyon	cLS	DS	BLN	GLMmS	X	X	AENs
Pinus elliottii slash pine	cSU	DSW	L	GLMmS	X	X	AsWwg

PLANT LIST	Plant Type	Env. Tol.	Aesthetic Value	Wildlife Value	Flower Color	Bloom Period	Landscape Uses
Pinus engelmannii Apache pine	cSC	S	L	GMmS	X	X	s
Pinus flexilis limber pine	cLS	DS	BL	GLMmS	X	X	Es
Pinus glabra spruce pine	cLC	FS	L	GMmS	X	X	BsT
Pinus jeffreyi Jeffrey pine	cLC	DSs	L	GLMmS	X	X	AIsT
Pinus lambertiana sugar pine	cLC	DS	BL	GLMmS	X	X	AIsT
Pinus leiophylla Chihuahuan pine	cSC	DS	—	GMmS	X	X	s
Pinus longaeva Intermountain bristle-cone pine	cSC	DS	BL	GMmS	X	X	ABs
Pinus monophylla single-leaf pinyon	cLS	DS	L	GLMmS	X	X	BNs
Pinus monticola western white pine	cLC	DSs	BL	GLMmS	X	X	BIsT
Pinus muricata Bishop pine	cLU	Na	L	GLMmS	X	X	Bs
Pinus palustris long-leaf pine	cLC	FS	L	GLMmS	X	X	BsT
Pinus ponderosa ponderosa pine	cLC	DSs	BL	GLMmS	X	X	BIsT
Pinus pungens Table Mountain pine	cSC	DS	L	GMmS	X	X	ABs
Pinus quadrifolia four-leaf pinyon	cSU	DS	L	GLMmS	X	X	BNs
Pinus radiata Monterey pine	cSC	D	L	GMmS	X	X	ABbsT
Pinus remota paper-shell pinyon	cLC	S	BfL	GLMmS	X	X	ABNsT
Pinus resinosa red pine	cLC	DSs	BL	GLMmS	X	X	BPsT
Pinus rigida pitch pine	cSC	DS	L	GLMmS	X	X	BEIs
Pinus sabiniana digger pine	cSC	DS	f	GLMmS	X	X	BNs
Pinus serotina pond pine	cLU	F	L	GLMmS	X	X	BsW
Pinus strobiformis southwestern white pine	cLC	DS	L	GMmS	X	X	BsT
Pinus strobus eastern white pine	cLC	Ss	BLf	GLMmS	X	X	ABCIsT
Pinus taeda loblolly pine	cLC	S	L	GLMmS	X	X	BCSsT
Pinus torreyana Torrey pine	cSU	DS	BfL	GMmS	X	X	ABs
Pinus virginiana Virginia pine	cSU	DS	L	GLMmS	X	X	BCs

PLANT LIST	Plant Type	Env. Tol.	Aesthetic Value	Wildlife Value	Flower Color	Bloom Period	Landscape Uses
Piscidia piscipula Florida fishpoison-tree	SC	S	—	U	PW	Su	ABs
Pisonia aculeata devil's-claw pisonia	SS	S	—	U	PuY	SpSu	B
Pisonia rotundata smooth devil's-claws	LS	DS	—	U	G	SpSu	I
Pithecellobium ebano ebony blackbead	eSU	DS	L	h	W	SpSuF	BHSs
Pithecellobium keyense Florida Keys blackbead	eSU	DS	F	U	PW	WSuSpF	Bs
Pithecellobium unguis-cati cat-claw blackbead	SU	FSs	F	U	P	SpF	BP
Planera aquatica planertree	SU	FSW	B	MW	GY	Sp	SWwg
Platanus occidentalis American sycamore	LC	DFS	BLf	LMS	yg	Sp	ABIST
Platanus racemosa California sycamore	LC	FS	BFL	LS	BrY	Sp	AST
Platanus wrightii Arizona sycamore	SC	FS	BFL	LS	R	Sp	AT
Pluchea sericea arrow-weed	SS	FNaS	BF	BhM	PuW	SpSu	gS
Poliomintha incana hoary rosemary-mint	SS	DNaS	L	U	PuW	SpSuF	AB
Populus angustifolia narrow-leaf cottonwood	LU	FS	L	BLM	Br	Sp	AT
Populus balsamifera balsam poplar	LC	FSW	L	BCGLMmS	G	Sp	SW
Populus deltoides eastern cottonwood	LC	DFS	ABL	BCGLMmSW	R	Sp	ABCESTW
Populus fremontii Fremont's cottonwood	LC	FS	BL	BGLMmS	G	Sp	AST
Populus grandidentata big-tooth aspen	SC	S	AL	BCGLMmS	G	Sp	AB
Populus heterophylla swamp cottonwood	LC	FSW	AL	BCGLMmS	G	Sp	ASWwg
Populus tremuloides quaking aspen	LU	S	AL	BCGLMmS	G	Sp	ABS
Prosopis glandulosa honey mesquite	SU	S	L	BGLMmS	yg	SuF	ABT
Prosopis pubescens American screw-bean	LU	SF	FL	BGLMmS	W	Su	ABT
Prosopis velutina velvet mesquite	SU	S	L	BGLMmS	yg	SuF	AB
Prunus americana American plum	SU	S	AFf	GLMmS	W	Sp	A
Prunus angustifolia chickasaw plum	LS	S	Ff	GLMmS	W	Sp	B
Prunus caroliniana Carolina laurel cherry	eLU	S	ABFfL	GLMmS	W	Sp	BFHs

PLANT LIST	Plant Type	Env. Tol.	Aesthetic Value	Wildlife Value	Flower Color	Bloom Period	Landscape Uses
Prunus emarginata bitter cherry	LS	FS	F	GLMmS	OW	Sp	BS
Prunus fasciculata desert almond	SS	DS	F	GLMmS	W	Sp	BM
Prunus ilicifolia holly-leaf cherry	eSU	DS	fL	GLMmS	W	Sp	ABgPs
Prunus mexicana bigtree plum	LU	FS	FL	GLMmS	W	Sp	AB
Prunus myrtifolia West Indian cherry	eSU	Ss	F	GLMmS	W	WSp	ABFIs
Prunus pensylvanica fire cherry	SU	DS	ABFfL	BGLMmS	W	Sp	BN
Prunus pumila Great Lakes sand cherry	SS	FW	AfL	BGLMmS	W	Sp	ENS
Prunus serotina black cherry	SC	DFS	ABFfL	BGLMmS	W	Sp	BNIW
Prunus texana peachbush	SS	—	FL	GLMmS	W	Sp	AB
Prunus virginiana choke cherry	LU	DSs	ABFfL	BGLMmS	W	Sp	BN
Pseudophoenix sargentii Florida cherry palm	SU	S	FL	mS	Y	WSpSuF	ABFI
Pseudotsuga macrocarpa big-cone douglas-fir	cLC	s	L	GLMmS	yg	—	ABsT
Pseudotsuga menziesii Douglas-fir	cLC	S	L	GLMmS	OR	SpSu	ABIsT
Psidium longipes mangroveberry	eLS	DSs	F	U	W	SuF	Is
Psorothamnus arborescens Mojave smokebush	MS	DSs	F	m	B	—	BMP
Psorothamnus emoryi Emory's smokebush	SS	DS	F	m	Pu	Sp	BP
Psorothamnus fremontii Fremont's smokebush	SS	DS	F	m	Pu	SpSu	BP
Psorothamnus schottii indigo bush	S	FSs	FL	m	B	Sp	BPS
Psorothamnus spinosus smokethorn	SU	FS	BF	m	BPu	Su	BPS
Psychotria ligustrifolia Bahama wild coffee	SU	s	f	U	W	SpSu	I
Psychotria nervosa Seminole balsamo	SS	s	L	U	W	WSpSuF	BFgI
Ptelea trifoliata common hoptree	SU	Ss	L	U	W	SpSu	AB
Purshia glandulosa antelope-brush	SS	DS	B	Mm	W	Su	AB
Purshia mexicana Mexican cliff-rose	eSS	DS	Ff	Mm	Y	Su	ABFR
Purshia tridentata bitterbrush	SS	DS	F	LMm	Y	Su	B

PLANT LIST	Plant Type	Env. Tol.	Aesthetic Value	Wildlife Value	Flower Color	Bloom Period	Landscape Uses
Pyrularia pubera buffalo-nut	MS	s	—	U	BrG	Sp	gI
Quercus agrifolia coastal live oak	eSC	DS	BL	GLMmS W	—	—	AFTs
Quercus alba northern white oak	LC	DSs	ABL	CGMmS	yg	Sp	AINT
Quercus arizonica Arizona white oak	SC	S	L	CGMmS	Y	—	AT
Quercus austrina bluff oak	LC	S	L	CGMmS W	Y	Sp	ABIT
Quercus bicolor swamp white oak	LC	DFSs W	AL	CGMmS W	yg	Sp	INSWwg
Quercus buckleyi Buckley's oak	SU	—	—	CGMmS	Y	Sp	—
Quercus chapmanii Chapman's oak	SC	DSs	BL	CGMmS W	yg	Sp	BIT
Quercus chrysolepis canyon live oak	eLC	DSs	BL	GLMmS W	RY	Su	ABIT
Quercus coccinea scarlet oak	SC	DS	AL	CGMmS	yg	Sp	ABIT
Quercus douglasii blue oak	SC	DS	BL	GLMmS W	yg	Sp	ABT
Quercus dumosa California scrub oak	eSS	DS	L	GLMmS W	yg	Sp	BM
Quercus durata leather oak	SS	S	L	GLMmS W	yg	Sp	B
Quercus ellipsoidalis northern pin oak	SC	DS	AL	CGMmS	yg	Sp	BT
Quercus emoryi Emory's oak	eSC	DS	BL	CGMmS	Y	Sp	ABNT
Quercus engelmannii Engelmann's oak	eLU	DS	L	GLMmS W	yg	Sp	ABET
Quercus falcata southern red oak	SC	DS	BL	CGMmS	R	Sp	ABT
Quercus fusiformis plateau oak	eLU	DS	L	GMmS	yg	Sp	AB
Quercus gambelii Gambel's oak	SU	DS	A	CGMmS W	Br	Sp	AB
Quercus garryana Oregon white oak	LC	FS	BL	CGMmS W	yg	Sp	BIST
Quercus geminata sand live oak	SC	DS	—	CGMmS W	Br	Sp	BI
Quercus gravesii Graves' oak	LU	S	AL	GMmS	Br	Sp	AT
Quercus grisea gray oak	SC	DS	L	CGMmS	yg	Sp	B
Quercus havardii Havard's oak	SS	DS	DS	GLS	yg	Sp	BM
Quercus hemisphaerica Darlington's oak	eSC	FS	L	CGMmS W	yg	Sp	ABSsTW

PLANT LIST	Plant Type	Env. Tol.	Aesthetic Value	Wildlife Value	Flower Color	Bloom Period	Landscape Uses
Quercus hypoleucoides silver-leaf oak	eSU	S	BL	GLS	RG	Sp	ABs
Quercus ilicifolia bear oak	LS	DS	AL	CGLmSW	yg	Sp	BEN
Quercus imbricaria shingle oak	LU	DFSs	AL	CGMmSW	yg	Sp	ABIT
Quercus incana bluejack oak	LU	DS	L	CGMmSW	RY	Sp	BT
Quercus inopina sandhill oak	LS	DS	L	CGMmSW	Br	Sp	BI
Quercus kelloggii California black oak	SC	DS	AL	GLMmSW	yg	Sp	ABT
Quercus laceyi Lacey's oak	LU	DS	AL	GLMmS	yg	Sp	ABT
Quercus laevis turkey oak	SC	DS	L	CGMmSW	R	Sp	BT
Quercus laurifolia laurel oak	SC	FSW	L	CMmSW	R	Sp	ASTW
Quercus lobata valley oak	LC	S	BL	GLMmSW	Y	Sp	ABT
Quercus lyrata overcup oak	SC	FSW	L	CMmSW	Y	Sp	SW
Quercus macrocarpa burr oak	LC	DFS	BL	CGmSW	yg	Sp	ABIT
Quercus margarettiae sand post oak	MS	DFS	BL	CGMmSW	yg	Sp	AI
Quercus marilandica blackjack oak	LU	DFS	L	CGMmSW	yg	Sp	AB
Quercus michauxii swamp chestnut oak	LC	S	L	CGMmSW	yg	Sp	AIT
Quercus minima dwarf live oak	MS	DS	L	CGMmSW	yg	Sp	ABIS
Quercus mohriana Mohr's oak	LS	DS	BL	CGMmS	yg	Sp	AB
Quercus muehlenbergii chinkapin oak	LC	DS	L	CGMmSW	yg	Sp	ABT
Quercus myrtifolia myrtle oak	eSU	DS	L	CGMmSW	R	Sp	BIs
Quercus nigra water oak	SC	FSW	L	CGMmSW	Br	Sp	ASTwg
Quercus oblongifolia Mexican blue oak	eSC	DS	B	CGMmS	yg	Sp	ABTs
Quercus pagoda cherry-bark oak	LC	FSs	BL	CGMmSW	yg	Sp	AIW
Quercus palustris pin oak	LC	FSW	AL	CGMmSW	yg	Sp	AITWwg
Quercus phellos willow oak	SC	DFSW	AL	CGMmSW	yg	Sp	AITwg
Quercus prinoides dwarf chinkapin oak	LS	DS	AL	CGLmSW	yg	Sp	B

PLANT LIST	Plant Type	Env. Tol.	Aesthetic Value	Wildlife Value	Flower Color	Bloom Period	Landscape Uses
Quercus prinus chestnut oak	SC	DSs	AL	CGLMm SW	yg	Sp	ABIT
Quercus pumila running oak	SS	DS	—	CGLMm SW	Y	Sp	BI
Quercus pungens sandpaper oak	eSU	DS	L	CGLMm S	yg	Sp	AT
Quercus rubra northern red oak	LC	Ss	ABL	CGMmS W	yg	Sp	ABIT
Quercus rugosa net-leaf oak	eSU	S	L	CGLMm S	yg	Sp	AB
Quercus sadleriana deer oak	eMS	Ss	L	CGLMm S	yg	Sp	AB
Quercus shumardii Shumard's oak	SC	DS	AL	CGmSW	yg	Sp	STW
Quercus sinuata bastard oak	MC	Fs	L	CGMm W	yg	Sp	S
Quercus stellata post oak	LU	DS	L	CGMmS W	yg	Sp	AB
Quercus texana Texas red oak	SU	DS	AL	CGLMm S	R	Sp	ABT
Quercus toumeyi Toumey's oak	eSU	DS	L	CGLMm S	yg	Sp	B
Quercus turbinella shrub live oak	eLS	DS	L	GLMmS W	Y	Sp	AB
Quercus vacciniifolia huckleberry oak	eSS	S	L	GLMmS W	yg	Sp	BM
Quercus velutina black oak	LC	DSs	AL	CGMmS W	yg	Sp	BT
Quercus virginiana live oak	eSC	DFS	BL	CGMmS W	Y	Sp	AITs
Quercus wislizenii interior live oak	eSC	S	BL	GLMmS W	yg	—	ABTs
Randia aculeata white indigo-berry	eSC	DFSs	f	U	W	Su	ABFIs
Reynosia septentrionalis darling-plum	eSU	DS	—	U	yg	SpSu	BFIs
Rhamnus alnifolia alder-leaf buckthorn	SS	FS	L	hS	GW	—	W
Rhamnus crocea holly-leaf buckthorn	eLS	DS	fL	GLMmS	GY	Sp	Bs
Rhapidophyllum hystrix needle palm	eMS	Fs	L	U	W	Sp	Ws
Rhizophora mangle red mangrove	eSC	SsW	L	W	Y	WSpSu F	SsW
Rhododendron albiflorum Cascade azalea	SS	Fs	FL	GLMm	W	—	ABIS
Rhododendron calendulaceum flame azalea	MS	Ss	AFL	BHGLM mS	O	SpSu	ABI
Rhododendron canescens mountain azalea	LS	Ss	FL	bGHMm S	PW	Sp	BIS

PLANT LIST	Plant Type	Env. Tol.	Aesthetic Value	Wildlife Value	Flower Color	Bloom Period	Landscape Uses
Rhododendron catawbiense catawba rosebay	eLS	DS	FL	GMmS	PPu	SpSu	ABs
Rhododendron macrophyllum California rhododendron	eSU	Ss	FL	GMmS	PPu	Sp	ABs
Rhododendron maximum great-laurel	eSU	FSs	FL	BHmS	P	Su	BIW
Rhododendron periclymenoides pink azalea	MS	DFSs	FL	BGHmS	PuW	Sp	BISW
Rhododendron vaseyi pink-shell azalea	LS	FSs	AFL	BGHLMmS	P	Sp	ABIS
Rhododendron viscosum clammy azalea	MS	FSs	FL	BGHLMmW	W	Su	ABW
Rhus aromatica fragrant sumac	MS	DS	AfL	GLMmS	Y	Sp	ABGN
Rhus copallinum winged sumac	SU	DS	AfL	GLmS	yg	Su	ABg
Rhus glabra smooth sumac	MS	S	AfL	GLMmS	G	Su	ABCEP
Rhus integrifolia lemonade sumac	eMS	NaS	AFL	GLMmS	W	Sp	B
Rhus microphylla little-leaf sumac	MS	DS	A	Mm	GW	Sp	AB
Rhus ovata sugar sumac	eMS	DS	L	GLMmS	—	Sp	E
Rhus trilobata ill-scented sumac	MS	S	AfL	mS	G	Sp	AH
Rhus virens evergreen sumac	eMS	S	FfL	GMmS	W	Su	bH
Ribes amarum bitter gooseberry	MS	DS	L	GLMmSW	Pu	Su	BNP
Ribes americanum wild black currant	SS	FSs	AL	GLMmSW	W	Sp	ABNW
Ribes aureum golden currant	MS	Fs	FfL	GLMmS	Y	Su	EISN
Ribes californicum California gooseberry	SS	DS	L	GLMmSW	W	Su	ABPN
Ribes cereum white squaw currant	SS	S	F	GLMmSW	PW	Su	ABN
Ribes cynosbati eastern prickly gooseberry	SS	DFSs	AL	GLMmSW	GW	Sp	ABHIPNW
Ribes glandulosum skunk currant	SS	FS	L	GLMmSW	—	Sp	AINW
Ribes hirtellum hairy-stem gooseberry	SS	FS	L	GLMmSW	—	Sp	IN
Ribes indecorum white-flower currant	S	—	—	GLMmSW	W	WSpF	—
Ribes lacustre bristly black gooseberry	SS	FS	L	GLMmSW	G	Sp	ANPW
Ribes malvaceum chaparral currant	SS	Ss	L	GLMmSW	P	WSp	ABINP

PLANT LIST	Plant Type	Env. Tol.	Aesthetic Value	Wildlife Value	Flower Color	Bloom Period	Landscape Uses
Ribes missouriense Missouri gooseberry	SS	Ss	f	GLMS	GW	Sp	AINP
Ribes montigenum western prickly gooseberry	SS	DS	f	GLMmS W	R	Su	ANP
Ribes pinetorum orange gooseberry	MS	s	f	GLMmS W	OR	Su	AFNP
Ribes roezlii Sierran gooseberry	SS	S	Ff	GLMmS W	R	Su	NP
Ribes velutinum desert gooseberry	SS	Ss	BfL	GLMmS	Y	Sp	ABIP
Ribes viburnifolium Santa Catalina currant	eSS	DS	L	GLMmS W	P	Su	N
Ribes viscosissimum sticky currant	SS	s	B	GLMmS W	W	Su	BN
Ribes wolfii winaha currant	MS	s	L	GLMmS	PW	Su	IN
Rivina humilis rougeplant	SS	s	L	M	PW	—	BI
Robinia hispida bristly locust	SS	DS	F	mS	Pu	Sp	ABE
Robinia neomexicana New Mexico locust	SU	S	F	GMm	PW	Su	AE
Robinia pseudo-acacia black locust	LC	S	L	mS	W	Su	E
Rosa arkansana prairie rose	SS	DS	F	BGLMmS	P	SpSu	bHM
Rosa blanda smooth rose	SS	S	Ff	BGLMmS	P	Sp	bHM
Rosa californica California rose	MS	S	Ff	BGLMmS	P	Sp	Bbg
Rosa carolina Carolina rose	SS	DS	Ff	BGLMmS	PW	Su	BbEFG
Rosa gymnocarpa wood rose	SS	s	F	GLMmS	P	SpSu	HIS
Rosa nutkana Nootka rose	MS	S	FL	BGLMmS	P	Sp	Bbg
Rosa palustris swamp rose	SS	FSW	F	BGLMmS	P	Su	SWwg
Rosa woodsii Woods' rose	SS	DS	F	BGLMmS	P	Su	AbHM
Roystonea elata Florida royal palm	eSC	S	L	mS	W	—	A
Rubus allegheniensis allegheny blackberry	MS	Ds	FfL	GLMmS	W	SpSu	NP
Rubus argutus saw-tooth blackberry	SS	D	FfL	GLMmS	W	Sp	NP
Rubus canadensis smooth blackberry	SS	S	FfL	GLMmS	W	Su	NP
Rubus chamaemorus cloudberry	SS	FS	FfL	BGLMmSW	W	SpSu	GNSW

PLANT LIST	Plant Type	Env. Tol.	Aesthetic Value	Wildlife Value	Flower Color	Bloom Period	Landscape Uses
Rubus cuneifolius sand blackberry	SS	F	FfL	BGLMmS	W	SpSu	NP
Rubus flagellaris whiplash dewberry	SS	D	FfL	BGLMmS	W	Su	NP
Rubus idaeus common red raspberry	MS	DFSs	AFfL	BCGLMmSW	W	SpSu	BgINW
Rubus lasiococcus hairy-fruit smooth dewberry	SS	DS	FfL	GLMmS	W	SpSu	BGIN
Rubus leucodermis white-stem raspberry	SS	Ss	FfL	GLMmS	W	Sp	BN
Rubus occidentalis black raspberry	SS	DFSs	FfL	BGLMmS	W	Sp	BIN
Rubus parviflorus western thimble-berry	SS	DSs	FfL	BGLMmS	W	SpSu	BN
Rubus pedatus strawberry-leaf raspberry	SS	Ss	FfL	BGLMmS	W	SpSu	GN
Rubus pubescens dwarf red raspberry	eSS	FS	FfL	BGLMmSW	W	SpSu	INW
Rubus spectabilis salmon raspberry	SS	FsW	BFfL	S	PR	Sp	BNSW
Rubus trivialis southern dewberry	MS	Ss	FfL	GLMmS	W	Sp	N
Rubus ursinis California Dewberry	SS	Ss	FfL	GLMmS	W	SpSu	MNP
Rubus vitifolius Pacific dewberry	eSS	FS	FfL	GLMmS	W	Sp	NS
Sabal etonia scrub palmetto	eMS	DS	L	mS	W	SpSu	AB
Sabal mexicana Rio Grande palmetto	eLU	S	L	mS	W	Su	AB
Sabal minor dwarf palmetto	LU	FSW	BL	LMS	W	Su	ABIW
Sabal palmetto cabbage palmetto	LC	DFSW	BL	LMS	W	SpSu	ABT
Sageretia wrightii Wright's mock buckthorn	MS	DS	—	U	W	SpSuF	B
Salazaria mexicana Mexican bladder-sage	SS	DFS	Ff	M	BPu	Sp	BgP
Salix amygdaloides peach-leaf willow	LU	FS	L	GLMmSW	yg	Sp	SW
Salix arctica stout arctic willow	SS	S	L	U	G	Sp	B
Salix bebbiana long-beak willow	SU	DFS	F	GLMmSW	yg	Sp	SW
Salix bonplandiana Bonpland's willow	eLS	FS	L	GMmS	GrY	—	S
Salix candida sage willow	MS	FSsW	BL	BGhLMmS	GW	SpSu	BSW
Salix caroliniana coastal-plain willow	SU	FSW	L	GLMmS	Y	Su	SW

PLANT LIST	Plant Type	Env. Tol.	Aesthetic Value	Wildlife Value	Flower Color	Bloom Period	Landscape Uses
Salix discolor tall pussy willow	SU	DFS	LF	BGhLM mSW	Gr	Sp	gSW
Salix drummondiana Drummond's willow	MS	FSW	L	GLMmS	yg	Sp	SW
Salix eriocephala Missouri willow	LS	FS	—	GLMmS	yg	Sp	SW
Salix exigua sandbar willow	SU	FNaS	L	GLMmS	Y	Sp	SW
Salix geyeriana Geyer's willow	LS	FSW	L	GLMmS	Y	Sp	SW
Salix gooddingii Goodding's willow	LU	FS	—	GLMmS	Y	Sp	ST
Salix hindsiana sandbar willow	LS	FS	—	GLMmS	yg	Sp	ES
Salix hookeriana coastal willow	LS	FNa	FL	GLMmS	Y	Sp	SW
Salix humilis small pussy willow	MS	DFS	L	CGhLM mSW	yg	Sp	gMW
Salix interior sandbar willow	LS	FSW	L	GLMmS W	Y	Sp	CESW
Salix lasiolepis arroyo willow	LS	FS	BL	GLMmS	—	WSp	S
Salix lucida shining willow	LS	DFS	L	GLMmS W	Gr	Sp	BSW
Salix lutea yellow willow	MS	DFS	—	GLMmS	—	Sp	S
Salix melanopsis dusky willow	LS	FS	—	GLMmS	—	Sp	S
Salix monticola mountain willow	LU	FW	L	GLMmS	Y	Sp	W
Salix nigra black willow	LU	DFSW	L	BCGhL MmSW	yg	Sp	ABCESW
Salix petiolaris meadow willow	LS	FS	L	GLMmS	—	Sp	SW
Salix scouleriana Scouler's willow	LU	s	L	GLMmS	—	Sp	I
Salix sericea silky willow	LS	FS	L	GLMmS	—	Sp	ABS
Salix sessilifolia sessile-leaf willow	SU	FS	L	GLMmS	Y	Sp	S
Salix sitchensis sitka willow	SU	FS	L	GLMmS	yg	SP	ST
Salvia apiana California white sage	MS	DS	FL	GS	W	—	BgM
Salvia dorrii gray ball sage	SS	DSs	FL	GS	BPu	SpSuF	BgM
Salvia funerea Death Valley sage	SS	DS	F	GHS	Pu	Sp	gM
Salvia leucophylla San Luis purple sage	SS	DS	F	GS	Pu	Sp	BgM

PLANT LIST	Plant Type	Env. Tol.	Aesthetic Value	Wildlife Value	Flower Color	Bloom Period	Landscape Uses
Salvia mellifera California black sage	SS	DS	F	GhS	W	Sp	BgM
Salvia mohavensis Mojave sage	SS	DS	F	GHS	B	SpSu	gM
Salvia pachyphylla rose sage	SS	DS	F	GHS	—	Su	E
Salvia vaseyi bristle sage	SS	DS	F	GHS	W	Sp	gM
Sambucus canadensis American elder	MS	DFSs	FfL	GLMmSW	W	SuF	BNW
Sambucus cerulea blue elder	SU	FS	FfL	GLMmS	W	SpSuF	A
Sambucus racemosa European red elder	LS	FSsW	FfL	GLMmSW	W	SuF	BSW
Sapindus saponaria wing-leaf soapberry	SC	FS	AF	S	W	SpSu	AST
Sapium biloculare Mexican jumping-bean	SU	DFS	F	U	G	SpSuF	MS
Sarcobatus vermiculatus greasewood	MS	NaS	B	LMm	G	SuF	EM
Sassafras albidum sassafras	LU	D	AL	GLMS	Y	Sp	AB
Savia bahamensis Bahama-maidenbush	SU	Ss	—	U	—	SpF	I
Scaevola plumieri gullfeed	eLS	S	F	U	W	WSpSuF	E
Schaefferia cuneifolia desert-yaupon	eSS	S	f	U	GY	Su	AH
Schaefferia frutescens Florida boxwood	SU	—	—	U	—	SpSu	I
Schoepfia chrysophylloides island beefwood	SU	Ss	—	U	GR	FW	I
Sclerocactus erectocentrus red-spine fish-hook cactus	C	DS	BF	mS	PPuWY	—	gM
Sclerocactus intertextus white fish-hook cactus	C	DS	BF	mS	PPuWY	—	gM
Sclerocactus scheeri Scheer's fish-hook cactus	C	DS	BF	mS	PPuWY	—	gM
Sclerocactus uncinatus Chihuahuan fish-hook cactus	C	DS	BF	mS	PPuWY	—	gM
Sebastiania fruticosa Gulf sebastian-bush	SS	F	A	U	yg	SpSuF	AS
Senna armata desert wild sensitive plant	SS	FS	F	U	PY	Su	MS
Sequoia sempervirens redwood	cLC	F	B	mS	X	X	BITs
Serenoa repens saw-palmetto	SS	DFNaS	—	hmS	W	SpSu	W
Serjania brachycarpa little-fruit slipplejack	WV	Ss	—	U	GWY	WSp	B

PLANT LIST	Plant Type	Env. Tol.	Aesthetic Value	Wildlife Value	Flower Color	Bloom Period	Landscape Uses
Shepherdia canadensis russet buffalo-berry	MS	DS	f	GLmS	Y	Sp	E
Sideroxylon celastrinum saffron-plum	SU	FS	—	U	yg	WF	W
Sideroxylon foetidissimum false mastic	eSC	DSs	B	U	Y	WSpSuF	ABIs
Sideroxylon lanuginosum gum bully	SU	DSs	—	U	Y	Su	ABI
Sideroxylon reclinatum Florida bully	LS	FS	—	U	—	SpSu	B
Sideroxylon salicifolium white bully	SC	DS	—	U	Y	WSp	B
Sideroxylon tenax tough bully	SU	DS	—	U	W	Sp	ABI
Simarouba glauca paradise-tree	SC	DSs	F	U	yg	Sp	ABI
Simmondsia chinensis jojoba	MS	DS	—	G	—	WSp	BM
Smilax auriculata ear-leaf greenbrier	WV	DS	—	GLMmS	yg	Su	I
Smilax bona-nox fringed greenbrier	WV	DFSW	—	GLMmSW	G	SpSu	BIW
Smilax glauca sawbrier	WV	Ss	—	GLMS	G	Su	BI
Smilax laurifolia laurel-leaf greenbrier	eWV	FSW	L	GLMmSW	GW	Su	BW
Smilax pumila sarsparilla-vine	WV	DS	—	GLMmSW	G	Su	BIM
Smilax rotundifolia horsebrier	WV	DFSs	fL	GLMmSW	Ryg	SpSu	BIPW
Smilax tamnoides chinaroot	WV	Ss	L	GLMmSW	G	Sp	BI
Smilax walteri coral greenbrier	WV	F	L	MmS	G	SpSu	BI
Solanum donianum mullien nightshade	SU	—	—	U	B	WSpSuF	—
Solanum dulcamara climbing nightshade	MS	S	F	GLmSW	G	SpSu	B
Solanum erianthum potato-tree	MS	Ss	FL	GLmS	W	SpSuF	B
Sophora affinis Eve's mecklacepod	LS	Ss	—	U	PW	—	—
Sophora secundiflora mescal-bean	eLS	Ss	FL	U	BPu	—	—
Sophora tomentosa yellow necklacepod	eMS	DFS	F	U	Y	WSpSuF	gMs
Sorbus americana American mountain-ash	SU	FS	FfL	GLMmS	W	Sp	W
Sorbus decora northern mountain-ash	SU	FS	ABfL	GLMmSW	W	Sp	BW

PLANT LIST	Plant Type	Env. Tol.	Aesthetic Value	Wildlife Value	Flower Color	Bloom Period	Landscape Uses
Sorbus scopulina Cascade Mountain-ash	SS	DS	fL	GLMmS	W	Su	B
Sorbus sitchensis Sitka mountain-ash	MS	DFSW	FfL	GLMmS	W	SU	BW
Sphaeralcea ambigua apricot globe-mallow	SS	DS	F	Mm	OR	SpSuF	BgM
Sphaeralcea incana soft globe-mallow	SS	DS	F	Mm	OR	SpSuF	BgM
Sphaeralcea rosea desert mallow	SS	DS	F	Mm	P	SpSu	BgM
Spiraea alba white meadowsweet	SS	SW	BF	BGM	W	SU	ABWwg
Spiraea douglasii Douglas' meadowsweet	MS	FS	B	GM	P	Su	A
Spiraea tomentosa steeplebush	SS	FSW	Ff	BGMmS W	PPu	Su	BWwg
Staphylea trifolia American bladdernut	MS	FSs	ABL	U	W	Sp	BISW
Stenocereus thurberi organ-pipe cactus	C	DS	BF	mS	PuR	Sp	AMPs
Stewartia malacodendron silky-camellia	SU	s	BF	U	W	SpSu	ABI
Styrax grandifolius big-leaf snowbell	LS	FSs	F	W	W	Sp	AgIS
Suaeda californica broom seepweed	SS	FNaS	B	W	—	F	ESW
Suaeda moquinii shrubby seepweed	SS	DNaS	—	W	—	—	M
Suriana maritima bay-cedar	MS	DNaS	F	U	Y	Sp	ES
Swietenia mahagoni West Indian mahogany	SC	—	Ff	U	yg	SpSu	AI
Symphoricarpos albus common snowberry	MS	DSs	f	GLMmS	P	SpSu	BEI
Symphoricarpos longiflorus desert snowberry	SS	DFS	f	GLMmS	P	Su	ES
Symphoricarpos mollis creeping snowberry	SS	Ss	f	GLMmS	RW	Sp	Bg
Symphoricarpos occidentalis western snowberry	SS	DS	FL	GhLMmS	P	Su	AB
Symphoricarpos orbiculatus coral-berry	SS	DSs	f	GLMmS	G	Su	BE
Symphoricarpos oreophilus mountain snowberry	SS	S	Ff	GLMmS	W	Sp	B
Symphoricarpos rotundifolius round-leaf snowberry	SS	S	Ff	GLMmS	P	Sp	AB
Symplocos tinctoria horsesugar	LS	FS	L	MS	Y	Sp	IW
Taxodium ascendens pond-cypress	LC	FS	L	W	—	Sp	ATW

PLANT LIST	Plant Type	Env. Tol.	Aesthetic Value	Wildlife Value	Flower Color	Bloom Period	Landscape Uses
Taxodium distichum southern bald-cypress	cLC	FSW	ABL	W	Pu	Sp	ATW
Taxus brevifolia Pacific yew	cSU	Ss	BfL	LS	Y	—	AI
Taxus canadensis American yew	cSS	s	L	GS	yg	Sp	EFGI
Tetradymia axillaris cottonthorn	SS	NaSs	F	U	Y	SuF	BM
Tetradymia canescens spineless horsebrush	SS	DS	F	U	Y	SuF	MR
Tetradymia glabrata little-leaf horsebrush	SS	DSs	F	U	Y	Su	BM
Tetradymia spinosa short-spine horsebrush	SS	DSs	F	U	Y	Su	BM
Tetrazygia bicolor Florida clover-ash	LS	DS	F	U	Y	SpSu	BFI
Thamnosma montana turpentine-broom	SS	DS	—	U	Pu	—	M
Thelocactus bicolor glory-of-Texas	C	DS	F	U	R	—	R
Thelocactus setispinus miniature-barrel cactus	C	DS	—	U	—	—	R
Thrinax morrisii Key thatch palm	SC	S	L	mS	Y	WSpSuF	ABI
Thrinax radiata Florida thatch palm	SC	S	L	mS	—	—	ABI
Thuja occidentalis eastern arborvitae	cSC	DFSs	L	MmS	X	X	AFHPSs
Thuja plicata western arborvitae	cLC	FSs	L	MmS	X	X	sT
Tilia americana American basswood	LC	Ss	AL	GhLMm	Y	Su	ABIT
Tiquilia canescens woody crinklemat	SS	Ss	FL	U	PR	—	BGgM
Tiquilia greggii plumed crinklemat	SS	DS	FL	U	PW	—	GgM
Torreya californica California-nutmeg	cLU	Fs	L	U	—	—	ABISs
Tournefortia hirsutissima chiggery-grapes	WV	Ss	F	U	W	WSpSuF	BI
Tournefortia volubilis twining soilderbush	WV	Ss	—	U	Wyg	WSpSu	BI
Toxicodendron diversilobum Pacific poison-oak	SS	Ss	AfL	GLMmS	G	—	BI
Toxicodendron pubescens Atlantic poison-oak	MS	FSs	L	GLMmS	Gr	—	BI
Toxicodendron radicans eastern poison-ivy	WV	DFSs	AL	GLMmS	GW	Sp	BI
Toxicodendron vernix poison-sumac	MS	FSs	AL	GLMmS	GW	—	W

PLANT LIST	Plant Type	Env. Tol.	Aesthetic Value	Wildlife Value	Flower Color	Bloom Period	Landscape Uses
Trema lamarckianum pain-in-back	eSU	DSs	B	S	PW	WSpSu F	BIS
Trema micranthum Jamaican nettletree	eSU	DSs	B	S	PW	—	BIS
Tsuga canadensis eastern hemlock	cLC	Ss	BL	MmS	X	X	ABFHIs
Tsuga caroliniana Carolina hemlock	cLC	Ss	BL	GMmS	X	X	ABHITS
Tsuga heterophylla western hemlock	cLC	Ss	BL	GMmS	X	X	ABIT
Tsuga mertensiana mountain hemlock	cLC	Ss	BL	GLMmS	X	X	ABIT
Ulmus alata winged elm	LU	Ss	L	GLMSW	GR	W	ABT
Ulmus americana American elm	LC	DFSs	AL	GLMmS W	BrG	Sp	ABITW
Ulmus crassifolia cedar elm	LC	DFS	L	GMmS	GR	Su	AT
Ulmus rubra slippery elm	LC	S	AL	GLMSW	yg	WSp	BIW
Ulmus thomasii rock elm	LC	Fs	L	GLMSW	BrR	Sp	BI
Umbellularia californica California-laurel	eSU	DFSs	L	mS	Y	WSp	BgRsT
Ungnadia speciosa Mexican-buckeye	LU	DFS	F	h	P	Sp	ABS
Vaccinium alaskense Alaska blueberry	MS	Ss	Ff	GLMmS	BrGP	SpSu	BGI
Vaccinium angustifolium late lowbush blueberry	SS	DFSs	AL	CGLmS W	W	Sp	N
Vaccinium arboreum tree sparkle-berry	eLS	DSs	BL	GLMmS W	W	Sp	BIg
Vaccinium cespitosum dwarf blueberry	SS	Ss	Lf	GLMmS	PW	SpSu	BMN
Vaccinium corymbosum highbush blueberry	MS	DFSs W	AL	CGLMm SW	W	Sp	BMN
Vaccinium crassifolium creeping blueberry	eSS	FSs	fL	GLMmS W	P	Sp	BGgsW
Vaccinium darrowii Darrow's blueberry	eSS	DS	L	LMmR	RW	Sp	BgIs
Vaccinium deliciosum rainier blueberry	SS	DSs	Lf	GLMmS	PW	Sp	BgI
Vaccinium elliottii Elliott's blueberry	SS	DFSs	L	GLMmS W	PW	Su	BISW
Vaccinium erythrocarpum southern mountain-cranberry	SS	Ss	Ff	GLMmS	PW	SpSu	BINR
Vaccinium fuscatum black blueberry	MS	FS	L	GLMmS W	W	Su	W
Vaccinium macrocarpon large cranberry	eSS	FSsW	AFL	CGLMm SW	WP	Su	MNsW

PLANT LIST	Plant Type	Env. Tol.	Aesthetic Value	Wildlife Value	Flower Color	Bloom Period	Landscape Uses
Vaccinium membranaceum square-twig blueberry	SS	Ss	L	GLMmS	GW	Sp	BIN
Vaccinium myrsinites shiny blueberry	eSS	DSs	FL	GLMmS	PW	Sp	BgM
Vaccinium myrtilloides velvet-leaf blueberry	SS	FSsW	F	GLMmS W	PW	Su	BW
Vaccinium myrtillus whortle-berry	SS	DSs	f	GLMS	P	Sp	BG
Vaccinium ovalifolium oval-leaf blueberry	MS	Ss	—	GLMmS	PW	Sp	BI
Vaccinium ovatum evergreen blueberry	eMS	Ss	Lf	GLMmS	PW	Sp	BgIN
Vaccinium oxycoccos small cranberry	eSS	FSsW	L	GLMMS W	PW	Su	W
Vaccinium pallidum early lowbush blueberry	SS	DSs	—	GLMmS W	PW	Sp	BI
Vaccinium parvifolium red blueberry	LS	Ss	f	GLMmS	G	Sp	BIN
Vaccinium scoparium grouseberry	SS	Ss	—	GLMmS W	W	Su	BI
Vaccinium stamineum deerberry	MS	DSs	Ff	GLMmS W	W	Sp	B
Vaccinium tenellum small black blueberry	SS	S	F	GLMmS W	PR	Sp	B
Vaccinium vitis-idaea northern mountain-cranberry	eSS	DFSW	L	GLMmS W	W	Sp	BINRW
Vallesia glabra pearlberry	MS	Ss	—	U	W	SpSuF	BI
Vauquelinia californica Arizona-rosewood	eSU	DS	FL	U	W	Su	Bgs
Viburnum acerifolium maple-leaf arrow-wood	SS	Ds	AFfL	BGLMm S	W	Sp	AbIN
Viburnum dentatum southern arrow-wood	MS	FfsW	AFfL	BGmS	W	Sp	AbgHIsW
Viburnum edule squashberry	SS	Ss	Ff	GLMmS	W	SpSu	Ig
Viburnum ellipticum western blackhaw	MS	DSs	FfL	BGLMm S	W	Sp	Bbg
Viburnum lantanoides hobblebush	MS	s	AL	BGLMm S	W	Sp	bGHIs
Viburnum lentago nanny-berry	LS	FSs	FfL	BS	W	Sp	ABbs
Viburnum nudum possumhaw	LS	FDSs W	AFfL	BGLmS W	W	SpSu	ABbgINW
Viburnum obovatum small-leaf arrow-wood	SU	s	Ff	BGLmS W	W	Sp	I
Viburnum prunifolium smooth blackhaw	SU	DS	AfL	BGMmS	W	Sp	ABbg
Viburnum rafinesquianum downy arrow-wood	SS	DSs	AFL	BGLMm S	W	Sp	ABbgI

PLANT LIST	Plant Type	Env. Tol.	Aesthetic Value	Wildlife Value	Flower Color	Bloom Period	Landscape Uses
Viburnum rufidulum rusty blackhaw	SU	DSs	AFfL	BGLSs	W	Sp	ABbg
Viguiera deltoidea triangle goldeneye	SS	DS	F	U	Y	SpSu	M
Viguiera stenoloba resinbush	SS	DS	F	BS	Y	Su	M
Vitis aestivalis summer grape	WV	DSs	fL	GLmSW	W	SpSu	BINP
Vitis arizonica canyon grape	WV	DSs	fL	GLMS	W	SpSu	BEN
Vitis californica California grape	WV	FS	FfL	GLMSW	W	Sp	BS
Vitis labrusca fox grape	WV	FSs	fL	GLMSW	GW	Su	BN
Vitis monticola sweet mountain grape	WV	S	fL	GLMmS	—	—	BN
Vitis munsoniana little muscádine	WV	DSs	f	GLMmS	G	SpSuF	BIN
Vitis mustangensis mustang grape	WV	DFS	L	GLMmS	R	Sp	BN
Vitis riparia river-bank grape	WV	DFSs	AL	GLmSW	yg	Sp	INS
Vitis rotundifolia muscadine	WV	FSs	L	MS	G	Su	BINS
Vitis shuttleworthii calloose grape	WV	Ss	f	GLMmS	G	—	BI
Vitis vulpina frost grape	WV	FSs	L	GLMmS	—	Su	N
Waltheria indica basora-prieta	SS	DS	F	Mm	Y	WSp	ABM
Washingtonia filifera California fan palm	eSU	FS	L	mS	W	Su	ABM
Wisteria frutescens American wisteria	WV	Ss	F	U	Pu	Su	AB
Xanthorhiza simplicissima shrub yellowroot	SS	FSs	L	U	BPu	Sp	GIS
Ximenia americana tallow-wood	SU	DSs	F	U	Y	SpF	BI
Yucca aloifolia aloe yucca	eMS	DS	—	mS	W	—	—
Yucca arkansana Arkansas yucca	eMS	DS	FL	HmS	W	—	AM
Yucca baccata banana yucca	eSS	DS	F	hm	W	Sp	AMP
Yucca brevifolia Joshua-tree	eSU	DS	BFL	mRT	yg	Sp	AM
Yucca campestris Plains yucca	eSS	DS	FL	HmS	PGW	Sp	AM
Yucca constricta Buckley's yucca	SS	DS	FL	HmS	GW	Sp	AM

PLANT LIST	Plant Type	Env. Tol.	Aesthetic Value	Wildlife Value	Flower Color	Bloom Period	Landscape Uses
Yucca elata soaptree yucca	eSU	DS	FL	HMmS	W	Su	AM
Yucca faxoniana Eve's-needle	eLS	S	FL	HmS	W	SpSu	AM
Yucca filamentosa Adam's-needle	eSS	DS	FL	HmS	W	Su	AbgM
Yucca glauca soapweed yucca	eMS	DS	FL	HmS	W	Su	AgM
Yucca schidigera Mojave yucca	eLS	DS	L	HmS	W	Sp	AM
Yucca schottii hoary yucca	eSU	DS	FL	HmS	W	SuF	AM
Yucca thompsoniana Thompson's yucca	eSU	DS	FL	HmS	W	Su	AM
Yucca torreyi Torrey's yucca	eSU	DS	FL	HmS	W	Sp	ABM
Yucca treculeana Don Quixote's-lace	eSU	DS	FL	HmS	W	WSp	ABM
Yucca whipplei Our-Lord's-candle	eMS	DS	FL	HmS	W	Sp	AM
Zamia pumila coontie	eSS	DS	LF	U	R	Su	BFg
Zanthoxylum americanum toothachetree	SU	DFS	AL	GmS	yg	Sp	BP
Zanthoxylum clava-herculis Hercules'-club	SU	Ss	—	GS	GW	Sp	I
Zanthoxylum fagara lime prickly-ash	eSU	Ss	—	GS	yg	WSpSu	BI
Zanthoxylum hirsutum Texas hercules'-club	LS	S	S	GS	yg	—	B
Zenobia pulverulenta honeycup	SS	FSs	AL	GmS	W	Su	BIW
Ziziphus obtusifolia lotebush	SS	DS	—	GmS	—	Su	BM

Herbaceous Plant Matrix

PLANT LIST	Plant Type	Env. Tol.	Aesthetic Value	Wildlife Value	Flower Color	Bloom Period	Landscape Uses
Acalypha gracilens slender three-seed mercury	SA	DS	—	GS	G	SuF	M
Acalypha virginica Virginia three-seed mercury	MA	DS	—	GS	G	SuF	CM
Achillea millefolium common yarrow	MP	DS	DFL	U	WP	SpSuF	BbCFgM
Achlys triphylla sweet-after-death	SP	Ss	DF	U	PuR	Su	BIS
Acleisanthes anisophylla oblique-leaf trumpets	SP	S	—	U	W	SpSuF	M
Acleisanthes longiflora angel's trumpets	SP	DS	F	B	PuW	SpSuF	bEM
Aconitum columbianum Columbian monkshood	MP	SF	FL	U	BW	SW	gSW
Acorus americanus sweetflag	MP	FSW	L	Lm	Y	Su	Wwg
Acourtia runcinata feather-leaf desert-peony	SP	Ss	F	U	PPu	SpSuF	BE
Acrostichum aureum golden leather fern	MP	FS	L	U	X	X	W
Acrostichum danaeifolium inland leather fern	MP	FS	L	U	X	X	W
Actaea pachypoda white baneberry	MP	s	FfL	Gm	W	Su	BgI
Actaea rubra red baneberry	SP	Ss	FfL	GM	W	SpSu	BgI
Adenocaulon bicolor American trailplant	MP	s	L	U	W	Su	I
Adiantum pedatum northern maidenhair	SP	s	L	M	X	X	BgI
Agalinis maritima salt marsh false foxglove	SA	FNaSW	FL	B	P	SpSu	bCSWwg
Agalinis purpurea purple false foxglove	LA	DFHaSsW	F	B	Ppu	SuF	BbgMW
Agalinis tenuifolia slender-leaf false foxglove	SA	FSs	F	B	PPu	SuF	WMBg
Agastache urticifolia nettle-leaf giant-hyssop	LP	Ds	F	U	LPu	Su	BgI
Ageratina altissima white snakeroot	MP	DSs	F	BGS	W	SuF	BIW
Agoseris glauca pale goat-chicory	SP	S	F	U	Y	SpSu	gM
Agrimonia gryposepala tall hairy grooveburr	LP	Ss	F	U	Y	Su	B

PLANT LIST	Plant Type	Env. Tol.	Aesthetic Value	Wildlife Value	Flower Color	Bloom Period	Landscape Uses
Agrostis diegoensis leafy bent	SPG	Ss	—	U	Gr	—	BM
Agrostis elliottiana Elliot's bent	SAG	S	—	M	G	Sp	M
Agrostis gigantea black bent	MPG	Fs	—	U	Gr	—	BIS
Agrostis hallii Hall's bent	MPG	s	—	U	Gr	—	BI
Agrostis hyemalis winter bent	MPG	DSs	—	U	Gr	—	M
Agrostis idahoensis Idaho bent	SPG	S	—	U	Gr	—	M
Agrostis scabra rough bent	MPG	DSs	—	U	Gr	—	M
Alisma plantago-aquatica water plantain	MP	FSW	FL	Lm	W	Su	Wwg
Alisma subcordatum American water-plantain	MP	FSW	FL	LM	W	Su	Wwg
Allionia incarnata trailing windmills	MP	DS	—	U	PPuW	SpSu	M
Allium acuminatum taper-tip onion	SP	DS	F	B	P	—	Bb
Allium canadense meadow garlic	SP	FSs	—	GHhS	PW	SpSu	BM
Allium cernuum nodding onion	MP	DS	F	B	P	Su	bM
Allium textile white wild onion	SP	DS	F	B	PW	SpSu	bM
Allium tricoccum ramp	MP	s	FL	B	G	Su	BgI
Alopecurus aequalis short-awn meadow-foxtail	MPG	FSW	F	GS	Gr	—	Wwg
Alopecurus borealis meadow-foxtail	PG	—	—	U	yg	SpSu	—
Alopecurus saccatus Pacific meadow-foxtail	SG	FSW	F	GS	X	X	MW
Alophia drummondii propeller-flower	SP	S	F	U	LPu	Sp	gM
Alternanthera flavescens yellow joyweed	P	—	—	U	YW	WSpSuF	—
Alternanthera philoxeroides alligator-weed	HVPAQ	FNaSW	—	U	W	SpSuF	W
Alyssum desertorum desert madwort	P	—	—	U	Y	Su	—
Amaranthus australis southern amaranth	LP	FNaS	—	GLMmSW	—	—	—

PLANT LIST	Plant Type	Env. Tol.	Aesthetic Value	Wildlife Value	Flower Color	Bloom Period	Landscape Uses
Amaranthus palmeri Palmer's amaranth	MA	DS	—	GLMmS W	—	—	—
Amblyolepis setigera huisache-daisy	SA	S	F	BC	Y	Sp	bCGgM
Amblyopappus pusillus dwarf coastweed	SA	FNaS	F	U	Y	Su	CSW
Ambrosia acanthicarpa flat-spine burr-ragweed	SA	S	—	GS	G	SuF	C
Ambrosia artemisiifolia annual ragweed	LA	DS	—	GS	G	SuF	CS
Ambrosia confertiflora weak-leaf burr-ragweed	LP	S	—	GMmS	G	SuF	EM
Ambrosia hispida coastal ragweed	MA	FS	—	GS	G	SuF	CS
Ambrosia psilostachya western ragweed	SAP	DS	—	GS	G	SuF	C
Ambrosia trifida great ragweed	LA	DFS	—	GS	G	SuF	CS
Amianthium muscitoxicum flypoison	MP	FS	F	U	W	Su	BWwg
Ammophila breviligulata American beach grass	MG	S	—	GS	Br	SuF	CES
Amphicarpaea bracteata American hog-peanut	HV	Ss	L	LS	Pu	SuF	CGI
Amphicarpum muhlenbergianum perennial goober grass	MPG	FS	—	U	Gr	SuF	W
Amsonia illustris Ozark bluestar	MP	FS	F	U	B	Sp	W
Amsonia tabernaemontana eastern bluestar	MP	Ss	F	U	B	SpSu	Bg
Andropogon floridanus Florida bluestem	MPG	DS	AL	GL	W	F	M
Andropogon gerardii big bluestem	LPG	DS	ABL	GL	GPu	SuF	EgM
Andropogon glomeratus bushy bluestem	LPG	FNaS	AFL	GL	GW	SuF	SW
Andropogon hallii sand bluestem	MPG	SD	AL	GL	GW	SuF	—
Andropogon ternarius split-beard bluestem	MPG	SF	AFL	GL	W	F	BM
Andropogon virginicus broom-sedge	MPG	S	ABFL	GL	W	SuF	EM
Androsace occidentalis western rock-jasmine	SA	DS	F	U	PW	Sp	Cg
Anemia adiantifolia pineland fern	SP	DS	F	U	—	SpSu	Bg

PLANT LIST	Plant Type	Env. Tol.	Aesthetic Value	Wildlife Value	Flower Color	Bloom Period	Landscape Uses
Anemone berlandieri ten-petal thimbleweed	P	—	—	U	PuW	Sp	—
Anemone canadensis round-leaf thimbleweed	SP	S	F	h	BW	Sp	BgM
Anemone caroliniana Carolina thimbleweed	SP	S	F	h	BW	Sp	BgM
Anemone cylindrica long-head thimbleweed	MP	DSs	FL	h	W	Sp	BgM
Anemone edwardsiana plateau thimbleweed	SP	Ss	FL	h	W	Sp	R
Anemone lyallii little mountain thimbleweed	P	—	—	U	BPW	Sp	—
Anemone piperi Piper's windflower	P	—	—	U	BPW	Sp	—
Anemone quinquefolia nightcaps	SP	S	FL	h	W	Sp	BgM
Anemone virginiana tall thimbleweed	MP	S	F	h	W	Su	BgM
Anemopsis californica yerba-mansa	SP	FNaS	FL	U	W	SpSu	gSW
Angelica atropurpurea purple-stem angelica	LP	FSW	FL	B	W	Su	gSW
Angelica pinnata small-leaf angelica	P	FW	—	U	PW	Su	W
Anisocoma acaulis scalebud	SA	DS	F	U	W	Sp	Cg
Antennaria dimorpha cushion pussytoes	SP	Ss	L	GLMm	—	SpSu	BG
Antennaria neglecta field pussytoes	SP	DS	L	GLMm	W	Sp	BGM
Antennaria parlinii Parlin's pussytoes	SP	DSs	FL	GMm	W	SpSu	BGgMR
Antennaria plantaginifolia woman's-tobacco	SP	Ds	AL	GLMm	W	Sp	BGR
Antennaria rosea rosy pussytoes	SP	S	F	GLMm	P	Su	G
Anthaenantia villosa green silkyscale	MPG	DSs	F	U	G	SuF	B
Aphanostephus ramosissimus plains dozedaisy	SA	S	F	B	LPPu	SpSu	CM
Aphanostephus skirrhobasis Arkansas dozedaisy	SA	DS	F	U	LW	SpSu	gM
Apios americana groundnut	HV	FSs	F	m	Br	SuF	GgI
Apocynum androsaemifolium spreading dogbane	MP	S	F	BC	P	Su	B

PLANT LIST	Plant Type	Env. Tol.	Aesthetic Value	Wildlife Value	Flower Color	Bloom Period	Landscape Uses
Aquilegia canadensis red columbine	SP	S	FL	Hh	RY	SpSu	Bbl
Aquilegia chrysantha golden columbine	MP	S	FL	h	Y	Su	BgR
Aquilegia coerulea Colorado blue columbine	MP	S	FL	h	B	Su	BbM
Aquilegia formosa crimson columbine	MP	S	FL	Hh	R	Su	BbM
Arabis breweri Brewer's rockcress	SP	S	—	CS	PuW	—	R
Arabis drummondii Canadian rockcress	MP	DSs	—	U	W	SpSu	B
Arabis holboellii Holboell's rockcress	SP	DS	—	CS	PW	Su	b
Arabis lignifera Owens Valley rockcress	P	—	—	U	PPu	Sp	—
Arabis lyrata lyre-leaf rockcress	SP	S	F	C	W	Sp	B
Arabis platysperma pioneer rockcress	SP	S	—	CS	PW	Su	B
Arabis repanda Yosemite rockcress	SB	S	—	CS	W	Su	B
Arabis sparsiflora elegant rockcress	BP	—	—	U	Pu	SpSu	—
Aralia californica California spikenard	LP	s	FfL	GMS	R	Sn	gIS
Aralia nudicaulis wild sarsaparilla	SP	S	L	U	GW	Su	BI
Aralia racemosa American spikenard	LP	Ss	fL	U	W	Su	BgI
Arenaria hookeri Hooker's sandwort	P	—	—	U	W	Su	—
Arenaria kingii King's sandwort	P	—	—	U	WP	Su	—
Arethusa bulbosa dragon's-mouth	SP	W	F	U	PPu	Su	W
Argentina anserina common silverweed	SP	FS	FL	U	Y	Su	GSW
Argentina egedii Pacific silverweed	P	—	—	U	Y	SpSu	—
Arisaema dracontium greendragon	MP	FS	LF	GS	yg	Sp	BI
Arisaema triphyllum jack-in-the-pulpit	SP	s	FfL	GS	GPu	Sp	BgI
Aristida californica California three-awn	SPG	DS	F	LMmS	Br	Su	M

PLANT LIST	Plant Type	Env. Tol.	Aesthetic Value	Wildlife Value	Flower Color	Bloom Period	Landscape Uses
Aristida divaricata poverty three-awn	SPG	DS	F	LMmS	Br	Su	M
Aristida longespica red three-awn	MAG	DS	F	LMmS	Pu	SuF	M
Aristida oligantha prairie three-awn	SAG	DS	F	LMmS	GPu	SuF	M
Aristida orcuttiana single-awn three-awn	PG	—	—	U	yg	SuF	M
Aristida purpurascens arrow-feather three-awn	MPG	DS	F	LMmS	Br	SuF	BM
Aristida purpurea purple three-awn	MPG	DS	F	LMmS	Br	SuF	BM
Aristida spiciformis bottlebrush three-awn	MPG	DS	F	LMmS	Br	F	B
Aristida stricta pineland three-awn	SPG	DS	F	LMmS	Br	SuF	BM
Aristida ternipes spider grass	SPG	DS	F	LMmS	Br	SuF	M
Aristolochia serpentaria Virginia-snakeroot	HV	Ss	F	U	PuR	SpSu	BI
Armeria maritima sea thrift	eSP	—	—	U	P	SuSp	—
Armoracia lacustris lakecress	APAQ	W	—	U	W	SpSu	wg
Arnica chamissonis leafy leopard-bane	P	FW	F	U	Y	Su	W
Arnica cordifolia heart-leaf leopard-bane	SP	Ds	F	U	Y	Su	B
Arnica latifolia daffodil leopard-bane	SP	Ss	F	U	Y	Su	B
Arnoglossum muehlenbergii great Indian-plantain	LP	S	—	U	W	Su	BI
Artemisia campestris Pacific wormwood	SBP	DS	—	U	WY	SuF	—
Artemisia dracunculus dragon wormwood	LP	—	L	GMm	WY	F	—
Artemisia ludoviciana white sagebrush	MP	SD	L	GMm	yg	SuF	M
Artemisia michauxiana Michaux's wormwood	P	—	—	GMm	Y	SpSu	—
Artemisia parryi Parry's wormwood	P	—	—	GMm	Y	Su	—
Artemisia tripartita three-tip sagebrush	eP	—	—	GMm	BrW	Su	—
Arthrocnemum subterminale Parish's glasswort	SP	NaS	—	W	—	—	S

PLANT LIST	Plant Type	Env. Tol.	Aesthetic Value	Wildlife Value	Flower Color	Bloom Period	Landscape Uses
Aruncus dioicus bride's-feathers	MP	Ss	FL	U	W	Su	BgI
Asarum canadense Canadian wild ginger	SP	s	L	U	Bu	Sp	BGgI
Asarum caudatum long-tail wild ginger	SP	s	L	U	Pu	Sp	BGgI
Asarum hartwegii Hartweg's wild ginger	SP	s	FL	U	Br	SpSu	GI
Asclepias amplexicaulis clasping milkweed	MP	DS	Ff	BCh	PPu	Su	BbgM
Asclepias arenaria sand milkweed	SP	S	F	B	GW	Su	bM
Asclepias asperula spider antelope-horns	SP	DSs	F	BC	Puyg	SpSu	BbgM
Asclepias curtissii Curtis' milkweed	MP	DS	F	B	W	SpSuF	b
Asclepias eriocarpa Indian milkweed	MP	DS	Ff	BM	WP	Su	bM
Asclepias incarnata swamp milkweed	MP	FS	Ff	BCh	P	Su	bWwg
Asclepias lanuginosa side-cluster milkweed	MP	S	Ff	BC	P	SpSu	M
Asclepias latifolia broad-leaf milkweed	MP	DS	Ff	BC	G	Su	M
Asclepias ovalifolia oval-leaf milkweed	SP	S	Ff	BCh	GW	Su	b
Asclepias perennis aquatic milkweed	MP	FSW	Ff	BCh	W	SpSu	blW
Asclepias speciosa showy milkweed	MP	S	Ff	BCh	P	SpSu	BbS
Asclepias stenophylla slim-leaf milkweed	MP	DS	Ff	BC	GW	SU	bMR
Asclepias subulata rush milkweed	LP	DFS	F	BM	WY	SpSu	bMW
Asclepias syriaca common milkweed	LP	DS	Ff	BCh	Pu	Su	bM
Asclepias tuberosa butterfly milkweed	MP	DS	Ff	BCh	O	SuF	bM
Asclepias verticillata whorled milkweed	MP	DS	F	BC	W	Su	bgM
Asclepias viridiflora green comet milkweed	MP	DS	F	BC	G	SU	bM
Ascyrum hypericoides St. Andrew's cross	HVP	D	FfL	U	Y	Su	BG
Asplenium platyneuron ebony spleenwort	eSP	s	L	U	X	X	BbI

PLANT LIST	Plant Type	Env. Tol.	Aesthetic Value	Wildlife Value	Flower Color	Bloom Period	Landscape Uses
Asplenium rhizophyllum walking fern	SP	s	L	U	X	X	BlR
Asplenium trichomanes maidenhair spleenwort	SP	s	L	U	X	X	Bgl
Ascyrum hypericoides St. Andrew's cross	HVP	D	FfL	U	Y	Su	BG
Aster acuminatus whorled wood aster	MP	S	F	BCGhL MmS	W	SuF	Bb
Aster alpigenus tundra aster	P	FW	F	GLMmS	BPu	SuF	bW
Aster anomalus many-ray aster	MP	Ds	F	BCGhL MmS	BP	F	BbEg
Aster breweri Brewer's aster	P	—	F	GLMmS ch	B	SuF	b
Aster ciliolatus Lindley's aster	MP	Ss	F	BCGhL MmS	B	SuF	b
Aster conspicuus eastern showy aster	P	—	F	U	Pu	Su	b
Aster cordifolius common blue wood aster	MP	S	F	CBh	Pu	F	Bb
Aster divaricatus white wood aster	MP	s	F	BGLMm CSh	W	SuF	BblW
Aster drummondii Drummond's aster	MP	DSs	F	BCGMm S	BP	SuF	BbgM
Aster dumosus rice button aster	MP	FSs	F	BCGhL MmS	BPuW	SuF	b
Aster ericoides white heath aster	MP	DS	F	BCGhL MmS	W	SuF	bM
Aster laevis smooth blue aster	MP	S	F	BCL	B	SuF	bM
Aster lanceolatus white panicle aster	MP	FS	Ch	BC	W	SuF	bW
Aster lateriflorus farewell-summer	LP	S	F	BCh	PuW	SuF	b
Aster linariifolius flax-leaf white-top aster	SP	S	F	BCh	B	SuF	bM
Aster linearifolius stiff-leaf aster	MP	DS	FL	BCGh	Bpu	F	Bbg
Aster macrophyllus large-leaf aster	LP	S	F	BCGLM mSh	Pu	SuF	b
Aster novae-angliae New England aster	LP	S	F	BS	Pu	F	bM
Aster occidentalis western mountain aster	P	FW	—	U	LPu	SuF	W
Aster oolentangiensis sky blue aster	MP	S	F	BCGhL MmS	B	SuF	bM

PLANT LIST	Plant Type	Env. Tol.	Aesthetic Value	Wildlife Value	Flower Color	Bloom Period	Landscape Uses
Aster paludosus southern swamp aster	SP	S	F	BCh	Pu	SuF	BbM
Aster patens late purple aster	MP	DSs	F	BCGhL MmS	BPu	F	BbEgM
Aster paternus toothed white-top aster	SP	S	F	BCGhL MmS	W	Su	Bl
Aster pilosus white oldfield aster	MP	DS	F	BS	W	F	bM
Aster puniceus purple-stem aster	LP	FSW	F	BCh	Pu	SuF	bW
Aster sericeus western silver aster	MP	DS	F	BS	Pu	F	bW
Aster solidagineus narrow-leaf white-top aster	MP	S	F	BCGHh LMmS	W	SuF	b
Aster tenuifolius perennial saltmarsh aster	SP	FSW	F	BGMmS	BNP	F	bW
Aster tortifolius dixie white-top aster	MP	DS	F	BCh	W	SuF	BbM
Aster undulatus waxy-leaf aster	MP	DSs	F	BCGhM mS	B	SuF	BbgM
Aster umbellatus parasol flat-top white aster	LP	FS	FL	BCGhL MmS	W	SuF	bW
Aster walteri Walter's aster	SP	DS	F	BCh	Pu	F	BbM
Astragalus adsurgens standing milk-vetch	SP	DS	F	GLMm	Pu	Su	BEM
Astragalus amphioxys Aladdin's-slippers	SAP	—	—	GLMm	PPu	SpSu	E
Astragalus arrectus palouse milk-vetch	P	—	—	GLMm	WY	SpSu	EM
Astragalus castaneiformis chestnut milk-vetch	P	—	—	GLMm	W	SpSu	E
Astragalus crassicarpus ground-plum	P	DS	F	GLMm	Pu	Sp	EM
Astragalus kentrophyta spiny milk-vetch	P	—	—	GLMm	PPuW	SuF	—
Astragalus miser timber milk-vetch	P	—	—	GLMm	Y	Sp	—
Astragalus missouriensis Missouri milk-vetch	SP	DS	F	U	PPu	Sp	GM
Astragalus mollissimus woolly milk-vetch	SP	DS	—	GLMm	Pu	SpSu	—
Astragalus nuttallianus turkey-pea	SA	DS	DF	GMm	PPu	Sp	GgM
Astragalus pectinatus narrow-leaf milk-vetch	P	—	—	GLMm	YW	SpSu	—

PLANT LIST	Plant Type	Env. Tol.	Aesthetic Value	Wildlife Value	Flower Color	Bloom Period	Landscape Uses
Astragalus platytropis broad-keel milk-vetch	SP	S	F	GLMm	PuY	Su	Fg
Astragalus purshii Pursh's milk-vetch	SP	DS	FL	GLMm	WPu	SpSu	EM
Astragalus spaldingii Spalding's milk-vetch	MP	—	L	GLMm	—	—	EM
Astragalus spatulatus tufted milk-vetch	P	—	—	GLMm	PPu	SpSu	—
Astragalus tener alkali milk-vetch	SP	DNaS	F	GLMm	Pu	Sp	EM
Astragalus tephrodes ashen milk-vetch	P	—	—	GLMm	PPu	SpSu	—
Astragalus troglodytus creeping milk-vetch	P	—	—	GLMm	PuR	SpSu	—
Astragalus wootonii Wooton's milk-vetch	SAB	DS	F	GMm	PRW	Sp	GgM
Athyrium filix-femina subarctic lady fern	MP	s	L	U	X	X	BgI
Atrichoseris platyphylla parachute-plant	A	—	—	U	PW	Sp	—
Atriplex argentea silverscale	SA	DS	—	GLMmS	G	SuF	C
Atriplex fruticulosa ball saltbush	SP	DNaS	—	GLMmS W	G	SuF	—
Atriplex patula halberd-leaf orache	SA	FNaS	L	GLMmS W	G	SuF	CSW
Atriplex pentandra crested saltbush	SAP	FNaS	—	GLMmS W	G	SuF	C
Atriplex phyllostegia arrow saltbush	SP	DS	—	GLMmS W	G	SuF	—
Atriplex powellii Powell's orache	SP	DNaS	L	GLMmS W	G	SuF	M
Atriplex tularensis Bakersfield saltbush	SP	DNaS	f	GLMmS W	—	SuF	M
Atriplex watsonii Watson's saltbush	SP	FSW	—	GLMmS W	G	Su	W
Aulacomnium palustre moss	m	FSs	—	U	X	X	IW
Aureolaria flava smooth yellow false-foxglove	LP	s	F	U	Y	Su	Bgl
Aureolaria laevigata entire-leaf yellow false-foxglove	MP	s	F	U	Y	Su	BI
Aureolaria pedicularia fern-leaf yellow false-foxglove	MA	DSs	F	U	Y	F	CBI
Aureolaria virginica downy yellow false-foxglove	MP	Ss	F	U	Y	SpSu	Bl

PLANT LIST	Plant Type	Env. Tol.	Aesthetic Value	Wildlife Value	Flower Color	Bloom Period	Landscape Uses
Azolla caroliniana Carolina mosquito fern	AQP	SW	LA	W	X	X	Wwg
Baileya multiradiata showy desert-marigold	SAP	DS	F	U	Y	Su	CgM
Balduina angustifolia coastal-plain honeycomb-head	MP	DS	—	U	Y	SuF	—
Balsamorhiza incana hoary balsamroot	SP	DS	F	U	Y	SpSu	BgM
Balsamorhiza sagittata arrow-leaf balsamroot	MP	DS	FL	U	Y	SpSu	BM
Baptisia alba white wild indigo	LP	DS	FL	B	W	SpSu	ABgM
Baptisia bracteata long-bract wild indigo	MP	DS	FL	B	WY	SpSu	ABgM
Baptisia lanceolata gopherweed	MP	S	FL	B	Y	Sp	ABgM
Barbarea orthoceras American yellow-rocket	S	FWs	F	GS	Y	Su	BSW
Batis maritima turtleweed	MP	NaS	—	U	W	Su	ES
Berlandiera subacaulis Florida greeneyes	SP	DS	F	BS	Y	SpSu	bM
Berlandiera texana Texas greeneyes	MP	FSs	F	U	OY	SpSu	BS
Berula erecta cat-leaf water-parsnip	MP	FSW	DL	U	W	SuF	SWwg
Besseya wyomingensis Wyoming coraldrops	P	—	—	U	Pu	Sp	—
Bidens aristosa bearded beggarticks	MP	DFS	FL	BGmSW	Y	SuF	BbCgMW
Bidens bipinnata Spanish-needles	MA	FSs	FL	BGmSW	Y	SuF	C
Bidens cernua nodding burr-marigold	MP	DFS	F	BS	Y	SuF	BbgMW
Bidens frondosa devil's-pitchfork	MP	FS	—	GmSW	yg	SuF	MW
Bifora americana prairie bishop	MA	S	—	U	W	SpSu	M
Blechnum spicant deerfern	eP	FW	—	U	X	X	W
Blennosperma nanum common stickyseed	AP	FW	—	U	YPuW	Sp	W
Blepharoneuron tricholepis pine-dropseed	PG	—	—	U	G	Su	—
Blephilia ciliata downy pagode-plant	SP	DS	FL	B	LPu	SpSu	BbgM

PLANT LIST	Plant Type	Env. Tol.	Aesthetic Value	Wildlife Value	Flower Color	Bloom Period	Landscape Uses
Boehmeria cylindrica small-spike false nettle	MP	FSs	—	U	G	SuF	IW
Boisduvalia glabella smooth spike-primrose	SA	S	FL	U	PuW	Sp	M
Boltonia asteroides white doll's-daisy	LP	FNaSW	F	B	PuW	SuF	bMW
Boltonia diffusa small-head doll's-daisy	MP	FSs	F	B	PuW	F	BM
Bothriochloa barbinodis cane beard grass	MPG	DS	F	GL	W	Su	AM
Bothriochloa saccharoides plumed beard grass	MPG	DS	F	GL	W	Su	AM
Botrychium virginianum rattlesnake fern	eSP	s	L	GLM	X	X	Bgl
Bouteloua breviseta gypsum grama	SPG	DS	F	GLMmS	G	Su	M
Bouteloua chondrosioides spruce-top grama	SPG	DS	F	GLMmS	G	Su	M
Bouteloua curtipendula side-oats grama	MG	DS	F	GLMmS	GPu	SuF	M
Bouteloua eriopoda black grama	SPG	DS	F	GLMmS	G	Su	M
Bouteloua gracilis blue grama	SPG	DS	F	GLMmS	Br	Su	M
Bouteloua hirsuta hairy grama	SPG	DS	F	GLMmS	Br	Su	M
Bouteloua pectinata tall grama	MPG	S	—	m	G	SpSuF	M
Bouteloua ramosa chino grama	MPG	DS	F	GLMmS	G	Si	MR
Bouteloua repens slender grama	SPG	DS	F	GLMmS	G	Su	MR
Bouteloua rigidiseta Texas grama	SPG	DS	F	GLMmS	G	Su	MR
Bouteloua trifida red grama	SPG	DS	F	GLMmS	G	Su	MR
Brachiaria ciliatissima fringed signal grass	LPG	S	F	U	G	SpSuF	EM
Brachyelytrum erectum bearded shorthusk	MPG	Ss	—	U	G	Su	BI
Brickellia cylindraceae gravel-bar brickellbush	MP	Ss	—	G	GW	SuF	M
Brickellia eupatarioides false boneset	LP	S	f	U	WY	SuF	gM
Brickellia oblongifolia narrow-leaf brickelbush	SP	DS	—	U	GPuW	SuF	M

PLANT LIST	Plant Type	Env. Tol.	Aesthetic Value	Wildlife Value	Flower Color	Bloom Period	Landscape Uses
Bromus anomalus nodding brome	SPG	DS	—	GLMmS	G	Su	EM
Bromus ciliatus fringed brome	MPG	Ss	—	GLMmS	G	Su	BES
Bromus laevipes woodland brome	PG	—	—	GLMmS	yg	SpSu	—
Bromus marginatus large mountain brome	MPG	DS	—	GLMmS	G	Su	BEM
Bromus orcuttianus Chinook brome	MPG	DSs	—	GLMmS	G	Su	BEI
Bromus pubescens hairy woodland brome	MPG	DSs	—	GLMmS	G	Su	BEI
Bromus vulgaris Columbian brome	MPG	Ss	—	GLMmS	G	Su	BE
Buchloë dactyloides buffalo grass	SPG	DS	F	LSW	G	Su	EM
Buchnera americana American bluehearts	MP	DFSsW	F	U	Pu	WSuF	BMWwg
Cabomba caroliniana Carolina fanwort	AQP	SsW	L	U	W	SpSu	Wwg
Cacalia atriplicifolia pale Indian-plantain	LP	Ss	L	h	WG	Su	BgM
Cacalia plantaginea groove-stem Indian-plantain	LP	DSW	L	h	WG	Su	BgM
Cakile edentula American searocket	SA	FNaS	L	U	Y	SpF	CEW
Calamagrostis breweri short-hair reed grass	SPG	S	L	LM	Pu	Su	M
Calamagrostis canadensis bluejoint	LPG	FSW	L	LM	Pu	Su	W
Calamagrostis nutkaensis nootka reed grass	LPG	Fs	L	LM	Pu	Su	W
Calamagrostis rubescens pinegrass	MPG	DS	L	LM	Pu	Su	BM
Calamagrostis stricta slim-stem reedgrass	MPG	FSW	L	LM	Pu	Su	W
Calamovilfa gigantea giant sand-reed	LPG	DS	—	LM	G	SuF	S
Calamovilfa longifolia prairie sand-reed	LPG	DS	—	LM	G	SuF	M
Calla palustris water-dragon	SP	SW	FfL	U	WY	SpSu	Wwg
Callirhoe involucrata purple poppy-mallow	SP	Ss	F	B	PuR	Sp	BgM
Calochortus ambiguus doubting mariposa-lily	P	—	—	U	GrP	SpSu	—

PLANT LIST	Plant Type	Env. Tol.	Aesthetic Value	Wildlife Value	Flower Color	Bloom Period	Landscape Uses
Calochortus invenustus plain mariposa-lily	SP	S	FL	U	Pu	Sp	gM
Calochortus luteus yellow mariposa-lily	SP	DS	F	U	OY	SpSu	gM
Calochortus nuttallii sego-lily	SP	DS	F	U	PuW	SpSu	gM
Calopogon tuberosus tubercus grass-pink	SP	FS	FL	U	P	Su	gW
Caltha leptosepala white marsh-marigold	SP	FS	F	hL	W	SpSu	SWwg
Caltha palustris yellow marsh-marigold	SP	FSsW	FL	hL	Y	Su	Wwg
Calycoseris parryi yellow tackstem	SA	DS	F	U	Y	Sp	C
Calylophus berlandieri Berlandier's sundrops	SAP	DS	F	U	Y	SpSu	gM
Calylophus hartwegii Hartweg's sundrops	SP	DS	F	U	Y	SpSu	gMR
Calylophus serrulatus yellow sundrops	MP	DS	—	U	Y	Su	M
Calystegia occidentalis chaparral false bindweed	PHV	—	—	U	P	SpSu	—
Calystegia sepium hedge false bindweed	HV	FS	FL	GLm	P	SpF	BS
Camassia quamash small camas	SP	FS	F	U	BW	Sp	MW
Camassia scilloides Atlantic camas	MP	s	F	U	B	Sp	BgI
Camissonia boothii shredding suncup	A	—	—	U	PW	Su	—
Camissonia brevipes golden suncup	A	—	—	U	Y	Sp	—
Camissonia chamaenerioides long-capsule suncup	SA	DS	—	U	W	Sp	—
Camissonia claviformis browneyes	A	—	—	U	RW	Sp	—
Camissonia scapoidea Paiute suncup	A	S	—	U	Y	Sp	—
Campanula aparinoides marsh bellflower	SP	FS	F	hS	BW	Su	gW
Campanula rotundifolia bluebell-of-Scotland	SP	S	F	hS	B	SuF	gM
Campanulastrum americanum American-bellflower	LA	s	F	U	B	Su	gI
Canavalia rosea bay-bean	MP	NaS	F	U	P	SpSuF	S

PLANT LIST	Plant Type	Env. Tol.	Aesthetic Value	Wildlife Value	Flower Color	Bloom Period	Landscape Uses
Canna flaccida bandanna-of-the-Everglades	MP	FSW	F	U	Y	SpSu	Wwg
Cardamine breweri Sierran bittercress	SP	S	F	U	W	Sp	M
Cardamine concatenata cut-leaf toothwort	SP	Fs	F	U	W	Sp	IS
Cardamine cordifolia large mountain bittercress	S	Fs	F	U	W	Sp	E
Cardamine diphylla crinkleroot	SP	Ss	L	U	W	Sp	SG
Cardamine longii Long's bittercress	SP	FNa SW	F	U	W	SuF	SW
Cardaria draba heart-pod hoarycress	SP	—	—	U	W	SP	—
Cardiospermum corindum faux-persil	HV	Ss	—	U	WY	SuF	BGM
Carex alata broad-wing sedge	MPG	SW	L	CGLMm SW	G	Su	W
Carex albicans white-tinge sedge	MPG	DsS	L	CGLMm S	G	SpSu	BI
Carex albursina white bear sedge	MPG	s	L	CGLMm S	G	Sp	I
Carex amphibola eastern narrow-leaf sedge	MPG	FSs	—	GLMmS	G	SpSu	BIMW
Carex aquatilis leafy tussock sedge	MPG	FSW	L	CGLMm SW	G	Su	Wwg
Carex arctata drooping woodland sedge	MPG	Ss	L	CGLMm S	G	Su	BI
Carex arkansana Arkansas sedge	MPG	S	L	CGLMm SW	BG	Su	MW
Carex atlantica prickly bog sedge	MPG	FSsW	L	GLMmS W	G	SpSu	BIW
Carex aurea golden-fruit sedge	SPG	FS	L	CGLMm S	G	Su	MS
Carex bebbii Bebb's sedge	MPG	S	L	CGLMm S	G	Su	M
Carex bicknellii Bicknell's sedge	SPG	DS	L	CGLMm SW	BG	SW	MW
Carex bolanderi Bolander's sedge	MPG	DS	L	CGLMm S	Br	Su	M
Carex brittoniana Britton's sedge	MPG	S	—	U	G	SpSu	EM
Carex bromoides brome-like sedge	MPG	FSs	L	CGLMm SW	G	SpSu	IMW
Carex canescens hoary sedge	MPG	FSW	L	CGLMm S	G	SpSu	Wwg

PLANT LIST	Plant Type	Env. Tol.	Aesthetic Value	Wildlife Value	Flower Color	Bloom Period	Landscape Uses
Carex caroliniana Carolina sedge	SPG	FSW	—	GLMmSW	G	Sp	SW
Carex cephalophora oval-leaf sedge	SPG	FSs	—	GLMmS	G	Sp	SW
Carex cherokeensis Cherokee sedge	MPG	Ss	—	U	G	Sp	BE
Carex comosa bearded sedge	MPG	WS	L	CGLMmSW	G	Su	W
Carex concinna low northern sedge	MPG	s	L	CGLMmS	G	Su	I
Carex concinnoides northwestern sedge	SPG	D	L	CGLMmSW	G	Su	M
Carex crinita fringed sedge	MPG	SW	FfL	CGLMmSW	G	Su	W
Carex debilis white-edge sedge	MPG	s	L	GLMmS	G	Su	BI
Carex deweyana round-fruit short-scale sedge	MPG	—	L	CGLMmS	G	Sp	BI
Carex diandra lesser tussock sedge	MPG	SW	L	CGLMmSW	G	SpSu	W
Carex disperma soft-leaf sedge	SPG	Ss	L	CGLMmS	G	SpSu	BI
Carex duriuscula spike-rush sedge	P	—	L	CGLMmS	G	Su	—
Carex eburnea bristle-leaf sedge	SPG	S	L	CGLMmS	G	SpSu	M
Carex emoryi Emory's sedge	MPG	FSs	—	GLMmS	G	Sp	SW
Carex exilis coastal sedge	MPG	FSW	L	CGLMmSW	G	SpSu	W
Carex exserta short-hair sedge	SPG	DS	L	CGLMmS	Br	Su	M
Carex festucacea fescue sedge	MPG	FSs	L	CGLMmS	G	SpSu	MW
Carex filifolia thread-leaf sedge	SPG	DS	L	CGLMmS	G	Su	M
Carex flaccosperma thin-fruit sedge	SPG	sW	—	GLMmSW	G	Sp	BIW
Carex flava yellow-green sedge	MPG	FS	L	CGLMmS	yg	Su	MS
Carex foenea dry-spike sedge	P	—	L	CGLMmS	G	Su	—
Carex folliculata northern long sedge	MPG	FSsW	L	CGLMmSW	G	Su	BIW
Carex frankii Frank's sedge	MPG	Ss	L	CGLMmS	G	Su	BIM

PLANT LIST	Plant Type	Env. Tol.	Aesthetic Value	Wildlife Value	Flower Color	Bloom Period	Landscape Uses
Carex geophila white mountain sedge	P	—	L	CGLMmS	G	Su	—
Carex geyeri Geyer's sedge	SPG	Ss	L	CGLMmS	G	Sp	BI
Carex gigantea giant sedge	MPG	FSW	L	CGLMmSW	G	SpF	W
Carex granularis limestone-meadow sedge	MPG	Ss	L	GLMmS	G	Su	BIM
Carex gravida heavy sedge	MPG	S	L	GLMmS	BrG	SpSu	M
Carex grayi Gray's sedge	MPG	Ss	L	GLMmS	G	SuF	BIM
Carex hassei Hasse's sedge	MPG	FS	L	GLMmSW	BrPu	Su	SW
Carex haydenii cloud sedge	MPG	SW	L	GLMmSW	G	SpSu	BIW
Carex heteroneura different-nerve sedge	MPG	S	L	GLMmS	G	Su	M
Carex hyalinolepis shoreline sedge	SPG	FSW	L	CGLMmSW	BrG	Su	MW
Carex hystericina porcupine sedge	MPG	FSW	L	GLMmSW	G	SuF	SW
Carex inops long-stolon sedge	SPG	D	L	GLMmS	G	Su	B
Carex interior inland sedge	SPG	FS	L	GLMmSW	G	SpSu	W
Carex intumescens great bladder sedge	MPG	Ss	L	GLMmS	G	SpF	BIM
Carex jonesii Jones' sedge	SPG	S	L	GLMmS	G	SpSu	M
Carex lacustris lakebank sedge	MPG	SW	L	GLMmSW	G	Su	W
Carex lanuginosa woolly sedge	MPG	FSW	L	GLMmSW	Br	SpSu	MWS
Carex lasiocarpa woolly-fruit sedge	MPG	FSW	L	GLMmSW	G	SpSu	SW
Carex laxiflora broad loose-flower sedge	MPG	s	L	GLMmS	G	SpSu	BI
Carex leptalea bristly-stalk sedge	SPG	FSs	L	GLMmS	G	Su	BIM
Carex limosa mud sedge	SPG	FSW	L	GLMmSW	G	SpSu	SW
Carex livida livid sedge	SPG	FSW	L	GLMmSW	G	SpSu	MW
Carex lupulina hop sedge	MPG	SW	L	GLMmSW	G	Su	W

PLANT LIST	Plant Type	Env. Tol.	Aesthetic Value	Wildlife Value	Flower Color	Bloom Period	Landscape Uses
Carex lurida shallow sedge	MPG	SW	L	GLMmSW	G	Su	W
Carex lyngbyei Lyngbye's sedge	MPG	DS	L	GLMmS	G	SpSu	M
Carex macloviana Falkland Island sedge	SPG	S	L	GLMmS	Br	Su	M
Carex magellanica boreal-bog sedge	MPG	SsW	L	GLMmS	G	Su	BW
Carex meadii Mead's sedge	MPG	DS	L	GLMmS	G	Su	M
Carex microdonta little-tooth sedge	SPG	DS	—	GLMmS	G	Su	M
Carex microptera small-wing sedge	P	FW	L	GLMmSW	G	Su	W
Carex muhlenbergii Muhlenberg's sedge	SPG	s	—	GLMms	G	Sp	EI
Carex multicaulis many-stem sedge	SPG	DFS	L	GLMmSW	G	Su	MS
Carex nebrascensis Nebraska sedge	MPG	FSW	L	GLMmSW	G	Su	MW
Carex nigromarginata black-edge sedge	SPG	Ss	L	GLMmS	PuG	Sp	BIM
Carex obnupta slough sedge	MPG	S	L	GLMmSW	Bl	Su	M
Carex obtusata blunt sedge	SPG	DS	L	GLMmS	G	Su	M
Carex occidentalis western sedge	P	—	L	GLMmS	G	Su	—
Carex oligosperma few-seed sedge	MPG	FSW	L	GLMmSW	G	Su	W
Carex pauciflora few-flower sedge	SPG	FSW	L	GLMmSW	G	SpSu	SW
Carex pedunculata long-stalk sedge	SPG	s	L	GLMmS	G	Sp	I
Carex pensylvanica Pennsylvania sedge	SPG	S	L	GLMmS	G	Sp	B
Carex planostachys cedar sedge	SPG	DSs	—	GLMmS	G	Sp	B
Carex plantaginea broad scale sedge	SPG	s	L	GLMmS	G	Sp	I
Carex platyphylla plantain-leaf sedge	SPG	s	L	GLMmS	G	Sp	I
Carex prairea prairie sedge	MPG	FSW	L	GLMmSW	G	SpSu	SW
Carex rosea rosy sedge	MPG	Ss	L	GLMmS	G	SpSu	BI

PLANT LIST	Plant Type	Env. Tol.	Aesthetic Value	Wildlife Value	Flower Color	Bloom Period	Landscape Uses
Carex rossii Ross' sedge	SPG	DS	L	GLMmS	G	SpSu	BM
Carex rostrata swollen beaked sedge	MPG	SW	L	GLMmS W	G	Su	W
Carex scabrata eastern rough sedge	MPG	FSs	L	GLMmS W	G	Su	BIMS
Carex schottii Schott's sedge	MPG	S	L	GLMmS	B	Su	M
Carex schweinitzii Schweinitz's sedge	MPG	FSsW	L	GLMmS W	G	SpSu	BIMW
Carex scoparia pointed broom sedge	MPG	NaSW	L	GLMmS W	G	Su	SW
Carex scopulorum Holm's Rocky Mountain sedge	SPG	S	L	GLMmS	G	Su	M
Carex senta western rough sedge	MPG	DS	L	GLMmS	G	Su	M
Carex spectabilis northwestern showy sedge	MPG	—	L	GLMmS	G	SpSu	—
Carex sprengelii long-beak sedge	P	FW	L	GLMmS W	G	Su	W
Carex squarrosa squarrose sedge	MPG	FSs	L	GLMmS W	G	SuF	BIMW
Carex sterilis dioecious sedge	SPG	FS	L	GLMmS W	G	SpSu	MW
Carex stricta uptight sedge	MPG	SW	L	GLMmS W	G	Su	W
Carex teneriformis Sierran slender sedge	SPG	DS	L	GLMmS	Br	Su	M
Carex tetanica rigid sedge	MPG	FSsW	L	GLMmS W	PuBr	SpSu	BIMW
Carex tribuloides blunt broom sedge	MPG	FSs	L	GLMmS W	G	SuF	BIMSW
Carex trichocarpa hairy-fruit sedge	MPG	FS	L	GLMmS W	Pu	Su	SW
Carex trisperma three-seed sedge	MPG	FSsW	L	GLMmS W	G	Su	BIW
Carex tumulicola foothill sedge	MPG	DSs	L	GLMmS W	Br	Su	BW
Carex typhina cat-tail sedge	MPG	FSsW	L	GLMmS W	G	SuF	IMW
Carex vesicaria lesser bladder sedge	MPG	SW	—	GLMmS W	GBr	Su	W
Carex viridula little green sedge	SPG	FS	L	GLMmS W	G	SuF	MW
Carex vulpinoidea common fox sedge	MPG	SsW	L	GLMmS W	G	Su	IW

PLANT LIST	Plant Type	Env. Tol.	Aesthetic Value	Wildlife Value	Flower Color	Bloom Period	Landscape Uses
Carpobrotus aequilateralus baby sun-rose	P	S	FL	U	Pu	SpF	EGg
Castilleja angustifolia northwestern Indian-paintbrush	MP	DS	F	U	R	Su	—
Castilleja applegatei wavy-leaf Indian-paintbrush	SP	DS	F	U	R	SpSu	MR
Castilleja coccinea scarlet Indian-paintbrush	SAB	SW	F	H	RY	SpSu	MW
Castilleja exilis small-flower annual Indian-paintbrush	MA	SW	F	U	R	SpSu	MW
Castilleja flava lemon-yellow Indian-paintbrush	SP	S	F	U	Y	SpSu	M
Castilleja indivisa entire-leaf Indian-paintbrush	SA	SW	F	BH	R	SpSu	bWwg
Castilleja integra squawfeather	SP	s	F	U	R	SpSu	I
Castilleja linariifolia Wyoming Indian-paintbrush	SP	Ss	F	HM	R	Su	BbM
Castilleja lutescens stiff yellow Indian-paintbrush	SP	—	F	HM	Y	Su	—
Castilleja martinii Martin's Indian-paintbrush	SP	DS	F	HM	—	—	—
Castilleja miniata great red Indian-paintbrush	MP	S	F	HM	R	SpF	BbM
Castilleja minor alkali Indian-paintbrush	SP	SW	F	HM	OPR	SpSu	MW
Castilleja rhexiifolia rosy Indian-paintbrush	P	FW	F	U	PR	Su	W
Castilleja sulphurea sulphur Indian-paintbrush	SP	S	P	HM	Y	SpSu	bM
Catabrosa aquatica water whorl grass	AQ SPG	SW	—	U	BrY	Su	W
Caulanthus amplexicaulis clasping-leaf wild cabbage	MP	S	—	U	Pu	SpSu	—
Caulophyllum thalictroides blue cohosh	MP	s	L	U	G	Su	Bl
Cayaponia quinqueloba five-lobe-cucumber	HV	FS	—	U	GW	Su	B
Cenchrus carolinianus coastal sand burr	MPG	FSs	—	MS	G	SuF	EMS
Cenchrus longispinus innocent-weed	MAG	DFS	—	MS	G	SuF	EMS
Cenchrus tribuloides sand-dune sandburr	SAG	S	—	SM	Br	SuF	BEM
Centaurea americana American star-thistle	LA	S	F	BGhS	PW	SpSu	bCgM

PLANT LIST	Plant Type	Env. Tol.	Aesthetic Value	Wildlife Value	Flower Color	Bloom Period	Landscape Uses
Centaurium exaltatum desert centaury	SA	S	F	U	P	Su	M
Centella asiatica spadeleaf	SP	FS	L	W	W	SpF	W
Centrosema virginianum spurred butterfly-pea	PHV	DSs	F	U	LPu	SpSuF	BgM
Cerastium arvense field mouse-ear chickweed	SP	S	—	U	W	SpSu	M
Cerastium beeringianum Bering Sea mouse-ear chickweed	SP	S	—	U	W	Su	M
Ceratophyllum demersum coontail	AQP	W	L	W	G	Su	wg
Chaenactis carphoclinia pebble pincushion	SA	DS	F	U	WY	Sp	M
Chaenactis fremontii morningbride	SA	DS	F	U	WY	Sp	M
Chaenactis santolinoides Santolina pincushion	SP	DSs	F	U	WY	Su	BM
Chaerophyllum procumbens spreading chevril	SA	s	—	U	W	Sp	I
Chaetopappa asteroides Arkansas leastdaisy	SA	DS	F	B	BP	SpSu	bCGM
Chaetopappa bellidifolia white-ray leastdaisy	SA	S	F	B	LVY	Sp	bGgM
Chaetopappa effusa spreading leastdaisy	MP	S	F	S	WY	SuF	M
Chamaecrista fasciculata sleepingplant	MP	DS	F	U	Y	SpSu	BM
Chamaesaracha sordida hairy five-eyes	P	—	—	U	W	SpSuF	—
Chamaesyce angusta blackfoot sandmat	SP	S	—	GMmS	W	SpSuF	BM
Chamaesyce bombensis dixie sandmat	SA	S	L	GMmS	G	SpF	CGM
Chamaesyce cordifolia heart-leaf sandmat	SA	DS	—	GMmS	W	SuF	CGM
Chamaesyce fendleri Fendler's sandmat	SP	DS	—	GMmS	W	SpSuF	M
Chamaesyce garberi Garber's sandmat	A	—	—	GMmS	W	WSpSuF	—
Chamaesyce geyeri Geyer's sandmat	SP	DS	—	GMmS	W	SuF	EM
Chamaesyce missurica prairie sandmat	SA	DS	—	GMmS	W	Su	CEG
Chamaesyce polygonifolia seaside sandmat	SA	S	L	GMmS	G	SpF	CS

PLANT LIST	Plant Type	Env. Tol.	Aesthetic Value	Wildlife Value	Flower Color	Bloom Period	Landscape Uses
Chamaesyce serpens matted sandmat	SA	S	—	GMmS	W	SpSuF	CM
Chamaesyce serpyllifolia thyme-leaf sandmat	SA	DS	—	GMmS	W	SuF	CGM
Chaptalia tomentosa woolly sunbonnets	SP	FS	—	U	W	Sp	W
Chasmanthium latifolium Indian wood-oats	MPG	FSs	DF	U	BrG	SuF	BElgs
Chasmanthium laxum slender wood-oats	MPG	FS	DF	U	BrG	SuF	SW
Cheilanthes lanosa hairy lip fern	SP	DS	L	M	X	X	R
Chelone glabra white turtlehead	MP	FSs	F	Bh	W	SuF	BbIW
Chelone lyonii pink turtlehead	MP	FSs	F	Bh	Pu	SuF	BbIW
Chenopodium atrovirens pinyon goosefoot	SA	—	—	GMmS	W	Su	C
Chenopodium desiccatum arid-land goosefoot	SA	DS	—	GMmS	W	SuF	CM
Chenopodium leptophyllum narrow-leaf goosefoot	MA	DS	—	GMmS	G	Su	CM
Chimaphila maculata striped prince's-pine	eSP	s	FL	U	W	Su	GI
Chimaphila menziesii little prince's-pine	eSP	s	FL	U	W	Su	GI
Chloracantha spinosa Mexican devilweed	P	—	—	U	W	SuF	—
Chloris ciliata fringed windmill grass	MPG	S	DF	U	G	SpSuF	M
Chloris cucullata hooded windmill grass	MPG	DS	—	U	G	SpSuF	EM
Chloris pluriflora multi-flower windmill grass	MPG	DSs	DF	M	GW	SuF	ABgM
Chloris verticillata tumble windmill grass	SPG	DS	F	U	G	SuF	EMR
Chorizanthe brevicornu brittle spineflower	SA	DS	F	U	W	—	CM
Chorizanthe rigida devil's spineflower	SA	DS	F	U	Y	Sp	CR
Chrysopsis gossypina cottony golden-aster	MP	S	F	BS	Y	F	BEM
Chrysopsis mariana Maryland golden-aster	SP	Ss	F	BS	Y	SuF	BbgM
Cicuta bulbifera bulblet-bearing water-hemlock	MP	FSW	FL	U	W	SuF	Wwg

PLANT LIST	Plant Type	Env. Tol.	Aesthetic Value	Wildlife Value	Flower Color	Bloom Period	Landscape Uses
Cicuta douglassii western water-hemlock	LP	FSW	FL	U	W	SuF	SW
Cicuta maculata spotted water-hemlock	LP	FSW	FL	U	W	SuF	W
Cimicifuga americana mountain bugbane	MP	s	FL	U	W	SuF	BgI
Cimicifuga racemosa black bugbane	LP	s	FL	U	W	Su	Bgl
Cinna arundinacea sweet wood-reed	LPG	s	FL	U	G	SuF	BI
Circaea alpina small enchanter's-nightshade	SP	s	—	U	W	SpSu	I
Circaea lutetiana broad-leaf enchanter's-nightshade	MP	s	—	U	W	Su	I
Cirsium canescens prairie thistle	SBP	SW	F	BHS	PuR	SuF	MW
Cirsium discolor field thistle	LB	S	F	BS	P	SuF	BM
Cirsium flodmanii Flodman's thistle	MB	S	F	BS	Pu	SuF	M
Cirsium horridulum yellow thistle	MP	FS	F	L	LY	SpSu	MW
Cirsium muticum swamp thistle	LP	FSW	F	BSW	P	SuF	W
Cistanthe umbellata Mt. Hood pussypaws	SP	DFSs	LF	U	PW	Su	IMR
Cladium mariscoides smooth saw-grass	MPG	FNaSW	—	U	Br	SuF	SWwg
Cladium mariscus swamp saw-grass	MPG	FNaSW	—	U	Br	SuF	W
Cladonia evansii lichen	m	DS	—	Mm	X	X	IR
Cladonia leporina lichen	m	DS	—	Mm	X	X	IR
Cladonia prostrata lichen	m	DS	—	Mm	X	X	IR
Cladonia subtenuis lichen	m	DS	—	Mm	X	X	IR
Clarkia rhomboidea diamond fairyfan	MA	DS	F	U	PPu	SpSu	CEM
Claytonia caroliniana Carolina springbeauty	SP	s	F	U	PW	Sp	BgI
Claytonia lanceolata lance-leaf springbeauty	SP	Ss	F	U	PWY	SpSu	BgIM
Claytonia perfoliata miner's-lettuce	SA	Ss	F	U	PW	SpSu	BgIM

PLANT LIST	Plant Type	Env. Tol.	Aesthetic Value	Wildlife Value	Flower Color	Bloom Period	Landscape Uses
Claytonia sibirica Siberian springbeauty	SA	Ss	F	U	PW	Sp	BgIM
Claytonia virginica Virginia springbeauty	SP	s	F	U	PW	Sp	BgI
Clematis drummondii Texas virgin's-bower	PHV	DSs	DF	U	W	SpSuF	Bs
Clematis fremontii Fremont's leather-flower	SP	DS	F	B	BPuY	Sp	bM
Clematis pitcheri bluebill	PHV	Ss	F	U	BPu	Su	B
Clintonia borealis yellow bluebead-lily	SP	Fs	Ff	U	yg	Sp	IW
Clintonia umbellulata white bluebead-lily	SP	s	FL	U	W	Sp	I
Clintonia uniflora bride's bonnet	SP	s	F	U	W	SpSu	gI
Clitoria mariana Atlantic pigeonwings	HV	DSs	F	U	B	Su	BGg
Cnidoscolus stimulosus finger-rot	MP	DSs	F	U	W	SpF	BM
Cnidoscolus texanus Texas bull-nettle	MP	DS	F	U	W	SpSuF	gM
Cocculus diverisifolius snailseed	LP	Ss	Ff	U	WY	Su	BM
Coelorachis cylindrica Carolina joint-tail grass	MPG	Ss	—	U	G	—	BM
Collinsia childii child's blue-eyed Mary	SA	S	F	G	BPu	SuSp	CM
Collinsia parviflora small-flower blue-eyed Mary	SA	S	L	G	BW	Sp	gR
Collinsia torreyi Torrey's blue-eyed Mary	SA	S	F	G	BY	SpSu	CM
Collinsonia canadensis richweed	MP	s	—	G	Y	SuF	CgI
Collomia grandiflora large-flower mountain-trumpet	MA	S	F	U	P	Su	CB
Collomia heterophylla variable-leaf mountain-trumpet	SA	s	F	U	PuR	Sp	CI
Collomia linearis narrow-leaf mountain-trumpet	SAB	DS	F	U	BP	SpSu	CMS
Comandra umbellata bastard-toadflax	SP	FSs	FL	U	W	SpSu	BIM
Comarum palustre purple marshlocks	SP	FSW	F	GLMmS	R	Su	Wwg
Commelina erecta white-mouth dayflower	SP	DS	F	U	BW	Su	BM

PLANT LIST	Plant Type	Env. Tol.	Aesthetic Value	Wildlife Value	Flower Color	Bloom Period	Landscape Uses
Commelina virginica Virginia dayflower	MP	FS	F	GMS	B	Su	BS
Conopholis americana American squawroot	SP	s	B	M	yg	Sp	I
Conyza canadensis Canadian horseweed	MP	S	—	U	WY	SuF	E
Coptis aspleniifolia fern-leaf goldthread	SP	s	fL	U	W	SpSu	GgI
Coptis laciniata Oregon goldthread	SP	s	fL	U	W	SpSu	GgI
Coptis occidentalis Idaho goldthread	SP	s	fL	U	W	SpSu	GgI
Coptis trifolia three-leaf goldthread	eSP	FSs	L	U	W	SpSu	GgIW
Corallorrhiza maculata summer coralroot	MP	s	F	U	BrRY	SpF	I
Cordia podocephala Texas manjack	SP	DSs	F	U	W	SpSuF	BM
Cordylanthus maritimus saltmarsh bird's-beak	SA	NaSW	—	U	W	Su	MSW
Cordylanthus nevinii Nevin's bird's-beak	SA	S	—	U	PuY	SuF	M
Cordylanthus ramosus bushy bird's-beak	SA	—	—	U	Y	Su	—
Cordylanthus rigidus stiff-branch bird's beak	MA	DS	—	U	PuWY	Su	—
Coreopsis basalis golden-mane tickseed	SA	S	F	BS	Y	Sp	bCM
Coreopsis grandiflora large-flower tickseed	MP	DS	F	BhS	Y	Su	BbgM
Coreopsis lanceolata lance-leaf tickseed	MP	DS	F	BS	Y	SpSu	bgM
Coreopsis major greater tickseed	MP	S	F	h	Y	Su	Bg
Coreopsis palmata stiff tickseed	MP	DS	F	h	Y	Su	gM
Coreopsis tripteris tall tickseed	LP	SD	F	BGmS	Y	Su	BbgM
Cornus canadensis Canadian bunchberry	SP	FSs	AFfL	BGS	WY	SpSu	BGI
Corydalis flavula yellow fumewort	SP	s	F	U	Y	Sp	BgI
Corydalis micrantha small-flower fumewort	SA	FSs	FL	U	Y	Sp	BC
Cotula coronopifolia common brassbuttons	SP	FNaSW	F	U	Y	WSp	Wwg

PLANT LIST	Plant Type	Env. Tol.	Aesthetic Value	Wildlife Value	Flower Color	Bloom Period	Landscape Uses
Crassula aquatica water pygmyweed	SA	FNaSW	—	U	GW	SuF	Wwg
Crepis acuminata long-leaf hawk's-beard	SP	DSs	FL	U	Y	SuSp	B
Crepis modocensis siskiyou hawk's-beard	P	—	—	U	Y	Su	—
Crepis occidentalis large-flower hawk's-beard	AP	—	—	U	Y	Su	—
Crepis runcinata fiddle-leaf hawk's-beard	SP	NaS	FL	U	W	Su	M
Cressa truxillensis spreading alkali-weed	SP	FS	—	U	W	SpSu	—
Crinum americanum seven-sisters	MP	FSsW	—	U	W	SuF	—
Croton argyranthemus healing croton	SP	S	—	GMmS	G	SuF	M
Croton capitatus hogwort	LA	DS	—	GMmS	W	SuF	CM
Croton dioicus grassland croton	SP	S	—	GMmS	W	SuF	M
Croton glandulosus vente-conmigo	SA	Ss	F	GMmS	G	SuF	BC
Croton punctatus Gulf croton	MP	S	L	GMmS	G	SpW	E
Crotonopsis elliptica egg-leaf rushfoil	SA	DFS	—	U	W	Su	CB
Cryptantha affinis quill cat's-eye	SA	S	—	U	W	Su	CBM
Cryptantha ambigua basin cat's-eye	SA	S	—	U	W	Su	CBM
Cryptantha cinerarius gray spring-parsley	S	DS	—	U	W	SpSu	—
Cryptantha cinerea James' cat's-eye	SBP	DS	—	U	WY	SpSu	—
Cryptantha circumscissa cushion cat's-eye	SA	—	—	U	W	SpSu	—
Cryptantha hoffmannii Hoffmann's cat's-eye	BP	—	—	S	W	Su	—
Cryptantha maritima Guadalupe cat's-eye	A	—	—	U	W	Sp	—
Cryptantha roosiorum bristle-cone cat's-eye	P	—	—	S	W	Su	—
Cryptantha watsonii Watson's cat's-eye	A	—	—	U	W	SpSu	—
Cryptogramma acrostichoides American rockbrake	SP	S	L	GLM	X	X	R

PLANT LIST	Plant Type	Env. Tol.	Aesthetic Value	Wildlife Value	Flower Color	Bloom Period	Landscape Uses
Cryptotaenia canadensis Canadian honewort	MP	s	L	U	W	SuF	BI
Ctenium aromaticum toothache grass	MPG	FSs	F	U	G	SuF	BW
Cunila marina common dittany	SP	DSs	FL	U	BPu	SuF	BI
Cymopterus bulbosus bulbous spring-parsley	SP	DS	—	U	PPu	Sp	M
Cymopterus cinerarius gray spring-parsley	SP	Ss	—	U	Pu	Sp	B
Cymopterus longipes long-stalk spring-parsley	P	S	—	U	YW	Sp	M
Cymopterus montanus mountain spring-parsley	SP	S	—	U	PuW	Sp	MW
Cynanchum angustifolium Gulf coast swallow-wort	HV	FNaSW	F	U	GW	SpSu	Wwg
Cyperus bipartitus shining flat sedge	SPG	FS	L	GSW	G	SuF	SWwg
Cyperus eragrostis tall flat sedge	MPG	S	L	GLmSW	G	Su	W
Cyperus esculentus chufa	SPG	S	L	GLmSW	G	Su	Wwg
Cyperus fendlerianus Fendler's flat sedge	MPG	S	L	GLmSW	Br	Su	M
Cyperus filiformis wiry flat sedge	P	S	L	GSW	G	Su	M
Cyperus schweinitzii sand flat sedge	SPG	DS	L	GLmS	G	Su	M
Cyperus squarrosus awned flat sedge	SAG	FS	L	GSW	G	Su	Wwg
Cypripedium acaule pink lady's-slipper	SP	s	FL	U	P	Su	gI
Cypripedium arietinum ram-head lady's-slipper	SP	FsW	FL	U	R	Sp	IW
Cypripedium calceolus showy lady's slipper	MP	FSs	FL	U	Y	SpSu	gIW
Cypripedium parviflorum lesser yellow lady's-slipper	MP	FSs	FL	U	Y	Sp	BgIW
Cypripedium pubescens greater yellow lady's-slipper	MP	s	F	U	Y	Sp	gI
Cyrtopodium punctatum cowhorn orchid	SP	FsW	F	U	W	SpSu	I
Cystopteris bulbifera bulblet bladder fern	SP	s	L	M	X	X	I
Cystopteris fragilis brittle bladder fern	SP	FSs	L	GLM	X	X	BIS

PLANT LIST	Plant Type	Env. Tol.	Aesthetic Value	Wildlife Value	Flower Color	Bloom Period	Landscape Uses
Dalea albiflora white-flower prairie-clover	P	DS	F	BGLM	W	Su	bM
Dalea aurea golden prairie-clover	SP	DS	F	B	Y	Su	gM
Dalea candida white prairie-clover	SP	DS	F	BGLM	W	Su	bM
Dalea compacta compact prairie-clover	MP	S	F	Bm	Y	Su	M
Dalea enneandra nine-anther prairie-clover	MP	DS	F	B	W	Su	Bg
Dalea multiflora round-head prairie-clover	SP	S	F	Bm	W	Su	M
Dalea pinnata summer-farewell	MP	DS	F	m	W	F	BgM
Dalea purpurea violet prairie-clover	MP	DS	F	BGLM	Pu	SuF	bM
Dalea versicolor oakwoods prairie-clover	P	—	F	BG	YW	SpSuF	M
Dalea villosa silky prairie-clover	MP	DS	F	BL	P	Su	bM
Dalibarda repens robin-run-away	SP	FSW	F	U	W	Su	gIW
Danthonia californica California wild oat grass	SPG	DSs	FL	M	G	Su	BM
Danthonia compressa flattened wild oat grass	MPG	S	FL	M	G	Su	BM
Danthonia intermedia timber wild oat grass	SPG	SW	FL	M	G	Su	MW
Danthonia spicata poverty wild oat grass	SPG	S	FL	M	G	Su	BM
Danthonia unispicata few-flower wild oat grass	SPG	DSs	—	U	—	Su	BgR
Delphinium andersonii desert larkspur	SP	DSs	F	U	B	—	BgI
Delphinium barbeyi subalpine larkspur	P	Ss	F	U	—	—	BgI
Delphinium bicolor flat-head larkspur	SP	Ss	F	U	BPu	SpSu	BgI
Delphinium carolinianum Carolina larkspur	MP	S	F	U	W	SpSu	BgM
Delphinium geyeri Geyer's larkspur	P	Ss	F	U	W	Su	BgI
Delphinium nuttallianum two-lobe larkspur	SP	DS	F	U	W	Su	M
Delphinium occidentale dunce-cap larkspur	LP	s	F	U	B	Su	gI

C-80

Landscape Restoration Handbook

PLANT LIST	Plant Type	Env. Tol.	Aesthetic Value	Wildlife Value	Flower Color	Bloom Period	Landscape Uses
Delphinium tricorne dwarf larkspur	SP	Ss	F	U	B	Sp	BgI
Deparia acrostichoides silver false spleenwort	LP	s	L	M	X	X	I
Deschampsia cespitosa tufted hair grass	MPG	FSW	L	M	G	Su	MWE
Deschampsia danthonioides annual hair grass	MAG	S	L	M	G	Su	M
Deschampsia flexuosa wavy hair grass	MPG	DSs	L	M	G	Su	BIM
Descurainia incana mountain tansy mustard	AB	—	—	U	Y	SpSu	—
Descurainia pinnata western tansy mustard	MA	DS	—	U	W	SpSu	CM
Desmanthus illinoensis prairie bundle-flower	MP	DFS	FfL	Gm	W	Su	CEgM
Desmodium canadense showy tick-trefoil	LP	S	F	GMm	P	Su	B
Desmodium ciliare hairy small-leaf tick-trefoil	SP	DSs	GMmS	GS	PuV	Su	BM
Desmodium glabellum Dillenius' tick-trefoil	LP	Ss	F	GMm	Pu	Su	BI
Desmodium glutinosum pointed-leaf tick-trefoil	MP	s	FL	GMmS	W	Su	I
Desmodium illinoense Illinois tick-trefoil	LP	DS	F	GMm	PuW	Su	M
Desmodium laevigatum smooth tick-trefoil	MP	DSs	F	GMm	Pu	Su	BIM
Desmodium nudiflorum naked-flower tick-trefoil	MP	s	F	GMm	PuW	Su	I
Desmodium obtusum stiff tick-trefoil	MP	DS	F	GMm	Pu	Su	—
Desmodium paniculatum panicled-leaf tick-trefoil	MP	DS	F	GMmS	Pu	Su	M
Desmodium perplexum perplexed tick-trefoil	MP	S	F	GMm	PPu	SuF	B
Desmodium psilophyllum simple-leaf tick-trefoil	MP	Ss	—	GMm	P	SpF	BM
Desmodium rotundifolium prostrate tick-trefoil	SP	Ds	F	GMm	PuW	SuF	I
Desmodium sessilifolium sessil-leaf tick-trefoil	MP	DS	F	GMmS	LP	SpSu	BM
Diarrhena americana American beakgrain	MG	s	BL	M	G	SuF	IS
Dicentra canadensis squirrel-corn	SP	Ss	FL	m	PW	Sp	BgI

PLANT LIST	Plant Type	Env. Tol.	Aesthetic Value	Wildlife Value	Flower Color	Bloom Period	Landscape Uses
Dicentra cucullaria Dutchman's-breeches	SP	Ss	FL	hm	W	Sp	Bgl
Dichanthelium acuminatum tapered rosette grass	SPG	Ss	—	GMS	G	Su	BM
Dichanthelium commutatum variable rosette grass	MPG	s	L	GLMS	G	Su	BI
Dichanthelium dichotomum cypress rosette grass	MPG	Ds	FL	GLMS	G	Su	BI
Dichanthelium laxiflorum open-flower rosette grass	MPG	s	—	GW	G	Su	EIS
Dichanthelium leibergii Leiberg's rosette grass	MPG	DS	L	GLMS	G	Su	M
Dichanthelium linearifolium slim-leaf rosette grass	SPG	Ds	—	GMS	G	Su	BI
Dichanthelium malacophyllum soft-leaf rosette grass	MPG	—	—	GLMS	G	Su	B
Dichanthelium oligosanthes Heller's rosette grass	SPG	s	L	GLMS	G	Su	BI
Dichanthelium pedicellatum cedar rosette grass	SPG	DS	—	GLMS	G	SpF	EM
Dichanthelium ravenelii Ravenel's rosette grass	MPG	DSs	—	G	G	Su	BEM
Dichanthelium sabulorum hemlock rosette grass	SPG	Ds	L	GLMSW	G	Su	BI
Dichanthelium sphaerocarpon round-seed rosette grass	MPG	FSsW	L	GLMSW	G	SuF	BIW
Dichelostemma pulchellum bluedicks	P	—	—	U	PuW	Sp	—
Dicliptera brachiata branched foldwing	MP	Fs	F	U	PuW	SuF	ISW
Dicranum polysetum moss	m	s	—	U	X	X	l
Dicranum undulatum feather moss	m	FsW	—	U	X	X	W
Digitaria californica California crab grass	MG	DS	—	U	G	Su	M
Digitaria texana Texas crab grass	MPG	S	—	GMS	G	WSpF	EM
Diodia teres poorjoe	SA	DS	F	GM	W	SuF	B
Dioscorea villosa wild yam	SP	FS	L	M	Y	Su	B
Disporum hookeri drops-of-gold	SP	Fs	FL	U	GW	SpSu	gIS
Disporum lanuginosum yellow fairybells	SP	s	Ff	U	Y	Sp	gI

PLANT LIST	Plant Type	Env. Tol.	Aesthetic Value	Wildlife Value	Flower Color	Bloom Period	Landscape Uses
Disporum smithii large-flower fairybells	P	—	F	U	W	Sp	—
Distichlis spicata coastal salt grass	MPG	DFNa SW	FL	MmW	G	SuF	SWwg
Dodecatheon jeffreyi tall mountain shootingstar	SP	FS	F	U	PR	Su	MS
Dodecatheon meadia eastern shootingstar	SP	DSs	FL	U	PW	Sp	gIM
Dodecatheon pulchellum dark-throated shootingstar	SP	Fs	F	U	P	Sp	M
Dodecatheon redolens scented shootingstar	SP	FS	FL	U	Pu	Sp	gM
Downingia bella Hoover's calico-flower	A	FW	—	U	B	SpSu	W
Draba albertina slender whitlow-grass	SBPG	FS	—	M	W	SpSu	MW
Draba cinera gray-leaf whitlow-grass	P	—	—	U	W	SpSu	—
Draba corrugata southern California whitlow-grass	SPG	Ss	FL	M	Y	SpSu	BIM
Draba verna spring whitlow-grass	SPG	DS	—	M	W	Sp	M
Dracocephalum parviflorum American dragonhead	MABP	DS	F	U	BL	SpSu	M
Drepanocladus revolvens moss	m	FSW	L	M	X	X	W
Drosera capillaris pink sundew	SP	FS	F	U	—	SpSu	W
Drosera intermedia spoon-leaf sundew	SP	FS	F	U	W	SpSu	W
Drosera rotundifolia round-leaf sundew	SP	FSW	F	U	W	Su	W
Dryopteris campyloptera mountain wood fern	MP	s	L	U	X	X	BgI
Dryopteris carthusiana spinulose wood fern	MP	FsW	L	U	X	X	BIW
Dryopteris cristata crested wood fern	eMP	FSs	L	U	X	X	BIW
Dryopteris expansa spreading wood fern	MP	s	L	U	X	X	BIW
Dryopteris goldiana Goldie's wood fern	MP	s	L	U	X	X	BgI
Dryopteris intermedia evergreen wood fern	MP	s	L	U	X	X	BgI
Dryopteris marginalis marginal wood fern	MP	s	L	U	X	X	BgI

PLANT LIST	Plant Type	Env. Tol.	Aesthetic Value	Wildlife Value	Flower Color	Bloom Period	Landscape Uses
Dudleya farinosa powdery live-forever	SP	DS	FL	U	W	SuF	gRM
Dugaldia hoopesii owl's-claws	LP	D	F	U	OY	Su	M
Dulichium arundinaceum three-way sedge	MPG	FS	L	U	G	Su	Wwg
Dyschoriste linearis polkadots	SP	DS	F	U	Pu	SpSu	GM
Dyschoriste oblongifolia oblong-leaf snakeherb	SP	DS	F	U	Pu	Su	—
Echinacea angustifolia blacksamson	MP	DS	F	BS	Pu	SpF	bM
Echinacea pallida pale purple-coneflower	MP	DS	F	BS	P	Su	bM
Echinacea purpurea eastern purple-coneflower	MP	DSs	F	BS	PPu	SuF	BbgM
Echinochloa crus-galli large barnyard grass	MAG	FSW	L	GLSW	G	SuF	Wwg
Echinochloa walteri long-awn cock's-spur grass	MAG	FSW	L	GLSW	G	SuF	Wwg
Echinocystis lobata wild cucumber	HV	FSs	FL	M	GW	SuF	S
Eleocharis acicularis neddle spike-rush	SPG	FS	L	LW	G	SuF	W
Eleocharis albida white spike-rush	MPG	FNaSW	L	LW	G	SuF	W
Eleocharis bella delicate spike-rush	MPG	SW	L	LW	G	SuF	MW
Eleocharis compressa flat-stem spike-rush	MPG	FS	L	LW	BrPu	SuF	Wwg
Eleocharis elliptica elliptic spike-rush	MPG	FS	L	LW	G	SpF	EW
Eleocharis elongata slim spike-rush	MPG	SW	L	LW	G	SuF	MW
Eleocharis equisetoides horsetail-spike-rush	SPG	FWS	L	LW	G	SuF	Wwg
Eleocharis erythropoda bald spike-rush	MPG	FS	L	LW	G	SuF	Wwg
Eleocharis fallax creeping spike-rush	MPG	FNaS	L	LW	G	Su	SW
Eleocharis flavescens yellow spike-rush	SPG	FS	L	LW	G	SuF	Wwg
Eleocharis intermedia matted spike-rush	SAG	FS	L	LW	G	SuF	CMSW
Eleocharis obtusa blunt spike-rush	MPG	FNaS	L	LW	G	Su	W

PLANT LIST	Plant Type	Env. Tol.	Aesthetic Value	Wildlife Value	Flower Color	Bloom Period	Landscape Uses
Eleocharis palustris pale spike-rush	MPG	FNaSW	L	LW	G	Su	Wwg
Eleocharis parvula little-head spike-rush	SAG	FSW	L	LW	G	SpSuF	CSW
Eleocharis quinqueflora few-flower spike-rush	MPG	SW	L	LW	G	Su	Wwg
Eleocharis rostellata beaked spike-rush	MPG	FNaS	L	LW	G	SuF	W
Eleocharis smallii Small's spike-rush	MPG	FNaS	L	LW	G	SuF	W
Eleocharis vivipara viviparous spike-rush	SPG	F	L	LW	G	SuF	Wwg
Elephantopus carolinianus Carolina elephant's-foot	SP	s	FL	B	PPu	SuF	I
Elephantopus nudatus smooth elephant's-foot	SP	FSs	FL	B	PPu	SuF	IW
Elymus canadensis nodding wild rye	LPG	DSs	FfL	LMS	G	Su	EMS
Elymus elymoides western bottle-brush grass	MPG	DS	L	LMS	G	Su	EM
Elymus glaucus blue wild rye	MPG	DFS	DFL	LMS	G	Su	BMS
Elymus hystrix eastern bottle-brush grass	LPG	DSs	DFL	LMS	G	Su	BIM
Elymus lanceolatus streamside wild rye	MPG	FS	L	LMS	G	Su	MS
Elymus trachycaulus slender wild rye	SPG	DS	L	LMS	G	Su	M
Elymus virginicus Virginia wild rye	MPG	FSs	DFL	LMS	G	SuF	BIMS
Elyonurus barbiculmis wool-spike grass	MPG	DS	FL	LMS	G	SuF	M
Elyonurus tripsacoides pan-American balsamscale	MPG	FS	FL	LMS	G	SuF	MS
Emmenanthe penduliflora yellow whispering-bells	SA	DS	FL	U	Y	Su	BgM
Encelia farinosa goldenhills	LP	DS	F	U	Y	Sp	M
Enemion biternatum eastern false rue-anemone	SP	s	FL	U	W	Sp	gI
Engelmannia pinnatifida Englemann's daisy	SB	S.	F	U	Y	WSpSu	M
Epifagus virginiana beechdrops	SP	s	B	Mm	Br	SuF	I
Epilobium angustifolium fireweed	LP	S	FL	Mm	P	SuF	g

PLANT LIST	Plant Type	Env. Tol.	Aesthetic Value	Wildlife Value	Flower Color	Bloom Period	Landscape Uses
Epilobium brachycarpum tall annual willowherb	SA	DS	F	Mm	PW	SuF	gM
Epilobium ciliatum fringed willowherb	SA	Fs	F	Mm	PuW	SuF	IW
Epilobium glaberrimum glaucous willowherb	MA	FS	F	Mm	PuW	SuF	BW
Epilobium palustre marsh willowherb	SP	DS	F	Mm	PW	Su	W
Epilobium strictum downy willowherb	SP	FSW	F	Mm	PR	SuF	W
Equisetum arvense field horsetail	MP	FS	L	LMW	X	X	SWwg
Equisetum fluviatile water horsetail	MP	FSW	L	LMW	X	X	SWwg
Equisetum hyemale tall scouring-rush	MP	SF	L	LMW	X	X	SWwg
Equisetum sylvaticum woodland horsetail	SP	Ss	L	LM	X	X	BI
Equisetum variegatum varigated scouring-rush	SP	FSW	L	LMW	X	X	SWwg
Eragrostis intermedia plains love grass	MPG	S	FL	MS	GPu	SuF	EM
Eragrostis secundiflora red love grass	MPG	DS	DF	U	G	Su	EM
Eragrostis spectabilis petticoat-climber	SPG	DS	FL	MS	GPu	SuF	EIM
Eragrostis trichodes sand love grass	MPG	DS	FL	MS	GPu	SuF	EM
Erigeron arizonicus Arizona fleabane	P	—	—	U	PW	Su	—
Erigeron bellidiastrum western daisy fleabane	MA	DS	F	B	PW	SpSu	BbM
Erigeron breweri Brewer's fleabane	SP	DS	F	B	Pu	Su	bGM
Erigeron caespitosus tufted fleabane	SP	SD	F	B	PW	SpSu	BbM
Erigeron compositus dwarf mountain fleabane	SP	DS	F	B	BPW	Su	bGgM
Erigeron concinnus Navajo fleabane	SP	DS	F	B	BW	Su	bEGM
Erigeron divergens spreading fleabane	SAP	DS	F	B	—	Su	bEM
Erigeron engelmannii Engelmann's fleabane	SP	DS	F	B	W	SpSu	BbM
Erigeron eximius spruce-fir fleabane	P	S	F	B	BP	Su	gM

PLANT LIST	Plant Type	Env. Tol.	Aesthetic Value	Wildlife Value	Flower Color	Bloom Period	Landscape Uses
Erigeron flagellaris trailing fleabane	SP	DS	F	B	W	Sp	bEGM
Erigeron formosissimus beautiful fleabane	SP	S	F	B	BPW	Su	M
Erigeron glaucus seaside fleabane	SP	DFS	F	B	LPu	Su	bGMS
Erigeron lemmonii Lemmon's fleabane	P	—	—	B	PW	SpSuF	—
Erigeron linearis desert yellow fleabane	SP	DS	F	B	Y	Su	bGM
Erigeron nematophyllus needle-leaf fleabane	SP	DS	F	B	W	Sp	BbM
Erigeron neomexicanus New Mexico fleabane	P	—	—	B	PW	SuF	—
Erigeron oreophilus chaparral fleabane	P	—	—	B	PW	SuF	—
Erigeron parishii Parish's fleabane	SP	DS	FL	U	PuV	Sp	gM
Erigeron platyphyllus broad-leaf fleabane	P	—	—	B	P	SuF	—
Erigeron poliospermus purple cushion fleabane	SP	DS	F	B	PuW	Su	bGM
Erigeron pumilus shaggy fleabane	SP	DS	F	B	PW	Sp	BbM
Erigeron strigosus prairie fleabane	MP	DS	F	B	W	SuSp	BbM
Erigeron ursinus Bear River fleabane	SP	Ss	F	B	PuW	Su	BbGM
Eriocaulon parkeri estuary pipewort	SP	FSW	B	U	W	SuF	Wwg
Eriogonum abertianum Abert's wild buckwheat	MA	DS	—	GMmS	WY	SpSuF	M
Eriogonum alatum winged wild buckwheat	SP	DS	—	U	Y	SuF	M
Eriogonum annuum annual wild buckwheat	SAB	DS	F	GLMmS	PW	SpSuF	GgMR
Eriogonum baileyi Bailey's wild buckwheat	SP	DS	F	GLMmS	WY	Su	GgMR
Eriogonum caespitosum matted wild buckwheat	SP	DS	—	U	Y	Su	—
Eriogonum cernuum nodding wild buckwheat	SA	DS	F	G	W	SpSu	—
Eriogonum compositum arrow-leaf wild buckwheat	SP	DS	F	GLMmS	WY	Su	GgMR
Eriogonum elongatum long-stem wild buckwheat	SS	DS	—	GLMmS	W	Su	BM

PLANT LIST	Plant Type	Env. Tol.	Aesthetic Value	Wildlife Value	Flower Color	Bloom Period	Landscape Uses
Eriogonum flavum alpine golden wild buckwheat	SP	DS	F	GLMmS	Y	Su	GgMR
Eriogonum heracleoides parsnip-flower wild buckwheat	SP	DS	F	GLMmS	RWY	Su	GgMR
Eriogonum incanum frosted wild buckwheat	SP	DS	F	GLMmS	RY	Su	GgMR
Eriogonum inflatum Indian-pipeweed	MAP	DNaS	—	GMmS	RY	SpSuF	M
Eriogonum kennedyi Kennedy's wild buckwheat	SP	DS	F	GLMmS	PW	Su	GgMR
Eriogonum nudum naked wild buckwheat	MP	DS	F	GLMmS	W	SuF	gMR
Eriogonum ovalifolium cushion wild buckwheat	SP	DS	F	GLMmS	Y	Su	GgMR
Eriogonum parishii mountainmist	SP	DS	F	GLMmS	P	Su	GgMR
Eriogonum rotundifolium round-leaf wild buckwheat	SA	DS	—	GMmS	W	SpSuF	M
Eriogonum saxatile hoary wild buckwheat	SP	S	F	GLMmS	WY	Su	GgM
Eriogonum tenellum tall wild buckwheat	SP	DS	—	GMmS	PW	SuF	MR
Eriogonum tomentosum dog-tongue wild buckwheat	MP	DS	F	GLMmS	W	SuF	GgM
Eriogonum trichopes little desert trumpet	SA	DS	F	U	Y	SpSu	—
Erioneuron pilosum hairy woolly grass	SGP	DS	—	U	GPu	Su	GgM
Erioneuron pulchellum low woolly grass	SGP	DS	—	U	GPu	SpSu	GgM
Eriophorum angustifolium tall cotton-grass	MPG	FSW	F	S	W	SpSu	Wwg
Eriophorum cringerum fringed cotton-grass	PG	—	—	S	G	Su	—
Eriophorum gracile slender cotton-grass	MPG	SW	F	S	W	SpSu	SWwg
Eriophorum vaginatum tussock cotton-grass	MPG	FSW	F	S	W	SpSu	Wwg
Eriophorum virginicum tawny cotton-grass	MPG	FSW	F	S	OY	SuF	Wwg
Eriophorum viridicarinatum tassel cotton-grass	SPG	FSW	F	S	W	SpSu	GWwg
Eriophyllum confertiflorum yellow-yarrow	SP	DS	F	U	Y	Su	GgM
Eriophyllum lanatum common woolly-sunflower	SP	DS	F	U	Y	SpSu	GgM

PLANT LIST	Plant Type	Env. Tol.	Aesthetic Value	Wildlife Value	Flower Color	Bloom Period	Landscape Uses
Eriophyllum stoechadifolium seaside woolly-sunflower	MP	DS	F	U	Y	Su	gM
Eryngium leavenworthii Leavenworth's eryngo	MP	DS	L	h	W	SuF	M
Eryngium vaseyi coyote-thistle	MP	DS	L	h	W	Su	W
Eryngium yuccifolium button eryngo	LP	S	DL	Bh	GW	Su	gM
Erysimum asperum plains wallflower	MP	DS	F	U	ORY	SpSu	gM
Erythronium albidum small white fawn-lily	SP	s	F	U	W	Sp	I
Erythronium americanum American trout-lily	SP	s	FL	U	Y	Sp	gl
Erythronium grandiflorum yellow avalanche-lily	SP	S	FL	U	Y	SpSu	BM
Erythronium montanum white avalanche-lily	P	—	F	U	W	Su	—
Erythronium umbilicatum dimpled trout-lily	SP	s	FL	U	Y	Sp	EGgI
Eschscholzia californica California-poppy	SP	DS	F	GmS	O	WSu	gM
Eupatorium betonicifolium betony-leaf thoroughwort	MP	SsW	F	B	B	SuF	Wwg
Eupatorium capillifolium dogfennel	MP	DSs	—	U	GPW	SuF	BM
Eupatorium coelestinum blue mistflower	MP	FS	F	B	B	SuF	BbMW
Eupatorium maculatum spotted joe-pye-weed	LP	FS	F	BS	P	Su	BbW
Eupatorium perfoliatum common boneset	MP	FS	F	BhS	W	SuF	bW
Eupatorium rotundifolium round-leaf thoroughwort	MP	FSs	—	S	W	SuF	BM
Eupatorium serotinum late-flowering thoroughwort	MP	Ss	F	Bh	W	SuF	BgIM
Eupatorium villosum Florida Keys thoroughwort	P	—	—	U	PW	WSpSuF	—
Euphorbia corollata flowering spurge	MP	S	F	GMmS	W	SuF	M
Euphorbia cyathophora fire-on-the-mountain	MA	s	L	GMmS	W	Su	I
Euphorbia dentata toothed spurge	SA	DS	L	U	G	SpSuF	CM
Euphorbia heterophylla Mexican-fireplant	MA	DS	L	GMmS	G	SuF	BM

PLANT LIST	Plant Type	Env. Tol.	Aesthetic Value	Wildlife Value	Flower Color	Bloom Period	Landscape Uses
Euphorbia hexagona six-angle spurge	SA	DS	—	U	W	SuF	MSW
Euphorbia ipecacuanhae American ipecac	SP	DSs	—	GMmS	G	SpSu	BM
Euphorbia roemeriana Roemer's spurge	SA	Ss	—	GMmS	W	Sp	BCM
Euphorbia tetrapora weak spurge	SA	DS	—	U	X	Sp	CM
Eustachys petraea pinewoods finger grass	MPG	Ds	F	U	Br	SuF	SW
Euthamia graminifolia flat-top goldentop	MP	S	F	BGLMmS	Y	SuF	bg
Evax caulescens involucrate pygmy-cudweed	SA	DFS	F	U	G	Su	CMS
Festuca arizonica Arizona fescue	SPG	Ss	—	GLMmS	G	Su	B
Festuca brachyphylla short-leaf fescue	SGP	S	—	GLMmS	G	Su	EGM
Festuca californica California fescue	LPG	DSs	—	GLMmS	G	Su	BM
Festuca campestris prairie fescue	MPG	S	—	GLMmS	G	Su	EM
Festuca idahoensis bluebunch fescue	MPG	Ss	—	GLMmS	G	Su	BM
Festuca kingii King's fescue	LPG	DS	—	GLMmS	G	Su	EM
Festuca occidentalis western fescue	MPG	DSs	—	GLMmS	G	Su	BI
Festuca rubra red fescue	MPG	FSW	—	GLMmS	Gr	Su	MW
Fimbristylis autumnalis slender fimbry	SAG	SW	—	U	G	SuF	CEGSW
Fimbristylis caroliniana Carolina fimbry	LPG	FNaS	F	U	G	SuF	BEgS
Fimbristylis castanea marsh fimbry	LPG	FSW	—	U	G	SpSuF	ESW
Fimbristylis puberula hairy fimbry	MG	DS	—	U	G	Su	M
Flaveria floridana Florida yellowtops	MP	DS	—	U	Y	F	M
Florestina tripteris sticky florestina	SA	Ss	—	U	PuW	SuF	BC
Fragaria chiloensis beach strawberry	SP	FS	F	B	W	SpSu	EGS
Fragaria vesca woodland strawberry	SP	Ss	Ff	GLMmS	W	SpSu	BM

PLANT LIST	Plant Type	Env. Tol.	Aesthetic Value	Wildlife Value	Flower Color	Bloom Period	Landscape Uses
Fragaria virginiana Virginia strawberry	SP	DSs	FN	BGLMmS	W	SpSu	bEG
Frasera albicaulis white-stem elkweed	SP	DS	F	U	W	—	gM
Frasera albomarginata desert elkweed	SP	DS	F	U	G	—	gM
Frasera caroliniensis American-columbo	MP	s	F	U	GY	Su	gIM
Frasera speciosa monument plant	LP	DS	F	M	GW	SuF	gM
Fremontodendron californicum California flannelbush	eLS	DSs	F	U	Y	—	ABF
Fritillaria pinetorum pinewoods missionbells	SP	s	F	U	Pu	Su	BI
Froelichia floridana plains snake-cotton	LA	DS	F	U	W	SuF	CM
Froelichia gracilis slender snake-cotton	SA	DS	—	U	W	SpSuF	M
Fuirena squarrosa hairy umbrella sedge	MPG	FSW	L	U	G	SuF	SW
Gaillardia aestivalis lance-leaf blanket-flower	MP	DSs	F	Bh	RY	SpSu	BbM
Gaillardia pulchella firewheel	SA	DS	F	B	PuRY	SpSu	bCgM
Gaillardia suavis perfumeballs	MP	S	F	Bh	RY	SpSu	bM
Galactia canescens hoary milk-pea	PHV	DS	F	U	P	SpSu	M
Galactia volubilis downy milk-pea	SP	DS	—	S	PPu	Su	B
Galax urceolata beetleweed	SP	s	L	M	W	SpSu	BIE
Galearis spectabilis showy orchid	SP	s	F	U	PW	Su	I
Galium aparine sticky-willy	MP	S	—	mW	W	SpSu	B
Galium asprellum rough bedstraw	MP	sW	—	M	W	Su	IW
Galium bifolium twin-leaf bedstraw	SA	S	—	m	W	Su	—
Galium boreale northern bedstraw	MP	FS	—	m	W	Su	S
Galium californicum California bedstraw	P	—	—	m	Y	SpSu	—
Galium circaezans bedstraw licorice	SP	s	L	m	W	Su	I

PLANT LIST	Plant Type	Env. Tol.	Aesthetic Value	Wildlife Value	Flower Color	Bloom Period	Landscape Uses
Galium concinnum shining bedstraw	SP	s	—	m	W	Su	I
Galium hypotrichium alpine bedstraw	P	—	—	m	yg	Su	—
Galium johnstonii Johnston's bedstraw	P	—	—	m	GW	Su	—
Galium obtusum blunt-leaf bedstraw	SP	FSsW	—	m	W	SpSu	W
Galium parishii Parish's bedstraw	SP	—	—	m	W	—	—
Galium pilosum hairy bedstraw	MP	s	—	m	W	Su	I
Galium sparsiflorum Sequoia bedstraw	SP	S	—	m	W	Su	M
Galium texense Texas bedstraw	SA	Ss	—	U	W	Sp	B
Galium tinctorium stiff marsh bedstraw	SP	FSW	—	m	W	SuF	W
Galium trifidum three-petal bedstraw	MP	s	—	m	W	Su	I
Galium triflorum fragrant bedstraw	SP	s	—	m	W	Su	I
Gaura coccinea scarlet beeblossom	SP	DS	F	B	R	SpSu	bM
Gaura parviflora velvetweed	MP	DS	F	B	W	Su	bM
Gaura villosa woolly beeblossom	P	—	—	B	W	SpSu	—
Gayophytum diffusum spreading groundsmoke	SA	DS	F	U	PW	Su	gM
Gayophytum heterozygum zigzag groundsmoke	SA	—	—	U	PW	Su	—
Gayophytum humile dwarf groundsmoke	SA	Ds	—	U	PW	Su	I
Gayophytum ramosissimum pinyon groundsmoke	A	DSs	—	U	PW	SpSu	B
Gentiana andrewsii closed bottle gentian	MP	FWSs	F	H	BW	SuF	BWwg
Gentiana newberryi alpine gentian	SP	S	F	H	B	Su	M
Gentiana puberulenta downy gentian	SP	DS	F	H	B	SuF	M
Gentiana saponaria harvestbells	MP	FSs	F	H	B	F	MW
Geocaulon lividum false toadflax	SP	Fs	—	U	Pu	SpSu	IW

PLANT LIST	Plant Type	Env. Tol.	Aesthetic Value	Wildlife Value	Flower Color	Bloom Period	Landscape Uses
Geranium caespitosum purple cluster crane's-bill	MP	Ss	F	U	PPuW	SpSu	BgI
Geranium maculatum spotted crane's-bill	MP	Ss	FL	GMmS	PPu	SpSu	BgI
Geranium richardsonii white crane's-bill	SP	Ss	F	GMmS	WL	SpSu	Bg
Geranium robertianum herbrobert	SP	S	FL	GMmS	P	SpF	BgI
Geranium viscosissimum sticky purple crane's-bill	MP	S	FL	GMmS	Pu	SpSu	B
Geum canadense white avens	SP	S	F	Gm	W	Su	B
Geum rivale purple avens	SP	FS	F	Gm	PuY	SpSu	W
Geum triflorum old-man's-whiskers	SP	DS	F	Gm	Pu	Sp	gM
Gilia clivorum purple-spot gily-flower	A	—	—	GMm	Y	Sp	—
Gilia interior inland gily-flower	SA	DS	—	GMm	B	SpSu	M
Gilia minor little gily-flower	SA	DS	—	GMm	B	SpSu	M
Gilia scopulorum rock gily-flower	SP	DS	—	GMm	LPY	SpSu	MR
Gilia splendens splendid gily-flower	A	—	—	HGMm	PR	SpSu	g
Gilia tricolor bird's-eyes	SA	S	F	GMm	BY	SpSu	gM
Glandularia bipinnatifida Dakota mock vervain	MP	—	F	U	PPu	WSp	gM
Glandularia canadensis rose mock vervain	MP	DS	F	U	PuW	WSpSuF	M
Glandularia wrightii Davis Mountain mock vervain	SP	DS	F	U	PPu	SpSuF	M
Glaux maritima sea-milkwort	SP	FNaSW	—	U	PW	Su	SW
Glyceria canadensis rattlesnake manna grass	MG	FS	L	W	—	Su	W
Glyceria elata tall manna grass	MG	FSs	L	W	—	Su	MIW
Glyceria grandis American manna grass	MG	FS	L	W	—	Su	SW
Glyceria melicaria melic manna grass	MG	FSs	L	W	—	Su	IW
Glyceria striata fowl manna grass	MG	SF	FL	W	—	Su	W

PLANT LIST	Plant Type	Env. Tol.	Aesthetic Value	Wildlife Value	Flower Color	Bloom Period	Landscape Uses
Glycyrrhiza lepidota American licorice	MP	S	—	U	W	SpSu	M
Goodyera oblongifolia green-leaf rattlesnake-plantain	SP	s	F	U	W	Su	I
Goodyera pubescens downy rattlesnake-plantain	SP	s	L	U	W	SpF	I
Gratiola ebracteata bractless hedge-hyssop	SA	FS	—	U	WY	Su	MW
Gratiola neglecta clammy hedge-hyssop	SP	FS	—	U	WY	SpF	W
Grayia spinosa spiny hop-sage	SS	DS	—	U	W	Sp	M
Grindelia camporum great valley gumweed	MP	DS	F	M	Y	SuF	M
Grindelia hirsutula hairy gumweed	P	—	F	M	Y	SpSu	—
Grindelia integrifolia Pudget Sound gumweed	MP	DS	F	U	Y	SuF	gMS
Grindelia paludosa Suisun Marsh gumweed	P	W	F	U	Y	SuSp	M
Grindelia squarrosa curly-cup gumweed	MBP	DS	F	U	Y	SuF	gMS
Gutierrezia sarothrae kindlingweed	SS	DS	F	GLMm	Y	SuF	—
Gymnocarpium dryopteris western oak fern	SP	Fs	L	U	X	X	GIS
Gymnopogon ambiguus bearded skeleton grass	SPG	S	L	m	G	SuF	M
Gymnopogon brevifolius short-leaf skeleton grass	SPG	S	L	m	G	SuF	M
Hackelia patens spotted stickseed	MP	DS	—	U	W	Su	M
Hackelia virginiana beggar's-lice	MP	s	—	U	BW	SuF	I
Hainardia cylindrica barb grass	AG	—	—	U	G	Su	—
Halodule beaudettei shoalweed	P	Na	—	U	BrG	Su	—
Harbouria trachypleura whiskybroom-parsley	SP	—	—	U	Y	—	—
Hedeoma acinoides slender false pennyroyal	SA	S	—	U	BPuW	Sp	GM
Hedeoma costatum ribbed false pennyroyal	SP	DS	—	U	—	SpSu	R
Hedeoma dentatum Arizona false pennyroyal	SAP	DS	—	U	P	Sp	M

PLANT LIST	Plant Type	Env. Tol.	Aesthetic Value	Wildlife Value	Flower Color	Bloom Period	Landscape Uses
Hedeoma drummondii Drummond's false pennyroyal	SAP	DS	—	U	B	SpSuF	R
Hedeoma hispidum rough false pennyroyal	SA	DS	—	U	B	SpSu	M
Hedeoma hyssopifolium aromatic false pennyroyal	P	—	—	U	P	F	—
Hedeoma oblongifolium oblong-leaf false pennyroyal	P	—	—	U	PPu	F	—
Hedyotis nigricans diamond-flowers	SP	DS	L	B	PPu	SuF	bM
Helenium amarum yellowdicks	SA	S	F	B	Y	SpSu	bgM
Helenium autumnale fall sneezeweed	MP	DFS	F	B	Y	Su	bMW
Helenium bigelovii Bigelow's sneezeweed	MP	FS	F	B	Y	SuF	BMW
Helianthella uniflora Rocky Mountain dwarf-sunflower	MP	DS	F	B	Y	Su	BbgM
Helianthemum canadense long-branch frostweed	SP	DS	F	B	Y	SpSu	BbM
Helianthemum georgianum Georgia frostweed	SP	DSs	F	U	Y	SpSu	BgM
Helianthus angustifolius swamp sunflower	LP	FS	F	BGLMmS	Y	SuF	BbgW
Helianthus annuus common sunflower	LA	DS	F	BGLmS	Y	SuF	M
Helianthus anomalus western sunflower	MA	DS	F	BGLMmS	Y	Su	BbgM
Helianthus argophyllus silver-leaf sunflower	LA	S	F	BmS	Pu	SuF	CM
Helianthus ciliaris Texas-blueweed	MP	FS	F	BGLmS	Y	SuF	bMS
Helianthus debilis cucumber-leaf sunflower	MP	DS	F	BGLMmS	Y	Su	BbgM
Helianthus divaricatus woodland sunflower	LP	DSs	F	BGLMmS	Y	SuF	Bg
Helianthus grosseserratus saw-tooth sunflower	LP	DS	F	BGLMmS	Y	SuF	BbgM
Helianthus hirsutus whiskered sunflower	LP	DS	F	BGLMmS	Y	SuF	BbGM
Helianthus maximiliani Michaelmas-daisy	LP	DS	F	BGLMmS	Y	SuF	BbgM
Helianthus mollis neglected sunflower	LP	DS	F	BGLMmS	Y	SuF	BbGM
Helianthus occidentalis few-leaf sunflower	LP	DS	F	BGLMmS	Y	SuF	BbgM

PLANT LIST	Plant Type	Env. Tol.	Aesthetic Value	Wildlife Value	Flower Color	Bloom Period	Landscape Uses
Helianthus pauciflorus stiff sunflower	LP	DS	F	BGLMmS	Y	SuF	BbM
Helianthus petiolaris prairie sunflower	MA	DS	F	BGLMm	Y	Su	BbM
Helianthus radula rayless sunflower	LP	FS	F	BGLMmS	Y	SuF	Bb
Helianthus strumosus pale-leaf woodland sunflower	LP	DS	F	BGLMmS	Y	SuF	Bb
Helianthus tuberosus Jerusalem-artichoke	LP	FSs	F	BGLMmS	Y	SuF	BEMSW
Heliomeris multiflora Nevada showy false goldeneye	MPA	NaS	F	BGS	Y	Su	BbMS
Heliopsis helianthoides smooth oxeye	LP	DS	F	BS	Y	SpSu	BbCEgM
Heliotropium convolvulaceum wide-flower heliotrope	SA	DS	—	G	W	SuF	M
Heliotropium curassavicum seaside heliotrope	SAP	DNaS	F	G	PuW	SpF	M
Heliotropium tenellum pasture heliotrope	SA	DS	—	U	W	Su	M
Hepatica nobilis liverwort	SP	s	F	U	WP	Sp	gI
Heracleum lanatum American cow-parsnip	LP	FW	FL	M	W	SuF	BIMSW
Heracleum maximum cow-parsnip	LP	FSs	F	M	W	WSuF	BIMW
Heracleum sphondylium eltrot	LP	FS	FL	U	W	Su	M
Heteranthera dubia grass-leaf mud-plantain	SP	FSW	F	W	Y	SuF	Wwg
Heteromeles arbutifolia California-Christmas-berry	eLS	Ss	FfL	GmS	W	Su	AB
Heteropogon contortus twisted tanglehead	PG	—	—	U	G	Su	—
Heterotheca bolanderi Bolander's false golden-aster	P	—	—	U	Y	SuF	—
Heterotheca subaxillaris camphorweed	MAB	DS	F	B	Y	SuF	CbgM
Heterotheca villosa hairy false golden-aster	MP	DS	F	B	Y	SpSuF	BgM
Heuchera abramsii San Gabriel alumroot	P	—	—	U	PPu	Sp	—
Heuchera americana American alumroot	MP	s	FL	U	yg	Sp	GI
Heuchera longiflora long-flower alumroot	SP	s	L	H	PPuW	SpSu	GI

PLANT LIST	Plant Type	Env. Tol.	Aesthetic Value	Wildlife Value	Flower Color	Bloom Period	Landscape Uses
Heuchera parviflora little-flower alumroot	SP	s	L	U	PW	SuF	GI
Heuchera richardsonii Richardson's alumroot	SP	DS	L	U	GW	Sp	GM
Heuchera rubescens pink alumroot	SP	DS	L	H	PuR	SpF	GR
Hexastylis arifolia little-brown-jug	eSP	s	L	U	Br	Sp	GI
Hexastylis virginica Virginia heartleaf	SP	s	L	U	Br	Sp	GI
Hibiscus laevis halberd-leaf rose-mallow	LP	FSW	F	H	PW	Su	SWwg
Hibiscus moscheutos crimson-eyed rose-mallow	LP	FSW	F	Hh	PW	Su	bWwg
Hieracium albiflorum white-flower hawkweed	MP	DS	F	GLM	W	Su	B
Hieracium cynoglossoides hound-tongue hawkweed	SP	DS	F	GLM	Y	Su	gM
Hieracium gronovii queendevil	LP	DSs	F	GLMm	Y	SuF	BlM
Hieracium venosum rattlesnake-weed	MP	SD	F	GLM	Y	SpSu	gM
Hilaria belangeri curly-mesquite	SPG	DS	—	M	G	SuF	EM
Hilaria jamesii James' galleta	SPG	DS	F	M	G	SuF	EM
Hilaria mutica tobosa grass	SPG	DS	—	M	G	SuF	EM
Hilaria rigida big galleta	MPG	DS	F	Mm	G	SuF	EM
Hoita orbicularis round-leaf leather-root	P	—	—	U	Y	SuF	—
Honkenya peploides seaside sandplant	HVP	DFNaS	L	U	W	Su	S
Hordeum brachyantherum meadow barley	MPG	FS	F	GLMmS	BrPu	Su	MW
Hordeum jubatum fox-tail barley	MPG	FS	f	GMSW	GPu	Su	MW
Hordeum pusillum little barley	SAG	DS	F	GLMmS	—	Su	M
Horkelia cuneata wedge-leaf honeydew	SP	S	—	U	W	—	M
Houstonia canadensis Canadian summer bluet	SP	DSs	F	U	BW	SpSuF	BgM
Houstonia longifolia long-leaf summer bluet	SP	S	F	S	PuW	Su	BgG

PLANT LIST	Plant Type	Env. Tol.	Aesthetic Value	Wildlife Value	Flower Color	Bloom Period	Landscape Uses
Houstonia procumbens round-leaf bluet	SP	DS	F	B	W	SpSu	BbEGg
Houstonia pusilla tiny bluet	SA	DS	F	U	BV	Sp	CgM
Hudsonia tomentosa sand golden-heather	SP	DS	F	U	Y	SpSu	E
Huperzia lucidula shining club-moss	eSP	s	L	U	X	X	GI
Hutchinsia procumbens ovalpurse	SA	W	—	U	W	SpSuF	W
Hybanthus concolor eastern green-violet	MP	FSs	—	U	G	Sp	I
Hydrastis canadensis goldenseal	MP	s	FfL	U	PuB	Sp	IN
Hydrocotyle bonariensis coastal marsh-pennywort	SP	FSW	L	U	GW	SpF	W
Hydrocotyle umbellata many-flower marsh-pennywort	SP	FSW	L	U	yg	SpF	W
Hydrocotyle verticillata whorled marsh-pennywort	SP	DFSW	L	U	yg	SpF	W
Hydrophyllum appendiculatum great waterleaf	SP	s	FL	U	P	SpSu	gI
Hydrophyllum canadense blunt-leaf waterleaf	SP	s	F	U	PW	SpSu	gI
Hydrophyllum capitatum cat's-breeches	SP	Ss	—	U	BW	SpSu	M
Hydrophyllum tenuipes Pacific waterleaf	MP	W	F	U	Bpu	SpSu	BIg
Hydrophyllum virginianum Shawnee-salad	SB	s	FL	U	B	SpSu	BIg
Hylocomium splendens feathermoss	m	Fs	L	U	—	—	GIW
Hymenocallis caroliniana Carolina spider-lily	SP	FSs	FL	U	W	Sp	gI
Hymenopappus artemisiifolius old-plains-man	MB	S	F	U	R	Sp	M
Hymenopappus filifolius fine-leaf woollywhite	MP	DS	—	U	WY	SpSuF	M
Hymenopappus scabiosaeus Carolina woollywhite	MB	S	F	B	W	SpSu	bM
Hymenoxys bigelovii Bigelow's rubberweed	MBP	DS	—	U	Y	Su	M
Hymenoxys cooperi Cooper's rubberweed	MBP	DS	—	U	Y	SuF	M
Hymenoxys quinquesquamata Ricon rubberweed	P	—	—	U	Y	SuF	—

PLANT LIST	Plant Type	Env. Tol.	Aesthetic Value	Wildlife Value	Flower Color	Bloom Period	Landscape Uses
Hymenoxys richardsonii Colorado rubberweed	MP	DS	—	U	Y	Su	M
Hypericum anagalloides tinker's penny	SP	FS	F	U	Y	Su	W
Hypericum drummondii nits-and-lice	SA	DS	—	U	OY	SuF	MR
Hypericum gentianoides orange-grass	SP	DS	F	U	Y	SuF	BgM
Hypericum majus greater Canadian St. John's-wort	SP	FS	F	U	Y	Su	SW
Hypnum circinale moss	m	—	—	U	X	X	GI
Hypoxis hirsuta eastern yellow star-grass	SP	DS	F	U	Y	SpSu	gM
Hyptis alata clustered bush-mint	LP	SsW	—	U	PuW	SpSu	gM
Impatiens capensis spotted touch-me-not	MP	FSsW	FL	BGHhL mS	O	SuF	bISWwg
Impatiens pallida pale touch-me-not	MP	FSsW	FL	BGHhL mS	Y	SuF	IbWwg
Ipomoea imperati beach morning-glory	HV	W	—	U	W	SpSuF	W
Ipomoea pes-caprae bay-hops	PV	DNaS	—	U	Pu	SuF	E
Ipomopsis aggregata scarlet skyrocket	MB	DS	F	U	BPRW	SuF	AgMR
Ipomopsis longiflora white-flower skyrocket	MP	DS	F	U	B	SpF	gM
Ipomopsis pumila low skyrocket	SA	DS	—	U	LW	Sp	R
Ipomopsis rubra standing-cypress	LB	DSs	F	BH	R	SuF	BbM
Ipomopsis spicata spiked skyrocket	SP	DS	F	U	WY	—	gM
Iris brevicaulis zigzag iris	SP	FSW	FL	GHLW	BPu	SpSu	AbWwg
Iris douglasiana mountain iris	P	—	FL	BHL	PuRW	Sp	—
Iris hartwegii rainbow iris	SP	S	FL	BHL	PuY	—	Bbg
Iris lacustris dwarf lake iris	SP	FS	FL	BHLW	B	Sp	AWwg
Iris missouriensis Rocky Mountain iris	SP	FS	FL	BHLW	BPu	SpSu	AbMSWwg
Iris tenax tough-leaf iris	MP	S	FL	BHL	LPu	Su	ABbM

PLANT LIST	Plant Type	Env. Tol.	Aesthetic Value	Wildlife Value	Flower Color	Bloom Period	Landscape Uses
Iris versicolor harlequin blueflag	MP	FSW	FL	BHLW	B	Su	AbWwg
Iris virginica Virginia blueflag	SP	FSW	FL	BHLW	B	Su	AbWwg
Isocoma drummondii Drummond's jimmyweed	MP	Ss	F	U	Y	SpSuF	BM
Isocoma pluriflora southern jimmyweed	MP	NaS	F	B	Y	SuF	gM
Isoetes howellii Howell's quillwort	MP	FS	—	U	X	X	SW
Iva axillaris deer-root	SP	DNaS	—	U	GW	SuF	MW
Iva imbricata seacoast marsh-elder	MP	DNaS	—	U	Y	SuF	E
Iva nevadensis Nevada marsh-elder	A	—	—	U	W	SuF	—
Ivesia purpurascens summit mousetail	P	—	FL	U	PuW	Su	R
Ivesia santolinoides Sierran mousetail	P	—	—	U	W	Su	—
Jaumea carnosa marsh jaumea	SP	FNaSW	F	U	Y	SuF	SW
Jeffersonia diphylla twinleaf	SP	s	FL	U	W	Sp	I
Juncus acuminatus knotty-leaf rush	MPG	FS	FL	L	Br	SpSu	W
Juncus acutus spiny rush	MPG	DS	—	L	Br	Su	M
Juncus alpinoarticulatus northern green rush	SAG	FSW	—	L	Br	Su	W
Juncus balticus Baltic rush	MPG	DFSW	—	L	BrG	SpF	Wwg
Juncus bufonius toad rush	SAG	FSW	—	GLm	G	SpF	W
Juncus canadensis Canadian rush	MPG	FSW	—	L	Br	Su	SW
Juncus cooperi Cooper's rush	MPG	NaSW	—	L	G	Su	W
Juncus covillei Coville's rush	MPG	—	—	L	Br	Su	—
Juncus dudleyi Dudley's rush	MPG	FS	—	L	Br	SuF	W
Juncus effusus lamp rush	MPG	FSW	FL	L	Br	SuF	Wwg
Juncus gerardii saltmarsh rush	MPG	FNaSW	—	L	Br	SuF	SW

PLANT LIST	Plant Type	Env. Tol.	Aesthetic Value	Wildlife Value	Flower Color	Bloom Period	Landscape Uses
Juncus interior inland rush	SPG	DS	D	L	BG	Su	W
Juncus lesueurii salt rush	PG	Na	—	L	BrG	SpSu	W
Juncus longistylis long-style rush	MPG	FSW	—	L	Br	Su	W
Juncus marginatus grass-leaf rush	MPG	FSW	—	L	Br	Su	W
Juncus mexicanus Mexican rush	MG	DS	—	L	Br	Su	M
Juncus nevadensis Sierran rush	SG	FS	—	L	Br	Su	MW
Juncus nodosus knotted rush	SPG	FSW	—	L	Br	Su	W
Juncus orthophyllus straight-leaf rush	PG	W	—	L	BrG	Su	W
Juncus patens spreading rush	SPG	FSW	—	L	BrG	Su	W
Juncus pelocarpus brown-fruit rush	SPG	FSW	—	L	G	SuF	W
Juncus roemerianus Roemer's rush	MPG	FNaSW	—	L	Br	SpF	W
Juncus stygius moor rush	SG	SW	—	L	W	Su	W
Juncus tenuis poverty rush	SPG	FSsW	—	LM	Br	SuF	BEGW
Juncus torreyi Torrey's rush	MG	S	—	L	Br	SuF	M
Keckiella antirrhinoides chaparral bush-beardtongue	P	—	—	U	Y	Sp	—
Kochia americana greenmolly	SP	DNaS	L	H	W	SuF	M
Kochia californica California summer-cypress	SP	Ss	BL	U	R	—	BM
Koeleria macrantha prairie Koeler's grass	SG	DS	fL	M	G	SpSu	AEM
Kosteletzkya virginica Virginia fen-rose	LP	DFNaSW	F	U	P	SuF	BgWwg
Krascheninnikovia lanata winterfat	SP	DNaS	L	mS	—	—	BEMS
Krigia biflora two-flowered dwarf-dandelion	SP	S	F	B	Y	SpSu	Bg
Lachnanthes caroliniana Carolina redroot	MP	FS	F	W	Y	Su	gW
Lactuca canadensis Florida blue lettuce	LP	S	F	GMmS	Y	SuF	B

PLANT LIST	Plant Type	Env. Tol.	Aesthetic Value	Wildlife Value	Flower Color	Bloom Period	Landscape Uses
Laennecia schiedeana pineland marshtail	A	—	—	U	W	F	—
Langloisia setosissma bristly-calico	SA	S	—	U	LPu	SpSu	BM
Laportea canadensis Canadian wood-nettle	MP	Fs	F	U	G	Su	IS
Lappula occidentalis flat-spine sheepburr	SA	DSs	—	U	B	Su	CBM
Lasthenia californica California goldfields	SA	S	F	U	Y	Sp	CgM
Lasthenia fremontii Fremont's goldfields	SA	S	F	U	Y	Sp	CgMW
Lasthenia glabrata yellow-ray goldfields	SA	FNaS	F	U	Y	Sp	CW
Lasthenia minor coastal goldfields	A	S	—	U	Y	Sp	—
Lathyrus graminifolius grass-leaf vetchling	SP	Ss	F	h	BPW	Sp	B
Lathyrus lanszwertii Nevada vetchling	HVP	—	—	U	BPu	SpSu	—
Lathyrus japonicus sea vetchling	HVP	DNaS	F	h	—	SpSu	ES
Lathyrus ochroleucus cream vetchling	MP	Ds	F	h	W	SpSu	BI
Lathyrus palustris marsh vetchling	MP	FSW	F	h	Pu	Su	W
Lathyrus polyphyllus leafy vetchling	MP	—	F	h	Pu	Su	—
Lathyrus venosus veiny vetchling	MP	FsW	F	h	Pu	SpSu	gISW
Layia chrysanthemoides smooth tidytips	SA	S	F	U	Y	SpSu	CgM
Layia fremontii Fremont's tidytips	SA	DS	F	U	YW	SpSu	M
Layia glandulosa white tidytips	SA	DS	F	U	W	SpSu	CgM
Lechea cernua nodding pinweed	SP	DS	—	S	R	Su	M
Lechea mucronata hairy pinweed	MP	DSs	—	S	R	Su	BM
Lechea san-sabeana San Saba pinwheel	SP	Ss	—	U	R	Sp	BM
Lechea tenuifolia narrow-leaf pinweed	SP	DSs	—	S	R	Su	BM
Leersia hexandra southern cut grass	MPG	SW	—	LSW	G	Su	ESW

PLANT LIST	Plant Type	Env. Tol.	Aesthetic Value	Wildlife Value	Flower Color	Bloom Period	Landscape Uses
Leersia lenticularis catchfly grass	LPG	FSW	F	LSW	G	F	EW
Leersia oryzoides rice cut grass	MPG	FSW	—	LSW	G	Su	EPSW
Leersia virginica white grass	MPG	FSs	F	LSW	G	SuF	IW
Lemna aequinoctialis lesser duckweed	AQ	FSW	—	W	X	X	Wwg
Lemna minor common duckweed	AQP	SW	L	W	G	Su	Wwg
Lemna perpusilla minute duckweed	AQP	SW	L	W	G	Su	Wwg
Lepidium densiflorum miner's pepperwort	SAB	DS	—	MmSW	GW	Su	M
Lepidium fremontii bush pepperwort	MP	DS	—	MmS	W	Su	M
Lepidium lasiocarpum hairy-pod pepperwort	SAB	DNaS	—	MmS	W	Sp	BM
Lepidium latipes San Diego pepperwort	SA	FNaS	—	MmSW	G	Sp	M
Leptochloa dubia green sprangletop	MPG	DS	—	U	G	SpF	MR
Leptodactylon pungens granite prickly-phlox	SP	DS	FL	B	P	Sp	BbGgMR
Lespedeza capitata round-head bush-clover	LP	DS	L	GhMS	W	Su	CEM
Lespedeza hirta hairy bush-clover	LP	S	L	GMS	YW	Su	CEM
Lespedeza procumbens trailing bush-clover	SP	DS	F	GMS	P	SuF	EB
Lespedeza repens creeping bush-clover	SP	DS	F	GMS	PPu	SpF	EB
Lespedeza stuevei tall bush-clover	MP	DS	F	GMS	Pu	SuF	EM
Lespedeza texana Texas bush-clover	SA	Fs	F	GhMS	BPu	SuF	CIS
Lespedeza violacea violet bush-clover	SP	DSs	F	GMS	V	Su	BgM
Lespedeza virginica slender bush-clover	SP	DS	F	GMS	PPu	SuF	BEM
Lesquerella gordonii Gordon's bladderpod	SA	DS	—	U	Y	Sp	M
Lesquerella ludoviciana Louisiana bladderpod	SP	DSs	F	U	Y	Sp	BI
Leucobryum glaucum white moss	em	s	L	U	X	X	GI

PLANT LIST	Plant Type	Env. Tol.	Aesthetic Value	Wildlife Value	Flower Color	Bloom Period	Landscape Uses
Leucocrinum montanum star-lily	MP	Ds	F	U	W	SpSu	BI
Lewisia nevadensis Nevada bitter-root	SP	S	F	S	PW	Su	BIM
Leymus cinereus Great Basin lyme grass	MPG	DSsW	DF	GS	G	SpSu	BEIMSW
Leymus condensatus giant lyme grass	MPG	DFS	—	GS	G	Su	EMSW
Leymus salinus salinas lyme grass	LPG	DSs	DF	GS	G	Su	BEIM
Leymus triticoides beardless lyme grass	MPG	FNaS	DF	GS	G	Su	ESW
Liatris aspera tall gayfeather	LP	DS	F	BHS	PPu	SuF	bgM
Liatris cylindracea Ontario gayfeather	SP	DS	DF	BHS	PPu	SuF	bgM
Liatris elegans pink-scale gayfeather	MP	S	DF	BS	PuW	SuF	bM
Liatris mucronata cusp gayfeather	MP	DS	F	BHS	Pu	SuF	bM
Liatris pauciflora few-flower gayfeather	MP	DS	F	BHS	Pu	SuF	bgM
Liatris punctata dotted gayfeather	MP	DS	F	BHS	PR	SuF	bgM
Liatris pycnostachya cat-tail gayfeather	MP	DS	F	BHS	Pu	SuF	M
Liatris scariosa devil's-bite	LP	S	F	BHS	P	SuF	bgM
Liatris spicata dense gayfeather	LP	FS	F	BHS	P	SuF	bgM
Liatris squarrosa scaly gayfeather	MP	DS	F	BHS	PPu	Su	bgM
Liatris tenuifolia short-leaf gayfeather	MP	DS	DF	BHS	Pu	SuF	bgM
Ligusticum filicinum fern-leaf wild lovage	MP	DSs	L	U	W	Su	BM
Ligusticum porteri Porter's wild lovage	MP	Ss	L	U	W	Su	BM
Lilaea scilloides flowering-quillwort	A	SW	L	U	G	SpSu	Wwg
Lilaeopsis chinensis eastern grasswort	SP	FS	L	U	W	SpF	W
Lilium catesbaei southern red lily	SP	FS	F	BHS	O	Su	bW
Lilium parryi lemon lily	MP	FS	FL	BS	Y	Su	bM

PLANT LIST	Plant Type	Env. Tol.	Aesthetic Value	Wildlife Value	Flower Color	Bloom Period	Landscape Uses
Lilium philadelphicum wood lily	MP	S	F	BHS	OR	Su	Bbgl
Lilium superbum Turk's-cap lily	LP	FS	F	BHS	OR	Su	BW
Limnanthes douglasii Douglas' meadowfoam	SP	FS	FL	U	WY	Sp	BgMW
Limnobium spongia American spongeplant	AQP	SW	L	U	G	SuF	Wwg
Limnodea arkansana Ozark grass	SAG	DS	—	U	G	Sp	M
Limonium californicum marsh-rosemary	SP	FNaS	F	U	Pu	SuW	SW
Limonium carolinianum Carolina sea-lavender	SP	NaSW	DF	B	Pu	SuF	SWwg
Limosella aquatica awl-leaf mudwort	SP	FS	—	U	PuW	Su	W
Limosella australis Welsh mudwort	SP	FNaS W	—	U	W	SuF	W
Linanthis nuttallii Nuttall's desert-trumpets	SP	Ss	F	U	PW	—	Bg
Linanthus ciliatus whiskerbrush	SA	S	F	U	PuR	—	Bg
Linnea borealis American twinflower	eHVP	DFsW	F	mSW	P	—	BbglMSW
Linum hudsonioides Texas flax	SA	DS	F	U	Y	SpSuF	gM
Linum lewisii prairie flay	SP	DS	F	B	B	SpSuF	bMR
Linum rigidum large-flower yellow flax	SA	DNaS	F	B	Y	Su	BbM
Liparis loeselii yellow wide-lip orchid	SP	FSs	FL	m	GY	SpSu	IW
Listera caurina northwestern twayblade	SP	Ss	F	U	GPu	SpSu	IM
Listera convallarioides broad-tip twayblade	SP	Fs	F	U	yg	SpSu	gI
Lithophragma glabrum bulbous woodlandstar	SP	Ss	F	U	P	Sp	BI
Lithophragma parviflorum prairie woodlandstar	SP	Ss	F	U	P	Sp	MR
Lithophragma tenellum slender woodlandstar	SP	Ss	F	U	P	Sp	MR
Lithospermum canescens hoary puccoon	SP	DS	F	U	OY	Sp	gM
Lithospermum caroliniense hairy puccoon	SP	Ss	F	U	OY	SpSu	BM

PLANT LIST	Plant Type	Env. Tol.	Aesthetic Value	Wildlife Value	Flower Color	Bloom Period	Landscape Uses
Lithospermum incisum fringed gromwell	SP	DS	F	U	OY	Sp	M
Lithospermum ruderale Columbian puccoon	SP	Ss	F	U	Y	Sp	BM
Lobelia cardinalis cardinal-flower	MP	FSs	F	Hh	R	SuF	ABbgIWS
Lobelia dortmanna water lobelia	SP	SW	F	B	Pu	SuF	bWwg
Lobelia kalmii brook lobelia	SP	FS	F	B	B	SuF	bWwy
Lobelia siphilitica great blue lobelia	MP	FSs	F	Bh	B	SuF	BbMSW
Lobelia spicata pale-spike lobelia	MP	DS	F	Bh	BW	SuF	BbM
Lomatium foeniculaceum carrot-leaf desert-parsley	SP	DS	—	U	Y	Su	BM
Lomatium triternatum nine-leaf desert-parsley	SP	S	—	U	Y	Su	M
Lotus nevadensis Nevada bird's-foot-trefoil	SP	S	F	GLMm	Y	SpSu	M
Lotus oblongifolius streambank bird's-foot-trefoil	SP	FS	F	GLMm	Y	SpF	S
Lotus plebus long-bract bird's-foot-trefoil	SP	FSs	F	GLMm	Y	Sp	B
Lotus rigidus broom bird's-foot-trefoil	SP	DS	F	GLMm	WY	Sp	M
Lotus salsuginosus coastal bird's-foot-trefoil	SP	DS	F	GLMm	Y	SpSu	M
Lotus wrightii scrub bird's-foot-trefoil	SP	DS	F	GLMm	Y	Sp	M
Ludwigia palustris marsh primrose-willow	MP	FSW	F	W	Y	Su	W
Ludwigia repens creeping primrose-willow	SP	FSsW	F	W	Y	Su	SW
Lupinus argenteus silver-stem lupine	MP	DSs	FL	GLMmS	B	Su	BgM
Lupinus arizonicus Arizona lupine	SA	S	FL	GLMmS	PuR	Sp	gM
Lupinus bicolor miniature annual lupine	SA	S	FL	GLMmS	BW	Sp	CgM
Lupinus brevicaulis short-stem lupine	SA	DNaSs	FL	GLMmS	B	SpSu	CBgM
Lupinus breweri matted lupine	SP	DS	FL	BCGMmS	B	Su	bM
Lupinus caudatus Kellogg's spurred lupine	SP	Ss	FL	GLMmS	B	Sp	BgM

PLANT LIST	Plant Type	Env. Tol.	Aesthetic Value	Wildlife Value	Flower Color	Bloom Period	Landscape Uses
Lupinus elatus tall silky lupine	SP	Ss	FL	GLMmS	PuW	SpSu	Bg
Lupinus excubitus interior bush lupine	MP	DFS	FL	GLMmS	B	SpSu	gM
Lupinus formosus summer lupine	MP	Ss	FL	GLMmS	BPuW	Su	BgM
Lupinus huachucanus Huachuca Mountain lupine	SP	Ss	F	GLMmS	BPu	Sp	BgM
Lupinus latifolius broad-leaf lupine	MP	S	F	CGHLMmS	Pu	SpSu	gM
Lupinus luteolus butter lupine	MA	DS	F	CGHLMmS	Y	SpSu	BgM
Lupinus neomexicanus New Mexico lupine	SP	S	F	CGHLMmS	L	SpSuF	gM
Lupinus palmeri Palmer's lupine	MP	S	F	CGHLMmS	Pu	SpSuF	gM
Lupinus peirsonii long lupine	P	—	F	BCH	Y	Sp	bg
Lupinus perennis sundial lupine	SP	DS	FL	BCGHLMmS	B	SpSu	bgM
Lupinus polyphyllus blue-pod lupine	MP	FS	F	CGLMmSH	BRY	SpSu	gMW
Lupinus sericeus Pursh's silky lupine	SP	S	FL	CGHLMmS	B	SpSu	gM
Lupinus subcarnosus Texas bluebonnet	SA	S	FL	GMmS	B	Sp	CbM
Lupinus texensis Texas lupine	SA	DS	F	BC	B	Sp	gM
Lupinus versicolor Lindley's varied lupine	SP	S	F	BCGH	BPPuY	SpSu	bM
Luziola fluitans southern water grass	G	W	—	U	—	—	wg
Luzula congesta heath wood-rush	SG	S	—	U	Br	—	M
Lycopodium clavatum running ground-pine	eSP	Ss	L	M	X	X	BGI
Lycopodium obscurum princess-pine	eSP	Ss	L	M	X	X	BGI
Lycopodium sabinifolium savin-leaf ground-pine	eSP	Ss	L	M	X	X	BGI
Lycopus americanus cut-leaf water-horehound	MP	FSW	—	U	W	SuF	W
Lycopus asper rough water-horehound	SP	FSW	—	U	W	SuF	W
Lycopus uniflorus northern water-horehound	SP	FS	—	U	W	SuF	W

PLANT LIST	Plant Type	Env. Tol.	Aesthetic Value	Wildlife Value	Flower Color	Bloom Period	Landscape Uses
Lycopus virginicus Virginia water-horehound	SP	FSW	—	U	W	SuF	W
Lycurus phleoides common wolf's-tail	SPG	DS	F	U	G	Su	MR
Lygodesmia juncea rush skeleton-plant	MP	DS	F	U	BPPu	Su	M
Lysichiton americanus yellow skunk-cabbage	MP	SW	FL	GW	Y	Su	Wwg
Lysimachia ciliata fringed yellow-loosestrife	MP	FSs	F	mW	Y	Su	BgSW
Lysimachia quadrifolia whorled yellow-loosestrife	MP	S	F	m	Y	Su	gM
Lysimachia thyrsiflora tufted yellow-loosestrife	MP	FW	F	U	Y	Su	Wwg
Lythrum alatum wing-angle loosestrife	MP	FSW	F	U	P	SuF	Wwg
Lythrum lineare saltmarsh loosestrife	MP	FNaSW	F	U	P	SuF	Wwg
Machaeranthera canescens hoary tansy-aster	MBP	DNaSs	F	B	PPuW	SuF	BbM
Machaeranthera phyllocephala camphor-daisy	A	FS	F	B	Y	—	bMS
Machaeranthera pinnatifida lacy tansy-aster	SP	DS	F	B	Y	SpSuF	bM
Machaeranthera tanacetifolia takhoka-daisy	SA	DS	F	B	BPPu	SpSu	CbM
Machaerocarpus californicus fringed-water-plantain	SP	FSW	F	U	PW	SpSu	Wwg
Madia exigua little tarplant	SA	S	F	GLmS	Y	Sp	CBM
Madia gracilis grassy tarplant	MA	S	F	GLmS	Y	Sp	CBM
Maianthemum canadense false lily-of-the-valley	SP	Ss	FL	GLm	W	Sp	BEGgI
Maianthemum dilatatum two-leaf false Solomon's-seal	SP	SW	Ff	G	W	Sp	IWwg
Maianthemum racemosum feathery false Solomon's-seal	MP	s	F	GmS	WG	Sp	BI
Maianthemum stellatum starry false Solomon's-seal	SP	DS	Ff	GLM	W	SpSu	B
Maianthemum trifolium three-leaf false Solomon's-seal	SP	sW	Ff	GLM	W	SpSu	IWwg
Malacothrix glabrata smooth desert-dandelion	SA	DS	F	U	Y	Sp	gM
Manfreda virginica false aloe	LP	DS	L	U	yg	Su	gM

PLANT LIST	Plant Type	Env. Tol.	Aesthetic Value	Wildlife Value	Flower Color	Bloom Period	Landscape Uses
Marshallia caespitosa puffballs	SP	DS	F	U	LWY	Sp	M
Marsilea vestita hairy water-clover	AQP	FSW	L	W	X	X	Wwg
Matelea edwardsensis plateau milkvine	PHV	DSs	—	U	G	Sp	BI
Matelea gonocarpa angular-fruit milkvine	HVP	s	F	BC	BrPu	Su	bI
Matelea reticulata netted milkvine	PHV	DSs	—	U	G	SpSuF	B
Matteuccia struthiopteris ostrich fern	LP	FSW	L	U	X	X	BSW
Mayaca fluviatilis stream bog-moss	AQ	SW	L	U	PW	SpSu	W
Medeola virginiana Indian cucumber-root	SP	s	FfL	S	yg	Sp	gI
Meehania cordata Meehan's-mint	SP	s	L	U	BL	SuF	I
Melampodium leucanthum plains blackfoot	SP	Ss	F	U	W	SpSuF	BM
Melampyrum lineare American cow-wheat	SP	DFS	F	U	W	Su	W
Melanthium parviflorum Appalachian bunchflower	LP	s	—	U	yg	SuF	I
Melica bulbosa onion grass	SPG	s	—	MS	G	Su	I
Melica californica California melic grass	MPG	Ss	—	MS	G	Su	BM
Melica harfordii Harford's melic grass	SPG	Ss	—	MS	G	Su	B
Melica imperfecta coast range melic grass	MPG	DSs	—	MS	G	Su	B
Melica stricta rock melic grass	SPG	S	—	MS	G	Su	MS
Melica sublata Alaska melic grass	SPG	S	—	MS	G	Su	BM
Melothria pendula Guadeloupe-cucumber	HPV	FSs	—	U	yg	SpF	B
Mentha arvensis American wild mint	SP	FS	F	U	BW	SuF	W
Mentzelia albicaulis white-stem blazingstar	SA	DS	BF	G	Y	Sp	CgMR
Mentzelia decapetala gumbo-lily	SAB	DS	BF	U	WY	SpSu	gM
Mentzelia involucrata white-bract blazingstar	SAB	DFS	BF	U	RWY	Sp	gMR

PLANT LIST	Plant Type	Env. Tol.	Aesthetic Value	Wildlife Value	Flower Color	Bloom Period	Landscape Uses
Mentzelia multiflora Adonia blazingstar	SAB	DS	BF	U	Y	Sp	gM
Mentzelia nuda goodmother	SAB	DS	BF	U	Y	SpSu	gM
Mentzelia oligosperma chickenthief	SAB	DS	BF	U	Y	SpSu	gM
Mentzelia pumila golden blazingstar	SAB	DS	BF	U	Y	SpSuF	gM
Menyanthes trifoliata buck-bean	SP	FSW	F	U	W	Su	SW
Mertensia arizonica aspen bluebells	SP	Ss	F	Hh	B	SpSu	Bbg
Mertensia ciliata tall fringe bluebells	MP	Ss	F	Hh	B	SpSu	Bbg
Mertensia lanceolata prairie bluebells	SP	Ss	F	Hh	B	SpSu	Bbg
Mertensia oblongifolia languid-lady	SP	Ss	F	Hh	B	SpSu	Bbg
Mertensia virginica Virginia bluebells	SP	Fs	F	Hh	B	Sp	bgIW
Mesembryanthemum nodiflorum slender-leaf iceplant	SA	S	F	U	W	SpSu	gM
Mikania scandens climbing hempvine	HV	FSsW	F	U	LPu	SuF	BSW
Mianthemum dilatatum two-leaf false Solomon's-seal	SP	sW	FfL	U	W	SpSu	GISW
Mimulus bigelovii yellow-throat monkey-flower	SA	Ss	F	U	BPu	—	B
Mimulus guttatus seep monkey-flower	MP	F	F	U	Y	Su	W
Mimulus moschatus muskflower	SP	FS	F	U	Y	—	SW
Mimulus primuloides yellow creeping monkey-flower	SP	S	FL	U	Y	Su	gM
Mimulus ringens Allegheny monkey-flower	SP	Fs	F	U	BW	SuF	BgI
Mimulus suksdorfii miniature monkey-flower	SP	SW	F	U	Y	SpSu	gMWwg
Mimulus tricolor tricolor monkey-flower	SP	FS	F	U	Pu	Sp	W
Minuartia californica California stitchwort	SA	DS	—	U	W	Sp	GR
Minuartia caroliniana pine-barren stitchwort	SP	Ss	L	U	Pw	SpSu	BMG
Minuartia nuttallii brittle stitchwort	SA	Ss	—	U	W	SpSu	BGE

PLANT LIST	Plant Type	Env. Tol.	Aesthetic Value	Wildlife Value	Flower Color	Bloom Period	Landscape Uses
Minuartia patula Pitcher's stitchwort	SA	DS	—	U	W	SpSu	GM
Mirabilis albida white four-o'clock	MP	DS	F	BHs	PW	SpSuF	bM
Mirabilis alipes winged four-o'clock	SP	NaSs	F	BHS	Ppu	SuF	BbM
Mirabilis bigelovii desert wishbonebush	SP	Ss	F	BHS	Ppu	SuF	Bb
Mirabilis hirsuta hairy four-o'clock	MP	DS	F	S	PPu	SuF	BgR
Mirabilis linearis narrow-leaf four-o'clock	LP	DS	F	S	PPu	SuF	BgR
Mitchella repens partridge-berry	SP	s	FL	GL	W	Sp	BGgI
Mitella diphylla two-leaf bishop's-cap	SP	s	FL	G	W	Sp	BGgI
Mitella nuda bare-stem bishop's-cap	SP	s	FL	U	BrG	SpSu	BGgI
Mitella pentandra five-stamen bishop's-cap	SP	SsW	—	U	G	SpSu	BISW
Mitella stauropetala side-flower bishop's-cap	SP	S	L	U	W	SpSu	GgI
Moehringia macrophylla big-leaf grove-sandwort	SP	SsW	—	U	W	SpSu	BW
Monanthochloe littoralis shore grass	SPG	FNaS	—	U	G	Sp	EW
Monarda citriodora lemon beebalm	MP	Ss	F	Bh	PW	SpSuF	BbM
Monarda clinopodioides basil beebalm	SA	S	F	Bh	PW	SpSu	BCM
Monarda didyma scarlet beebalm	MP	FSs	F	H	PuR	SuF	BbSW
Monarda fistulosa Oswego-tea	MP	DSs	DF	BHL	L	Su	BbM
Monarda punctata spotted beebalm	MP	DS	DF	Bh	LPu	SuF	bM
Monardella cinerea gray mountainbalm	SP	S	F	U	PuW	Su	M
Monardella linoides flax-leaf mountainbalm	SP	S	F	U	PuW	Su	gM
Monardella odoratissima alpine mountainbalm	SP	DS	F	U	PuW	SuF	M
Monolepis nuttalliana Nuttall's poverty-weed	SA	DS	—	U	GR	W	M
Monotropa hypopithys many-flower Indian-pipe	SP	s	F	U	P	SpSu	I

PLANT LIST	Plant Type	Env. Tol.	Aesthetic Value	Wildlife Value	Flower Color	Bloom Period	Landscape Uses
Monotropa uniflora one-flower Indian-pipe	SP	s	F	U	W	SpSu	I
Monroa squarrosa false buffalo grass	G	—	—	U	G	Su	—
Montia linearis linear-leaf candy-flower	SA	FS	L	mGS	W	SuF	Swg
Muhlenbergia capillaris hair-awn muhly	MPG	Ds	—	GLM	G	SuF	BI
Muhlenbergia cuspidata stony-hills muhly	MPG	DS	—	GLM	G	SuF	M
Muhlenbergia emersleyi bull grass	SPG	S	—	GLM	G	SuF	R
Muhlenbergia filiculmis slim-stem muhly	PG	—	—	GLM	G	Su	—
Muhlenbergia filiformis pullup muhly	SPG	S	—	GM	G	SuF	BM
Muhlenbergia frondosa wire-stem muhly	MPG	DS	—	GLM	G	SuF	M
Muhlenbergia glomerata spiked muhly	MPG	FSW	—	GLM	G	SuF	W
Muhlenbergia lindheimeri Lindheimer's muhly	LPG	FSs	—	GM	G	F	BES
Muhlenbergia minutissima least muhly	SAG	DS	—	GLM	G	SuF	R
Muhlenbergia montana mountain muhly	SPG	DSs	—	GLM	G	F	BR
Muhlenbergia porteri bush muhly	SPG	DSs	—	GLM	G	SuF	BM
Muhlenbergia pungens sandhill muhly	SPG	DSs	—	GLM	G	Su	BM
Muhlenbergia racemosa green muhly	MPG	DFSs	—	GLM	G	SuF	BEMSW
Muhlenbergia reverchonii seep muhly	SPG	DS	—	GLM	G	SuF	M
Muhlenbergia richardsonis matted muhly	SG	DSNa	—	GLM	G	SuF	EM
Muhlenbergia schreberi nimblewill	SPG	sW	—	GM	G	F	BEI
Muhlenbergia torreyi ringed muhly	MPG	DSs	—	GLM	G	SuF	BM
Muhlenbergia utilis aparejo grass	SPG	FSs	—	GM	G	F	SW
Muhlenbergia virescens screw leaf muhly	SPG	DSs	—	GLM	G	SuF	BM
Muhlenbergia wrightii Wright's muhly	SPG	S	—	GLM	G	SuF	M

PLANT LIST	Plant Type	Env. Tol.	Aesthetic Value	Wildlife Value	Flower Color	Bloom Period	Landscape Uses
Musineon divaricatum leafy wild parsley	SP	DNaS	—	C	Y	—	M
Myosotis verna spring forget-me-not	SA	FSs	—	U	W	Sp	BCMS
Myosurus minimus tiny mousetail	SA	FS	—	U	WY	SpSu	CM
Myriophyllum aquaticum parrot's feather	AQP	FSW	L	W	W	Su	Wwg
Myriophyllum heterophyllum two-leaf water-milfoil	AQP	FSW	L	W	W	Su	Wwg
Najas marina holly-leaf waternymph	AQP	W	L	W	Br	—	wg
Nassella cernua tussock grass	MPG	S	L	LMmS	—	—	M
Nassella lepida tussock grass	MPG	S	L	LMmS	—	—	M
Nassella leucotricha Texas wintergrass	MPG	DS	L	LMmS	Br	—	M
Nassella pulchra tussock grass	MPG	DS	L	LMmS	Pu	—	M
Nassella viridula green tussock grass	MPG	DS	L	LMmS	BrG	—	M
Navarretia intertexta needle-leaf pincushion-plant	SA	S	F	U	PuW	Sp	M
Navarretia leucocephala white-flower pincushion-plant	SA	FS	F	U	W	Su	CGgMW
Nelumbo lutea American lotus	AQP	SW	FL	LMW	Y	Su	wg
Nemastylis geminiflora prairie pleatleaf	SP	S	F	U	B	Sp	gM
Nemophila breviflora Great Basin baby-blue-eyes	SA	Ss	F	U	W	SpSu	BCGI
Nemophila phacelioides large-flower baby-blue-eyes	A	Ss	F	U	BPu	Sp	Bg
Neostapfia colusana colusa grass	SAG	DFs	—	U	G	Su	MS
Neptunia lutea yellow puff	SP	DS	F	U	Y	SpSuF	gM
Nitrophila occidentalis boraxweed	SP	DNaS	—	U	PW	Su	EM
Nolina bigelovii bigelow's bear-grass	LP	DS	—	U	—	Sp	M
Nothoscordum bivalve cowpoison	SP	DS	F	U	W	SpSuF	BgMR
Nuphar lutea yellow pond-lily	AQP	SW	FL	LW	Y	Su	Wwg

PLANT LIST	Plant Type	Env. Tol.	Aesthetic Value	Wildlife Value	Flower Color	Bloom Period	Landscape Uses
Nuttalanthus texanus Texas-toadflax	AB	—	—	U	BL	SpSu	—
Nyctaginia capitata devil's-bouquet	SP	S	F	H	PR	SpSuF	bGM
Nymphaea mexicana banana water-lily	AQP	SW	FL	LW	Y	Su	Wwg
Nymphaea odorata American white water-lily	AQP	SW	FL	LMW	W	Su	Wwg
Nymphoides aquatica big floatingheart	AQP	SW	FL	LMW	W	Su	Wwg
Nymphoides cordata little floatingheart	AQP	SW	FL	LMW	W	Su	Wwg
Oenanthe sarmentosa Pacific water-dropwort	AQP	W	—	U	W	SuF	W
Oenothera albicaulis white-stem evening-primrose	MA	Ss	F	HMS	PW	SpSu	Bb
Oenothera biennis king's-cureall	LP	DS	F	MS	Y	SuF	bM
Oenothera cespitosa tufted evening-primrose	SP	DS	F	HMS	PW	Su	bEMR
Oenothera deltoides devil's lantern	SA	Ss	F	HMS	PW	SpSu	Bb
Oenothera drummondii beach evening-primrose	MP	S	F	B	Y	WSpSu	bMS
Oenothera elata Hooker's evening-primrose	SB	FSs	F	HMS	Y	SpSu	BbI
Oenothera humifusa seaside evening-primrose	SP	DNaS	FL	MS	Y	SpF	bE
Oenothera pallida white-pole evening-primrose	SPA	Ss	F	MS	LPW	SpSuF	BI
Oenothera rhombipetala greater four-point evening-primrose	MA	S	F	MS	Y	Su	M
Oenothera speciosa pinkladies	SP	DS	F	B	P	SpSu	bgM
Okenia hypogaea burrowing-four-o'clock	MP	DS	—	U	—	SuF	E
Onoclea sensibilis sensitive fern	MP	SsW	fL	U	X	X	BCWI
Ophioglossum engelmannii limestone adder's tongue	SP	FSs	L	U	X	X	BMR
Oplismenus setarius short-leaf basket grass	SPG	Fs	—	U	G	SuF	IS
Oreonana vestita woolly mountain-parsley	SP	S	—	U	W	SpSuF	M
Orobanche fasciculata clustered broom-rape	SP	SsF	F	U	Y	Sp	IM

PLANT LIST	Plant Type	Env. Tol.	Aesthetic Value	Wildlife Value	Flower Color	Bloom Period	Landscape Uses
Orobanche ludoviciana Louisiana broom-rape	SP	FS	F	U	Pu	SuF	S
Orontium aquaticum goldenclub	AQP	SW	FL	U	Y	Sp	Wwg
Orthilia secunda sidebells	eSP	s	LF	G	W	Su	Gl
Orthocarpus attenuatus valley-tassels	SAP	DS	F	U	WY	Sp	gM
Orthocarpus campestris field owl-clover	SA	DS	F	U	W	Sp	gM
Orthocarpus erianthus Johnnytuck	SA	DS	F	U	Y	Sp	gM
Orthocarpus linearilobus pale owl-clover	SA	DS	F	U	W	Sp	gM
Orthocarpus luteus golden-tongue osa-clover	SA	DS	F	U	Y	SuF	gM
Orthocarpus purpurascens red owl-clover	SP	S	F	U	PY	Sp	BM
Orthocarpus tolmiei Tolmie's owl-clover	A	—	—	U	G	Su	—
Oryzopsis asperifolia white-grain mountain-rice grass	MPG	Ds	—	GMmS	G	Su	BlE
Oryzopsis hymenoides Indian mountain-rice grass	MG	DS	—	GMmS	G	SpSu	EM
Oryzopsis pungens short-awn mountain-rice grass	SPG	DS	—	GMmS	G	Su	EM
Oryzopsis racemosa black-seed mountain-rice grass	MPG	s	—	GMmS	G	Su	EI
Osmorhiza chilensis mountain sweet-cicely	MP	Ss	FL	U	W	Su	BglS
Osmorhiza claytonii hairy sweet-cicely	MP	s	FL	U	W	Su	Bgl
Osmorhiza depauperata blunt-fruit sweet-cicely	MP	FSs	FL	U	W	Su	BglS
Osmorhiza occidentalis Sierran sweet-cicely	MP	FSs	F	U	GW	Su	BIS
Osmunda cinnamomea cinnamon fern	LP	FSsW	fL	m	X	X	ABglWwg
Osmunda claytoniana interrupted fern	MP	s	L	m	X	X	Bgl
Osmunda regalis royal fern	MP	FsW	fL	m	X	X	ABglSWwg
Oxalis dichondraefolia peony-leaf wood-sorrel	SP	Ss	FL	CGMS	Y	WSpSuF	G
Oxalis montana sleeping-beauty	SP	Ss	FL	GLMS	W	SpSu	Bgl

PLANT LIST	Plant Type	Env. Tol.	Aesthetic Value	Wildlife Value	Flower Color	Bloom Period	Landscape Uses
Oxalis oregana redwood-sorrel	P	—	—	GLMS	PuW	WSpSu	—
Oxalis violacea violet wood-sorrel	SP	Ss	FL	GLMS	Pu	Sp	gM
Oxytropis lagopus hare-foot locoweed	P	—	—	U	PPu	Su	—
Oxytropis lamberti stemless locoweed	SP	DSs	F	U	PPuW	SpSu	BM
Oxytropis parryi Parry's locoweed	SP	S	F	U	PPu	SpSu	M
Palafoxia feayi Feay's palafox	MP	DS	F	U	P	F	BI
Palafoxia rosea rosy palafox	SA	S	—	U	P	SuF	M
Palafoxia texana Texas palafox	SA	S	F	U	P	SpSuF	CGM
Panax quinquefolius American ginseng	SP	s	FfL	U	W	Su	BgI
Panax trifolius dwarf ginseng	SP	FSs	FL	U	W	SpSu	BgI
Panicum abscissum cut-throat grass	MPG	FS	—	GLMS	G	SuSpF	W
Panicum amarum bitter panic grass	MPG	DNaS	—	GLMS	G	SuF	ES
Panicum anceps beaked panic grass	MPG	DSs	—	GMS	G	SuF	BM
Panicum bulbosum bulb panic grass	SPG	Ss	—	GLMS	G	SuF	BM
Panicum flexile wiry panic grass	SPG	S	—	GLMS	G	SuF	BM
Panicum hallii Hall's panic grass	SPG	DFSs	—	GLMmSW	G	—	BEM
Panicum hemitomon maiden-cane	MPG	FSW	—	GLMSW	G	SuF	Wwg
Panicum obtusum blunt panic grass	SPG	FS	—	GLMSW	G	SuF	EMS
Panicum philadelphicum Philadelphia panic grass	SPG	DS	—	GLMS	G	SuF	M
Panicum repens torpedo grass	MPG	FSW	—	GLMS	G	SpSuF	S
Panicum rigidulum red-top panic grass	MPG	FS	—	GLMS	G	WSpSuF	W
Panicum virgatum wand panic grass	LPG	DFNaSsW	A	GLMSW	G	SuF	EMW
Pappophorum bicolor pink pappus grass	MPG	S	—	U	GP	SpSu	M

PLANT LIST	Plant Type	Env. Tol.	Aesthetic Value	Wildlife Value	Flower Color	Bloom Period	Landscape Uses
Pappophorum vaginatum whiplash pappus grass	G	—	—	U	G	Su	—
Parietaria pensylvanica Pennsylvania pellitory	SA	s	—	U	G	Sp	IRS
Parnassia glauca fen grass-of-Parnassus	SP	FS	FL	U	W	SuF	MgSW
Paronychia drummondii Drummond's nailwort	SAB	Ds	—	U	W	SpSuF	CBI
Paronychia jamesii James' nailwort	SP	DFS	—	U	—	SpSuF	EMRS
Paronychia sessiliflora low nailwort	S	DSs	—	U	—	SpSu	GMR
Parthenium hysterophorus Santa Maria feverfew	MA	S	F	U	W	SuF	M
Parthenium integrifolium wild quinine	MP	DS	FD	U	W	Su	BgM
Pascopyrum smithii western-wheat grass	MG	DS	—	U	G	Su	M
Paspalum bifidum pitchfork crown grass	MPG	Ss	—	GLMSW	G	F	BI
Paspalum dilatatum golden crown grass	LPG	SW	—	MS	G	Su	BMW
Paspalum distichum jointed crown grass	MPG	FS	—	GLMSW	G	SpSuF	s
Paspalum floridanum Florida crown grass	LPG	S	—	GLMSW	G	SpSuF	s
Paspalum monostachyum Gulf dune crown grass	LPG	FS	—	GLMSW	G	SuF	BIs
Paspalum plicatulum brown-seed crown grass	MPG	DS	—	GLMSW	G	SpSuF	M
Paspalum pubiflorum hairy-seed crown grass	MPG	SsW	—	MS	G	SpSu	BEMW
Paspalum setaceum slender crown grass	MPG	FS	—	GLMSW	G	WSpSuF	sW
Paspalum vaginatum talquezal	SPG	FNaSW	—	GLMSW	G	WSuF	W
Passiflora affinis bracted passion flower	PHV	Ss	F	U	yg	Su	Bg
Passiflora foetida scalet-fruit passion-flower	PHV	NaSs	F	C	PuW	SpSuF	BbM
Passiflora incarnata purple passion-flower	HV	Ss	FL	H	L	Su	bB
Passiflora lutea yellow passion-flower	HV	Ss	FL	H	Y	Su	bBI
Passiflora tenuiloba bird-wing passion-flower	PHV	Ss	F	U	G	SpF	Bg

PLANT LIST	Plant Type	Env. Tol.	Aesthetic Value	Wildlife Value	Flower Color	Bloom Period	Landscape Uses
Pedicularis canadensis Canadian lousewort	SP	Ss	F	U	RY	Sp	BI
Pedicularis groenlandica bull elephant's-head	MP	SsW	F	U	BPu	Sp	BgMI
Pedicularis procera giant lousewort	MP	Ss	F	U	Y	Sp	BgM
Pedicularis racemosa parrot's-beak	SP	Ss	F	U	PuWY	Sp	BgI
Pedicularis semibarbata pinewoods lousewort	SP	Ds	L	H	Y	SpSu	bGI
Pediocactus simpsonii snowball cactus	C	DS	B	U	Pu	Sp	gM
Pediomelum argophyllum silver-leaf Indian-breadroot	SP	S	FL	U	B	Su	M
Pediomelum cuspidatum large-bract Indian-breadroot	SP	DS	FL	U	B	Su	M
Pediomelum esculentum large-Indian-breadroot	SP	DS	L	B	B	SpSu	b
Pellaea breweri Brewer's cliffbrake	eSP	s	L	U	X	X	gIR
Pellaea mucronata bird-foot cliffbrake	eSP	DS	L	U	X	X	gIR
Peltandra virginica green arrow-arum	SP	FSW	L	W	G	SpSu	Wwg
Penstemon albidus red-line beardtongue	MP	DS	F	Hh	W	SpSu	bgM
Penstemon angustifolius broad-beard beardtongue	SP	S	F	Hh	BWY	SpSu	bgM
Penstemon arenicola red desert beardtongue	SP	DS	F	Hh	BPu	Su	bgM
Penstemon australis Eustis Lake beardtongue	MP	DSs	F	MmS	PuRY	SpSu	Bg
Penstemon barbatus beard-lip beardtongue	MP	DS	F	HMmS	R	SuF	bgM
Penstemon caesius San Bernardino beardtongue	SP	S	F	HMmS	B	Su	bgM
Penstemon cobaea cobaea beardtongue	MP	S	F	MmS	Pu	SpSu	BgM
Penstemon davidsonii timberline beardtongue	SP	S	FL	HMmS	B	Su	bGgR
Penstemon digitalis foxglove beardtongue	LP	DS	F	HhS	W	Su	bM
Penstemon eatonii Eaton's beardtongue	SP	DS	F	HhMmS	R	Su	bgM
Penstemon fremontii Fremont's beardtongue	SP	DSs	F	HMmS	BW	Su	BbgM

PLANT LIST	Plant Type	Env. Tol.	Aesthetic Value	Wildlife Value	Flower Color	Bloom Period	Landscape Uses
Penstemon gracilis lilac beardtongue	SP	S	F	HMmS	BPu	Su	bgM
Penstemon grinnellii Grinnell's beardtongue	P	—	F	HMmS	BPuW	SpSu	bg
Penstemon hirsutus hairy beardtongue	MP	DSs	F	BHMmS	BW	SpSu	BbgR
Penstemon humilis low beardtongue	SP	Ss	F	BMmS	B	Su	bBg
Penstemon labrosus San Gabriel beardtongue	MP	DSs	F	HMmS	R	Su	Bbgl
Penstemon linarioides toadflax beardtongue	SP	S	F	HhMmS	Pu	Su	bgM
Penstemon palmeri scented beardtongue	MP	DSs	F	HhMmS	WY	Su	BbgM
Penstemon rostriflorus beaked beardtongue	MP	Ss	F	HMmS	ORY	Su	BbgM
Penstemon rydbergii meadow beardtongue	SP	Ss	F	HhMmS	Pu	Su	BbgM
Penstemon speciosus royal beardtongue	SP	S	F	HhMmS	BPu	SpSu	bgM
Penstemon subglaber Utah smooth beardtongue	SP	Ss	F	HhMmS	PW	Su	BbgM
Penstemon tubiflorus white wand beardtongue	LP	S	F	MmS	W	SpSu	gM
Penstemon virgatus upright blue beardtongue	SP	Ss	F	HhMmS	BLPu	Su	BbgM
Penstemon watsonii Watson's beardtongue	SP	DS	F	HhMmS	BPu	Su	bgM
Penstemon whippleanus dark beardtongue	MP	Ss	F	HhMmS	BPuW	Su	Bbgm
Pentalinon luteum hammock vipertail	HVP	—	—	U	Y	WSpSuF	—
Perideridia parishii Parish's yampah	P	—	—	U	PW	SuF	—
Perityle emoryi Emory's rockdaisy	SAB	DSs	F	B	WY	SpSu	bCgMR
Phacelia austromontana southern sierran scorpion-weed	SA	Ss	F	Mm	BL	SpSu	CGgI
Phacelia bipinnatifida fern-leaf scorpion-weed	SB	Ss	FL	Bm	BL	Sp	BbCGgI
Phacelia crenulata notch-leaf scorpion-weed	SA	S	F	m	Pu	WSp	CR
Phacelia distans distant scorpion-weed	SP	Ss	F	m	BPuW	SpSu	Bg
Phacelia imbricata imbricate scorpion-weed	P	—	—	m	W	SpSu	—

PLANT LIST	Plant Type	Env. Tol.	Aesthetic Value	Wildlife Value	Flower Color	Bloom Period	Landscape Uses
Phacelia patuliflora sand scorpion-weed	SA	Ss	F	U	LV	Sp	BGg
Phacelia purshii Miami-mist	SA	Fs	F	m	B	SpSu	BCgS
Phacelia sericea purplefringe	SP	Ss	F	m	BPu	Su	BgI
Phegopteris connectilis narrow beech fern	SP	s	L	m	X	X	I
Phleum alpinum mountain timothy	SPG	FSW	—	MS	G	SuF	W
Phlox caespitosa clustered phlox	SP	S	F	HMS	PuW	Sp	bgM
Phlox covillei Coville's phlox	P	—	—	HMS	PW	SuF	—
Phlox divaricata wild blue phlox	SP	s	FL	BHMS	B	Sp	BbGI
Phlox drummondii annual phlox	MA	DS	F	BHS	R	Su	CbgM
Phlox gracilis slender phlox	SP	s	F	HMS	PuW	Sp	bGgI
Phlox hoodii carpet phlox	SP	DSs	F	BHMS	LW	SpSu	BbGgM
Phlox longifolia long-leaf phlox	SP	DS	F	HMS	PPuW	SpSu	bGgM
Phlox multiflora Rocky Mountain phlox	SP	DSs	F	BHMS	PW	SpSu	BbGIM
Phlox pilosa downy phlox	SP	DS	F	BhMS	BP	Su	bGgM
Phragmites australis common reed	LPG	FSW	FL	U	GrR	SuF	W
Phryma leptostachya American lopseed	MP	S	F	U	PuW	SuF	BI
Phyllanthus abnormis Drummond's leaf-flower	SA	DS	—	U	W	SpSuF	M
Phyllanthus polygonoides smartweed leaf-flower	SP	DS	—	U	G	SpSuF	M
Physalis angustifolia coastal ground-cherry	SP	DS	—	GLm	W	WSpSuF	M
Physalis virginiana Virginia ground-cherry	SP	Ss	f	GLm	Y	Su	BI
Physalis walteri Walter's ground-cherry	SP	DNaS	—	GLm	W	SpSu	—
Physostegia intermedia slender false dragonhead	LP	FSW	F	h	L	SpSu	Wwg
Physostegia virginiana obedient-plant	MP	FS	FL	hS	P	SuF	bgMW

PLANT LIST	Plant Type	Env. Tol.	Aesthetic Value	Wildlife Value	Flower Color	Bloom Period	Landscape Uses
Pilea pumila Canadian clearweed	SP	Fs	L	U	W	SuF	IS
Pilularia americana American pillwort	SP	FSW	—	U	X	X	RWwg
Pinaropappus roseus white rock-lettuce	SP	DS	—	U	LRW	Sp	MR
Piptochaetium avenaceum black-seed spear grass	MPG	s	F	MmS	Br	Sp	BI
Piptochaetium fimbriatum pinyon spear grass	MPG	S	F	MmS	G	SuF	MR
Piptochaetium pringlei pringle's spear grass	MPG	s	F	MmS	Br	Sp	BI
Pistia stratiotes water-lettuce	AQP	SW	L	U	Y	WSpSu F	wg
Pityopsis graminifolia golden-aster	MP	DS	—	U	Y	SpSu	M
Plagiobothrys acanthocarpus adobe popcorn-flower	SA	—	—	U	W	—	C
Plagiobothrys distantiflorus California popcorn-flower	A	—	—	U	YW	SpSu	—
Plagiobothrys humistratus dwarf popcorn-flower	SA	—	—	U	W	—	—
Plagiobothrys hystriculus bearded popcorn-flower	A	—	—	U	YW	Sp	—
Plagiobothrys jonesii Mojave popcorn-flower	SA	DS	—	U	—	—	CM
Plagiobothrys kingii Great Basin popcorn-flower	SA	DS	—	U	W	Sp	CM
Plagiobothrys leptocladus alkali popcorn-flower	SA	SW	—	U	W	—	CW
Plagiobothrys nothofulvus rusty popcorn-flower	SP	S	—	U	W	Sp	M
Plagiobothrys stipitatus stalked popcorn-flower	SA	S	—	U	W	Sp	M
Plantago decipiens seaside plantain	SP	FNaS W	—	GLMmS	W	SuF	W
Plantago eriopoda red-woolly plantain	SP	FNaS	—	GLMmS	W	SuF	W
Plantago maritima goosetongue	SP	FS	—	GLMmS	W	SuF	W
Plantago ovata blond plantain	A	—	—	GLMmS	G	Sp	—
Plantago patagonica woolly plantain	SA	DS	—	GLMmS	PuW	SuF	M
Platanthera blephariglottis white fringed orchid	SP	FSW	F	M	W	SuF	Wwg

PLANT LIST	Plant Type	Env. Tol.	Aesthetic Value	Wildlife Value	Flower Color	Bloom Period	Landscape Uses
Platanthera clavellata green woodland orchid	SP	FSW	F	M	WGY	Su	W
Platanthera dilatata scentbottle	MP	FS	F	M	W	SpF	W
Platanthera grandiflora greater purple fringed orchid	MP	FS	F	M	Pu	Su	Wwg
Platanthera leucophaea prairie white fringed orchid	MP	FS	F	M	W	Su	W
Platanthera leucostachys sierra rein orchid	MP	FS	F	M	W	Su	BMW
Platanthera praeclara Great Plains white fringed orchid	MP	FSW	F	M	W	Su	W
Platanthera sparsiflora canyon bog orchid	SP	FS	F	M	G	Su	W
Pleurozium schreberi feather moss	m	FSsW	—	U	X	X	BGIW
Pluchea camphorata plowman's wort	LP	FS	F	U	P	SuF	W
Pluchea foetida stinking camphorweed	MP	FSW	F	m	W	SuF	Wwg
Pluchea rosea rosy camphorweed	MP	FSW	F	m	P	SuF	Wwg
Poa alpina alpine blue grass	SPG	FS	—	GLMmS	G	Sp	EM
Poa arachnifera Texas blue grass	MPG	S	—	U	G	F	M
Poa bolanderi Bolander's blue grass	SG	Ss	—	GLMmS	G	Sp	BEM
Poa fendleriana mutton grass	SPG	DSs	—	GLMmS	G	Sp	BEM
Poa glauca white blue grass	SG	S	—	GLMmS	G	SuF	EMR
Poa nervosa Hooker's blue grass	SPG	Ss	—	GLMmS	G	SpSu	BE
Poa palustris fowl blue grass	MG	S	—	GLMmS	G	SpSu	EM
Poa reflexa nodding blue grass	SPG	S	—	GLMmS	G	SpSu	EM
Poa secunda curly blue grass	MPG	DSs	—	GLMmS	G	SpSu	EBM
Podophyllum peltatum may-apple	SP	Ss	FfL	U	W	Sp	I
Pogogyne ziziphoroides Sacramento mesa-mint	SA	FS	—	U	Pu	Sp	M
Pogonia ophioglossoides snake-mouth orchid	SP	FSW	FL	U	P	SpSu	W

PLANT LIST	Plant Type	Env. Tol.	Aesthetic Value	Wildlife Value	Flower Color	Bloom Period	Landscape Uses
Polanisia tenuifolia slender-leaf clammeyweed	MP	DS	—	U	W	Su	B
Polemonium foliosissimum towering Jacob's-ladder	MP	Ss	FL	U	BPuW	Sp	Bg
Polemonium pulcherrimum beautiful Jacob's-ladder	SP	Ss	FL	U	BPuW	Sp	BgM
Polemonium reptans Greek-valerian	SP	Ss	F	U	L	Sp	BgIM
Polygala alba white milkwort	SA	DS	F	U	W	SpSu	gM
Polygala lindheimeri shrubby milkwort	SP	DSs	—	U	P	SpF	BM
Polygala nuttallii Nuttall's milkwort	SP	DS	F	U	Pu	SuF	gM
Polygala ovatifolia egg-leaf milkwort	SP	DSs	F	U	PW	SpSu	BM
Polygala paucifolia gaywings	SP	s	FL	U	P	Sp	BgI
Polygala verticillata whorled milkwort	SA	DS	F	U	P	Su	GM
Polygonatum biflorum King Solomon's-seal	LP	FSs	FL	G	W	Sp	BgIS
Polygonatum pubescens hairy Solomon's-seal	MP	s	F	G	GW	Sp	gI
Polygonella articulata coastal jointweed	SA	DS	F	U	PW	SuF	gM
Polygonella polygama October-flower	SP	DSs	—	U	PY	F	BM
Polygonum amphibium water smartweed	MP	FSW	BF	GLmSW	PR	SuF	Wwg
Polygonum bistortoides American bistort	MP	Ss	F	GmS	PW	Su	BM
Polygonum douglasii Douglas' knotweed	SA	Ss	—	GLmS	GPW	Su	B
Polygonum glaucum seaside knotweed	MA	DSNa	L	GLmSW	PW	Su	S
Polygonum hydropiperoides swamp smartweed	SP	SW	—	GMmS W	PW	Su	Wwg
Polygonum pensylvanicum pinkweed	MP	FSW	F	GLmSW	P	SpF	Wwg
Polygonum punctatum dotted smartweed	MP	FSW	F	GLmSW	W	SuF	Wwg
Polygonum saggittatum arrow-leaf tearthumb	SP	FNaS W	FL	GLmSW	W	SuF	Wwg
Polygonum virginianum jumpseed	MP	s	—	GLmS	P	Su	I

PLANT LIST	Plant Type	Env. Tol.	Aesthetic Value	Wildlife Value	Flower Color	Bloom Period	Landscape Uses
Polypodium hesperium western polypody	eSP	s	L	GLM	X	X	gIR
Polypodium virginianum rock polypody	SP	Ds	L	GLM	X	X	gIR
Polypremum procumbens juniper-leaf	SAP	DS	—	U	W	SpSuF	BM
Polystichum acrostichoides Christmas fern	eSP	s	L	GLM	X	X	BgI
Polystichum munitum pineland sword fern	eMP	s	L	GLM	X	X	BgI
Polytaenia nuttallii Nuttall's prairie-parsley	MP	DS	L	U	Y	SpSu	BM
Polytaenia texana Texas false-parsley	MP	S	DF	U	Y	SpSu	M
Polytrichum strictum feather moss	m	FSW	—	U	X	X	W
Pontederia cordata pickerelweed	MP	FSW	FL	LW	B	SuF	Wwg
Potentilla arguta tall cinquefoil	MP	DS	F	GLMmS	WY	Su	gM
Potentilla breweri Sierran cinquefoil	SP	S	F	GLMmS	Y	Su	gM
Potentilla crinita bearded cinquefoil	SP	Ss	F	BGLMmS	Y	Su	BbgM
Potentilla glandulosa sticky cinquefoil	SP	DS	F	BGLMmS	Y	SpSu	bgM
Potentilla gracilis graceful cinquefoil	SP	DS	F	BGLMmS	WY	Su	bgM
Potentilla hippiana woolly cinquefoil	SP	DS	F	BGLMmS	Y	Su	bgM
Potentilla paradoxa bushy cinquefoil	SAP	FSW	FL	BGLMmS	Y	SuF	bgMW
Potentilla simplex oldfield cinquefoil	SP	S	F	BGmLMS	Y	Sp	BbgM
Potentilla thurberi scarlet cinquefoil	P	W	F	BGLMmS	R	SuF	W
Prenanthella exigua brightwhite	SA	S	F	U	PW	SpSu	BM
Prenanthes alba white rattlesnake-root	LP	Ss	F	U	W	Su	BI
Prenanthes altissima tall rattlesnake-root	LP	s	—	U	W	SuF	BI
Prenanthes racemosa purple rattlesnake-root	LP	FSW	F	U	P	SuF	MWS
Prenanthes trifoliolata gall-of-the-earth	LP	DSs	F	U	W	F	B

PLANT LIST	Plant Type	Env. Tol.	Aesthetic Value	Wildlife Value	Flower Color	Bloom Period	Landscape Uses
Primula parryi brook primrose	SP	Ss	F	U	PuR	—	gM
Proboscidea louisianica ram's-horn	MA	S	f	U	WY	Su	BgM
Pseudocymopterus montanus alpine false mountain-parsley	SP	DS	—	U	Y	Su	BM
Pseudoroegneria spicata bluebunch-wheat grass	MPG	DS	—	U	Gr	Su	M
Pseudostellaria jamesiana sticky-starwort	P	—	—	U	W	SpSu	—
Psilocarphus tenellus slender woollyheads	SA	DS	—	U	—	SpSu	MW
Psoralidium tenuiflorum slender-flower lemonweed	MP	DS	—	U	Pu	Su	BM
Psorothamnus fremontii Fremont's smokebush	MS	DS	FL	U	B	Su	BgM
Pteridium aquilinum northern bracken fern	MP	DSs	L	GLM	X	X	BgI
Pterocaulon virgatum wand blackroot	SP	S	L	U	yg	Sp	M
Pterospora andromedea pine drops	MA	s	F	U	W	SuF	gI
Ptilagrostis kingii Sierran false needle grass	PG	—	—	U	G	Su	—
Ptilium crista-castrensis feathermoss	m	s	L	U	X	X	GI
Puccinellia kurilensis dwarf alkali grass	PG	—	—	U	G	Su	—
Puccinellia nuttalliana Nuttall's alkali grass	SPG	S	—	U	G	—	EM
Pulsatilla patens American pasqueflower	SP	DS	F	h	BPuW	Sp	bgM
Pycnanthemum californicum California mountain-mint	MP	DS	FL	B	W	Su	bgM
Pycnanthemum flexuosum Appalachian mountain-mint	MP	DFS	FL	B	W	SuF	BbgSW
Pycnanthemum tenuifolium narrow-leaf mountain-mint	SP	DSs	DF	B	W	Su	BbgM
Pycnanthemum virginianum Virginia mountain-mint	MP	Ss	F	Bh	PuW	Su	Bbg
Pyrola asarifolia pink wintergreen	eSP	sW	FL	G	PPu	Su	GgIWwg
Pyrola chlorantha green-flower wintergreen	eSP	sW	FL	G	PW	Su	GgIWwg
Pyrola elliptica shinleaf	eSP	s	FL	G	GW	Su	GgI

PLANT LIST	Plant Type	Env. Tol.	Aesthetic Value	Wildlife Value	Flower Color	Bloom Period	Landscape Uses
Pyrola picta white vein wintergreen	eSP	s	FL	G	GWY	Su	GgI
Pyrrocoma recemosa clustered goldenweed	SP	NaS	F	u	Y	Su	gM
Rafinesquia neomexicana New Mexico plumseed	SA	DSs	F	U	PPuW	Sp	BCG
Ranunculus abortivus kidney-leaf buttercup	SP	S	—	GLMmS	Y	SpSu	M
Ranunculus californicus California buttercup	SP	S	F	GLMmS	Y	SpSu	gM
Ranunculus cymbalaria alkali buttercup	SP	NaS	FL	GLMmS	Y	SpF	gMS
Ranunculus fascicularis early buttercup	SP	Ss	F	GLmS	Y	Sp	BgIM
Ranunculus glaberrimus sagebrush buttercup	SP	S	FL	GLMmS	Y	Sp	BgM
Ranunculus hispidus bristly buttercut	SP	Ss	F	U	Y	Sp	BgI
Ranunculus longirostris long-beak water-crowfoot	P	SsW	F	GLMmS W	W	SpSu	Wwg
Ranunculus macranthus large buttercup	MP	FSsW	F	GMmW	Y	Sp	IW
Ranunculus recurvatus blisterwort	SP	s	F	GLMmS	Y	Sp	gI
Ranunculus reptans lesser creeping spearwort	SP	FS	F	GLMmS W	Y	Su	gMS
Ranunculus sceleratus cursed buttercup	MAP	SW	F	GLMmS W	Y	SpSu	Wwg
Ratibida columnifera red-spike Mexican-hat	LP	DS	F	BS	Y	SuF	bgM
Ratibida pinnata gray-head Mexican-hat	LP	DS	F	BS	Y	Su	bgM
Redfieldia flexuosa blowout grass	PG	—	—	U	G	Su	—
Reimarochloa oligostachya Florida reimar grass	MPG	DS	—	U	Gr	SpSu	W
Rhexia mariana Maryland meadow-beauty	SP	FSW	F	U	PuR	Su	MW
Rhynchosia americana American snout-bean	PHV	Ss	F	U	Y	SpSuF	B
Rhynchosia tomentosa twining snout-bean	MP	DSs	—	U	O	SpSu	BM
Rhynchospora alba white beak sedge	LPG	Ss	—	GLMSW	W	Su	W
Rhynchospora capillacea needle beak sedge	SPG	FSW	—	GLMSW	G	SuF	W

PLANT LIST	Plant Type	Env. Tol.	Aesthetic Value	Wildlife Value	Flower Color	Bloom Period	Landscape Uses
Rhynchospora colorata narrow-leaf whitetop	SPG	FS	—	GLMSW	G	Su	W
Rhynchospora corniculata short-bristle horned beak sedge	MPG	FNaSW	—	GLMSW	G	SuF	W
Rhynchospora harveyi Harvey's beak sedge	MPG	FSs	—	GMmSW	G	SpSu	BEMW
Rhynchospora inundata narrow-fruit horned beak sedge	SAPG	FS	—	GLMSW	G	Su	W
Rhynchospora megalocarpa sandy-field beak sedge	MPG	DS	—	GLMS	G	Su	M
Rhynchospora microcarpa southern beak sedge	MPG	FSsW	—	GLMSW	G	Su	W
Rhynchospora tracyi Tracy's beak sedge	MPG	FSW	—	GLMSW	G	SuF	W
Rivina humilis rougeplant	LP	FSs	Ff	U	PW	SpF	B
Rorippa teres southern marsh yellowcress	SA	SsW	—	U	Y	WSp	W
Rudbeckia fulgida orange coneflower	MP	Ss	F	BGS	OY	Su	BgM
Rudbeckia hirta black-eyed-Susan	MBP	DS	F	BCS	Y	SuF	BbgM
Rudbeckia laciniata green-head coneflower	LP	FS	F	BS	Y	SuF	bMW
Rudbeckia occidentalis western coneflower	LP	Fs	F	BS	Y	Su	BbIS
Ruellia caroliniensis Carolina wild petunia	MP	Ss	F	GS	B	SpSuF	BbgM
Ruellia drummondiana Drummond's wild petunia	MP	Ss	F	U	L	SuF	BRS
Ruellia humilis fringe-leaf wild petunia	MP	Ss	F	GS	BL	Su	BgM
Ruellia nudiflora violet wild petunia	MP	Ss	F	GS	BPu	SpSuF	BbgM
Rumex maritimus golden dock	SA	FNaS	—	CGLMmSW	G	Su	C
Rumex orbiculatus greater water dock	LP	SW	L	CGLMmSW	G	SuF	BWwg
Rumex verticillatus swamp dock	LP	FSW	—	CGMmSW	BrG	Su	W
Ruppia maritima beaked ditch-grass	AQPG	NaW	—	U	—	WSpSuF	Wwg
Sabatia angularis rose-pink	MBP	DS	F	U	P	SuF	BgM
Sabatia arenicola sand rose-gentian	SA	Ss	F	B	P	SpSu	BbGM

PLANT LIST	Plant Type	Env. Tol.	Aesthetic Value	Wildlife Value	Flower Color	Bloom Period	Landscape Uses
Sabatia campestris Texas-star	SP	S	F	U	Y	SuF	M
Saccharum giganteum giant plume grass	PG	W	—	U	G	Su	W
Sacciolepis striata American glenwood grass	LPG	FSsW	—	U	G	SuF	ESW
Sagittaria calycina hooded arrowhead	P	SsW	FL	LW	W	Su	Wwg
Sagittaria graminea grass-leaf arrowhead	SP	SW	FL	LW	W	SuF	Wwg
Sagittaria lancifolia bull-tongue arrowhead	P	FW	FL	MW	W	SpSuF	Wwg
Sagittaria latifolia duck-potato	P	Ss	FL	LW	W	Su	Wwg
Sagittaria papillosa nipple-bract arrowhead	P	FNaW	FL	MW	W	SpSuF	Wwg
Salicornia bigelovii dwarf saltwort	SA	DNaS	—	W	W	SuF	W
Salicornia maritima sea saltwort	A	NaSW	—	W	G	SuF	W
Salicornia rubra red saltwort	SA	DNaS W	—	W	G	SuF	W
Salicornia virginica woody saltwort	SP	FNaS W	—	W	G	SuF	W
Salvia azurea azure-blue sage	LP	DS	F	BH	B	SuF	bgM
Salvia coccinea blood sage	MAP	Ss	F	BHhS	R	SpSuF	BbgM
Salvia columbariae California sage	SA	S	F	U	BPu	Sp	M
Salvia lyrata lyre-leaf sage	SP	DS	F	U	BPu	SpSu	BgM
Salvia roemeriana cedar sedge	MP	Ss	F	HhS	R	SpSu	Bb
Salvia spathacea hummingbird sage	MA	S	F	GHS	R	Sp	bg
Samolus ebracteatus limewater brookweed	SP	FNaS W	F	U	PW	SpSuF	MW
Samolus valerandi seaside brookweed	SP	SF	—	U	W	SpSu	W
Sanguinaria canadensis bloodroot	SP	s	FL	ANTS	W	Sp	gI
Sanicula arctopoides footsteps-of-spring	SP	S	—	U	Y	Su	GM
Sanicula bipinnatifida purple black-snakeroot	SP	S	L	U	Y	Su	M

PLANT LIST	Plant Type	Env. Tol.	Aesthetic Value	Wildlife Value	Flower Color	Bloom Period	Landscape Uses
Sanicula canadensis Canadian black-snakeroot	MBP	DSs	L	U	W	SpSu	BI
Sanicula crassicaulis Pacific black-snakeroot	MP	—	—	U	Ypu	Su	BI
Sanicula marilandica Maryland black-snakeroot	MP	s	L	U	GW	Su	BI
Sanicula odorata clustered black-snakeroot	MP	W	—	U	W	SpSu	W
Sarcocornia pacifica Pacific swampfire	SP	NaSW	—	U	—	—	EW
Sarcodes sanguinea snowplant	SP	s	FL	U	R	SpSu	I
Sarracenia flava yellow pitcherplant	MP	SW	FL	U	Y	Sp	Wwg
Sarracenia purpurea purple pitcherplant	SP	SW	FL	U	R	SpSu	W
Sarracenia rubra sweet pitcherplant	SP	FS	FL	U	R	Sp	W
Satureja douglasii Oregon-tea	P	—	—	U	PuW	SpSuF	—
Saururus cernuus lizard's-tail	LP	FSW	FL	U	W	SuF	SWwg
Saxifraga odontoloma streambank saxifrage	MP	FSW	FL	U	Pu	SpSu	SWwg
Saxifraga pensylvanica eastern swamp saxifrage	MP	FSW	F	m	yg	Sp	Wwg
Saxifraga virginiensis early saxifrage	SP	s	FL	U	W	Sp	BgIR
Scheuchzeria palustris rannoch-rush	SPG	Fs	—	U	W	Sp	EW
Schizachne purpurascens false melic grass	MPG	DSs	F	U	—	Su	BgRI
Schizachyrium cirratum Texas false bluestem	MPG	S	B	GMmS	W	SuF	EM
Schizachyrium rhizomatum Florida false bluestem	MPG	S	FL	GMmS	G	Su	M
Schizachyrium sanguineum crimson false bluestem	MPG	DS	FL	GMmS	W	SuF	EM
Schizachyrium scoparium little false bluestem	MPG	DS	AFL	GMmS	W	SuF	BEgM
Schizachyrium tenerum slender false bluestem	MPG	s	FL	U	W	SuF	EI
Schoenocrambe linifolia Salmon River plains-mustard	P	—	—	U	Y	SpSu	—
Schoenus nigricans black bog-rush	SPG	DFS	—	U	Br	SpSu	SW

PLANT LIST	Plant Type	Env. Tol.	Aesthetic Value	Wildlife Value	Flower Color	Bloom Period	Landscape Uses
Schrankia microphylla little-leaf sensitive-briar	SP	DS	FL	U	P	SpF	BM
Scirpus acutus hard-stem bulrush	LPG	FNaSW	L	GLmSW	G	SuF	Wwg
Scirpus americanus chairmaker's bulrush	LPG	FNaSW	L	GLmSW	BrG	SuF	Wwg
Scirpus atrovirens dark-green bulrush	LPG	FSW	DL	GLmSW	G	Su	MW
Scirpus californicus California bulrush	LPG	FNaSW	L	GLmSW	Br	Su	Wwg
Scirpus cespitosus tufted bulrush	MPG	FSW	L	GLmSW	G	Su	Wwg
Scirpus congdonii Congdon's bulrush	MG	FSW	L	GLMsW	G	Su	SW
Scirpus cyperinus cottongrass bulrush	LPG	FSW	DL	GLmSW	G	SuF	SWwg
Scirpus fluviatilis river bulrush	LPG	FSW	L	GLmSW	G	SuF	Wwg
Scirpus heterochaetus pale great bulrush	LPG	FSW	L	GLmSW	G	SuF	Wwg
Scirpus lineatus drooping bulrush	MPG	FS	L	GLmSW	G	Su	MW
Scirpus maritimus saltmarsh bulrush	LPG	FS	L	GLmSW	G	SuF	W
Scirpus microcarpus red-tinge bulrush	MPG	FSW	L	GLmSW	G	Su	MsW
Scirpus pallidus pale bulrush	MPG	SW	L	GLmSW	G	Su	W
Scirpus pungens three-square	MPG	FNaSW	L	GLmSW	Br	SuF	Wwg
Scirpus robustus seaside bulrush	LPG	FNaSW	L	GLmSW	Br	SuF	Wwg
Scirpus tabernaemontani soft-stem bulrush	LPG	FNaSW	DL	GLmSW	BrG	SuF	SWwg
Scleria oligantha little-head nut-rush	MPG	FSs	—	GSW	Br	SpSu	BIM
Scleria triglomerata whip nut-rush	MPG	Ss	—	GSW	Br	SpF	BI
Scleria verticillata low nut-rush	SPG	FSW	—	GSW	Br	SuF	W
Scleropogon brevifolius burro grass	SPG	Ss	FL	M	G	—	BEM
Scorpidium scorpioides moss	m	SW	L	U	X	X	Wwg
Scrophularia lanceolata lance-leaf figwort	LP	Ss	—	U	yg	Su	B

PLANT LIST	Plant Type	Env. Tol.	Aesthetic Value	Wildlife Value	Flower Color	Bloom Period	Landscape Uses
Scrophularia parviflora pineland figwort	P	—	—	U	RG	SuF	—
Scutellaria drummondii Drummond's skullcap	SA	DSs	F	H	BPu	SpSu	BMR
Scutellaria galericulata hooded skullcap	MP	FS	F	U	B	Su	BW
Scutellaria lateriflora mad dog skullcap	MP	FS	F	U	BW	SuF	W
Scutellaria ovata heart-leaf skullcap	MP	s	F	U	B	Su	I
Scutellaria resinosa resin-dot skullcap	SP	DS	F	U	BPu	SpSuF	MR
Sedum niveum Davidson's stonecrop	P	—	—	U	W	Su	—
Sedum nuttallianum yellow stonecrop	SA	DS	FL	BC	Y	SpSu	GgMR
Sedum pulchellum widow's cross	SABP	DSs	FL	U	PW	SpSu	GgR
Sedum rhodanthum queen's-crown	SP	Ss	FL	U	PW	SpSu	GgM
Sedum stenopetalum worm-leaf stonecrop	SP	Ss	F	U	Y	—	Gg
Sedum ternatum woodland stonecrop	SP	DSs	FL	U	W	Sp	EGgIR
Selaginella arenicola sand spike moss	SPm	DS	L	U	X	X	Gg
Senecio ampullaceus Texas ragwort	MA	DS	F	B	Y	Sp	M
Senecio anonymus Small's ragwort	MP	S	FL	B	Y	SpSu	bM
Senecio aureus golden ragwort	SP	FSW	FL	B	Y	SpSu	bW
Senecio canus silver-woolly ragwort	SP	DSs	FL	B	Y	SpSu	BbM
Senecio cynthioides white mountain ragwort	P	—	FL	B	Y	Su	b
Senecio eremophilus desert ragwort	MP	Ss	FL	B	Y	SpSu	BbEm
Senecio flaccidus thread-leaf ragwort	P	—	FL	B	Y	SpSu	b
Senecio glabellus cress-leaf ragwort	MP	FS	F	B	Y	Sp	bSW
Senecio multilobatus lobe-leaf ragwort	SBP	Ss	FL	B	Y	SpSu	Bb
Senecio parryi mountain ragwort	SA	—	—	B	Y	F	—

PLANT LIST	Plant Type	Env. Tol.	Aesthetic Value	Wildlife Value	Flower Color	Bloom Period	Landscape Uses
Senecio pauperculus balsam ragwort	MP	DS	F	B	Y	Su	BbW
Senecio plattensis prairie ragwort	MP	DSs	F	B	Y	SpSu	bMR
Senecio riddellii Riddell's ragwort	MP	S	F	B	Y	SuF	bM
Senecio scorzonella Sierran ragwort	SP	S	FL	BM	Y	—	bM
Senecio spartioides broom-like ragwort	MP	FNaS	FL	B	Y	SuF	BbS
Senecio triangularis arrow-leaf ragwort	LP	FS	FL	B	Y	SuF	bSW
Senna bauhinioides shrubby wild sensitive-plant	SP	S	F	GMm	Y	Sp	M
Senna hirsuta woolly wild sensitive-plant	P	—	—	U	Y	Su	—
Senna roemeriana two-leaved wild sensitive-plant	SP	Ss	L	GmS	Y	Su	BgM
Sesuvium portulacastrum shoreline sea-purslane	SP	DNaS	F	U	PPu	WSpSuF	GgM
Setaria corrugata coastal bristle grass	MAG	FSs	—	GLmS	G	Su	CS
Setaria leucopila strembed bristle grass	MPG	DFS	—	U	G	SuF	MS
Setaria macrostachya plains bristle grass	MPG	S	—	GLmS	G	SuF	EM
Setaria magna giant bristle grass	LAG	FNaSW	—	GLmSW	G	SuF	EW
Setaria ramiseta Rio Grande bristle grass	SPG	DS	—	GLmS	G	SpSuF	EM
Setaria scheelei southwestern bristle grass	MPG	s	DF	U	G	SpSuF	IS
Sibbaldia procumbens creeping-glow-wort	SP	Ss	—	U	Y	Su	BGM
Sibbaldiopsis tridentata shrubby-fivefingers	MP	FSs	—	mS	W	Su	gR
Sicyos angulatus one-seed burr-cucumber	HV	FS	L	U	W	SuF	B
Silene caroliniana sticky catchfly	SP	s	F	HmS	PW	Sp	BGgI
Silene douglasii seabluff catchfly	SP	Ss	F	U	GPW	SpSu	BM
Silene lemmonii Lemmon's catchfly	SP	S	F	mS	W	—	BGgM
Silene parishii Parish's catchfly	SP	Ss	F	mS	WY	—	BGgI

PLANT LIST	Plant Type	Env. Tol.	Aesthetic Value	Wildlife Value	Flower Color	Bloom Period	Landscape Uses
Silene regia royal catchfly	MP	DSs	F	HS	R	Su	BbgM
Silene stellata widow's-frill	MP	SD	FL	BmS	W	SuF	BbM
Silene verecunda San Francisco catchfly	P	S	G	mS	P	—	gM
Silene virginica fire-pink	SP	s	F	H	R	SpSu	BgI
Silphium asteriscus starry rosinweed	LP	DSs	F	BGS	Y	Su	Bbg
Silphium compositum kidney-leaf rosinweed	LP	DS	F	BS	Y	SuF	bM
Silphium integrifolium entire-leaf rosinweed	LP	DS	F	BS	Y	Su	bM
Silphium laciniatum compassplant	LP	DS	FL	BS	Y	Su	bM
Silphium radula rough stem rosinweed	P	DS	F	BS	Y	Su	—
Silphium terebinthinaceum prairie rosinweed	LP	DS	FL	BS	Y	Su	bM
Simsia calva awnless bush-sunflower	SP	S	F	U	Y	SpSuF	BM
Siphonoglossa pilosella hairy tubetongue	SP	DFSs	F	U	BPuW	SpSuF	BMS
Sisyrinchium albidum white blue-eyed-grass	SP	DS	F	U	VW	Sp	BgM
Sisyrinchium angustifolium narrow-leaf blue-eyed-grass	SP	Ss	FL	U	B	Sp	gM
Sisyrinchium bellum California blue-eyed-grass	SP	S	FL	U	B	Sp	gM
Sisyrinchium douglasii grasswindows	SP	DS	F	U	PuW	Sp	BM
Sium suave hemlock water-parsnip	LP	FSW	F	U	W	Su	W
Smilax ecirrata upright carrion-flower	HPV	s	—	MS	G	Sp	I
Smilax herbacea smooth carrion-flower	HPV	s	—	MS	G	Sp	I
Solanum bahamense Bahama nightshade	P	—	—	U	BW	WSpSuF	—
Solanum elaeagnifolium silver-leaf nightshade	MP	DSs	—	GS	BPuW	SpSuF	BM
Solanum xantii chaparral nightshade	SP	DS	F	GLmSW	B	—	GgM
Solidago bicolor white goldenrod	MP	DS	F	BGLMmS	W	F	BgIR

PLANT LIST	Plant Type	Env. Tol.	Aesthetic Value	Wildlife Value	Flower Color	Bloom Period	Landscape Uses
Solidago buckleyi Buckley's goldenrod	MP	Ss	F	BCS	Y	F	BbI
Solidago caesia wreath goldenrod	MP	s	F	BGLMmS	Y	F	BbI
Solidago californica northern California goldenrod	MP	DS	F	GLMmS	Y	F	M
Solidago canadensis Canadian goldenrod	LP	FS	F	BGLMmS	Y	SuF	bMW
Solidago flexicaulis zigzag goldenrod	MP	s	F	BGLMmS	Y	SuF	bI
Solidago gigantea late goldenrod	LP	S	F	BGLMmS	Y	SuF	Bb
Solidago glomerata clustered goldenrod	MP	FS	F	GLMmS	Y	F	W
Solidago hispida hairy goldenrod	MP	S	F	BGLMmS	Y	SuF	bM
Solidago juncea early goldenrod	MP	DSs	F	BGLMmS	Y	SuF	BbM
Solidago macrophylla large-leaf goldenrod	MP	s	F	BGLMmS	Y	SuF	Bb
Solidago missouriensis Missouri goldenrod	MP	S	F	BGLMmS	Y	SpF	BbM
Solidago mollis velvet goldenrod	SP	DS	F	BGLMmS	Y	SuF	bM
Solidago multiradiata Rocky Mountain goldenrod	MP	S	F	BGLMmS	Y	SuF	bM
Solidago nemoralis gray goldenrod	SP	S	F	BGLMmS	Y	SuF	bM
Solidago odora anise-scented goldenrod	MP	S	F	BGLMmS	Y	SuF	bM
Solidago ohioensis Ohio goldenrod	MP	FS	F	BGLMmS	Y	SuF	bMW
Solidago patula round-leaf goldenrod	MP	FS	F	BGLMmS	Y	F	bMW
Solidago ptarmicoides prairie goldenrod	MP	DS	F	BGLMmS	Y	F	bM
Solidago rigida hard-leaf goldenrod	LP	DS	F	BGLMmS	Y	SuF	bM
Solidago rugosa wrinkle-leaf goldenrod	MP	FSW	F	BGMmS	Y	F	bMW
Solidago sempervirens seaside goldenrod	LP	FNaS	F	BGLMmS	Y	SuF	bW
Solidago spathulata coastal-dune goldenrod	SP	Ss	F	BGLMmS	Y	F	BbG
Solidago speciosa showy goldenrod	LP	DS	F	BGLMmS	Y	SuF	BbM

PLANT LIST	Plant Type	Env. Tol.	Aesthetic Value	Wildlife Value	Flower Color	Bloom Period	Landscape Uses
Solidago uliginosa bog goldenrod	LP	FS	F	BGLMmS	Y	SuF	bMW
Solidago velutina three-nerve goldenrod	P	—	F	BGLMmS	Y	SuF	—
Solidago wrightii Wright's goldenrod	MP	DSs	F	BGLMmS	Y	SuF	BbM
Sorghastrum nutans yellow Indian grass	LPG	DFS	L	L	Y	SuF	M
Sparganium americanum American burr-reed	MPG	SW	L	LmSW	G	SpSu	Wwg
Sparganium eurycarpum broad-fruit burr-reed	LP	FSs	DL	LW	G	Su	SWwg
Spartina alterniflora saltwater cord grass	LPG	FNaSW	L	LMSW	G	SuF	W
Spartina bakeri bunch cord grass	LPG	FS	L	LMSW	G	WSpF	W
Spartina cynosuroides big cord grass	LG	FNaSW	L	LMSW	G	SuF	Wwg
Spartina foliosa California cord grass	MPG	FNaSW	L	MmSW	G	SuF	Wwg
Spartina gracilis alkali cord grass	MG	FS	FL	LMSW	G	SuF	Wwg
Spartina patens salt-meadow cord grass	MG	FNaSW	L	LMSW	G	SuF	Wwg
Spartina pectinata freshwater cord grass	LG	FNaSW	L	LMSW	G	SuF	W
Spartina spartinae gulf cord grass	LG	FNaSW	L	LMSW	G	SuF	W
Spergularia salina sandspurrey	SA	DFNaSW	—	U	PuW	SpSu	CW
Sphaeralcea angustifolia copper globe-mallow	SP	DS	DF	U	LPuR	WSpSuF	GgRM
Sphaeralcea coccinea scarlet globe-mallow	SP	DS	DF	U	LR	SpSu	GgM
Sphaeralcea emoryi Emory's globe-mallow	P	—	DF	U	O	SpSuF	—
Sphaeralcea fendleri thicket globe-mallow	SP	DS	DF	U	LPuR	SpSuF	GgMR
Sphaeralcea grossulariifolia currant-leaf globe-mallow	MP	Ss	FL	U	OP	—	BgM
Sphaeralcea incana soft globe-mallow	LP	DS	F	U	P	SuF	gMR
Sphaeralcea leptophylla scaly globe-mallow	SP	Ss	DF	U	R	SpSu	BGgM
Sphaeromeria cana gray chicken-sage	SS	DS	FL	Mm	OR	SuF	BgM

PLANT LIST	Plant Type	Env. Tol.	Aesthetic Value	Wildlife Value	Flower Color	Bloom Period	Landscape Uses
Sphagnum angustifolim moss	m	SsW	L	mR	X	X	W
Sphagnum capillifolium moss	m	FSs	L	mR	X	X	SW
Sphagnum centrale moss	m	FSs	L	GW	X	X	IW
Sphagnum contortum moss	m	FSsW	L	mR	X	X	WM
Sphagnum cuspidatum moss	m	FSW	L	mR	X	X	Wwg
Sphagnum fallax moss	m	FSW	L	mR	X	X	W
Sphagnum fimbriatum moss	m	FSsW	L	mR	X	X	W
Sphagnum fuscum moss	m	FSsW	L	mR	X	X	W
Sphagnum magellanicum moss	m	FSsW	L	mR	X	X	W
Sphagnum majus moss	m	FSsW	L	mR	X	X	Wwg
Sphagnum nemoreum moss	m	FSsW	L	mR	X	X	W
Sphagnum papillosum moss	m	FSsW	L	mR	X	X	W
Sphagnum recurvum moss	m	FSsW	L	mR	X	X	BW
Sphagnum rubellum moss	m	FSsW	L	mR	X	X	W
Sphagnum russowii moss	m	FSsW	L	mR	X	X	BW
Sphagnum subsecundum moss	m	FSsW	L	mR	X	X	MWwg
Sphagnum teres moss	m	FSsW	L	mR	X	X	Wwg
Sphagnum warnstorfii moss	m	FSsW	L	mR	X	X	BW
Sphenosciadium capitellatum swamp whiteheads	LP	FSW	F	U	W	Su	MW
Spiranthes cernua white nodding ladies'-tresses	SP	FS	F	M	W	SuF	gMW
Spiranthes lucida shining ladies'-tresses	SP	S	F	M	Y	Su	BgM
Spiranthes romanzoffiana hooded ladies'-tresses	SP	FSsW	F	M	W	SuF	gMW
Sporobolus airoides alkali-sacaton	MPG	FNaS	FL	GMmS	G	SuF	EM

PLANT LIST	Plant Type	Env. Tol.	Aesthetic Value	Wildlife Value	Flower Color	Bloom Period	Landscape Uses
Sporobolus asper tall dropseed	MPG	S	L	GMmS	G	SuF	EM
Sporobolus contractus narrow-spike dropseed	MPG	DFNaSs	L	GMmS	G	SuF	BM
Sporobolus cryptandrus sand dropseed	MPG	DS	L	GMmS	G	SuF	M
Sporobolus flexuosus mesa dropseed	MPG	S	L	GMmS	G	SuF	M
Sporobolus giganteus giant dropseed	LPG	DSs	L	GMmS	G	Su	BEM
Sporobolus heterolepis prairie dropseed	MPG	DS	L	GMmS	G	SuF	EM
Sporobolus indicus smut grass	MPG	DS	L	GMmS	G	Su	EM
Sporobolus junceus wire grass	MPG	DSs	L	GMmS	G	SuF	BEM
Sporobolus pyramidatus whorled dropseed	MPG	NaS	—	GMmS	G	SuF	EM
Sporobolus silveanus Silvens' dropseed	MPG	Ss	L	GMmS	G	SuF	BE
Sporobolus vaginiflorus poverty dropseed	SAG	DS	L	GMmS	G	SuF	EM
Sporobolus virginicus seashore dropseed	SPG	FNaS	L	GMmS	G	SuF	ESW
Sporobolus wrightii Wright's dropseed	MPG	DNaSs	FL	GMmS	G	SuF	BEM
Stachys drummondii Drummond's hedge-nettle	MAB	Ss	F	U	LP	WSpSu	B
Stachys mexicana Mexican hedge-nettle	P	W	—	U	Pu	Su	W
Stachys palustris woundwort	SP	S	F	U	Pu	Su	M
Stachys tenuifolia smooth hedge-nettle	LP	FS	F	U	P	SuF	W
Stellaria nitens shiny starwort	SA	FSs	—	m	W	Su	BM
Stellaria prostrata prostrate starwort	SA	Fs	—	U	W	SpSuF	GIS
Stellaria pubera great chickweed	SP	s	—	GMmS	W	Sp	BI
Stenosiphon linifolius false gaura	LP	DS	—	U	W	SuF	R
Stenotaphrum secundatum St. Augustine grass	SPG	FS	—	U	R	SuF	MS
Stenotus acaulis stemless mock goldenweed	eSS	DS	FL	U	Y	SpSu	GgMR

PLANT LIST	Plant Type	Env. Tol.	Aesthetic Value	Wildlife Value	Flower Color	Bloom Period	Landscape Uses
Stephanomeria pauciflora brown-plume wire-lettuce	SP	DNaSs	—	U	PW	SuF	BEM
Stillingia sylvatica queen's-delight	MP	S	L	G	Y	SpSu	B
Stillingia texana Texas toothleaf	SP	S	—	G	G	Sp	BM
Stipa comata needle-and-thread	SPG	DS	F	MmS	G	Sp	GM
Stipa coronata giant needle grass	LPG	DS	F	MmS	Pu	SpSu	M
Stipa lettermanii Letterman's needle grass	SPG	DS	—	LMmS	G	Su	GM
Stipa nelsonii Nelson's needle grass	P	—	—	MmS	G	Su	—
Stipa neomexicana New Mexico needle grass	MPG	DSs	—	MmS	G	SuF	BM
Stipa occidentalis western needle grass	SPG	DS	F	MmS	G	Su	BEM
Stipa pinetorum pine-forest needle grass	SPG	Ss	—	MmS	G	Su	BE
Stipa spartea porcupine grass	MPG	DS	F	LMmS	G	Sp	EM
Stipa speciosa desert needle grass	SPG	DS	F	MmS	G	Sp	EGM
Stipa thurberiana Thurber's needle grass	SPG	DS	F	MmS	G	Su	EM
Stipulicida setacea pineland scaly-pink	SP	Ss	F	U	—	SpSuF	BgM
Streptanthus bernardinus Laguna Mountain jewelflower	MP	Ss	F	U	PuW	—	Bg
Streptopus amplexifolius clasping twistedstalk	MP	s	FL	U	GW	SpSu	gI
Streptopus roseus rosy twistedstalk	MP	s	FL	U	P	SpSu	gI
Streptopus streptopoides small twistedstalk	MP	s	FL	U	PuG	SpSu	gI
Strophostyles helvula trailing fuzzy-bean	HVA	FS	F	G	PPu	SuF	BEGS
Strophostyles leiosperma slick-seed fuzzy-bean	HVA	DS	F	G	PPu	SuF	BEG
Strophostyles umbellata pink fuzzy-bean	HVP	S	F	G	P	SuF	BG
Stylisma pickeringii Pickering's dawnflower	PHV	S	F	U	W	SuF	gM
Stylocline micropoides woolly-head neststraw	SA	DS	B	U	U	Sp	—

PLANT LIST	Plant Type	Env. Tol.	Aesthetic Value	Wildlife Value	Flower Color	Bloom Period	Landscape Uses
Stylophorum diphyllum celandine-poppy	SP	s	FL	U	Y	Sp	BgI
Stylosanthes biflora side-beak pencil-flower	SP	DS	F	U	Y	SpF	gRM
Suaeda californica broom seepweed	MP	FNaSW	F	U	—	F	W
Suaeda linearis annual seepweed	MA	FNaSW	F	U	—	SuF	SW
Suaeda maritima herbaceous seepweed	SA	FNaS	F	U	W	SuF	—
Symplocarpus foetidus skunk-cabbage	SP	SW	FL	GW	GPu	WSp	W
Synthyris reniformis snowqueen	P	—	—	U	BPu	Sp	—
Talinum parviflorum sunbright	SP	DS	FL	U	PW	Su	gR
Talinum teretifolium quill fameflower	SP	DS	FL	U	PPu	Su	gR
Tellima grandiflora fragrant fringecup	MP	s	FL	U	RW	SP	EgI
Tephrosia lindheimeri Lindheimer's hoary-pea	MP	S	FL	BG	Pu	SpSuF	EGM
Tephrosia spicata spiked hoary-pea	SP	S	FL	BG	WP	SpSu	EBG
Tephrosia virginiana goat's-rue	SP	S	FL	BG	PY	SpSu	EBG
Tetragonotheca ludoviciana Louisiana neveray	LP	S	F	U	Y	SuF	gI
Tetraneuris acaulis stemless four-nerve-daisy	SP	DS	F	B	Y	SpSu	MR
Teucrium canadense American germander	MP	FSs	F	U	P	SuF	BS
Thalia geniculata bent alligator-flag	AQP	FSW	L	U	W	SuF	Wwg
Thalictrum clavatum mountain meadow-rue	MP	FSs	L	B	W	SpSu	BIS
Thalictrum dasycarpum purple meadow-rue	LP	FS	FL	B	Wyg	Su	BgS
Thalictrum dioicum early meadow-rue	LP	Ss	FL	B	W	Sp	Bb
Thalictrum fendleri Fendler's meadow-rue	MP	Ss	FL	B	W	SpSu	Bb
Thalictrum occidentale western meadow-rue	MP	s	FL	B	W	Su	BbI
Thalictrum pubescens king-of-the-meadow	LP	Fs	FL	B	W	Su	BbIW

PLANT LIST	Plant Type	Env. Tol.	Aesthetic Value	Wildlife Value	Flower Color	Bloom Period	Landscape Uses
Thalictrum thalictroides rue-anemone	SP	s	FL	B	PW	Sp	BbI
Thamnosma texanum rue-of-the-mountains	MP	S	FL	U	PuY	SpF	gM
Thaspium trifoliatum purple meadow-parsnip	MP	FSs	F	B	Y	SpSu	BbMW
Thelesperma megapotamicum Hopi-tea	MP	S	F	U	Y	SpSu	M
Thelesperma nuecense Rio Grand greenthread	MA	S	F	B	Y	SpSuF	C
Thelesperma simplicifolium slender greenthread	MP	DS	F	U	Y	Su	EgM
Thelespermia filifolium stiff greenthread	MA	S	F	U	Y	SpSu	CgM
Thelypodiopsis elegans westwater tumble-mustard	MAB	Ss	F	U	LPW	—	BC
Thelypodium sagittatum arrowhead thelypody	SB	NaS	L	U	PuW	—	GM
Thelypteris noveboracensis New York fern	MP	Fs	L	M	X	X	BgIS
Thelypteris palustris eastern marsh fern	MP	FSsW	L	M	X	X	BgW
Thermopsis rhombifolia prairie golden-banner	SP	Ss	F	U	Y	—	BEGg
Thymophylla acerosa American pricklyleaf	SP	Ss	F	U	Y	—	BGg
Thymophylla pentachaeta five-needle pricklyleaf	SP	Ss	F	U	Y	—	BGg
Tiarella cordifolia heart-leaf foamflower	SP	s	FL	G	W	SpSu	GgI
Tiarella trifoliata three-leaf foamflower	SP	s	FL	G	W	SpSu	GgI
Tinantia anomala widow's-tears	MA	Ss	F	U	BL	SpSu	BCR
Tipularia discolor crippled-cranefly	SP	s	FL	U	G	Su	GgI
Tiquilia canescens woody crinklemat	SP	S	F	U	P	SpSu	M
Tolmiea menziesii piggyback-plant	MP	Ss	L	U	BrGPu	—	BgIS
Tomenthypnum nitens moss	m	FSW	—	GMmW	X	X	W
Tortula brevipes moss	m	—	—	GMmW	X	X	—
Tortula princeps moss	m	—	—	GMmW	X	X	—

PLANT LIST	Plant Type	Env. Tol.	Aesthetic Value	Wildlife Value	Flower Color	Bloom Period	Landscape Uses
Tortula ruralis moss	m	DS	L	GMmW	X	X	ERS
Trachypogon secundus one-sided crinkle-awn grass	MPG	Ss	—	U	G	SuF	BEM
Tradescantia edwardsiana plateau spiderwort	MP	s	F	B	PW	Sp	gI
Tradescantia gigantea giant spiderwort	MP	S	F	U	BP	Sp	gM
Tradescantia occidentalis prairie spiderwort	SP	DS	F	U	BP	SpSu	gM
Tradescantia ohiensis bluejacket	SP	S	LF	U	BP	Su	gM
Tradescantia virginiana Virginia spiderwort	SP	Ss	FL	U	B	SpSu	BgI
Tragia ramosa branched noseburn	SP	Ss	—	U	G	SpSuF	BP
Tragia urticifolia nettle-leaf noseburn	MP	DS	—	U	—	Su	BM
Triadenum virginicum Virginia marsh-St. John's-wort	MP	SW	F	W	P	Su	W
Trichocolea tomentella leafy liverwort	m	—	—	U	X	X	—
Trichoneura elegans Silveus' grass	MAG	DS	—	U	G	SpF	EM
Trichoptilium incisum yellowdome	SA	DS	L	U	W	Su	M
Trichostema dichotomum forked bluecurls	SA	DS	F	G	BP	SuF	CgM
Tridens eragrostoides love fluff grass	MPG	Ss	—	U	G	SpSuF	E
Tridens flavus tall redtop	LPG	DSs	DF	U	GPu	SuF	BEM
Tridens muticus awnless fluff grass	SPG	DS	—	U	GPu	SuF	M
Tridens strictus long-spike fluff grass	MG	Ss	—	U	Pu	Su	BIW
Trientalis borealis American starflower	SP	sW	F	U	W	Sp	gI
Trifolium albopurpureum rancheria clover	SP	DS	F	GLMmSW	Pu	Sp	GM
Trifolium barbigerum bearded clover	SP	S	FL	GLMmSW	P	Sp	GM
Trifolium cyathiferum bowl clover	SP	DS	FL	GLMmSW	PW	SpSu	GM
Trifolium depauperatum balloon sack clover	MP	S	FL	GLMmSW	WPY	Sp	M

PLANT LIST	Plant Type	Env. Tol.	Aesthetic Value	Wildlife Value	Flower Color	Bloom Period	Landscape Uses
Trifolium fucatum sour clover	MP	FNaS	F	GLMmS W	PW	Sp	MS
Trifolium gracilentum pin-point clover	SP	S	FL	GLMmS W	P	Sp	GM
Trifolium gymnocarpon holly-leaf clover	SP	Ss	F	GLMmS W	LPPu	Su	BGM
Trifolium microdon valparaiso clover	SP	S	L	GLMmS W	PW	Sp	GM
Trifolium monanthum mountain carpet clover	SP	FS	FL	GLMmS W	PW	Sp	GMS
Trifolium olivaceum olive clover	SP	S	FL	GLMmS W	PuW	Sp	GM
Trifolium variegatum white-tip clover	SP	FSs	FL	GLMmS W	Pu	Su	MS
Triglochin concinnum slender arrow-grass	SPG	NaSW	—	W	G	SpSu	W
Triglochin maritimum seaside arrow-grass	SPG	FNaS W	—	W	G	SpSu	W
Triglochin palustre marsh arrow-grass	SPG	FNaS W	—	W	G	SpSu	SW
Trillium catesbaei bashful wakerobin	SP	s	FL	U	P	Sp	gI
Trillium cernuum whip-poor-will-flower	SP	Fs	F	U	W	SpSu	gIW
Trillium cuneatum little-sweet-Betsy	SP	s	F	U	R	Sp	gI
Trillium erectum stinking-Benjamin	SP	s	F	U	R	Sp	gI
Trillium flexipes nodding wakerobin	SP	s	F	U	W	SpSu	gI
Trillium grandiflorum large-flower wakerobin	SP	s	F	U	W	Sp	gI
Trillium luteum yellow wakerobin	SP	s	FL	U	Y	Sp	gI
Trillium nivale dwarf white wakerobin	SP	s	F	U	W	Sp	gI
Trillium ovatum western wakerobin	SP	s	F	U	RW	Sp	gI
Trillium recurvatum bloody-butcher	SP	s	F	U	R	Sp	gI
Trillium sessile toadshade	SP	s	FL	U	Br	SpSu	gI
Trillium undulatum painted wakerobin	SP	s	F	U	W	Sp	gI
Triplasis purpurea purple sand grass	MPG	DS	B	S	G	SuF	BEM

PLANT LIST	Plant Type	Env. Tol.	Aesthetic Value	Wildlife Value	Flower Color	Bloom Period	Landscape Uses
Tripsacum dactyloides eastern mock grama	LPG	FS	—	U	G	SpF	SW
Trisetum cernuum nodding false oat	LPG	—	FL	U	G	Su	BEMS
Trisetum spicatum narrow false oat	SPG	S	F	U	Pu	Su	M
Triteleia grandiflora large-flower triplet-lily	MP	Ss	F	U	B	SpSu	B
Trollius laxus American globeflower	SP	FS	F	U	Y	SpSu	gW
Typha angustifolia narrow-leaf cat-tail	LPG	FSW	FfL	LW	Br	Su	Wwg
Typha domingensis southern cat-tail	LPG	SF	FfL	LW	Br	SpSu	Wwg
Typha latifolia broad-leaf cat-tail	LPG	FSW	FfL	LW	Br	Su	Wwg
Uniola paniculata sea-oats	LG	S	DFfL	U	G	Su	ECS
Urtica chamaedryoides heart-leaf nettle	MA	s	—	U	G	SpSu	CI
Urtica dioica stinging nettle	MP	DS	—	U	G	SuF	MP
Utricularia cornuta horned bladderwort	AQ	W	—	U	Y	Su	Wwg
Utricularia foliosa leafy bladderwort	AQP	SW	—	U	Y	WSpSuF	Wwg
Utricularia minor lesser bladderwort	AQP	Ss	FL	U	Y	Su	Wwg
Utricularia subulata zigzag bladderwort	AQP	SW	F	U	Y	SpF	Wwg
Uvularia grandiflora large-flower bellwort	SP	Ss	FL	U	Y	Sp	gI
Uvularia perfoliata perfoliate bellwort	SP	s	FL	U	Y	Sp	gI
Uvularia puberula mountain bellwort	SP	s	FL	U	Y	Sp	BgI
Uvularia sessilifolia sessile leaf bellwort	MP	s	FL	U	Y	Sp	BgI
Valeriana arizonica Arizona valerian	SP	Ss	FL	U	PW	Su	BgM
Valeriana occidentalis small-flower valerian	MP	Ss	—	U	W	SpSu	BM
Valeriana sitchensis Sitka valerian	MP	S	F	U	W	Su	gM
Valerianella stenocarpa narrow-cell cornsalad	SA	S	—	U	W	Sp	MRS

PLANT LIST	Plant Type	Env. Tol.	Aesthetic Value	Wildlife Value	Flower Color	Bloom Period	Landscape Uses
Valerianella texana Edwards plateau cornsalad	SA	FSs	—	U	W	Sp	BS
Vancouveria hexandra white inside-out-flower	SP	Ss	FL	U	W	SpSu	BGI
Vaseychloa multinervosa Texas grass	MPG	s	—	U	G	SpSuF	EI
Veratrum californicum California false hellebore	LP	FSsW	FL	U	GW	Su	Wwg
Veratrum fimbriatum fringed false hellebore	LP	S	F	U	GW	Su	BgM
Veratrum viride American false hellebore	LP	SsW	F	U	W	Su	Wwg
Verbena bonariensis purple-top vervain	LP	DS	F	LS	Pu	Su	gM
Verbena bracteata carpet vervain	HVAP	DS	FL	LMS	BLPu	SpSuF	MR
Verbena halei Texas vervain	LP	Ss	F	MS	BPu	SpSuF	gM
Verbena hastata simpler's-joy	LP	FS	F	LSW	B	SuF	gMW
Verbena stricta hoary vervain	MP	S	F	mS	BPu	Su	BgM
Verbesina alternifolia wingstem	LP	Fs	F	B	Y	SuF	MS
Verbesina chapmanii Chapman's crownbeard	MP	FS	F	B	Y	Su	bMW
Verbesina encelioides shrubby seepweed	SA	DS	F	GMS	Y	SpSuF	gMS
Verbesina microptera Texas crownbeard	LP	S	F	BGmS	WY	SuF	bM
Verbesina virginica white crownbeard	LP	FSs	F	B	W	SuF	BMS
Vernonia baldwinii western ironweed	LP	DS	F	B	PuW	SuF	M
Vernonia fasciculata prairie ironweed	LP	DS	F	B	Pu	SuF	M
Vernonia gigantea grant ironweed	LP	DS	F	B	Pu	Su	M
Vernonia texana Texas ironweed	MP	S	F	B	Pu	Su	bM
Veronica americana American-brooklime	SP	FsW	F	h	B	SpSu	W
Veronica peregrina neckweed	SP	FS	F	h	W	SpSu	MW
Veronicastrum virginicum culver's-root	MP	DS	F	H	PW	Su	bM

PLANT LIST	Plant Type	Env. Tol.	Aesthetic Value	Wildlife Value	Flower Color	Bloom Period	Landscape Uses
Vicia americana American purple vetch	MP	FS	F	BCGSm	Pu	SpSu	bMS
Vigna luteola piedmont cow-pea	SP	SW	F	U	Y	SpSuF	W
Viola adunca hook-spur violet	SP	Ss	FL	CGLmS	Pu	SpSu	BGgI
Viola affinis sand violet	SP	FSs	FL	CGLmS	W	Sp	BGgIMS
Viola beckwithii western pansy	SP	Ss	FL	CGLmS	RPu	SpSu	BGgM
Viola blanda sweet white violet	SP	s	FL	CGLmS	W	Sp	bGgI
Viola canadensis Canadian white violet	SP	s	FL	CGLmS	W	Sp	BGgI
Viola conspersa American dog violet	SP	Ss	F	CGLmS	Pu	SpSu	BGgI
Viola cucullata marsh blue violet	SP	FS	FL	CGLmS	B	Sp	Wwg
Viola glabella pioneer violet	SP	Fs	FL	CGLmS	Y	SpSu	GgIS
Viola hastata halberd-leaf yellow violet	SP	s	F	GmS	Y	Sp	gI
Viola lanceolata bog white violet	SP	FS	FL	CGLmS	W	Sp	GgSWwg
Viola lobata moose-horn violet	SP	S	FL	GLmS	Y	SpSu	BGg
Viola macloskeyi smooth white violet	SP	FsW	FL	GmMS	W	SpSu	GgW
Viola missouriensis Missouri violet	SP	Fs	L	GCmS	B	Sp	EgIS
Viola nephrophylla northern bog violet	SP	FW	FL	GMmS	B	SpSu	GWwg
Viola orbiculata evergreen yellow violet	SP	s	FL	CGLmS	Y	Sp	GgI
Viola palmata early blue violet	SP	s	FL	CGLmS	Pu	Sp	BGgI
Viola pedata bird foot violet	SP	SD	FL	BCGhLmS	B	Sp	GgbR
Viola pedatifida crow-foot violet	SP	S	FL	CGLmS	Pu	Sp	GgM
Viola pubescens downy yellow violet	SP	s	FL	CGLmS	Y	Sp	GgI
Viola purpurea goose-foot yellow violet	SP	Ss	FL	GLmS	Y	SpSu	BGgM
Viola rostrata long-spur violet	SP	s	FL	GmMS	L	Sp	BGgI

PLANT LIST	Plant Type	Env. Tol.	Aesthetic Value	Wildlife Value	Flower Color	Bloom Period	Landscape Uses
Viola rotundifolia round-leaf yellow violet	SP	s	FL	CGLmS	Y	Sp	BGgI
Viola sempervirens redwood violet	SP	s	FL	CGLmS	Y	Sp	GgI
Viola sororia hooded blue violet	SP	s	FL	CGLmS	Pu	SpSu	GgI
Viola umbraticola ponderosa violet	SP	—	FL	CGLmS	—	—	GgI
Vulpia microstachys small six-weeks grass	SAG	S	—	GLMmS	G	Su	EM
Vulpia myuros rat-tail six-weeks grass	SPG	S	—	MS	G	SuF	M
Vulpia octoflora eight-flower six-weeks grass	SAG	S	FL	GLMmS	G	Sp	EM
Waldsteinia fragarioides Appalachian barren-strawberry	SP	s	F	GLMmS	Y	Sp	BG
Wedelia hispida hairy creeping-oxeye	P	—	—	U	Y	SuF	—
Whipplea modesta modesty	SP	Ss	—	U	W	SpSu	B
Woodsia oregana Oregon cliff fern	SP	s	L	GLM	X	X	GI
Woodsia scopulina Rocky Mountain cliff fern	SP	s	L	GLM	X	X	GI
Woodwardia areolata netted chain fern	MP	SsW	L	GLM	X	X	gIW
Woodwardia virginica Virginian chain fern	MP	SsW	L	GLM	X	X	gIW
Wyethia amplexicaulis northern mule's-ears	SP	S	FL	Gm	Y	SpSu	GM
Wyethia mollis woolly mule's-ears	MP	DS	FL	Gm	Y	Su	BM
Xanthisma texanum Texas sleepy-daisy	MA	DS	F	—	Y	SpSuF	gM
Xerophyllum asphodeloides eastern turkeybeard	MP	Ss	F	U	W	Su	BI
Xerophyllum tenax western turkeybeard	MP	DS	—	U	W	Su	BM
Xylorhiza glabriuscula smooth woody-aster	SP	NaS	F	U	PW	Su	gM
Xyris montana northern yellow-eye-grass	SP	FSW	L	U	Y	SuF	W
Zephyranthes atamasca atamasco-lily	SP	Ss	F	U	W	Sp	BgI
Zigadenus elegans mountain deathcamas	MP	S	F	U	GWY	Su	gM

PLANT LIST	Plant Type	Env. Tol.	Aesthetic Value	Wildlife Value	Flower Color	Bloom Period	Landscape Uses
Zigadenus paniculatus sand-corn	MP	Ss	F	U	WY	Su	B
Zigadenus venenosus meadow deathcamas	MP	FSs	F	U	WY	Su	BgMS
Zinnia acerosa white zinnia	SP	DS	F	BS	W	SuF	bgM
Zinnia grandiflora little golden zinnia	SP	DS	F	BS	GRY	SuSpF	bgM
Zizania aquatica Indian wild rice	MA	SW	L	SW	G	Su	W
Zizaniopsis miliacea marsh-millet	MG	FS	L	W	—	Su	SW
Zizia aptera heart-leaf alexanders	SP	DSs	F	U	Y	SpSu	BgI

APPENDIX D

Nursery Sources for Native Plants and Seeds

Introduction

The nurseries in this appendix responded to a survey that was sent to over 500 nurseries throughout the United States. The addresses were gathered from a number of published sources that gave the address of nurseries interested in native plants and seeds or listed a nursery as a general reference for information. We received over 250 responses. The survey asked what percent of the species offered are native and to which region they are native. This information can be found listed under each nursery in this appendix. We believe that the percent of native stock offered will grow as the demand grows. An attempt was made to include only sources that sell nursery–propagated material.

Each source is listed by state in alphabetical order. All listings include an address and telephone number. The italicized type corresponds to a question on the survey followed by the nursery's respective answer. If there was no response to a question on the survey, the code N/A indicates this or, in some cases, a nursery catalog was sent without the survey. Every effort was made to provide accurate and complete information.

This list is not inclusive of all native-plant nurseries or of all nurseries offering native species. The intention is to help you get in touch with sources offering native-plant material or information. Use Appendix A to identify species native to your region. Appendix C, the plant matrices, will give you an idea of native species that will best suit your landscaping needs. Many of the nurseries listed in Appendix D offer consultation services. Additional consultants are listed in Appendix E. Use this list to get started on naturalizing your area, whether through restoration or natural landscaping. Additional sources of information can be found in the National Wildflower Research Center's *Widlflower Handbook*.

Alabama

Byers Nursery Company, Inc.
P. O. Box 560
Meridianville, AL 35759 (205) 828-0625
Sales type: wholesale
Stock type: trees, shrubs, seeds
Native stock: 50%+
Regional focus: southeastern U.S.
Services: N/A
Comments: We offer excellent landscape trees and shrubs.

Arkansas

Holland Wildflower Farm
P. O. Box 328
Elkins, AR 72727 (501) 643-2622
Sales type: mail order, retail, wholesale
Stock type: seeds, seed mixes, herbaceous plants
Native stock: 80%
Regional focus: eastern U.S.
Services: consultation
Comments: We specialize in seeds or plants that require
little care from east of the Rockies.

Pittman Wholesale Nursery
P. O. Box 606
Magnolia, AR 71754 (800) 553-6661
Sales type: wholesale
Stock type: trees, shrubs, herbaceous plants
Native stock: 25%
Regional focus: N/A
Services: N/A
Comments: We offer a large selection of perennials.

Arizona

Cactus World
2955 E. Chula Vista Dr.
Tucson, AZ 85716 (520) 795-6028
Sales type: wholesale
Stock type: cacti seeds
Native stock: N/A
Regional focus: southwest
Services: N/A
Comments: N/A

Mountain States Wholesale Nursery
P. O. Box 2500
Phoenix, AZ 85340 (800) 840-8509
Sales type: wholesale only
Stock type: trees, shrubs, groundcover, perennials
Native stock: 60%
Regional focus: southwest
Services: contract growing
Comments: We grow desert-adapted plants from one
gallon to box trees.

Plant for the Southwest
50 E. Blacklidge
Tucson, AZ 85705 (602) 628-8773
Sales type: mail order, retail, wholesale
Stock type: wildflower seeds, cacti, and succulents
Native stock: 60%
Regional focus: southwest U.S.
Services: N/A
Comments: Our specialities are specimen-size cacti
and succulents.

Silver Bell Nursery
2730 N. Silverbell Rd.
Tucson, AZ 85745 (520) 622-3894
Sales type: retail, wholesale
Stock type: trees, shrubs, herbaceous plants, seeds,
seed mixes
Native stock: 50%+
Regional focus: southwest U.S. and Sonoran desert
Services: landscaping, contract growing
Comments: We propagate a large variety of native and
nonnative groundcovers, shrubs, and trees.

Wild Seed, Inc.
P. O. Box 27751
Tempe, AZ 85285 (602) 276-3536
Sales type: mail order, retail, wholesale
Stock type: seeds, seed mixes
Native stock: 95%
Regional focus: southwest U.S.
Services: N/A
Comments: We have a large selection of native
western wildflower, grass, shrub, and tree seeds.

California

Albright Seed Company
487 Dawson Dr., Bay 5S
Camarillo, CA 93012 (805) 484-0551
Sales type: wholesale
Stock type: seeds, seed mixes, fertilizer, mulch
Native stock: 65%
Regional focus: southwest U.S., Pacific coast
Services: habitat and wetland restoration
Comments: We maintain our inventory and do custom collections.

Albright Seed Company
189-A Arthur Rd.
Martinez, CA 94553 (510) 372-8245
Sales type: wholesale
Stock type: seeds, seed mixes
Native stock: 50%
Regional focus: western U.S.
Services: vegetation consulting, erosion control, landscaping golf courses
Comments: We specialize in a full range of services for erosion control and revegetation.

Anderson Valley Nursery
P. O. Box 504, 18151 Mt. View Rd.
Boonville, CA 95415 (707) 895-3853
Sales type: retail, wholesale
Stock type: trees, herbaceous plants, shrubs
Native stock: 60%
Regional focus: California
Services: consultation
Comments: We provide California native and Mediterranean type shrubs or perennials for water conservation landscaping.

Appleton Forestry Native Plant Nursery
1369 Tilton Rd.
Sebastopol, CA 95472 (707) 823-3776
Sales type: wholesale
Stock type: trees, herbaceous plants
Native stock: 100%
Regional focus: Pacific northwest
Services: habitat and wetland restoration
Comments: We specialize in riparian habitat species.

C. H. Baccus
900 Boynton Ave.
San Jose, CA 95117 (408) 244-2923
Sales type: mail order, retail
Stock type: herbaceous plants and bulbs
Native stock: 100%
Regional focus: western U. S.
Services: contract growing
Comments: We are a small mail order grower selling 100% seed grown stock.

Calaveras Nursery
1622 Highway 12
Valley Springs, CA 95252 (209) 772-1823
Sales type: retail, wholesale
Stock type: trees
Native stock: 60%
Regional focus: California
Services: N\A
Comments: We are specimen tree growers.

California Flora Nursery
P. O. Box 3
Fulton, CA 95439 (707) 528-8813
Sales type: retail, wholesale
Stock type: perennials, grasses, vines, shrubs
Native stock: 40%
Regional focus: San Francisco Bay area
Services: N/A
Comments: We have small numbers of many species produced from local seeds and cuttings for riparian, chaparral, and coastal communities.

Carter Seeds
475 Mar Vista Dr.
Vista, CA 92083 (619) 724-5931
Sales type: wholesale
Stock type: seeds
Native stock: 40%
Regional focus: all regions
Services: habitat restoration
Comments: We provide seeds for trees, shrubs, flowers, and wildflowers.

Central Coast Wilds
114 Liberty St.
Santa Cruz, CA 95060 (408) 459-0656
Sales type: retail, wholesale
Stock type: seed, seed mixes, trees, herbaceous plants
Native stock: 100%
Regional focus: west coast
Services: habitat restoration, landscaping, wetland restoration
Comments: We specialize in California native grasses, site-specific seed collections, and custom orders for seeds.

Christensen's Nursery Company
16000 Sanborn Rd.
Saratoga, CA 95070-9707 (408) 867-4181
Sales type: wholesale
Stock type: trees, shrubs, herbaceous plants, seeds, seed mixes
Native stock: N/A
Regional focus: N/A
Services: N/A
Comments: N/A

Circuit Rider Productions
9619 Old Redwood Hwy.
Windsor, CA 95492 (707) 838-6641
Sales type: wholesale
Stock type: trees, shrubs
Native stock: 100%
Regional focus: northern California
Services: wetland and habitat restoration
Comments: We specialize in northern California
woody plants for restoration and revegetation work.

Clyde Robin Seed Company, Inc.
P. O. Box 2366
Castro Valley, CA 94546 (510) 785-0425
Sales type: wholesale
Stock type: seeds, seed mixes
Native stock: 65%
Regional focus: western, midwestern
Services: N/A
Comments: We specialize in native California species
and provide both regionalized mixtures and individual
species.

Clyde Robin Seeds Company, Inc.
3670 Enterprise Ave.
Hayward, CA 94545 (510) 785-0425
Sales type: mail order, wholesale
Stock type: seeds, seed mixes
Native stock: 65%
Regional focus: west, midwestern
Services: N/A
Comments: We specialize in native California species
and provide both regionalized mixtures and individual
species.

Cornflower Farms
P. O. Box 896
Elk Grove, CA 95759 (916) 689-1015
Sales type: mail order, wholesale
Stock type: trees, herbaceous plants
Native stock: 90%
Regional focus: California
Services: on-site collection, consultation
Comments: We provide restoration and revegetation
plant materials for California including site-specific
collections.

Endangered Species
P. O. Box 1830
Tustin, CA 92781 (909) 940-0043
Sales type: mail order, retail, wholesale
Stock type: palms, bamboo, cycads
Native stock: 30%
Regional focus: all U.S.
Services: habitat and wetland restoration, landscaping
Comments: We have a huge selection of hardy, tropical
bamboo, native palms, and rare ornamentals.

Environmental Seed Producers, Inc.
P. O. Box 2709
Lompoc, CA 93438-2709 (805) 735-8888
Sales type: wholesale
Stock type: seeds and custom wildflower seed mixes
Native stock: 50%
Regional focus: all U.S.
Services: custom wildflower seed mixes
Comments: We offer over 150 species of wildflowers.

Forest Seeds of California
1100 Indian Hill Rd.
Placerville, CA 95667 (916) 621-1551
Sales type: mail order, wholesale
Stock type: seeds
Native stock: 95%
Regional focus: west coast
Services: habitat restoration, weed control, forestry
services
Comments: We offer tree and shrub seed collection
and processing and are a licensed pest control operator.

Freshwater Farms, Inc.
5851 Myrtle Ave.
Eureka, CA 95503 (707) 444-8261
Sales type: mail order, retail, wholesale
Stock type: seeds, seed mixes, trees, herbaceous plants
Native stock: 90%
Regional focus: west coast
Services: habitat and wetland restoration, landscaping,
contract growing
Comments: We provide wetland and riparian plants
and seeds with a wide selection of redwood understory
plants.

Habitat Restoration
3234 "H" Asford St.
San Diego, CA 92111 (619) 279-8769
Sales type: wholesale
Stock type: herbaceous plants, seeds, seed mixes
Native stock: 98%
Regional focus: southwest
Services: habitat and wetland restoration, mitigation
landscaping
Comments: Habitat restoration specializing in the
endangered habitats of coastal San Diego County.

Hardscrabble Seed Company
P. O. Box 60
Peirpoint Springs, CA 93208
(209) 542-3208
Sales type: mail order, wholesale
Stock type: seeds
Native stock: 100%
Regional focus: Sierra Nevada
Services: N/A
Comments: We specialize in giant sequoia seed and
handle some drought-resistant flowering native trees
and shrubs.

Hedgerow Farms
21740 County Rd. 88
Winters, CA 95694 (530) 662-4570
Sales type: retail, wholesale
Stock type: grass seeds
Native stock: 100%
Regional focus: California
Services: restoration consultation
Comments: We specialize in grassland restoration.

J. L. Hudson, Seedsman
Star route 2, P. O. Box 337
La Honda, CA 94020
Sales type: mail order, retail, wholesale
Stock type: seeds
Native stock: 100%
Regional focus: all regions
Services: ecological enrichment
Comments: We carry a large selection of species.

KSA Jojoba
19025 - EF Parthenia
Northridge, CA 91324 (818) 701-1534
Sales type: mail order, wholesale
Stock type: Jojoba seeds
Native stock: 100%
Regional focus: Sonoran Desert area of California,
Arizona, Mexico
Services: N/A
Comments: We provide a free catalog for sending a
self-addressed stamped (2 stamps) envelope.

Larner Seeds
P. O. Box 407
Bolinas, CA 94924 (415) 868-9407
Sales type: mail order, retail
Stock type: seeds, seed mixes, herbaceous plants,
books
Native stock: 100%
Regional focus: California
Services: habitat restoration, landscaping
Comments: We specialize in California coastal plants.

Las Pilitas Nursery
3232 Las Pilitas Rd.
Santa Margarita, CA 93453
(805) 438-5992
Sales type: mail order, retail, wholesale
Stock type: trees, herbaceous plants
Native stock: 100%
Regional focus: western U. S.
Services: habitat and wetland restoration, landscaping,
biosurveys
Comments: We grow over 800 species of California
plants and provide expertise about native plants and
their placement in native systems.

Louisiana-Pacific Forest Tree Nursery
1508 Crannell Rd.
Trinidad, CA 95570 (707) 443-7511
Sales type: mail order, retail, wholesale
Stock type: containerized seedlings, plugs
Native stock: 90%
Regional focus: northwest U. S.
Services: reforestation following logging
Comments: Our primary focus is reforestation seed-
lings and habitat restoration for northwest California
and southwest Oregon.

LSA Associates, Inc.
157 Park Place
Pt. Richmond, CA 94801 (510) 236-6810
Sales type: N/A
Stock type: trees, shrubs, herbaceous plants
Native stock: N/A
Regional focus: California
Services: installation, landscape design
Comments: We contract grow.

Mockingbird Nurseries, Inc.
1670 Jackson St.
Riverside, CA 92504 (909) 780-3571
Sales type: wholesale
Stock type: trees, herbaceous plants
Native stock: 85%
Regional focus: southern California
Services: habitat restoration
Comments: We do contract growing with on-site
collection when necessary.

Mostly Natives Nursery
P. O. Box 258, 27215 Hwy. 1
Tomales, CA 94971 (707) 878-2009
Sales type: mail order, retail, wholesale
Stock type: trees, shrubs, herbaceous plants
Native stock: 50%
Regional focus: west coast
Services: N/A
Comments: We specialize in coastal native plants.

Native Sons Nursery, Inc.
379 W. El Campo Rd.
Arroyo Grande, CA 93420 805-481-5996
Sales type: wholesale
Stock type: trees, herbaceous plants
Native stock: 40%
Regional focus: southwest U. S.
Services: habitat restoration
Comments: We limit our restoration work to the
central coast of California.

Northcoast Native Nursery
2710 Chileno Valley Rd.
Petaluma, CA 94952 (707) 769-1213
Sales type: retail, wholesale
Stock type: trees, herbaceous plants, native grasses
Native stock: 100%
Regional focus: Pacific northwest and California
Services: habitat and wetland restoration, planning
and design
Comments: N/A

Pacific Coast Seed, Inc.
6144 Industrial Way, Bldg. A
Livermore, CA 94550 (925) 373-4417
Sales, type: wholesale
Stock type: seeds, custom seed mixes
Native stock: 60%
Regional focus: northern California, western Nevada
Services: habitat and wetland restoration, landscaping
Comments: We specialize in California native grasses,
native shrubs, trees, and wildflowers.

Redwood City Seed Company
P. O. Box 361
Redwood City, CA 94064 (415) 325-7333
Sales type: mail order, retail, wholesale
Stock type: grass plugs of California native perennial
bunch grass
Native stock: 100%
Regional focus: California
Services: habitat restoration, landscaping, bunch
grasses for lawns
Comments: We use 100% native grass plugs in habitat
restoration and landscaping.

S & S Seeds
P. O. Box 1275
Carpinteria, CA 93014 (805) 684-0436
Sales type: wholesale
Stock type: seeds, seed mixes
Native stock: 80%
Regional focus: southwest U. S. and California
Services: custom collect seeds
Comments: We specialize in wildflower, tree, shrub,
grasses, and native plant seeds used for erosion
control, revegetation, and landscaping.

Santa Barbara Botanic Garden
1212 Mission Canyon Rd.
Santa Barbara, CA 93105 (805) 682-4726
Sales type: retail, wholesale
Stock type: trees, herbaceous plants, shrubs
Native stock: 50%
Regional focus: California, central coast
Services: contract growing
Comments: We have a wide array of natives and will
contract grow plants for restoration projects.

Saratoga Horticultural Foundation
15185 Murphy Ave.
San Martin, CA 95046 (408) 779-3303
Sales type: retail, wholesale
Stock type: trees, herbaceous plants, shrubs
Native stock: 34%
Regional focus: California
Services: contract growing
Comments: We are primarily a research foundation
finding and introducing new plants to the landscaping
nursery trade.

Sunset Coast Nursery
P. O. Box 221, 2745 Tierra Way
Watsonville, CA 95076 (408) 726-1672
Sales type: retail, wholesale
Stock type: trees, shrubs, herbaceous plants, bare root,
grasses
Native stock: 100%
Regional focus: California, southwest
Services: habitat and wetland restoration, site-specific
services
Comments: We specialize in coastal plants native to
the Monterey Bay area.

The Living Desert
47900 Portola Ave
Palm Desert, CA 92260 (760) 346-5694
Sales type: retail, wholesale
Stock type: trees, shrubs, herbaceous plants
Native stock: 90%
Regional focus: southwest U. S.
Services: N/A
Comments: We have a zoological garden with a small
retail nursery.

The Reveg Edge
P. O. Box 609
Redwood City, CA 94064 (415) 325-7333
Sales type: mail order, retail, wholesale
Stock type: bare root seedlings
Native stock: 100%
Regional focus: all U. S.
Services: habitat restoration
Comments: We specialize in growing native seeds
supplied by customer.

Theodore Payne Foundation
10459 Tuxford St.
Sun Valley, CA 91352 (818) 768-1802
Sales type: mail order, retail
Stock type: seeds, seed mixes, trees, herbaceous plants
Native stock: 100%
Regional focus: California
Services: consultation
Comments: We offer many wildflowers as well as
shrubs, trees, and perennials from coast to high
mountain sites.

Tree of Life Nursery
P. O. Box 635
San Juan Capistrano, CA 92693
(714) 728-0685
Sales type: wholesale
Stock type: seed mixes, trees, herbaceous plants
Native stock: 100%
Regional focus: California
Services: contract growing
Comments: We are the largest landscaping supplier of
California native plants for landscape and restoration.

Village Nurseries
1589 No. Main St.
Orange, California 92825　(714) 279-3100
Sales type: retail, wholesale
Stock type: trees, shrubs, herbaceous plants
Native stock: 25%
Regional focus: western U.S.
Services: habitat and wetland restoration
Comments: Our 300-acre nursery offers native and
drought-tolerant plants in sizes from liners to 48"
boxes.

Wapumne Native Plant Nursery Company
3807 Mt. Pleasant Rd.
Lincoln, CA 95648　(916) 645-9737
Sales type: retail, wholesale
Stock type: trees, shrubs, herbaceous plants, seeds,
grasses
Native stock: 100%
Regional focus: northern California
Services: consultation
Comments: We are a small but experienced (23 yrs.)
nursery; knowledgeable of natural environments.

Wildflowers International, Inc.
967 Hwy. 128
Philo, CA 95466 (707) 895-3500
Sales type: wholesale
Stock type: seeds
Native stock: 50%
Regional focus: U. S.
Services: N/A
Comments: We have bulk and custom mixes.

Wildwood Farm
10300 Sonoma Hwy.
Kenwood, CA 95452 (707) 833-1161
Sales type: mail order, retail
Stock type: trees, herbaceous plants
Native stock: 50%
Regional focus: all U.S.
Services: landscaping
Comments: We focus on unusual plants for shade as
well as sunny locations.

Colorado

Applewood Seed Company
5380 Vivian St.
Arvada, CO 80002　(303) 431-7333
Sales type: mail order, wholesale
Stock type: seeds, seed mixes
Native stock: 50%
Regional focus: all U.S.
Services: consultation, custom mixes
Comments: We have bird, butterfly, native, regional,
and golf course mixes.

Arkansas Valley Seed Company
P. O. Box 16025, 4652 Colorado Blvd.
Denver, CO 80216　(303) 320-7500
Sales type: wholesale
Stock type: seeds
Native stock: 20%
Regional focus: intermountain, semiarid regions
Services: N/A
Comments: We are a wholesale supplier of rangeland
grasses, turfgrass, and pasture grass seed.

Southwest Seed, Inc.
13260 Ct. Rd. 29
Dolores, CO 81323 (970) 565-8722
Sales type: mail order, retail, wholesale
Stock type: seeds, seed mixes
Native stock: 30%
Regional focus: western U.S.
Services: consulting, custom growing
Comments: We custom mix seeds for the individual
needs of sites.

Western Native Seed
P. O. Box 1463
Salida, CO 81201　(719) 539-1071
Sales type: mail order, retail, wholesale
Stock type: seeds, seed mixes
Native stock: 100%
Regional focus: Rocky Mountains and western Great
Plains
Services: N/A
Comments: We have a large selection of wildflowers,
grasses, wetland species, trees, and shrubs.

Wild Things
218 Quincy
Pueblo, CO 81004　(719) 543-2722
Sales type: wholesale
Stock type: herbaceous plants, shrubs
Native stock: 95%
Regional focus: southwest
Services: N/A
Comments: We specialize in native prairie plants.

Connecticut

Broken Arrow Nursery
13 Broken Arrow Rd.
Hamden, CT 06518 (203) 288-1026
Sales type: mail order, retail, wholesale
Stock type: trees, shrubs
Native stock: 30%
Regional focus: eastern U.S.
Services: N/A
Comments: We specialize in varieties of mountain laurel and have a large selection of natives and exotics.

Oliver Nurseries
1159 Bronson Rd.
Fairfield, CT 06430 (203) 259-5609
Sales type: retail
Stock type: trees, shrubs, herbaceous plants
Regional focus: northeastern U.S.
Services: landscaping
Comments: We specialize in dwarf conifers, alpines, and unusual cultivars.

Select Seeds
180 Stickney Hill Rd.
Union, CT 06076 (860) 684-9310
Sales type: mail order
Stock type: seeds, tools, books
Native stock: 10%
Regional focus: N/A
Services: N/A
Comments: We have rare old cultivars of garden flowers, heirlooms, fragrant varieties, flowering vines, and cutting garden favorites.

Florida

Breezy Oaks Nursery
23602 S.E. Hawthorn Rd.
Hawthorn, FL 32640 (352) 481-3795
Sales type: retail, wholesale
Stock type: trees, shrubs
Native stock: 65%
Regional focus: Florida
Services: N/A
Comments: We specialize in native azaleas and red anise.

Bullbay Creek Farm
1033 Old Bumpy Rd.
Tallahassee, FL 32311 (850) 878-3989
Sales type: wholesale
Stock type: herbaceous plants, palms
Native stock: 100%
Regional focus: southeast U.S.
Services: N/A
Comments: We have a large selection of native ferns and wetland palms.

Central Florida Lands & Timber, Inc.
Rt. 1, Box 899
Mayo, FL 32066 (904) 294-1211
Sales type: wholesale
Stock type: trees
Native stock: 95%
Regional focus: southeastern U. S.
Services: N/A
Comments: We have bare-root seedlings and containerized trees from one to twenty gallons.

Central Florida Native Flora, Inc.
P. O. Box 1045
San Antonio, FL 33576 (352) 588-3687
Sales type: wholesale
Stock type: trees, herbaceous plants, shrubs, aquatic plants
Native stock: 100%
Regional focus: southeast U.S.
Services: habitat and wetland restoration, landscaping, contract growing
Comments: We have forty acres of native plants in containers from one to sixty-five gallons.

Chiappini Farm Native Nursery
P. O. Box 436
Melrose, Fl 32666 (800) 293-5413
Sales types: wholesale
Stock type: trees, shrubs
Native stock: 100%
Regional focus: north Florida, south Georgia, coastal South Carolina
Services: habitat and wetland restoration, landscaping
Comments: We have native trees and shrubs for north Florida.

Florida Environmental
P. O. Box 321
Palmetto, FL 34221 (941)-729-5015
Sales type: wholesale
Stock type: herbaceous plants
Native stock: 100%
Regional focus: southeast U.S.
Services: landscaping, mitigation
Comments: We specialize in mangroves and seaoats.

Florida Keys Native Nursery, Inc.
102 Mohawk St.
Tavernier, FL 33070 (305) 852-2636
Sales type: mail order, retail, wholesale
Stock type: trees, shrubs, ground cover
Native stock: 100%
Regional focus: south Florida, Bahamas, Cuba, Puerto Rico, Caribbean basin
Services: habitat restoration, landscaping
Comments: Over 100 native Florida trees and shrubs, all containerized from small seedlings to large trees.

Florida Native Plants, Inc.
730 Myakka Rd.
Sarasota, FL 34240 (441) 322 -1915
Sales type: retail, wholesale
Stock type: trees, shrubs, native grasses, wildflowers
Native stock: 100%
Regional focus: central and southwest Florida
Services: landscaping
Comments: We have a full spectrum of native regional plants (140 species) and have been in business since 1982.

Gann's NativeTropical Greenery
22140 SW 152nd Ave.
Goulds, FL 33170 (305) 248-5529
Sales type: retail, wholesale
Stock type: trees, herbaceous plants, shrubs, wildflowers, grasses
Native stock: 99%
Regional focus: south Florida
Services: contract growing habitat-specific plants
Comments: We have a large selection of plants for Florida-specific tropical habitats.

Horticultural Systems, Inc.
P. O. Box 70, Golf Course Rd.
Parrish, FL 34219 (800) 771-4141
Sales type: mail order, retail, wholesale
Stock type: trees, herbaceous plants
Native stock: 100%
Regional focus: southeast U.S.
Services: habitat and wetland restoration, landscaping, nursery setup consulting
Comments: We have 65 acres with over 135 varieties of native plants including native plant tissue cultures established in 1974.

Joseph L. Gilio
P. O. Box 1122
Jensen Beach, FL 34958 (861) 283-3646
Sales type: wholesale
Stock type: trees, herbaceous plants
Native stock: 100%
Regional focus: southeast
Services: habitat and wetland restoration, lake creation
Comments: We have a large selection of wetland and southeast lake plants.

Liner Farm, Inc.
P. O. Box 701369
St. Cloud, FL 34770 (800) 330-1484
Sales types: mail order, wholesale
Stock type: trees, herbaceous plants
Native stock: 10%
Regional focus: southeast, southwest
Services: N/A
Comments: We have over 2900 types of starter plants for restoration and nurseries.

Mesozoic Landscape, Inc.
7667 Park Lane West
Lake Worth, FL 33467 (561) 967-2630
Sales type: retail, wholesale
Stock type: trees, herbaceous plants, shrubs
Native stock: 98%
Regional focus: south Florida
Services: landscaping
Comments: We have container trees, shrubs, and herbaceous plants grown from local stock.

Native Green Cay
12750 Hagen Ranch Rd.
Boynton Beach, FL 33437 (561) 496-1415
Sales type: wholesale
Stock type: trees
Native stock: 100%
Regional focus: southern Florida
Services: N/A
Comments: We have forty species of field-grown, Florida native trees.

Native Nurseries
1661 Centerville Rd.
Tallahassee, FL 32308 (850) 386-8882
Sales type: mail order, retail
Stock type: trees, shrubs, herbaceous plants
Native stock: 80%
Regional focus: southeast U.S.
Services: landscaping
Comments: We specialize in bird and butterfly gardens.

Native Southeastern Trees, Inc.
P. O. Box 780
Osteen, FL 32764 (407) 322-5133
Sales type: wholesale
Stock type: trees
Native stock: 75%
Regional focus: southeast U.S.
Services: N/A
Comments: N/A

Native Tree Nursery, Inc.
17250 SW 232rd St.
Homestead, FL 33030 (305) 247-4499
Sales type: wholesale
Stock type: trees, shrubs
Native stock: 90%
Regional focus: Florida
Services: N/A
Comments: N/A

Slocum Water Gardens
1101 Cypress Gardens Blvd.
Winter Haven, FL 33884 (941) 293-7151
Sales type: mail order, retail, wholesale
Stock type: aquatic
Native stock: 10%
Regional focus: southeast U.S.
Services: N/A
Comments: We have everything for the garden pool.

Superior Trees, Inc.
P. O. Box 9325
Lee, FL 32059 (904) 971-5159
Sales type: wholesale
Stock type: trees
Native stock: 95%
Regional focus: southeast U.S.
Services: N/A
Comments: We have bare-root and containerized
seedlings.

The Natives
2929 J. B. Carter Rd.
Davenport, FL 33837 (941)-422-6664
Sales type: retail, wholesale
Stock type: trees, shrubs, herbaceous plants, grasses
Native stock: 100%
Regional focus: Florida
Services: Upland habitat design and restoration
Comments: We offer sand scrub and sandhill plants
and many wildflowers.

The Wetlands Company, Inc.
1785 Southwood St.
Sarasota, FL 34231 (941) 921-6609
Sales type: retail, wholesale
Stock type: wetland herbaceous plants
Native stock: 100%
Regional focus: southeast U.S.
Services: consultants
Comments: We specialize in wetlands.

Wild Azalea Nursery
17970 SW 111th St.
Brooker, FL 32622 (352) 485-3556
Sales type: mail order, retail
Stock type: seeds, seed mixes, trees, herbaceous plants,
mosses
Native stock: 90%
Regional focus: southeast Florida
Services: N/A
Comments: We have aquatic species.

Georgia

Goodness Grows, Inc.
P. O. Box 576
Crawford, GA 30630 (706) 743-5055
Sales type: mail order, retail, wholesale
Stock type: trees, shrubs, herbaceous plants
Native stock: 50%
Regional focus: U.S.
Services: N/A
Comments: We specialize in herbaceous plants.

Transplant Nursery, Inc.
1586 Parkertown Rd.
Lavonia, GA 30553 (706) 356-8947
Sales type: mail order, retail, wholesale
Stock type: shade loving and ericaceous plants
Native stock: 30%
Regional focus: southeast U.S.
Services: N/A
Comments: We have native deciduous and evergreen
azaleas, rhododendrons, and other companion plants.

Twisted Oaks Nursery
P. O. Box 10
Waynesboro, GA 30830 (706) 554-3040
Sales type: wholesale
Stock type: trees, shrubs, seeds
Native stock: 20%
Regional focus: southeast
Services: landscaping, wetland restoration
Comments: We develop new cultivars of native plant
material, like *Myrica pumila* Fairfax.

Idaho

High Altitude Gardens
P. O. Box 1048
Hailey, ID 83333 (208) 788-4363
Sales type: mail order
Stock type: seeds, seed mixes, herbaceous plants
Native stock: 100%
Regional focus: high elevations in all regions
Services: N/A
Comments: We provide many appropriate species never before available for reclamation.

Native Seed Foundation
Star Route
Moyle Springs, ID 83845 (208) 267-7938
Sales type: wholesale
Stock type: seeds
Native stock: 99%
Regional focus: intermountain northwest, some maritime northwest
Services: N/A
Comments: We primarily offer seeds of woody shrubs and ground covers, but have some trees, forbs, and flowers.

Northplain/Mountain Seed
P. O. Box 9107
Moscow, ID 83843 (208) 882-8040
Sales type: mail order, retail, wholesale
Stock type: seeds, seed mixes
Native stock: 99%
Regional focus: inland and pacific northwest, eastern U. S.
Services: N/A
Comments: N/A

Reggear Tree Farm
1500 Loseth Rd.
Orofino, ID 83544 (208) 476-7364
Sales type: wholesale
Stock type: trees
Native stock: 50%
Regional focus: intermountain west
Services: N/A
Comments: We sell only conifers. We have large, landscape trees to 50 feet.

Winterfield Ranch Seed
P. O. Box 97
Swan Valley, ID 83449 (208) 483-3683
Sales type: wholesale
Stock type: seeds, seed mixes
Native stock: 95%
Regional focus: northwest
Services: habitat restoration, landscaping
Comments: We provide quality native seeds for grasses and wildflowers.

Illinois

Bluestem Prairie Nursery
13197 E. 13th Rd.
Hillsboro, IL 62049 (217) 532-6344
Sales type: mail order
Stock type: seeds, seed mixes, herbaceous plants
Native stock: 100%
Regional focus: Midwest
Services: Answer questions on plant propagation and restoration
Comments: We have a large selection of seed packets and bare-root plants of prairie and woodland species.

Country Road Greenhouse, Inc.
RR1, Box 62
Malta, IL 60150 (815) 825-2305
Sales type: wholesale
Stock type: herbaceous plants
Native stock: 100%
Regional focus: Midwest
Services: contract growing, consultation
Comments: We grow large quantities of savanna, prairie, and wetland species.

Genesis Nursery
23200 Hurd Rd.
Tampico, IL 61283 (815) 438-2220
Sales type: mail order, retail, wholesale
Stock type: herbaceous plants, seeds, seed mixes
Native stock: 100%
Regional focus: Midwest
Services: habitat and wetland restoration, landscaping, consulting, monitoring
Comments: We offer a large variety of savanna, wetland, and prairie species.

LaFayette Home Nursery, Inc.
Rt. 1, Box 1A
LaFayette, IL 61449 (309) 995-3311
Sales type: mail order, retail, wholesale
Stock type: seeds, seed mixes, trees, herbaceous plants, tools, straw mulch
Native stock: 95%
Regional focus: Midwest
Services: habitat and wetland restoration, landscaping, prairie and woodland restoration
Comments: We supply species for, install, and maintain prairie, woodland, and wetland aquatic systems.

Midwest Wildflowers
P. O. Box 64
Rockton, IL 61072 (815) 624-7040
Sales type: mail order, retail
Stock type: seeds
Native stock: 85%
Regional focus: Midwest
Services: N/A
Comments: We have seed packets including 125
species.

Purple Prairie Nursery
Rt. 2, Box 176
Wyoming, IL 61491 (309) 286-7356
Sales type: retail, wholesale
Stock type: seeds, herbaceous plants
Native stock: 100%
Regional focus: Illinois
Services: N/A
Comments: N/A

The Natural Garden
38 W 443 Hwy. 64
St. Charles, IL 60175 (630) 584-0150
Sales type: mail order, retail, wholesale
Stock type: seeds, seed mixes, native shrubs, vines,
ground cover, grasses, perennials
Native stock: 25%
Regional focus: Chicago region
Services: habitat and wetland restoration, landscaping,
contract growing
Comments: We have a large selection of species of
native plants for open shade or prairie and can provide
large numbers of plants.

Indiana

Heartland Restoration Services
349 Airport North Office Park
Fort Wayne, IN 46825 (219) 489-8511
Sales type: mail order, retail
Stock type: seeds
Native stock: 100%
Regional focus: Midwest
Services: N/A
Comments: We have local genotype native seeds.

Iowa

Allendan Seed
1966 175th Ln.
Winterset, IA 50273 (515) 462-1241
Sales type: mail order, retail, wholesale
Stock type: herbaceous, mostly seeds
Native stock: 100%
Regional focus: Midwest U.S.
Services: N/A
Comments: Most of our grasses are certified Iowa
ecotypes.

Cascade Forestry Nursery
22033 Fillmore Rd.
Cascade, IA 52033 (319) 852-3042
Sales type: N/A
Stock type: seeds, trees, shrubs, tools
Native stock: 95%
Regional focus: Midwest U.S.
Services: habitat restoration, reforestation
Comments: N/A

Ion Exchange
1878 Old Mission Dr.
Harpers Ferry, IA 52146 (319) 535-7231
Sales type: mail order, retail, wholesale
Stock type: seeds, containerized herbaceous plants
Native stock: 100%
Regional focus: upper Midwest
Services: N/A
Comments: We have 250 riverine, savanna, and
wetland species.

Iowa Prairie Seed Company
1740 220th St.
Sheffield, IA 50475 (515) 892-4111
Sales type: mail order, wholesale
Stock type: herbaceous, seed
Native stock: 100%
Regional focus: Midwest
Services: N/A
Comments: We have local ecotype seeds for prairie
plantings.

Osenbaugh Grass Seeds
RR 1, Box 44
Lucas, IA 50151 (515) 766-6476
Sales type: mail order, retail, wholesale
Stock type: seeds, seed mixes
Native stock: 99%
Regional focus: Midwest, northeast
Services: planting
Comments: We specialize in bird habitat mixes.

Smith's Nursery Company
P. O. 515
Charles City, IA   50616 (515) 228-3239
Sales type: mail order, retail, wholesale
Stock type: seeds, shrubs, tree
Native stock: 50%
Regional focus: N/A
Services: N/A
Comments: We specialize in rare and unusual plants.

Kansas

Sharp Brothers Seed Company
P. O. Box 140
Healy, KS 67850 (316) 398-2231
Sales type: retail, wholesale
Stock type: seeds, seed mixes
Native stock: N/A
Regional focus: Midwest, high plains
Services: habitat restoration
Comments: We specialize in native grasses.

Kentucky

Akinback Farm
2501 Hwy. 53 South
LaGrange, KY 40031 (502) 222-5791
Sales type: retail
Stock type: herbaceous perennials
Native stock: 40%
Regional focus: Midwest, southeast
Services: display gardens available for viewing
Comments: We carry over 1000 perennials and 350
varieties of native and ornamental herbs.

Caudill Seed
1201 Story Ave.
Louisville, KY 40206 (502) 583-4402
Sales type: mail order, retail, wholesale
Stock type: seeds, seed mixes
Native stock: N/A
Regional focus: Midwest
Services: custom orders
Comments: We specialize in rare and exotic species for
larger size projects (several pounds).

Dabney Herbs
P. O. Box 22061
Louisville, KY 40252 (502) 893-5198
Sales type: mail order, wholesale
Stock type: trees, shrubs, herbaceous plants, seeds
Native stock: 45%
Regional focus: Midwest, Appalachia
Services: habitat restoration, landscaping information
Comments: We offer nursery propagated plants,
medicinal herbs, and shade plants.

Jane's Jungle
635 Sawdridge Creek W.
Monterey, KY 40359 (502) 484-2044
Sales type: retail, wholesale
Stock type: seeds of wetlands, some upland species,
trees, shrubs, herbaceous plants
Native stock: 100%
Regional focus: Kentucky
Services: landscape consulting
Comments: N/A

Nolin River Nut Tree Nursery
797 Port Wooden Rd.
Upton, KY 42784 (502) 369-8551
Sales type: mail order
Stock type: trees
Native stock: 75%
Regional focus: N/A
Services: N/A
Comments: We have over 200 varieties of grafted nut
trees, persimmons, and pawpaws.

Shooting Star Nursery
444 Bates Rd.
Frankfort, KY 40601 (502) 223-1679
Sales type: mail order, retail
Stock type: seeds, seed mixes, trees, herbaceous plants,
shrubs
Native stock: 99%
Regional focus: eastern U. S.
Services: habitat restoration, landscaping, environmen-
tal stewardship
Comments: We have a large selection of nursery-
propagated wildflowers, trees, shrubs, and vines native
to the eastern U. S.

Stinson Rhododendron Nursery
10400 Florian Rd.
Louisville, KY 40223 (502) 722-0514
Sales type: wholesale
Stock type: trees, shrubs
Native stock: 75%
Regional focus: Appalachian Mountain Range
Services: N/A
Comments: We offer flowering shrubs.

Louisiana

Louisiana Nature and Science Center
P. O. Box 870610
New Orleans, LA 70187 (504) 246-5672
Sales type: mail order, retail, wholesale
Stock type: trees, shrubs, herbaceous plants, seeds, seed mixes
Native stock: N/A
Regional focus: N/A
Services: N/A
Comments: N/A

Louisiana Nursery
5853 Hwy. 182
Opelousas, LA 70570 (318) 948-3696
Sales type: mail order, retail
Stock type: trees, herbaceous plants, shrubs, vines, grasses, aquatic plants
Native stock: 25%
Regional focus: southeast U.S.
Services: N/A
Comments: We have over 5000 different taxa and specialize in the unique, rare, and unusual.

Natives Nurseries
320 N. Theard St.
Covington, LA 70433 (504) 892-5424
Sales type: retail, wholesale
Stock type: trees, herbaceous plants
Native stock: 80%
Regional focus: southeast U.S.
Services: habitat and wetland restoration
Comments: We have a range of plants for wetland, prairie, and woodland situations.

Prairie Basse Native Plants
Rt. 2, Box 491F
Carencro, LA 70520 (318) 896-9187
Sales type: mail order, retail
Stock type: trees, shrubs, herbaceous plants
Native stock: 95%
Regional focus: Gulf Coastal Plain
Services: habitat and wetland restoration, landscaping
Comments: We specialize in habitat enrichment/restoration and do ecological consultation.

Sherwood Akin's Greenhouses
P. O. Box 6
Sibley, LA 71073 (318) 377-3653
Sales type: mail order, retail
Stock type: trees, shrubs, herbaceous plants
Native stock: N/A
Regional focus: N/A
Services: grow superior fruited mayhaws
Comments: We are a small nursery specializing in unusual, edible landscaping plants.

Maine

Fieldstone Gardens
620 Quaker Ln.
Vassalboro, ME 04989 (207) 923-3836
Sales type: mail order, retail
Stock type: herbaceous plants
Native stock: 10–15%
Regional focus: cold climate states
Services: N/A
Comments: We offer over 600 species of rock garden, border, shade, and full sun plants.

Maryland

Bluemount Nurseries, Inc.
2103 Blue Mount Rd.
Monkton, MD 21111 (410) 329-6226
Sales type: wholesale
Stock type: herbaceous plants
Native stock: 10%
Regional focus: mid-Atlantic, northeast
Services: custom growing
Comments: We have a broad selection of herbaceous perennials, vines, ferns, and ornamental grasses.

Carroll Gardens, Inc.
444 E. Main St.,
Westminster, MD 21157 (410) 848-5422
Sales type: mail order, retail, wholesale
Stock type: trees, shrubs, herbaceous plants
Native stock: N/A
Regional focus: N/A
Services: N/A
Comments: N/A

Environmental Concern, Inc.
P. O. Box P, 210 W. Chew Ave.
St. Michaels, MD 21663　(301) 745-9620
Sales type: wholesale
Stock type: trees, shrubs, herbaceous plants, seeds
Native stock: 100% (herbaceous), 30% (woody)
Regional focus: northeast, coastal Maryland, Delaware, Virginia
Services: habitat, wetland restoration, landscaping
Comments: The nursery specializes in herbaceous and woody wetland species. Consulting specialists in wetland, stream, and upland forest restoration.

Kurt Bluemel, Inc.
2740 Greene Ln.
Baldwin, MD 21013　(410) 557-7229
Sales type: mail order, retail, wholesale
Stock type: herbaceous plants
Native stock: 20%
Regional focus: U.S.
Services: N/A
Comments: We specialize in ornamental grasses.

Maryland Aquatic Nurseries
3427 N. Furnace Rd.
Jarrettsville, MD 21084　(410) 557-7615
Sales type: mail order, wholesale
Stock type: aquatic plants, moisture-loving perennials and grasses
Native stock: 25%
Regional focus: eastern seaboard
Services: landscaping, specializing in pond design
Comments: We offer a broad selection of aquatic plants as well as exclusive M.A.N.-made products.

Native Seeds, Inc.
14590 Triadelphia Mill Rd.
Dayton, MD 21036　(301) 596-9818
Sales type: mail order, retail, wholesale
Stock type: seeds, seed mixes
Native stock: 90%
Regional focus: all U. S.
Services: N/A
Comments: We offer a selection of wildflower seeds in bulk and different-sized packets.

Massachusetts

Bestmann Green Systems
53 Mason St.
Salem, MA 01970　(508) 741-1166
Sales type: retail, wholesale
Stock type: seeds, seed mixes, shrubs, herbaceous plants, tools, restoration supplies
Native stock: 99%
Regional focus: northeast, east central, and Midwest
Services: habitat and wetland restoration, erosion control
Comments: We are one of the largest and oldest wetland nurseries in northeast and offer over 40 wetland species.

Botanicals, Inc.
P. O. Box 65
London, MA 02173　(508) 358-1029
Sales type: wholesale
Stock type: trees, herbaceous plants, water plant software
Native stock: 85%
Regional focus: northeast, mid-Atlantic, southeast
Services: landscaping for low maintenance, low water use
Comments: N/A

Brairwood Gardens
14 Gully Ln., Rt. 3
East Sandwich, MA 02537 (508) 888-2146
Sales type: mail order, retail, wholesale
Stock type: trees, shrubs, rhododendrons
Native stock: 5%
Regional focus: east coast
Services: expert advice, landscape design
Comments: Experienced wholesaler and dealer in landscaping plants with a large selection of rhododendrons.

Donaroma's Nursery
P. O. Box 2189
Edgartown, MA 02539　(508) 627-8366
Sales type: mail order, retail
Stock type: herbaceous plants
Native stock: 20%
Regional focus: northeast
Services: wetland and grassland restoration, landscaping
Comments: We grow herbaceous perennials and design, construct, and maintain landscapes.

F. W. Schumacher Company, Inc.
P. O. Box 1023, 36 Spring Hill Rd.
Sandwich, MA 02563 (508) 888-0659
Sales type: mail order, retail, wholesale
Stock type: seeds
Native stock: 70%
Regional focus: all regions
Services: habitat and wetland restoration, landscaping
Comments: We offer ornamental landscaping and sell seeds of trees and shrubs.

New England Wetland Plants, Inc.
800 Main St.
Amherst, MA 01002 (413) 256-1752
Sales type: wholesale
Stock type: trees, shrubs, herbaceous plants
Native stock: 100%
Regional focus: New England
Services: N/A
Comments: We carry erosion control material.

New England Wildflower Society
Garden in the Woods
Hemenway Rd.
Framingham, MA 01701 (508) 877-7630
Sales type: retail
Stock type: trees, shrubs, herbaceous plants, seeds, seed mixes
Native stock: 99%
Regional focus: eastern U.S.
Services: N/A
Comments: We are nonprofit and offer field trips and botany and horticulture courses.

Rock Spray Nursery, Inc.
P. O. Box 693
Truro, MA 02666 (508) 349-6769
Sales type: mail order, retail, wholesale (send $1.00 for catalog)
Stock type: heath, heather
Native stock: 100%
Regional focus: U.S., Canada
Services: growers of heather
Comments: We grow over 120 varieties of heath and heather, hardy to zone 4.

Tripple Brook Farm
37 Middle Rd.
Southampton, MA 01073 (413) 527-4626
Sales type: mail order, retail
Stock type: trees, shrubs, herbaceous plants
Native stock: 50%
Regional focus: eastern U.S.
Services: N/A
Comments: We are a small nursery propogating a wide range of woody plants and herbaceous perennials including 200 species of eastern U.S. natives.

Weston Nurseries, Inc,
E. Main St., P. O. Box 186
Hopkinton, MA 01748 (508) 435-3414
Sales type: retail, wholesale
Stock type: trees, shrubs, herbaceous plants
Native stock: 10-20%
Regional focus: N/A
Services: landscaping
Comments: New England's largest selection of landscape-size plants, trees, and shrubs.

Michigan

Armintrout's West Michigan Farms, Inc.
1156 Lincoln Rd.
Allegan, MI 49010 (616) 673-6627
Sales type: mail order, wholesale
Stock type: trees
Native stock: 90%
Regional focus: northern U.S.
Services: N/A
Comments: We specialize in evergreen trees.

BioEnt
14174 Hoffman Rd.
Three Rivers, MI 49093 (616) 278-3975
Sales type: retail, wholesale
Stock type: trees, shrubs, herbaceous plants, seeds, seed mixes
Native stock: 80%
Regional focus: Great Lakes region
Services: habitat and wetland restoration, landscaping, contract growing
Comments: We offer both wetland and dry land native species.

Hortech
P. O. Box 533
Spring Lake, MI 49456 (616) 842-1392
Sales type: wholesale
Stock type: herbaceous perennial
Native stock: 20%
Regional focus: Michigan, northern Indiana, and northwest Ohio
Services: container grower
Comments: We specialize in ground cover.

Nesta Prairie Perennials
1019 Miller Rd.
Kalamazoo, MI 49001 (800) 233-5025
Sales type: retail, wholesale
Stock type: herbaceous plants
Native stock: 100%
Regional focus: upper Midwest
Services: N/A
Comments: We specialize in native perennials of the prairie, savannah, and wetland edge in plugs, quarts, and gallons.

Oikos Tree Crops
P. O. Box 19425
Kalamazoo, MI 49065 (616) 624-6233
Sales type: mail order, retail, wholesale
Stock type: trees
Native stock: N/A
Regional focus: northeast and Midwest
Services: biodegradable-paper pot-grown native fruits
Comments: We have native fruits and oaks.

The Michigan Wildflower Farm
11770 Cutler Rd.
Portland, MI 48875 (517) 647-6010
Sales type: mail order, retail
Stock type: seeds, seed mixes, herbaceous plants
Native stock: 99%
Regional focus: Michigan, Wisconsin, Minnesota
Services: habitat restoration, landscaping, consultation
Comments: We have primarily prairie and wetland species seeds with some plants.

Vans Pines Nursery
7550 144th Ave.
West Mt. Olive, MI 49460 (800) 888-7337
Sales type: mail order, wholesale
Stock type: seeds, trees, herbaceous plants
Native stock: 75%
Regional focus: Great Lakes
Services: site consultation
Comments: We have over 200 species of native and ornamental evergreen and deciduous trees and shrubs.

Wavecrest Nursery
2509 Lakeshore Dr.
Fennville, MI 49408 (616) 543-4175
Sales type: mail order, retail, wholesale
Stock type: trees, shrubs, herbaceous plants
Native stock: 10%
Regional focus: Midwest
Services: habitat and wetland restoration, landscaping
Comments: We offer a large selection of shade-loving woodland plants and many prairie grasses.

Wetlands Nursery
P. O. Box 14553
Saginaw, MI 48601 (517) 752-3492
Sales type: retail, wholesale
Stock type: seeds, herbaceous plants
Native stock: 95%
Regional focus: Michigan
Services: custom plantings, landscaping
Comments: We specialize in native wetland species with Michigan genotypes.

Minnesota

Bailey Nurseries, Inc.
1325 Bailey Rd.
St. Paul, MN 55119 (612) 459-9744
Sales type: wholesale only
Stock type: trees, shrubs, herbaceous plants
Native stock: 50%
Regional focus: northern U.S.
Services: N/A
Comments: We specialize in bare root material.

Feder's Prairie Seed Company
12871 3380th Ave.
Blue Earth, MN 56013 (507) 526-3049
Sales type: mail order
Stock type: seeds, seed mixes
Native stock: 100%
Regional focus: upper Midwest
Services: habitat restoration, landscaping
Comments: We have local ecotype prairie seeds at competitive prices.

Landscape Alternatives, Inc.
1705 St. Albans St. No.
Lewisville, MN 55113 (612) 488-3142
Sales type: mail order, retail, wholesale
Stock type: herbaceous plants, seeds
Native stock: 95%
Regional focus: upper Midwest, Minnesota
Services: habitat restoration, landscape design
Comments: We offer over 100 species of prairie wildflowers and grasses as well as many woodland species.

Mohn Frontier Seed and Nursery
Rt. 1, P. O. Box 152
Cottonwood, MN 56229 (507) 423-6482
Sales type: retail, wholesale
Stock type: seeds, seed mixes
Native stock: 95%
Regional focus: upper Midwest
Services: planting information for native grass or wildflower seeds
Comments: We provide native prairie seeds or seed mixes.

Orchid Gardens
2232 139th Ave. NW
Andover, MN 55304 (612) 755-0205
Sales type: mail order
Stock type: trees, shrubs, herbaceous plants
Native stock: N/A
Regional focus: Midwest
Services: habitat restoration, landscaping
Comments: We deal exclusively with plants native to Minnesota, mostly wildflowers.

Prairie Hill Wildflowers
Rt. 1, Box 191-A
Ellendale, MN 56026 (507) 451-7791
Sales type: retail, wholesale
Stock type: herbaceous plants, seeds, custom seed mixes
Native stock: 100%
Regional focus: upper Midwest
Services: habitat, wetland, and prairie restoration
Comments: We offer a good selection of prairie and wetland species.

Prairie Moon Nursery
Rt. 3, Box 163
Winona, MN 55987 (507) 452-1362
Sales type: mail order, retail
Stock type: seeds, seed mixes, trees, herbaceous plants, books
Native stock: 100%
Regional focus: upper Midwest
Services: consulting
Comments: We have 400 species of wetland, prairie, savanna, and woodland species.

Prairie Restorations, Inc.
P. O. Box 327
Princeton, MN 55371 (612) 389-4342
Sales type: mail order, retail, wholesale
Stock type: seeds, seed mixes, trees, herbaceous plants, tools, books
Native stock: 100%
Regional focus: Minnesota
Services: consultation, installation, management
Comments: We design, manage, and restore prairie and other native plant communities.

Rice Creek Gardens, Inc.
11506 Hwy. 65
Blaine, MN 55434 (612) 754-8090
Sales type: retail
Stock type: shrubs, herbaceous plants
Native stock: 10%
Regional focus: Midwest
Services: N/A
Comments: N/A

The Environmental Collaborative
P. O. Box 539
Osseo, MN 55369
Sales type: mail order
Stock type: trees, shrubs
Native stock: 95%
Regional focus: northeast, Midwest, north central
Services: N/S
Comments: We offer over fifty species of hardwoods, conifers, and shrubs.

Wildlife Habitat
5114 NE 46th St.
Owatonna, MN 55060 (507) 451-6771
Sales type: mail order, retail, wholesale
Stock type: seeds, seed mixes, ground prairie grass mulch
Native stock: 100%
Regional focus: Minnesota, Iowa, Wisconsin, South and North Dakota
Services: N/A
Comments: We grow and condition six species of warm season native prairie grasses.

Missouri

Elixir Farm Botanicals
General Delivery
Brixey, MO 65618 (417) 261-2393
Sales type: mail order
Stock type: seeds, books
Native stock: 50%
Regional focus: eastern U.S.
Services: habitat restoration, indigenous medicinal plant seeds
Comments: We have a variety of indigenous medicinal plant seeds.

Forrest Keeling Nursery, Inc.
P. O. Box 135, 80 Keeling Ln.
Elsberry, MO 63343 (800) 356-2401
Sales type: retail, wholesale
Stock type: trees, herbaceous plants
Native stock: 50%
Regional focus: Midwest and east central U.S.
Services: N/A
Comments: We specialize in native trees and shrubs including twenty-six different types of oaks and provide bare root, container, and field grown specimens.

Hamilton Seeds
16786 Brown Rd.
Elk Creek, MO 65464 (417) 967-2190
Sales type: mail order, retail, wholesale
Stock type: seeds, seed mixes, herbaceous plants
Native stock: 100%
Regional focus: Midwest, eastern U.S.
Services: N/A
Comments: We provide prairie, woodland, and
wetland species.

J & J Seeds
Rt. 3
Gallatin, MO 64640 (660) 663-3165
Sales type: retail, wholesale
Stock type: seeds
Native stock: 100%
Regional focus: upper Midwest
Services: N/A
Comments: We have Conservation Reserve Program
mixes, warm season grasses.

Missouri Wildflowers Nursery
9814 Pleasant Hill Rd.
Jefferson City, MO 65109 (314) 496-3492
Sales type: mail order, retail, wholesale
Stock type: seeds, seed mixes, herbaceous plants
Native stock: 100%
Regional focus: central U.S.
Services: N/A
Comments: We specialize in prairie and woodland
herbaceous species (99% Missouri genetic origin) with
some trees and shrubs.

Sharp Brothers Seed Company
Rt. 4, P. O. Box 237A
Clinton, MO 64735 (800) 451-3779
Sales type: mail order, retail, wholesale
Stock type: seeds, seed mixes
Native stock: 90%
Regional focus: central, Great Plains
Services: N/A
Comments: We have specialized in native grasses and
wildflowers for over thirty years.

Montana

Bitterroot Restoration, Inc.
445 Quast Ln.
Corvallis, MT 59828 (406) 961-4991
Sales type: wholesale
Stock type: trees, herbaceous plants, wetland and riparian
plants
Native stock: 100%
Regional focus: northwest, southwest, Rocky Mts., west.
Services: wetland and habitat restoration, landscaping,
erosion control, consulting
Comments: We are a full-service revegetation company and
native plant nursery.

Four Winds Nursery
5853 East Shore Route
Polson, MT 59860 (406) 887-2215
Sales type: mail order, retail
Stock type: trees, shrubs, herbaceous plants, seeds, seed
mixes
Native stock: 20%
Regional focus: N. Rocky Mts., northern plains
Services: landscaping of alpine, rock, bog gardens
Comments: We are a small mom-and-pop business and carry
a large variety of hardy, regional plants.

Lawyer Nursery, Inc.
950 Hwy. 200 West
Plains, MT 59859 (406) 826-3881
Sales type: mail order, wholesale
Stock type: trees, shrubs, seeds
Native stock: 50%
Regional focus: U.S.
Services: N/A
Comments: We have been in business since 1959.

Valley Nursery
P. O. Box 4845
Helena, MT 59604 (406) 458-3992
Sales type: mail order, retail, wholesale
Stock type: trees, shrubs, herbaceous plants
Native stock: 50%
Regional focus: northern U.S.
Services: N/A
Comments: N/A

Nebraska

Bluebird Nursery, Inc.
P. O. Box 460
Clarkson, NE 68629 (402) 892-3457
Sales type: wholesale
Stock type: herbaceous plants
Native stock: 33%
Regional focus: all U.S.
Services: N/A
Comments: We are a forty-year old company with 2000 varieties, specializing in perennials.

Paul E. Allen Farm Supply
Rt. 2, Box 8
Bristow, NE 68719 (402) 583-9924
Sales type: mail order, retail, wholesale
Stock type: seeds, seed mixes
Native stock: 90%
Regional focus: Midwest
Services: N/A
Comments: We specialize in native wildflowers and prairie grasses.

Stock Seed Company
28008 Mill Rd.
Murdock, NE 68407 (402) 867-3771
Sales type: mail order, retail, wholesale
Stock type: seeds, seed mixes
Native stock: N/A
Regional focus: all U. S.
Services: N/A
Comments: We have specialized in production of seeds for prairie restoration for forty years.

New Jersey

Pinelands Nursery
323 Island Rd.
Columbus, NJ 08022 (609) 291-9486
Sales type: wholesale
Stock type: trees, shrubs, herbaceous plants
Native stock: 100%
Regional focus: eastern, northeast U.S.
Services: N/A
Comments: We carry erosion control materials.

Princeton Nurseries
P. O. Box 185, Ellisdale Rd.
Allentown, NJ 08501 (609) 254-7671
Sales type: wholesale
Stock type: trees, shrubs
Native stock: 25%
Regional focus: northeast U.S.
Services: N/A
Comments: We offer native woody plants, bare root, in containers or B&B.

Well-Sweep Herb Farm
205 Mt. Bethel Rd.
Port Murray, NJ 07865 (908) 852-5390
Sales type: mail order, retail
Stock type: seeds, herbaceous plants
Native stock: 5%
Regional focus: N/A
Services: N/A
Comments: We carry woodland plants.

Wild Earth Native Plant Nursery
P. O. Box 7258
Freehold, NJ 07728 (732) 780-5661
Sales type: mail order, retail, wholesale
Stock type: trees, shrubs, herbaceous plants
Native stock: 99%
Regional focus: east coast
Services: custom growing, landscape design
Comments: We specialize in propagated native plants from cells to gallon containers.

New Mexico

Agua Fria Nursery, Inc.
1409 Agua Fria
Santa Fe, NM 87501 (505) 983-4831
Sales type: mail order, retail
Stock type: herbaceous plants, shrubs, small evergreens
Native stock: 70%
Regional focus: southwest
Services: N/A
Comments: We specialize in native perennials of the southwest.

Bernardo Beach Native Plant Farm
3729 Arno St. NE
Albuquerque, NM 87107 (505) 345-6248
Sales type: retail, local sales only
Stock type: trees, shrubs, herbaceous plants, grasses
Native stock: 70%
Regional focus: southwest, high desert
Services: landscape design
Comments: We design landscapes with native plants that are ecologically adapted.

Curtis and Curtis Seed
Star Route Box 8A
Clovis, NM 88101 (505) 762-4759
Sales type: wholesale
Stock type: seed
Native stock: 85%
Regional focus: southwest
Services: N/A
Comments: We have a large selection of grasses and some wildflowers and shrubs.

Desert Moon Nursery
P. O. Box 600
Veguita, NM 87062 (505) 864-0614
Sales type: mail order, retail, wholesale
Stock type: seeds, trees, herbaceous plants, cacti, succulents
Native stock: 99%
Regional focus: southwest
Services: landscape design, consultation
Comments: We specialize in native plants of the Chihuahuan desert.

Mesa Garden
P. O. Box 72
Belen, NM 87002 (505) 864-3131
Sales type: mail order only
Stock type: cacti and succulents only
Native stock: 100%
Regional focus: western U.S.
Services: N/A
Comments: We offer many types of cacti and succulents for the mountain regions.

Plants of the Southwest
Rt. 6, Box 11-A
Santa Fe, NM 87501 (505) 438-8888
Sales type: mail order, retail, wholesale
Stock type: trees, shrubs, herbaceous plants, seeds, seed mixes
Native stock: 90%
Regional focus: southwest
Services: landscaping, habitat and wetland restoration
Comments: We offer native seeds, wildflowers, shrubs, and grasses.

New York

John H. Gordon Nursery
1385 Campbell Blvd.
Amherst, NY 14228 (716) 691-9371
Sales type: mail order
Stock type: seeds, trees, books
Native stock: 90%
Regional focus: northeast
Services: habitat and wetland restoration
Comments: We have nuts, pawpaw, and persimmons for western New York.

Panfield Nurseries, Inc.
322 Southdown Rd.
Huntington, NY 11743 (516) 427-0112
Sales type: retail, wholesale
Stock type: meadow wildflowers (100% propagated), woodland plants and ferns
Native stock: 20%
Regional focus: northeast
Services: habitat restoration, landscaping, consultation
Comments: Established in 1931, we have a large assortment of perennials, native and woodland plants.

S. Scherer & Sons
104 Waterside Rd.
Northport, NY 11768 (516) 261-7432
Sales type: mail order, retail, wholesale
Stock type: trees, shrubs, herbaceous plants, seeds, seed mixes
Native stock: 30%
Regional focus: all U.S.
Services: growers of aquatic plants
Comments: We offer aquatic plants and water garden accessories.

Southern Tier Consulting, Inc.
P. O. Box 30
West Clarksville, NY 14786
(716) 968-3120
Sales type: wholesale
Stock type: trees, shrubs, herbaceous plants, seeds, seed mixes
Native stock: 95%
Regional focus: eastern U.S.
Services: habitat and wetland restoration, bioengineering
Comments: We focus on wetland plants, restoration, and mitigation.

North Carolina

Argura Nurseries, Inc.
7000 Canada Rd.
Tuckasegee, NC 28783 (704) 293-5550
Sales type: mail order, wholesale
Stock type: trees, shrubs, ground covers
Native stock: 90%
Regional focus: eastern U.S.
Services: habitat and wetland restoration
Comments: We offer conifer seedlings, transplants, field trees, and deciduous liners.

Boothe Hill Wildflowers
921 Boothe Hill
Chapel Hill, NC 27514 (919) 967-4091
Sales type: mail order
Stock type: seeds, seed mixes
Native stock: 100%
Regional focus: southeast and other areas of U.S.
Services: N/A
Comments: We specialize in native wildflowers for the southeast. Seeds collected under field conditions.

Carolina Seacoast Beach Plants
P. O. Box 418
Salterpath, NC 28575 (252) 240-2415
Sales type: retail, wholesale
Stock type: beach grasses
Native stock: 100%
Regional focus: eastern seaboard
Services: custom design, stabilization project, custom plant, and install sand fences.
Comments: We specialize in stabilization of sand dunes.

Humphries Nursery
Rt. 7, Box 202 C
Durham, NC 27707 (919) 489-0952
Sales type: retail, wholesale
Stock type: herbaceous plants
Native stock: 25%
Regional focus: North Carolina Piedmont
Services: landscape design
Comments: We specialize in perennials.

Lamtree Farm
2323 Copeland Rd.
Warrensville, NC 28693 (919) 385-6144
Sales type: mail order, retail, wholesale
Stock type: seeds, trees, shrubs
Native stock: 90%
Regional focus: eastern U. S.
Services: landscaping
Comments: We have a large selection of native azaleas and rhododendrons.

Niche Gardens
1111 Dawson Rd.
Chapel Hill, NC 27516 (919) 967-0078
Sales type: mail order, retail
Stock type: trees, herbaceous plants, shrubs
Native stock: 65%
Regional focus: all U S.
Services: landscaping, garden design, contract growing
Comments: We have a large selection of sun and shade natives.

TakeRoot
220 Blakes Dr.
Pittsboro, NC 27312 (919) 967-9515
Sales type: retail, wholesale
Stock type: ferns
Native stock: 80%
Regional focus: southeast
Services: free site analysis
Comments: We grow all ferns from spores.

We-Du Nurseries
Rt. 5, Box 724
Marion, NC 28752 (704) 738-8300
Sales type: mail order
Stock type: trees, herbaceous plants, shrubs
Native stock: 50%
Regional focus: Midwest, southeast, east
Services: N/A
Comments: We offer native plants of woody species and perennials.

Ohio

Bekatha's Garden
P. O. Box 615, 3354 Lebanon Rd.
Lebanon, OH 45036 (513) 932-1070
Sales type: mail order, retail, wholesale
Stock type: trees, shrubs, herbaceous plants, seeds, seed mixes
Native stock: N/A
Regional focus: N/A
Services: habitat and wetland restoration, landscaping
Comments: N/A

Envirotech Nursery
462 S. Ludlow Alley
Columbus, OH 43215 (614) 224-1920
Sales type: retail, mail order, wholesale
Stock type: herbaceous plants
Native stock: 90%
Regional focus: Midwest
Services: installation
Comments: We specialize in aquatic and prairie plants and constructed wetlands.

Garden Place
P. O. Box 388
Menton, OH 44061-0388 (216) 255-3705
Sales type: mail order and retail only
Stock type: herbaceous perennials
Native stock: N/A
Regional focus: all of U.S.
Services: N/A
Comments: We specialize in perennials.

Jones Fish and Lake Management
3433 Church St.
Newtown, OH 45244 (800) 662-3474
Sales type: retail, wholesale
Stock type: herbaceous plants
Native stock: 90%
Regional focus: Midwest
Services: lake and pond management
Comments: We specialize in water gardens, ponds
and provide perennials that are potted and bare rooted.

Mellingers
2310 W. So. Range Rd.
North Lima, OH 44452 (330) 549-9861
Sales type: retail, wholesale
Stock type: trees, shrubs, herbaceous plants
Native stock: 60%
Regional focus: all of U.S.
Services: N/A
Comments: We have a full line of garden supplies and
plants.

Oklahoma

Grasslander
Rt. 1, Box 56
Hennessey, OK 73742 (405) 853-2607
Sales type: retail, wholesale
Stock type: seeds, seed mixes, grassland seeders
Native stock: 25%
Regional focus: central U. S.
Services: wetland restoration, golf course consultation
Comments: We provide grasses for prairies and
wetlands as well as grassland seeders.

Guy's Seed Company
2520 Main
Woodward, OK 73801 (405) 254-2926
Sales type: wholesale
Stock type: seeds
Native stock: 100%
Regional focus: Oklahoma, north Texas, southern
Kansas
Services: N/A
Comments: We do contract seed collecting and
cleaning.

Johnston Seed Company
319 West Chestnut
Enid, OK 73701 (405) 233-5800
Sales type: mail order, retail, wholesale
Stock type: seeds, seed mixes
Native stock: 90%
Regional focus: Midwest, southwest, southeast, central
plains
Services: N/A
Comments: We specialize in native prairie grasses and
forbs.

Oregon

Balance Restoration Nursery
27995 Chambers Mill Rd.
Lorane, OR 97451 (514) 942-5530
Sales type: mail order, wholesale
Stock type: trees, shrubs, herbaceous plants, emergents
Native stock: 94%
Regional focus: northwest
Services: consultation
Comments: We specialize in native wetland plants.

Callahan Seeds
6045 Foley Ln.
Central Point, OR 97502 (503) 855-1164
Sales type: mail order, retail, wholesale
Stock type: seeds
Native stock: 80%
Regional focus: western
Services: N/A
Comments: We specialize in native tree and shrub
seeds.

Forestfarm
990 Tetherwild Rd.
Williams, OR 95744 (541) 846-6963
Sales type: mail order
Stock type: trees, shrubs, herbaceous plants
Native stock: 20%
Regional focus: all regions, especially the west
Services: N/A
Comments: We offer extensive selections of American
native plants for sunny meadows, shady woodlands,
and wet and dry conditions.

Great Western Seed Company
P. O. Box 387
Albany , OR 97321 (541) 928-3100
Sales type: wholesale
Stock type: seeds
Native stock: 60%
Regional focus: all of U.S.
Services: N/A
Comments: We provide turfgrasses, wildflowers, and
some native grasses.

Greer Gardens
1280 Goodpasture Island Rd.
Eugene, OR 97401 (503) 686-8266
Sales type: mail order, retail
Stock type: trees, herbaceous plants, tools
Native stock: N/A
Regional focus: all U. S.
Services: N/A
Comments: We have many native, rare, and unusual
plants.

Nature's Garden
40611 Hwy. 226
Scio, OR 93734 (503) 394-3217
Sales type: mail order, retail, wholesale
Stock type: trees, shrubs, herbaceous plants
Native stock: 10%
Regional focus: western
Services: commercial agriculture
Comments: We grow mostly Iris, Campanulas, and
Sisyrinchium, but do have some woodland plants.

Russell Graham
4030 Eagle Crest Rd. NW
Salem, OR 97304 (503) 362-1135
Sales type: mail order
Stock type: herbaceous plants
Native stock: 80%
Regional focus: northwest
Services: N/A
Comments: We specialize in native perennials of the
northwest. Catalog is $2.00.

Sevenoaks Native Nursery
29730 Harvest Dr.
Albany, OR 97321 (514) 757-6520
Sales type: wholesale
Stock type: trees, shrubs (all propagated)
Native stock: 90%
Regional focus: north and southwest, Rocky Mts..
Services: custom growing not rooted cutting
Comments: N/A

Siskiyou Rare Plant Nursery
2825 Cummings Rd.
Medford, OR 97501 (514) 772-6846
Sales type: mail order, retail
Stock type: herbaceous plants, shrubs
Native stock: 35%
Regional focus: all of U.S.
Services: N/A
 Comments: We specialize in plants for borders and
rock gardens.

Turf Seed, Inc.
P. O. Box 250
Hubbard, OR 97032 (503) 651-2130
Sales type: wholesale
Stock type: seeds, seed mixes
Native stock: N/A
Regional focus: northern tier states
Services: seed production
Comments: We offer a large selection of turfgrass seed
for golf course use, cool season species. Many of our
varieties contain "endophytes" for natural insect
resistance without herbicides.

Pennsylvania

Appalachian Wildflower Nursery
723 Honey Creek Rd.
Reedsville, PA 17084 (717) 667-6998
Sales type: mail order, retail
Stock type: trees, shrubs, herbaceous plants
Native stock: 60%
Regional focus: southeast coast and mountains
Services: N/A
Comments: We focus on Ranunculaceae and mono-
cots.

Carino Nurseries
P. O. Box 538
Indiana, PA 15701 (724) 463-3350
Sales type: mail order, retail, wholesale (now shipping to customers in forty-six states)
Stock type: trees, shrubs
Native stock: 75%
Regional focus: most of U.S.
Services: habitat and wetland restoration, landscaping, soil conservation
Comments: We offer an abundance of bare root seedlings and transplants, conifers and deciduous.

England's Herb Farm
33 Todd Rd.
Honey Brook, PA 19344 (610) 273-2863
Sales type: mail order, retail
Stock type: herbaceous plants
Native stock: 30%
Regional focus: mid-Atlantic
Services: habitat and wetland restoration, landscaping
Comments: We have a selection of herbaceous perennials for wetlands and meadows for mid-Atlantic region.

Ernst Conservation Seeds
9006 Mercer Pike
Meadville, PA 16335 (800) 873-3321
Sales type: wholesale
Stock type: seeds, seed mixes, trees, herbaceous plants
Native stock: 75%
Regional focus: northeast, east coast, Midwest
Services: habitat and wetland restoration, landfill reclamation
Comments: We provide quality native seeds and plants.

Flickingers' Nursery
P. O. Box 245
Sagamore, PA 16250 (412) 783-6528
Sales type: mail order, wholesale
Stock type: trees, shrubs, ground cover, Christmas tree seedlings
Native stock: 90%
Regional focus: northeast, Midwest
Services: habitat restoration
Comments: We offer a complete selection of bare root pines, spruce, fir, and Canadian hemlock.

Johnston Nurseries
Rt. 1, Box 100
Creekside, PA 15732 (724) 463-8456
Sales type: mail order
Stock type: trees, shrubs, Christmas trees (all bare root nursery stock)
Native stock: 50%
Regional focus: northeast, southeast, Midwest
Services: N/A
Comments: We specialize in reforestation and conservation planting.

Musser Forests, Inc.
Dept. EF, Box 340
Indiana, PA 15701 (412) 465-5685
Sales type: mail order, retail, wholesale
Stock type: trees, shrubs, herbaceous plants
Native stock: 10%
Regional focus: northeast, Midwest
Services: N/A
Comments: N/A

Natural Landscapes
354 N. Jennersville Rd.
West Grove, PA 19390 (610) 869-3788
Sales type: wholesale
Stock type: trees, shrubs
Native stock: 100%
Regional focus: mid-Atlantic
Services: N/A
Comments: We specialize in native azaleas.

North Creek Nurseries, Inc.
R. R. 2, Box 33, 33 North Creek Rd.
Landenberg, PA 19350 (610) 255-0100
Sales type: wholesale
Stock type: herbaceous plants
Native stock: 90%
Regional focus: eastern U.S.
Services: N/A
Comments: N/A

Octoraro Wetland Nurseries
6126 Steet Rd.
Kirkwood, PA 17536 (717) 529-3160
Sales type: wholesale
Stock type: seeds, seed mixes, trees, herbaceous plants, shrubs, aquatics plants
Native stock: 98%
Regional focus: mid-Atlantic, northeast
Services: habitat and wetland restoration, stream bank stabilization, tidal restoration
Comments: We provide native plants for restoration.

Pine Grove Nursery, Inc.
R.D. #3, Box 146
Clearfield, PA 16830 (814) 765-2363
Sales type: mail order, retail, wholesale
Stock type: trees, shrubs
Native stock: 50%
Regional focus: all of U.S.
Services: N/A
Comments: We have deciduous bareroot seedlings
grown from seeds.

Silva Native Nursery and Seed Company
1683 Sieling Farm Rd.
New Freedom, PA 17349 (717) 227-0486
Sales type: retail, wholesale
Stock type: seeds, seed mixes, trees, herbaceous plants
Native stock: 100%
Regional focus: eastern U.S.
Services: contract growing
Comments: We have woody and herbaceous species
for woodlands, wetlands, grasslands, dunes, and urban
habitats from Maine to Georgia.

The Primrose Path
Rt. 2, Box 110
Scottdale, PA 15683 (724) 887-6756
Sales type: mail order, retail
Stock type: herbaceous plants (all plants are nursery
propagated)
Native stock: 60%
Regional focus: northern two-thirds of the U.S.
Services: garden installation and design
Comments: N/A

The Rosemary House
120 S. Market St.
Mechanicsburg, PA 17055 (717) 697-5111
Sales type: mail order, retail
Stock type: herbaceous plants, herb seeds
Native stock: 10%
Regional focus: northern U.S.
Services: garden design
Comments: We specialize in herbs for all purposes.

South Carolina

Coastal Gardens & Nursery
4611 Socastee Blvd.
Myrtle Beach, SC 29575 (803) 293-2000
Sales type: mail order, retail, wholesale
Stock type: shrubs, herbaceous plants
Native stock: 10%
Regional focus: southeast, southern U.S.
Services: landscaping, wetland restoration
Comments: We offer shade and ornamental wetland
plants.

Oak Hill Farm
204 Pressly St.
Clover, SC 29710-1233 (803) 222-4245
Sales type: wholesale
Stock type: shrubs
Native stock: 1%
Regional focus: all U.S.
Services: N/A
Comments: We specialize in cold hardy azaleas. We
have no catalog.

Park Seed Company, Inc.,
P. O. Box 31, Cokesburg Rd.
Greenwood, SC 29647 (864) 223-8555
Sales type: mail order
Stock type: herbaceous plants, seeds
Native stock: 100%
Regional focus: N/A
Services: N/A
Comments: N/A

Wayside Gardens
1 Garden Ln.
Hodges, SC 29695 (800) 845-1124
Sales type: mail order
Stock type: trees, shrubs, herbaceous plants
Native stock: 100%
Regional focus: N/A
Services: N/A
Comments: N/A

Woodlanders, Inc.
1128 Colleton Ave.
Aiken, SC 29801 (803) 648-7522
Sales type: mail order
Stock type: trees, herbaceous plants, vines, shrubs,
perennials
Native stock: 65%
Regional focus: southeast
Services: N/A
Comments: We specialize in rare and hard to find
plants.

South Dakota

Northern Plains Seed Company
P. O. Box 964
Sioux Falls, SD 57101 (605) 336-0623
Sales type: retail, wholesale
Stock type: grass seeds
Native stock: N/A
Regional focus: northern U.S.
Services: N/A
Comments: We specialize in grass seeds.

Tennessee

Boyd's Nursery
P. O. Box 71
McMinnville, TN 37110 (931) 668-9898
Sales type: wholesale
Stock type: trees, shrubs, herbaceous plants
Native stock: 100%
Regional focus: all of U.S.
Services: N/A
Comments: We have plants which are hardy for zones 5, 6, and 7.

Hidden Springs Nursery
170 Hidden Springs Ln.
Cookeville, TN 38501 (615) 268-2592
Sales type: mail order, retail
Stock type: trees
Native stock: 30%
Regional focus: all U. S.
Services: N/A
Comments: We are a small organic operation with rare and unusual food-producing trees.

Mountain Gardens and Music
4227 Ft. Henry Dr.
Kingsport, TN 37663 (423) 239-6257
Sales type: retail
Stock type: herbaceous plants
Native stock: 100%
Regional focus: eastern U.S.
Services: N/A
Comments: We specialize in butterfly gardening.

Native Gardens
5737 Fisher Ln.
Greenback, TN 37742 (423) 856-0220
Sales type: mail order, retail, wholesale
Stock type: seeds, herbaceous plants
Native stock: 99%
Regional focus: eastern U.S.
Services: design consulting
Comments: We have over 200 species of wildflowers, shrubs, vines, ferns, and grasses for a wide variety of habitats.

Scott Brothers Nursery Company
P. O. Box 581
McMinnville, TN 37111 (931) 473-2954
Sales type: wholesale
Stock type: herbaceous plants
Native stock: 10%
Regional focus: all of U.S.
Services: N/A
Comments: We have a large selection of perennials.

Sunlight Gardens, Inc.
174 Golden Ln.
Andersonville, TN 37705 (423) 494-8237
Sales type: mail order, retail, wholesale
Stock type: herbaceous plants
Native stock: 75%
Regional focus: eastern U.S.
Services: habitat restoration, landscaping, custom growing
Comments: We provide entirely nursery-propagated native perennials, ferns, and forbs.

Triangle Nursery
Rt. 2, Box 229
McMinnville, TN 37110 (931) 668-8022
Sales type: wholesale
Stock type: trees, shrubs
Native stock: 75%
Regional focus: eastern U.S.
Services: N/A
Comments: N/A

Texas

Aldridge Nursery, Inc.
P. O. Box 1299
Van Ormy, TX 78073 (210) 622-3491
Sales type: wholesale
Stock type: trees, shrubs
Native stock: 30%
Regional focus: south and central U.S.
Services: N/A
Comments: We specialize in shade and ornamental trees.

Anderson Landscape and Nursery
2222 Pech
Houston, TX 77055 (713) 984-1342
Sales type: retail
Stock type: trees, shrubs, herbaceous plants
Native stock: 85%
Regional focus: Texas
Services: landscaping
Comments: N/A

Antique Rose Emporium
9300 Lueckemyer
Brenham, TX 77833 (409) 836-9051
Sales type: mail order, retail, wholesale
Stock type: roses
Native stock: 5%
Regional focus: N/A
Services: N/A
Comments: We are reintroducing old varieties of roses.

Bamert Seed Company
Rt. 3, Box 1120
Muleshoe, TX 79347 (806) 272-5506
Sales type: retail, wholesale
Stock type: seeds, seed mixes
Native stock: 100%
Regional focus: N/A
Services: N/A
Comments: N/A

Barton Springs Nursery
3601 Bee Cave Rd.
Austin, TX 78746 (512) 328-6655
Sales type: retail
Stock type: trees, shrubs, herbaceous plants, seeds
Native stock: 73%
Regional focus: Texas
Services: N/A
Comments: We specialize in native plants.

Breed & Company
718 W. 29th St.
Austin, TX 78705 (512) 474-7058
Sales type: retail
Stock type: shrubs, herbaceous plants
Native stock: 50%
Regional focus: central Texas
Services: questions and answers for landscaping
Comments: N/A

Browning Seed, Inc.
Box 1836, South IH 27
Plainview, TX 79073 (806) 293-5271
Sales type: mail order, wholesale
Stock type: seed
Native stock: 50%
Regional focus: southwest, Colorado, Oklahoma
Services: restoration
Comments: We harvest and condition plants, native grasses, and forbs.

Buchanan's Native Plants
611 East 11th St.
Houston , TX 77008 (713) 861-5702
Sales type: retail
Stock type: trees, shrubs, herbaceous plants
Native stock: 75%
Regional focus: Texas
Services: N/A
Comments: We have native, drought-tolerant plants.

Callahan's General Store
501 Bastrop Hwy.
Austin, TX 78741 (512) 385-3452
Sales type: retail
Stock type: seed
Native stock: 10%
Regional focus: Texas
Services: N/A
Comments: N/A

Color Spot Nurseries
7960 Cagnon Rd.
San Antonio, TX 78252 (512) 677-8020
Sales type: wholesale
Stock type: trees, shrubs, herbaceous plants
Native stock: 10%
Regional focus: southwestern and southern U.S.
Services: N/A
Comments: We specialize in container-grown southwestern and southern prairie natives.

Dallas Nature Center/Native Plant Nursery
7171 Mountain Creek Parkway
Dallas, TX 75249 (972) 296-1955
Sales type: retail
Stock type: trees, shrubs, herbaceous plants, seeds
Native stock: 100%
Regional focus: Blacklands
Services: N/A
Comments: We sell only what we grow.

Dodds Family Tree Nursery
515 W. Main
Fredericksburg, TX 78624
(830) 997-9571
Sales type: retail
Stock type: trees, shrubs, herbaceous plants
Native stock: 50%
Regional focus: Texas
Services: landscaping, design
Comments: We have organic natives and perennials.

Doremus Wholesale Nursery
P. O. Box 750, Rt. 2
Warren, TX 77664 (409) 547-3536
Sales type: wholesale
Stock type: trees, shrubs, herbaceous plants
Native stock: 60%
Regional focus: south and southeastern Texas
Services: N/A
Comments: We offer a large selection of uncommon native trees.

Douglass W. King Company, Inc.
P. O. Box 200320
San Antonio, TX 78220 (210) 661-4191
Sales type: wholesale
Stock type: seed
Native stock: 59%
Regional focus: southwest
Services: N/A
Comments: We offer warm season grasses.

Foster Rambie Grass Seed
P. O. Box 85386
Uvalde, TX 78802 (830) 278-2711
Sales type: retail, wholesale
Stock type: seeds
Native stock: 50%
Regional focus: Edwards Plateau, south, central, and north Texas
Services: N/A
Comments: We harvest and clean grass seeds.

Gardens
1818 West 35th St.
Austin, TX 78703 (512) 451-5490
Sales type: retail
Stock type: trees, shrubs, herbaceous plants
Native stock: 60%
Regional focus: Texas
Services: landscaping, landscape design
Comments: N/A

Green 'n Growing
P. O. Box 855
Pflugerville, TX 78691 (512) 251-3262
Sales type: retail
Stock type: trees, shrubs, herbaceous plants, seeds
Native stock: 50%
Regional focus: central Texas
Services: N/A
Comments: N/A

Gunsight Mountain Ranch & Nursery
P. O. Box 86
Tarpley, TX 78883 (830) 562-3225
Sales type: retail, wholesale
Stock type: trees, herbaceous plants, shrubs
Native stock: 98%
Regional focus: central and south Texas
Services: habitat restoration, consultation
Comments: We specialize in xeric native Texas trees and shrubs.

Harpool Farm and Garden
420 E. McKinney
Denton, TX 76201 (940) 387-0541
Sales type: retail
Stock type: trees, shrubs, herbaceous plants, seeds
Native stock: 25%
Regional focus: North Texas
Services: landscaping
Comments: We provide special orders.

Heep's Nursery
1705 Jason #1
Edinburg, TX 78539 (512)381-8813
Sales type: retail, wholesale
Stock type: trees, shrubs, herbaceous plants, grasses
Native stock: 100%
Regional focus: south Texas
Services: habitat restoration, landscaping
Comments: We offer over 100 species native to south Texas. All plants are propagated by seed or cuttings.

Hill Country Landscape Garden Center
13561 Pond Springs Rd.
Austin, TX 78720 (512) 258-0093
Sales type: retail
Stock type: trees, shrubs, herbaceous plants
Native stock: 25%
Regional focus: central and south Texas
Services: landscaping
Comments: N/A

Island Botanics Environmental Consultants
3734 Flour Bluff Dr.
Corpus Christi, TX 78418 (512) 937-4873
Sales type: contract grower only, consulting
Stock type: grasses, dune species
Native stock: 100%
Regional focus: Texas Gulf Coast
Services: habitat and wetland restoration, consulting,
contract growing
Comments: N/A

J'Don Seeds International
Box 10998-533
Austin, TX 78766 (800) 848-1641
Sales type: mail order, wholesale
Stock type: seeds, seed mixes
Native stock: 40%
Regional focus: all regions
Services: N/A
Comments: We offer a large selection of native
wildflower seed, regional mixes, and custom mixes.

Jenco Wholesale Nurseries
1445 MacArthur Dr., Suite 264
Carrollton, TX 75007 (972) 446-1820
Sales type: wholesale
Stock type: trees, shrubs, herbaceous plants, seeds
Native stock: 30%
Regional focus: Texas, Oklahoma, Kansas, New
Mexico, Missouri
Services: N/A
Comments: We have stores in Dallas, Houston, and
Lubbock, Texas, in Tulsa, Oklahoma, and Oklahoma
City and Wichita, Kansas.

Kings Creek Gardens
813 Straus Rd
Cedar Hill, TX 75104 (972) 291-7650
Sales type: retail
Stock type: trees, shrubs, herbaceous plants, seeds
Native stock: 25%
Regional focus: south-central Texas
Services: landscaping
Comments: N/A

Madrone Nursery
2318 Hilliard Rd.
San Marcos, TX 78666 (512) 353-3944
Sales type: wholesale
Stock type: trees, herbaceous plants
Native stock: 98%
Regional focus: southwest U. S.
Services: habitat restoration and assessment
Comments: We specialize in rare and hard to find
plants native to Texas.

Native Ornamentals
P. O. Box 997
Mertzon, TX 76941 (915) 835-2021
Sales type: retail, wholesale
Stock type: trees, herbaceous plants
Native stock: 99%
Regional focus: Chihuahuan desert
Services: landscaping, xeriscaping
Comments: We have a variety of perennial wildflow-
ers, shrubs, and trees of west Texas.

Native Son
7400 McNeil Dr.
Austin, TX 78729 (512) 444-2610
Sales type: retail, wholesale
Stock type: trees, shrubs, herbaceous
Native stock: 85%
Regional focus: central Texas
Services: landscaping, consulting
Comments: We have offered woody plants of the
Edwards Plateau since 1950.

Native Texas Nursery
1141 Penion Dr.
Austin, TX 78748 (512) 280-2824
Sales type: wholesale
Stock type: trees, shrubs
Native stock: 80%
Regional focus: central Texas
Services: N/A
Comments: N/A

Native Tree Farm
Hwy. 29 E., FM 1660
Georgetown, TX 78628 (512) 635-0103
Sales type: retail, wholesale
Stock type: trees, shrubs
Native stock: 100%
Regional focus: central and northern Texas
Services: installation and tree removal
Comments: We specialize in special-order plants and
tree care.

Powers Wholesale Nursery
7310 Sherwood Rd.
Austin, TX 78745 (512) 444-5511
Sales type: wholesale
Stock type: trees, shrubs, herbaceous plants
Native stock: 25%
Regional focus: central Texas
Services: N/A
Comments: N/A

Scherz Landscape Company
P. O. Box 60087
San Angelo, TX 76906 (915) 944-0511
Sales type: retail
Stock type: trees, shrubs, herbaceous plants, seeds
Native stock: 25%
Regional focus: Texas
Services: landscaping and maintenance
Comments: N/A

Shades of Green
334 W. Sunset Rd.
San Antonio, TX 78209 (210) 824-3772
Sales type: retail
Stock type: trees, shrubs, herbaceous plants
Native stock: 50%
Regional focus: south Texas
Services: N/A
Comments: We specialize in xeriscape plants, organic gardening, and garden supplies.

Spring Creek Nursery
2702 Co. Rd. 202
Mertzon, TX 76941 (915) 632-3203
Sales type: retail, wholesale
Stock type: trees, herbaceous plants, grasses
Native stock: 98%
Regional focus: Texas
Services: landscaping
Comments: We grow perennial bunch grasses, woody shrubs, and small trees for habitat restoration.

Sunbelt Trees, Inc.
16008 Boss Gaston
Richmond, TX 77469 (800) 635-4313
Sales type: wholesale
Stock type: trees
Native stock: 40%
Regional focus: central Texas
Services: N/A
Comments: We specialize in west Texas oaks.

Texas Native Trees
P. O. Box 817
Leander, TX 78646 (512) 260-1697
Sales type: mail order, wholesale
Stock type: seeds, trees, shrubs
Native stock: 85%
Regional focus: Texas
Services: habitat and wetland restoration, landscaping, environmental control
Comments: We specialize in Texas native species.

Texas Star Gardens
P. O. Box 663
Abilene, TX 79604 (915) 692-2733
Sales type: wholesale
Stock type: trees, shrubs, herbaceous plants
Native stock: 100%
Regional focus: central and western Texas
Services: N/A
Comments: We have xeric plants.

Turner Seed Company, Inc.
Rt. 1, Box 292
Breckenridge, TX 76024 (254) 559-2065
Sales type: retail, wholesale
Stock type: seeds
Native stock: 80%
Regional focus: Kansas, Texas, Oklahoma, New Mexico
Services: N/A
Comments: N/A

W. H. Anton Seed Company, Inc.
P. O. Box 667
Lockhart, TX 78644 (512) 398-2433
Sales type: mail order, retail, wholesale
Stock type: seeds
Native stock: 5%
Regional focus: Great Plains, Texas
Services: N/A
Comments: N/A

Weston Gardens in Bloom, Inc.
8101 Anglin Dr.
Fort Worth, TX 76140 (817) 572-0549
Sales type: retail
Stock type: trees, shrubs, herbaceous plants, seeds
Native stock: 60%
Regional focus: northern Texas
Services: landscaping and design
Comments: We provide educational seminars and ornamental native grasses.

Wildseed Farms, Inc.
P. O. Box 308
Eagle Lake, TX 77434 (409) 234-7353
Sales type: mail order, retail
Stock type: seeds, seed mixes
Native stock: 60%
Regional focus: all U. S.
Services: N/A
Comments: We are the south's largest grower and producer of wildflower seeds.

Utah

Granite Seed Company
1697 West 2100 North
Lehi, UT 84043 (801) 768-4422
Sales type: wholesale
Stock type: seeds, seed mixes
Native stock: 80%
Regional focus: west, plains
Services: N/A
Comments: We have a large selection of seed for land reclamation.

Intermountain Cactus
1478 No. 750 East
Kaysville, UT 84037 (801) 546-2006
Sales type: mail order
Stock type: herbaceous plants, cactus
Native stock: 100%
Regional focus: western U.S.
Services: N/A
Comments: We offer a large selection of winter hardy cactus.

Maple Leaf Industries
480 South 50 East
Ephraim, UT 84627 (801) 283-4701
Sales type: mail order, retail
Stock type: seeds
Native stock: 60%
Regional focus: west
Services: N/A
Comments: We specialize in native shrub, forb, and grass seeds.

Stevenson Intermountain Seed
P. O. Box 2
Ephraim, UT 84627 (435) 283-6639
Sales type: retail
Stock type: shrubs, herbaceous plants, seeds
Native stock: 60%
Regional focus: intermountain area
Services: N/A
Comments: N/A

Vermont

American Meadows, Inc.
P. O. Box 5
Charlotte, VT 05445 (802) 425-3500
Sales type: wholesale in bulk only
Stock type: seeds
Native stock: N/A
Regional focus: all of U.S.
Services: consulting with orders
Comments: N/A

Vermont Wildflower Farm
Rt. 7, P. O. Box 5
Charlotte, VT 05445 (802) 425-3500
Sales type: mail order, retail
Stock type: seeds, seed mixes
Native stock: N/A
Regional focus: all of U.S.
Services: consulting
Comments: N/A

Virginia

Edible Landscaping
P. O. Box 77
Afton, VA 22920 (804) 561-0234
Sales type: mail order, retail
Stock type: trees, herbaceous plants
Native stock: 20%
Regional focus: eastern and southern U.S.
Services: N/A
Comments: We have plans with ornamental and edible attributes which require less care.

Green Gardens
7247 Baldwin Ridge Rd.
Warrenton, VA 20187 (540) 347-7663
Sales type: retail, wholesale
Stock type: trees, shrubs, herbs
Native stock: 50%
Regional focus: Virginia
Services: landscape design, installation, and maintenance
Comments: We specialize in woody plants.

Virginia Natives
P. O. Box 18
Hume, VA 22639 (540) 364-1001
Sales type: retail, wholesale
Stock type: herbaceous plants
Native stock: N/A
Regional focus: mid-Atlantic, southeast
Services: habitat restoration, landscaping
Comments: N/A

Washington

Abundant Life Seed Foundation
P. O. Box 772
Port Townsend, WA 98368
(360) 385-5660
Sales type: retail
Stock type: seeds
Native stock: 10%
Regional focus: N/A
Services: rare seed preservation
Comments: We have a small offering of native seeds
for trees, shrubs, vines, and berries.

Aldrich Berry Farm & Nursery, Inc.
190 Aldrich Rd.
Mossyrock, WA 98564 (360) 983-3138
Sales type: mail order, retail, wholesale
Stock type: trees, shrubs
Native stock: 80%
Regional focus: Pacific northwest
Services: contract growing
Comments: We grow reforestation seedings and
specialize in conifers.

Foliage Gardens
2003 128 Ave. SE
Bellevue, WA 98005 (425) 747-2998
Sales type: mail order, retail
Stock type: trees, herbaceous plants
Native stock: 20%
Regional focus: all U. S.
Services: N/A
Comments: We carry hardy and exotic ferns.

Frosty Hollow Ecological Restoration
P. O. Box 53
Langley, WA 98260 (360) 579-2332
Sales type: mail order, retail, wholesale
Stock type: seeds
Native stock: 100%
Regional focus: Pacific northwest
Services: habitat and wetland restoration, consultation
Comments: We provide consultation in ecological
restoration and supplies of Pacific Northwest native
seeds.

Henry's Plant Farm
4522 132nd St. SE
Snohomish, WA 98290 (425) 337-8120
Sales type: mail order, wholesale
Stock type: herbaceous plants, ferns
Native stock: N/A
Regional focus: all of U.S.
Services: contract growing, liners
Comments: We specialize in ferns grown from spores
and young plants for nurseries.

Inside Passage Seeds
P. O. Box 639
Port Townsend, WA 98368
(360) 385-6114
Sales type: mail order, retail, wholesale
Stock type: seeds, books
Native stock: 95%
Regional focus: northwest coast
Services: consultation on landscaping and habitat
restoration
Comments: N/A

McLaughlin's Seeds
Buttercup's Acre, Box 550
Mead, WA 99021-0550
FAX (509) 466-0230
Sales type: mail order, wholesale
Stock type: seeds
Native stock: N/A
Regional focus: Pacific Northwest, N. Rockies, N.
Cascades, Olympia
Services: N/A
Comments: Established in 1929, we have been a
pioneer in the production and marketing of authentic
native wildflower, alpine, and rock plant seeds in
consignment packets and by mail order.

Plants of the Wild
P. O. Box 866
Tekoa, WA 99033 (509) 284-2848
Sales type: mail order, retail, wholesale
Stock type: seeds, seed mixes, trees, herbaceous plants
Native stock: 90%
Regional focus: Pacific northwest
Services: N/A
Comments: We have container-grown trees, shrubs, grasses, and forbs native to the northwest.

Rainier Seeds, Inc.
1404 4th St.
Davenport , WA 99122 (800) 828-8873
Sales type: mail order, retail, wholesale
Stock type: seeds
Native stock: N/A
Regional focus: all of U.S.
Services: habitat restoration, erosion control
Comments: We produce and clean reclamation and native grass species.

Silvaseed Company
P. O. Box 118
Roy, WA 98580 (253) 843-2246
Sales type: retail, wholesale
Stock type: native conifers
Native stock: 100%
Regional focus: Pacific northwest
Services: conifer seed processing
Comments: We have plugs and plug 1s selling by the bag.

Wisconsin

Applied Ecological Services Inc.
17921 Smith Rd.
Broadhead, WI 53520 (608) 897-8641
Sales type: mail order, retail, wholesale
Stock type: seeds, herbaceous plants
Native stock: 100%
Regional focus: Wisconsin and Illinois
Services: habitat and wetland restoration, wetland mitigation
Comments: We are a full service restoration/ mitigation corporation.

Hauser's Superior View Farm
Rt. 1, Box 199
Bayfield, WI 54814 (715) 779-5404
Sales type: retail
Stock type: herbaceous plants
Native stock: 50%
Regional focus: Midwest, northeast
Services: The best plants at low prices.
Comments: Established in 1908, Hauser's sells only one- to two-year old, field-grown plants.

Hild and Associates
326 South Glover Rd.
River Falls, WI 54022 (715) 426-5131
Sales type: retail, wholesale
Stock type: shrubs, herbaceous plants, seeds
Native stock: 100%
Regional focus: N/A
Services: installation
Comments: We contract grow wetland and prairie species.

J & J Tranzplant Aquatic Nursery Inc.
P. O. Box 227
Wild Rose, WI 54984 (920) 622-3552
Sales type: mail order, retail, wholesale
Stock type: seeds, seed mixes, trees, herbaceous plants, aquatic plants
Native stock: 90%
Regional focus: Midwest
Services: wetland restoration, consultation
Comments: We provide potted and bare root plants.

J. W. Jung Seed Company
335 S. High St.
Randolph, WI 53957 (920) 326-3121
Sales type: mail order, retail
Stock type: trees, shrubs, herbaceous plants, seeds, seed mixes, bulbs
Native stock: 85%
Regional focus: Midwest
Services: landscaping, home gardens
Comments: We offer species well adapted for the Midwest.

K. F. Evergreen
Rt. 4, Box 141
Sparta, WI 54656 (800) 458-7275
Sales type: mail order
Stock type: trees, shrubs, Christmas trees
Native stock: 90%
Regional focus: northern U.S.
Services: N/A
Comments: We specialize in native reforestation and wildlife habitat species.

Kester's Wild Game Food Nursery, Inc.
P. O. Box 516
Omro, WI 54963 (920) 685-2929
Sales type: retail, mail order, wholesale
Stock type: herbaceous plants
Native stock: 100%
Regional focus: upper Midwest
Services: installation
Comments: We consult for wild game and small pond
needs.

Kettle Moraine Natural Landscaping
W 996 Birchwood Dr.
Campbellsport, WI 53010
(920) 533-8939
Sales type: mail order, retail
Stock type: seeds, mixed seeds
Native stock: 100%
Regional focus: southeast Wisconsin
Services: habitat restoration, landscaping
Comments: We specialize in locally gathered prairie
seed from southeast Wisconsin.

Little Valley Farm
Rt. 3, Box 544
Spring Green, WI 53588 (608) 935-3324
Sales type: mail order, retail
Stock type: shrubs, herbaceous plants, some seeds
Native stock: 100%
Regional focus: Midwest
Services: landscaping
Comments: We offer landscaping with native shrubs,
and a select group of woodland and prairie species.

Milaeger's Gardens
4838 Douglas Ave.
Racine, WI 53402 (414) 639-2372
Sales type: mail order, retail
Stock type: herbaceous plants
Native stock: 15%
Regional focus: Midwest
Services: N/A
Comments: We have over 1000 perennials.

Prairie Future Seed Company, LLC
P. O. Box 644
Menomonee Falls, WI 53052
(414) 820-0221
Sales type: mail order, retail, wholesale
Stock type: herbaceous plants, seeds, seed mixes
Native stock: 93%
Regional focus: Midwest
Services: landscaping, consultation
Comments: We offer over 200 native prairie species.

Prairie Nursery
P. O. Box 306
Westfield, WI 53964 (608) 296-3679
Sales type: mail order, wholesale
Stock type: seeds, seed mixes, herbaceous plants
Native stock: 99%
Regional focus: upper Midwest, mid-Atlantic, New
England
Services: habitat and wetland restoration, landscaping,
species inventory
Comments: One hundred percent nursery-propagated
prairie, wetland, and woodland native plants and seeds.

Prairie Ridge Nursery
9783 Overland Rd.
Mt Horeb, WI 53572 (608) 437-5245
Sales type: mail order, retail, wholesale
Stock type: herbaceous plants, seeds, seed mixes
Native stock: 100%
Regional focus: Midwest
Services: habitat, wetland, and woodland restoration
Comments: We offer complete consulting, planting,
and management services for prairie, wetland and
woodland restorations, and gardens. All plants are
propagated at our nursery.

Prairie Seed Source
P. O. Box 83
North Lake, WI 53064 (414) 673-7166
Sales type: mail order, retail
Stock type: seeds
Native stock: 100%
Regional focus: southeast Wisconsin
Services: On-site consultation and custom mixes for
large sites (1\4 acre and larger).
Comments: We have over 170 species of grasses,
forbs, and shrubs and specialize in southeast Wiscon-
sin genotypes and diversity.

Reeseville Ridge Nursery
P. O. Box 171, 309 S. Main St.
Reeseville, WI 53579 (920) 927-3291
Sales type: mail order, retail, wholesale
Stock type: seeds, trees, shrubs, ground cover
Native stock: 50%
Regional focus: northeast and Midwest
Services: N/A
Comments: We have a wide selection of native woody
plants.

Superior View Farm
Rt. 1, Box 199
Bayfield, WI 54814 (715) 779-5404
Sales type: mail order
Stock type: herbaceous plants
Native stock: 10%
Regional focus: all of U. S
Services: N/A
Comments: We specialize in perennials.

Taylor Creek Restoration Nurseries
17921 Smith Rd.
Brodhead, WI 53520 (608) 897-8641
Sales type: mail order, retail, wholesale
Stock type: herbaceous plants, grasses
Native stock: 100%
Regional focus: upper Midwest
Services: consulting, habitat restoration, custom
growing
Comments: We are a full spectrum nursery with over
5000 species of native plants.

Wehr Nature Center
9701 W. College Ave.
Franklin, WI 53132 (414) 425-8550
Sales type: retail;
Stock type: seed mixes
Native stock: 100%
Regional focus: upper Midwest, Wisconsin
Services: N/A
Comments: Our mix includes native prairie forb
species harvested from prairies in southeast Wisconsin.

Wyoming

Wind River Seed
3075 Lane 51
Manderson, WY 82432 (307) 568-3361
Sales type: mail order, retail, wholesale
Stock type: seeds, seed mixes
Native stock: 95%
Regional focus: northern Great Plains, Rocky Mts.,
Midwest, northwest
Services: N/A
Comments: We have a large variety of seeds of
grasses, wildflowers, wetland shrubs, and trees.

APPENDIX E

Ecological Restoration Resources

Table of Contents

This is a very preliminary list of self reported consultants. The quality of their work is unknown to the authors.

State natural heritage programs have ecological community and local species information. These programs have staff expertise and may know local consultants and experts.

The Internet is a good way to obtain current information about an organization. The sites selected have many links to other sites.

Restoration Consultants

Arkansas

Holland Wildflower Farm
P. O. Box 328
Elkins, AR 72727 (501) 643-2622
Contact person: Bob and Julie Holland
Consulting: environmental, native plant landscaping, degraded land restoration
Community type: N/A
Regional focus: eastern U. S. and Canada

California

Albringt Seed Company
487 Dawson Dr. #5S
Camarillo, CA 93012 (805) 484-0551
Contact person: Paul Albright
Consulting: native plant landscaping, bioengineering, degraded land restoration, exotic species management and eradication
Community type: wetland, upland, estuarine
Regional focus: Pacific coast, California, southwest

All Season's Landscape Maintenance
P. O. Box 1256
Alberta, CA 95626 (916) 991-1974
Contact person: Robert Andrade
Consulting: native plant landscaping
Community type: N/A
Regional focus: N/A

Anderson Valley Nursery
P. O. Box 504
Boonville, CA 95415 (707) 895-3853
Contact person: Ken Montgomery
Consulting: environmental, native plant landscaping
Community type: N/A
Regional focus: California, esp. north coastal

Center for Natural Lands Management
425 E Alvarado, Suite H
Fallbrook, CA 92028 (760) 731-7790
Contact person: Sherry Teresa
Consulting: degraded land restoration, exotic species management and eradication
Community type: wetland, upland, estuarine
Regional focus: all of U. S.

Central Coast Wilds
114 Liberty St.
Santa Cruz, CA 95060 (408) 459-0656
Contact person: Josh Fodor
Consulting: environmental consulting, native plant landscaping, degraded land restoration, exotic species management and eradication
Community type: wetland, upland, estuarine
Regional focus: west coast

Circuit River Productions
9619 Old Redwood Hwy.
Windsor, CA 95492 (707) 838-6641
Contact person: Rocky Thompson
Consulting: environmental, bioengineering, degraded land restoration, exotic species management and eradication
Community type: wetland, upland, estuarine
Regional focus: northern California

EIP Associates
601 Montgomery St., Suite 500
San Francisco, CA 94111 (415) 362-1500
Contact person: Richard Nichols
Consulting: environmental, bioengineering, degraded land restoration, exotic species management and eradication
Community type: wetland, upland, estuarine
Regional focus: California, Nevada

Freshwater Farms
5851 Myrtle Ave.
Eureka, CA 95503 (707) 444-8261
Contact person: Rick Storre
Consulting: environmental, native plant landscaping, bioengineering
Community type: wetland, upland, estuarine
Regional focus: west coast to Montana and Colorado

Glenn Lukos Assoicates, Inc.
23441 South Pointe Dr., Suite 150
Laguna Hills, CA 92653 (714) 837-0404
Contact person: Greg Prettyman
Consulting: environmental, degraded land restoration, exotic species management and eradication
Community type: wetland, upland, estuarine
Regional focus: California, Nevada

Jones & Stokes Associates
2600 V St.
Sacramento, CA 95818 (916) 737-3000
Contact person: Greg Sutter
Consulting: environemntal consulting, native plant landscaping, degraded land restoration, exotic species management and eradication
Community type: wetland, upland, estuarine
Regional focus: western U. S.

Las Pilitas Nursery
3232 Las Pilitas Rd.
Santa Margarita, CA 93453
(805) 438-5992
Contact person: Bert or Celest Wilson
Consulting: environmental, native plant landscaping,
bioengineering, degraded land restoration, exotic
species management and eradication
Communtiy type: wetland, upland, estuarine
Regional focus: California

LSA Associates, Inc.
157 Park Place
Point Richmond, CA 94801
(510) 236-6810
Contact person: Malcolm Sproul
Consulting: environmental, bioengineering, native
plant landscaping, degraded land restoration, exotic
species management and eradication
Community type: upland, wetland, estuarine
Regional focus: California

Natures Image, Inc.
26741 Portola Pkwy., Suite 1E-437
Foothill Ranch, CA 92610
(714) 589-1254
Contact person: John Caruana
Consulting: native plant landscaping, degraded land
restoration, exotic species management and eradication
Community type: wetland, upland, estuarine
Regional focus: southwest, southern California

Pacific OpenSpace, Inc.
P. O. Box 744
Petaluma, CA 94953   (707) 769-1213
Contact person: Dave Kaplow
Consulting: environmental, native plant landscaping,
degraded land restoration, exotic species management
and eradication
Community type: wetland, upland, estuarine
Regional focus: California and Pacific northwest

Pacific Southwest Biological Services
P. O. Box 985
National City, CA 91951 (619) 447-5333
Contact person: Mitch Beecham
Consulting: environmental, native plant landscaping,
bioengineering, degraded land restoration, exotic
species management and eradication
Community type: wetland, upland, estuarine
Regional focus: southwest

Santa Barbara Botanic Garden
1212 Mission Canyon Rd.
Santa Barbara CA 93105
(805) 682-4726 ext. 114
Contact person: Carol Bornstein
Consulting: environmental
Community type: upland
Regional focus: California's cental coast region

Stephen Dreher
298 Mariposa Ave.
Sierra Madre, CA 91024   (818) 355-4706
Contact person: Stephen Dreher
Consulting: environmental, native plant landscaping,
exotic species management and eradication
Community type: wetland, upland
Regional focus: California

Stivers and Associates
150 El Camino Real #120
Tustin, CA 92686   (714) 838-9811
Contact person: Guy Stivers
Consulting: environmental, native plant landscaping,
degraded land restoration, exotic species management
and eradication
Community type: wetland, upland, esturarine
Regional focus: southern California

The Reveg Edge
P. O. Box 609
Redwood City, CA 94064   (650) 325-7333
Contact person: Craig C. Dremann
Consulting: environmental, bioengineering, degraded
land restoration, exotic species managment and
eradication
Community type: upland
Regional focus: western U. S., specializing in ultra-
arid grassland communitites

Wapumne Native Plant Nursery Company
3807 Mt. Pleasant Rd.
Lincoln, CA 95648 (916) 645-9737
Contact person: Everett Butts
Consulting: native plant landscaping
Community type. upland
Regional focus: Sierras, northern California

Zentner and Zentner
4240 Hollis St., Suite 360
Emeryville, CA 94608 (510) 596-2690
Contact person: John Zentner
Consulting: environmental, native plant landscaping
Community type: wetland, upland, estuarine
Regional focus: California, Bay area, Sacramento
Valley

Colorado

Aquatic and Wetland Company
1655 Walnut St., Suite 205
Boulder, CO 80302   (303) 442-5770
Contact person: Laurie Rink
Consulting: environmental, native plant landscaping, bioengineering, degraded land restoration
Community type: wetland
Regional focus: Rocky Mountains, southwest, California

Phillips Seeding and Reclamation, Inc.
1183 Billings Ave.
Lafayette, CO 80026 (303) 665-2610
Contact person: Robin or Mark Phillips
Consulting: degraded land restoration
Community type: upland
Regional focus: Colorado and surrounding states

Western Native Seed
P. O. Box 1463
Salida, CO 81201 (714) 539-1071
Contact person: Alex Tonnesen
Consulting: native plant landscaping, degraded land restoration
Community type: wetland, upland
Regional focus: Rocky Mountains and western Great Plains

Connecticut

Dr. Jennifer H. Mattei
Dept. of Biology
Sacred Heart University
5151 Park Ave.
Fairfield, CT 06432   (203) 365-7577
Contact person: Dr. Jennifer H. Mattei
Consulting: environmental, native plant landscaping, degraded land restoration, exotic species management and eradication
Community type: upland
Regional focus: northeast

Delaware

Keith Clancy
Delaware Natural Heritage Program
4876 Hay Point Landing Rd.
Smyrna, DE 19977 (302) 653-2880
Contact person: Keith Clancy
Consulting: native plant landscaping
Community type: wetland, upland, estuarine
Regional focus: mid-Atlantic

Florida

Central Florida Native Flora, Inc.
P. O. Box 1045
San Antonio, FL 33576   (352) 588-3687
Contact person: Brightman S. Logan
Consulting: environmental, native plant landscaping
Community type: wetland, upland
Regional focus: southeast, mainly central Florida

Chiappini Farm Native Nursery
P. O. Box 436
Melrose, FL 32666 (800) 293-5413
Contact person: David Chiappini
Consulting: native plant landscaping, degraded land restoration
Community type: wetland, upland, estuarine
Regional focus: northeast Florida

Ecogroup International Corp.
P. O. Box 792, Golf Course Rd.
Parrish, FL 34219   (800) 771-4114
Contact person: Michael M. Bundy
Consulting: environmental, native plant landscaping, degraded land restoration, exotic species management and eradication
Community type: wetland, upland, esturaine
Regional focus: southeast U. S.

Florida Keys Native Nursery, Inc.
102 Mohawk St.
Tavernier, FL 33070 (305) 852-2636
Contact person: Donna N. Sprunt
Consulting: environmental, native plant landscaping, exotic species management and eradication
Community type: wetland, upland
Regional focus: Florida, Caribbean

Lotspeich and Associates, Inc.
422 W. Fairbanks Ave., Suite 201
Winter Park, FL 32789   (407) 740-8482
Contact person: Renee L. Thomas
Consulting: environmental, native plant landscaping,
degraded land restoration, exotic species management
and eradication
Community type: wetland, upland, estuarine
Regional focus: southeast

Mesozoic Landscapes, Inc.
7667 Park Lane West
Lake Worth, FL 33467   (561) 967-2630
Contact person: Richard Moyroud
Consulting: native plant landscaping, exotic species
management and eradication
Community type: estuarine (maritime forests)
Regional focus: south Florida

Nautilus Environmental Services, Inc.
P. O. Box 497
Parrish, FL 34219   (941) 776-2524
Contact person: Otto M. Bundy
Consulting: environmental consulting
Community type: wetland, estuarine
Regional focus: southeast

The Wetlands Company, Inc.
7650 So. Tamiami Trail #10
Sarasota, FL 34231   (813) 921-6609
Contact person: Michael Reilly
Consulting: native plant landscaping, exotic species
eradication
Community type: wetland, upland, estuarine
Regional focus: southeast U. S.

Wetlands Management, Inc.
P. O. Box 7122
Jensen Beach, FL 34958   (561) 283-8420
Contact person: Joseph L. Gilio
Consulting: environmental, bioengineering, exotic
species management and eradication
Community type: wetland, upland, estuarine
Regional focus: southeast U. S.

Wild Azalea Nursery
17970 SW 111st
Brooker, FL 32622   (352) 485-3556
Contact person: Don Graham
Consulting: native plant landscaping
Community type: N/A
Regional focus: north central Florida

Georgia

Eco-South, Inc.
144 College Ave.
Covington, GA 30209   (404) 385-1849
Contact person: Butch Register or Rick Larsen
Consulting: environmental, bioengineering
Community type: wetland, upland, estuarine
Regional focus: southeast

The Jaeger Company
119 Washington St.
Gainesville, GA 30501 (770) 534-0506
Contact person: Dale Jaeger
Consulting: environmental, native plant landscaping,
bioengineering, exotic species management and
eradication
Community type: wetland, upland, estuarine
Regional focus: southeast

Idaho

Winterfeld Ranch Seed
P. O. Box 97
Swan Valley, ID 83449   (208) 483-3683
Contact person: Delbert F. Winterfeld
Consulting: environmental, native plant landscaping,
degraded land restoration
Community type: wetland, upland
Regional focus: west and southeast Idaho

Illinois

Bluestem Prairie Nursery
13197 E. 13th Rd.
Hillsboro, IL 62049 (217) 532-6344
Contact person: Ken Schaal
Consulting: restoration
Community type: upland
Regional focus: midwest

Christopher B. Burke Engineering, Ltd.
9575 W. Higgins Rd., Suite 600
Rosemont, IL 60018   (847) 823-0500
Contact person: Jedd Anderson
Consulting: environmental, native plant landscaping,
bioengineering, exotic species management and
eradication
Community type: wetland, upland
Regional focus: Illinois, Indiana, Wisconsin

Genesis Nursery
RR 1, Box 32
Walnut, IL 61376 (815) 379-9060
Contact person: anyone
Consulting: native plant landscaping
Community type: wetland, upland
Regional focus: Chicago region

Karin Wisiol & Associates
1411 Wild Iris Ln.
Grayslake, IL 60030   (847) 548-1650
Contact person: Karin Wisiol, Ph.D.
Consulting: environmental, native plant landscaping
Community type: wetland, upland
Regional focus: midwest and national

LaFayette Home Nursery, Inc.
RR#1 Box 1A
LaFayette, IL 61449   (309) 995-3311
Contact person: Dave Lahr
Consulting: native plant landscaping, bioengineering,
degraded land restoration, exotic species managment
and eradication
Community type: wetland, upland
Regional focus: midwest

Prairie Sun Consultants
612 Staunton Rd.
Naperville, IL 60565   (630) 983-8404
Contact person: Patricia Armstrong
Consulting: environmental, native plant landscaping
Community type: upland
Regional focus: midwest

Ted Gray and Associates
1024 Laurie Ln.
Burr Ridge, IL 60521 (630) 734-0270
Contact person: Ted Gray
Consulting: environmental, bioengineering, native
plant restoration
Community type: wetlands
Regional focus: midwest

The Natural Garden
38 W. Hwy. 64
St. Charles, IL 60175   (630) 584-0150
Contact person: Vallari Talapatra
Consulting: environmental, native plant landscaping,
exotic species managment and eradication
Community type: wetland, upland
Regional focus: Chicago region, midwest

The S/E Group
2000 York Rd., Suite 112
Oak Brook, IL 60523   (630) 990-0005
Contact person: Tom Slowiuski
Consulting: environmental, bioengineering, degraded
land restoration, exotic species management and
eradication
Community type: wetland
Regional focus: Great Lakes states

Indiana

Earthsource, Inc.
349 Airport North Office Park
Fort Wayne, IN 46825   (219) 489-8511
Contact person: Eric Ellingson
Consulting: environmental, bioengineering, native
plant landscaping, degraded land restoration, exotic
species management and eradication
Community type: upland, wetland, estuarine
Regional focus: Midwest

J. F. New and Associates, Inc.
708 Roosevelt Rd.
Walkerton, IN 46574   (219) 586-3400
Contact person: Will Ditzler
Consulting: native plant landscaping, bioengineering,
exotic species management and eradication
Community type: wetland, upland
Regional focus: Great Lakes and midwest

Spence Landscaping and Nursery, Inc.
2220 E. Fuson Rd.
Muncie, IN 47302   (765) 286-7154
Contact person: Kevin Tungesvick
Consulting: native plant landscaping, exotic species
management and eradication
Community type: wetland, upland
Regional focus: Indiana, Ohio

Iowa

Cascade Forestry Service
22033 Fillmore Rd.
Cascade, IA 52033   (319) 852-3042
Contact person: Leo Frueh, Sid Munford, Ed
Vandermillen
Consulting: environmental, native plant landscaping,
degraded land restoration
Community type: wetland, upland
Regional focus: Midwest

Kansas

Sharp Bros. Seed Company
P. O. Box 140
Healy, KS 67850 (316) 398-2231
Contact person: Art Armbrust
Consulting: degraded land restoration
Community type: upland
Regional focus: U. S.

Kentucky

Eco-Tech, Inc.
1003 E. Main St.
Frankfort, KY 40601 (502) 695-8060
Contact person: Hal Bryan, Gary Libby, or James Kiser
Consulting: environmental, bioengineering, degraded land restoration, exotic species management and eradication
Community type: wetland, upland, estuarine
Regional focus: southeast

Jane's Jungle
635 Sawdridge Creek W.
Monterey, KY 40359 (502) 484-2044
Contact person: Jane Harrod
Consulting: native plant landscaping, exotic species eradication
Community type: wetland, upland
Regional focus: Kentucky

Nolin River Nut Tree Nursery
797 Port Wooden Rd.
Upton, KY 42784 (502) 369-8551
Contact person: John or Lisa Brittain
Consulting: native plant landscaping
Community type: N/A
Regional focus: N/A

Louisiana

Natives Landscape Corp.
320 N.Theard St.
Covington, LA 70433 (504) 892-5424
Contact person: John Mayronne
Consulting: environmental, native plant landscaping, exotic species management and eradication
Community type: wetland, upland
Regional focus: southeast

Maryland

Ecosystem Recovery Institute
P. O. Box 249
Freeland, MD 21053   (717) 235-8426
Contact person: Michael Hollins
Consulting: environmental education, native plant landscaping, bioengineering, exotic species management and eradication
Community type: wetland, upland, estuarine
Regional focus: all of U. S.

Envirens, Inc.
P. O. Box 299
Freeland, MD 21053   (717) 227-0073
Contact person: Sheri Early
Consulting: environmental, native plant landscaping, bioengineering, degraded land restoration, exotic species management and eradication
Community type: wetland, upland, estuarine
Regional focus: eastern and central U. S.

Environmental Concern, Inc.
P. O. Box P., 210 W. Chew Ave.
St. Michaels, MD 21663   (301) 745-9620
Contact person: F. Albert McCullough, III
Consulting: native plant landscaping, degraded land restoration
Community type: wetland, upland, estuarine
Regional focus: eastern U. S.

Massachusetts

B & C Associates, Inc.
2 Rice St.
Hudson, MA 01749   (508) 568-0135
Contact person: David Crossman
Consulting: wetland
Community type: wetland
Regional focus: northeast

Botanicals
P. O. Box 65
London, MA 02173   (508) 358-1029
Contact person: Jeff Licht
Consulting: environmental, native plant landscaping
Community type: wetland, upland, estuarine
Regional focus: northeast

Carr Research Laboratory, Inc.
17 Waban St.
Wellesley, MA 02181   (508) 651-7027
Contact person: Dr. Jerome B. Carr
Consulting: environmental, bioengineering, degraded
land restoration
Community type: wetland, upland, estuarine
Regional focus: New England

Donaroma's Nursery
P. O. Box 2189
Edgartown, MA 02539   (508) 627-8366
Contact person: Cathy Wood
Consulting: native plant landscaping, exotic species
management and eradication
Community type: upland
Regional focus: northeast

LEC Environmental Consultants, Inc.
#3 Otis Park Dr.
Bourne, MA 02532   (508) 759-0050
Contact person: Paul Lezito
Consulting: environmental, native plant landscaping,
bioengineering, degraded land restoration, exotic
species management and eradication
Community type: wetland, estuarine
Regional focus: northeast

Metcalf and Eddy, Inc.
30 Harvard Mill Square, P. O. Box 4071
Wakefield, MA 01880   (617) 224-6238
Contact person: Margaret McBrien
Consulting: environmental, exotic species management
and eradication
Community type: wetland, estuarine
Regional focus: all regions, primarily Atlantic states

New England Environmental
800 Main St.
Amherst, MA 01002   (413) 256-0202
Contact person: Michael Marcus
Consulting: environmental, native plant landscaping,
bioengineering, degraded land restoration, exotic
species management and eradication
Community type: wetland, upland, estuarine
Regional focus: New England

The Bioengineering Group, Inc.
53 Mason St.
Salem, MA   01970   (508) 740-0096
Contact person: Wendi Goldsmith
Consulting: environmental, native plant landscaping,
bioengineering, degraded land restoration, exotic
species management and eradication
Community type: wetland, estuarine
Regional focus: northeast, southeast

Michigan

Bioent
14174 Hoffman Rd.
Three Rivers, MI 49093   (616) 278-3975
Contact person: Gary Manley
Consulting: environmental, native plant landscaping,
bioengineering
Community type: wetland, upland
Regional focus: Midwest, Michigan, Indiana

JJR Inc.
110 Miller Ave.
Ann Arbor, MI 48104   (313) 669-2663
Contact person: Steve Ott
Consulting: environmental, native plant landscaping,
bioengineering
Community type: wetland, upland
Regional focus: Midwest, east

The Michigan Wildflower Farm
11770 Cutler Rd.
Portland, MI 48875   (517) 647-6010
Contact person: Esther Durnwald
Consulting: native plant landscaping
Community type: N/A
Regional focus: Michigan

Minnesota

Feder's Prairie Seed Company
12871 380th Ave.
Blue Earth, MN 56013   (507) 526-3049
Contact person: Wayne Feder
Consulting: environmental, native plant landscaping
Community type: upland
Regional focus: upper Midwest

Mohn Frontier Seed and Nursery
RR 1, Box 152
Cottonwood, MN 56229   (507) 423-6482
Contact person: Robert or Lydia
Consulting: environmental, native plant landscaping,
degraded land restoration
Community type: wetland, upland
Regional focus: upper Midwest

Prairie Moon Nursery
Rt. 3, Box 163
Winona, MN 55987   (507) 452-1362
Contact person: Alan Wade
Consulting: environmental
Community type: wetland, upland
Regional focus: upper Midwest

Prairie Restorations, Inc.
P. O. Box 327
Princeton, MN 55371   (612) 389-4342
Contact person: Robin Straka
Consulting: environmental, native plant landscaping
Community type: wetland, upland
Regional focus: primarily Minnesota

The Kestrel Design Group, Inc.
5140 Hankerson Ave., Suite 1
Edina, MN 55436   (612) 928-9600
Contact person: L. Peter MacDonagh
Consulting: environmental, bioengineering, degraded
land restoration
Community type: wetland, upland
Regional focus: Midwest, upper Midwest

Missouri

Elixir Farm Botanicals, LLC
General Delivery
Brixey, MO 65618   (417) 261-2393
Contact person: Vinnie McKinney
Consulting: environmental, native plant landscaping,
degraded land restoration
Community type: wetland, upland
Regional focus: eastern indigenous forest

Forrest Keeling Nursery, Inc.
P. O. Box 135
Elsberry, MO 63343   (573) 898-5571
Contact person: Hugh K. Stearenson
Consulting: environmental, native plant landscaping
Community type: wetland
Regional focus: Midwest

Hamilton Seeds
16786 Brown Rd.
Elk Creek, MO 65464   (417) 967-2190
Contact person: Rex and Amy Carrie
Consulting: N/A
Community type: N/A
Regional focus: N/A

Montana

Bitterroot Restoration, Inc.
445 Quast Ln.
Corvallis, MT 59828   (406) 961-4991
Contact person: Tim Miekle
Consulting: environmental, native plant landscaping,
bioengineering, degraded land restoration, exotic
species management and eradication
Community type: wetland, upland, estuarine
Regional focus: northwest, southwest, Rocky Moun-
tains, west

Wildlands Center for Preventing Roads
P. O. Box 7516
Missoula, MT 59807   (406) 543-9551
Contact person: Bethanie Walder
Consulting: environmental, degraded land restoration
Community type: wetland, upland
Regional focus: U. S.

Nebraska

Stock Seed Company
28008 Mill Rd.
Murdock, NE 68407   (402) 867-3771
Contact person: Rod Fritz or Dave Stock
Consulting: native plant landscaping, degraded land
restoration
Community type: upland
Regional focus: U. S.

New Jersey

Environmental Liability Management, Inc.
218 Wall St., Research Park
Princeton, NJ 08540   (609) 683-4848
Contact person: Phil Sandine
Consulting: environmental, bioengineering
Community type: wetland, upland, estuarine
Regional focus: northeast

New Mexico

Desert Moon Nursery
P. O. Box 600
Veguita, NM 87062　(505) 864-0614
Contact person: Ted Hodoba
Consulting: native plant landscape design
Community type: N/A
Regional focus: southwest

New York

John Gordon Nursery
1385 Campbell Blvd.
Amherst, NY 14228　(716) 691-9371
Contact person: John Gordon
Consulting: N/A
Community type: upland
Regional focus: western New York

Southern Tier Consulting, Inc.
2701-A Rt. 305
West Clarksville, NY 14786
(716) 968-3120
Contact person: Gary Pierce
Consulting: native plant landscaping, bioengineering
Community type: wetland, estuarine
Regional focus: eastern U. S.

North Carolina

Boothe Hill Wildflowers
921 Boothe Hill
Chapel Hill, NC 27514　(919) 967-4091
Contact person: Nancy Easterling
Consulting: native plant landscaping
Community type: N/A
Regional focus: southeast

Carolina Seacoast Beach Plants
P. O. Box 418
Salterpath, NC 28575 (252) 240-2415
Contact person: Don Carpenter
Consulting: degraded dune restoration
Community type: dunes
Regional focus: eastern seaboard and Gulf Coast

Lamtree Farm
2323 Copeland Rd.
Warrensville, NC 28693　(910) 385-6144
Contact person: Lee A. Morrison
Consulting: native plant landscaping
Community type: N/A
Regional focus: eastern U. S.

Niche Gardens
1111 Dawson Rd.
Chapel Hill, NC 27516　(919) 967-0078
Contact person: Dianne Ford
Consulting: native plant landscaping
Community type: N/A
Regional focus: southeast

North Dakota

Prairie Habitats, Inc.
Box 1, Argyle, MB
Canada ROC OBO　(204) 467-5007
Contact Person: C. or J. Morgan
Consulting: environmental consulting, native plant landscaping, degraded land restoration, exotic species management and eradication
Community type: wetland, upland
Regional focus: North Dakota, Minnesota, South Dakota

Ohio

Envirotech Consultants, Inc.
462 South Ludlow Alley
Columbus, OH 43215　(614) 224-1920
Contact person: John M. Kiertscher
Consulting: environmental, native plant landscaping
Community type: wetland, upland
Regional focus: all U. S., primarily Midwest

Oklahoma

Grasslander
Rt. 1, Box 56
Hennessey, OK 73742　(405) 853-2607
Contact person: Chuck Grimes
Consulting: environmental, native plant landscaping
Community type: wetland, upland
Regional focus: south-central U. S.

Oregon

Mark Griswold Wilson
1123 SE Harney St.
Portland, OR 97202   (503) 234-2233
Contact person: Mark Griswold Wilson
Consulting: environmental, native plant landscaping,
degraded land restoration, exotic species management
and eradication
Community type: wetland, upland
Regional focus: Oregon, Washington

Pennsylvania

Ecological Restoration, Inc.
1633 Gilmar Rd.
Apollo, Pa 15613 (724) 733-9969
Contact person: Dave Hails
Consulting: environmental, native plants landscaping,
bioengineering, degraded land restoration, exotic
species management and eradication
Community type: wetland, upland, estuarine
Regional focus: east of Mississippi

England's Herb Farm
33 Todd Rd.
Honey Brook, PA, 19344   (610) 273-2863
Contact person: Yvonne England
Consulting: native plant landscaping
Community type: wetland
Regional focus: Chester County Pennsylvania

Landstudies, Inc.
6 S. Broad St.
Lititz, PA 17543   (717) 627-4440
Contact person: Mark Gutshall
Consulting: environmental, native plant landscaping,
bioengineering, exotic species management and
eradication
Community type: wetland, upland, estuarine
Regional focus: mid-Atlantic, northeast

McTish, Kunkel & Associates
2402 Sunshine Rd.
Allentown, PA 18103   (610) 791-2700
Contact person: Robert H. Hosking, Jr.
Consulting: environmental, native plant landscaping,
bioengineering, degraded land restoration, exotic
species management and eradication
Community type: wetland, upland
Regional focus: Appalachian Mountains, Great Valley,
Glaciated Plateau

Munro Ecological Services, Inc.
990 Old Sumneytown Pike
Harleysville, PA 19438   (610) 287-9671
Contact person: John W. Munro
Consulting: environmental, native plant landscaping,
bioengineering, degraded land restoration, exotic
species management and eradication
Community type: wetland, upland, estuarine
Regional focus: eastern U. S.

Nemeth Landscaping
1469 Oakridge Dr.
South Fork, PA 15956   (814) 266-4663
Contact person: Charlie
Consulting: native plant landscaping
Community type: wetland, upland
Regional focus: Pennsylvania, West Virginia, Maryland

Normandeau Associates
3450 Schuylkill Rd.
Spring City, PA 19475   (610) 948-4700
Contact person: Margaret P. O'Malley
Consulting: environmental
Community type: wetland, upland, estuarine
Regional focus: New England, mid-Atlantic, southeast

Schmid and Company, Inc.
1201 Cedar Grove Rd.
Media, PA 19063   (610) 356-0903
Contact person: James Schmid
Consulting: environmental
Community type: wetland, estuarine
Regional focus: mid-Atlantic

Tennessee

Native Gardens
5737 Fisher Ln.
Greenback, TN 37742   (423) 856-0220
Contact person: Edward and Meredith Clebsch
Consulting: environmental, native plant landscaping
Community type: upland
Regional focus: southeast

Sunlight Gardens
174 Golden Ln.
Andersonville, TN 37705   (423) 496-8237
Contact person: Andrea Sessions
Consulting: native plant landscaping
Community type: N/A
Regional focus: southeast

Texas

Bamert Seed Company
Rt. 3, Box 1120
Muleshoe, TX 79347 (806) 272-5506
Contact person: Nick Bamert
Consulting: N/A
Community type: upland
Regional focus: all U. S.

Browning Seed, Inc
P. O. Box 1836, South 1H 27
Plainview, TX 29073 (806) 293-2571
Contact person: John Browning
Consulting: environmental consulting, native plant
landscaping, degraded land restoration, exotic species
management and eradication
Community type: wetland, upland
Regional focus: Nebraska, Oklahoma, Texas, Kansas,
New Mexico, Colorado

Gunsight Mt. Ranch and Nursery
P. O. Box 86
Tarpley, TX 78883 (830) 562-3225
Contact person: Patricia Wilkie or Dorothy B. Mattiza
Consulting: environmental consulting, native plant
landscaping, degraded land restoration
Community type: wetland, upland
Regional focus: central and south Texas

Island Botanics Environmental Consultants
3734 Flour Bluff Dr.
Corpus Christi, TX 78418 (512) 937-4873
Contact person: Paul Carangelo
Consulting: environmental consulting, bioengineering,
degraded land restoration, exotic species management
and eradication
Community type: wetland, upland, estuarine
Regional focus: New Mexico, Texas

Madrone Nursery
2318 Hilliard Rd.
San Marcos, TX 78666 (512) 353-3944
Contact person: Dan Hosage, Jr.
Consulting: environmental consulting, native plant
landscaping, bioengineering, degraded land restoration
Community type: wetland, upland
Regional focus: Edwards Plateau, Trans-Pecos, greater
southwest U. S. and northern Mexico

Native American Seed
610 Main St.
Junction, TX 76849
Contact person: Jan Neiman
Consulting: environmental consulting, degraded land
restoration, exotic species management and eradication
Community type: upland
Regional focus: Texas

Native Ornamentals
P. O. Box 997
Mertzon, TX 76941 (915) 835-2021
Contact person: Steve or Valorie Lewis
Consulting: environmental consulting, native plant
landscaping
Community type: upland
Regional focus: Chihuahuan desert region of west Texas

Native Sons Plant Nursery
507 Lockhart Dr.
Austin, TX 78704 (512) 343-1448
Contact person: Sheryl McLaughlin
Consulting: native plant landscaping, degraded land
restoration, exotic species management and eradication
Community type: upland
Regional focus: central Texas

Neiman Environments, Inc.
P. O. Box 185
Junction, TX 76849 (915) 446-3600
Contact person: Bill or Jan Neiman
Consulting: environmental consulting, degraded land
restoration, exotic species management and eradication
Community type: upland
Regional focus: Texas

Spring Creek Nursery
2702 Co. Rd. 202
Mertzon, TX 76941 (915) 632-3203
Contact person: Terry K. Tate
Consulting: native plant landscaping
Community type: N/A
Regional focus: west Texas, southeast New Mexico

Texas Native Trees
P. O. Box 817
Leander, TX 78646 (512) 260-1697
Contact person: Ray Rodriguez
Consulting: environmental consulting, native plant
landscaping, exotic species management and eradication
Community type: wetland, upland, estuarine
Regional focus: all U. S.

W. H. Anton Seed Company, Inc.
P. O. Box 667
Lockhart, TX 78644 (512) 398-2433
Contact person: Jim Priddy
Consulting: N/A
Community type: upland
Regional focus: Texas and Great Plains

Utah

REVEG Environmental Consulting, Inc.
1697 W. 2100 North
Lehi, UT 84043 (801) 768-4422
Contact: William Agnew
Consulting: environmental, native plant landscaping, bioengineering, degraded land restoration, exotic species management and eradication
Community type: wetland, upland
Regional focus: Arizona, Utah, Colorado, Wyoming, Nevada, and New Mexico

Virginia

Perennium
841 Oak Leaf Ct.
Warrenton, VA 20186 (540) 347-0027
Contact person: Regina Yurkonis
Consulting: native plant landscaping
Community type: wetland, upland
Regional focus: Virginia and mid-Atlantic

Virginia Natives
P. O. Box 18
Hume, VA 22639 (540) 364-1001
Contact person: Mary Painter
Consulting: environmental consulting, native plant landscaping
Community type: wetland, upland
Regional focus: mid-Atlantic, southeast

Washington

Agua Tierra Environmental Consulting, Inc.
506 Edison St. SE, Suite 100
Olympia, WA 98501 (360) 754-3755
Contact person: Christian Fromuth
Consulting: environmental consulting, native plant landscaping, bioengineering, degraded land restoration, exotic species management and eradication
Community type: wetland, estuarine
Regional focus: Pacific northwest, Rocky Mountain states, Basin and Range Province

Biologic Environmental Consulting
2505 Richardson St.
Fitchburg, WA 53711 (608) 277-9960
Contact person: Michael Anderson
Consulting: environmental, native plant landscaping, degraded land restoration, exotic species management and eradication
Community type: wetland, upland
Regional focus: upper Midwest

East West Landscape
3325 McNeill St.
Port Townsend, WA 98368
(360) 379-8908
Contact person: John Barr
Consulting: native plant landscaping
Community type: wetland, upland
Regional focus: maritime northwest

Frosty Hollow Ecological Restoration
P. O. Box 53
Langley, WA 98260 (360) 579-2332
Contact person: Marianne Edain
Consulting: environmental consulting, degraded land restoration, exotic species management and eradication
Community type: wetland, upland, estuarine
Regional focus: Pacific northwest

Inside Passage Seeds
P. O. Box 639
Port Townsend, WA 98368
(360) 385-6114
Contact person: Forest Shomer
Consulting: native plant landscaping
Community type: wetland, upland, estuarine
Regional focus: northwest coast

Ridolfi Engineers and Assoc., Inc.
1001 Fourth Ave., Suite 2720
Seattle, WA 98154 (206) 682-7294
Contact person: Sue Alvarez
Consulting: environmental consulting, bioengineering,
degraded land restoration
Community type: N/A
Regional focus: northwest, Idaho, Oregon, Washington, Alaska

The Watershed Company
P. O. Box 1180
Kirkland, WA 98033 (425) 822-5242
Contact person: Bill Way
Consulting: environmental consulting, bioengineering,
degraded land restoration, exotic species management
and eradication
Community type: wetland, upland
Regional focus: northwest

Wisconsin

Applied Ecological Services, Inc.
17921 Smith Rd.
Broadhead, WI 53520 (608) 897-8641
Contact person: Gary Cole or John Larison
Consulting: environmental consulting, native plant
landscaping, degraded land restoration, exotic species
management and eradication
Community type: wetland, upland, estuarine
Regional focus: upper Midwest

CRM Ecosystems, Inc.
9738 Overland Rd.
Mt. Hored, WI 53572 (608) 437-5245
Contact person: N/A
Consulting: environmental consulting, native plant
landscaping, degraded land restoration, exotic species
management and eradication
Community type: wetland, upland
Regional focus: upper Midwest

Kettle Moraine Natural Landscaping
W. 996 Birchwood Dr.
Campbellsport, WI 53010 (920) 533-8939
Contact person: Connie Ramthun
Consulting: native plant landscaping, exotic species
management and eradication
Community type: upland
Regional focus: southeast Wisconsin

Mead and Hunt, Inc.
6501 Watts Rd.
Madison, WI 53719 (608) 273-6380
Contact person: Dennis Geary
Consulting: environmental, bioengineering, degraded
land restoration
Community type: wetland, upland
Regional focus: Midwest, northeast, and California

Prairie Nursery
P. O. Box 306
Westfield, WI 53964 (608) 296-3679
Contact person: Neil Diboll or Bob Van Abel
Consulting: native plant landscaping, degraded land
restoration, exotic species management and eradication
Community type: wetland, upland
Regional focus: upper Midwest, mid-Atlantic and New
England

Prairie Seed Source
P. O. Box 83
North Lake, WI 53064
Contact person: R. Ahrenheorster
Consulting: environmental consulting
Community type: upland
Regional focus: southeast Wisconsin

Natural Heritage Programs

Alabama Natural Heritage Program
Huntingdon College
Massey Hall
1500 Fairview Ave.
Montgomery, AL 36106-2148
(334) 834-4519 *Fax:* (334) 834-5439

Arizona Heritage Data Management System
Arizona Game and Fish Department
WM-H
2221 W. Greenway Rd.
Phoenix, AZ 85028
(602) 789-3612 *Fax:* (602) 789-3928

Arkansas Natural Heritage Commission
Suite 1500, Tower Building
323 Center St.
Little Rock, AR 72201
(501) 324-9150 *Fax:* (501) 324-9618

California Natural Heritage Division
Department of Fish and Game
1220 S St.
Sacramento, CA 95814
(916) 322-2493 *Fax:* (916) 324-0475

Colorado Natural Heritage Program
Colorado State University
254 General Services Building
Fort Collins, CO 80523
(970) 491-1309 *Fax:* (970) 491-3349

Connecticut Natural Diversity Database
Natural Resources Center
Department of Environmental Protection
79 Elm St., Store Level
Hartford, CT 06106-5127
(860) 424-3540 *Fax:* (860) 424-4058

Delaware Natural Heritage Program
Division of Fish and Wildlife
Department of Natural Resources and Environmental
Control
4876 Hay Point Landing Rd.
Smyrna, DE 19977
(302) 653-2880 *Fax:* (302) 653-3431

District of Columbia Natural Heritage Program
13025 Riley's Lock Rd.
Poolesville, MD 20837
(301) 427-1302 *Fax:* (301) 427-1355

Florida Natural Areas Inventory
1018 Thomasville Rd., Suite 200-C
Tallahassee, FL 32303
(904) 224-8207 *Fax:* (904) 681-9364

Georgia Natural Heritage Program
Wildlife Resources Division
Georgia Department of Natural Resources
2117 U. S. Highway 278 S.E.
Social Circle, GA 30279
(706) 557-3032 *Fax:* (706) 557-3033

Idaho Conservation Data Center
Department of Fish and Game
600 S. Walnut St., Box 25
Boise, ID 83707-0025
(208) 334-3402 *Fax:* (208) 334-2114

Illinois Natural Heritage Division
Department of Natural Resources
Division of Natural Heritage
524 S. Second St.
Springfield, IL 62701-1787
(217) 785-8774 *Fax:* (217) 785-8277

Indiana Natural Heritage Data Center
Division of Nature Preserves
Department of Natural Resources
402 W. Washington St., Room W267
Indianapolis, IN 46204
(317) 232-4052 *Fax:* (317) 233-0133

Iowa Natural Areas Inventory
Department of Natural Resources
Wallace State Office Building
Des Moines, IA 50319-0034
(515) 281-8524 *Fax:*(515) 281-6794

Kansas Natural Heritage Inventory
Kansas Biological Survey
2041 Constant Ave.
Lawrence, KS 66047-2906
(913) 864-3453 *Fax:*(913) 864-5093

Kentucky Natural Heritage Program
Kentucky State Nature Preserves Commission
801 Schenkel Ln.
Frankfort, KY 40601
(502) 573-2886 *Fax:*(502) 573-2355

Louisiana Natural Heritage Program
Department of Wildlife and Fisheries
P. O. Box 98000
Baton Rouge, LA 70898-9000
(504) 765-2821 *Fax:* (504) 765-2607

Maine Natural Areas Program
Department of Conservation
93 State House Station
Augusta, ME 04333-0093
(207) 287-8044 *Fax:* (207) 287-8040

Maryland Heritage and Biodiversity Conservation Programs
Department of Natural Resources
Tawes State Office Building, E-1
Annapolis, MD 21401
(410) 260-8540 *Fax:* (410) 260-8595

Massachusetts Natural Heritage and Endangered Species Program
Division of Fisheries and Wildlife
Rte. 135
Westborough, MA 01581
(508) 792-7270 *Fax:* (508) 792-7275

Michigan Natural Features Inventory
Mason Building, 5th Floor
Box 30444
Lansing, MI 48909-7944
(517) 373-1552 *Fax:* (517) 373-6705

Minnesota Natural Heritage and Nongame Research
Department of Natural Resources
500 Lafayette Rd., Box 7
St. Paul, MN 55155
(612) 297-4964 *Fax:* (612) 297-4961

Mississippi Natural Heritage Program
Museum of Natural Science
111 N. Jefferson St.
Jackson, MS 39201-2897
(601) 354-7303 *Fax:* (601) 354-7227

Missouri Natural Heritage Database
Missouri Department of Conservation
P. O. Box 180
Jefferson City, MO 65102-0180
(573) 751-4115 *Fax:* (573) 526-5582

Montana Natural Heritage Program
State Library Building
1515 E. 6th Ave.
Helena, MT 59620
(406) 444-3009 *Fax:* (406) 444-0581

Nebraska Natural Heritage Program
Game and Parks Commission
2200 N. 33rd St.
Lincoln, NE 68503
(402) 471-5421 *Fax:* (402) 471-5528

Nevada Natural Heritage Program
Department of Conservation and Natural Resources
1550 E. College Parkway, Suite 145
Carson City, NV 89706-7921
(702) 687-4245 *Fax:* (702) 885-0868

New Hampshire Natural Heritage Inventory
Department of Resources and Economic Development
172 Pembroke St.
P. O. Box 1856
Concord, NH 03302
(603) 271-3623 *Fax:* (603) 271-2629

New Jersey Natural Heritage Program
Office of Natural Lands Management
22 S. Clinton Ave., CN404
Trenton, NJ 08625-0404
(609) 984-1339 *Fax:* (609) 984-1427

New York Natural Heritage Program
Department of Environmental Conservation
700 Troy-Schenectady Rd.
Latham, NY 12110-2400
(518) 783-3932 *Fax:* (518) 783-3916

North Carolina Heritage Program
NC Department of Environment, Health and Natural Resources
Division of Parks and Recreation
P. O. Box 27687
Raleigh, NC 27611-7687
(919) 733-4181 *Fax:* (919) 715-3085

North Dakota Natural Heritage Inventory
North Dakota Parks and Recreation Department
1835 Bismarck Expressway
Bismarck, ND 58504
(701) 328-5357 *Fax:* (701) 328-5363

Ohio Natural Heritage Data Base
Division of Natural Areas and Preserves
Department of Natural Resources
1889 Fountain Square, Building F-1
Columbus, OH 43224
(614) 265-6453 *Fax:* (614) 267-3096

Oklahoma Natural Heritage Inventory
Oklahoma Biological Survey
111 E. Chesapeake St.
University of Oklahoma
Norman, OK 73019-0575
(405) 325-1985 *Fax:* (405) 325-7702

Oregon Natural Heritage Program
Oregon Field Office
821 SE 14th Ave.
Portland, OR 97214
(503) 731-3070 *Fax:* (503) 230-9639

Pennsylvania Natural Diversity Inventory - East
The Nature Conservancy
34 Airport Dr.
Middletown, PA 17057
(717) 948-3962 *Fax:* (717) 948-3957

Pennsylvania Natural Diversity Inventory - West
Western Pennsylvania Conservancy
209 4th Ave.
Pittsburgh, PA 15222
(412) 288-2777 *Fax:* (412) 281-1792

Pennsylvania Natural Diversity Inventory - Central
Bureau of Forestry
P. O. Box 8552
Harrisburg, PA 17105-8552
(717) 783-0388 *Fax:* (717) 783-5109

Rhode Island Natural Heritage Program
Department of Environmental Management
Division of Planning and Development
83 Park St.
Providence, RI 02903
(401) 277-2776 *Fax:* (401) 277-2069

South Carolina Heritage Trust
South Carolina Department of Natural Resources
P. O. Box 167
Columbia, SC 29202
(803) 734-3893 *Fax:* (803) 734-6310

South Dakota Natural Heritage Data Base
South Dakota Department of Game, Fish and Parks
Wildlife Division
523 E. Capitol Ave.
Pierre, SD 57501-3182
(605) 773-4227 *Fax:* (605) 773-6245

Tennessee Division of Natural Heritage
Department of Environment and Conservation
401 Church St.
Life and Casualty Tower, 8th Floor
Nashville, TN 37243-0447
(615) 532-0431 *Fax:* (615) 532-0614

Texas Biological and Conservation Data System
3000 S. IH-35, Suite 100
Austin, TX 78704
(512) 912-7011 *Fax:* (512) 912-7058

Utah Natural Heritage Program
Division of Wildlife Resources
1596 W. North Temple
Salt Lake City, UT 84116
(801) 538-4761 *Fax:* (801) 538-4709

Vermont Nongame and Natural Heritage Program
Vermont Fish and Wildlife Department
103 S. Main St., 10 South
Waterbury, VT 05671-0501
(801) 241-3700 *Fax:* (802) 241-3295

Virginia Division of Natural Heritage
Department of Conservation and Recreation
Division of Natural Heritage
217 Governor St., 3rd Floor
Richmond, VA 23219
(804) 786-7951 *Fax:* (804) 371-2674

Washington Natural Heritage Program
Department of Natural Resources
P. O. Box 47016
Olympia, WA 98504-7016
(360) 902-1340 *Fax:* (360) 902-1783

West Virginia Natural Heritage Program
Department of Natural Resources Operations Center
Ward Rd., P. O. Box 67
Elkins, WV 26241
(304) 637-0245 *Fax:* (304) 637-0250

Wisconsin Natural Heritage Program
Endangered Resources
Department of Natural Resources
101 S. Webster St., Box 7921
Madison, WE 53707
(608) 266-7012 *Fax:* (608) 266-2925

Wyoming Natural Diversity Database
1604 Grand Ave., Suite 2
Laramie, WY 82070
(307) 745-5026 *Fax:* (307) 745-5026

Organizations and Internet Resources

Primary Sources for Restoration

Society for Ecological Restoration
http://nabalu.flas.ufl.edu/ser/SERhome.html

> The Society for Ecological Restoration (SER) is the single most important contact point for anyone seriously interested in restoration. A significant amount of good information can be found on line. SER has a newsletter and two important journals—*Restoration and Management Notes* (see next listing) and *Restoration Ecology: The Journal of the Society for Ecological Restoration*.

> They will also provide the names of individuals who are members in a particular region. This could be helpful for making contacts with consultants and locally knowledgeable professionals.

> The Society for Ecological Restoration
> 1207 Seminole Highway
> Madison, WI 53711
> 608-262-9547

Restoration & Management Notes
http://wiscinfo.doit.wisc.edu/arboretum/homepage.html

The Association of State Wetlands Managers
http://members.aol.com/aswmi/homepage.html

> The Association is a nonprofit membership organization open to people and organizations interested in wetland management. The goal is to help public and private wetland decision makers use the "best possible scientific information and techniques in wetland delineation, assessment, mapping, planning, regulation, acquisition, restoration, and other management."

> At this site you can find the National Registry of Wetland Professionals. This registry gives name, address, phone number, expertise, and whether they are willing to consult.

> Hard copy of the Registry can be ordered from ASWM Registry, P.O. Box 269, Berne, NY 12023, phone number 518-872-1804.

American Society of Landscape Architects (ALSA)
http://www.asla.org

> This web site has a search engine for ASLA firms. Nonmembers of ASLA can search by firm name, type, city, and/or state. This is a good way to get some potentially helpful contacts for restoration work.

> They can also be reached at ASLA, 636 Eye Street, NW, Washington, D.C., 20001, phone number 202-898-2444.

Additional Sources

The following web sites are not in any particular order of importance. Many of the sites have links to other web sites. Both the Internet and the field of ecological restoration are changing so fast it is hard to maintain any type of current list.

Soil Ecology and Restoration Group
http://www.rohan.sdsu.edu/dept/SERG/serg.html

> This site contains information about the group, restoration techniques, links to other sites, resources, and equipment suppliers.
>
> Biology Department
> San Diego State University
> San Diego, CA 92182
> 619-594-2883

Society for Wetland Scientists
http://www.sws.org

> This site describes and offers information related to the organization. It includes links to other sites and allows browsing and searching of *Wetlands*, the society's journal.
>
> P.O. Box 1897
> Lawrence, KS 66044
> 913-843-1221

Stream Corridor Restoration
http://www2.hqnet.usda.gov/stream-restoration

> This site provides an online draft copy of the handbook *Stream Corridor Restoration*.

The Coastal Program
http://www.fws.gov/cep/cepcode.html

> This site contains information on the U.S. Fish and Wildlife Services coastal program.

Ecological Society of America
http://esa.sdsc.edu

> This site contains many links to other Internet sites and information about "An Initiative for a Standardized Classification of Vegetation in the U.S."

Native Plant Conservation Initiative
http://www.aqd.nps.gov/npci

> The NPCI is a consortium of nine federal government member agencies and over 100 nonfederal cooperators representing various disciplines within the conservation field: biologists, botanists, habitat preservationists, horticulturists, resources management consultants, soil scientists, special interest clubs, nonprofit organizations, concerned citizens, nature lovers, gardeners. NPCI members and cooperators work collectively to solve the problems of native plant extinction and native habitat restoration, ensuring the preservation of our ecosystem.

U. S. Environmental Protection Agency
http://www.epa.gov

National Wetlands Inventory
http://www.nwi.fws.gov

Soil and Water Conservation Society
http://www.swcs.org

Tallgrass Prairie in Illinois (Illinois Natural History Survey)
http://www.prairienet.org/tallgrass

This site contains information on prairie restoration and many links.

The Desert Lands Restoration Task Force
http://www.serg.sdsu.edu/SERG/dlrtf.html

This site contains a lot of good information, resources, links, and an online guide to desert land restoration.

Arid Lands Newsletter
http://ag.arizona.edu/OALS/ALN/ALNHome.html

This online newsletter is published by the Office of Arid Lands Studies at the University of Arizona.

The Nature Conservancy
http://www.tnc.org

This site contains information about The Nature Conservancy and good links to conservation science information.

Natural Heritage Network Central Server
http://www.abi.org

This site provides access to Natural Heritage Programs and other information related to them.

Audubon International
http://www.audubonintl.org

Audubon International has numerous programs designed to improve environmental decision making, especially the Cooperative Sanctuary System and the Audubon Signature Program. This site also has links to other related sites.

United States Golf Association
http://www.usga.org

This site gives information on USGA's Environmental Education Program and Turfgrass and Environmental Research Program. They also list the golf courses in the Audubon Cooperative Sanctuary Program for Golf Courses and provide information on their Wildlife Links program.

Society for Conservation Biology
http://conbio.rice.edu/scb

This site gives information about the society. It also allows the table of contents for the journal *Conservation Biology* and the latest newsletter of the society to be reviewed.

Natural Areas Association
http://www.natareas.org

This site allows a review of the table of contents of the *Natural Areas Journal* and access to abstracts of the articles. It also includes links to related web sites.

U.S. Army Corps of Engineers
http://www.usace.army.mil/inet/functions/cw/index.htm

California Exotic Pest Plant Council
http://www.igc.apc.org/ceppc

Florida Exotic Pest Plant Council
http://www.fleppc.org

Tennessee Exotic Pest Plant Council
http://www.webriver.com/tn-eppc

Weed Science Society of America
http://ext.agn.uiuc.edu/wssa/index.html

The Wetlands Regulation Center
http://www.wetlands.com

> Online text of federal wetlands regulations.

Iowa Prairie Network Homepage
http://www.netins.net/showcase/bluestem/ipnapp.htm

> This site contains an online newsletter and lots of good links.

Northern Prairie Wildlife Research Center
http://www.npwrc.usgs.gov

> This site contains a lot of information and links related to the northern prairies region.

International Erosion Control Association
http://www.ieca.org

Land and Water
http://www.landandwater.com

> This is a commercial magazine on natural resource management and restoration. Some articles are available online.

National Biological Service
http://www.its.nbs.gov

Sierra Club
http://www.sierraclub.org

USDA Plant database
http://plants.usda.gov

U.S. Geological Survey
http://info.er.usgs.gov

UNESCO World Heritage List
http://www.cco.caltech.edu/~salmon/world.heritage.html

International Institute for Ecological Economics
http://kabir.cbl.cees.edu/miiee/miiee.html

Directories for Internet Sites and Other Resources

Botany - Internet Directory for Botany
http://herb.biol.uregina.ca/liu/bio/botany.shtml

> This site contains an alphabetical listing of Internet sites related to botany.

National Library for the Environment
http://www.cnie.org/nle

National Biological Information Infrastructure
http://www.nbii.gov

Environmental Organizations Web Directory
http://www.webdirectory.com

Directory of Web Sites
http://www.mindspring.com/~lshull/wetlands.html

The Need to Know—Ecology and Environment Page
http://www.peak.org/~mageet/tkm/ecolenv.htm

California Environmental Resources Evaluation System
http://ceres.ca.gov

APPENDIX F

Regulatory Considerations for Restoring Wetlands

Various state and federal laws are used to protect wetlands. The primary federal laws for protecting wetlands are Section 404 of the 1977 Clean Water Act and the Swampbuster Provision of the Food Security Act. Section 404 of the 1977 Clean Water Act is jointly administered by the U.S. Army Corps of Engineers and the U.S. Environmental Protection Agency with consultation from the U.S. Fish and Wildlife Service and states. The Swampbuster Provision of the Food Security Act denies federal subsidies to any farm owner who knowingly converts wetlands to farmland. A wetland subject to regulation is called a *jurisdictional wetland*. The U.S. Army Corps of Engineers developed a regulatory definition of wetlands used in the Section 404 permitting process. The legal definition included in that regulation is

> The term "wetlands" means those areas that are inundated or saturated by surface or ground water at a frequency and duration sufficient to support, and that under normal circumstances do support, a prevalence of vegetation typically adapted for life in saturated soil conditions. Wetlands generally include swamps, marshes, bogs, and similar areas.

Delineation of wetland boundaries is a critical part of the regulatory process. The delineation methods include three criteria: hydrology, soils, and vegetation. Wetland hydrology includes the presence of water, either at the surface or within the root zone. Hydrologic conditions can vary throughout the year from standing water to dry. Wetland soils have characteristics distinct from upland soils. Soil scientists have identified a series of hydric soils. Wetland plants (hydrophytes) are specifically adapted to wet conditions. They include facultative wetland plants that can live in either wet or dry conditions and obligate wetland plants that are able to live only in a wet environment. Experts should be consulted for the technical, regulatory delineation of wetlands. The government's approach to delineation is contained in a single manual *Corp of Engineers Wetland Delineation Manual* (Environmental Laboratory 1988).

The "no net loss" concept has become an important part of wetland conservation in the United States. This concept was formulated by the National Wetlands Policy Forum in 1987. The one overall objective recommended by the forum was

> to achieve no overall net loss of the nation's remaining wetlands base and to create and restore wetlands, where feasible, to increase the quantity and quality of the nation's wetland resource base.

In addition to the Federal 404 permit, states have the power to restrict the discharges of "dredged or fill material" into wetlands. This state process is called the 401 certification process. A 404 permit cannot be issued unless the state certifies it meets water quality standards through the 401 process. It is during the 404 permit review process that the 401 certification review occurs by the state. Figure F.1 shows the 404 permitting process. Figure F.2 shows the U.S. Army Corps of Engineers regions and Table F.1 provides addresses for Corps division offices.

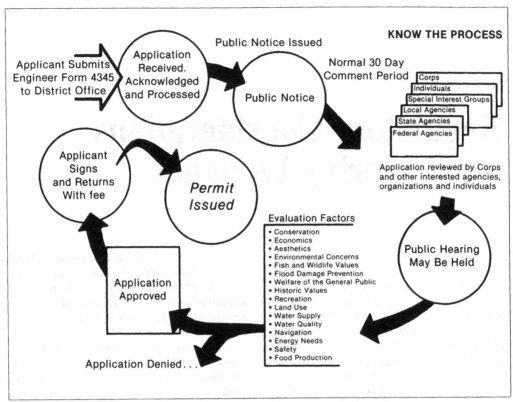

KNOW THE PROCESS

Figure F.1. Corps of Engineers Permit Review Process (source: U.S. Army Corps of Engineers)

Table F.1. Regional Offices of the U.S. Army Corps of Engineers

Missouri River Division
P.O. Box 103 Downtown Station
Omaha, NE 68101

Ohio River Division
P.O. Box 1159
Cincinnati, OH 45201

New England Division
424 Trapelo Road
Waltham, MA 02254

Pacific Ocean Division
Building 230, Fort Shafter
Honolulu, HI 96858

North Atlantic Division
90 Church Street
New York, NY 10007

South Atlantic Division
77 Forsyth Street, SW
Atlanta, GA 30303

North Central Division
536 S. Clark Street
Chicago, IL 60605

South Pacific Division
630 Sansome Street, Room 1218
San Francisco, CA 94111

Southwestern Division
1114 Commerce Street
Dallas, TX 75242

Lower Mississippi Valley Division
P.O. Box 80
Vicksburg, MS 39180

North Pacific Division
P.O. Box 2870
Portland, OR 97208

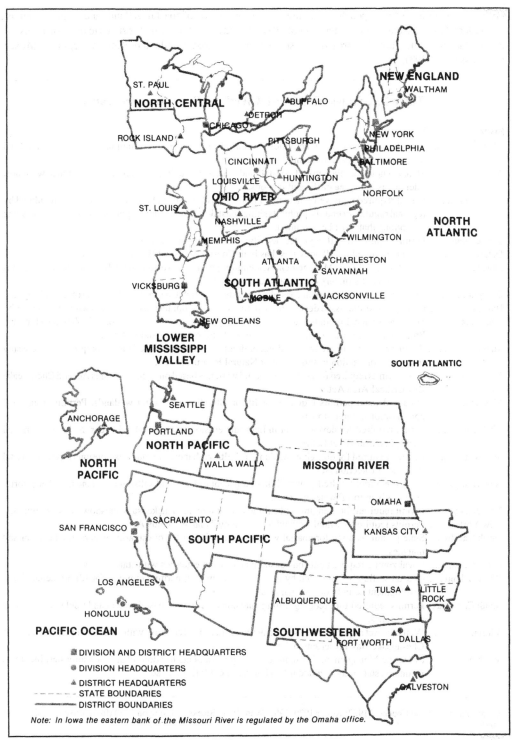

Figure F.2. Map of U.S. Army Corps of Engineers Regions

Individual states have developed their own laws for the protection of coastal wetlands and inland wetlands. The federal Coastal Zone Management Act of 1972 created a grant program for states to develop plans for coastal management. Table F.2 provides a list of states with coastal wetland protection programs (Mitsch and Gosselink 1993).

Table F.2. States with Coastal Wetland Protection Programs

State	Program
Alabama	Permits are required for activities in coastal zone (dredging, dumping, etc.) that alter tidal movement or damage flora and fauna.
Alaska	State agencies regulate use of coastal land waters, including offshore areas, estuaries, wetlands, tideflats, islands, sea cliffs, and lagoons.
California	Permit required for development up to 1,000 yards from mean high tide; coastal zone regulated by regional regulatory boards; prohibits siting coastal-dependent developments in wetlands with some exceptions that must be permitted.
Connecticut	Permit required for all regulated activity; state inventory required.
Delaware	Permits required for all activities; has both Coastal Zone and Beach Protection Acts.
Florida	Florida Coastal Zone Management Act requires permit for erosion-control devices and excavations or erections of structures in coastal environment.
Georgia	Permits required for work in coastal saltmarshes through Coastal Marshlands Protection Program.
Hawaii	County authorities issue development permits for development of coastal area with state oversight.
Louisiana	State and/or local permits required for activity in coastal wetlands. Coastal Wetland Planning, Protection, and Restoration Act passed in 1990 to restore coastal wetlands.
Maine	Permits required for dredging, filling, or dumping into coastal wetlands. Comprehensive coastal/freshwater protection in Protection of Natural Resources Act.
Maryland	State permits required for activity in coastal wetlands based on Tidal Wetlands Act and Chesapeake Bay Critical Area Act.
Massachusetts	State and local permits required for fill or alteration of coastal wetlands. Permits from local conservation commissioners.
Michigan	Permit required for development in high rule erosion areas, flood risk areas, and environmental areas of coastal Great Lakes.
Mississippi	Permits required for dredging and dumping, although there are many exemptions through Coastal Wetlands Protection Act.
New Hampshire	Permit required for dredge and fill in or adjacent to fresh and saltwater wetlands; higher priority usually given to saltwater marshes.
New Jersey	Permit required for dredging and filling; agriculture and Hackensack meadowlands exempted.
New York	Permits required for tidal wetland alteration by Tidal Wetlands Act.
North Carolina	State permit required for coastal wetland excavation or fill of estuarine waters, tidelands, or salt marshes.
Oregon	Local zoning requirements on coastal marshes and estuaries with state review.
Rhode Island	Coastal wetlands designated by order and use limited; permits required for filling, aquaculture, development activity on salt marshes.
South Carolina	Permits required for dredging, filling, and construction in coastal waters and tidelands including salt marshes.
Virginia	Wetlands Act requires permits for all activities in coastal counties with some exemptions; also 1988 Chesapeake Bay Preservation Act.
Washington	Shoreline Management Act requires local governments to adopt plans for shorelines, including wetlands; state may regulate if local government fails to do so.

Source: After Zinn and Copeland 1982; Kusler 1979, 1983; Want 1990; Meeks and Runyon 1990 (from Mitsch and Gosselink 1993)

State programs relating to protecting inland wetlands vary condsiderbly. Table F.3 lists states that have comprehensive wetland laws for inland waters (Mitsch and Gosselink 1993). It is recommended that the U.S. Army Corps of Engineers and the state agency responsible for water be contacted about current wetland regulations.

Table F.3. States that Have Comprehensive Wetland Laws for Inland Waters

State	Law
Connecticut	Inland Wetlands and Watercourses Act
Delaware	The Wetlands Act
Florida	Henderson Wetlands Protection Act of 1984
Maine	Protection of Natural Resources Act
Maryland	Chesapeake Bay Critical Area Act
Massachusetts	Wetland Protection Act
Michigan	Goemaere-Anderson Wetland Protection Act
Minnesota	The Wetland Conservation Act of 1991
New Hampshire	Fill and Dredge in Wetlands Act
New Jersey	Freshwater Wetlands Protection Act of 1987
New York	Freshwater Wetlands Act
North Dakota	No Net Wetlands Loss Bill of 1987
Oregon	Fill and Removal Act Comprehensive Land Use Planning Coordination Act
Rhode Island	Freshwater Wetlands Act
Vermont	Water Resources Management Act
Wisconsin	Water Resources Development Act Shoreland Management Program

Source: Want 1990; Meeks and Runyon 1990 (from Mitsch and Gosselink 1993)

The most comprehensive definition of wetlands was adopted by the U.S. Fish and Wildlife Service. This is an important ecological definition and is widely accepted by wetland scientists. The U.S. Army Corps of Engineers definition is the regulatory definition. The definition and a full classification of wetlands is presented in *Classification of Wetlands and Deepwater Habitats of the United States* by Cowardin et al. (1979).

> *Wetlands are lands transitional between terrestrial and aquatic systems where the water table is usually at or near the surface or the land is covered by shallow water.... Wetlands must have one or more of the following three attributes: (1) at least periodically, the land supports predominantly hydrophytes, (2) the substrate is predominantly undrained hydric soil, and (3) the substrate is nonsoil and is saturated with water or covered by shallow water at some time during the growing season of each year. The term Wetland includes a variety of areas that fall into one of five categories: (1) areas with hydrophytes and hydric soils, such as those commonly known as marshes, swamps, and bogs; (2) areas without hydrophytes but with hydric soils— for example, flats where drastic fluctuation in water level, wave action, turbidity, or high concentration of salts may prevent the growth of hydrophytes; (3) areas with hydrophytes but nonhydric soils, such as margins of impoundments or excavations where hydrophytes have become established but hydric soils have not yet developed; (4) areas without soils but with hydrophytes such as the seaweed-covered portion of rocky shores; and (5) wetlands without soil and without hydrophytes, such as gravel beaches or rocky shores without vegetation.*

The definitions of wetlands do not include drained hydric soils that have the hydrology so altered that they will not support wetland plants. These areas of hydric soils are good candidates for restoration if the wetland hydrology is reestablished.

If you are impacting a wetland, contact the U.S. Army Corps of Engineers regional office. Do not use the information in this appendix to secure compliance.

Map of the Natural Regions of the United States and Küchler Codes List

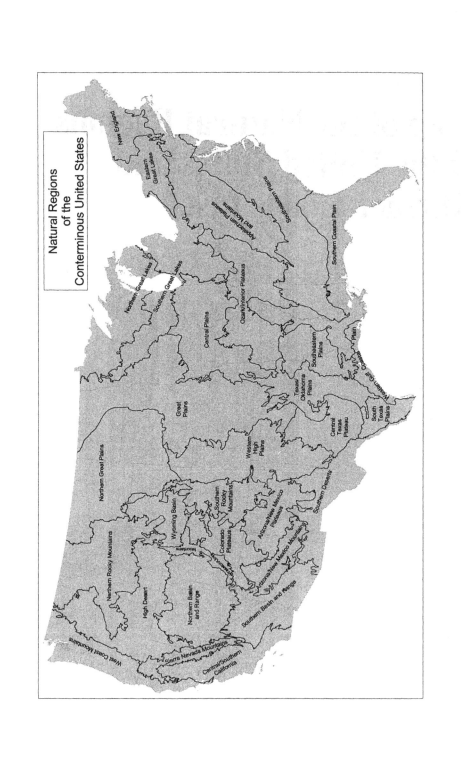

Natural Regions of the Conterminous United States

New England
Eastern Great Lakes
Southwestern Plains
Appalachian Plateaus
Southern Coastal Plain
Northern Great Lakes
Southern Great Lakes
Central Plains
Ozark/Interior Plateaus
Southeastern Plains
Central Plain
Texas/Oklahoma Plains
South Texas Plains
Great Plains
Central Texas Plateau
Western High Plains
Northern Great Plains
Southern Rocky Mountains
Wyoming Basin
Arizona/New Mexico Plateaus
Southern Deserts
Colorado Plateaus
Northern Rocky Mountains
Arizona/New Mexico Mountains
High Desert
Northern Basin and Range
Southern Basin and Range
West Coast Mountains
Sierra Nevada Mountains
Central/Southern California

CENTRAL AND EASTERN GRASSLANDS

GRASSLANDS

63 Foothills prairie
64 Grama-needlegrass-wheatgrass
65 Grama-buffalo grass
66 Wheatgrass-needlegrass
67 Wheatgrass-bluestem-needlegrass
68 Wheatgrass-grama-buffalo grass
69 Bluestem-grama prairie
70 Sandsage-bluestem prairie
71 Shinnery
72 Sea oats prairie
73 Northern cordgrass prairie
74 Bluestem prairie
75 Nebraska Sandhills prairie
76 Blackland prairie
77 Bluestem-sacahuista prairie
78 Southern cordgrass prairie
79 Palmetto prairie

GRASSLAND AND FOREST COMBINATIONS

80 Marl-Everglades
81 Oak savanna
82 Mosaic of 74 and 100
83 Cedar glades
84 Cross timbers
85 Mesquite-buffalo grass
86 Juniper-oak savanna
87 Mesquite-oak savanna
88 Fayette prairie
89 Blackbelt
90 Live oak-sea oats
91 Cypress savanna
92 Everglades

EASTERN FORESTS

NEEDLELEAF FORESTS

93 Great Lakes spruce-fir forest
94 Conifer bog
95 Great Lakes pine forest
96 Northeastern spruce-fir forest
97 Southeastern spruce-fir forest

BROADLEAF FORESTS

98 Northern floodplain forest
99 Maple-basswood forest
100 Oak-hickory forest
101 Elm-ash forest
102 Beech-maple forest
103 Mixed mesophytic forest
104 Appalachian oak forest
105 Mangrove

BROADLEAF AND NEEDLELEAF FORESTS

106 Northern hardwoods
107 Northern hardwoods-fir forest
108 Northern hardwoods-spruce forest
109 Transition between 104 and 106
110 Northeastern oak-pine forest
111 Oak-hickory-pine forest
112 Southern mixed forest
113 Southern floodplain forest
114 Pocosin
115 Sand pine scrub
116 Subtropical pine forest

WESTERN FORESTS

NEEDLELEAF FORESTS

1 Spruce-cedar hemlock forest

2 Cedar-hemlock-Douglas fir forest

3 Silver fir-Douglas fir forest

4 Fir-hemlock forest

5 Mixed conifer forest

6 Redwood Forest

7 Red fir forest

8 Lodgepole pine-subalpine forest

9 Pine-cypress forest

10 Ponderosa shrub forest

11 Western Pondersa Forest

12 Douglas fir forest

13 Cedar-hemlock-pine forest

14 Grand fir-Douglas fir forest

15 Western spruce-fir forest

16 Eastern ponderosa forest

17 Black Hills pine forest

18 Pine-Douglas fir forest

19 Arizona pine forest

20 Spruce-fir-Douglas fir forest

21 Southwestern spruce-fir forest

22 Great Basin pine forest

23 Juniper-pinyon woodland

24 Juniper steppe woodland

BROADLEAF FORESTS

25 Alder-ash forest

26 Oregon oakwoods

27 Mesquite bosques

BROADLEAF AND NEEDLELEAF FORESTS

28 Mosaic of 2 and 26

29 California mixed evergreen forest

30 California oakwoods

31 Oak-juniper woodland

32 Transition between 31 and 37

WESTERN SHRUB AND GRASSLANDS

SHRUB

33 Chaparral

34 Montane chaparral

35 Coastal sagebrush

36 Mosaic of 30 and 35

37 Mountain mahogany-oak scrub

38 Great Basin sagebrush

39 Blackbrush

40 Saltbush-greasewood

41 Creosote bush

42 Creosote bush-bur sage

43 Palo verde-cactus shrub

44 Creosote bush-tarbush

45 Ceniza shrub

46 Desert: vegetation largely absent

GRASSLANDS

47 Fescue-oatgrass

48 California steppe

49 Tule marshes

50 Fescue-wheatgrass

51 Wheatgrass-bluegrass

52 Alpine meadows and barren

53 Grama-galleta steppe

54 Grama-tobosa prairie

SHRUB AND GRASSLANDS COMBINATIONS

55 Sagebrush steppe

56 Wheatgrass-needlegrass shrubsteppe

57 Galleta-three awn shrubsteppe

58 Grama-tobosa shrubsteppe

59 Trans-Pecos shrub savanna

60 Mesquite savanna

61 Mesquite-acacia savanna

62 Mesquite-live oak savanna